Handbook
of
Sensory Physiology

Volume VII/2

Physiology of
Photoreceptor Organs

By

I. Abramov · C. G. Bernhard · P. O'Bryan · A. I. Cohen
M. G. F. Fuortes · G. Gemne · P. Gouras · H. K. Hartline
A. Kropf · W. R. Levick · J. Z. Levinson · D. Mauzerall
F. Ratliff · G. Seitz · W. Sickel · W. K. Stell
T. Tomita · O. Trujillo-Cenoz · G. Westheimer

Edited by

M. G. F. Fuortes

With 342 Figures

Springer-Verlag Berlin · Heidelberg · New York 1972

QP
351
.H34
nr.7
pt.2

ISBN 3-540-05743-9 Springer-Verlag Berlin · Heidelberg · New York
ISBN 0-387-05743-9 Springer-Verlag New York · Heidelberg · Berlin

Typesetting, printing and binding: Brühlsche Universitätsdruckerei Gießen

Preface

This volume is a collection of essays which attempts to summarize the recent progress in the field of photoreceptor and retinal physiology.

Reflecting the way in which research is organized, each author reports on the studies performed with the techniques with which he is most familiar: morphological, chemical or physiological. The first chapters describe the structure of visual cells and the histological architecture of the retina. Next comes a summary of the laws governing photochemical reactions and a report on the biochemistry of photopigments. Four articles cover the optical properties of invertebrate eyes and the electrophysiology and the interactions of their photoreceptors. These are followed by a discussion of the properties of vertebrate eyes, including chapters on optics, on the electrical responses of rods and cones and on the functional organization of the retina. The final chapter provides an extensive review of retinal biochemistry and metabolism.

Even though the experimental approach differs, all studies are directed toward the solution of two basic problems: transduction in the photoreceptors and organization (often called "information processing") in the retina.

The central problem of photoreceptor cells is to determine how light produces a response. We know that illumination evokes electrical changes and we have recently learned a great deal about the features of these changes. The evidence indicates however that elaborate processes must be interposed between the absorption of photons by the pigment and the production of electric currents through the membrane. These intermediary events remain to be unraveled.

The final aim of the studies on the retina is to find how the different cells interact in order to bring about the messages which are transmitted along the optic nerve. In the *Limulus*, inhibitory interactions between ommatidia have been discovered and carefully analyzed. They seem to be appropriate for sharpening the resolution of spatial contrast. In the vertebrate retina, systematic investigations have been started only recently, but they have already disclosed important, (and often unfamiliar) properties of responses, of synaptic mechanisms and of interactions.

Reading the reviews collected in this volume, it will be evident that remarkable progress has been made, both in the studies on photoreceptors and in those on the retina. And yet it will be realized that the same advances which have clarified a number of important details have also opened up new problems, often unsuspected and puzzling. No unifying generalization has yet been reached to permit a simple description of the principles governing the functions of visual organs.

The electron-microscope can now resolve the structure of cells with great detail but we do not yet understand what order is hidden under the convoluted outlines of the histological sections. Similarly, the responses recorded with micro-electrodes give evidence of astonishing subtleties of function, which we do not know how to analyze, and comparative studies show a multiplicity of action which is probably related to a disappointing diversity of mechanisms.

It seems that much painstaking work will still have to be performed. But perhaps the profitable path to follow has been traced and we can now rely for further progress on the tenacity and talent of the young investigators who have recently joined the field of vision.

M. G. F. FUORTES

Contents

List of Contributors

ABRAMOV, Israel
The Rockefeller University, New York, New York 10021, USA

BERNHARD, C. G.
Karolinska Institutet, Fysiologiska Institutionen II, Solnavägen 1, Stockholm, Sweden

COHEN, Adolph
Ophthalmology Department, Washington University Medical School, 660 South Euclid Avenue, St. Louis, Missouri 63110, USA

FUORTES, M. G. F.
Room 2C02, Building 36, National Institutes of Health, Bethesda, Maryland 20014, USA

GEMNE, Gösta
Department of Physiology, Karolinska Institutet, Solnavägen 1, S-10401 Stockholm 60, Sweden

GOURAS, Peter
Laboratory of Vision Research, National Eye Institute National Institutes of Health, Bethesda, Maryland 20014, USA

HARTLINE, Haldan Keffer
The Rockefeller University, New York, New York 10021, USA

KROPF, Allen
Dept. of Chemistry, Amherst College, Amherst, Mass. 01002, USA

LEVICK, W. R.
Department of Physiology, John Curtin School of Medical Research Australian National University, P. O. Box 334, Canberra City, A.C.T. 2601 Australia

LEVINSON, John Z.
Dept. of Psychology, University of Maryland, College Park, Md. 20740, USA

MAUZERALL, David
66th Street and York Ave, Rockefeller University, New York, New York 10021, USA

O'BRYAN, Paul, M., Jr.
Room 2C02, Building 36, National Institutes of Health, Bethesda, Maryland 20014, USA

RATLIFF, Floyd
The Rockefeller University, New York, New York 10021, USA

SEITZ, Georg
2. Zoologisches Institut der Universität, D-852 Erlangen, Bismarckstr. 10, Germany

SICKEL, Werner
University of Cologne, Dept. of Physiology,
D-5000 Köln 41, Robert-Koch-Str. 39, Germany

STELL, William K.
Jules Stein Eye Institute, The Center for the Health Sciences,
University of California, Los Angeles, California 90024, USA

TOMITA, Tsuneo
Department of Physiology, Keio University School of Medicine
Shinanomachi, Shinjuku-ku, Tokyo, Japan
and
Department of Ophthalmology and Visual Science,
Yale University School of Medicine, New Haven, Connecticut 06510, USA

TRUJILLO-CENOZ, Omar
Instituto de Investigation de Ciencias Biologicas
Departamento de Ultrastructura Cellular, Avda. Italia 3318, Montevideo, Uruguay

WESTHEIMER, Gerald
Department of Physiology-Anatomy, University of California
Berkeley, California 94720, USA

Introduction

By

HALDAN K. HARTLINE, New York, New York (USA)

Visual science has undergone remarkable development in the past two decades. A generation of competent, vigorous workers, skilled in the use of the latest biophysical and biochemical techniques has built a structure of new understanding on older foundations that puts this branch of science in the very forefront of Neurophysiology.

The interplay of diverse disciplines and techniques in the visual sciences has borne rich fruits. Biochemistry, spectrophotometry, and electrophysiology probe the individual receptors. Electronmicroscopy together with newer developments of conventional histology and cytology elucidate the structures whose functions we seek to understand. The electronic amplifier is now a sophisticated, versatile instrument far easier to use than heretofore; with it we exploit the unitary analysis of the activity of visual neurons and their intricate interactions. With ultrafine micropipette electrodes, and the circuitry required by them, we elucidate basic cellular mechanisms. Computers, still scorned by a few, are valuable laboratory tools for many; they will be increasingly necessary as the formidable complexities of organized nervous functions are faced.

Vision is almost universal among all but the lowest animals. Comparative methods in physiology and in behavioral science, and especially in their combination, have come into increasing prominence, and give to visual science the biological breadth it deserves. Invertebrate photoreceptors and optic ganglia have furnished valuable preparations for the physiologist. The technical difficulties that slowed the study of vertebrate visual systems have now been largely overcome and the physiology of the vertebrate retina and higher visual systems has advanced rapidly in the last few years.

Valuable and essential as choice of preparation and supporting technology is, the more powerful tools are the intellectual. We understand much more of the biochemistry of the visual pigments than we did even a few years ago, we see much more of the elemental intracellular structures of receptors and visual neurons, and of their interconnections in optic tissue. The fundamental mechanisms of the nerve fibers and the synapse seem now to be on firm foundations. We can speak with more confidence of the antagonistic processes in organized neural action: the "interplay of excitation and inhibition" has been documented by experiment, and is more than an interplay of words. "Subjective" experience — in its exact form, psychophysics — comes closer now to physiology: to understand how we see we must first know what we see.

In the articles which follow this introduction the above comments will be well illustrated. The subjects covered, however, have been restricted to the peripheral visual processes in the receptors and retina. Equally important advances in the physiology of higher visual centers will be treated elsewhere in this Handbook, as will the advances in visual psychophysics.

Several specific topics suggest themselves for brief comment. We have always thought — or hoped — that the electrical responses of visual receptor cells would turn out to be a crucial link in the transduction of light to nervous action. Not many years ago invertebrate photoreceptors furnished the only preparations in which electrical responses could be studied readily. Those first studied respond to illumination by electrical depolarization. The local currents, as a result, flow in the direction expected for excitation of the neural pathways they serve. With the advent of ultra-fine intracellular electrodes, the individual rods and cones of vertebrate retinas have at last been brought to study. They dealt us a surprise; these receptors undergo an *increase* in electrical polarization, in response to illumination. Such hyperpolarization, usually associated with inhibition, seems strange as the response of a primary photoreceptor — although some invertebrate photoreceptors are known which initiate shadow responses by eliciting trains of nerve impulses — "off" responses — at the "rebound" from the receptor hyperpolarization induced by illumination. Perhaps the vertebrate receptors release synaptic transmitter in a manner different from that with which we are more familiar. Perhaps, as has been suggested, transmitter is released spontaneously in the dark, and its cessation is the signal that means "light". It seems a strange way to build a quantum-catcher.

However this may be, the more familiar forms of receptor potentials are certainly several steps removed from the very first events set off by the absorption of photons. The recently discovered electrical event, the "early receptor potential", appears to be more closely related to the initial biochemical events in light sensitive structures. There is hope that in it we now have — at last — a link between the very first photochemical events and receptor action. But how the linkage of this to later events takes place is still unknown.

The vertebrate retina has produced another surprise. A microelectrode in the ganglion cell layer gives noisy evidence of the discharges of bursts and trains of familiar nerve impulses transmitting information over the optic nerve to the higher visual centers in the brain. But the other layers of the retina are strangely silent. This could mean no more than a technical deficiency — although similar penetration of, say, the optic ganglia of insects rarely fails to elicit cascades of crackles, pops and buzzes in the monitoring loud speaker. Recently, elegant techniques of ionophoretic staining of the penetrated cells have been developed. With them, we can now identify specific responses of specific cell types. Such staining provides assurance as to the location of the microelectrode and it now seems fairly certain that, except for the ganglion cells (and possibly the amacrines), vertebrate retinal neurons and receptors generate no nerve impulses. The graded electrical changes, spreading electrotonically over the copious and short cellular branches seem to be sufficient to liberate the necessary transmitter at the chemical synapses, or to act directly through electrical ones. Interaction within the vertebrate retina, serving the first steps of the complex processing of visual information, may well be accomplished without benefit of the trains of nerve impulses that are required for rapid communication over long pathways.

In no part of visual physiology has progress been more gratifying than in the study of color vision. We have had many years of background of very exact psychophysical studies, to which more recently the technique of reflection densitometry in human subjects has been added. And now the absorption spectra and action spectra of individual receptors in the vertebrate retina, from fish to man, have been measured. The retinal cones have sensitivities which peak in different regions of the visible spectrum; in many of the forms studied the cones fall into three groups with sensitivities peaking in the red or orange, the green and the blue. This is the receptor scheme first postulated long ago by THOMAS YOUNG, and taken up in detail by HELMHOLTZ. But electrophysiological analysis reveals that at the very next stage in the retina this chromatic information is processed into antagonistic responses, much as was postulated by HERING. Thus the century-old controversy between HELMHOLTZ and HERING now seems to be resolved: both were right.

Visual information — spatial, temporal, chromatic — undergoes extensive processing as it is transmitted, stage by stage, from receptors to higher visual centers. Excitatory and inhibitory influences, acting in various combinations over direct pathways, or laterally, or by recurrent feed-back, generate complex patterns of activity. In these patterns of nervous activity various features of the patterns of light and shade and color in the visual image are accentuated, or suppressed, or recombined to serve the needs of recognizing and responding appropriately to significant features in the animal's lighted surroundings. It is with the formidable task of describing and analyzing these various patterns that much of the work in the following articles is concerned.

Chapter 1

The Structural Organization of the Compound Eye in Insects

By

Omar Trujillo-Cenóz, Montevideo (Uruguay)

With 36 Figures

Contents

Introduction

Ever since the first use of the light microscope for examining biological specimens, the compound eye of arthropods has attracted the attention of scholars. More recently, the compound eye has been found to be a particularly suitable model for studying the general problems of the flow and processing of sensory information in the nervous centers. At the same time, for complete understanding of the new data resulting from these novel lines of investigation, a solid base of detailed anatomical knowledge is needed.

Early workers, in spite of the limitations of their rather simple histological techniques, were able to describe excellently the dioptric apparatus and associated photoreceptor cells found in the retina of several kinds of arthropods (GRENACHER, 1879). Nevertheless, accurate histological observations of the optic centers were possible only after the development of the embedding media and histological stains. Several fundamental papers were published by VIALLANES (1887a, b), who introduced the serial-section technique for the study of the brain and optic centers of insects. He discovered (VIALLANES, 1887a) the existence of discrete units in the lamina ganglionaris of insects and also found them in the lamina of *Palinurus* (Crustacea). VIALLANES (1892) coined the term "neuroommatides" for describing these units of the lamina.

The exclusive use of the aniline stains for studying the compound eye resulted in misconceptions concerning the structural organization of the optic centers. The synaptic neuropiles were considered to be composed of a fine punctuate substance transversed by bundles of fine fibrils. This conceptual fog was dissipated by the introduction of the Golgi and Ehrlich methods. The pioneer publication by KENYON (1896) was followed by that of VIGIER (1908), who successfully applied the Golgi method for studying the interconnections between the first and second order nerve fibers in the lamina of the fly.

The publication of VIGIER's paper prompted CAJAL to report his own observations on the visual system of muscoid flies (CAJAL, 1909, 1910). He later published a more complete study of the retina and visual centers in several species of insects (CAJAL and SÁNCHEZ, 1915) and this study remains an unexhausted source of anatomical information for those interested in insect vision. Concomitantly, ZAWARZIN (1914) achieved brilliant results using the Ehrlich method. He gave a detailed account of the neuronal interconnections in the main optic centers of the dragonfly larva.

Another important advance in our knowledge of the structural organization of the compound eye was achieved by employment of the electron microscope. This new era was initiated by the works of FERNÁNDEZ-MORÁN (1956—1958), GOLD-SMITH and PHILPOTT (1957), MILLER (1957) and WOLKEN et al. (1957). The work of MILLER (1957) demonstrated that the rhabdom ("das Stäbchen" of GRENACHER) is composed of tubular structures derived from the plasma membrane of the photoreceptor cells. Further refinements of the electron-microscope technique (SJÖ-STRAND, 1958) have made possible the three-dimensional reconstruction of small portions of the ommatidia and visual neuropiles (TRUJILLO-CENÓZ, 1965a; MELAMED and TRUJILLO-CENÓZ, 1968).

The identification of the cell types and cell connections in the synaptic fields has been greatly facilitated by the combination of the Golgi and electron microscope techniques (BLACKSTAD, 1965). Using this technical approach, Golgi-stained neurons of the visual centers of blowflies were followed and studied, taking advantage of the high resolution offered by the electron microscope (TRUJILLO-CENÓZ and MELAMED, 1970).

Functional and anatomical data have been obtained from vertebrate retinas by the intracellular injection of fluorescent dyes, (KANEKO, 1970). This technique introduced by STRETTON and KRAVITZ (1968) allows the recording of the bioelectrical activity of the cell and also permits its subsequent morphological study under a fluorescence microscope. Preliminary studies by MOTE and GOLDSMITH (1971) indicate that this method will yield valuable information about the structure and function of the arthropod visual system.

The present chapter is mainly concerned with the compound eye of insects, and special emphasis has been given to the electronmicroscopic findings. However, interpretation of the electronmicrographs depends upon data that often can be obtained only from light microscopic investigations. Therefore, correlation of both technical approaches has been attempted in the text and figures. The bibliography cited is by no means exhaustive and a more complete list of references can be found in the excellent reviews by GOLDSMITH (1964) and HORRIDGE (1965).

I. General Anatomy of the Compound Eye and Associated Nervous Centers — Axes of Reference

Unlike vertebrates, in which the photoreceptor cells and the first synaptic neuropiles are included in a single retina, insects have similar elements distributed within three distinct regions: the retina, the lamina and the medulla (CAJAL and SÁNCHEZ, 1915; VARELA, 1971). They also have a large optic lobe or lobula which in lepidopterans and dipterans is divided in two unequal neuropilic masses (Fig. 1).

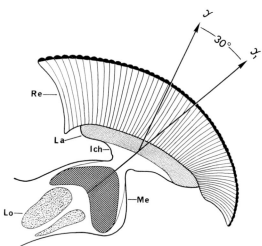

Fig. 1. Schematic drawing showing the retina (Re) and main optic centers in muscoid flies. Just below the retina lies a thin lamina (La) which is connected to the medulla (Me) by the intermediate chiasm (Ich) or optic nerve. The lobula (Lo) is divided in two unequal neuropilic masses termed by CAJAL, ovoidal and laminar ganglia. The γ axis is defined as an axis perpendicular to the external surface of the eye passing through the medial region of the retina-lamina complex. The γ_1 axis is defined as an axis passing through the central region of the medulla. Note that these two axes depart 30° from one another.

Predators and fast flying insects usually have large retinas composed of thousands of small units or ommatidia. Conversely, the retina in specimens of more primitive groups (Thysanura, Collembola) consists of a few ommatidia.

The regular array of the ommatidia allows, at least in dipterans, a reliable correlation between the gross anatomy of the eye and the spatial arrangement of the photoreceptor cells and fibers. Then, it is important to define some geometrical terms and reference axes which will assist understanding of the structural organization of the eye. Considering that the compound eye of muscoid flies is one of the best known from an anatomical point of view, it will be used as a model. As shown in Fig. 2 the antero-medial area of the compound eye of *Lucilia* is characterized by the hexagonal array of the ommatidia. The same figure shows the two axes, α and β, which are used as reference for defining the terms "horizontal" and "vertical" when applied to the ommatidial rows. The horizontal rows of ommatidia are those parallel to the α axis, whereas the vertical rows are those parallel to the β axis. The horizontal axis of the eye is here defined as an axis coincident with the longest horizontal ommatidial row.

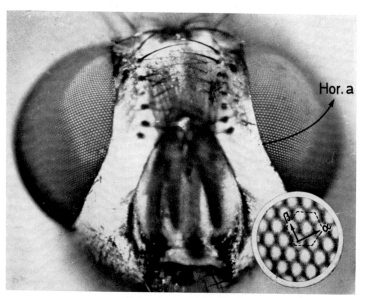

Fig. 2. Low power micrograph (25 ×) of the two compound eyes of *Lucilia*. The horizontal axis of the eye (Hor. a) has been indicated in the left eye by a heavy line. The inset shows, at higher magnification (approximately 80 ×), the hexagonal array of the ommatidia and the two axes (α and β) used as reference for defining the terms horizontal and vertical when applied to the ommatidial rows.

The lamina also shows a geometrical hexagonal array of its constituent units (the "neuroommatides" or optical cartridges). Therefore the axes and terms as defined for the retina are valid at this level too. A line perpendicular to the external surface of the eye passing through the center of the retina and the lamina has been named the γ axis.

The main axes of the medulla are shown in Fig. 3. The α_1 axis departs approximately 15° from the horizontal axis of the retina, the β_1 axis is defined as an axis perpendicular to the α_1 axis. The γ_1 axis is coincident with a straight line passing through the central region of the medulla. This axis is not coincident with the γ axis of the retina-lamina complex but departs approximately 30° from it (Fig. 1).

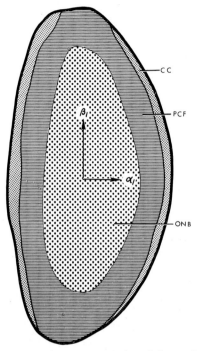

Fig. 3. Schematic drawing representing a cross section of the medulla (section perpendicular to the γ_1 axis) to show its two main axes (α_1 and β_1). CC, cellular cortex; PCF, post chiasmatic fibers; ONB, optic nerve bundles or medullary units.

II. The Retina

From an anatomical point of view the retina can be considered as the sum of its elementary units or ommatidia. Each ommatidium consists of an optical system showing different degrees of structural complexity and a group of photoreceptor cells (eight to eleven) exhibiting dissimilar morphology in the various groups of insects.

1. The Optical System of the Ommatidium

Since GRENACHER's work (GRENACHER, 1879) it is classical to distinguish in insects three main types of eyes which differ in the subcorneal components of the dioptric system. Usually, the optical system of the ommatidium consists of a corneal lens and a crystalline cone. Some insect eyes have in addition, long crystalline tracts which extend from the apex of the cone to the tips of the photo receptor cells. These are the so-called "clear-zone" eyes, (HORRIDGE, 1971).

The eyes having a well-developed solid cone, have been termed *eucone eyes*. Other groups (diptera sub-orders Brachycera and Cyclorrapha), lack a solid cone but its place is occupied by a cup-shaped organ filled with a gelatinous substance. These are the *pseudocone eyes* of GRENACHER. *Acone eyes* have beneath the cornea, neither a solid cone nor a soft pseudocone, but only four small, poorly differentiated cells. This type of eye occurs in primitive dipterans *(Tipula)*, Dermaptera, some Hemiptera and several species of Coleoptera. There are finally, some Coleoptera *(Lampyris, Photuris, Phausis)* in which the inner side of the cornea projects towards the photoreceptor cells forming a long, solid chitinous cone (EXNER, 1891; HORRIDGE, 1968 a; SEITZ, 1969).

The compound eye of dark-adapted moths shows a bright glow when observed under reflected light. This glow or eye-shine is caused by a reflecting layer — the tapetum — derived from the tracheolar system. Most butterflies have a more elaborated kind of tapetum which behaves as a colored mirror, tinting the light acting on the photoreceptor structures. This type of tapetum consists of periodic sets of cytoplasmic plates that alternate with air spaces (MILLER and BERNARD, 1968; BERNARD and MILLER, 1970).

Considering that the pigment granules contained in the pigment cells absorb light and therefore have a screening function, these cells will be described here together with the other components of the optical system.

a) The Corneal Lenses. The corneal lenses are modified transparent portions of the chitinous exoskeleton. Usually they have hexagonal outlines but shape deviations are common in different species, and even in different areas of a single eye.

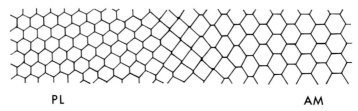

PL   AM

Fig. 4. The corneal lenses in the compound eyes of muscoid flies show important variations in size and shape. The antero-medial region of the eye (AM) contains large hexagonal lenses oriented with a side upwards, whereas the postero-lateral region (PL) consists of smaller hexagonal facets oriented with an edge upwards. Between these two zones there is a narrow band of rhomboidal lenses which extends from the dorsal to the ventral region of the eye (*Musca*, approximately 200 ×)

There are dipterans, like *Eristalis*, in which the majority of the ommatidial lenses exhibit a square or rhomboidal form. In muscoid flies each compound eye has three distinct areas which differ in the size, shape and orientation of the corneal lenses (Fig. 4) (KUIPER, 1966). The antero-medial portion of the eye contains large, hexagonal lenses (up to 40 μ in *Lucilia, Calliphora* and *Sarcophaga*) oriented with one side upwards. A narrow zone composed of rhomboidal lenses extends from the dorsal to the ventral region of the eye. This dorso-ventral band separates the antero-medial zone composed of large ommatidia from the postero-lateral region

containing small hexagonal facets oriented with an edge upwards. Still more striking differences have been found in the so-called "divided type" of eye. May-flies
(order Ephemeroptera) and some aquatic coleoptera have their compound eye
divided in two main portions, each of which is composed of ommatidia dissimilar
in size and shape. The dragonflies of the genus *Aeschna* have large hexagonal lenses
in the dorsal portion of the eye and much smaller lenses in the ventral region
(Fig. 5).

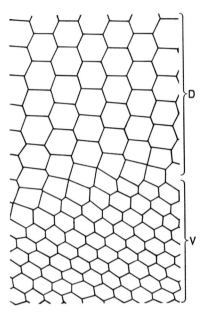

Fig. 5. In the dragon-fly *Aeschna* each compound eye is divided in two zones. There is a dorsal
zone (D) containing large hexagonal lenses and a ventral zone (V) composed of smaller corneal
lenses (250 ×)

KUIPER (1966) stated that in *Calliphora* the larger lenses have external convex
surfaces whereas their inner surfaces are practically flat. In this species, biconvex
lenses are found only at the ventral and lateral parts of the eye. Studies made on
Musca by VOWLES (1966) demonstrated that the ommatidial lenses are not
spherical; each lens has a different radius of curvature in the vertical and horizontal
planes. There are, therefore, two focal lengths, one for the vertical plane (approximately $65\,\mu$ from the front surface of the lens) and another for the horizontal plane
(approximately $75\,\mu$ from the front surface of the lens).

SEITZ (1968) and VARELA and WIITANEN (1970) have studied the optical properties of the corneal lenses in *Calliphora* and *Apis* respectively. In these two
insects each corneal lens consists of three concentric layers exhibiting slightly different refractive indexes. The external layer shows the higher refractive index
(1.490 for the bee and 1.473 for the fly) while the inner layer, contacting the cone
or the pseudocone has the lower one (1.435 for the bee and 1.415 for the fly). From
SEITZ and VARELA's data it is clear that the corneas of these two species do not act

as a lens cylinder, the positive lens function is explained by the convex external surface of the cornea. Working on the firefly *Phausis*, SEITZ (1969) found that the corneal lens is a homogeneous spherical lens while the processus corneae is a lens cylinder.

Additional information about the morphology of the corneal lenses has been provided by the electron microscope. One of the most interesting findings was reported by BERNHARD and MILLER (1962). They discovered that the corneal lenses of certain insects have, at the interface between air and chitin, small conical protuberances measuring approximately 0.2 μ from top to base. These structures were named "corneal nipples" and are particularly well developed in nocturnal Lepidoptera (Fig. 6) and in some species of the order Collembola (BARRA, 1971).

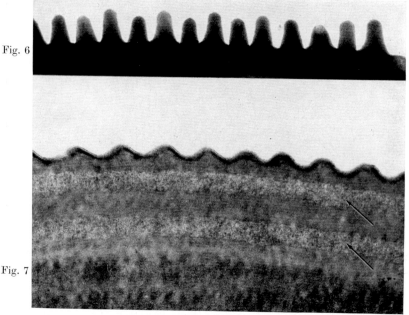

Fig. 6

Fig. 7

Figs. 6 and 7. In many insects the external surface of the corneal lens is covered by thousands of small conical protuberances known as the "corneal nipples". This electronmicrograph shows a longitudinal section through the cornea of a nocturnal moth in which the corneal nipples are particularly well developed (45000 ×). Unlike moths, dipterans do not show typical corneal nipples but only small round elevations. Note the dense and rare alternating bands occurring in the corneal lens (arrows). These bands are the morphological substratum of a system of interference filters tinting the light reaching the photoreceptors. Such system occurs in many dipterans. *Sympycnus* — Approx: × 45000

Muscoid flies do not have typical corneal nipples (nipple-in-air array) but only short, round elevations. According to MILLER et al. (1966) the corneal nipple array is found in flies not at the corneal-air interface, but partially embedded in the outer layers of the lens chitin. During development of the lepidopteran eye the "nipple anlage" seems to be completed before deposition of the inner, thicker corneal layers (GEMNE, 1966).

The corneal nipples act as an antireflection coating increasing transmission of light through the ommatidial lens (BERNHARD et al., 1965). Three possible biological functions of the corneal nipples have been postulated by MILLER et al. (1968): 1) the corneal nipples may be used for camouflage by reducing the eye reflections, 2) in nocturnal animals they may play a rôle in increasing sensitivity near absolute threshold of vision, 3) they may prevent that light, reflected from the tapetum, can reach the rhabdom after a second reflection at the front corneal surface. Such reflected spurious light could interfere with visual perception.

A recent paper by BERNHARD et al. (1970) contains an assay of interpretation, both phylogenetic and functional, of the corneal nipples. (See also Chapter 9 in this volume.)

An elaborated system of corneal layers was discovered by BERNARD and MILLER (1968) in tabanid and dolichopodid eyes. Electron micrographs taken from sections normal to the eye surface show a set of dense and clear alternating bands located just beneath the air-chitin interface (Fig. 7). These bands have different refractive indexes and according to BERNARD and MILLER (1968—1970) they are the origin of the structural colors typical of the compound eyes in different families of diptera. Furthermore, it has been postulated by the same authors that this peculiar arrangement of thin layers acts as interference filters which may represent the material substratum for a color contrast enhancement system. The measurement of the reflectance spectra from single facets has been made by BERNARD (1971). His complete study, covering four families of Diptera, showed that "in localized regions of an eye that contains a mixture of filters colors, the reflectance characteristics often fit in the sense that one type of filters is maximally reflecting at a wave-length where another type is minimally reflecting". These findings support the contrast-enhancement hypothesis, but do not exclude the possibility that the corneal filters could be part of a color-vision system.

b) The Cone and Pseudocone. In the so-called *eucone eyes* there is a solid crystalline cone beneath the corneal lens. Commonly, the crystalline cone consists of four specialized cells — the cone cells — which unite to form a conical or cylindro-conical body with its apex directed inward. Butterflies for example, have large cones which, when observed in fresh, appear as highly refractile bodies (Fig. 8). The electronmicrographs clearly show that the crystalline substance is an intracellular differentiation of the cone cells. Each of these cells has a very electron-dense core surrounded by two or three layers of a lighter cytoplasm (FISCHER and HORSTMANN, 1971) (Fig. 9). The electron-dense core is composed of minute tightly packed granules and filaments whereas the external, clear layers contain fine fibrils and a few mitochondria. The cell nuclei are located in the most distal portion of the cone, near the corneal lens. In certain species of Lepidoptera, Neuroptera and Coleoptera each cone cell sends out a proximal process which, together with the processes arising from the three other neighbor cells, form a single crystalline tract. The crystalline cones in beetles of the genus *Repsimus* show a coat of 12 to 15 layers of rough endoplasmic reticulum. Microtubules are found at the peripheral zone of the cone just beneath the outer membrane of the cone cells (HORRIDGE and GIDDINGS, 1971).

In the bee, (VARELA and PORTER, 1969) the crystalline cone is composed of four cells containing small granules and microtubules which begin to appear in the

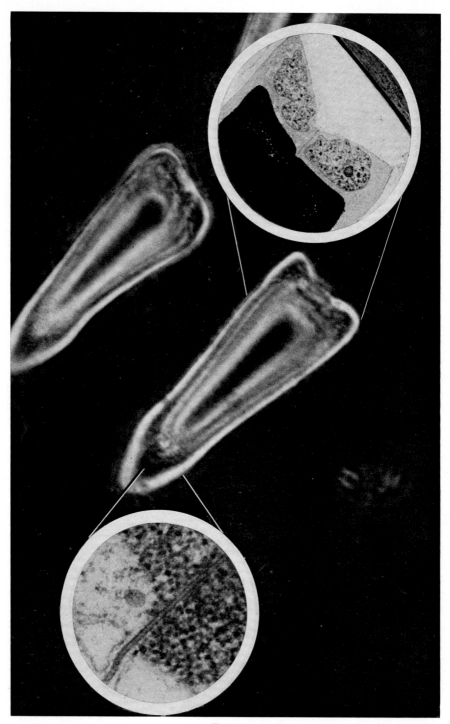

Fig. 8

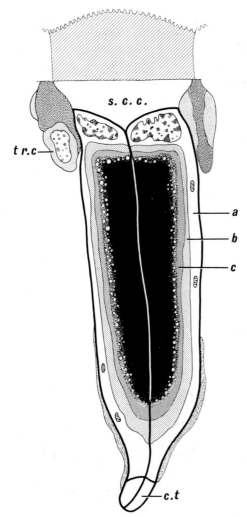

Fig. 9. This semischematic drawing represents a longitudinal section through the crystalline cone of a moth. As usual in insects the crystalline cone consists of four cells which unite to form a single structure (this longitudinal section shows only two of them). In the cytoplasm of these cells there are zones of different electron density: 1) there is an outer light zone (a) which contains the nucleus and a few mitochrondria, 2) there are two intermediate zones (b and c) which are denser and 3) there is a central core of a very high electron density. Note the vacuoles occurring at the periphery of the electron dense core. S.C.C., sub-corneal cavity; tr.c, trichogen cell; c.t., crystalline tract. × 4000

Fig. 8. Phase contrast photomicrograph (1500 ×) showing two unfixed crystalline cones of a nocturnal moth. The cones appear in fresh, as highly refractive structures. The two insets are electronmicrographs showing at higher magnification, structural details of the crystalline cones. The one located in the upper portion of the picture (2500 ×) depicts the nuclei of the cone cells. The nuclei are located in the peripheral zone of the cone facing the inner surface of the corneal lens. Note that the peripheral cytoplasm consists of a light fine granular matrix which contrasts with the dark aspect of the inner portion of the cone cells. The structural organization of the inner portion is shown at higher magnification (50000 ×) in the other inset. The inner portion of the cone consists of small dense filaments and granules. Complementary information concerning the fine structure of the crystalline cone can be obtained from Fig. 9.

basal third of the cone cells. In many species, each cone cell sends out a long process which runs parallel to the rhabdom and terminates near the basement membrane of the retina. Profiles of such "cone roots" (HORRIDGE, 1966) are usually found in cross sections of ommatidia, as clear oval elements lying between adjacent photoreceptors.

The crystalline cone of the bee is an isotropic structure and its refractive index is, according to VARELA and WIITANEN (1970), 1.3477 (— + 0.0004 sp). Working on the moth *Ephestia*, KUNZE and HAUSEN (1971) found an inhomogeneous refractive index which decreases from the axis to the periphery of the crystalline cone.

A different anatomical plan occurs in onany dipterans. As mentioned above, these insects have a pseudocone eye. Just beneath the cornea is a cup-shaped cavity containing a semi-fluid substance. In muscoid flies the lateral wall of the pseudocone consists of two pigmented cells, here referred to as the *pseudocone cells*. The pseudocone cells contain minute granules of an orange-brownish pigment. The outer surface of each pseudocone cell contacts several apical prolongations of the large pigment cells whereas the inner surface is bathed by the semi-fluid material filling the pseudocone cavity (Fig. 10). It is clear that this gelatinous

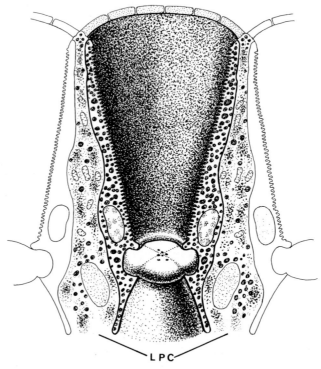

Fig. 10. In muscoid flies, below the corneal lens, there is a cup-shaped cavity — *the pseudocone cavity* — containing an amorphous viscous substance. The lateral walls of this cavity are composed of two cells — *the pseudocone cells* — containing pigment granules. The floor of the pseudocone cavity consists of four triangular cells which form a transparent plate coupling the dioptric structures with the photoreceptor cells. L.P.C., large pigment cells

substance is not an intracellular product, since there is a conspicuous membrane separating it from the pseudocone-cell cytoplasm. Moreover, short unequal micro-villi can be seen arising from the inner side of the cells and projecting into the pseudocone cavity. Desmosomes occur at the regions in which the two pseudocone cells come in contact. According to SEITZ (1968) the pseudocone of *Calliphora* is homogeneous and isotropic in all directions. The refractive index is 1.337.

The pseudocone cavity is closed from below by four transparent flat cells which unite to form a circular or trapezoidal plate. The inferior surface of this plate has a central excavation which receives the apical extracellular prolongations of the rhabdomeres (TRUJILLO-CENÓZ, 1965b). This plate insures the mechanical and optical coupling between the dioptric structures and the photoreceptor cells.

The crystalline cone and tracts of *Archichauliodes* (Megaloptera) show striking changes during dark and light adaptation (WALCOTT and HORRIDGE, 1971). The images recorded from light-adapted eyes, show the existence of crystalline tracts ending between the distal processes of the photoreceptor cells. At the regions in which the photoreceptor cell membranes meet the crystalline tract components, there are desmosome-like structures. In dark-adapted state there is no apparent crystalline tract, the distal photoreceptor cells are found contacting directly the crystalline cones. Similar movements have been observed in *Ephestia* (Lepidoptera), *Repsimus* (Coleoptera) and *Dytiscus* (Coleoptera). HORRIDGE and GIDDINGS (1971) have suggested that in these species the crystalline tracts are formed by the outflow of the peripheral cytoplasm of the cone cells. However, the mechanism responsible for these dramatic changes still remains unknown.

c) The Crystalline Tracts. In some groups of insects there is a relatively long distance $(20-1000\,\mu)$ between the proximal portion of the cone and the distal segment of the rhabdom. Such space is occupied by slender cylindrical structures connecting the rhabdoms with the peripheral dioptric structures. The crystalline tracts derive, either from the four cone cells or from the retinular cells (MILLER et al., 1968). The electron microscope has allowed a more detailed analysis of these tiny structures. In *Chrysopa*, for example, each crystalline tract is composed of four cytoplasmic pro-cesses, each of which represents the distal prolongation of one of the four cone cells (Fig. 11). In the species in which the crystalline tracts are not cone derivatives, but extensions of the retinular cells, a more complicated structural organization can be found. In *Ceratomia* (MILLER, personal communication) each crystalline tract consists of eight twisted processes containing minute granules and microtubules. Desmosome-like structures occur near the longitudinal axis of the tract. Observations by HORRIDGE (1968a), by DÖVING and MILLER (1969) and by MILLER et al. (1968) strongly suggest that the crystalline tracts act as wave guides trapping and conducting light towards the rhabdoms. Their conclusions are in conflict with the classical superposition theory as proposed by EXNER (1891). However, the question still remains open as proved by the papers published by KUNZE (1970). His behavioral experiments made on the moth *Ephestia*, suggest that visual information which is significant to the animal "is processed via a superposition dioptric apparatus". An up to date review and discussion of the superposition theory can be found in a recent paper by HORRIDGE (1971).

He states that the existence of a clear zone between the cone tips and the rhabdom layer, is a basic anatomical requisite for the summation mechanisms which enhance sensitivity of a dark-adapted eye. Light, scattered at the cone tip can reach photoreceptors belonging to neighbor ommatidia. In addition to the scattering phenomena, light is carried towards the rhabdoms along the different types of crystalline tracts. They provide separate optical pathways for each individual ommatidium. Such light-guide systems operate not only at light- but also at dark-adapted states. When dealing with KUNZE's optical and behavioral experiments, HORRIDGE proposes a different alternative interpretation.

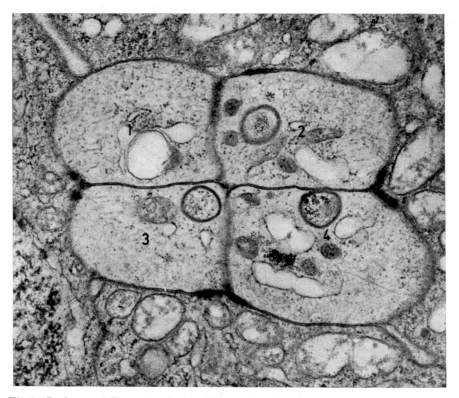

Fig. 11. In the eye of *Chrysopa* each crystalline tract consists of four cytoplasmic prolongations (1—4) derived from the four cells forming a single cone. This electronmicrograph shows the transitional zone between the crystalline cone and the tract — × 25 000

d) Tapetal Microstructure. Studying the anatomical substrate for the colored eye-glow exhibited by different species of diurnal lepidoptera, MILLER and BERNARD (1968) discovered a multilayered device occurring at the most proximal end of the rhabdom. It derives from tracheolar cells and consists of modified taenidial ridges regularly separated by air spaces (there are forty taenidia in the monarch butterfly *Danaus*).

These tracheolar microcomponents function optically, as a reflection interference filter (a quarter wave-length interference filter). Such filters cause the colored glow observed in the eye. Variations of butterfly glow color at different regions of the eye is explained by a corresponding variation in the filter layer thickness. Butterfly filters tint the light passing through the rhabdom and this effect will be superimposed upon the absorption properties of the photosensitive pigment. Tapetum filters are thought to improve contrast by enhancing brightness differences. In nocturnal moths the tapetum consists of bundles of small tracheae — the tracheole bush — lying around the rhabdoms.

e) **The Pigment Cells.** Available information concerning the morphology of the pigment cells is in many respects insufficient; correspondingly, the terminology is profuse and confused (a similar statement was made forty five years ago by SNODGRASS, 1926). Based mainly on studies made on lepidoptera three main types of pigment cells (JOHNAS, 1911) are classically described:

a) the primary pigment cells which lie between the cones and crystalline tracts;
b) the accessory pigment cells which lie between the photoreceptor elements; and
c) the basal pigment cells whose nuclei are located below the basement membrane.

SNODGRASS (1926) reported the occurrence of two sets of pigment cells; the distal or *iris pigment cells* which invest the cone and corneal cells and the proximal or *retinal pigment cells* which surround and separate the ommatidia. WIGGLESWORTH (1965) used a still different nomenclature. He distinguished the *primary iris cells* covering the crystalline cones and the *secondary iris cells* ensheathing both the primary iris cells and the ommatidia.

In spite of this variety of names, when facing a specific biological specimen it is difficult to make a valid correlation between the classical terms and the anatomical observations. Therefore, it seems advisable, at least in dipterans, to use a simpler nomenclature based on the main anatomical characteristics of the pigment cells.

The procedure of squashing a fresh slice from a fly eye (Musca) and observing it under the light microscope allows one to distinguish two main classes of pigment cells. One class is represented by large elongated cells containing a purple pigment whereas the other class includes small basal cells, containing a yellow-brownish pigment. The large cells, here referred to as the *large pigment cells*, extend from the inner surface of the corneal lenses to the basement membrane. GRENACHER (1879) and HICKSON (1885) observed that the nuclei of the large pigment cells lie in the peripheral portion of the retina, approximately at the level of the photoreceptor cell nuclei. Cross sections of the retina show six or seven of these cells surrounding each ommatidium (Fig. 12). However, the regular arrangement of the retinal elements allows a single pigment cell to be in contact with photoreceptor cells belonging to different but adjacent ommatidia. The electronmicrographs clearly show that the large pigment cells send out irregular cytoplasmic processes which run between the photoreceptor cells and penetrate into invaginations of their plasma membrane. Besides mitochondria and the typical pigment granules, the cytoplasm of these cells contains small dense particles with the aspect of glycogen (Fig. 12, inset). It is interesting to note that glycogen-like particles occur only in the large pigment cells. This peculiar characteristic facilitates recognition of their intraommatidial processes.

The cells containing a yellow pigment (here referred to as the *small pigment cells*) give origin to thick cytoplasmic prolongations which penetrate into the

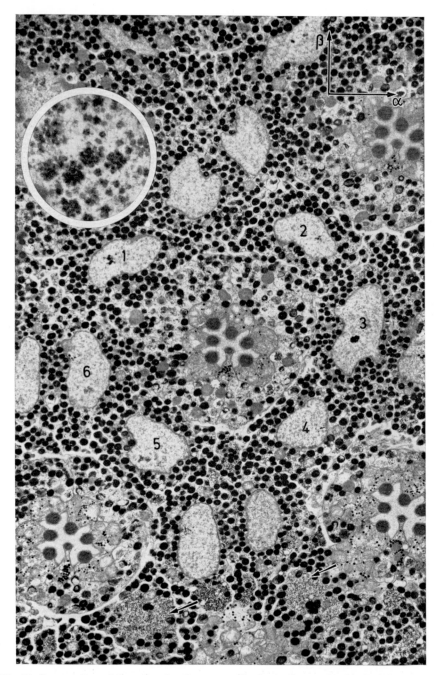

Fig. 12. Cross section of the retina (section perpendicular to the γ axis) showing the crown of *large pigment cells nuclei*. (1—6) surrounding the ommatidial units. The large pigment cells are characterized by the presence of minute dense particles with the aspect of glycogen (inset). These particles appear scattered in the cytoplasm or concentrated in well-delimited areas (arrows). *Sarcophaga,* 3000 × ; inset 60000 ×

ommatidial cavity. Usually there are four of these prolongations at the base of each ommatidium. At this level the photoreceptor cells have lost their rhabdomeres and have an axon-like appearance (Fig. 13). Serial section studies demonstrate that the small pigment cells ensheath incompletely the photoreceptor axons at their passage through the basement membrane.

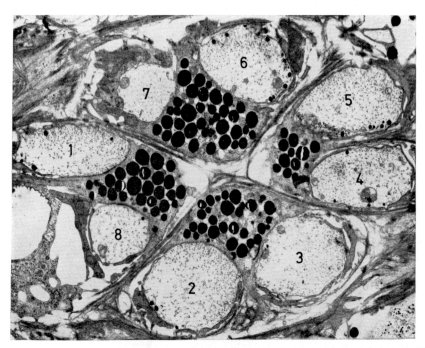

Fig. 13. Near the basement membrane separating the retina from the lamina, the ommatidial cavity is occupied by four large processes derived from the *small pigment cells*. As shown in this electronmicrograph the pigment cell cytoplasm partially ensheath the photoreceptor axons (1—8), *Lucilia*, 20000 ×

The occurrence of pigment movement correlated with changes in illumination has been recognized in several species since EXNER's classical work.

This pigment migration is particularly dramatic in nocturnal insects having the so-called superposition type of eye. In the light, the pigment granules are scattered all along the ommatidium. After dark adaptation the pigment withdraws upwards covering only the cones and the distal portion of the crystalline tracts.

As stated by WIGGLESWORTH (1965) there is no change in the cell morphology but only concentration or dispersion of the granules in the cell cytoplasm. Pigment migration occurs mainly in the secondary iris cells. A useful table containing the photomechanical responses produced by light adaptation in different orders of insects was published by GOLDSMITH (1964).

Electrophysiological and histological studies by BERNHARD (BERNHARD et al., 1963; BERNHARD and OTTOSON, 1960—1964) allowed a correlation between the changes in threshold occurring during dark adaptation and the position of the pigment granules. In nocturnal moths the sensitivity change of the photoreceptors show two phases: a first rapid phase and a second slow one, maintained up to ten to twenty minutes. This second phase has been related with the positional changes

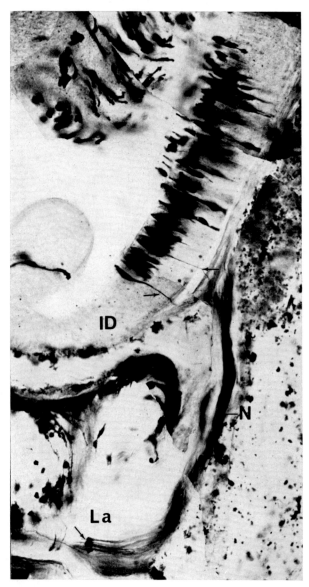

Fig. 14. In diptera the photoreceptor elements differentiate from epithelial cells composing of the imaginal discs (I.D.). At early pupal stages the embryonic photoreceptors can be stained using the Golgi technique. Each photoreceptor sends out a single axon (arrows) which enters, via a relatively thick nerve (N), into the central nervous system. Some of the photoreceptor axons terminate by means of club-shaped masses (arrow) within the lamina anlage. Other photoreceptors (the future central cells) do not terminate within the lamina anlage but in deeper visual centers (medulla). *Lucilia*, approx. × 400

of the pigment granules from light to dark condition. In moths the pigment granules consist mainly of ommin (LINZEN, 1967). Their spectral absorptions were studied by HÖGLUND (HÖGLUND et al., 1970; HÖGLUND and STRUWE, 1970) in the moth *Celerio euphorbiae* and in the wasp *Vespa*. There is an absorption maximum between 540 and 550 nm. However, the effect of pigment movements on the spectral sensitivity of the photoreceptor cells is small (HÖGLUND and STRUWE, 1970).

2. The Photoreceptor Cells

It is generally agreed (JOHANNSEN and BUTT, 1941; BODESTEIN, 1950) that the photoreceptor cells of insects derive from ectodermal cells. VIALLANES (1890) stated that in *Mantis* the development of the compound eyes is brought about by the formation of an eye disc in the ectodermal layer. In diptera the photoreceptors differentiate from some of the cells forming the antennal-eye imaginal discs which appear as ectodermal evaginations from the pharyngeal cavity. The earliest modifications of the ectodermal cells in the eye "anlage" of *Pieris* were described by SÁNCHEZ Y SÁNCHEZ (1919) using the Golgi technique. The information obtained from Golgi-impregnated imaginal discs of *Lucilia* (TRUJILLO-CENÓZ and MELAMED, unpublished) proved that typical axons originate from the immature photoreceptors and connect them to the nerve centers (Fig. 14). Nevertheless, classical [see BODESTEIN (1953) for complete references] and modern transplantation experiments (EICHENBAUM and GOLDSMITH, 1968) demonstrate that the establishment of central connection is not necessary for the development of normal ommatidial receptors in insects.

It is convenient for descriptive purposes to consider two main portions in the photoreceptor cells: a) the somata, lying in the retina and b) the axonal segment, which forms part of the first synaptic neuropile or lamina.

The most typical structure of the photoreceptor cell is the rhabdomere (the terms rhabdom and rhabdomere are used following the definitions given by BULLOCK and HORRIDGE, 1965). In the ommatidia of diptera (VIGIER, 1907; CAJAL, 1909; DIETRICH, 1909), certain aquatic hemiptera (LÜDTKE, 1953; BURTON and STOCKHAMMER, 1969), dermaptera and cerambicid beetles (mentioned by HORRIDGE and GIDDINGS, 1971) the rhabdomeres of the different photoreceptors can be seen as independent structures (Fig. 15). In most species however, the rhabdomeres fuse to form a single central rhabdom (Fig. 16).

A functional consequence of the existence of independent rhabdomeres is that their optical axes do not coincide. This was conclusively proved in muscoid flies in

---

Fig. 15. The retina of muscoid flies consists of thousands of ommatidia distributed according to an hexagonal pattern. Each ommatidium contains eight photoreceptor cells. The rhabdomeres of six of these cells (1—6) are distributed asymmetrically around a central cavity. Near the ommatidium axis there is a central rhabdom (7) composed of two unequal rhabdomeres derived from the superior and the inferior central cells (the section shown here passes through the apical portion of the ommatidia, therefore only superior central cells can be seen). Note that the trapezoid figure resulting from the arrangement of the rhabdomere is, in the three superior ommatidia the mirror image of that resulting from the distribution of rhabdomeres in the three inferior ones (equatorial region). The inset shows "in vivo" the rhabdomeres of one ommatidium. These structures act as light guides and appear, under the light microscope, as seven bright spots. Hor. a, Horizontal axis of the eye. *Lucilia*, approx × 4000

Fig. 15 see page 24

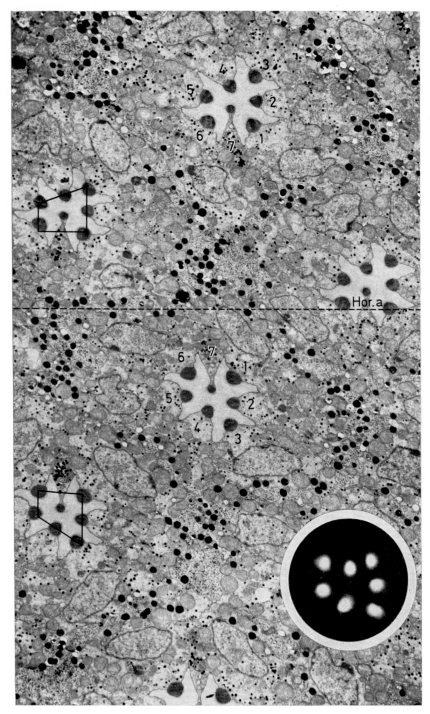

Fig. 15. Legend see page 23

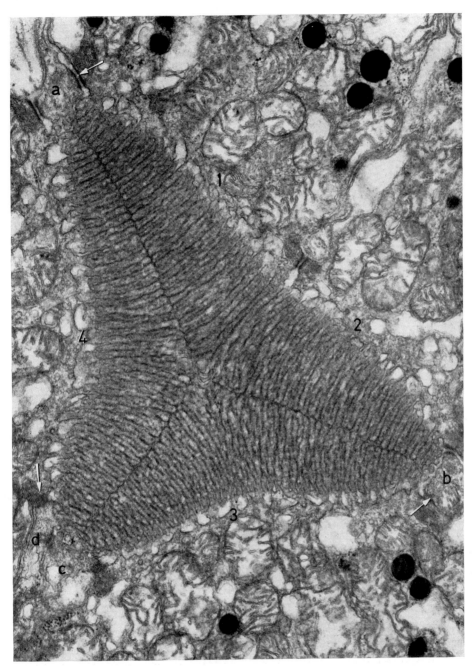

Fig. 16. The ommatidium of the dragon-fly *Aeschna* shows at this level four large (1—4) and four small photoreceptor cells (*a—d*). Unlike dipterans the rhabdomeres of the different photoreceptor cells are fused in a common central rhabdom. The rhabdom is mainly formed by the four large photoreceptors since the small ones only bear short scarce microvilli. Typical desmosomes can be seen at the points where adjacent photoreceptors meet (arrows). Approximately 50000 ×

which each one of the seven rhabdomeres "looks" in a different direction (Autrum and Wiedemann, 1962; Kuiper, 1962; Wiedemann, 1965; Kirschfeld, 1967).

There are insects like the bee (Varela and Porter, 1970) in which the fused rhabdom is formed by a roughly uniform contribution of the light common photoreceptors. Sections cut at different levels only show small variations in size of the rhabdomeres. In other species however (damsel-flies, Ninomiya et al., 1969; cockroach, Trujillo-Cenóz and Melamed, 1971, dragon-flies, Eguchi, 1971) rhabdomeres derived from different photoreceptors attain their maximum development at different ommatidial levels. Working on different species of beetles (Fam. Dytiscidae, Gyrinidae, Hydrophilidae and Carabidae) Horridge et al. (1970) and Horridge and Giddings (1971) found a peculiar "tiered" type of rhabdom. In these insects each ommatidium has one distal independent rhabdomere contacting the crystalline tract. At a deeper level is the large, fused type of rhabdom derived from six photoreceptors. Finally, near the basement membrane lies one small basal photoreceptor cell bearing its own rhabdomere. Such stratification of the rhabdomeres in two or three layers may be functionally important to retain both acuity and sensitivity. The peripheral rhabdomere could maintain acuity by excluding all but a pencil of light, whereas the proximal rhabdomeres could receive light from more than one facet (Horridge et al., 1970). The fused rhabdom or the independent rhabdomeres appear in fresh preparations as refractive, homogeneous rods which in general are very resistant to cellular desintegration (when immersed in water the rhabdomeres of the fly preserve their normal appearance after complete disruption of the remainder of the photoreceptor cells).

Measurement of the refractive index of the rhabdom in Diptera and Hymenoptera (de Vries, 1956; Kuiper, 1962; Seitz, 1968; Varela and Wiitanen, 1970) shows that it is higher (1.349—1.347) than that of the surrounding media. This indicates that the rhabdom or the independent rhabdomeres may act as wave guides capturing light and conducting it along a path coincident with the rhabdom longitudinal axis (Exner, 1891).

The light-conducting property of the rhabdomeres is easy to observe in freshly cut slices of the fly eye. If a small eye slice is suspended in a drop of saline and the photoreceptor cells are illuminated through the corneal lenses the independent rhabdomeres appear as seven bright, light spots (Fig. 15, inset).

The electronmicroscope investigations made by several workers Fernández-Morán (1956—1958), Goldsmith and Philpott (1957), Miller (1957) and Wolken et al. (1957) have demonstrated that the rhabdomere consists of tightly packed microvilli lying in planes perpendicular to the long axis of the photoreceptor cell.

The development of the rhabdomere was studied by Waddington and Perry (1960) in the fruit-fly Drosophila. According to them the rhabdomeric microvilli arise from small, sac-like invaginations of the plasma membrane. They also pointed out the contribution of multivesicular bodies to the rhabdomere formation. Further studies by Trujillo-Cenóz and Melamed (unpublished) in Sarcophaga and Lucilia have partially confirmed the observations mentioned above. It is interesting to note that, at early pupal stages, the photoreceptors of muscoid flies are grouped together without any evidence of the central cavity which characterizes the fully developed ommatidia (Fig. 17).

The development of the rhabdom and the appearance of the electrical responses of the compound eye were studied by EGUCHI et al. (1962) in the silk-worm *Bombyx mori*. No electrical responses to photic stimulation were recorded from the pupal eye which had non-differentiated rhabdoms. Parallel to the appearance of the rhabdomeric microvilli, the first electrical responses to photic stimulation could be recorded.

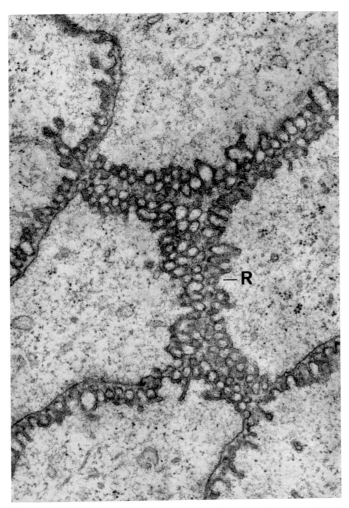

Fig. 17. During development, the fly ommatidia do not show the photoreceptor cells separated by a central cavity such as occurs in adult animals. On the contrary, the photoreceptors meet forming some sort of embryonic fused rhabdom (R). At this stage the rhabdomeric microvilli are uneven in size and shape. *Sarcophaga*. × 30000

In *Lucilia* as in most other species of dipterans, each microvillus measures about 0.75 μ in length and has a diameter of 500 Å. However the microvilli do not show a uniform diameter, but have short neck-like constrictions at the sites of insertion

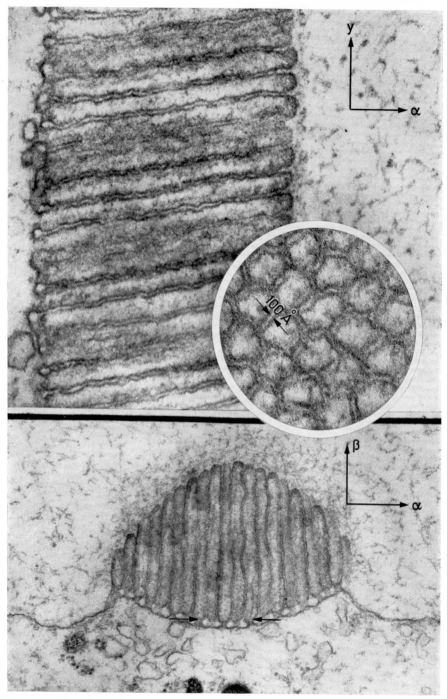

Fig. 18 and 19. The rhabdomeres consist of closely packed microvilli oriented perpendicular to the pathway of light. When viewed in cross section (inset), the microvilli appear separated by a distance of approximately 100 Å. Sections perpendicular to the photoreceptor main axis (Fig. 19) show that the whole rhabdomere has an oval or triangular outline. It is also possible to observe the existence of a short neck connecting each microvillus to the mother cell (arrows).

Fig. 18 approximately 140000 × ; Fig. 19 approximately 60000 ×).

in the mother cell. These constrictions form some sort of boundary between the undifferentiated cell cytoplasm and the rhabdomere. Such boundary is more evident in transverse sections of the photoreceptor cells. Cross sections of the microvilli show a typical honeycomb pattern (Figs. 18—19). In osmium fixed, Araldite-embedded material, the bounding membranes of adjacent microvilli are separated by a gap 100—120 Å wide which is continuous with the extra-cellular space. This gap provides space for ion flow during the photoreceptor stimulation (LASANSKY, 1970). Therefore, the rhabdomere, like the outer segment of vertebrate rods and cones, represents a multiple-membrane system oriented perpendicular to the pathway of light. Before having direct experimental evidence, the rhabdomere was considered the organelle containing the photosensitive pigments. More recently, microspectrophotometric measurements have made possible the tentative identification of two visual pigments in the fly rhabdomeres (LANGER and THORELL, 1966). The six peripheral rhabdomeres (1—6) of *Calliphora* (chalky mutant) have a photosensitive pigment with two maxima of absorption at 500 nm and 350—380 nm. The central rhabdom may have a different pigment with the main maximum at 470 nm. However, as stated by KIRSCHFELD (1971) the observation of two dissimilar extinction spectra does not necessary implys the existence of two different photopigments. A shift of the extinction maximum to shorter wave-lenghts can be explained too, by the smaller diameter of the central rhabdom (its diameter is approximately one half of that of the other six peripheral rhabdomeres).

Long, membrane-bounded channels and pinocytotic vesicles are commonly found at the base of the microvilli. The long channels communicate with the ommatidial central cavity and penetrate deeply in the cytoplasm. HORRIDGE and BARNARD (1965) have found that the endoplasmic reticulum cisternae are scattered in the photoreceptor cell cytoplasm in the light adapted eye, but form a palisade around the rhabdom in the dark adapted eye. Mitochondria are numerous in the photoreceptor cell somata. In *Lucilia*, *Sarcophaga* and *Musca* they are more abundant in the apical region of the cell. Changes in the distribution of mitochondria have been reported to occur in *Locusta* in relation to light or dark adaptation (HORRIDGE, 1966).

Using the freeze-etching method for electron microscopy and the light microscope, SEITZ (1970) studied the changes occurring in the photoreceptors of *Calliphora*, during light and dark adaptation. In the dark-adapted eye there is a layer of vesicles located close to the rhabdomeres of the six common photoreceptors. In the light-adapted state, only a few vesicles can be found near the rhabdomeres. No differences could be observed between the two states of adaptation in the cytoplasm of the superior and inferior central cells.

Microvesicular bodies (Mvb) similar to those described by SOTELO and PORTER (1959) in the cytoplasm of rat oocytes have been also found in arthropod photoreceptor cells (MELAMED and TRUJILLO-CENÓZ, 1966; EGUCHI and WATERMAN, 1967). In dipteran photoreceptors these bodies show a wide variation in size and number. Each Mvb consists of a single bounding membrane enclosing a variable number of small vesicles. EGUCHI and WATERMAN (1967) working on *Libinia* (Crustacea) have made statistical estimates of the number of Mvb in the photoreceptor cells when exposed to different light conditions. According to these authors, marked reductions occur on dark adaptation and a significant increase at the beginning of light adaptation.

The photoreceptor cell cytoplasm also contains minute granules of pigment. These granules are scattered all along the photoreceptor cell somata (concentrations of these granules can be found at the axon terminals). When compared with the pigment granules of the pigment cells they are smaller (0.1—0.2 μ).

KIRSCHFELD and FRANCESCHINI (1969) have reported the occurrence of positional changes in the photoreceptor pigment granules related to different conditions of illumination. Working on *Musca* eyes, they saw the granules scattered over the whole area of the photoreceptor as long as the cell is dark-adapted. During light adaptation the pigment granules migrate towards the rhabdomeres in the six common photoreceptor cells but not in the superior central cell (No. 7 of DIETRICH). It is assumed, that during light adaptation the pigment granules lie so close to the rhabdomeres that "total reflection is frustrated" and light conduction along the rhabdomere diminished. Movement of the photoreceptor cells including the rhabdom and the cell nucleus has been described several years ago (see summary and table by GOLDSMITH (1964). WALCOTT (1969), using boiling water as main fixative, has confirmed and extended previous observations. In coleopteran eyes, in the dark-adapted state the distal portions of the photoreceptors lie close to the crystalline cone. During light adaptation the photoreceptor cell cytoplasm moves away from the cone. Concomitantly an intervening crystalline tract is formed.

It is known that not all photoreceptor cells of a single ommatidium are morphologically alike. DIETRICH (1909) and CAJAL (1909) observed the existence of a slender more central rhabdomere in the fly ommatidium. This rhabdomere belongs to a flat cell (No. 7 following DIETRICH's nomenclature) which occupies a fixed position in relation to the other thicker more peripheral photoreceptors. DIETRICH also indicated the existence of an eighth basal "rudimentary" retinula cell. The first electron microscope studies (GOLDSMITH and PHILPOTT, 1957; FERNÁNDEZ-MORÁN, 1958) confirmed the existence of the seventh "eccentric cell" in the fly ommatidium. Later investigations (TRUJILLO-CENÓZ and MELAMED, 1966a; MELAMED and TRUJILLO-CENÓZ, 1968) have demonstrated that the central rhabdom actually consists of two unequal rhabdomeres. There is a long superior rhabdomere originating from the *superior central cell* (cell No. 7 of DIETRICH) and an inferior shorter rhabdomere derived from a smaller *inferior central cell* (cell No. 8 of DIETRICH). Both rhabdomeres have a common longitudinal axis which practically coincides with the ommatidium axis. However, the microvilli which form the two rhabdomeres lie in mutually perpendicular planes (Fig. 20). The somata of the superior and inferior central cells are also oriented at right angles to one another. The anatomical arrangement of the rhabdomeres, together with the long, independent pathway of the axons suggests that the central may represent an intraommatidial polarized light analyzer. Recent experiments by KIRSCHFELD and REICHARDT (1970) tend to support this hypothesis. It seems probable that only the superior and inferior central cells can transduce and convey polarized light information to mediate an optomotor response. Since the six common photoreceptors are also sensitive to polarized light, but cannot mediate a reaction to this kind of illumination (KIRSCHFELD and REICHARDT, 1970) it is possible that such information would be lost in the optic cartridges (MELAMED and TRUJILLO-CENÓZ, 1968; KIRSCHFELD, 1971; STRAUS-FELD, 1971a).

A variation of the orthogonal arrangement of microvilli of the central cells, occurs in dolichopodid flies (TRUJILLO-CENÓZ and BERNARD, 1972), In this small insect half of the ommatidia (Type B) show the orthogonal pattern described in *Lucilia*, *Calliphora* and *Musca* while in the other half (Type A) the rhabdomeric microvilli of the superior and inferior central cells are oriented parallel to one another. A regular row by row distribution of the two kinds of ommatidia was found in the eye.

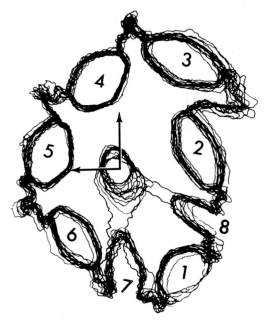

Fig. 20. This diagram was made photographically superposing the outlines of thirty serial electronmicrographs. It shows the position of the two central cells within the ommatidium. The arrows indicate the direction of the microvilli in the superior (vertical arrow) and inferior central cells (horizontal arrow). 1—6, rhabdomeres of the six common photoreceptors; 7, 8, superior and inferior central cells. (*Lucilia*). From MELAMED and TRUJILLO-CENÓZ, J. Ultrastructure Research **21**, 313—334 (1968). Academic Press Inc.

Rhabdomeres are dichroic, with maximal absorption of linearly polarized light when its electric vector is parallel to the axes of the rhabdomeric microvilli (WATERMAN et al., 1969). Therefore, the vertical orientation of the type A central cells is appropriate for diminishing absorption of the horizontally polarized light reflected from water or leaf surfaces. Such cells would be relatively blind to surface glare and thereby may improve prey detection through the water surface or against glossy leaves, grass or mud. A similar structural organization has been described by SCHNEIDER and LANGER (1969) in the water strider *Gerris lacustris*. In this aquatic hemiptera, ommatidia of the ventral portion of the eye show in both central cells the rhabdomeric microvilli oriented in dorsoventral direction (the same exhibited by type A ommatidia in dolichopodid flies). Their structure is assumed to be differentiated to obtain a better view through the water. Measurements of the dichroic absorption in the seventh central rhabdom of *Musca* (KIRSCHFELD, 1969) may suggest a different functional interpretation. In KIRSCHFELD's experiments the seventh rhabdom exhibited the maximum absorption of polarized light, when the *e*-vector was oriented perpendicular to the main axes of the rhabdomeric microvilli (it appears as an atypical rhabdomere, since the remainder six ones show their maximum when the *e*-vector is parallel to the rhabdomeric microvilli). If this observation is confirmed and extended to other groups (dolichopodids for example) the type A ommatidium should be considered not partially blind, but particularly sensitive to surface glare.

HORRIDGE (1966) working on *Locusta* reported the existence of an eccentric cell which occupies a fixed angular position relative to the other common photoreceptors. Three small retinular cells have been described in the drone (PERRELET and BAUMANN, 1969). Two of them extend along the entire length of the ommatidium while the third one is shorter and confined to the proximal part of the retina. A small cell devoid of rhabdomere occurs in the ommatidia of the dragonfly *Aeschna* (EGUCHI, 1971). Nevertheless, such atypical photoreceptor also sends its axon to the lamina.

III. The Lamina Ganglionaris

The gross anatomy of the lamina shows some variations in the different species of insects. In flies and dragon-flies it appears as a thin layer (90—150 μ thick) closely apposed to the inner surface of the retina. Parallelly, its histological architecture is relatively simple. In other insects like bees, butterflies and locusts, the lamina is a thicker more complex region.

When stained with the common histological dyes the lamina of insects shows three main zones (Fig. 21): a) an external zone composed of trachea and bundles of photoreceptor axons, b) an intermediate zone characterized by the presence of the neuron somata and c) a neuropile zone consisting of thousands of cylindrical units — the optical cartridges. Each one of these concentric zones contains nervous and non-nervous components.

It is important to emphasize that the optical cartridges are distinct units in Diptera, Odonata and some Lepidoptera but in other groups they cannot be clearly defined.

Together with the typical nervous components, the lamina also contains glial cells which invest neurons and nerve fibers. A peculiar type of glial element "the epithelial cells" of CAJAL (1909) represents one of the most conspicuous elements of the dipteran lamina. These "epithelial cells" ensheath the optical cartridges, giving the lamina a typical multimodular aspect.

1. The Nervous Components of the Lamina

Useful information concerning the nerve elements of the lamina resulted from the works of VIGIER (1908), CAJAL (1909), CAJAL and SÁNCHEZ (1915). Using the Golgi technique they were able to recognize that the lamina is the site of synaptic

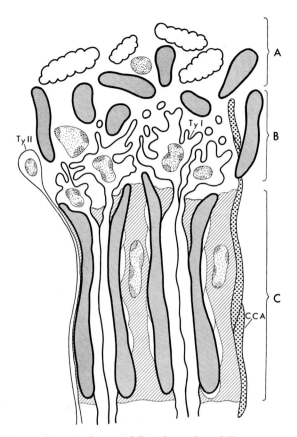

Fig. 21. The lamina ganglionaris of muscoid flies shows three different zones: an external zone (A) composed of trachea and bundles of photoreceptor axons, an intermediate zone (B) characterized by the presence of the neuron somata and a neuropile zone (C) consisting of thousands of cylindrical units, – the optical cartridges. There are two main types of neurons (Ty. I and Ty. II) and three main types of neuroglia cells. Long photoreceptor axons (CCA) can be seen travelling through the lamina and entering the optic nerve. These are the axons of the superior and inferior central cells.

connection between the photoreceptor axons and the second order neurons of the visual pathway. Moreover, they saw that in flies these contacts occur at the level of peculiar daisy-like units termed by CAJAL and SÁNCHEZ "optical cartridges". More recently, STRAUSFELD and BLEST (1970) and STRAUSFELD (1970–1971a) using the Golgi technique too, have added new valuable data about the histological organization of the lamina in Lepidoptera and Diptera.

To facilitate understanding of the basic nerve circuits of this region it is desirable to begin with the description of the optical cartridges.

a) The Optical Cartridge. Cross sections of the lamina of dipterans illustrate about the structural organization of the optical cartridges. As shown in Fig. 22 each cartridge basically consists of six peripheral photoreceptor axon terminals and two central second-order fibers. As described by CAJAL (1909) and VIGIER

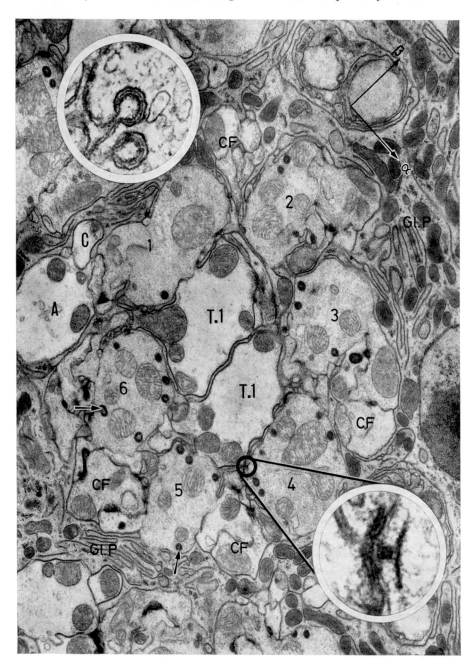

Fig. 22. Each optical cartridge consists of six photoreceptor axon endings (1—6) and two second-order fibers (T_1). These two second-order fibers establish synaptic connections with each one of the six photoreceptor endings. The synaptic points are characterized by the presence of T-shaped pre-synaptic ribbons. One of these ribbons is shown at higher magnification (approx. 70000 ×) in the inset at the right inferior angle of the picture. The plasma membrane of the photoreceptor axons shows, at cartridge level, small invaginations (arrows) which are occupied by the "capitate projections". The capitate projections are specialized glial processes. Two of

(1908) and confirmed in modern light and electron microscope studies (TRUJILLO-CENÓZ and MELAMED (1963), TRUJILLO-CENÓZ (1965a), STRAUSFELD (1970—1971a, b), BOSCHEK (1971) the second-order fibers give off numerous short collateral prolongations which penetrate the narrow spaces existing between adjacent photoreceptor endings. Three axons of smaller diameters can be seen associated to each optical cartridge (STRAUSFELD, 1971a). They arise from three distinct sub-types of intralaminar neurons (more details are given under the heading "The neurons of the lamina").

The typical pattern of organization — 2 central fibers + 6 peripheral photoreceptor endings — has been found so far in the non-equatorial regions of the eye of calliphorids, sarcophagids, muscids, syrphids and dolichopodids. Cartridges with eight and seven photoreceptor endings occur normally in the equatorial region of the lamina (HORRIDGE and MEINERTZHAGEN, 1970a; BOSCHEK, 1971). In other insect groups, the number and arrangement of the photoreceptor endings as well as that of the second-order fibers is different from the one described in Diptera Fig. 23 (NINOMIYA et al., 1969, VARELA, 1970).

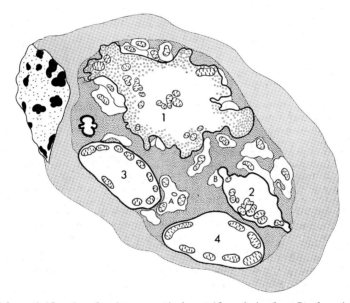

Fig. 23. Schematic drawing showing an optical cartridge of *Aeschna*. In these insects each cartridge consists of four large photoreceptor endings (1—4) and two thin second order fibers (A and B). The second-order fibers ramify profusely within the cartridge. A loose glial sheath derived from large neuroglia cells surrounds each cartridge (light dotted frame). Dense glial prolongations (heavy dotted frame) can be found at the center of the cartridges running between the nerve fibers

these processes are shown at higher magnification (60000 ×) in the inset at the superior left angle of the picture. Nerve elements (CF) considered to form part of some of the systems of centrifugal fibers described in the lamina can be seen in contact with the photoreceptor endings. The optical cartridges are ensheathed by cytoplasmic prolongations (Gl.P.) derived from large neuroglia cells. A and C, asymmetric and connecting cells. *Lucilia,* × 12000

The electronmicrographs usually show the two central fibers as clear, axon-like elements which contain large mitochondria. In some instances however, clusters of small vesicles can be observed near the plasma membrane. Another interesting feature revealed by the electron microscope is that the two second-order fibers maintain, in great part of their intra-laminar pathway a close apposition. At the cartridge, the plasma membranes are separated by a distance of 100 Å, so that the overall width of the two membranes and intervening cleft is approximately 250 Å. A regular periodicity of about 100 Å can be observed in the interfiber cleft (Tru-jillo-Cenóz, 1965a). Boschek (1971) also described septate junctions between other elements of the cartridge.

The most striking morphological characteristic of the photoreceptor axon terminals is the presence of numerous small invaginations of the plasma membrane which occur all along their course in the cartridges. These invaginations are oc-cupied by club-shaped processes. Pedler and Goodland (1965) considered pre-sumptively, these club-shaped processes as nerve elements establishing synaptic contacts with the photoreceptor axon terminals (button synapses). However, this point of view cannot be maintained, since serial section electronmicroscope studies (Trujillo-Cenóz, 1965a) demonstrated that these processes actually are specializ-ed glial prolongations (capitate projections).

Besides numerous mitochondria and pigment granules the photoreceptor terminals contain small pre-synaptic ribbons. They appear as T-shaped electron-dense structures, lying close to the photoreceptor axon membrane (Trujillo-Cenóz, 1965a; Boschek, 1971) (Fig. 22 inset). Commonly, synapses occur between the photoreceptor endings and the collateral branches of the second order fibers. However, direct contacts with the mother trunks are not rare. Very frequently one sees that two second-order fibers establish synaptic connections on the same patch of the photoreceptor axon membrane; in such cases the tips of the two post-synaptic fibers lie in close contact. Post-synaptic, plate-like structures have been found by Boschek (1971) in Musca. In gluta-osmium fixed material embedded in Araldite, the synaptic cleft measures approximately 150 Å. Microvesicles are scarce and when present they are concentrated near the synaptic loci.

The three dimensional reconstructions of short segments of the optical cartridges have shed some light on the spatial distribution of the synaptic loci. It has been possible to demonstrate by this method, that *each one of the two second-order fibers establishes several synaptic contacts with each one of the six photoreceptor axons* (Trujillo-Cenóz, 1965a). This fact is of particular importance when considering the functional significance of the optical cartridges.

A complex system of efferent fibers has been described in the lamina of several species of insects (Cajal and Sánchez, 1915; Strausfeld and Blest, 1970; Strausfeld, 1970—1971a). There are in this region: 1) short fibers originating from neurons lying in the optic nerve, 2) long fibers poorly ramified, and 3) long fibers which branch profusely and form basket-shaped arborizations at the peri-phery of the optical cartridge (the "nervous bags" of Cajal and Sánchez). These last two types originate from neurons located in deeper visual centers.

The electron microscope study of the optical cartridges has revealed a set of two fibers (α and β; Trujillo-Cenóz, 1965a) climbing on the surface of the photo-receptor axon endings (Figs. 22—24). These fibers frequently change place and it is

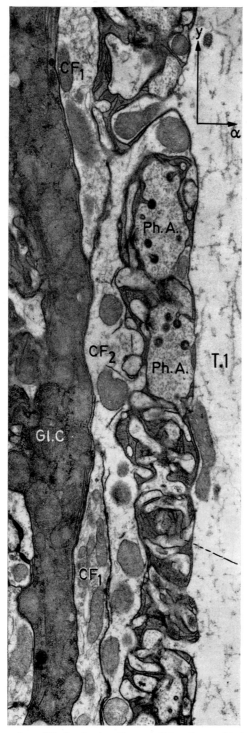

Fig. 24. Longitudinal section of the lamina (section parallel to the γ axis) showing two centrifugal fibers (CF_1 and CF_2) climbing on the surface of the photoreceptor axon endings (Ph.A). T_1, second-order fiber (Type I neuron axon) Gl.C. neuroglia cell. *Lucilia* (approx. $\times$ 30000)

common to see one of them running over the other. When followed in the series of sections one sees that these elements jump from one photoreceptor to another, usually to one of their two closest neighbors.

The electronmicrographs also show the occurrence of typical synaptic ribbons within these fibers thus suggesting a centrifugal flow of activity from deeper centers to the lamina. Combining the Golgi and the electronmicroscope techniques it has been possible to identify accurately (TRUJILLO-CENÓZ and MELAMED, 1970) one of these systems of climbing fibers with the "nervous bags" described by CAJAL and SÁNCHEZ (1915) in blow flies. Each nervous bag originates from a thin long nerve fiber which crosses the optic nerve or intermediate chiasm. This long fiber represents the peripheral division of a T-shaped trunk stemming from a small neuron soma located in the cell rind of the medulla. The peripheral division terminates around an optical cartridge forming a network of varicose branches. The short central division of the T-shaped trunks terminates by means of an oval or pyramidal mass close to the endings of the Type I monopolar cells (the central fibers of the cartridges) within the first synaptic stratum of the medulla. Electronmicroscope observations of intramedullar portions of the centrifugal fibers suggest that they may be the morphological substratum for a regulatory mechanism controlling the cartridge input by feeding back its output to the crown of photoreceptor axon endings. In his work on Musca, BOSCHEK (1971) described a similar kind of fibers but considering them as post-synaptic elements relative to the photoreceptor endings.

One point still remains obscure. It concerns the nature of the other "mimetic" fibers that always follow the ramifications of the nervous bag fibers. According to STRAUSFELD (1971 a) the mimetic fibers may represent a different kind of bag or "basket" (T_1 in his nomenclature) also originating from medullar neurons.

Simpler centrifugal fibers also occur in the lamina. They were observed by CAJAL and SÁNCHEZ (1915) but properly described by STRAUSFELD (1970—1971 a).

The so-called "short centrifugal fibers" of CAJAL, which originate from neurons located within the optic nerve have been also found by STRAUSFELD (1970—1971 a) and described as "amacrine cells". These neurons are particularly difficult to impregnate with the silver salts and therefore their synaptic connections are unknown. In addition to the vertical columnar organization represented by the optical cartridges there are two sites in which horizontal interconnections may occur. Just below the neuron somata there is a plexus consisting of extremely thin varicose fibers (TRUJILLO-CENÓZ, 1965 a). STRAUSFELD (1970—1971 a) has shown the existence of long thin fibers running at this level ("the tangential fibers") which probably are components of this external plexus. MELAMED (unpublished) has seen the branches of the midget neurons entering and synapsing with undetermined elements of this plexus. At the inner margin of the lamina there is a second plexus or network of horizontally running fibers. It was discovered and analyzed by STRAUSFELD and BRAITENBERG (1970) who demonstrated that it contains the lateral processes of the connecting neurons (L_4 or tripartite cell; further details will be given when dealing with the neurons of the lamina).

b) The Neurons of the Lamina. In his paper on the visual centers of insects CAJAL (CAJAL and SÁNCHEZ, 1915) described two main classes of neurons in the lamina: the "giant unipolar cells" and the "small unipolar cells". This last group contains, according to CAJAL, not less than three sub-types of neurons. CAJAL's work covered different insect groups (Diptera, Hymenoptera, Odonata and Orthoptera) and his analogies between neurons in such diversity of insects are sometimes difficult to accept. Moreover, the term "giant" introduced by CAJAL is not a fortunate one since the somata of the intra-laminar neurons rarely exceed 10 μ.

STRAUSFELD and BLEST (1970) preferred to restrict the term "giant" to those cells having lateral processes spreading through more than one group of photo-receptor endings. Conversely, the small unipolar cells are for them those neurons whose lateral processes are confined to one optical cartridge. Therefore, the "giant" neurons of *Calliphora* are for these authors "small" unipolar cells.

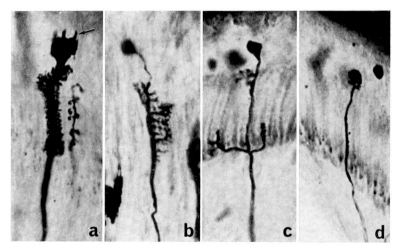

Fig. 25. The lamina of *Lucilia* contains two main types of neurons (Type I and Type II). Type I neurons (a) are characterized by the presence of dendritic like processes arising from the cell body (arrow), by a short neck or axon-hillock and by short numerous collateral branches which synapse with the six photoreceptor axon endings. The main prolongations of two Type I neurons occupy the center of each optical cartridge. Three subtypes can be recognized within the Type II neurons: the asymmetrical or brush cell (b), the connecting cell (c) and the midget cell (d). The synaptic connections of the Type II neurons are still partially known. Golgi technique, *Lucilia*, approx. × 600

Our light and electron microscope investigations made on muscoid flies, have confirmed the occurrence of two main classes of neurons in the lamina: Type I and Type II. Type I neurons have been identified as the giant unipolar cells of CAJAL and SÁNCHEZ. Within the second type (Type II) it is possible to recognize three subtypes which show dissimilar morphology and different anatomical connections.

α) *The Type I Neurons.* In muscoid flies the Type I neurons are characterized by the following morphological characteristics: 1) short, thick intercalar segment or "neck" connecting the soma with the main axonal trunk; 2) numerous short lateral processes which occur all along the intracartridge portion of the main trunk: 3) in *Sarcophaga*, *Lucilia* and *Calliphora* the soma have conspicuous dendritic-like processes. All these features can be observed in properly Golgi-stained specimens (Fig. 25a). However, in our opinion the most important characteristic which individualizes the Type I neurons concerns the position and connections of their main trunks in the optical cartridges. As clearly stated by CAJAL and SÁNCHEZ (1915) and confirmed by electronmicroscopy (TRUJILLO-CENÓZ, 1965a) the main prolongations of two of these neurons occupy the center of each optical cartridge.

As described when dealing with the optical cartridges the lateral processes arising from the main trunk of the Type I neurons establish synaptic connections with each one of the six photoreceptor axon endings forming the cartridge. Type I neurons correspond to neurons L_1 and L_2 of BRAITENBERG (1967) and BOSCHEK (1971) and to the so-called radial bistratified diffuse monopolar cell and bilateral diffuse monopolar cell of STRAUSFELD (1970—1971a). Light and electronmicroscope studies carried out in our laboratory do not support STRAUSFELD's discrimination of two kinds of Type I neurons. In spite of minor morphological differences it seems well proved that the two central fibers of the cartridges have, at laminar level, similar synaptic relations. As described previously (TRUJILLO-CENÓZ, 1969) and confirmed recently (STRAUSFELD, 1971a), the axons of the Type I neurons terminate within two different synaptic strata of the medulla.

β) The Type II Neurons. All of them exhibit the following common characteristics: a) Smooth type of soma (they never show dendritic-like prolongations) b) a slender neck connecting the soma with the main intra-laminar prolongation; c) the collateral processes are always less abundant than in the Type I neurons. In spite of these common features three distinct sub-types can be recognized in muscoid flies.

1) The Asymmetrical or Brush Cell. The occurrence of a third fiber running close or within the optical cartridges was mentioned first by CAJAL and SÁNCHEZ (1915) and more recently by BRAITENBERG (1967) (his L_3 fiber); STRAUSFELD and BRAITENBERG (1970), BOSCHEK (1971) and STRAUSFELD (1971a).

Studying under the electron microscope transverse series of sections cut from Golgi-stained eyes, MELAMED (unpublished) was able to correlate the Golgi and the electronmicroscope images. The L_3 fiber originates from a relatively small neuron soma lying close to the basement membrane. It has a long slender "neck" or "intercalar segment" which enters in the synaptic field of the lamina together with the prolongations of the Type I neurons. In the outer regions of the lamina the "third fiber" runs between the photoreceptor axons of the cartridges but at deeper regions it is found outside the cartridges. In *Lucilia* (Fig. 25b) a variable number of collateral processes (not more than 20) spread out from the main trunk and penetrate between the photoreceptor axons and the central fibers (Type I main trunks) of the cartridge. It is important to note that these processes arise only from one side of the main trunk giving this neuron its typical "asymmetrical" appearance.

2) The Connecting Cell. This interesting type of laminar neuron was discovered by STRAUSFELD and BRAITENBERG (1970) in *Musca* and confirmed in Calliphorinae and Syrphidae by STRAUSFELD (1971a). It has been also found by MELAMED (unpublished) in *Lucilia* (Fig. 25c) and by TRUJILLO-CENÓZ (unpublished) in dolichopodid flies. The neuron somata such as the ones of the asymmetrical cells, occupy an external position near the basement membrane. Two types of processes arise from the main trunk of the connecting cell: a few short collateral branches at the outer margin of the synaptic field of the lamina and three or four longer collaterals which contribute to compose a network of fibers at the inner margin of the lamina. These inner branches interconnect three or four optical cartridges. The connecting neuron was christened "tripartite cell" by STRAUSFELD (1971a). However, taking into account the variability in the number of branches it seems advisable to use a noncommittal name as the one proposed here.

3) The Midget Cell. The lamina contains a third class of Type II neurons whose somata are located near the outer limit of the external plexiform layer. Considering that these are the smallest cells of the lamina, the term midget coined by STRAUSFELD (1970—1971a) seems to be appropriate for describing them. The main prolongation of the midget cell shows only few collateral processes at the beginning of its pathway through the lamina (Fig. 25d). Studies by MELAMED (unpublished) suggest that the midget cell collaterals form part of the thin plexus occurring just below the cell bodies layer. However, STRAUSFELD (1971a) described the collateral branches penetrating the cartridge between photoreceptors 4 and 5. The main prolongation runs outside the cartridge.

Investigations made in our laboratory (MELAMED and TRUJILLO-CENÓZ, unpublished) have allowed a precise identification of the different neuronal fibers in cross sections of the lamina (Fig. 26).

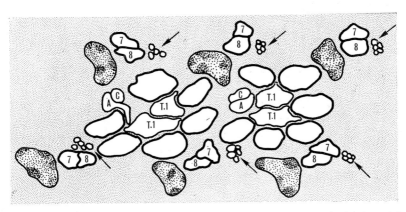

Fig. 26. This schematic drawing based upon electronmicrographs from the lamina of *Lucilia* complements the information provided by the precedent figure. It shows the position of the axons originating from the four different types of laminar neurons. T_1, Type I neurons; A, asymmetric cell; C. connecting cell; 7—8, axons of the superior and inferior central cells. The arrows indicate the groups of fibers containing the axons of the midget cells

c) The Retina-Lamina Projection. The analysis of the mechanisms involved in motion perception (HASSENSTEIN, 1951; REICHARDT, 1962; McCANN and MacGINITIE, 1965; THORSON, 1964—1966a, b) have increased the interest of the studies dealing with possible anatomical sites for interommatidial integration. In spite of the studies by HORRIDGE (1968) and SHAW (1968) on *Schistocerca* and that of HORRIDGE and MEINERTZHAGEN (1970b) on *Schistocerca*, *Apis*, *Notonecta* and

Aeschna detailed knowledge on this subject is still practically limited to investigations made on dipterans (VIGIER, 1908; CAJAL and SÁNCHEZ, 1915; TRUJILLO-CENÓZ and MELAMED, 1966 a—b; BRAITENBERG, 1966—1967; KIRSCHFELD, 1967; MELAMED and TRUJILLO-CENÓZ, 1968; HORRIDGE and MEINERTZHAGEN, 1970a; STRAUSFELD, 1971).

As stated in preceding pages the ommatidium of dipterans contains six peripheral or common photoreceptors and two central ones termed the superior and inferior central cells.

Serial section studies have demonstrated that after crossing the basement membrane photoreceptor axons belonging to different but neighboring ommatidia join to form common thick bundles. However, the ommatidial units reappear at a deeper level represented by the so-called *pseudo-cartridges* (TRUJILLO-CENÓZ and

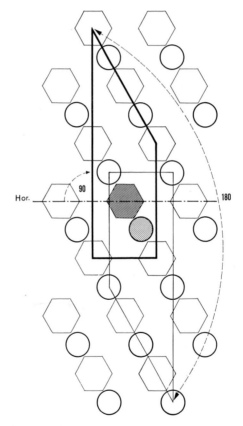

Fig. 27. This drawing shows the projection of an ommatidium (hatched hexagon) upon the layer of optical cartridges (open circles joined by light lines) and also the projection of an optical cartridge (dotted circle) upon the ommatidial field (open hexagons joined by heavy lines). The geometrical figure resulting from such projection is a trapezoid whose longest dimension forms an angle of 90° with a horizontal row of ommatidia. The pattern of projection of the ommatidium is rotated 180° from that of the cartridge. This drawing also shows that the arrays of ommatidia and optical cartridges are not coincident. [From TRUJILLO-CENÓZ and MELAMED, J. Ultrastructure Research **16**, 395—398 (1966). Academic Press Inc.]

MELAMED, 1966a). In each pseudo-cartridge the axons of the six peripheral photo-receptors form a crown which surrounds the two axons of the superior and inferior central cells. The topographical distribution of the pseudo-cartridges in the lamina is exactly the same as that of the ommatidia in the retina. A new rearrangement of fibers occurs near the outer margin of the neuropile zone: the peripheral photoreceptor axons crisscross in such a way that those stemming from one ommatidium participate in the formation of six optical cartridges. These are distributed according to a trapezoidal pattern, which is identical to that exhibited by the ommatidia included in the "pseudopupil". The first report of the asymmetrical projection of one ommatidium upon six optical cartrigdes was made by TRU-JILLO-CENÓZ and MELAMED in 1965 (Symp. Wennergren Center, listed in the references as TRUJILLO-CENÓZ and MELAMED 1966a). Such report was followed by a paper of BRAITENBERG (1966) in which he described, using reduced silver, the same asymmetrical projection. He also attempted to correlate the projection patterns with the gross anatomy of the eye. This correlation however, was not correct since the geometrical figures resulting from the projection patterns appear oriented at 30 degrees in respect to the horizontal axis of the eye (x axis in his former axes of reference). A second more complete electron microscope study made in our laboratory established the proper correlation between the projection patterns and the main axes of the eye (TRUJILLO-CENÓZ and MELAMED, 1966b). As shown in Fig. 27 the longest dimension of the trapezoid resulting from the projection of one ommatidium upon the lamina is perpendicular to the horizontal axis of the eye. As expected the pattern of projection of one cartridge upon the ommatidia is identical to the pattern of projection of one ommatidium upon the cartridges but it is rotated 180°. This study also revealed that the six photoreceptor axons forming one optical cartridge originate from photoreceptor cells occupying sequential angular positions in the six ommatidia (Fig. 28). These findings were confirmed later by BRAITENBERG (1967).

The functional significance of this peculiar neural relationship between ommatidia and optical cartridges depends critically on the direction in space of the optical axes of the photoreceptor cells. DE VRIES (1956) studying fresh slices of fly eyes, saw that adjacent photoreceptors in one ommatidium have different visual fields. Further studies by KUIPER (1962) showed that in blow-flies the optical axes of the rhabdomeres of a single ommatidium diverge by several degrees. We owe to KIRSCHFELD (1967) a complete understanding of the optical and anatomical features of these eyes. His study demonstrated that each optical cartridge gathers together the axons of six photoreceptors looking at the same point in space. This means that the optical aperture for such a point is six facets rather than one and allows to consider the fly eye as a "neural superposition eye" (KIRSCHFELD, 1967 — 1971).

An important conclusion derived from these studies is that the so-called outer chiasm (the zone containing the crisscrossing axons) cannot provide the anatomical basis for motion-perceiving interactions. Finally it must be emphasized that the axons of the two central cells bypass the lamina and end directly in the medulla (MELAMED and TRUJILLO-CENÓZ, 1968). The patterns of connections of the photoreceptors and the laminar neurons were also studied by HORRIDGE and MEINERTZHAGEN (1970a) in *Calliphora*. They found the same asymmetrical projection pattern described by TRUJILLO-CENÓZ and MELAMED (1966b) and by BRAITENBERG (1967). Moreover they described the occurrence of optical cartridges with more than six photoreceptor axon endings. Cartridges with seven or eight photoreceptor endings can be found at the equatorial region of the eye. Four rows of lamina cartridges

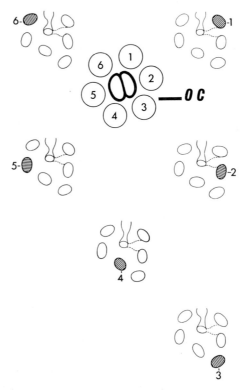

Fig. 28. Schematic representation of the projection of one optical cartridge (OC) over the ommatidial field. The cartridge gathers together axons 1—6, originating from photoreceptor cells occupying sequential angular positions, 1—6. [From MELAMED and TRUJILLO-CENÓZ, J. Ultrastructure Research **21**, 313—334 (1968). Academic Press Inc.]

contain eight photoreceptor endings and two other rows flanking them, contain seven. Even in these atypical optical cartridges the photoreceptor endings terminate in well-defined places. Such findings are in full agreement with the "atypical pseudo-pupils" described by KIRSCHFELD (1967) in the equatorial region of the eye. A complication is introduced at this region because the dorsal half of the retina has an ommatidial pattern that is the mirror image of that appearing in the ventral half (DIETRICH, 1909).

A different problem is posed by the eyes composed of ommatidia having the rhabdomeres fused in a single central rhabdom. In *Schistocerca* for example, the photoreceptor cells have the same visual axis (SHAW, 1967). Parallelly, Golgi studies have revealed that at least four photoreceptor axons from each ommatidium converge on the same optical cartridge (HORRIDGE, 1968b).

A similar situation seems to occur in the eye of the dragon-fly *Aeschna* (HOR-RIDGE and MEINERTZHAGEN, 1970 b). Our preliminary studies (TRUJILLO-CENÓZ and MELAMED, unpublished) on *Aeschna* have failed to depict any crossing of the photoreceptor axons. However, the electronmicroscope investigations by NINOMIYA et al. (1969) on damsel flies suggest the occurrence of crisscrossing of axons between neighbor ommatidia.

2. The Non-Nervous Components of the Lamina

a) The Neuroglia Cells. In the lamina of dipterans there are three types of neuroglia cells. At the level of the neuron somata there are numerous satellite glial cells which ensheath the cell bodies and the dendrite-like prolongations of Type I neurons. The cytoplasm of these cells shows scarce profiles of agranular endoplasmic reticulum. Tubular structures of undefined length and diameter ranging from 200—400 Å are also present in the cytoplasm. The nuclei of the satellite glial cells lie near the small groups of neurons scattered in the intermediate zone of the lamina. Desmosomes can be seen at the sites of glial prolongation contacts.

A row of elongated nuclei is found in the middle of the neuropile zone when longitudinal sections of the lamina are stained with the common aniline dyes. These nuclei belong to the so-called "epithelial cells" (CAJAL, 1909) which extend from the inner margin of the lamina to the upper limit of the neuropile zone. These epithelial cells of CAJAL actually are a special type of neuroglia cell, similar to the Müller cells found in the retina of vertebrates. BOSCHEK (1971) found, working on *Musca*, a precise pattern of organization of the epithelial cells throughout the optic cartridge zone. Each epithelial cell lies in contact with three optic cartridges. Conversely, each optic cartridge is surrounded by three epithelial cells.

Thin processes derived from the main prolongations of the epithelial cells penetrate between the photoreceptor axons and the second-order fibers. These processes were designated "capitate projections" since they end by means of club-shaped masses within small invaginations of the photoreceptor axon membrane. Each of these terminal masses has a diameter of approximately 1000 Å and is separated from the photoreceptor axon membrane by a constant distance of 250 Å. Interposed between the glial and nerve membrane there is an amorphous substance 100 Å thick which appears in the electronmicrographs as a third membrane (Fig. 22, superior inset). The cytoplasmic stalk connecting the terminal mass with its mother trunk has a constant diameter of about 500 Å. It shows, however, wide variations in length. The presence of capitate projections has to be considered in the lamina an exclusive and characteristic feature of the photoreceptor axon terminals [they also occur in the terminal portion of the ocellar photoreceptors, TRUJILLO-CENÓZ and MELAMED (unpublished), and according to CAMPOS-ORTEGA and STRAUSFELD (1972) in the intramedullar terminals of the central cells]. These structures are absent both from the axons of the superior and inferior central cells and from the other nerve elements of the cartridges. The epithelial cells end close to the optic nerve by means of a rounded foot which envelope the most proximal segment of the photoreceptor axon endings.

Just at the limit between the lamina and the optic nerve there is a layer composed of small glial cells aligned in a single row. These are the marginal glial cells

(Trujillo-Cenóz, 1965a) which ensheath the second order fibers as soon as they leave the optic cartridges. The cytoplasmic prolongations of these cells interlock forming a continuous layer perforated by the nerve-fibers which enter or leave the lamina.

b) The Tracheoblasts. The tracheal tubes are always covered by a layer of cytoplasm derived from large cells known as tracheoblasts. In some areas of the lamina, such as the tracheal or fenestrated zone, they are particularly abundant ensheathing the nerve elements. Tubular structures 200 Å wide occur in the cytoplasm of these cells.

IV. The Intermediate Chiasm or Optic Nerve

The intermediate chiasm or optic nerve is a compact mass of fibers connecting the lamina to the medulla. Its main constituents are first- and second-order fibers which decussate in the chiasm before their termination in the medulla. It is known after the investigations of Cajal (1909) that the optic nerve also contains centrifugal fibers which, as described in precedent pages, terminate in the lamina. A precise anatomical knowledge of this region is essential for understanding the projection of the lamina upon the medulla. Some basic points are now clear:

1) Decussation of fibers only occurs between the antero-medial and postero-lateral regions of the lamina. Therefore, there is no dorso-ventral intercrossing of

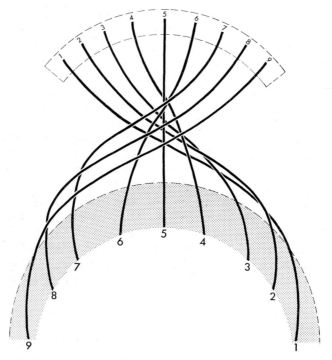

Fig. 29. Schematic drawing showing the pathway of the fibers in the intermediate chiasm or optic nerve. Dotted area: first synaptic stratum of the medulla. (Based on Sánchez y Sánchez's studies on the retina and optic centers of *Pieris*)

fibers (CAJAL and SÁNCHEZ, 1915; SÁNCHEZ Y SÁNCHEZ, 1919; LARSEN, 1966; TRUJILLO-CENÓZ, 1969; STRAUSFELD, 1971 b).

2) The two second-order fibers arising from each optical cartridge (axons of the Type I neurons) do not separate at chiasmatic level but run together up to their termination in the synaptic strata of the medulla (TRUJILLO-CENÓZ, 1969).

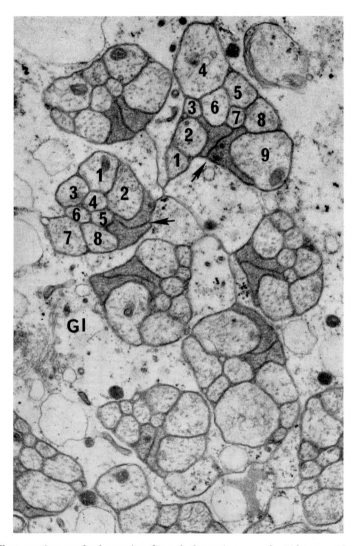

Fig. 30. Electronmicrograph of a section through the optic nerve of a 72 h. pupa *(Paralucilia)*. During development the optic nerve has a loose modular structure contrasting with the compact organization of the same region at adult stage. Fiber bundles containing 10 or 11 axons can be seen separated by layers of glial cytoplasm (Gl.). Two dense axons (arrows) are easily recognized in each fiber group; they represent the centripetal prolongations of the superior and inferior central cells. Five of the clear fibers arise from intralaminar neurons whereas the three or four remainder ones originate from neurons located in deeper centers ("centrifugal fibers"). Approx. × 20000

3) As shown in Fig. 29 crisscrossing of fibers occurs in such a way that a horizontal row of fibers numbered 1—9 (from left to right) will be projected upon the medulla in a horizontal row of nerve endings numbered 9—1 (from left to right) (SÁNCHEZ Y SÁNCHEZ, 1919; STRAUSFELD, 1971b); the uncrossed fibers described by CAJAL and SÁNCHEZ (1915) and by TRUJILLO-CENÓZ (1969) have not been found by STRAUSFELD (1971b).

4) The centrifugal fibers for a determined cartridge run together with the second-order and central cells fibers leaving such a cartridge; all of them arrive at the same area of the medulla (STRAUSFELD, 1971 b).

When studying the intermediate chiasm of adult flies, only the serial section technique allows the accurate identification of the different types of fibers leaving and entering the lamina. A clearer schematic situation occurs at pupal stage, since during development the optic nerve has a loose modular organization. Electron-microscope studies at this stage (TRUJILLO-CENÓZ and MELAMED, unpublished) show discrete nerve bundles containing 10 or 11 fibers (Fig. 30). This observation corroborates previous reports (TRUJILLO-CENÓZ, 1969; TRUJILLO-CENÓZ and ME-LAMED, 1970; STRAUSFELD, 1971b) describing fiber bundles with ten or more elements in the outer portion of the medulla. Serial section studies of the lamina "anlage" have made possible the identification of five of these fibers as prolongations of intralaminar neurons. The two denser axons are the centripetal prolongations of the two central cells while the three or four remainder ones represent centrifugal fibers whose mother cell bodies lie in the medulla.

Three kinds of neurons can be found in the optic nerve. One type is represented by the neuron somata giving origin to the "short centrifugal fibers". These are the "amacrine cells" of STRAUSFELD (1971a). A second type corresponds to the cell bodies from which the tangential fibers arise (STRAUSFELD, 1970—1971a). There are finally, displaced third-order neurons. Such elements normally occur in the cell rind of the medulla but it seems that during the lamina migration some of them move out towards the optic nerve.

The electronmicrographs show that along their course in the intermediate chiasm the nerve fibers are covered by thin glial processes.

V. The Medulla

The medulla is a large optic center interposed between the lamina and the lobula. The convex external surface receives the optic nerve fibers whereas its internal concave surface sends out numerous bundles of thin fibers towards the lobula. The histological preparations stained with the common aniline dyes show two main portions in the medulla: a peripheral portion containing the neuron somata and an inner portion consisting of several strata of interwoven fibers (nine according to CAJAL and SÁNCHEZ, 1915). The stratification of the neuropile core was observed by VIALLANES (1887a, b) and CUCCATI (1888) but only after the studies of KENYON (1896), CAJAL (1909), ZAWARZIN (1914), STRAUSFELD and BLEST (1970) and STRAUSFELD (1970) we know something about the nerve connections occurring at this level. CAJAL and SÁNCHEZ (1915) stated that the medulla of insects can be homologated to the inner plexiform layer of the vertebrate retina. Like this layer it is the site of termination of the second-order neurons of the visual pathway. We know however, that at least in some dipterans, other two types of nerve fibers leave the lamina, decussate at chiasmatic level and terminate in the medulla. These are the axons of the two central cells of the ommatidia and the axons of the different kinds of Type II neurons. The Golgi studies have also shown three or four systems of horizontal fibers which run between different synaptic strata. The synaptic strata contain in addition, the arborization of several types of inter-neurons (the amacrine cells of CAJAL and SÁNCHEZ, 1915) and that of important contingents of nerve fibers originating from deeper nerve centers (the centrifugal fibers for the deep retina — CAJAL and SÁNCHEZ, 1915).

It is impossible at present to give a detailed, complete description of the different nerve components of the medulla since the electron microscope investigation of this region has just begun (TRUJILLO-CENÓZ, 1969; CAMPOS-ORTEGA and STRAUS-FELD, 1972). Considering that these studies are still limited to muscoid flies the information reported in the following lines concerns this group of insects.

1. Nervous Components of the Medulla

Before its penetration of the neuropile zone the optic nerve divides in numerous small bundles or columns which form several rows parallel to the horizontal axis of the medulla (Fig. 31). When those bundles are observed at a higher magnification one sees that each of them contains two thick fibers (the axons of two Type I neurons) and eight to ten thinner ones (TRUJILLO-CENÓZ, 1969; TRUJILLO-CENÓZ and MELAMED, 1970; STRAUSFELD, 1971 b) (Fig. 31 inset). At this superficial level each bundle is completely covered by a glial sheath which becomes less evident at inner levels of the first synaptic stratum.

Using the serial section technique the two thick fibers of six optic bundles were followed to their terminations in the neuropile zone. This method of study demonstrates that the thick fibers (T-1) undergo gradual morphological changes transforming them in typical synaptic endings (Fig. 32). Concurrently with their entrance in the first synaptic stratum the pair of fibers undergoes a clockwise rotation which will determine the final position of the synaptic terminals in relation to the horizontal axis of the medulla (TRUJILLO-CENÓZ, 1969). STRAUSFELD (1971 b) demonstrated in *Calliphora* that each pair of Type I fibers leaving a cartridge undergoes a total rotation of about 180°.

It is interesting to note that one of the fibers precedes the other in showing synaptic characteristics. However, this fiber will end at a deeper level. As observed in other areas of the insect nervous system (TRUJILLO-CENÓZ and MELAMED, 1962; SMITH, 1965) the pre-terminal and terminal segments of the fibers contain an increased number of mitochondria and numerous small vesicles. At this level, the fibers lose their smooth oval outlines and show irregular invaginations and folds. In the middle of the first synaptic stratum one of the thick fibers has completely lost its axonal characteristics appearing as a flattened nerve ending oriented perpendicular to the horizontal axis of the medulla. At an inner region one sees that it divides in two terminal lobulated masses (Fig. 32). The other fiber (the one showing first the synaptic structures) undergoes similar morphological changes but divides and ends at a deeper level. To obtain more data about the thick fibers one must recourse to Golgi specimens. CAJAL (1909) was the first to recognize that the second order fibers of *Calliphora* terminate in the first synaptic stratum of the medulla by means of big club-shaped masses. Furthermore, he saw that "in frontal sections the terminal mass appears narrower than in horizontal sections, thus revealing that

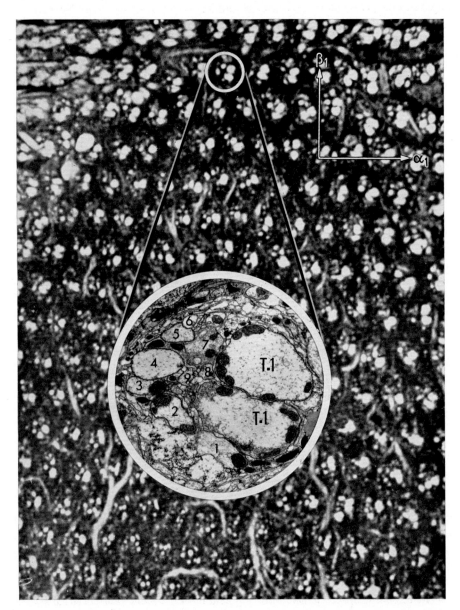

Fig. 31. Cross sections of the medulla (perpendicular to the γ_1 axis) show the optic nerve divided in numerous small bundles arranged in horizontal rows. When one of these bundles is observed at higher magnification under the electron microscope it appears composed of two thick fibers (T_1) and several thinner ones (usually 8 or 10). The two thick fibers correspond to the two second order fibers lying in the center of each optical cartridge. *Lucilia*. Photomicrograph, × 2000, inset approx. × 7000

such endings are more or less flattened in antero-posterior direction''. CAJAL's observations were confirmed by our own Golgi impregnations of the medulla in *Lucilia*. In this species the second-order fibers (the axons of the Type I neurons) terminate in large bi-lobulated masses similar to the ones found in *Tabanus* (CAJAL and SÁNCHEZ, 1915). Therefore, the electron microscope observations complemented by the Golgi studies have made possible the accurate identification of

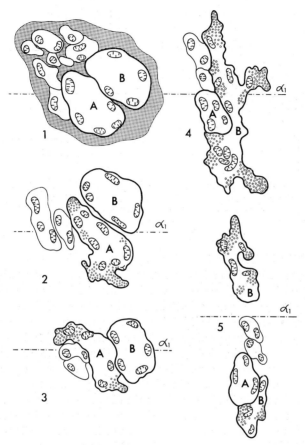

Fig. 32. When followed in long series of sections one sees that the two thick fibers (A and B, T_1 axons) undergo morphological changes transforming them into typical synaptic endings. This schematic drawing represents five sections (sect. No. 1—40—80—120—200) belonging to a series of more than 200 sections. Before termination each fiber divides in two lobulated masses.

the synaptic terminals of the Type I neurons. All these synaptic terminals are flattened in parallel planes perpendicular to the horizontal axis (α_1) of the medulla.

Our knowledge about the synaptic connections of the second-order endings is far from complete. The synaptic points can be distinguished by the existence of synaptic ribbons similar to the ones found in the lamina. However, more detailed studies made on these synaptic contacts have shown the three-dimensional aspect of the *T*-shaped synaptic ribbons (Fig. 33). Usually three or four thin post-synaptic fibers can be seen converging on the same synaptic locus. These post-synaptic fibers contain a dense osmiophilic substance concentrated near the plasma membrane (Trujillo-Cenóz, 1969). The synaptic cleft (250–300 Å wide) is occupied by a fine granular substance which appears heavily stained by the phosphotungstic acid (technique by Bloom and Aghajanian, 1966).

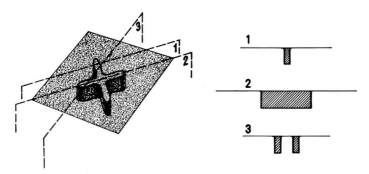

Fig. 33. Three dimensional representation of a pre-synaptic ribbon. Sections cut at different angles (1—2—3) show the pre-synaptic ribbons as having different aspects. [From Tujillo-Cenóz, J. Ultrastructure Research, **27**, 533—553 (1969). Academic Press Inc.]

In spite of the high resolution offered by the electron microscope and the new possibilities opened by the serial section technique it is particularly difficult to identify precisely the post-synaptic fibers. As shown in Golgi preparations, some of the post-synaptic elements are collateral branches of the third-order fibers. However, the electron microscopic identification of the third-order collaterals is handicapped by the presence of the so-called "amacrine cells" which differ from the third-order elements only by their site of termination. Even though our knowledge is still fragmentary there is a peculiar type of post synaptic fiber deserving a special description. Using the Golgi technique it is possible to stain a system of horizontal fibers running in the outer margin of the neuropile zone. As shown in Fig. 34 these fibers send out short vertical processes arranged like hooks in a fish-line. The electron microscope shows that these vertical branches establish synaptic connections with the second-order endings. However, these connections occur only with one member of each pair of second-order fibers. It is possible that these horizontal fibers may provide anatomical support for some of the motion detection systems found in the medulla-lobula region (Bishop et al., 1968).

As stated previously, each optic nerve bundle has in addition to the thick fibers, eight to ten smaller fibers which include the axons of the three sub-types of Type II neurons. Their terminations in the medulla have not been securely identified and therefore their synaptic relations are unclear yet. Something is known however about the pathway and termination of the axons of the superior and inferior central cells. By studying under the electron microscope large areas of the medulla (sections parallel to γ_1 axis) it was possible to discover a special type of nerve ending containing pigment granules. Taking into account that the granules of pigment normally occurring in the photoreceptor cells, concentrate in the terminal segments of the axons, these endings were identified as the synaptic terminals of the central cells (Melamed and Trujillo-Cenóz, 1968). Typical synaptic ribbons and vesicles have been found within these photoreceptor axon terminals. Combining different histological techniques, Campos-Ortega and Strausfeld (1972) have established

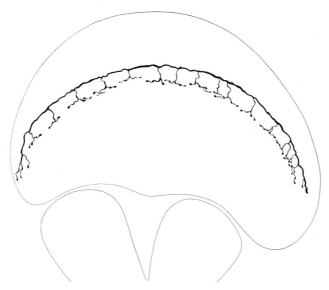

Fig. 34. Drawing of one of the fibers composing the first plexus of horizontal fibers (Golgi method). Note that the main trunk gives off vertical branches arranged like hooks in a fish-line. Parallel electron microscope studies have shown that these collateral branches establish synaptic connections with the second order endings (Type I neuron axon terminals).

accurately that the axon endings of the central cells occupy two different strata of the medulla. The superior central cell axons terminate beneath the deepest layer of the Type I neuron endings. The axons of the inferior central cells follow a shorter pathway and their endings are found just between the two layers of Type I neuron endings.

The intramedullar terminations of the centrifugal fibers forming the "nervous bags" in the lamina have also been described and studied combining the Golgi and the electronmicroscope techniques (Trujillo-Cenóz and Melamed, 1970). As described in precedent pages when dealing with the lamina, such nerve terminations lie in the first synaptic stratum of medulla. The thin collateral branches of these synaptic masses establish a close relationship with the synaptic endings of the Type I neurons.

2. Non-Nervous Components of the Medulla. — The Neuroglia Cells

At present only two defined types of neuroglia cells have been recognized in the medulla. One type is shown in Fig. 35. It is represented by large neuroglia cells which extend from the outer surface of the medulla to the inner limit of the cellular

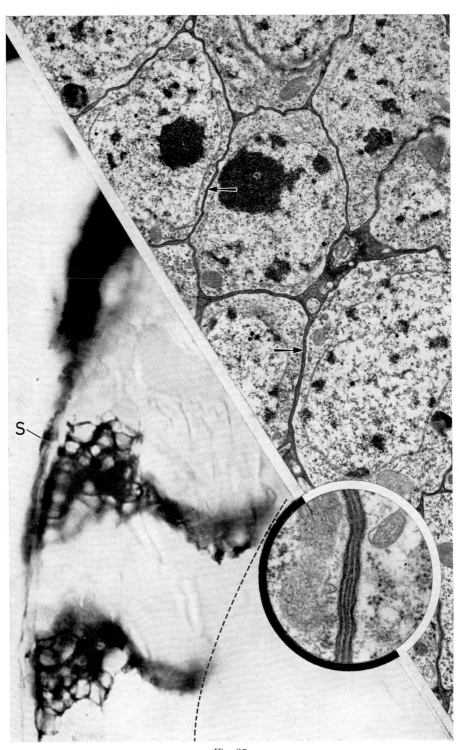

Fig. 35

cortex. When impregnated with silver chromate they show numerous laminar processes which invest the neuronal cell bodies. Complementary information about these cells can be obtained from the electron microscope studies. In gluta-osmium fixed material the neuroglia processes appear as dense cytoplasmic lamellae which cover completely the neuron somata.

The other type of neuroglia cell occurs in the outer limit of the neuropile zone. These cells have large eccentric nuclei which lie close to the optic nerve bundles. Their cytoplasm ensheath these bundles like typical oligodendrocytes.

VI. The Lobula or Optic Lobe

The lobula (HORRIDGE, 1965) or optic lobe (CAJAL and SÁNCHEZ, 1915) is the site of termination of the third order neurons. Its main histological characteristics were described by CAJAL and SÁNCHEZ (1915) who studied this optic center mainly in flies and bees. In Hymenoptera the lobula appears as a single neuropilic mass but in Diptera, Odonata and Lepidoptera it is divided in two or three distinct zones.

Horizontal sections of the lobula of muscoid flies shows two masses of interwoven fibers separated by a wedge-shaped region composed of numerous thin fibers. The anterior neuropile mass corresponds to the ovoid ganglion of CAJAL whereas the posterior narrower mass corresponds to the laminar ganglion of the same author (the laminar ganglion has been termed more recently lobular plate by HORRIDGE, 1965).

Little is known about the fine structure of the lobula. There are, however, some morphological details worth mentioning here. The great majority of the nerve fibers occurring in the lobula are extremely thin elements (below 0.5 μ). However, as shown in Fig. 36 the lobular plate contains several very large fibers which run just in the limit between the cellular cortex and the neuropile zone. Some of these fibers reach 15—20 μ in diameter. It is possible that this system of large fibers may correspond to the "posterior tract of fibers" mentioned by LARSEN (1966) in *Phormia*. When followed in serial sections one sees that they give origin to collateral branches which penetrate the neuropile establishing synaptic connections with undetermined number of fibers. As in other optic centers of the fly the synaptic points show typical *T*-shaped pre-synaptic ribbons.

A second interesting type of nerve element is represented by thick varicose fibers interconnecting at regular intervals the ovoid and laminar ganglia. These features suggest that they may correspond to the "colossal centrifugal fibers" described by CAJAL and SÁNCHEZ in *Tabanus*.

Fig. 35. The cortex of the medulla consists of thousands of small neurons (4—5 μ in diameter) which are covered by thin, dense glial processes (arrows). These processes originate from large neuroglia cells which also lie in the peripheral zone of the medulla. When impregnated with silver chromate (Cajal's rapid method) it is possible to observe that they extend from the external surface of the medulla (S) to the outer margin of the neuropile core (dotted line). Deposition of silver salts in the thinnest glial prolongations permits to see clearly, the outlimits of the small unstained neurons. The inset shows that in some areas two thin glial processes are interposed between adjacent neurons. Photomicrograph, × 800; low power electronmicrograph, × 7500; inset, × 30000 — *Lucilia*

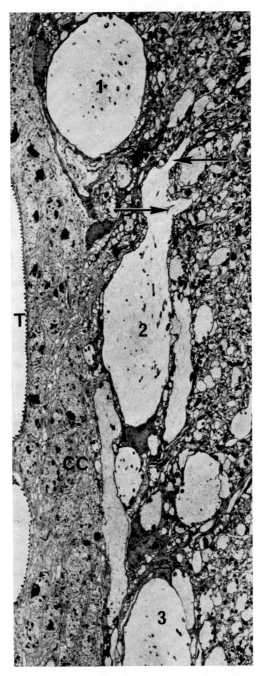

Fig. 36. Low power electronmicrograph showing some components of the system of giant fibers occurring in the lobular plate (1—2—3). Note that one of these giant fibers give off thinner branches which penetrate the neuropile zone (arrows). CC, cellular cortex of the lobula; T, trachea. Approx. ×2500, *Lucilia*.

Acknowledgements

The research work of the author was supported by Grants 618-64, 618-66, 618-67 from the Air Force Office of Scientific Research, Office of Aerospace Research, United States Air Force and by Grants NSO 8669-01-02-03 from the National Institutes of Health, U.S.A.

The author is indebted to Dr. J. R. SOTELO, Dr. J. W. RENFREW and Dr. J. MELAMED for reading the manuscript.

Acknowledgements are also due to Miss S. PAIS for technical assistance and to Misses M. D'ABBISOGNO and ELSA TRINKLE for typing the manuscript.

References

AUTRUM, H., WIEDEMANN, I.: Versuche über den Strahlengang im Insektenauge (Appositionsauge). Z. Naturforsch. **17**, 480—482 (1962).

BARRA, I. A.: Les photorécepteurs des collemboles, étude ultrastructurale. I. L'appareil dioptrique. Z. Zellforsch. **117**, 322—353 (1971).

BERNARD, G. D.: Evidence for visual function of corneal interference filters. J. Insect Physiol. **17**, 2287—2300 (1971).

— MILLER, W. H.: Interference filters in the cornea of Diptera. Invest. Ophth. **7**, 416—434 (1968).

— — What does antenna engineering have to do with insect eyes? IEEE Student Journal. Jan.-Feb., 1—8 (1970).

BERNHARD, C. G., GEMNE, G., SÄLLSTRÖM, J.: Comparative ultrastructure of corneal surface topography in insects with aspects on phylogenesis and function. Z. vergl. Physiol. **67**, 1—25 (1970).

— HÖGLUND, G., OTTOSON, D.: On the relation between pigment position and light sensitivity of the compound eye in different nocturnal insects. J. insect Physiol. **9**, 573—586 (1963).

— MILLER, W. H.: A corneal nipple pattern in insect compound eyes. Acta physiol. scand. **56**, 385—386 (1962).

— — MØLLER, A. R.: The insect corneal nipple array. Acta physiol. scand. **63**, suppl. 243, 1—79 (1965).

— OTTOSON, D.: Studies on the relation between the pigment migration and the sensitivity change during dark adaptation in diurnal and nocturnal Lepidoptera. J. gen. Physiol. **44**, 205—215 (1960).

— — Quantitative studies on pigment migration and light sensitivity in the compound eye at different light intensities. J. gen. Physiol. **47**, 465—478 (1964).

BISHOP, L. G., KEEHN, D. C., McCANN, G. D.: Motion detection by interneurons of optic lobes and brain of the flies Calliphora phaenicia and Musca comestica. J. Neurophysiol. **31**, 509—525 (1968).

BLACKSTAD, T. W.: Mapping of experimental axon degeneration by electronmicroscopy of Golgi preparations. Z. Zellforsch. **67**, 819—834 (1965).

BLOOM, F. E., AGHAJANIAN, C. K.: Cytochemistry of synapses: Selective staining for electron microscopy. Science **154**, 1575—1577 (1966).

BODESTEIN, D.: The postembryonic Development of Drosophila. In: DEMEREC, M. (Ed.): Biology of Drosophila. New York: Wiley 1950.

— Postembryonic Development. In: ROEDER, K. (Ed.): Insect Physiology. New York: Wiley 1953.

BOSCHEK, C.: On the fine structure of the peripheral retina and lamina ganglionaris of the fly, Musca domestica. Z. Zellforsch. **118**, 369—409 (1971).

BRAITENBERG, V.: Unsymmetrische Projektion der Retinulazellen auf die Lamina ganglionaris bei der Fliege Musca demestica. Z. vergl. Physiol. **52**, 212—214 (1966).

— Patterns of projection in the visual system of the fly. I. Retina-lamina projections. Exp. Brain Res. **3**, 271—290 (1967).

Bullock, T. H., Horridge, G. A.: Structure and Function in the Nervous Systems of Invertebrates. San Francisco: Freeman 1965.

Burton, P., Stockhammer, R.: Electron microscopic studies of the compound eye of the Toadbug, Gelastocoris oculatus. J. Morph. **127**, 233—258 (1969).

Cajal, S. R.: Nota sobre la estructura de la retina de la mosca (M. vomitoria L.). Trab. Lab. Invest. Biol. (Madrid) **7**, 217—257 (1909).

— Nota sobre la retina de los múscidos. Bol. R. Soc. Esp. Hist. Nat. **10**, 92—95 (1910).

— and Sánchez, D.: Contribución al conocimiento de los centros nerviosos de los insectos. Trab. Lab. Invest. Biol. Univ. Madrid. **13**, 1—164 (1915).

Campos-Ortega, J. A., Strausfeld, N. J.: The columnar Organization of the Second Synaptic Region of the Visual System of *Musca domestica* L. I. Receptor terminals in the Medulla. Z. Zellforsch. **124**, 561—585 (1972).

Cuccati, J.: Über die Organisation des Gehirns der Somomya erithrocephala. Z. wiss. Zool. **46**, 240—269 (1888).

de Vries, H. C.: Physical aspects of the sense organs. Progr. Biophys. **6**, 207—264 (1956).

Dietrich, W.: Die Facettenaugen der Dipteren. Z. wiss. Zool. **92**, 465—539 (1909).

Döving, K., Miller, W.: Function of insect compound eyes containing crystalline tracts. J. gen. Physiol. **54**, 250—267 (1969).

Eguchi, E.: Fine Structure and Spectral Sensitivities of Retinular cells in the Dorsal Sector of Compound Eyes in the Dragonfly Aeshna. Z. vergl. Physiologie **71**, 204—218 (1971).

— Naka, K., Kuwabara, M.: The development of the rhabdom and the appearance of the electrical response in the insect eye. J. gen. Physiol. **46**, 143—157 (1962).

— Waterman, T.: Changes in retinal fine structure induced in the crab Libinia by light and dark adaptation. Z. Zellforsch. **79**, 209—229 (1967).

Eichenbaum, D., Goldsmith, T. H.: Properties of intact photoreceptor cells lacking synapses. J. exp. Zool. **169**, 15—32 (1968).

Exner, S.: Die Physiologie der facettierten Augen von Krebsen und Insekten. Leipzig: Deuticke 1891.

Fernández-Morán, H.: Fine structure of the insect retinula as revealed by electron microscopy. Nature (Lond.) **177**, 742—743 (1956).

— Fine structure of the light receptors in the compound eyes of insects. Exp. Cell Res. Suppl. **5**, 586—644 (1958).

Fischer, A., Horstmann, G.: Der Feinbau des Auges der Mehlmotte, Ephestia kuehniella Zeller (Lepidoptera, Pyralididae). Z. Zellforsch. **116**, 275—304 (1971).

Gemne, G.: Ultrastructural ontogenesis of cornea and corneal Nipples in the compound eye of insects. Acta physiol. scand. **66**, 511—512 (1966).

Goldsmith, T. H.: The visual system of insects. In: The Physiology of Insecta, Vol. I. New York: Academic Press 1964.

— Philpott, D. E.: The microstructure of the compound eye of insects. J. biophys. biochem. Cytol. **3**, 429—440 (1957).

Grenacher, G. H.: Untersuchungen über das Sehorgan der Arthropoden insbesondere der Spinnen, Insecten und Crustacen. Göttingen: Vandenhoeck u. Ruprecht 1879.

Hassenstein, B.: Ommatidienraster und afferente Bewegungsintegration (Versuche an dem Rüssel-Käfer Chlorophanus). Z. vergl. Physiol. **33**, 301—326 (1951).

Hickson, S.: The eye and optic tract of insects. Quart. J. Microbiol. Sci. **25**, 215—251 (1885).

Höglund, G., Langer, H., Struwe, G., Thorell, B.: Spectral absorption by screening pigment granules in the compound eyes of a moth and a wasp. Z. vergl. Physiol. **67**, 238—242 (1970).

— Struwe, G.: Pigment migration and spectral sensitivity in the compound eye of moths. Z. vergl. Physiol. **67**, 229—237 (1970).

Horridge, G. A.: Arthropoda. Receptors for light and optic lobe. In: Bullock, T. H., Horridge, G. A. (Eds.): Structure and Function in the Nervous Systems of Invertebrates. San Francisco: Freeman 1965.

— The retina of the locust. In: The Functional Organization of the Compound Eye. Symp. Werner-Gren Center (1965). London: Pergamon 1966.

HORRIDGE, G. A.: Pigment movement and the crystalline threads of the firefly eye. Nature (Lond.) **218**, 778—779 (1968a).

— Affinity of neurons in regeneration. Nature (Lond.) **219**, 737—740 (1968b).

— Alternatives to superposition images in clear-zone eyes. Pro. roy. Soc. B **179**, 97—124 (1971).

— BARNARD, P. B. T.: Movement of palisade in locust retinula cells when illuminated. Quart. J. micr. Sci. **106**, 131—135 (1965).

— GIDDINGS, C.: Movement on dark-light adaptation in beetles eyes of the Neuropteran type. Proc. roy. Soc. B **179**, 73—85 (1971).

— MEINERTZHAGEN, I.: The accuracy of the patterns of connexions of the first- and second-order neurons of the visual system of Calliphora. Proc. roy. Soc. London B **175**, 69—82 (1970a).

— MEINERTZHAGEN, I. A.: The exact neural projection of the visual fields upon the first and second ganglia of the insect eye. Z. vergl. Physiol. **66**, 369—378 (1970b).

— WALCOTT, B., IOANNIDES, A. C.: The tiered retina of Dytiscus: a new type of compound eye. Proc. roy. Soc. B **175**, 83—94 (1970).

JOHANNSEN, O., BUTT, F.: Embryology of Insects and Myriapods. New York: McGraw-Hill 1941.

JOHNAS, W.: Das Facettenauge der Lepidopteren. Z. wiss. Zool. **97**, 218—261 (1911).

KANEKO, A.: Physiological and morphological identification of horizontal, bipolar, and amacrine cells in the goldfish retina. J. Physiol. (Lond.) **207**, 623—633 (1970).

KENYON, P. C.: The brain of the bee. J. comp. Neurol. **6**, 133—210 (1896).

KIRSCHFELD, K.: Die Projektion der optischen Umwelt auf das Raster der Rhabdomere im Komplexauge von Musca. Exp. Brain Res. **3**, 248—270 (1967).

— Absorption Properties of Photopigments in Single Rods, Cones and Rhabdomeres. In: REICHARDT, W. (Ed.): Processing of Optical Data by Organisms and by Machines. New York: Academic Press 1969.

— Aufnahme und Verarbeitung optischer Daten im Komplexauge von Insekten. Naturwissenschaften **58**, 201—209 (1971).

— FRANCESCHINI, W.: Ein Mechanismus zur Steuerung des Sichtflusses in den Rhabdomeren des Komplexauges von Musca. Kybernetik **6**, 13—22 (1969).

— REICHARDT, W.: Optomotorische Versuche an Musca mit linear polarisiertem Licht. Z. Naturforsch. B **25**, 228 (1970).

KUIPER, J. W.: The optics of the compound eye. Symp. Soc. exp. Biol. **16**, 58—71 (1962).

— On the image formation in a single ommatidium of the compound eye in Diptera. In: The Functional Organization of the Compound Eye. Symp. Wenner-Gren Center (1965). London: Pergamon 1966.

KUNZE, P.: Verhaltensphysiologische und optische Experimente zur Superpositionstheorie der Bildentstehung in Komplexaugen. Deutsch. Zool. 64. Tagung, 234—238. Stuttgart: Fischer 1970.

— HAUSEN, K.: Inhomogeneous refractive index in the crystalline cone of a moth eye. Nature (Lond.) **231**, 392—393 (1971).

LANGER, H., THORELL, B.: Microspectrophotometric assay of visual pigments in single rhabdomeres of the insect eye. In: Functional Organization of the Compound Eye. Symp. Wenner-Gren Center (1965). London: Pergamon 1966.

LARSEN, J. R.: The relationship of the optic fibers to the compound eye and centers of integration in the blow fly Phormia regina. In: The Functional Organization of the Compound Eye. Symp. Wenner-Gren Center (1965). London: Pergamon 1966.

LASANSKY, A.: Ultrastructural features of and [sic] extracellular ionic pathways in the Limulus ommatidium. Neurosci. Res. Prog. Bull. 8, 467—469 (1970).

LINZEN, B.: Zur Biochemie der Ommochrome. Naturwissenschaften **54**, 259—267 (1967).

LÜDTKE, H.: Retinomotorik und Adaptationsvorgänge im Auge des Rückenschwimmers (Notonecta glauca, L.). Z. vergl. Physiol. **35**, 129—152 (1953).

MCCANN, G. D., MACGINITIE, G. F.: Optomotor response studies of insect vision. Proc. roy. Soc. B **163**, 369—401 (1965).

Melamed, J., Trujillo-Cenóz, O.: The fine structure of visual system of Lycosa (Araneae: Lycosidae). Part I. Retina and optic nerve. Z. Zellforsch. **74**, 12—31 (1966).
— — The fine structure of the central cells in the ommatidia of Dipterans. J. Ultrastruct. Res. **21**, 313—334 (1968).
Miller, W. H.: Morphology of the compound eye of Limulus. J. biophys. biochem. Cytol. **3**, 421—428 (1957).
— Bernard, G.-D.: Butterfly glow. J. Ultrastruct. Res. **24**, 286—294 (1968).
— — Allen, J. L.: The optics of insects compound eyes. Science **162**, 760—767 (1968).
— Møller, A. R., Bernhard, G.-D.: The corneal nipple array. In: The functional organization of the compound eye. Symp. Wenner-Gren Center (1965). London: Pergamon 1966.
Mote, M., Goldsmith, T.: Compound Eyes: Localization of two color Receptors in the same Ommatidium. Science **171**, 1254—1255 (1971).
Ninomiya, W., Tominaga, Y., Kuwabara, M.: The fine structure of the compound eye of a Damsel-fly. Z. Zellforsch. **98**, 17—32 (1969).
Pedler, C., Goodland, H.: The compound eye and first optic ganglion of the fly. J. roy. micr. Soc. **84**, 161—179 (1965).
Perrelet, A., Baumann, F.: Presence of three small retinula cells in the ommatidium of the honey-bee drone eye. J. Micr. **8**, 497—502 (1969).
Reichardt, W.: Nervous integration in the facet eye. Biophys. J. **2**, 121—143 (1962).
Sánchez y Sánchez, D.: Sobre el desarrollo de los elementos nerviosos en la retina del Pieris brassicae, L. Trab. Lab. Invest. Biol. Univ. Madrid **17**, 1—63 (1919).
Schneider, L., Langer, H.: Die Struktur des Rhabdoms im Doppelauge des Wasserläufers Gerris lacustris. Z. Zellforsch. **99**, 538—559 (1969).
Seitz, G.: Der Strahlengang im Appositionsauge von Calliphora erythrocephala (Meig). Z. vergl. Physiol. **59**, 205—231 (1968).
— Untersuchungen am dioptrischen Apparat des Leuchtkäferauges. Z. vergl. Physiol. **62**, 61—74 (1969).
— Nachweis einer Pupillenreaktion im Auge der Schmeißfliege. Z. vergl. Physiol. **69**, 169—185 (1970).
Shaw, S. R.: Organization of the Locust retina. Symp. Zool. Soc. Lond. **23**, 135—163 (1968).
— Simultaneous recording from two cells in the locust retina. Z. vergl. Physiol. **55**, 183—194 (1967).
Sjöstrand, F. S.: Ultrastructure of retinal rod synapses of the guinea pig as revealed by three-dimensional reconstruction from serial sections. J. Ultrastruct. Res. **2**, 122—170 (1958).
Smith, D. S.: Synapses in the insect nervous system. In: The Physiology of the Insect Central Nervous System. New York: Academic Press 1965.
Snodgrass, R.: The morphology of insect sense organs and the sensory nervous system. Smithsonian Miscell. Collect. **77**, 1—80 (1926).
Sotelo, J. R., Porter, K.: An electron microscope study of the rat ovum. J. biophys. biochem. Cytol. **5**, 327—342 (1959).
Strausfeld, N. J.: Golgi stuides on insects. Part II. The optic lobes of Diptera. Phil. Trans. roy. Soc. London B **258**, 135—223 (1970).
— The organization of the insect visual system (light microscopy) I. Projections and arrangements of neurons in the lamina ganglionaris. Z. Zellforsch. **121**, 377—441 (1971 a).
— The organization of the insect visual system (light microscopy). II. The projection of fibres across the first optic chiasma. Z. Zellforsch. **121**, 442—454 (1971 b).
— Blest, A.: Golgi studies on insects. Part. I. The optic lobes of Lepidoptera. Phil. Trans. roy. Soc. Lond. B **258**, 81—134 (1970).
— Braitenberg, V.: The compound eye of the fly (Musca domestica): connections between the cartridges of the lamina ganglionaris. Z. vergl. Physiol. **70**, 95—104 (1970).
Stretton, A. O., Kravitz, E. A.: Neuronal geometry: determination with a technique of intracellular dye injection. Science **162**, 132—134 (1968).

Thorson, I.: Dynamics of motion perception in the desert locust. Science **145**, 69—71 (1964).

— Small-signal analysis of a visual reflex in the locust. I. Input parameters. Kybernetik **3**, 41—53 (1966a).

— Small-signal analysis of a visual reflex in the locust. II. Frequency dependence. Kybernetik **3**, 53—66 (1966b).

Trujillo-Cenóz, O.: Some aspects of the structural organization of the intermediate retina of Dipterans. J. Ultrastruct. Res. **13**, 1—33 (1965a).

— Some aspects of the structural organization of the arthropod eye. Cold Spr. Harb. Symp. quant. Biol. **30**, 371—382 (1965b).

— Some aspects of the structural organization of the medulla in muscoid flies. J. Ultrastruct. Res. **27**, 533—553 (1969).

— Bernard, G.-D.: Some aspects of the retinal organization of Sympycnus lineatus Loew (Diptera, Dolichopodidae). J. Ultrastruct. Res. **38**, 149—160 (1972).

— Melamed, J.: Electron microscope observations on the calyces of the insect brain. J. Ultrastruct. Res. **7**, 389—398 (1962).

— — On the fine structure of the photoreceptor — second optical neuron synapse in the insect retina. Z. Zellforsch. **59**, 71—77 (1963).

— — Electron microscope observations on the peripheral and intermediate retinas of Dipterans. In: The Functional Organization of the Compound Eye. Symp. Wenner-Gren Center (1965). London: Pergamon Press 1966a.

— — Compound eye of Dipterans: Anatomical basis for integration. An electron microscope study. J. Ultrastruct. Res. **16**, 395—398 (1966b).

— — Light and electronmicroscope study of one of the systems of centrifugal fibers found in the lamina of muscoid flies. Z. Zellforsch. **110**, 336—349 (1970).

— — Spatial distribution of photoreceptor cells in the ommatidia of Periplaneta americana. J. Ultrastruct. Res. **34**, 397—400 (1971).

Varela, F.: The fine structure of the visual system of the honeybee II. The lamina. J. Ultrastruct. Res. **31**, 178—194 (1970).

— The Vertebrate and the Insect Compound Eye in Evolutionary perspective. Vision Res. Supplement **3**, 201—209 (1971).

— Porter, K.: Fine structure of the visual system of the honeybee (Apis mellifera). I. The retina. J. Ultrastruct. Res. **29**, 236—259 (1969).

— Wiitanen, W.: The optics of the compound eye of the honey bee (Apis mellifera). J. gen. Physiol. **55**, 336—358 (1970).

Viallanes, H.: Études histologiques et organologiques sur les centres nerveux et les organes des sens des animaux articulés — quatrième mémoire — le cerveau de la guêpe (Vespa cabro et vespa vulgaris). Ann. Sci. Natur. 7e. Série. **2**, 5—100 (1887a).

— Études histologiques et organologiques sur les centres nerveux et les organes de sens des animaux articulés — cinquième mémoire — I. Le cerveau du criquet. Ann. Sci. Natur. 7e. Série. **4**, 1—120 (1887b).

— Sur quelques points de l'histoire du dévelopement embryonnaire de la Mante religieuse. Rev. Biol. Nord. France **2**, 479—488 (1890).

— Contribution à l'histologie du système nerveux des invertébrés. La lame ganglionnaire de la langouste. Ann. Sci. Natur. 7e. Série. **13**, 385—398 (1892).

Vigier, P.: Mécanisme de la synthèse des impressions lumineuses recueillis par les yeux composés des Diptères. C. R. Acad. Sci. (Paris) **145**, 122—124 (1907).

— Sur l'existence réèlle et le roles des appendices piriformes des neurones. Le neurone perioptique des diptères. C. R. Soc. Biol. **64**, 959—961 (1908).

Vowles, D. M.: The receptive fields of cells in the retina of the housefly (Musca domestica). Proc. roy. Soc. B **164**, 552—576 (1966).

Waddington, C. H., Perry, M.: The ultra-structure of the developing eye of Drosophila. Proc. roy. Soc. B **153**, 155—178 (1960).

Walcott, B.: Movement of retinula cells in insect eyes on light adaptation. Nature (Lond.) **223**, 971—972 (1969).

— Horridge, G. A.: The compound eye of Archichauliodes (Megaloptera). Proc. roy. Soc. B **179**, 65—72 (1971).

Waterman, T., Fernandez, H., Goldsmith, T.: Dichroism of photosensitive pigment in rhabdoms of the crayfish Orocnetes. J. gen. Physiol. **54**, 415—432 (1969).

Wiedemann, I.: Versuche über den Strahlengang im Insektenauge (Appositionsauge). Z. vergl. Physiol. **44**, 526—542 (1965).

Wigglesworth, V.: The Principles of Insect Physiology. London: Methuen 1965.

Wolken, J. J., Capenos, J., Turano, A.: Photoreceptor structures. III. Drosophila melanogaster. J. biophys. biochem. Cytol. **3**, 441—448 (1957).

Zawarzin, A.: Histologische Studien über Insekten. XV. Die optischen Ganglien der Aeschna-Larven. Z. wiss. Zool. **108**, 175—257 (1914).

Chapter 2

Rods and Cones*

By

Adolph I. Cohen, St. Louis, Missouri (USA)

With 10 Figures

Contents

A. Origin and Environment

The photoreceptor cells of vertebrates, the rods and cones, detect and convey information and may thus be classified as neuronal cells. Their form and position suggests an evolution related to that of ependymal cells, as suggested by Krause (1892, 1893, 1894). The ependymal cells constitute the lining cells of the hollow, fluid filled ventricular systems of vertebrate central nervous systems. The vertebrate eye, in a very real sense, is an externalized portion of the brain. During embryonic development the eyes form from hollow side chambers off the axial neural tube, and the lumen of the tube is continuous with that of the side chambers, known as optic vesicles. Ciliated ependymal cells eventually line these lumina. At a later stage of development, the lateral wall of each growing optic vesicle collapses against the medial wall, thus forming a double walled optic cup, whose resultant concavity is directed laterally. This concavity will form the vitreal chamber of the

* This work has been supported in part by Research Grant Number NB-04816-06, and Career Development Award Number 5-K3-NB-3170 of the National Institute of Neurological Diseases and Blindness, National Institutes of Health, Bethesda, Maryland.

mature eye. The thin space between the two walls of the optic cup persists as a
ventricular space between the retina (formed from cells of the lateral cup wall) and
the pigment epithelium (formed by cells of the medial cup wall). The ventricular
space becomes cut off from the main system of brain ventricles, yet retains many
of its functional attributes. The cells of the pigment epithelium remain as a sheet
of cells only one cell thick and become backed by a rich capillary plexus. This combination of a capillary plexus backing a thin sheet of ependymal cells constitutes
a choroid plexus as seen anywhere in the central nervous system. The specialized
fine structural features of the capillaries behind the retinal pigment cells, and those
of the pigment epithelial cells themselves, are essentially similar to those seen in
choroid capillaries and choroid plexus ependymal cells in other brain regions
(MAXWELL and PEASE, 1956). Thus, the capillaries exhibit "fenestrated" endothelial cells (WISLOCKI and LADMAN, 1955). The pigment cells, apart from the
presence of pigment, are like all choroid plexus ependyma in having highly infolded
cell membrane surfaces facing the capillaries and at the opposite surface which
faces the retina, there are found at the cell margins terminal bars with completely
girdling tight junctions (COHEN, 1965c; LASANSKY and deFISCH, 1965). There
are also microvillous projections from pigment cell surfaces facing the retina. The
ventricular space between the pigment cells and the retina persists in the adult
retina (Fig. 1).

The photoreceptor cells themselves develop from the primitive neuroepithelium
of the lateral wall of the optic cup and fit into the mosaic of ependymal cells on its
ventricular surface. The surface on the retinal side of the ventricle resembles in
significant aspects ependymal surfaces of ventricles of the mature central nervous
system in non-choroid plexus regions (BRIGHTMAN and PALAY, 1963). The cell
margins are joined by terminal bars, but these lack girdling tight junctions (COHEN,
1965c). However, at least in certain species, small tight patches are sometimes
seen between membranes of the glial cells of Müller at this surface. When fairly
large materials (ferritin, horseradish peroxidase, toluidine blue etc.) are injected
into brain ventricles, they readily pass between ependymal cells and enter the
neural tissue (BRIGHTMAN, 1965, 1967). While such tests are not readily performed
in retinas, the similar morphology predicts a similar outcome. The intercellular
junctions defined by terminal bars on the retinal side of the ventricle, occur precisely at the surface level. By virtue of staining heavily, these produce in thick
sections of retina, viewed by light microscopy, a line designating the retinal
surface. This surface is termed the external limiting membrane (Fig. 1).

The photoreceptor cells are bipolar in form and have their cell bodies (somas)
within the retina. Their axons erupt from the somas and head in the vitread direction to their terminations with second order neurons. (The terms *vitread*, towards
the vitreous chamber, and *sclerad*, towards the sclera, are useful in this spherical
organization.) The sclerad pole of the photoreceptor is constituted by a lengthy
portion of the cell which protrudes from the soma into the ventricle. This ventricular portion is structurally divided into an outer segment, lying closest to the
pigment epithelium, and an inner segment, lying closest to the soma. The two
segments are linked by a modified ciliary process (DeROBERTIS, 1956b; SJÖSTRAND
and ELFVIN, 1956) erupting somewhat below the apex of the inner segment but whose
microtubular components cross a short ciliary bridge and then course eccentrically

as a sort of a "backbone" running beneath the cell membrane of the outer segment at one region of its circumference (Fig. 1). Occasional receptors may exhibit a secondary cytoplasmic connection of the inner and outer segments (RICHARDSON, 1969), but it is not yet clear whether this is a normal or cyclical variation.

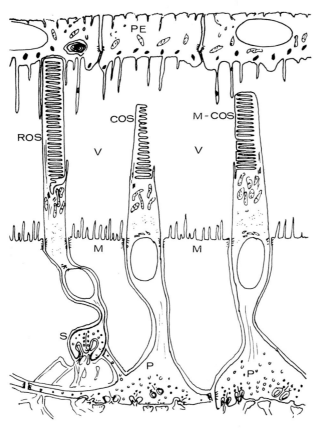

Fig. 1. A diagram of the vertebrate photoreceptor and its environment. PE = Pigment epithelium; ROS = Rod outer segment; COS = Cone outer segment; M-COS = Mammalian cone outer segment; IS = Inner segment; V = Ventricle; M = Müller cells; S= Spherule; P = Pedicle

Other cells of the primitive neuroepithelium develop into ependymoglial cells, the so-called Müller cells. These resemble astrocytes in cytoplasmic detail and span the full thickness of the retina and indeed, constitute the chief glial component of the retina. At the ventricular surface, with certain non-random exceptions to be noted later in this article, the bases of the ventricular portions of each photo-receptor are separated from each other by surrounding Müller cell cytoplasm. Thus the cell junctions on the ventricle involve either junctions between Müller cells, or between a receptor and a Müller cell, but typically there are no junctions between photoreceptors.

B. The Light Pathway through the Photoreceptor

Light normally enters the retina at its vitreal surface, passes through its thickness, and thus passes through photoreceptors from their synaptic (vitread) poles to and through their outer segments at the cell's sclerad poles. The illumination is then largely absorbed by melanotic pigment found to a minor extent in cells of the pigment epithelium and to a major extent within melanophores of the uveal coat of the eye. However, some light is measurably backscattered through the photoreceptors, retina, and out of the eye. This is a serious functional complication in albino animals, and an experimental complication when parameters related to visual pigments are studied, as in addition to possible differential scattering of certain wave-lengths by small particles of melanin pigment (BÜLOW, 1968), some light is absorbed by hemoglobin within blood cells.

In the portions of the photoreceptor located within the retina proper, there appears to be less precision in the spatial relations of the photoreceptor somas, axons, and surrounding glia. In general, where the axons erupt from the cell somas they are well separated by glial cells. In animals with many rods, such as mice, photoreceptor somas may sometimes be contiguous (COHEN, 1960). It is likely that there is no significant difference between the refractive indices of receptor and glial cytoplasm, and the intercellular spaces are only of the order of 200 Å. Thus, it is probable that this zone is optically homogenous. However, this homogeneity must break down when the ventricular border is reached. VAN HARREVELD and KHATTAB (1968) have estimated that the extracellular space in the ventricular region of the mouse retina is about 38 %. One might guess that the underlying ionic composition of this space, in view of its homologies, is probably akin to that of cerebrospinal fluid. However, it is known (FINE and ZIMMERMAN, 1963) to be rich in an acid mucopolysaccharide, possibly produced within inner segments of the photoreceptors (FINE and ZIMMERMAN, 1963; OCUMPAUGH and YOUNG, 1966). SIDMAN (1957, 1958) has studied the photoreceptors by interferometry and noted that by comparison to general cytoplasm, outer segments have an exceptionally high concentration of solids, 40—43 %, and a refractive index of 1.41. Thus the refractive step between the cell and ventricular fluid is probably highly significant. In some photoreceptors, e.g. those of the rat and mouse, the ciliary bridge between the inner and outer segments extends over 1—2 microns so that light passing up the cell would be affected by a refractive step, but in other photoreceptors, e.g. those of the frog, the ciliary bridge is short and the light transmission likely not to be affected. Cones, in general, exhibit minimal gaps at this junction between inner and outer segments. LATIES et al. (1968) have recently advanced evidence to the effect that the axis formed by the inner and outer segment of a photoreceptor may not be perpendicular to the retinal surface at the ventricle, but rather, is variously tipped in such directions in different retinal regions as to possibly obtain a better coincidence with the light path.

Another complication stems from the fact that the diameters of photoreceptors may be only a few times the wave-lengths of light in water. The outer segment of a human rod has a diameter of about 1.3 μ, that of a flying squirrel, 1.1 μ. TORALDO DI FRANCHIA (1949) and JEAN and O'BRIEN (1949) have pointed out that portions of the receptor can therefore act as dielectric antennas with the radiation undergoing complex reflections depending on the internal angles of incidence and some-times passing in and out of the ventricular portions of the cells. Enoch (1960,1961) has obtained evidence for the actual existence of these phenomena inasmuch as when receptors were viewed with the light passing through in the normal direction, modes indicating wave guiding were observed, and even color separations were seen when white light was employed. Although there is good evidence that color vision is not based on such phenomena (Brindley and Rushton, 1959) it does pose technical problems at least, in assessing the flux passing through receptors and in the assessment of the spectra of visual pigments and the pigment densities since some light may "escape" from one receptor to another.

A related aspect of receptor optics, is the possible "funnelling" or concentration on outer segments of the luminous flux passing through inner segments when the latter are of significantly greater diameter (c.f. Rushton, 1962). This is even possible in the "rod-like" foveal cones of primates. Here Dowling (1965) found cone inner segments to taper but even at their narrow apical region to be wider ($1.5\,\mu$ versus $0.9\,\mu$) than the outer segments. The latter did not seem to taper although it is hard to rule out a slight tapering of outer segments.

Receptor shape has also entered into the consideration of possible mechanisms of the Stiles-Crawford effect (Stiles and Crawford, 1933). This phenomenon con-sists of retinal responses based on cone activity exhibiting an efficiency markedly dependent on the axis of illumination, in contrast to rod initiated responses where the effect, if present, is much less. Unfortunately there has been but one study taking advantage of the variety of receptor shapes in different species, to attempt to distinguish how the shape of receptors relates to the phenomenon. Westheimer (1967) found that the "rod-like" foveal cones of the human retina exhibited less of the effect than did the more obviously conical parafoveal cones. This seems to implicate receptor shape in the origin of the phenomenon.

Although as just noted, photoreceptors in general are not contiguous to one another, in the retinas of certain fish, amphibians, reptiles and birds there exist variations in the mosaic of receptors in which paired receptors occur with their inner segments in broad contiguity, the two cells appearing as if compressed to yield a circular cross-section for the pair. These pairings generally involve cones. When the two members of a pair are morphologically similar, the combination is termed a "twin receptor," if dissimilar, "double receptor." Since there is no refractive break between the paired receptive elements, light should pass through them as a unit. It seems likely that this situation permits two dissimilar cells to sample similar if not identical regions of the visual field (Cohen, 1963a). In the case of the double cones, one cell generally has a colored oil drop in the apex of its inner segment, so that this in itself would mean that two dissimilar instruments were sampling the same visual field region. In the case of the morphologically similar "twin" receptors, it seems a reasonable prediction that the outer seg-ments will prove to have different visual pigments. Marks (1965) has already shown this to be the case for twin cones of goldfish. The mosaic patterns are described for various fish (Lyall, 1957; Engström, 1963), a bird (Engström, 1958; Morris 1970), in reptiles (Underwood, 1951; 1954), and in the leopard frog (Nilsson, 1964a, b).

C. Rods and Cones, Classical Definition

The observations of Schultze (1866) that photoreceptors could be classified by differences of form into two groups, rods and cones, and that cones predominated in diurnal animals, is a landmark in visual physiology. A summary discussion of the basis of this division with reference to modern views of receptor form and function will follow the physical description of the photoreceptors. In the classical view, typical rods have cylindrical inner and outer segments whose diameters are similar. Typical cones on the other hand have a conically formed outer segment and an inner segment whose diameter is larger than the outer.

D. Outer Segments and Their Membranous Discs

The photoreceptor outer segments are objects of special interest in view of the high probability that this is the region of the cell wherein occurs the interaction of visual pigment and light that initiates the visual process. All such visual pigments as have been studied appear to be combinations of the chromophores, 11-*cis* retinal$_1$, or 11-*cis* retinal$_2$, and various proteins called opsins. The latter have been described both as lipoproteins (Krinsky, 1958; Poincelot et al., 1969) and as lipid free glycoproteins (Heller, 1968a, b, 1969). They prove to be insoluble in the usual saline solutions but may be brought into colloidal suspension with detergents. This type of data has suggested that the visual pigments are built into the membranous structures of the cell. Because of the use of detergents in solubilizing these pigments, it would seem to be possible to trap lipid with the protein in the detergent micelles, thus the claim that opsin is a lipoprotein is the more difficult to establish critically.

When photoreceptors were under intensive investigation in the latter part of the 19th century, Kühne (1878) noticed that when outer segments were placed in solutions of bile salts, they came apart as if they consisted of a pile of discs or coins. With the advent of electron microscopy, Sjöstrand (1949, 1953) was able to show that the basic architecture of the outer segment was that of a bell jar-like cell membrane enclosing a pile of double membrane discs or flattened saccules (Fig. 1, 2). Since polarized light studies suggested (Schmidt, 1938; Denton, 1959; Wald et al., 1962) that the chromophores were perpendicular to the axes of the outer segments, they were therefore parallel to the discs and in view of the insolubility of the molecule, it seemed likely that the photopigment was part of the disc structure.

Sjöstrand (1959, 1961), Moody and Robertson (1960), Yamada (1960b), Okuda (1961), and Cohen (1961a, b, 1963) noted that in cones of a variety of species many of the double membrane discs appeared to be topological continuations of the cell membrane (Fig. 3) as if they developed by an infolding or ingrowth of the cell membrane of the outer segment. The openings into the discs tended to be widest in the basal discs and could be quite narrow at the apex (Cohen, 1963a) of the outer segment. The openings were not in perfect register in successive discs (Sjöstrand, 1961; Cohen, 1963a). While most of the double membrane discs of rods lacked such connections, at the base of the outer segments of rods similar connections of discs with the cell membrane were occasionally

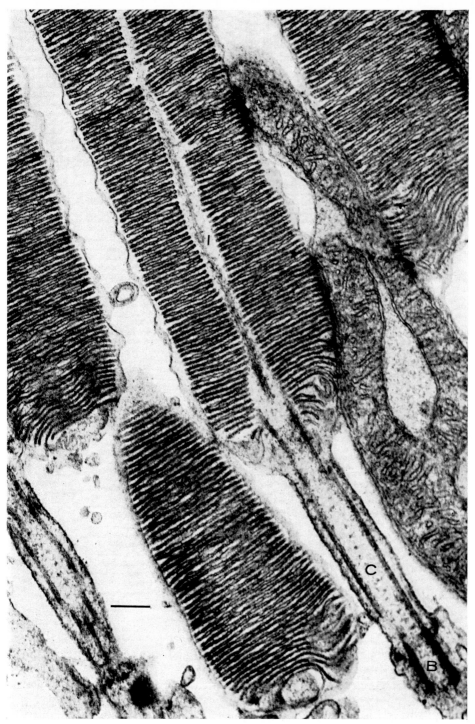

Fig. 2. A slightly oblique section through the outer segments of the rods of the flying squirrel *(Glaucomys volans)* to illustrate the outer segment discs and the alignment of the ciliary connective (C) with the disc incisure (I). The basal centriole (B) of the cilium is also evident. Bar designates 0.25 microns

observed (Moody and Robertson, 1960; Cohen, 1961a, 1963a). Mammalian
cones have been studied in monkeys (Cohen, 1961a, b; Ordy and Samorajski,
1968), the rabbit (DeRobertis and Lasansky, 1958), the ground squirrel (Hollenberg and Bernstein, 1966), and the cat (Cohen, unpublished observations).

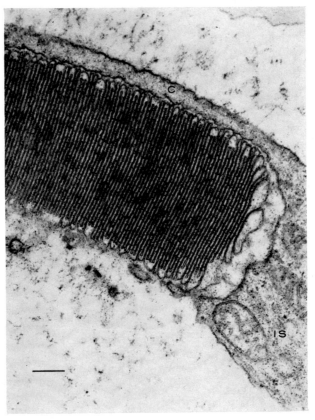

Fig. 3. An electron micrograph of the base of the outer segment of a foveal cone of a Rhesus
monkey. Note the infoldings of the cell membrane to form the discs. The section has missed
the infolding zones of the apparently isolated discs and also the connecting cilium. A calycal
process (C) from the inner segment (IS) is evident. Bar designates 0.25 microns

In the monkeys, ground squirrels, and cat, discs were seen connected to extracellular space in the inner third of the outer segment, but not in the outer remainder. Sjöstrand and Nilsson (1964) did not find in the rabbit the cone outer
segments described by DeRobertis and Lasansky. Indeed, while they identify
a type II photoreceptor, they do not call it a cone. This cell lacked open discs,
had a short outer segment, and was relatively infrequent.

While cross-sections of the outer segments of rods have all seemed to indicate
that the discs were not in continuity with the cell membrane (Cohen, 1960, 1963,
1965a; Dowling and Gibbons,1961; Nilsson, 1964, 1965) there is always a
possibility that a small connection might be artifactually severed or hard to

discern. This also applies to the apparently isolated discs in the distal portions of the outer segments of mammalian cones. This point is of considerable importance. If the visual pigment is located in discs or flattened saccules which are isolated from the cell membrane but the visual signal consists of a potential change across the cell membrane, then one must consider mechanisms whereby changes in disc pigment can bridge the gap and affect cell membrane. If a disc or flattened saccule is still attached to, and a continuum of, the cell membrane, then its apparent lumen is a potential extension of extracellular space. If one now had a colloidal marker which could not cross cell membranes but which was fine enough to infiltrate *apparently* intracellular compartments that are *actually* connected to extracellular space, then a possible means of demonstrating non-isolation of rod saccules would exist. In such experiments a failure to infiltrate the saccules of rod outer segments is only suggestive because a finer colloid or one with a different charge might do so. With this in mind, COHEN (1968) produced precipitates of lanthanum salts in the vicinity of receptors of glutaraldehyde fixed retinas of frogs and found that while cone saccules did infiltrate, those of rods did not except for rare, most basal saccules in the region where rod saccules are occasionally seen to be open to extracellular space. However, in local zones where rod outer segments were damaged, the precipitate entered the cytoplasm of the outer segments and also could infiltrate the lumina of presumptively damaged rod saccules in these regions. Fine lanthanum precipitates seem confined to extracellular space when used to infiltrate other tissues (REVEL and KARNOVSKY, 1968, BRIGHTMAN and REESE, 1969). In later studies (COHEN, 1970), the same procedure resulted in the infiltration of many cone discs in Rhesus monkeys and pigeons, but not of rod discs of monkeys, pigeons, rats, or mice. Infiltration with a visualizable protein such as horseradish peroxidase or ferritin is a preferable test because the infiltration can be carried out prior to fixation. BONTING and BANGHAM (1968) have briefly reported infiltrating outer segments of cattle and rabbits with ferritin. The particles they illustrate were found free in the cytoplasm of the inner and outer segments, and in the cilium of the same cells. When intact cells take up ferritin it is found in vacuoles. Some of the particles in their illustrations seem to be in saccules. Free cytoplasmic ferritin strongly indicates breakage of cell membrane (BRIGHTMAN, 1965 b). It is difficult to avoid damaging some outer segments during experimental manipulations. Ferritin in rod saccules could then mean that the saccule membrane was torn at some point. I have been unable to repeat their observations but have recently infiltrated some unfixed saccules of frog cones with ferritin under conditions of partial disruption of the saccules. DOGGENWEILER and FRENK (1965), in a study directed to another purpose, also noticed the occasional infiltration of rod saccules of frogs with lanthanum colloid. However, this occurred in less than 10 % of the rods (DOGGENWEILER, personal communication), in which the colloid was also seen free in the cytoplasm of the outer segment. Recently, LATIES and LIEBMAN (1970) injected the fluorescent dye Procian Yellow into the vitreous chamber of eyes of living frogs and mudpuppies and subsequently observed by fluorescence microscopy of frozen-dried material that cone outer segments alone exhibited fluorescence. LATIES (personal communication) reports that more recent work with this small molecule (molecular weight about 500) has given results identical to those observed with lanthanum in the monkey and rat.

These results suggest that cone discs are often open to extracellular space. Even in monkeys, where such connections are not usually evident in the sclerad two-thirds of the outer segments of cones by conventional electron microscopy, the lanthanum test shows that some discs at every level are open to this infiltrate. The non-infiltrating cone discs may represent discs which are artifactually or actually separated from the membrane. As for rod discs, current evidence of morphology

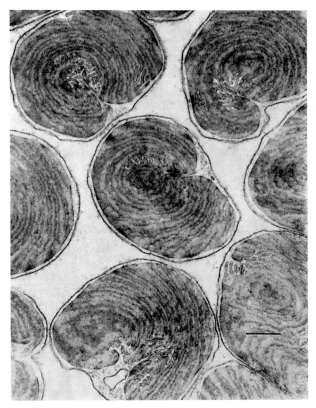

Fig. 4. A cross-section of the outer segment of a rod of a flying squirrel to indicate the apparent isolation of the discs from the cell membrane. Bar designates 0.25 microns

still favors their isolation from the cell membrane and from each other. Arden and Ernst (1969) have recently detected physiological differences between rods and cones which they believe could be based upon the isolation of rod discs.

Another difference between rod and cone outer segments is seen when these are examined in cross-section (Fig. 4–6). The perimeters of the discs of cones are circular, except at points where they may be seen to attach to the cell membrane (Cohen, 1961a, b, 1963a, 1968; Nilsson, 1965). On the other hand, the rod discs show a lobulated perimeter (Sjöstrand, 1949, 1953a; Porter, 1956; Wald et al., 1963; Cohen, 1960, 1961b, 1963a, b, 1965a; Nilsson, 1965). In case of fish or amphibia the disc perimeters show 8–12 deep incisures forming the lobulations, whereas in rods of the monkey, pigeon, and man (COHEN, 1961 a, b, 1963 a, b, 1965 a), the incisures are relatively superficial. In the guinea pig (SJÖSTRAND, 1949, 1953 a), mouse (COHEN, 1960), rat (DOWLING and GIBBONS, 1961) and flying squirrel (COHEN, unpublished observations) there is a single, deep disc incisure. This lies facing the bundle of ciliary microtubules lying under the outer segment membrane but persists above the termination of the microtubules.

Fig. 5. A cross-section of the outer segment of a rod of a frog. The striping effect is due to disc curvature causing many to be included in the section plane. Note the deep incisures. Processes (P) from pigment cells are also evident. Bar designates 0.5 microns

The incisures of successive discs tend to be in register. In amphibia a given group of registered incisures may only extend over a limited portion of the outer segment length, i.e. there is considerable irregularity in the lobulations of the discs so that both in number and location discs incisures may not match at two distant outer segment levels. The amphibian incisures are so deep that a given disc lobe may only possess a narrow connection to the body of the disc. Under conditions where $BaSO_4$ precipitate infiltrated the cytoplasm of outer segments of frog rods, COHEN (1968) observed that some lobules were completely surrounded by precipitate, suggesting a vertical connection to the remainder of the disc.

The significance of the lobulation of rod discs is not clear. The phenomenon produces an increase in the length of the disc perimeter, slightly reduces the surface area, and in rods of large diameter, increases cytoplasmic access to the central region of the discs.

There has been some suggestion that the edges of discs possess a special character. Thus, following certain fixation procedures where the saccule lumina expands, the disc margins seem to resist the deformations accompanying the expansions (DEROBERTIS and LASANSKY, 1961). The edges may also possess a slightly different appearance when viewed in cross-section (BROWN et al., 1963). When COHEN (1968) exposed unfixed, detached frog retinas to barium sulfate suspensions, this

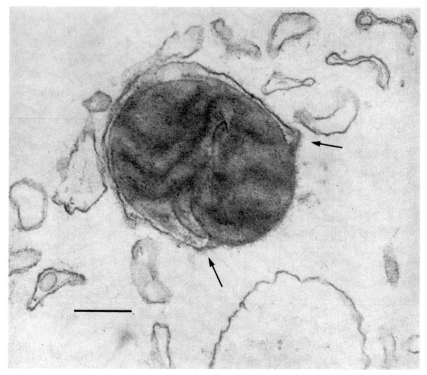

Fig. 6. A cross-section of the outer segment of a cone of a frog. The discs have apparently formed by infoldings between the arrows. Bar designates 0.25 microns

fine precipitate occasionally entered local regions of a few outer segments of rods at sites where the outer cell membranes were apparently torn. Subsequent fixation and microscopy revealed that while the precipiate could infiltrate around the discs, and into their incisures, it apparently had difficulty in penetrating between discs. This suggests the presence of a low density inter-disc material which possibly holds discs together, perhaps at their edges. FALK and FATT (1969) have recently found that when isolated outer segments of frog receptors were fixed in buffered osmium for one hour and then subjected to prolonged washing (24 hours) with Tris buffer [Tris (hydroxymethyl) amino-methane], the centers of the discs of rods eroded and washed away, leaving the disc rims adherent to each other and to the cell membrane. It is of interest to recall the old observation of KÜHNE (1878) that after one hour's fixation in the dark with 1 % osmium tetroxide, outer segments seemed reddish brown but readily bleached on exposure to light. I have confirmed this observation with both osmium tetroxide and potassium dichromate (COHEN, 1963). As the ability of visual pigment to bleach seems largely intact, this probably means that the retinal is in a highly hydrophobic zone where it is not attacked by these oxidants. As fixative diffusion proceeds from the edges of a disc towards its center, the central region could contain protein that is more poorly cross-linked by one hour's exposure to this fixative than the edges and may therefore dissolve more readily with prolonged washing. FALK and FATT, however, believe that a specific effect of Tris on osmicated protein is implicated. The discs of the thinner and much smaller outer segments of cones did not erode after the same treatment. This suggests either that the open saccules of cones are better fixed or that the membrane is intrinsically different.

DeROBERTIS (1956 b), TOKUYASU and YAMADA (1959). and CARASSO (1959) have studied the development of the outer segment. TOKUYASU and YAMADA (1959), and DeROBERTIS and LASANSKY (1961) indicated that discs seemed to form by an ingrowth or infolding of the plasma membrane of the outer segment. Recent studies (OLNEY, 1968 b) yield views consistent with this idea. EAKIN (1965) found that photoreceptors differentiate normally in the absence of light.

In 1961 and 1963, DROZ initiated studies of the renewal of protein within photoreceptors by combining electron microscopy with radioautography. He obtained evidence for protein synthesis in the inner segments of rats (confirming NOVER and SCHULTZE, 1960) and evidence that this protein moved into the outer segment. This raised the possibility of the persistence of saccule formation during life (COHEN, 1963 b), and could explain the occasional open rod discs at the base of rod outer segments. That this seems to be the case, at least in rods, was es- tablished by the elegant work of YOUNG and his collaborators (YOUNG and DROZ, 1968; YOUNG, 1965–1969; YOUNG and BOK, 1969). When a pulse of radioactive amino acid was given to rats or frogs, the activity first appeared in the inner seg- ments of the photoreceptors, it later could be located within the cilium (YOUNG, 1968) and finally was seen as a band at the base of the rod outer segment (YOUNG and DROZ, 1968). This band, with time, moved up the outer segment of the rod until it reached the tip. At this point the outer segment activity was lost but what seemed to be radioactive fragments of outer segment (phagosomes) now appeared in the cells of the pigment epithelium (YOUNG and BOK, 1969). Such objects within vacuoles of cells of the pigment epithelium have previously been attributed to outer segment fragments (COHEN, 1960, 1963 b) by way of distinguishing them from the non-vacuolar but superficially similar myeloid bodies which may occur in the same pigment cells in amphibians, reptiles, and birds. Complimenting these observations, recent studies indicate that the opsin of rhodopsin is among the proteins incorporating radioactive label in this procedure (HALL et al., 1968; BARGOOT et al., 1969). The suggestion then is that rhodopsin is intrinsically rather stable but largely turns over because of the attrition of outer segment tips and their continuous growth. Indeed, DOWLING and SIDMAN (1962) had described a rat strain with receptor dystrophy in which the pathology appeared to take the initial form of an excessive rhodopsin content and length of outer segment. This could now be viewed as suggesting an imbalance of production and attrition of outer segments and HERRON et al. (1969) have recently obtained evidence suggest- ing that the defect is in the pigment epithelium.

One study (Spitznas and Hogan, 1970) on the shedding of rod discs in the human retina suggests that the pigment epithelium may play an active role in the cutting off of groups of discs, but another study on Rhesus rods (Young, 1971 b) suggests that the cutting off of the discs does not involve the pigment epithelium. Both studies implicate the latter as actively phagocytosing the fragments, once formed.

Further complimenting these observations, Kroll and Machemer (1969 a) have recently examined the fate of retinas of monkeys which were first detached (i.e. enlarging the ventricular space with fluid to force the retina away from its close approximation to the pigment epithelium), and then reattached by experimental means. The separation of the retina from the pigment epithelium has long been known (Kühne, 1878) to inhibit rhodopsin regeneration and has suggested a necessary metabolic interaction of photoreceptor and pigment cells. These investigators noted that the outer segments of rods degenerated on detachment but could regenerate on reattachment. They have recently (Kroll and Machemer, 1969 b) noted that cone outer-segments do the same, but that regeneration is somewhat slower. Much of this was foretold by the finding of Dowling (1964) that rat outer segments, caused to degenerate by vitamin A deficiency, would regenerate on refeeding vitamin A.

Returning to the observations of Young (1969), he noted that in the outer segment of frog cones the radioactivity was not incorporated as a basal band moving up the outer segment but rather that the incorporation of radioactivity was diffuse. He points out that special problems are posed by the conicity of the outer segment of many cones since the more sclerad discs are progressively of smaller diameter. While in the initial development of a cone outer segment or in the course of a total regeneration of an outer segment, the first discs made could be smaller than those next produced, in a situation of continuous replacement of discs from below, the large discs would have to lose mass and diameter as they moved up a conical outer segment. There is as yet no chemical evidence of cone pigment turnover and animals like ground squirrels, said to have all cone retinas, have not yet been studied to ascertain the generality of these observations on frog cones. It would be interesting to learn whether the apical attrition phenomenon is only seen where discs are not attached to the cell membrane. Studies on cones of the ground squirrel would also be of interest because the outer segments are either cylindrical or minimally tapered. Recent work by Young (1971 a) on the rods and cones of the Rhesus monkey does confirm the differences seen in frogs.

There has been a good deal of attention to the cell membrane comprising the double-walled discs, both by those interested in membranes *per se*, and by those seeking more information on the natural situation of the photopigments. Useful early reviews are those of Sjöstrand (1959b, 1963), Robertson (1960), Moody and Robertson (1960), Fernandez-Morán (1961), and Moody (1964). More recent views of the general molecular architecture of cell membranes are those of Robertson (1966), Korn (1966), Lucy (1968), and Cherry (1968). It is clear from these reviews that the field is in considerable flux and that our early models of cell membranes (Danielli and Davson, 1935; "unit membrane" of Robertson, 1961), will probably undergo marked revisions. In electron micrographs, the usual black lines designating cell membranes, at least in part, may be critically resolved and viewed at higher magnifications to reveal two black lines separated by a lighter space, the so-called "unit membrane." In some cases, the two outer black lines are irregular and send spurs across the light gap to yield an overall globular appearance (SJÖSTRAND, 1963; NILSSON, 1964c). It should be said at once that "membranes" prepared for electron microscopy have been the subjects of rather violent treatments. They usually have been subjected to chemical attack by cross-linking denaturants (i.e. fixatives), and then to the violent effects of removing both free and bound water and finally embedding agents polymerized by further cross-linking. Moreover, when viewed in electron microscopes, the possibility of optical and photographic artifact is ever present. Membranes often possess curvature and the depth of focus of electron microscopes is such that all elements in the usual sections are in focus at the same time. Thus a report by ROBERTSON (1965) purporting to show a single pleating membrane forming the rod discs, was later shown by him (ROBERTSON, 1966) to be an optical artifact based upon a moiré pattern.

The fixatives mainly employed for electron microscopy (glutaraldehyde, formaldehyde gas in solution, and osmium tetroxide) are soluble to some extent both in aqueous and lipid phases and probably penetrate cells to some degree by both routes. It is a classical observation that as a first-order approximation, cell membranes tend to be penetrated by materials in direct proportion to their fat solubility and of the common fixatives, glutaraldehyde would seem to be a likely candidate for a high degree of solubility in lipid. Potassium permanganate is sometimes employed as a membrane fixative (MOODY and ROBERTSON, 1960). It is readily decomposed by many organic materials including lipids. All these reagents potentially react with certain groupings in carbohydrates, lipids and proteins. Thus, it is somewhat precarious to attempt to deduce the native molecular architecture of either the cell or disc membranes from their residual appearances in electron micrographs. This is not to argue that these attempts are not worthwhile, but rather that the data be treated cautiously.

The outer segments have an extraordinarily high lipid content, possibly 40 % by weight (SJÖSTRAND and GIERER, in SJÖSTRAND, 1959a) and phosphatidyl-ethanolamine, lecithin and acidic phospholipids have been identified (FLEISCHER and MCCONNELL, 1966). Recent studies by POINCELOT et al. (1969) and by DE PONT, DAEMEN and BONTING (1968) have in fact suggested that the 11-*cis* retinal might be only indirectly coupled to the protein opsin via a Schiff base linkage to phosphatidyl-ethanolamine, assuming the protein to be a lipoprotein. However, many aldehydes can form Schiff base linkages to phosphatidyl-ethanolamine, and the fact that retinal can do so does not prove that it is is so linked in native rhodopsin. ANDERSON (1970) has presented quantitative evidence bearing upon the possible association of phosphatidyl-ethanolamine and rhodopsin. In keeping with the idea of a weak linkage, a Schiff base linkage was first suggested by PITT et al. (1955), and evidence for it obtained by BOWNDS and WALD (1965), and AKHTAR et al. (1965). BOWNDS (1967) and AKHTAR et al. (1968), after illuminating receptors in the presence of sodium borohydride, found they could then isolate an N-retinyl decapeptide fraction – suggesting that the chromophore, at least at some point after illumination, was linked to protein. HELLER (1968a, b, 1969) has presented evidence indicating that opsin is a glycoprotein, rather than a lipoprotein. His isolated "native" rhodopsin had no lipid ($< 2 \%$) and in his view the linkage of retinaldehyde to opsin may be via a substituted aldimine bond (HELLER, 1968 b). HELLER (1968 a) notes that native opsin is rich in non-polar amino acid residues and that this likely accounts for its insolubility.

Studies on the outer segments employing polarized light have been of interest since the pioneering works of SCHMIDT (1938), and more recent work of DENTON (1959), WALD et al. (1963), and WEALE (1968). There seems to be a general agreement that the chromophore in the dark adapted receptor lies parallel to the disc planes. DENTON believed that the Vitamin A eventually formed after bleaching was oriented perpendicular to the membrane but no such change was observed by WALD's group. On the other hand, when they bleached in the presence of hydroxylamine, their data suggested that the resultant oxime was oriented perpendicular to the membrane. They suggest that DENTON was not, in fact, observing Vitamin A but offer no suggestion as to what other intermediate he might have been observing. WEALE (1968) has recently made the interesting observation that certain fixatives more rapidly alter the axis of polarization in rods, rather than cones, despite the apparent connection of the lumina of the cone discs to extracellular space. He suggests that there may be very significant differences in the permeability of the membranes of the two types of outer segments. Because of the fact that the agents he employed are to a degree both water and lipid soluble, it would be of interest in terms of considering permeability to observe and compare the effects of heat, ionizing radiation, and a larger range of chemical agents of differing water and lipid solubilities to see whether the same order of reaction of the receptor types is obtained.

As polarized light studies tell us something about chromophore orientation, they can be used to estimate the amount of chromophore oriented in a given way relative to the axial or disc orientation. Unfortunately, variations in disc planarity and receptor axes will impose errors in such studies.

Where osmium tetroxide is employed as the fixative, disc membranes in both rods and cones, usually appear in electron micrographs as two $25-30$ Å black lines enclosing an empty disc "lumen" of some $70-80$ Å. SJÖSTRAND noted that the black lines at the disc edge seemed thicker, about 60 Å. About 100 Å separates the discs. A useful table summarizing dimensions of discs, interspaces and repeat periods, as taken from the literature, is given by NILSSON (1965). With permanganate fixation, each of the two membranes infolding to form the walls of a disc readily shows "unit" membrane architecture, i.e. a 75 Å grouping of two 25 Å black lines with a light interval. The double-membrane discs appear solid with three dark lines and two light intervals, as the opposing dark lines of the two inner membrane surfaces either fuse, or the dark components interdigitate, i.e. the trilaminate membranes combine to form a pentalaminate disc. No disc "lumen" is evident. The total disc thickness is about the same as after osmium fixation, i.e. about 140 Å. However, inasmuch as the lanthanum technique (COHEN, 1968) suggests that discs of frog cones can be infiltrated and that even those of frog rods have potential lumina, it is likely that the central dark line of permanganate fixation represents an optical or photographic blending of the outer unit membrane lines of permanganate fixation, and that we are dealing with a so-called "gap" junction where a narrow $20-30$ Å space, penetrable by lanthanum salts, persists between the membrane surfaces as in some other CNS membrane junctions (REVEL and KARNOVSKY, 1967; BRIGHTMAN and REESE, 1969). It is quite possible that the gap normally contains a hydrated material. An amorphous material on the outer segment surface could be seen at least superficially entering the invaginations forming the discs of monkey cones (COHEN, 1961 b). BRIGHTMAN and REESE (1969) have advanced reasonable arguments to show that when gaps of only 20–30 Å separate nerve cell membranes, they do not preclude electrical coupling.

As an indication of the variability in disc dimensions after different technical procedures, one may review studies on the outer segments of frog photoreceptors, the latter being a frequent object of investigation. After osmium fixation (NILSSON, 1965) both rods and cones had double membrane discs of 125 Å thickness. The absolute interdisc spacing was markedly affected by whether the knife edge employed in sectioning was parallel or perpendicular to the axis of the receptor, but was 50 % greater in cones. With a similar fixative, WOLKEN (1957) obtained a value of 150 Å for the disc thickness of frog rods. FERNANDEZ-MORÁN (1961a) found the same discs to be 110–150 Å thick in standard osmium preparations but 160–200 Å thick when low temperature, freeze substitution and sub-zero dehydration methods were employed. BOROVIAGIN (1962a, b) obtained a disc thickness for rods of 110–115 Å and for cones of 115–120 Å after conventional osmium fixation, but a 150 Å thickness for either disc after potassium permanganate. MOODY and ROBERTSON (1960) likewise found a disc thickness of 150 Å for either frog rod or cone discs after permanganate fixation. Recent X-ray studies of fresh rods by BLAUROCK and WILKINS (1969) gave results consistent with the above orders of dimension.

One may also view membranes when osmication and metal staining *follow* dehydration and lipid extraction. Such procedures have been employed in studies of myelin (NAPOLITANO et al., 1967), and mitochondria (FLEISCHER et al., 1967). I have recently removed water and phospholipid from frog retinas with acetone or ammoniacal acetone with or without prior glutaraldehyde fixation and only then osmicated with 1 % osmium tetroxide in carbon tetrachloride. Despite the absence of most phospholipid, the double membrane discs of outer segments still exhibit an essentially pentalaminar organization, seen in cones to be derived from the coming together of trilaminate membranes. With glutaraldehyde prefixation, cone discs were about 135 Å thick, and those of rods about 115 Å. Without glutaraldehyde prefixation cone discs were about 125 Å and those of rods about 100 Å.

In certain species, for a certain fixation procedure, the discs of rods may exhibit an obvious difference in thickness from those of cones and this may aid in their distinction. Thus, after osmium fixation, cone discs are thinner in primates, rod discs thinner in frogs.

It is clear that while such investigations establish orders of magnitude, the data are too subject to artifact and too variable for precise molecular modelling and can accommodate a number of membrane models. Attempts have been made to discern differences in the dimensions of discs of rabbit rods as related to pigment bleaching (PATEL, 1967a, b). The differences reported are unconvincing. FERNAN-DEZ-MORÁN (1961a) has also claimed to see distinctions in light and dark adapted outer segments. He indicates that these could have preserved differently. KRINSKY (1958) pointed out that more lipid could be extracted from light adapted rods.

Fernandez-Morán (1962) has pointed out that bound water could be an important part of the structure of membranous systems, such as those in outer segments. Recent techniques employing nuclear magnetic resonance measurements (Clifford et al., 1968; Cope, 1969) confirm the presence in certain membranes of considerable levels of bound water, stable over biological temperature ranges. While such studies have not yet been applied to the packed membranous discs of outer segments, it is probable that these too contain significant "bound" water. Proton mobility in bound water exceeds that in solution (Hechter, 1965). Such considerations are relevant to speculations regarding charge migrations in discs within outer segments, or in the segments themselves. Rosenberg et al. (1961) have presented evidence for photoconduction and semiconduction in dried photoreceptors of sheep (however, see Abrahamson and Ostroy, 1967; Leslie, 1968; Cherry, 1968). Rosenberg and Jendrasiak (1968) have also discussed the great influence of hydration on lipid conductivity. Falk and Fatt (1963) have studied photoconductance in suspensions of outer segments.

Another means of studying the physical nature of discs and outer segments has been to expose them to osmotic shock. De Robertis and Lasansky (1961) reported that the discs within some outer segments of toad rods were swollen after fixation in a hypotonic fixative, but this was an incidental observation and not a controlled study. If the swelling seen was indeed osmotic, then it is apparent that their fixation process did not swiftly render the cell membrane highly permeable to water and ions and that evidence of the swollen state can survive processing for electron microscopy, although probably not in a quantitative manner.

It is interesting that fragments of isolated bovine outer segments behaved in experiments like ideal osmometers (Brierley et al., 1968). In isolating an outer segment, the ciliary stalk connecting it to the inner segments is necessarily severed and the mechanics of washing and separating the outer segments might well damage the cell membranes. This is a possible problem in flux studies with outer segments (Falk and Fatt, 1966; Bonting and Bangham, 1968). Possibly the ciliary patency (ca. 0.25μ) seals or the leak is inconsequential or, if the outer membrane is in fact damaged and patent, then these experiments reflect the osmotic and permeance properties of persisting intact saccules, not those of the outer membrane of the outer segment. If it could be critically demonstrated that rod discs or saccules are osmotically responsive to dilution of the cytoplasm about them, it would support the view that rod saccules are not connected to extracellular space and that the apparent isolation of the rod saccule is real. If cone saccules are open to extracellular space, then they should not be subject to osmotic swelling although they might be pulled open by forces acting on the cell membrane to which the are joined.

If discs are indeed isolated from the cell membranes of outer segments of rods, it becomes important to attempt to distinguish whether the properties of the saccule membranes differ from those of the cell membrane. This applies in particular to properties which influence ion movements. How do ion fluxes across saccule membranes influence local ion concentrations and potentials across adjacent outer membrane ? To use osmotic stress as a tool in this regard, isolated, detached retinas of rats or frogs were exposed to various dilutions of appropriate balanced salt solutions of a type in which electrical activity is ordinarily maintained. They were incubated for brief periods in the dark in these solutions under conditions in which pH control and oxygenation were assured and were then fixed in solutions matching the incubation solutions but which contained 1 % osmium tetroxide (an addition of 33 milliosmols). Outer segments of rats and frogs (rods and cones), thus handled, were observed (COHEN, 1971) to have swollen to a spherical configuration when exposed to hypotonic solutions but the saccules within them exhibited minimal swelling, if any. Moreover, identical results were obtained with hypotonic sucrose incubations and with incubations in isosmotic solutions of potassium chloride or potassium acetate but not potassium methylsulfate. Less swelling was seen with isosmotic solutions where the cation was sodium and the anions chloride or acetate. No swelling was seen with sodium methylsulfate. While these results might largely or partially reflect water movements through parts of the cell other than the still attached outer segments, there are sufficient differences between these results and the preliminary data of BRIERLY et al. (1968) to suggest there is an experimental opportunity here to distinguish the behavior of the outer and disc membranes. The swelling with fragments seen by BRIERLY et al., suggests that under conditions where the outer membrane is not intact, the saccules may more readily swell in hypotonic solutions. HELLER et al. (1971) have recently obtained physical and electron microscopic evidence on isolated outer segments of frogs to support this view.

Turning to other aspects of the membranes, CLARK and BRANTON (1968) have recently investigated the appearance of outer segments of rods of guinea pig retinas which had been swiftly frozen in liquid Freon and subjected to the freeze etching technique. Some retinas had been briefly exposed to hypotonic solutions of unstated osmolarity prior to freezing and this caused some disc swelling. Two distinctive fracture faces of the discs were seen, and the authors discuss where they might occur in the complex structure. They also indicate that the method is not free of artifact. Along these lines it should be noted that VILLERMET and WEALE (1969) found that even quick freezing in liquid nitrogen altered the polarization properties and lengths of outer segments from that of the native state.

The manner in which the visual protein is built into the discs is important to understanding the transduction process. However, it is hard to rule out that a small amount of visual pigment may be elsewhere. A small amount of newly synthesized visual pigment may be in transit and not yet in the discs. Another possible site for visual pigment is the cell membrane of the outer segment. In polarized light studies there is an indication that some fraction of the chromophore does not conform to the major portion of chromophore. To a large extent this could be accounted for by structural deviations. There may be deviations from true planarity in photoreceptor discs, as in fact is seen in fixed preparations, particularly in rods of large diameter. There may also be deviations of the average disc plane from a perpendicular to the receptor axis. Where individual receptors are not being studied, there may be slight deviations of individual receptor axes within a group. However, some of the deviant chromophore could be in the cell membrane itself.

One approach to the localization of the visual protein is immunochemical. DEWEY et al. (1966, 1969) prepared antibodies to a purified frog rhodopsin and the fluorescent tagged antibody stained isolated outer segments. When the electron microscope was used to examine isolated, unfixed, outer segments which had been exposed to antibody, BLASIE et al. (1966, 1969) found evidence for adhesion of antibody in a square array at 70 Å intervals. The antibody employed stained outer segments of both red and green rods as well as cones. The cell membrane of the outer segment stained as did isolated discs and there was some inner segment staining as well. DEWEY et al. (1969) also believe that the antibody entered intact outer segments to stain the discs but they call their evidence inconclusive and their illustrations are unconvincing. BLASIE et al. (1965) have also presented X-ray diffraction and electron microscopic evidence suggesting the presence of a 40 Å particle in outer segment membrane. MCCONNELL (1965) has also reported the presence of sub-units in this membrane.

A recently discovered phenomenon, the early receptor potential (BROWN and MURAKAMI, 1964; CONE, 1964; BROWN et al., 1965) has been attributed to a capacitive current resulting from a transient charge movement in visual pigment molecules that have captured quanta. In retinas containing both rods and cones such as the monkey (BROWN et al., 1965; GOLDSTEIN, 1969) man (GOLDSTEIN and BERSON, 1970) or the frog (GOLDSTEIN, 1967, 1968; TAYLOR, 1969), the response is dominated by the relatively small amount of cone pigments. However, rods do contribute to the response and the response is readily obtained from retinas with few, if any, cones (CONE, 1964). MURAKAMI and PAK (1970) have recently recorded this potential both intra- and extracellularly in the nocturnal Gecko and indicate that it is a potential of the cell membrane. According to PEDLER and TILLY (1964) all geckos have basically one type of receptor. The one outer segment of a nocturnal Gecko that they illustrate is typical of cones in having all discs connected to the cell membrane and COHEN (1970) has also noted some outer segments in Gecko gecko with a broad basal zone of open saccules.

LETTVIN (1965) and BRINDLEY and GARDNER-MEDWIN (1966) have argued that if the rod discs are indeed isolated and if the visual pigment is symmetrically located in both membranous walls of a disc, then the electrical displacements in each disc wall should have opposite directions and cancel. It is argued that as the potential is, in fact, detected, then either the idea of rod disc isolation or pigment symmetry must be wrong. LETTVIN (1965) states that if the discs were still in continuity with the cell membrane, then the disc lumen is a continuum of extracellular space and the charge translation in each membrane wall is actually across the outer cell membrane and that these would be electrically in parallel and detectable.

However, it is possible that some visual pigment resides in the cell membrane of outer segments of rods. It is also clear that the most basal saccules of rod outer segments are sometimes seen connected to the cell membrane. It is therefore possible that the detected portion of rod responses could originate in one or both of these sites. This would explain the domination of the response by the less numerous but open cone discs when these are present. However, it should also be emphasized that a similar photoelectric potential can be obtained from pigment epithelia, with action spectra and light stability suggesting that its source is melanin (BROWN and CRAWFORD, 1967a, b). This potential can be observed both intra- and extracellularly. Since there is no indication of melanin in the cell membrane or in saccules connected to extracellular space then there must be some means (heat ?, charge migration ?) by which the illumination of pigment cell melanin influences pigment cell membrane potential despite the apparent isolation of the melanin from the membrane. The same mechanism could apply to photopigment except that very much less heat would be generated.

As for the above mentioned possibility that visual pigment in the plasma membrane of rods could be the source of their contribution to the ERP (see also Cohen, 1969) one notes the consistent observation of Dewey et al. (1969) that fluorescent antibody to frog rhodopsin stained the surface of outer segments of frog rods. To further explore the quantitative feasibility of this idea, at the suggestion of H. Ripps, he and I used the data of Nilsson (1965) for the total surface area of discs of an individual frog cone and calculated from the same source the total plasma membrane area (without discs) of outer segments of an average red and green rod and cone of the frog. The value for the cone (with discs) and the values for the rods (without discs) were weighted by multiplying them by the frequency of red rods, green rods, and cones in a given retinal area as obtained from Nilsson (1964 a), and the results summed. The rod plasma membrane proved to be 22% of the total. While the presence of some cones with rhodopsin-like pigment (Liebman and Entine, 1968) and green rods complicate matters, one notes that Taylor (1969) suggests that 15% of the ERP in frogs is from rods and that Goldstein (1968) states that 20—25% of the frog ERP derives from short-wavelength pigments. Thus, for frogs at least, the possibility of pigment in the rod plasma membrane contributing to the ERP is worthy of serious consideration. Two further factors are that the visual pigment in the plasma membrane would not be optimally oriented for light absorption and that the area of any basal rod discs, still attached to the plasma membrane, should be added to its area.

The aforementioned studies by Young (1968, 1969), and Young and Droz (1968) suggest that the molecules of frog and rat rhodopsin are stable until they reach the apices of the outer segments. Here, by virtue of the fact that tip portions containing groups of saccules are pinched off and digested in the pigment epithelium (Young, 1969), the rhodopsin molecule is broken down. Thus there seems to be little intrinsic instability of this molecule. However, Noell et al. (1966a) noted that albino rats exposed to continuous illumination from fluorescent lights for 24 hours showed signs of retinal damage and with intense light a 1-hour exposure could produce damage. The action spectrum for the damage roughly approximates that of rhodopsin. Noell et al. (1966b) noted that electron microscopic changes were seen early in the pigment epithelium and outer segment. Kuwabara (1966) independently observed these changes and by light and electron microscopy. Kuwabara and Gorn (1967) believed the first signs to occur in the apices of the rod outer segments. Grignolo et al. (1969) have also confirmed these observations. Further questions about opsin stability are raised by the following experiments. Dowling and Gibbons (1961) placed rats on a Vitamin A deficient diet but supplemented this diet with Vitamin A acid. This regime prevents the general development of signs of Vitamin A deficiency, but inasmuch as the acid does not serve as a precursor for retinal, the visual symptoms of Vitamin A deficiency do appear. Under these circumstances, the outer segments deteriorate suggesting that in the absence of retinal, either opsin synthesis or disc synthesis fails or that opsin in the absence of retinal is unstable and that the discs deteriorate in its absence. The effects of Vitamin A deficiency on the cones of a lizard have been studied by EAKIN (1964) with similar findings. On the other hand, BONTING, CARAVAGGIO, and GOURAS (1961) found that during development rhodopsin could more than double while outer segment length (and mass?) remained the same, suggesting that the content of rhodopsin in discs may vary and that disc formation was not stoichiometrically tied to rhodopsin production.

WALD (1954) has indicated that some mechanism of signal amplification must exist in the rod. He suggested as possibilities a local alteration of membrane permeability for some ion or the activation of some enzyme with a large turnover number. Another possibility is a light modulated membrane "pump" for specific ions, as may occur in invertebrates (SMITH et al., 1968). McCONNELL and SCARPELLI (1963) reported that rhodopsin might have light influenced ATP-ase activity but two subsequent studies (BONTING et al., 1963; FRANK and GOLDSMITH, 1965) failed to confirm this. SCARPELLI and CRAIG (1963) presented histochemical evidence for nucleosidase activity in outer segment discs but the distribution of the precipitate suggests the possibility of localization artifact (COHEN, 1968).

MATSUSAKA (1967b) has illustrated apparent fusions of outer segments and pigment cell processes, but these may be artifactual.

E. The Ciliary Connective and Ciliary Microtubules

The ciliary junction between the inner and outer segments was first described by SJÖSTRAND (1953a, b) and by DE ROBERTIS (1956a). DE ROBERTIS (1956a), COHEN (1960), and DOWLING and GIBBONS (1961) noted that in contra-distinction to motile cilia, while possessing the usual 9 pairs of circumferential tubules, the central pair was absent. GRIGNOLO and ORZALESI (1963) and LIEB and KNAUF (1964) have confirmed these observations. The ciliary cross-section is about $0.25\,\mu$ and the central region appears markedly less dense than the outer cytoplasmic zone in which are embedded the ciliary microtubules (COHEN, 1965a) (Fig. 7). As noted earlier, YOUNG (1968) has obtained evidence for diffusion of materials through the cilium.

The fibrils within the outer segment occur eccentrically and run no more than half the outer segment length in the few species in which this has been studied. In the Rhesus monkey the nine microtubular pairs, quickly become nine single tubules and show no precise positioning (COHEN, 1965a). When lanthanum precipitate artifactually entered torn outer segments of rat rods, the precipitate while readily penetrating between the discs, failed to penetrate a zone containing the microtubules (COHEN, unpublished data). This suggests that the tubules may be embedded in a gel.

In the apex of the inner segment the pairs of microtubules feed into a basal body identical in all known respects to a centriole and an accessory centriole is likewise present. Fibrillar rootlet systems erupt from both centrioles (COHEN, 1960). In addition, TOKUYASU and YAMADA (1959) found evidence of radial connections of the cell membrane to the centriolar apparatus and structures consistent with these have been seen by others (SJÖSTRAND, 1961a; COHEN, 1961c, 1963, 1965a).

The ciliary region besides representing a cytoplasmic connection between the two ventricular portions of the photoreceptors, also must be included in the signalling path between the inner and outer segment. Current evidence holds (Brown and Watanabe, 1962a, b; Bortoff, 1964; Bortoff and Norton, 1965; Tomita, 1965; Werblin and Dowling, 1969) that photoreceptors transmit electrotonically rather than in all or non-fashion. The fact that this 0.25 μ diameter, the narrowest

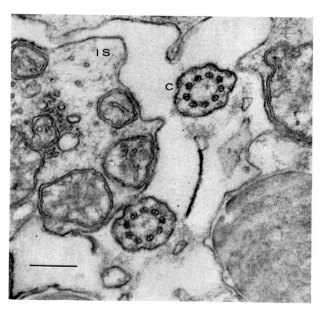

Fig. 7. A cross-section through the apex of an inner segment (IS) and two connecting ciliary stalks in the flying squirrel. Bar designates 0.25 microns

in the photoreceptor length, is interposed between the photopigment bearing region and the rest of the cell, must be taken into account in considering the cable properties of this cell.

Munk and Andersen (1962) have reported that certain fish possess a ciliumlike structure erupting from the inner segment alongside the conventional outer segment.

F. Inner Segments and Their Organelles

The inner segment of the photoreceptor (Fig. 8) shows a complex linear differentiation (Sjöstrand, 1953b). Its apical end may be well separated from the outer segment by a ciliary stalk that is 1—2 microns in length, as in some rodents, or the stalk may be very short and the inner segment may directly about on the outer. As the attachment of the cilium in the inner segment is eccentric, in some species the remaining apex of inner segment often extends along the ciliary stalk and abuts on or fits into a concavity in the base of the outer segment. In instances described as abnormal (Tokuyasu and Yamada, 1960; Pedler and Tilly, 1965), the outer

and inner segments are continuous, with the outer segment discs apparently forming at the apex of the inner segment. Richardson (1969) believes that, in mammals, inner segments have a separate, non-ciliary connection to the outer segment,
but while such instances do occur, he has not established their normality and
frequency by valid criteria. Perhaps these secondary junctions are intermittent
phenomena.

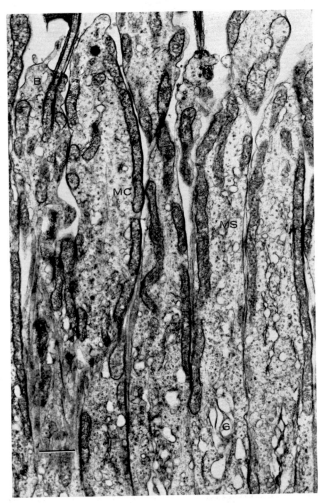

Fig. 8. A longitudinal section through the inner segment of a rat. Note at the apex the basal
centriole (B) with emerging rootlets, mitochondria (MC), fine microsomes (MS) and a Golgi
region (G). Bar designates 0.5 microns

From the circumference of the apex of the inner segment arise microvillous
processes which have been termed calycal processes inasmuch as they form a ruff
around the base of the outer segment. These number 9—12 in the macaque
(Cohen, 1961 b, 1965 a), and about 27 in Necturus where Brown et al. (1963), who

call them "dendrites", observe that in this species, they fit into the vertical surface folds of the cell membrane of the outer segment, reflecting the vertical grooves of the underlying disc incisures in register. These processes have no known function and their relation to the membrane of the outer segment is one of contiguity without evidence of specialized attachment, the membranous gap being 200 Å or more.

The rootlets descending from the centrioles are filamentous bundles exhibiting cross-striations (SJÖSTRAND, 1953b; TOKUYASU and YAMADA, 1959; MISSOTTEN, 1960a; COHEN, 1960, 1961b), and are in all known respects similar to those erupting from basal bodies of conventional cilia. MATSUSAKA (1967a) has reported cytochemical studies at the electron microscopic level which indicate ATP-ase activity in the ciliary rootlets of human rods. In the mouse (COHEN, 1960), the rootlets are typically "sandwiched" between long intracellular vacuoles. Rootlet function is unknown. They seem to be shorter in the inner segments of primate cones.

In many amphibians, reptiles, and birds a common feature of certain cone inner segments is a large oil droplet occurring at the apex of the inner segment. Sometimes multiple fine droplets are seen (PEDLER and BOYLE, 1969). The oil drop is often colored by the presence of light stable carotenoids. The mosaic of colored or clear droplets may not be random. Thus in pigeons there occurs a retinal area dominated by red droplets (known as the red field) surrounded by retina with yellow and orange oil droplets. In the double cones of the pigeon, one cone, the "chief" cone, possesses the oil droplet whereas its "accessory" cone partner lacks this object. The absorption spectra of the various colored droplets in the chicken have been studied by MEYER, COOPER and GERNEZ (1965), and related to types of cones (MEYER and COOPER, 1966). COOPER and MEYER (1968) have also studied the manner of first appearance of the droplets in the chick embryo. LIEBMAN (personal communication) has recently begun to study the absorption spectra of visual pigments of cones containing oil droplets and finds that the spectra are typical visual pigment spectra and vary in their absorption maxima. While in some cases the colored oil droplets cut off wavelengths outside the significant absorption range of the visual pigments, in other cases the masking absorption of a droplet overlaps the absorption range of the visual pigment in the same receptor. BERGER (1966) suggests that oil drops may originate from mitochondria.

In all inner segments, starting at the apical end or just below the oil droplet, if such is present, is a zone containing a large aggregation of mitochondria (Fig. 8). This mass has come to be known as the "ellipsoid." Its proximity to the outer segment suggests a role in the special metabolism of this region. BERGER (1964) has studied the origin of the mitochondria. KÜHNE (1876) suggested that the outer segment might serve to trigger a more definitive neuronal response by the rest of the cell and SJÖSTRAND (1963b) has stated that the outer segment might set off a generator potential in the inner. However, both these conjectures are only of historical interest in view of modern ideas of the electrophysiology of the photoreceptor (TOMITA, 1970; HAGINS et al., 1970).

It is also of interest that, in the Rhesus monkey at least, the mass of mitochondria in cones seems to have some positive correlation with the diameter of the inner segment and the extensiveness of the synaptic terminal of the receptor

(Cohen, 1961 b). It is likely that ongoing mitochondrial activity maintains certain metabolic pools and is not directly tied to the visual impulse of the photoreceptor. Enoch (1963, 1966) found that there were staining differences with neo-tetra-zolium-B in ellipsoids of cells which had been bleached as compared to unbleached cells. Report in the literature on staining differences in rods and various cones may in part be based on differential activities of the cells prior to fixation. Webster and Ames (1965), utilizing electron microscopy of rabbit retinas, believed that they could observe changes related to experimental anoxia of 3 minutes duration. These were reversible for 20 minutes.

The color of the mitochondrial mass, particularly if the cytochromes are in the reduced state, could modify the spectral distribution of the light passing through it.

Another occasional organelle of the inner segment is the so-called "paraboloid," found in inner segments of some receptors of certain fish, amphibians, birds, and reptiles. While there is considerable variation in the details of its structure in different species, in general the paraboloid is a region of many intracellular vacuoles between which are found masses of presumptive glycogen granules (Yamada, 1960a; Cohen, 1963a; Matsusaka, 1963, 1966). Rabinovitch et al. (1954) have presented histochemical evidence for glycogen in paraboloids of chicken rods and cones, but also for some in the ellipsoid region. Ishikawa and Pei (1965) have presented electron microscopic evidence for the presence of glycogen in mitochondria of the inner segment and synaptic terminal of the rod of the rat.

Walls (1942) has suggested that both the ellipsoid and paraboloid regions are refractile in the living state and may affect the passage of illumination. The cytological evidence suggests that these regions have other functions but cannot rule out optical effects as well. Nguyen H. Anh (1968, 1969a, b) has made a comparative study of inner segments, and detailed studies on this region in receptors of frogs and diurnal lizards.

The region between the cell nucleus and the ellipsoid is sometimes called the "myoid" region. This term derives from the well documented observation (Walls, 1942) that the ventricular portions of photoreceptors of certain species are capable of radial movements in response to light and that the "myoid" region becomes thin or thick as a photoreceptor is extended or contracted. In general, during the light adapted state the cone outer segments are brought closer to the retina proper. Rods may not move or become somewhat extended. The converse is true in the dark adapted state. Processes from the cells of the pigment epithelium interdigitate with the receptors in all retinas and in many species the pigment granules flow into these processes in response to light and out of them in the dark adapted state. These phenomena are most common in diurnal frogs, reptiles, and birds, and in certain fish such as in catfish.

Microtubules are another cytoplasmic feature of inner segments (Kuwabara, 1965) and are found throughout their length. The mechanism of photomechanical movement of receptors is unknown. A contraction of the ciliary rootlets is a possibility or the microtubular elements may be involved, as these are often associated with motile processes in diverse sorts of cells (Satir, 1967), as well as in the transport of materials (Rudzinska, 1967). Glicstein et al. (1969) have recently presented evidence for the selective translatory movement of a sub-class of the cone population of the goldfish by controlling the wavelengths initiating photomechanical movement. BERGER (1967) has studied subsurface membranes in inner segments of cones of fish.

The most vitread portion of the inner segment is rich in ribosomes and above the cell nucleus is found a typical Golgi complex of somewhat flattened vacuoles. In cones of sharks and teleosts (STELL, personal communication), and birds (CARASSO, 1956; YAMADA, 1960b; YASUZUMI et al., 1958; OKUDA, 1961; COHEN, 1963; NGUYEN H. ANH, 1968), the basal portion of the inner segments have regular vertical ridges so that in cross-section they appear like cog-wheels or gears. These are only lacking at the zone of contiguity of double or twin receptors. They have no known function.

UGA et al. (1970) have reported dense inter-receptor contacts (140 Å) of the inner segments of rods, and cones and rods in the human retina. LOCKET (1969, 1970a, b) studying retinas of deep sea fish, has reported (1970a) that there are tight junctions between the myoid regions of the receptors. However, his illustrations are not adequate to support the description of tight junctions in the specialized sense in which this term is employed by neurocytologists. Since receptors are well separated by Müller cells at the external limiting membrane, their diameters would have to increase in the region of the inner segment to permit inter-receptor contacts and it seems odd that this has not been previously noted.

G. External Limiting Membrane

The above term applies to the outer surface of the retina as demarked by a densely staining system of terminal bars, i.e. intercellular adhesions joining the mosaic of photoreceptor cells and glial cells of Müller. The latter send microvillous protrusions into the ventricular space and these form the classical "fiber baskets" seen around the base of the inner segments of the photoreceptors. As noted earlier, while the terminal bars at the junctions of the anterior faces of the cells of the pigment epithelium include a continuously girdling tight junction, in contradistinction such a tight element has been lacking in species studied thus far at the homologous junction of photoreceptors with Müller cells and was usually minimal, intermittent or lacking at the junction of two Müller cells (COHEN, 1965c; LASANSKY and DE FISCH, 1965). It has been suggested (COHEN, 1965c) that the tight junctions at the anterior face of the pigment cells could constitute the resistance barrier known as the "R" membrane as defined by electrophysiologists studying the retina (BRINDLEY, 1960). In those retinas where Müller cell junctions may include a tight zone, as in some amphibians (LASANSKY and DE FISCH, 1965), the external limiting membrane may constitute a second "R" membrane, but one of lesser resistance. Evidence for two resistance barriers has been advanced by BYSOV and HANITZSCH (1964).

The absence of a tight junction between photoreceptors and adjoining cells permits an unimpeded path for the flow of external currents between different regions of the photoreceptor. In the case of twin or double receptors, the junction between the members of the pair is unspecialized, 200 Å wide, and does not include any features of a terminal bar.

H. Cell Somas and Axons

There are no apparent specializations of the region of the soma. In duplex retinas, the nuclei of cones tend to occur close to the plane of the external limiting membrane whereas those of the more numerous rods occur vitread to this level. Microtubular elements are evident within the soma (KUWABARA, 1965). In general, extensions of the glial cells of Müller tend to separate the cells.

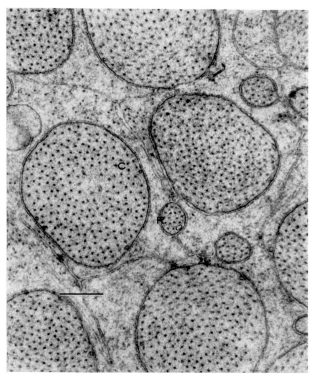

Fig. 9. Cross-sections of axons of cones (C) and rods (R) from the macular region of a Rhesus monkey. Müller cell processes separate the axons. Bar designates 1.0 micron

The photoreceptor axon (defined morphologically) erupts at the vitread pole of the soma and proceeds to the level of the cell terminal. During this course it is usually well separated from other axons by Müller cell extensions. In cross-section (Fig. 9) the axon shows numerous microtubular elements similar to those seen in axons and dendrites of conventional nerve cells. KUWABARA (1965) notes that in cones of the human retina about 250 such tubules were present and that the micro-tubules were well spaced from one another. This regularity of arrangement seems more striking than in axons elsewhere. In addition, at intervals along the axons, generally transverse arrangements of the system of cytoplasmic tubules known as endoplasmic reticulum were seen by KUWABARA to cross the axons. There are sug-gestions of connections of these irregular tubules to the surface. Well separated microtubules have also been seen in photoreceptor axons of the Rhesus monkey (COHEN, unpublished data).

A special problem in intracellular communication is posed by the distance between the terminals of foveal cones and the site of origin of their axons. This distance may exceed 125 μ. The axon diameter of a Rhesus foveal cone is about 1.3 μ (COHEN, unpublished data). BROWN and WATANABE (1962), recording extra-cellularly from the Rhesus fovea, failed to obtain any evidence of spike activity. Recent intracellular recordings from photoreceptor inner segments (TOMITA, 1965; TOYODA et al., 1969; WERBLIN, personal communication) likewise yielded no sign of "all or none" propagating transmission. While in Purkinje cells of the cerebellum the distance from the dendritic tips to the axon hillock is about 300 μ and the communication is probably electrotonic, these cells receive many excitatory inputs over their entire dendritic expanse and the stimulus may build in time. One must therefore imagine that if spike activity is absent in photoreceptors of the fovea, then the axon must have features which favor current spread, such as possessing an axonal membrane of high resistance. In many photoreceptors of extrafoveal areas or in retinas lacking foveas, the axons are so very short that propagated transmission is probably unnecessary. Indeed in the mouse and guinea pig, the synaptic sites with second order neurons are often at the level of the soma.

I. Receptor Terminals: Inter-Receptor Contacts and Synaptic Contacts

The synaptic bases of the photoreceptors generally lend themselves to the description "spherules" if they are small and round or "pedicles" if they are somewhat larger and present a flat base to the second order neurons (Fig. 10). The usual endings of mammalian rods are "spherules" and cones "pedicles." While this is also true of some forms like the alligator (KALBERER amd PEDLER, 1963), and catfish (COHEN, unpublished data), all the terminals in the pigeon (COHEN, 1963a) and frog (NILSSON, 1964a; DOWLING, 1968) qualify as pedicles despite the pre-sence of both rods and cones. However, NILSSON (1964a) indicates that frog rods have smaller pedicles than cones and this could prove to be generally true for rod pedicles. DOWLING and WERBLIN (1969) describe the rod terminals in Necturus as non-spherular. These only possess superficial synapses. SJÖSTRAND (1958) was the first to note two terminal classes in a rodent. The guinea pig possesses spherular and pedicular terminals which he termed alpha and beta terminals respectively. Similar terminal distinctions have been reported in such nocturnal rodents as the mouse (COHEN, 1960, 1967), rat (DOWLING, 1967b) and flying squirrel (COHEN, unpublished observations).

The terminals generally contain elements of the endoplasmic reticulum, small numbers of mitochondria, at times filamentous components (MOUNTFORD, 1964; OLNEY, 1968b), and occasional crystalloids (RADNOT and LOVAS, 1967). In the gecko (PEDLER and TANSLEY, 1963) and gray squirrel (COHEN, 1964) were also found large membranous aggregations formed from flattened vacuoles. There are also numerous small vesicles of about 300—500 Å diameter of the type now com-monly termed "synaptic vesicles." While these are suspected to serve as carriers

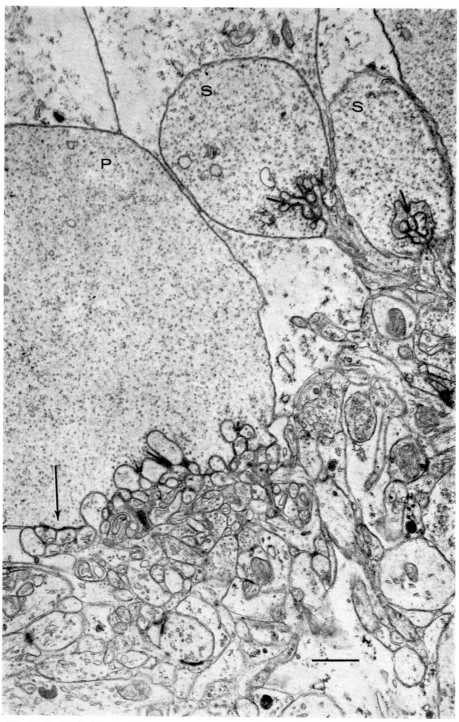

Fig. 10. Rod spherules (S) and a portion of a cone pedicle (P) of a Rhesus monkey. Note the invaginated synapses in both varieties of terminal and superficial synapses on the pedicle (arrow). Bar designates 1.0 micron

for neurohumors, it is not clear whether the vesicles break down as part of the release mechanism. Coated vesicles have also been observed (COHEN, 1964). No photoreceptor neurohumor has as yet been identified. OHMAN and SHELLEY (1968) have described a chemical procedure which causes the appearance of a fluorescent material in the somal region of only the photoreceptors among the retinal neurons. As their chemical procedure relates to substances resembling known neurohumors, this finding may represent a significant clue. In any event, MOUNTFORD (1963) could find no statistical evidence for changes in vesicle size after various periods in the light or dark but because of numerous factors affecting vesicle distribution which could not be readily controlled, no attempt was made to see whether the vesicle concentration or amount has changed. DE ROBERTIS and FRANCHI (1956) had previously reported size and distribution changes of vesicles in response to light. CRAGG (1969) found that upon the first exposure of rats to light there was a reduction in the width of rod terminals, and a reduced dispersion of the synaptic vesicles but no obvious change in their number.

Before dealing with the possible synaptic relations of receptors with the neurites of bipolar and horizontal cells, an interesting relationship among receptors must be considered. In SJÖSTRAND's (1958) study of the guinea pig, he noted that the beta (pedicular) receptors sent out processes which terminated on the alpha (spherular) terminals and that the contacts exhibited increased membrane density. Further work by SJÖSTRAND and MOUNTFORD (SJÖSTRAND, 1965) revealed that while short receptor processes may end on nearby receptor terminals, longer processes may bypass several nearby terminals to reach a more distant one. The synaptic bases of pigeon photoreceptors likewise give rise to long processes which run horizontally (COHEN, 1963a) and similar processes have been observed in the alligator (KALBERER and PEDLER, 1963). Processes from receptor terminals which contact terminals of other receptors have also been noted in the gray squirrel (COHEN, 1964), mouse (COHEN, 1967), human (MISSOTTEN et al., 1963), and in the human and macaque (COHEN, 1965b). Broadside contiguities of receptor terminals, lacking membrane densification, have been seen in the frog (NILSSON, 1964b) and in Necturus (DOWLING and WERBLIN, 1969). It is doubtful that this is an equivalent arrangement to the contacts exhibiting membrane densities.

In studies on the human retina, these contacts were seen to exhibit a highly organized pattern (COHEN, unpublished data). For example, in a horizontal retinal section passing through the pedicles of human foveal cones, each such pedicle was linked to all adjacent pedicles. Serial sections revealed that the contacts were likewise not confined to a single plane. The contacts were established by short processes from the pedicles. In a slightly more peripheral region, the short pedicular processes also terminated on all nearby spherules of rods. In the retinal periphery, cone to cone contacts were readily found and a serial reconstruction of any single pedicle revealed that processes emerged to contact all nearby rod spherules. Rod spherules may occasionally receive more than one contact but do not themselves give rise to contacting processes. The intercellular space at a contact site was not tight, although by manipulation of the fixation and tonicity (COHEN, 1965b) it could be reduced well below 200 Å. However, these tightened junctions never achieved the crisp, "crystalline" appearance of well accepted tight junctions, as in the terminal bars of the cells of the pigment epithelium in the same preparations.

The latter tight zones did not require any special manipulation of the fixation medium for their demonstration. However, MISSOTTEN and VAN DEN DOOREN (1966) call the junctions tight, and DOWLING and BOYCOTT (1966) see them as tight "occasionally."

The fact that synaptic vesicles may be present in these processes is of little value in deciding whether the contacts are of synaptic significance because of the presence of such vesicles everywhere in the terminal. The fact that the vesicles do not cluster at the contact points is likewise not a strong argument against their synaptic nature because vesicles do not cluster against presumptive presynaptic membrane anywhere in the terminal (COHEN, 1965b).

The most powerful arguments against receptors interacting at this level are physiological, such as that cones containing a particular pigment are said to only inhibit cones with the same pigment (ALPERN and RUSHTON, 1965), and that the best resolution of the foveal area (PIRENNE, 1962), measured with grating techniques which bypass the optics of the eye, tend to approximate the distances between individual cones. Perhaps the right physiological questions have not been asked. A negative anatomical argument is that the contacts are less frequent in the macaque than in man. It could be argued that the system is of mechanical significance in terms of resisting horizontal physical stresses on the retina. While clearly possible, the junctions at the external limiting membrane would seem to make such a system redundant. An open mind on these matters is clearly in order.

The definitive synaptic junctions with photoreceptors fall into two groups, i.e. superficial and invaginated. Superficial contacts of presumptive synaptic significance and dendritic origin were observed on pedicle terminals in the Rhesus monkey (COHEN, 1961b, Fig. 19; DOWLING and BOYCOTT, 1966), in the alligator (KALBERER and PEDLER, 1963), in man (MISSOTTEN et al., 1964), in the gray squirrel and mouse (COHEN, 1964, 1967), in the frog (DOWLING, 1968), and in the turtle (LASANSKY, 1969). Apart from the absence of clustered vesicles, the junctions appear to be synaptic contacts with a 200 Å separation and with membrane densities on both sides. In the alligator, KALBERER and PEDLER (1963) described what they termed "sawtooth" striations of some material in the gap and LASANSKY (1969), referring to similar findings in the turtle, has pointed to their similarity to the septate junctions of invertebrates which are said to facilitate electrical coupling of neurons. LASANSKY also believes that he has seen suggestions of these in toad and rat retinas.

All the synapses of spherules and the remaining pedicular synapses can be termed invaginated synapses in that the synapsing neurites infold the terminal membrane of the photoreceptor. This is most striking in the case of spherules.

The deep insertion of neurites into the spherule bases was noted by DE ROBERTIS and FRANCHI (1956) and CARASSO (1957) and studied in more detail by LADMAN (1958), SJÖSTRAND (1958), LANZAVECCHIA (1960) and COHEN (1960). In general these early studies showed that in the receptor cytoplasm, opposite to the tips of the invaginated neurites, there was to be found a lunate plate. This was oriented perpendicular to the receptor membrane, with the lunate concavity facing the membrane. LADMAN (1958) named this object a synaptic lamella, a name aptly corresponding to its three-dimensional structure, but other workers, impressed by the appearance of sections of this plate, have termed this a synaptic ribbon, a term now in common use. It is typical of the lamella or ribbon, that it tends to lie in a plane which passes between the often paired invaginated neurites. LANZAVEC-CHIA (1960) first noted that the lamella had a pentalaminate appearance in cross-section. This organelle seems to be surrounded by a clear zone of about 200 Å and this, in turn, by a halo of synaptic vesicles. However this apparatus may not directly abut on the receptor membrane opposite the contacting neurites. For as LADMAN (1960) first observed, the lamella is separated from the membrane by a ribbon-like density which he termed the arciform density. This paralleled the lunate concavity of the overlying synaptic lamella and was "U" shaped in cross-section. The identity of the entering neurites is beyond the subject of this article but the invagination they enter is not a simple concavity. Serial reconstructions (COHEN, unpublished data) show that evaginated protrusions of the receptor base weave among the neurites within the concavity of the spherules so that all the invaginated processes become adjacent to receptor membrane.

In the case of the pedicles, the sites of invaginations are more numerous, less deep and the synaptic lamellae or ribbons somewhat smaller. More recent investigations of these regions include the work of MISSOTTEN et al. (1963, 1964), MISSOTTEN (1965), SJÖSTRAND and MOUNTFORD (in SJÖSTRAND, 1965), and the work of STELL (1965, 1967).

The deep insertion of the invaginated neurites lends itself to the speculation that it is important to confine the site of interaction (COHEN, 1963 b). This might mean that the diffusion of neurohumor was restricted or that there was low shunting of current. Indeed, while in most chemical synapses there is no apparent electrical coupling of the pre- and post-synaptic elements (FATT and KATZ, 1951) something of this type does occur in the chick ciliary ganglion (MARTIN and PILAR, 1963) where the pre-synaptic element makes a chalice or cup which tends to envelop a substantial portion of the post-synaptic element. It is not clear whether any class of neurite at the synapse is electrically stimulated or facilitated. However, there are many missing elements for a rational analysis of the synaptic area of receptors, such as a knowledge of the resistances of the membrane involved.

It should also be noted that organelles resembling the synaptic lamella or ribbon occur in other receptors such as in cochlear hair cells (SMITH and SJÖSTRAND, 1961); lateral line receptors (BARETS and SZABO, 1962); and in some invertebrate eyes (TRUJILLO-CENOZ, 1965). Their function is unclear.

While photoreceptors are not the first neuronal type which can be differentiated in the development of the retina, the synapses of the photoreceptor are the first to develop to their mature appearance, location, and numbers (OLNEY, 1968; WEIDMAN and KUWABARA, 1968; NILSSON and CRESCITELLI, 1969). Correlations with the appearance of electrical activities are of interest in such studies and NILSSON and CRESCITELLI found some light-evoked electrical activity in the bullfrog retina preceding the maturation of the receptor terminals.

J. Rods and Cones: A Contemporary View

The terms "rod" and "cone" basically refer to the morphology of that portion of the photoreceptor which protrudes into the ventricular space between the retina and pigment epithelium. Although originally referring to the shape of the outer segment, the term "cone" now seems to be applied to cells with an overall tapered shape of coupled inner and outer segments, due to an inner segment possessing an obviously larger diameter than a relatively short outer segment, whether the latter is clearly conical or not.

Walls (1942) believed that receptor form is but one of a family of receptor properties having evolved in relation to whether the receptor participates in a scotopic or photopic retinal system. If evolution seems to have resulted in the possible adaptation of receptor form to a more efficient role in such systems, it becomes reasonable to inquire as to whether other receptor properties are likewise so modified. Even if there are some unusual environmental situations affecting the quality of available illumination, such as life in the depths of the sea, the lack of clarity in muddy streams, etc., for most animals the visual habitat includes some segments of the alternating pattern of night and day and it is this line of reasoning that makes the duplicity theory of retinal function most plausible.

The other properties of receptors that may contribute to retinal duplicity are: the visual pigment, the electrical responses of the various receptor regions including their modification by light level and recent activity, ultrastructural details, biochemistry, and the organization of the synaptic terminals. One must consider how these various properties separately and collectively relate to the receptors communicative behavior and how this, in turn, fits into various functional systems of retinas operating at scotopic, mesopic and photopic levels.

Unfortunately, as soon as some instances of correlations with receptor form have been shown to exist, there has been a premature tendency to extend the definition of "rod" or "cone" so that they may be identified by the presence of these other properties. In certain instances these correlations are based on very few cases or for a few related species of animals. Moreover, as there is an enormous variety in visual environments and the properties are, at least in some cases, potentially independent variables, exceptions to the correlations are quite likely to occur. It seems reasonable therefore to retain a morphological definition of "rod" and "cone" and to regard the form as a probabilistic indicator only for the presence of members of a certain loose grouping of functionally associated properties, not totally identical in different species.

In some species the form of the receptors may be too ambiguous to permit initially classifying them as "rod" or "cone" but the later exploration of other properties may indicate that it is both non-arbitrary and useful to extend the definition and do so. Thus the ultrastructure and physiology of the "rod-like" cones of the primate fovea clearly places them at one end of the spectrum of the more conventional cones of this retina. In a relatively few instances, a summary view of the properties of a receptor type may not resolve an ambiguity and such receptors may not usefully be termed a "rod" or "cone." One may speculate, as does Walls (1942), that these ambiguous varieties are in evolutionary transition between rods and cones but untestable speculations are of little value. In other instances, it may be useful to precede "rod" or "cone" by "morphological" to emphasize an uncertainty about other properties.

The presence of rods and cones still predicts a duplex retinal property. The work of Westheimer (1967), suggesting that cones of differing shape may differentially contribute to the Stiles-Crawford effect (1933), shows the potentialities of form as an optical adaptation. Further indications of physiological divergences of rods and cones are seen in the developing case of cones having many more discs open to extracellular space than rods, and in the rod discs possessing scalloped perimeters. Do the open discs correlate with the faster decay of cone potentials, as in the monkey (BROWN et al., 1965), or with faster pigment regeneration in cones or other metabolic interchanges involving the discs? The behavioral distinctions between rods and cones of certain species, as seen in opposing photomechanical movements (WALLS, 1942), are an obvious physiological divergence. Differential responses to drugs and X-ray by the rods and cones of certain retinas are also well known (NOELL, 1965).

Difficulties can also arise if one attempts to regard the mechanism of the duplex retinal property as entirely residing in the photoreceptors themselves. Thus WALLS (1942) proposed calling rods, low threshold "scotocytes" and cones, high threshold "photocytes" despite the fact that, as WEALE (1961) has emphasized, it is by no means established that scotopic versus photopic *retinal* responses are paralleled in scotopic versus photopic *receptor* responses. The approach of WALLS was premature. The new ability to place microelectrodes in photoreceptors will go a long way towards clarifying this point and others but it can be said *a priori* that differential sensitivity of the retina as a function of illuminance level does not require scotopic and photopic receptor sensitivities, although these certainly might enhance the function of the system. If one has two receptor classes based on differing visual pigments, then a shift in retinal spectral sensitivity with illumin- ance will depend in part on the absorptions of the pigments and their concen- trations, the relative numbers of receptors in the two classes, the degrees of neu- ronal convergence of signals from each class based upon retinal "wiring," the resultants of the inhibitory interactions of signals in the two systems, etc. Indeed, differential retinal sensitivity in the scotopic and photopic ranges could be achieved with two receptor classes containing the same pigment, provided they were wired with different degrees of convergence and some adaptive "switching" mechanism was present. Similarly, one must distinguish the later neural contributions to flicker fusion from those of rods or cones.

It is also unsatisfactory to define "rods" and "cones" in terms of the absorption spectrum of the contained photopigment, although it is clear that evolutionary pressures will tend to favor the presence of visual pigments in particular photo- receptors so that the spectral sensitivity of the cell will best fit its visual function. The new tool of microspectrophotometry of single photoreceptors will reveal whether one photoreceptor always has one pigment and what the absorption characteristics of that pigment happens to be. Since we already know of red and green rods in the frog and of various cone pigments of man it is clear that form is not an absolute indicator of pigment. While it is likely that a morphological rod, in a retina also possessing cones, will prove to possess a rhodopsin or porphyropsin, there is no necessity for this to be true. There is also no *a priori* reason why one particular pigment, defined in terms of absorption characteristics, might not func- tion in a photoreceptor feeding information to a scotopic system in one organism and to a photopic system in another. Indeed, LIEBMAN and ENTINE (1968) have found pigments in accessory cones of the frog and turtle with absorption spectra which qualify them as rhodopsins. DOWLING (1967 b) has suggested the possible existence of rhodopsin containing cones in the rat, alongside the rhodopsin containing rods (but see GREEN, 1971).

As an example of an ambiguous situation, consider the American gray squirrel. This diurnal animal was said to possess an all cone retina (WALLS, 1942) and its photopic character and absence of a Purkinje shift seemed supported by electroretinography (ARDEN and TANSLEY, 1955). Yet DARTNALL (1960) had extracted retinal pigment and found a typical rhodopsin whose peak absorption did not correspond to any ERG sensitivity peak nor to absorption spectra as deduced from reflection densitometry studies (WEALE, 1955). However, a behavior study by ARDEN and SILVER (1962) supported DARTNALL in that a Purkinje shift was seen and at the lower range of illuminance the action spectrum peaked at 502 Å. In the electron and light microscopes two receptor classes were then discerned (COHEN, 1964) and these occurred in roughly equal numbers. In terms of overall shape, both could be called cones despite possessing essentially cylindrical outer segments, because both had inner segments whose diameter was greater than the outer. Two terminal types were seen. Light microscopy showed that one class had a much longer outer segment. Electron microscopy revealed that this receptor type had considerably fewer open discs at the base of the outer segment than did the other. The nuclei of the two varieties were also distinctive and COHEN (1964) suggested that as one cell type possessed certain details often seen in rods, it might be tied to a scotopic system. About the same time GOURAS (1964) reported a Purkinje shift in an electroretinographic study, thus strengthening the case for a duplex retina.

Another possibly ambiguous situation exists for the lizards known as Geckos. WALLS (1942) describes the nocturnal geckos as possessing all rod retinas. PEDLER and TILLY (1964) have studied both nocturnal and diurnal geckos and DUNN (1966a, b, c, 1969) nocturnal geckos only by electron microscopy. PEDLER and TILLY illustrate one outer segment, that of the nocturnal form, *Gecko gecko*, and it is seen to have discs infolding from the cell membrane, as is the usual finding in outer segments of cones. But they do not comment on this point. DUNN on the other hand, while illustrating and enumerating various varieties of rods, photographed at low magnification, does not appear to have concentrated on details of the ultrastructure of the outer segments, and regards the retinas as all rod retinas. Both investigations show at least some cells with inner segments decidedly larger than outer segments and all cells end in pedicles. DODT and JESSEN (1961) and CRESCITELLI (1965) have found duplex function in the retinas of nocturnal Geckos. CRESCITELLI moreover found two pigments adsorbing at 521 nm and 478 nm respectively, in *Gecko gecko*. DENTON's (1956) study of the pupillary action spectrum in this form gave a result far from that of a rhodopsin dominated system. TOYODA et al. (1969) obtained intracellular recordings from receptors of the nocturnal Gecko and compared them to similar records from the salamander, Necturus. They regarded the Gecko responses as being rod-like in exhibiting a slow decay and a saturating level 2-log units below those of Necturus. The latter's responses seemed conelike despite the fact that about half the receptors are said to be rods.

It is not clear therefore whether the two pigments (at least) are segregated in rod classes or whether a more detailed ultrastructural analysis would show that one of the classes being termed "rods" actually possessed features more commonly seen in cones.

KALBERER and PEDLER (1963) note that while in sections over half the receptors in the duplex retina of the alligator end in pedicles as opposed to spherules, only one in ten of the photoreceptors, classified by outer segments is cone-like. One must be careful here because pedicles have a greater diameter than outer segments, and would appear in more sections. GRANIT (1962) has proposed that pedicles may be useful hallmarks of cones. In particular species, such as man, this is clearly true. But the commonly accepted rods and cones of frogs, all end in pedicles (NILSSON, 1964; DOWLING, 1968). DOWLING and WERBLIN (1969) found the likely rod terminals in *Necturus* to be non-spherular with superficial contacts only, but with larger ribbons than the cone pedicles. Thus, while PEDLER (1965) correctly points out that the complexity and connectivity of the terminal is an important variable in the classification of photoreceptors, it clearly should not determine their classification as "rod" or "cone".

When rods end in pedicles, as in the frog, each terminal contacts more neurites than does a spherule of a rat rod. In each form, rod bipolars receive inputs from more than one rod. It would be of some interest to compare the totality of cross-section for quantal capture represented by the receptor disc area and pigment concentration of all rods signalling an individual rod bipolar in these two species. This might help indicate the functional distinctions, if any, between having many fine rods converging on individual bipolars versus relatively fewer large rods, similarly converging.

Thus far electron microscopists have failed to find a retina that is free of pedicles. Few species have been studied, but this may suggest that spherules are more recent in the evolutionary sense. Animals that are reputedly nocturnal, like the mouse (COHEN, 1960), rat (DOWLING, 1967b), fruit bat (PEDLER and TILLY, 1969), little brown bat and flying squirrel (COHEN, unpublished observations), while having most receptors terminating in spherules, yet have significant numbers terminating in pedicles. Certain duplex functions are probably present in all these species. Indeed they may also have morphological cones which are poorly preserved or hard to find. DOWLING (1967b) has reported indications of these in the rat and COHEN (unpublished observations) has seen suggestions of cones in the flying squirrel. HAMASAKI (1967) has found morphological cones among the rods in the nocturnal owl monkey and scotopic and photopic limbs in their dark adaptation and flicker fusion curves. JONES and JACOBS (1963) observed a Purkinje shift in this animal. DOWLING (1967b) failed to find scotopic and photopic limbs in the usual ERG adaptation curves of the rat, nor did HELLNER (1966) find them in the mouse. But DODT and ECHTE (1961) and DOWLING (1967b) did find photopic and scotopic limbs in studying adaptational effects on flicker fusion electroretinograms in rats. DOWLING believes that a small Purkinke shift seen in the rat by DODT and ECHTE might be due to reflectance. In his studies the same spectral sensitivity was seen at both the scotopic and photopic levels and he suggests that rats may have rhodopsin cones alongside rhodopsin rods.

A most valuable experimental animal would be one with but one variety of photoreceptor in terms of form, pigment, and position in the neurological organization of the retina as evidenced by the absence of duplex physiology and behavior. Should such exist, it would be most useful for exploring subsequent retinal connections such as what receptors would interact via horizontal neurons, and on the relations of bipolars and horizontal cells in organizing visual fields. While RIPPS and DOWLING (personal communication) have some good evidence suggesting that the retina of a skate may meet the criteria of a truly all rod retina, most other claims of retinas with but one receptor type rest on insufficient or equivocal evidence.

Clearly the remarkable evolutionary diversity of photoreceptors, retinas, and visual systems, affords many opportunities for analyzing the significant variables in visual function.

References

ABRAHAMSON, E. W., OSTROY, S. E.: The photochemical and macromolecular aspects of vision. Progr. Biophys. Molec. Biol. **17**, 181—215 (1967).

AKHTAR, M., BLOSSE, P. T., DEWHURST, P. B.: The reduction of a rhodopsin derivative. Life Sci. **4**, 1221—1226 (1965).

— — — Studies on vision: The nature of the retinal-opsin linkage. Biochem. J. **110**, 693—702 (1968).

ALPERN, M., RUSHTON, W. A. H.: The specificity of the cone interaction in the after flash effect. J. Physiol. (Lond.) **176**, 473—482 (1965).

ANDERSON, R. E.: Is retinal-phosphatidyl ethanolamine in chromophore of rhodopsin? Nature (Lond.) **227**, 954—955 (1970).

ARDEN, G. B., ERNST, W.: Mechanism of current production found in pigeon cones but not in pigeon or rat rods. Nature (Lond.) **223**, 528—531 (1969).

— SILVER, P. H.: Visual thresholds and spectral sensitivities of the gray squirrel (Sciurus carolinensis leucotis). J. Physiol. (Lond.) **163**, 540—557 (1962).

— TANSLEY, K.: The spectral sensitivity of the pure cone retina of the gray squirrel (Sciurus carolinensis leucotis). J. Physiol. (Lond.) **127**, 592—602 (1955).

BARETS, A., SZABO, T.: Appareil synaptique des cellules sensorielles de l'ampoule de Lorenzini chez la Torpille, *Torpedo marmorata*. J. Microscopie **1**, 47—54 (1962).

BERGER, E. R.: Mitochondria genesis in the retinal photoreceptor inner segment. J. Ultrastruc. Res. **11**, 90—111 (1964).

— On the mitochondrial origin of oil drops in the retinal double cone inner segments. J. Ultrastruc. Res. **14**, 143 (1966).

— Subsurface membranes in paired cone photoreceptor inner segments of adult and Neonatal Lebistes retinae. J. Ultrastruc. Res. **17**, 220—232 (1967).

BLASIE, J. K., DEWEY, M. M., BLAUROCK, A. E., WORTHINGTON, C. R.: Electron microscope and low-angle x-ray diffraction studies on outer segment membranes from the retina of the frog. J. molec. Biol. **14**, 143—152 (1965).

— — WORTHINGTON, C. R.: Molecular localization of the photopigment in the outer segment membranes of frog retinal receptors. J. Histochem. Cytochem. **14**, 789 (1966).

— WORTHINGTON, C. R., DEWEY, M. M.: Molecular localization of frog retinal receptor photopigment by electron microscopy and low-angle x-ray diffraction. J. molec. Biol. **39**, 407—416 (1969).

BLAUROCK, A. E., WILKINS, M. H. F.: Structure of frog photoreceptor membrane. Nature (Lond.) **223**, 906—909 (1969).

BONTING, S. L., CARAVAGGIO, L., CANADY, M. R.: Studies of sodium potassium activated adenosine triphosphatase. X. Occurrence in retinal rods and relation to rhodopsin. Exp. Eye Res. **3**, 47—56 (1963).

— BANGHAM, A. D.: On the biochemical mechanism of the visual process. In: Biochemistry of the Eye. Basle: Karger 1968.

BOROVIAGIN, V. L.: Electron microscopic examination of retinal cones in the frog. Biofizika **7**, 154—164 (1962a).

— On the submicroscopic structure of retinal rods in frogs. Biofizika **7**, 734—740 (1962b).

BORTOFF, A.: Localization of slow potential responses in the Necturus retina. Vision Res. **4**, 627—635 (1964).

— NORTON, A.: Positive and negative potential responses associated with vertebrate photo-receptor cells. Nature **206**, 626—627 (1965).

BOWNDS, D.: Site of attachment of retinal in rhodopsin. Nature (Lond.) **216**, 1178—1181 (1967).

— WALD, G.: Reaction of the rhodopsin chromophore with sodium borohydride. Nature (Lond.) **205**, 254—257 (1965).

BRIERLEY, G. P., FLEISHMAN, D., HUGHES, S. D., HUNTER, G. R., McCONNELL, D. G.: On the permeability of isolated bovine retinal receptor outer segment fragments. Biochim. Biophys. Acta **163**, 117—120 (1968).

BRIGHTMAN, M. W.: The distribution within the brain of ferritin injected into cerebrospinal fluid compartments. I. Ependymal distribution. J. Cell Biol. **26**, 99—123 (1965a).

— The distribution within the brain of ferritin injected into cerebrospinal fluid compartments. II. Parenchymal distribution. Amer. J. Anat. **117**, 193—220 (1965b).

— The intracerebral movement of proteins injected into blood and cerebrospinal fluid of mice. Progr. Brain Res. **29**, 19—37 (1967).

— PALAY, W. L.: The fine structure of ependyma in the brain of the rat. J. Cell Biol. **19**, 415—439 (1963).

— REESE, T. S.: Junctions between intimately apposed cell membranes in the vertebrate brain. J. Cell Biol. **40**, 648—677 (1969).

BRINDLEY, G. S.: Physiology of the visual pathway. London: Arnold 1960.

— RUSHTON, W. A. H.: The colour of monochromatic light when passed into the human retina from behind. J. Physiol. (Lond.) **147**, 204—208 (1959).

— GARDNER-MEDWIN, A. R.: The origin of the early receptor potential of the retina. J. Physiol. (Lond.) **182**, 185—194 (1966).

BROWN, K. T., WATANABE, K.: Isolation and identification of a receptor potential from the pure cone fovea of the monkey retina. Nature (Lond.) **193**, 958—960 (1962a).

— — Rod receptor potential from the retina of the night monkey. Nature (Lond.) **193**, 547—550 (1962b).

— MURAKAMI, M.: Early receptor potential of the vertebrate retina. Nature (Lond.) **204**, 736—740 (1964).

— WATANABE, K., MURAKAMI, M.: The early and late receptor potential of monkey cones and rods. C. S. H. Symp. Quant. Biol. **30**, 457—482 (1965).

— CRAWFORD, J. M.: Intracellular recording of rapid light-evoked responses from pigment epithelium cells of the frog eye. Vision Res. **7**, 149—163 (1967a).

— — Melanin and the rapid light-evoked responses from pigment epithelium cells of the frog eye. Vision Res. **7**, 165—178 (1967b).

BROWN, P. K., GIBBONS, I. R., WALD, G.: The visual cells and visual pigment of the mudpuppy, *Necturus*. J. Cell Biol. **19**, 79—106 (1963).

BÜLOW, N.: Light scattering by pigment granules in the human retina. Acta ophthal. (Kbh.) **46**, 1048—1053 (1968).

BYSOV, A. V., HANITZSCH:, R.: Die elektrischen Eigenschaften und die Struktur der R-Membran. Vision Res. **4**, 483—492 (1964).

CARASSO, N.: Mise en évidence de prolongements cytoplasmiques inframicroscopiques au niveau du segment interne des cellules visuelles du Gecko (Reptile). C. R. Acad. Sci. (Paris) **242**, 2988—2991 (1956).

— Etude au microscope électronique des synapses chez le têtard *d'Alytes obstetricans*. C. R. Acad. Sci. (Paris) **245**, 216—219 (1957).

— Etude au microscopique électronique de la morphogenèse du segment externe des cellules visuelles chez le Pleurodèle. C. R. Acad. Sci. (Paris) **248**, 3058—3060 (1959).

CHERRY, R. J.: Semiconduction and photoconduction of biological pigments. Quart. Rev. Chem. Soc. (Lond.) **22**, 160—178 (1968).

CLARK, A. W., BRANTON, D.: Fracture faces of frozen outer segments from Guinea Pig retina. Z. Zellforsch. **91**, 586—603 (1968).

Clifford, J., Pethica, B. A., Smith, E. G.: A nuclear magnetic resonance investigation of motion in erythrocyte membranes. In: Membrane models and the formation of biological membranes. New York: Wiley 1968.

Cohen, A. I.: The ultrastructure of the rods of the mouse retina. Amer. J. Anat. 107, 23—48 (1960).

— Some preliminary electron microscopic observations of the outer receptor segments of the retina of the *Macaca rhesus*. In: The structure of the eye. New York: Academic Press 1961a.

— The fine structure of the extra-foveal receptors of the Rhesus monkey. Exp. Eye Res. 1, 128—136 (1961b).

— The fine structure of the visual receptors of the pigeon. Exp. Eye Res. 2, 88—97 (1963a).

— Vertebrate retinal cells and their organization. Biol. Rev. 38, 427—459 (1963b).

— Some observations on the fine structure of the retinal receptors of the American Gray squirrel. Invest. Ophthal. 3, 198—216 (1964).

— New details of the ultrastructure of the outer segments and ciliary connectives of the rods of human and macaque retinas. Anat. Rec. 152, 63—80 (1965a).

— Some electron microscopic observations on inter-receptor contacts in the human and macaque retinae. J. Anat. (Lond.) 99, 595—610 (1965b).

— A possible cytological basis for the "R" membrane in the vertebrate eye. Nature (Lond.) 205, 1222—1223 (1965c).

— An electron microscope study of the modification by monosodium glutamate of the retinas of normal and "rodless" mice. Amer. J. Anat. 120, 319—356 (1967).

— New evidence supporting the linkage to extracellular space of outer segment saccules of frog cones but not rods. J. Cell Biol. 37, 424—444 (1968).

— Rods and cones and the problem of visual excitation. In: The retina: Morphology, function and clinical characteristics (Jules Stein Symposium 1966). U.C.L.A. Forum in Med. Sci., 1969.

— Further studies on the question of the patency of saccules in outer segments of vertebrate photoreceptors. Vision Res. 10, 445—454 (1970).

— Electron microscopic observations on form changes in photoreceptor outer segments and their saccules in response to osmotic stress. J. Cell Biol. 48, 547—565 (1971).

Cone, R. A.: Early receptor potential of the vertebrate retina. Nature (Lond.) 204, 736 (1964).

Cooper, T. G., Meyer, D. B.: Ontogeny of retinal oil droplets in the chick embryo. Exp. Eye Res. 7, 434—442 (1968).

Cope, F. W.: Nuclear magnetic resonance evidence using D_2O for structured water in muscle and brain. Biophys. J. 9, 303—319 (1969).

Cragg, B. G.: Structure changes in naive retinal synapses detectable within minutes of first exposure to daylight. Brain Res. 15, 79—96 (1969).

Crescitelli, F.: The spectral sensitivity and visual pigment content of the retina of Gecko gecko. In: Color vision, physiology and experimental psychology. Ciba symposium. Boston: Little Brown (1965).

Danielli, J. F., Davson, H. A.: A contribution to the theory of the permeability of thin films. J. cell. comp. Physiol. 5, 495—508 (1935).

Dartnall, H. J. A.: Visual pigment from a pure cone retina. Nature (Lond.) 188, 475—479 (1960).

Denton, E. J.: The responses of the pupil of the Gecko gecko to external light stimulus. J. gen. Physiol. 40, 201—215 (1956).

— The contribution of the oriented photosensitive and other molecules to the absorption of the whole retina. Proc. roy. Soc. B. 150, 78—94 (1959).

Depont, J. J. H. H. M., Daemen, F. J. M., Bonting, S. L.: Biochemical aspects of the visual process. II. Schiff base formation in phosphatidylethanolamine monolayers upon penetration by retinaldehyde. Biochim. biophys. Acta (Amst.) 163, 204—211 (1968).

De Robertis, E.: Electron microscope observations on the submicroscopic organization of the retinal rods. J. biophys. biochem. Cytol. 2, (Suppl.) 319—330 (1956a).

— Morphogenesis of the retinal rods. J. biophys. biochem. Cytol. (Suppl.) 2, 209—218 (1956b).

— Franchi, C. M.: Electron microscope observations on synaptic vesicles in synapses of the retinal rods and cones. J. biophys. biochem. Cytol. 2, 307—318 (1956).

De Robertis E., Lasansky, A.: Submicroscopic organization of the retinal cones of the rabbit. J. biophys. biochem. Cytol. **4**, 743—746 (1958).
— — Ultrastructure and chemical organization of photoreceptors. In: The structure of the eye. New York: Academic Press 1961.
Dewey, M.M., Davis, P., Blasie, J.K.: Immunofluorescent localization of rhodopsin in the retina of the frog. J. Histochem. Cytochem. **14**, 789 (1966).
— Davis, P.K., Blasie, J.K., Barr, L.: Localization of rhodopsin antibody in the retina of the frog. J. molec. Biol. **39**, 395—405 (1969).
Dodt, E., Echte, K.: Dark and light adaptation in pigmented and white rat as measured by electroretinogram threshold. J. Neurophysiol. **24**, 427—445 (1961).
— Jessen, J.K.: The duplex nature of the retina of the nocturnal gecko as reflected in the electroretinogram. J. gen. Physiol. **44**, 1143—1158 (1961).
Doggenweiler, C.F., Frenk, S.: Staining properties of lanthanum on cell membranes. Proc. nat. Acad. Sci. (Wash.) **53**, 425—430 (1965).
Dowling, J.E.: Foveal receptors of the monkey retina: Fine structure. Science **147**, 57—59 (1965). Letter. Science **157**, 584—585 (1967).
— Synaptic organization of the frog retina: An electron microscopic analysis comparing the retinas of frogs and primates. Proc. roy. Soc. B **170**, 205—228 (1968).
— Gibbons, I.R.: The effect of vitamin A deficiency of the fine structure of the retina. In: The structure of the eye. New York: Academic Press 1961.
— Sidman, R.L.: Inherited retinal dystrophy in the rat. J. Cell Biol. **14**, 73—109 (1962).
— Boycott, B.B.: Organization of the primate retina: Electron microscopy. Proc. roy. Soc., B **166**, 80—111 (1966).
— Werblin, F.S.: Organization of the retina of the Mudpuppy, *Necturus maculosus*. I. Synaptic structure. J. Neurophysiol. **32**, 315—338 (1969).
Droz, B.: Synthesis and migration of proteins in the visual cells of rats and mice. Anat. Rec. **139**, 221 (1961).
— Dynamic condition of proteins in the visual cells of rats and mice as shown by radio-autography with labeled amino acids. Anat. Rec. **145**, 157—168 (1963).
Dunn, R.F.: Studies on the retina of the gecko, *Coleonyx variegatus*. I. The visual cell classification. J. Ultrastruct. Res. **16**, 651—671 (1966a).
— II. The rectilinear visual cell mosaic. J. Ultrastruct. Res. **16**, 672—684 (1966b).
— III. Photoreceptor cross-sectional area relationships. J. Ultrastruct. Res. **16**, 685—692 (1966c).
— The dimensions of rod outer segments related to light absorption in the gecko retina. Vision Res. **9**, 603—609 (1969).
Eakin, R.M.: The effect of vitamin A deficiency on photoreceptors in the lizard *Sceloporus occidentalis*. Vision Res. **4**, 17—22 (1964).
— Differentiation of rods and cones in total darkness. J. Cell Biol. **25**, 62—65 (1965).
Engström, K.: On the cone mosaic in the retina of *Parlis major* (Great Tit). Acta Zool. (Stockh.) **39**, 65—69 (1958).
— Cone types and cone arrangements in the retina of some Gadids. Acta Zool. (Stockh.) **42**, 228—243 (1961).
— Cone arrangements in Teleost retinae. Acta Zool. (Stockh.) **44**, 179—243 (1963).
Enoch, J.M.: Wave guide modes: Are they present and what is their role in the visual mechanism? J. opt. Soc. Amer. **50**, 1025—1026 (1960).
— Nature of the transmission of energy in the retinal receptors. J. opt. Soc. Amer. **51**, 1122—1126 (1961).
— The use of tetrazolium to distinguish between retinal receptors exposed and not exposed to light. Invest. Ophthal. **2**, 16—23 (1963).
— Validation of an indicator of mammalian retinal receptor response: Density of stain as a function of stimulus magnitude. J. opt. Soc. Amer. **56**, 116—123 (1966).
Falk, G., Fatt, P.: Photoconductive changes in a suspension of rod outer segments. J. Physiol. (Lond.) **167**, 36P—37P (1963).
— — Rapid hydrogen ion uptake of rod outer segments and rhodopsin solutions on illumination. J. Physiol. (Lond.) **183**, 211—224 (1966).

Falk, G., Fatt, P., Distinctive properties of the lamellar and disk edge structures of the rod outer segment. J. Ultrastruct. Res. **28**, 41—60 (1969).

Fatt, P., Katz, B.: An analysis of the end plate potential recorded with an extracellular electrode. J. Physiol. (Lond.) **115**, 320—369 (1951).

Fernandez-Morán, H.: The fine structure of vertebrate and invertebrate receptors as revealed by low temperature electron microscopy. In: The structure of the eye. New York: Ronald 1961 a.

— Lamellar systems in myelin and photoreceptors as revealed by high resolution electron microscopy. In: Macromolecular complexes. New York: Ronald 1961 b.

— Cell membrane ultrastructure, low temperature electron microscopy, and x-ray diffraction studies of lipoprotein components in lamellar systems. Circulation **26**, 1039—1065 (1962).

Fine, B. S., Zimmerman, L. E.: Observations on the rod and cone layer of the human retina. Invest. Ophthal. **2**, 446—459 (1963).

Fleischer, S., McConnell, D. G.: Preliminary observations on the lipids of bovine retinal outer segment discs. Nature (Lond.) **212**, 1366—1367 (1966).

— Fleischer, B., Stoekenius, W.: Fine structure of lipid depleted mitochondria. J. Cell Biol. **32**, 193—208 (1967).

Frank, R. N., Goldsmith, T. H.: Adenosine triphosphatase activity in the rod outer segments of the pig's retina. Arch. biochem. Biophys. **110**, 517—525 (1965).

Glicstein, M., Labossiere, E., Yager, D.: Cone position in partially light adapted goldfish retina. Anat. Rec. **163**, 189 (1969).

Goldstein, E. B.: Early receptor potential of the isolated frog (Rana pipiens) retina. Vision Res. **7**, 837—845 (1967).

— Visual pigments and the early receptor potential of the isolated frog retina. Vision Res. **8**, 953—963 (1968).

— Visual pigments and the early receptor potential of frog and monkey retinas. Exp. Eye Res. **8**, 247—248 (1969).

— Berson, E. L.: Rod and cone contributions to the human early receptor potential. Vision Res. **10**, 207—218 (1970).

Gouras, P.: Duplex function in the gray squirrel's electroretinogram. Nature (Lond.) **203**, 767—768 (1964).

Granit, R.: Retina and optic nerve. In: The eye, Vol. 2. New York: Academic Press 1962.

Green, D. G.: Light adaptation in the rat retina: Evidence for two receptor mechanisms. Science **174**, 598—600 (1971).

Grignolo, A., Orzalesi, N.: La ultrastruttura dei fotorecettori della retina humana. Soc. Oftal. Ital. **21**, 3—7 (1963).

— — Castellazzo, R., Vittone, P.: Retinal damage by visible light in albino rats. An electron microscope study. Ophthalmologica (Basel) **157**, 43—59 (1969).

Hagins, W. A., Penn, R. D., Yoshikami, S.: Dark current and photocurrent in retinal rods. Biophys. J. **10**, 380—412 (1970).

Hall, M. O., Bok, D., Bacharach, A. D. E.: Visual pigment renewal in the mature frog retina. Science **161**, 787—789 (1968).

Hamasaki, D. I.: An anatomical and electrophysiological study of the retina of the Owl Monkey, Aotes trivirgatus. J. comp. Neurol. **130**, 163—174 (1967).

Hechter, O.: Role of water structure in the molecular organization of cell membranes. Fed. Proc. **24**, S91—S102 (1965).

Heller, J.: Structure of visual pigments. I. Purification, molecular weight, and composition of bovine visual pigment$_{500}$. Biochemistry **7**, 2906—2913 (1968a).

— Structure of visual pigments. II. Binding of retinal and conformational changes on light exposure in bovine visual pigment$_{500}$. Biochem. **7**, 2914—2920 (1968 b).

— Comparative study of a membrane protein. Characterization of bovine, rat, and frog visual pigments$_{500}$. Biochemistry **8**, 675—679 (1969).

— Ostwald, T. J., Bok, D.: The osmotic behaviour of rod photoreceptor outer segment discs. J. Cell Biol. **48**, 633—694 (1971).

Hellner, K. A.: Das adaptive Verhalten der Mausnetzhaut. Arch. klin. exp. Ophthal. **169**, 166—179 (1966).

HOLLENBERG, M. J., BERNSTEIN, M. H.: Fine structure of the photoreceptor cells of the ground squirrel (Citellus tridecemlineatus). Amer. J. Anat. **118**, 359—373 (1966).

ISHIKAWA, T., PEI, Y. F.: Intramitochondrial glycogen particles in rat retinal receptor cells. J. Cell Biol. **25**, 402—407 (1965).

JEAN, J., O'BRIEN, B.: Microwave test of a theory of the Stiles-Crawford effect. J. opt. Soc. Amer. **39**, 1057 (1949).

JONES, A. E., JACOBS, G. H.: Electroretinographic luminosity functions of (Aotes trivirgatus) the owl monkey. Amer. J. Physiol. **204**, 47—50 (1963).

KALBERER, M., PEDLER, C.: The visual cells of the alligator: An electron microscopic study. Vision Res. **3**, 323—329 (1963).

KORN, E. D.: Structure of biological membranes. Science **153**, 1491—1498 (1966).

KRAUSE, W.: Die Retina. Int. Mschr. Anat. Physiol. **9**, 150—236 (1892).

— Die Retina. Int. Mschr. Anat. Physiol. **10**, 12—62 (1893).

— Die Retina. Int. Mschr. Anat. Physiol. **11**, 1—122 (1894).

KRINSKY, N. I.: The lipoprotein nature of rhodopsin. Arch. Ophthal. **60**, 688—694 (1958).

KROLL, A. J., MACHEMER, R.: Experimental retinal detachment in the owl monkey. V. Electron microscopy of the reattached retinal. Amer. J. Ophthal. **67**, 117—130 (1969a).

— — Experimental retinal detachment and reattachment in the Rhesus monkey: Electron microscopic comparison of rods and cones. Amer. J. Ophthal. **68**, 58—77 (1969 b).

KÜHNE, W.: On the photochemistry of the retina and on visual purple. Edition with notes by M. FOSTER. London: Macmillan 1878.

KUWABARA, T.: Microtubules in the retina. In: Structure of the eye. II. Stuttgart: Schattauer 1965.

— Membranous transformation of photoreceptic organ by light. Proc. 6th Int. Cong. Electron Micros. (Kyoto) Tokyo: Maruzen 1966.

— GORN, R. A.: Retinal damage by visible light. Arch. Ophthal. **79**, 69—78 (1968).

LADMAN, A. J.: The fine structure of the rod-bipolar synapse in the retina of the albino rat. J. biophys. biochem. Cytol. **4**, 459—466 (1958).

LANZAVECCHIA, G.: Ultrastruttura dei coni e dei bastoncelli della retina di *Xenopus laevis*. Arch. ital. Anat. Embriol. **65**, 417—435 (1960).

LASANSKY, A.: Basal junctions at synaptic endings of turtle visual cells. J. Cell Biol. **40**, 577—581 (1969).

— DEFISCH, F. W.: Studies on the function of the pigment epithelium in relation to ionic movement between retina and choroid. In: The structure of the eye. II. Stuttgart: Schattauer 1965.

LATIES, A. M., LIEBMAN, P. A., CAMPBELL, C. E. M.: Photoreceptor orientation in the primate eye. Nature (Lond.) **218**, 172—173 (1968).

LESLIE, R. B.: Membranes and bioenergetics. In: Biological membranes: Physical fact and function. New York: Academic 1968.

LETTVIN, J. Y.: General discussion: Early receptor potential. C.S.H. Symp. Quant. Biol. **30**, 501—502 (1965).

LIEB, W. A., KNAUF, H.: Über die Ultrastruktur der Photorezeptoren der Netzhaut. Mbl. Augenheilk. **145**, 657—676 (1964).

LIEBMAN, P. A., ENTINE, G.: Visual pigments of frog and tadpole (Rana pipiens). Vision Res. 8, 761—776 (1968).

LOCKET, N. A.: The retina of Poromitra nigrofulvus (Garman). An optical and electron microscope study. Exp. Eye Res. 8, 265—275 (1969).

— Deep-sea fish retinas. Brit. med. Bull. **26**, 107—111 (1970a).

— Retinal structure in a deep-sea fish, Ternoptyx diaphana Hermann. Exp. Eye Res. **9**, 22—27 (1970b).

LUCY, J. A.: Theoretical and experimental models for biological membranes. In: Biological membranes: Physical fact and function. New York: Academic Press 1968.

MARKS, W. B.: Visual pigments of single cones. In: Colour vision, physiology and experimental psychology. Ciba Symposium. Boston: Little Brown 1965.

MARTIN, A. R., PILAR, G.: Dual mode of synaptic transmission in the avian ciliary ganglion. J. Physiol. (Lond.) **168**, 443—463 (1963).

Matsusaka,T.: Electron microscopic observations on cytology and cytochemistry of the paraboloid glycogen of chick retina. Jap. J. Ophthal. **7**, 238—253 (1963).
— Some observations on the inner segment of the accessory cone in the chick retina as revealed by electron microscopy. Jap. J. Ophthal. **10**, 266—281 (1966).
— ATPase activity in the ciliary rootlet of human retinal rods. J. Cell Biol. **33**, C203—C208 (1967a).
— The intracytoplasmic channel in pigment epithelial cells of the chick retina. Z. Zellforsch. **81**, 100—113 (1967b).
Maxwell,D.S., Pease,D.C.: The electron microscopy of the choroid plexus. J. biophys. biochem. Cytol. **2**, 467—475 (1956).
McConnell,D.G.: The isolation of retinal outer segment fragments. J. Cell Biol. **27**, 459—473 (1965).
— Scarpelli,D.G.: Rhodopsin: An enzyme. Science **139**, 848 (1963).
Meyer,D.B., Cooper,T.G., Gernez,C.: Retinal oil droplets. In: The structure of the eye. II. Stuttgart: Schattauer 1965.
— — The visual cells of the chicken as revealed by phase contrast microscopy. Amer. J. Anat. **118**, 723—734 (1966).
Missotten,L.: Etude des batonnets de la rétine humaine au microscope électronique. Ophthalmologica (Basel) **140**, 200—214 (1960a).
— Etude des synapses de la rétine humaine au microscope électronique. Proc. Europ. Reg. Conf. on electron microscopy (Delft) **2**, 818—821 (1960b).
— L'ultrastructure des cones de la rétine humaine. Bull. Soc. belge Ophtal. **132**, 472—502 (1963).
— The synapses in the human retina. In: The structure of the eye. II. Stuttgart: Schattauer 1965.
— Appelmans,M., Michiels,J.: L'ultra-structure des synapses des cellules visuelles de la rétine humaine. Bull. Mem. Soc. franç. Ophtal. **76**, 59—82 (1963).
— Dehauwere,E., Guzik,A.: L'ultrastructure de la rétine humaine. A propos des cellules bipolaires et leur synapses. Bull. Soc. belge Ophtal. **137**, 277—293 (1964).
— Dooren,E.Van Den: L'ultrastructure de la retine humaine. Les contacts lateraux des pédoncules de cones de la fovea. Bull. Soc. belge Opthal. **144**, 800—805 (1966).
Moody,M.F.: Photoreceptor organelles in animals. Biol. Rev. **39**, 43—86 (1964).
— Robertson,J.D.: The fine structure of some retinal photoreceptors. J. biophys. biochem. Cytol. **7**, 87—92 (1960).
Morris,V.B.: Symmetry in a receptor mosaic demonstrated in the chick from the frequencies, spacing, and arrangement of the types of retinal receptor. J. comp. Neurol. **140**, 359—398 (1970).
Mountford,S.: Effects of light and dark adaptation on the vesicle populations of receptor-bipolar synapses. J. Ultrastruct. Res. **9**, 403—418 (1963).
— Filamentous organelles in receptor bipolar synapses of the retina. J. Ultrastruct. Res. **10**, 207—216 (1964).
Munk,O., Andersen,S.R.: Accessory outer segment, a re-discovered cilium-like structure in the layer of rods and cones of the human retina. Acta Ophthal. **40**, 526—531 (1962).
Murakami,M., Pak,W.L.: Intracellularly recorded early receptor potential of the vertebrate photoreceptors. Vision Res. **10**, 965—975 (1970).
Napolitano,L., Le Baron,F.N., Scaletti,J.V.: Preservation of myelin lamellar structure in the absence of lipid: A correlated chemical and morphological study. J. Cell Biol. **34**, 817—826 (1967).
Nguyen H. Anh,J.: Ultrastructure des récepteurs visuels de la rétine de Lacerta viridis. Bull. Ass. Anat. **53**, 1247—1259 (1968).
— Organisation particuliére du reticulum endoplasmique dans le segment interne des batonnets de la rétine chez les batraciens. J. Microscop. **8**, 145—148 (1969a).
— Ultrastructure des récepteurs visuels chez les vertébrés. Arch. Ophthal. (Paris) **29**, 795—822 (1969b).
Nilsson,S.E.G.: An electron microscopic classification of the retinal receptors of the leopard frog (Rana pipiens). J. Ultrastruct. Res. **10**, 390—410 (1964a).

— Inter-receptor contacts in the retina of the frog (Rana pipiens). J. Ultrasttruc. Res. **11**, 207—216 (1964b).

— A globular substructure of the retinal receptor outer segment membranes and some other cell membranes in the tadpole. Nature (Lond.) **202**, 509—510 (1964c).

— The ultrastructure of the receptor outer segment in the retina of the leopard frog (Rana pipiens). J. Ultrastruct. Res. **12**, 207—231 (1965).

— CRESCITELLI, F.: Changes in ultrastructure and electroretinogram of bullfrog retina during development. J. Ultrastruct. Res. **27**, 45—62 (1969).

NOELL, W. K.: Aspects of experimental and hereditary retinal degeneration. In: Biochemistry of the retina. New York: Academic Press 1965.

— THEMANN, H., KANG, B. S., WALKER, V. S.: Functional and structural manifestations of a damaging effect of light on the retina. Fed. Proc. **25**, 329 (1966a).

— WALKER, V. S., KANG, B. S., BERMAN, S.: Retinal damage by light in rats. Invest. Ophthal. **5**, 450—473 (1966b).

NOVER, A., SCHULTZE, B.: Autoradiographische Untersuchung über den Eiweißstoffwechsel in den Geweben und Zellen des Auges. (Untersucht mit S³⁵-thio-aminosäuren, C¹⁴-aminosäuren, H³-leucin an Maus, Ratte und Kaninchen.) Arch. Ophthal. **161**, 554—578 (1960).

OCUMPAUGH, D. E., YOUNG, R. W.: Distribution and synthesis of sulfated mucopolysaccharides in the retina of the rat. Invest. Ophthal. **5**, 196—203 (1966).

OHMAN, S., SHELLEY, W. B.: Induction of a unique fluorescence in the photoreceptor of the retina. Nature (Lond.) **220**, 378—379 (1968).

OKUDA, K.: Electron microscopic observations of the vertebrate retina. Acta Soc. Ophthal. Jap. **65**, 2126—2151 (1961).

OLNEY, J.: Centripetal sequence of appearance of receptor-bipolar synaptic structures in developing mouse retina. Nature (Lond.) **218**, 281—282 (1968a).

— An electron microscopic study of synapse formation, receptor outer segment development, and other aspects of developing mouse retina. Invest. Ophthal. **7**, 250—268 (1968b).

ORDY, J. M., SAMORAJSKI, T.: Visual acuity and ERG-CFF in relation to the morphologic organization of the retina among diurnal and nocturnal primates. Vision Res. 8, 1205—1225 (1968).

PATEL, S. C.: Determination of the sites of rhodopsin in the ultrastructure of the mammalian rod photoreceptor using a selective fixation technique. Part I. Brit. J. Physiol. Opt. **24**, 1—22 (1967).

— Determination of the sites of rhodopsin in the ultrastructure of the mammalian rod photoreceptor using a selective fixation technique. Part II. Brit. J. Physiol. Opt. **24**, 61—102 (1967).

PEDLER, C.: Rods and cones — a fresh approach. In: Colour vision, physiology and experimental psychology. Ciba symposium. Boston: Little Brown 1965.

— TANSLEY, K.: The fine structure of the cone of a diurnal gecko (Phelsuma inunguis). Exp. Eye Res. **2**, 39—47 (1963).

— TILLY, R.: The nature of the Gecko visual cell. A light and electron microscopic study. Vision Res. **4**, 499—510 (1964).

— — Ultrastructural variations in the photoreceptors of the macaque. Exp. Eye Res. **4**, 370—373 (1965).

— — The retina of a fruit bat (*Pteropus giganteus* Brünnich). Vision Res. **9**, 909—922 (1969).

— BOYLE, M.: Multiple oil droplets in the photoreceptors of the pigeon. Vision Res. **9**, 525—528 (1969).

PIRENNE, M. H.: Visual acuity. In: The eye. II. New York: Academic Press 1962.

PITT, G. A. J., COLLINS, F. D., MORTON, R. A., STOK, P.: Studies on rhodopsin. VIII. Retinylidenemethylamine, an indicator yellow analogue. Biochem. J. **59**, 122—128 (1955).

POINCELOT, R. P., MILLAR, P. G., KIMBEL, R. L., JR, ABRAHAMSON, E. W.: Lipid to protein chromophore transfer in the photolysis of visual pigments. Nature (Lond.) **221**, 256—257 (1969).

PORTER, K. R.: The submicroscopic morphology of protoplasm. Harvey Lect. **51**, 175—228 (1956).

Rabinovitch, M., Mota, J., Yoneda, S.: Note on the histochemical localization of glycogen and pentose nucleotides in the visual cells of the chick (Gallus gallus). Quart. J. micr. Sci. **95**, 5—9 (1954).

Radnot, M., Lovas, B.: Die Ultrastruktur der Photoreceptor-Synapsen in einem Falle von Sehnervenatrophie. Arch. klin. exp. Ophthal. **173**, 56—63 (1967).

Revel, J. P., Karnovsky, M. J.: Hexagonal array of subunits of intercellular junctions of the mouse heart and liver. J. Cell Biol. **33**, C7—C12 (1967).

Richardson, T. M.: Cytoplasmic and ciliary connections between the inner and outer segments of mammalian visual receptors. Vision Res. **9**, 727—732 (1969).

Ripps, H., Dowling, J. E.: Personal communication.

Robertson, J. D.: The molecular structure and contact relationships of cell membranes. Prog. Biophys. **10**, 343—418 (1960).

— A possible ultrastructural correlate of function in the frog retinal rod. Proc. nat. Acad. Sci. (Wash.) **53**, 860—866 (1965).

— Granulo-fibrillar and globular structure in unit membranes. Ann. N. Y. Acad. Sci. **137**, 421—440 (1966).

Rosenberg, B., Orlando, R. A., Orlando, J. M.: Photoconduction and semiconduction in dried receptors of sheep eyes. Arch. biochem. Biophys. **93**, 395—398 (1961).

— Jendrasiak, G. L.: Semiconductive properties of lipids and their possible relation to lipid bilayer conductivity. Chem. Phys. Lipids **2**, 47—54 (1968).

Rudzinska, M. A.: Ultrastructure involved in the feeding mechanism of Suctoria. Trans. N. Y. Acad. Sci. **29**, 512—525 (1967).

Rushton, W. A. H.: Visual pigments in man. Springfield, Ill.: Thomas 1962.

Satir, P.: Morphological aspects of ciliary motility. In: The contractile process. Boston: Little Brown 1967.

Scarpelli, D. G., Craig, E. L.: The fine localization of nucleoside triphosphatase in the retina of the frog. J. Cell Biol. **17**, 279—288 (1963).

Schmidt, W. J.: Polarisationsoptische Analyse eines Eiweiß-Lipoid-Systems, erläutert am Außenglied der Sehzellen. Kolloid-Z. **85**, 137—148 (1938).

Schultze, M.: Anatomie und Physiologie der Netzhaut. Arch. mikr. Anat. **2**, 175—286 (1866).

Sidman, R. L.: The structure and concentration of solids in photoreceptor cells studied by refractometry and interference microscopy. J. biophys. biochem. Cytol. **3**, 15—30 (1957).

— Histochemical studies of photoreceptor cells. Ann. N. Y. Acad. Sci. **74**, 182—195 (1958).

Sjöstrand, F. S.: Electron microscopy of retinal rods. J. cell. comp. Physiol. **33**, 383—405 (1949).

— The ultrastructure of the outer segments of rods and cones of the eye as revealed by the electron microscope. J. cell. comp. Physiol. **42**, 15—44 (1953a).

— The ultrastructure of the inner segments of the retinal rods of the guinea pig eye as revealed by electron microscopy. J. cell. comp. Physiol. **42**, 45—70 (1953b).

— Ultrastructure of retinal rod synapses of the guinea pig eye as revealed by three dimensional reconstructions from serial sections. J. Ultrastruct. Res. **2**, 122—170 (1958).

— The ultrastructure of the retinal receptors of the vertebrate eye. Ergebn. Biol. **21**, 128—160 (1959a).

— Fine structure of cytoplasm: The organization of membranous layers. Rev. Mod. Phys. **31**, 301—318 (1959b).

— Electron microscopy of the retina. In: The structure of the eye. New York: Academic Press 1961.

— A new ultrastructural element of the membranes in mitochondria and of some cytoplasmic membranes. J. Ultrastruct. Res. **9**, 340—361 (1963).

— The synaptology of the retina. In: Colour vision. Physiology and experimental psychology. Ciba symposium. Boston: Little Brown 1965.

— Elfvin, L. E. G.: Some observations on the retinal receptors of the toad's eye as revealed by the electron microscope. Proc. Stockholm Conf. Electr. Micros., (1956) 194—196. New York: Academic Press 1957.

— Nilsson, S. E. G.: Electron microscopy of the retina. In: The rabbit in eye research. Springfield: Thomas 1964.

SMITH, C. A., SJÖSTRAND, F. S.: A synaptic structure in the hair cells of the guinea pig cochlea. J. Ultrastruct. Res. **5**, 184—192 (1961).

SMITH, T. G., STELL, W. K., BROWN, J. E., FREEMAN, J. A., MURRAY, G. C.: A role for the sodium pump in photoreception in Limulus. Science **162**, 456—458 (1968).

SPITZNAS, M., HOGAN, M. J.: Outer segments of photoreceptors and the retinal pigment epithelium. Arch. Ophthal. **84**, 810—819 (1970).

STELL, W. K.: Correlation of retinal cytoarchitecture and ultrastructure in Golgi preparations. Anat. Rec. **153**, 389—398 (1965).

— The structure and relationships of horizontal cells and photoreceptorbipolar synaptic complexes in goldfish retina. Amer. J. Anat. **120**, 401—423 (1967).

STILES, W. S., CRAWFORD, B. H.: The luminous efficiency of rays entering the eye pupil at different points. Proc. roy. Soc. B **112**, 428—450 (1933).

TAYLOR, J. W.: Cone and possible rod components of the fast photovoltage in the frog eye: A new method of measuring cone regeneration rates *in vivo*. Vision Res. **9**, 443—452 (1969).

TOKUYASU, K., YAMADA, E.: The fine structure of the retina studied with the electron microscope. IV. Morphogenesis of outer segments of retinal rods. J. biophys. biochem. Cytol. **6**, 225—230 (1959).

— The fine structure of retina. V. Abnormal retinal rods and their morphogenesis. J. biophys. biochem. Cytol. **7**, 187—190 (1960).

TOMITA, T.: Electrophysiological study of the mechanisms subserving color coding in the fish retina. C.S.H. Quant. Biol. **30**, 559—566 (1965).

— Electrical activity of vertebrate photoreceptors. Quart. Rev. Biophys. **3**, 179—222 (1970).

TOYODA, JUN-ICHI, HIROSHI, N., TOMITA, T.: Light-induced resistance changes in single photoreceptors of *Necturus* and *Gecko*. Vision Res. **9**, 453—463 (1969).

TORALDO DI FRANCHIA, G.: Retinal cones as dielectric antennas. J. Opt. Soc. Amer. **39**, 324 (1949).

TRUJILLO-CENOZ, O.: Some aspects of the structural organization of the arthropod eye. C.S.H. Symp. Quant. Biol. **30**, 371—382 (1965).

UGA, S., NAKAO, F., MIMWIA, M., IKUI, H.: Some new findings on the fine structure of the human photoreceptor cells. J. Electronmicroscop. (Japan) **19**, 71—84 (1970).

UNDERWOOD, G.: Reptilian retinas. Nature (Lond.) **167**, 183—185 (1951).

— On the evolution and classification of the Geckoes. Proc. Zool. Soc. (Lond.) **124**, 465—492 (1954).

VAN HARREVELD, A., KHATTAB, F. I.: Electron microscopy of mouse retina prepared by freeze-substitution. Anat. Rec. **161**, 125—140 (1968).

VILLERMET, G. M., WEALE, R. A.: The optical activity of bleached photoreceptors. J. Physiol. (Lond.) **201**, 425—435 (1969).

WALD, G.: Mechanisms of vision. Fourth conference on nerve impulse. Josiah Macy Foundation, 11—57 (1954).

— BROWN, P. K., GIBBONS, I. R.: Visual excitation: A chemoanatomical study. Symp. Soc. exp. Biol. **16**, 32—57 (1962).

WALLS, G. L.: The vertebrate eye and its adaptive radiation. Bloomfield Hills: Cranbrook 1942.

WEALE, R. A.: Bleaching experiments of eyes of living gray squirrels (Sciurus carolinensis leucotis). J. Physiol. (Lond.) **127**, 587—591 (1955).

— The duplicity theory if vision. Ann. Roy. Coll. Surgeons **28**, 16—35 (1961).

— Optical activity and the fixation of rods and cones. Nature (Lond.) **220**, 583 (1968).

WEBSTER, H. DE F., AMES, A.: Reversible and irreversible changes in the fine structure of nervous tissue during oxygen and glucose deprivation. J. Cell Biol. **26**, 885—909 (1965).

WEIDMAN, T. A., KUWABARA, T.: Postnatal development of the rat retina. Arch. Ophthal. **79**, 470—484 (1968).

WERBLIN, F. S., DOWLING, J. E.: Organization of the retina of the Mudpuppy *(Necturus maculosus)*. II. Intracellular recording. J. Neurophysiol. **32**, 339—355 (1969).

WESTHEIMER, G.: Dependence of the magnitude of the Stiles-Crawford effect on retinal location. J. Physiol. (Lond.) **192**, 309—315 (1967).

WISLOCKI, G. B., LADMAN, A. J.: The demonstration of a blood-ocular barrier in the albino rat by means of the intravitam deposition of silver. J. biophys. biochem. Cytol. **1**, 501—510 (1955).

WOLKEN, J.J.: A comparative study of photoreceptors. Trans. N. Y. Acad. Sci. **19**, 315—327 (1957).

YAMADA, E.: The fine structure of the paraboloid of the turtle retina as revealed by electron microscopy. Anat. Rec. **136**, 352 (1960a).

— Observations on the fine structure of photoreceptive elements in the vertebrate eye. J. Electronmicroscop. **9**, 1—14 (1960b).

YASUZUMI, G., TEZUKA, O., IKEDA, T.: The submicroscopic structure of the inner segments of rods and cones in the retina of Uroloncha striata var. domestica Flower. J. Ultrastruct. Res. **1**, 295—306 (1958).

YOUNG, R.W.: Renewal of photoreceptor outer segments. Anat. Rec. **151**, 484 (1965).

— Further studies on the renewal of photoreceptor outer segments. Anat. Rec. **154**, 446 (1966).

— Protein renewal in rods and cones studied by electron microscopy radioautography. J. Cell Abstr. Pap, Amer. Soc. Cell Biol. Paper 306 (1967)

— Passage of newly formed protein through the connecting cilium of retinal rods in the frog. J. Ultrastruct. Res. **23**, 462—473 (1968).

— A difference between rods and cones in the renewal of outer segment protein. Invest. Ophthal. 8, 222—231 (1969).

— The renewal of rod and cone outer segments in the Rhesus monkey. J. Cell Biol. **49**, 303—318 (1971a).

— Shedding of discs from rod outer segments in the Rhesus monkey. J. Ultrastruc. Res. **34**, 190—203 (1971b).

— BOK, D.: Participation of the retinal pigment epithelium in the rod outer segment renewal process. J. Cell Biol. **42** 392—403 (1969).

— DROZ, B.: The renewal of protein in retinal rods and cones. J. Cell Biol. **39**, 169—184 (1968).

Chapter 3

The Morphological Organization
of the Vertebrate Retina

By

William K. Stell, Bethesda, Maryland (USA)

With 28 Figures

Contents

I. Introduction

The photoreceptor cells are transducers which absorb incident photons and transmit a proportional signal to other cells. More complex retinal functions such as the discrimination of colors and the detection of contrast and movement are accomplished by networks of many cells. This chapter is a critical historical review of attempts to establish the spatial relationships of the structural, functional, and chemical units in the retina which correspond to such networks.

The general features of retinal histology (Fig. 1 and 2) are well known from classical works (e.g. CAJAL, 1892; POLYAK, 1941). I shall describe these features primarily in terms of the direct visual pathway, proceeding toward the brain from *distal* (external, outer, sclerad) to *proximal* (internal, inner, vitread). The neurons are segregated along this pathway into three layers, which I shall designate the *layer of photoreceptor cells*, including the *layers of outer and inner segments* and the *layer of photoreceptor cell bodies* (or *outer nuclear layer*); the *layer of intermediate neurons* (or *inner nuclear layer*); and the *layer of ganglion cells*. Processes of photoreceptors and intermediate neurons are interconnected in the *first synaptic layer* (or *outer plexiform layer*) and processes of intermediate neurons and ganglion cells are interconnected in the *second synaptic layer* (or *inner plexiform layer*). Sublayering, especially of the intermediate neuronal and second synaptic layers, is observed in some animals. The perikarya of the neurons are more or less sheathed in neuroglia; the distal parts of the photoreceptors traverse the ventricle of the optic vesicle freely to contact cells of the often pigmented *retinal epithelium*. In all chordates the retina receives at least one blood supply by means of the choriocapillaris, at the base of these epithelial cells. In some species this is the sole vascular supply; in others, a second supply may be provided from the vitreal side in the form of an intraretinal or a hyaloretinal capillary network or special intravitreal structures (e.g.: pecten, conus papillaris).

The retina is inhomogeneous *horizontally* (in a plane perpendicular to the visual pathway) as well as vertically (along the distal-proximal pathway). Regional (horizontal) variations in cell form, size, and distribution generally reflect the specialization of those regions for adapting permanently to specific visual tasks. Although the *fovea* and *area centralis* are the best known of these, other regional modifications may be observed in many retinas (WALLS, 1942). In a more microscopic domain, regular geometric horizontal patterns or *mosaics* of nearby cells, especially photoreceptors, are found in many animals.

In the sections which follow, the retina is seen from the point of view of the microscopist, the histochemist, and the correlative physiologist. The accomplishments of CAJAL and POLYAK are so great that I cannot begin to review their contribution in a work such as this. To slight them thus is a significant deficiency. I acknowledge a special debt to the recent reviews of COHEN (1963a), ROHEN (1964), and DOWLING (1970). The astounding activity in the field considered here may be better appreciated by consulting COHEN's brief but nevertheless comprehensive treatment of it only a few years ago. Finally, I am indebted particularly to many friends who have graciously contributed illustrations, unpublished data, stimulating discussions, and valuable criticisms during the preparation of this review.

This work was accomplished in part during a visit to the Instituto Venezolano de Investigaciones Científicas, Caracas, Venezuela. I am especially grateful to Dr. MIGUEL LAUFER and the members of the Departamento de Neurobiologia for their hospitality and assistance.

II. The Structure of Cells and Their Relationships as Revealed by Optical Microscopy

A. Methods

The oldest means for the investigation of retinal micro-structure was the examination of fresh tissue under an optical magnifier. Although HANNOVER (1840, 1844) was able to discern mosaic patterns of teleostean cones by this method, its use is limited to the observation of large and refractile surface structures. Sectioning or maceration and dissection of chemically fixed tissue, which made possible more than a century ago the astounding observations (cf. Fig. 1 and 3) of HEINRICH MÜLLER (1857), and general tissue stains such as carmine were useful but did not reveal the finer neuronal processes and their relationships.

Special neurohistological procedures, specifically the silver-chromate method of GOLGI and the intravital methylene blue method of EHRLICH (see introductory sections of CAJAL, 1909—1911; and POLYAK, 1941), have increased vastly our knowledge of the fine details of neuronal structure (for outstanding recent reviews of neuroanatomical methods, see NAUTA and EBBESSON, 1970). The better of these procedures is Golgi's several variations of which

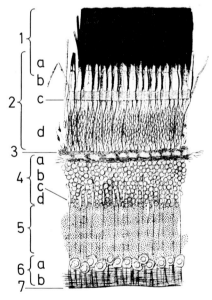

Fig. 1. Drawing of the retina of a teleost (*Perca*) fixed in chromic acid, sectioned freehand, and examined without staining. *1* epithelium (pigmented). *2* photoreceptor layer: *a* outer segments; *b* inner segments; *c* external junctional layer (limiting membrane); *d* receptor cell bodies (outer nuclear layer). *3* first synaptic (outer plexiform) layer. *4* layer of intermediate neurons (inner nuclear layer): *a* horizontal cells; *b* bipolar cells; *c* glial cells of Müller; *d* amacrine cells. *5* second synaptic (inner plexiform) layer. *6* ganglion cell layer: *a* layer of ganglion cell bodies; *b* layer of ganglion cell axons (optic fiber layer). *7* internal limiting membrane (reproduced from H. MÜLLER, 1857; by permission of Akademische Verlagsgesellschaft)

are in use. Most common are the *rapid Golgi* method, in which fixation in an osmium-dichromate mixture is followed by treatment in silver nitrate, and modifications in which a different fixative is employed, such as Golgi-Kopsch (formaldehyde-dichromate) and Golgi-Colonnier (glutaraldehyde-dichromate). In the Golgi-Cox method, the tissue is treated with a single solution containing mercuric chloride and chromate/dichromate. When successful, these methods combine superior chemical fixation with impregnation of even the finest neuronal processes by a very dense material, probably silver chromate (Stell, 1965a; Blackstad, 1965 and personal communication), which allows visualization of neurons in their entirety (cf. Fig. 2, 6, 25). Drawbacks of the Golgi methods include: (1) the possibility of incomplete impregnation, not always obvious, which may result in the failure to visualize some cell processes; (2) the capriciousness of impregnation — one cannot elect to demonstrate cells of a given type nor ascertain what types, if any, have *not* been impregnated; (3) a limitation upon the ability to see details of cellular relationships, imposed by the limits of optical resolution and the thickness (usually $50-100 \, \mu m$) of the sections cut so as to include entire cells. Because of the third limitation, it is impossible to know from optical microscopy alone whether two cells are in contact with one another. Although this problem may be overcome in part by optical examination of semithin ($0.5-2 \, \mu m$) sections of plastic-embedded Golgi preparations (see e.g. Boycott and Dowling, 1969), such an approach has been taken infrequently. Despite these difficulties it must be said that we owe more of our knowledge of retinal cells and probable patterns of interconnections to the Golgi methods than to any others. Intravital staining with methylene blue, which may provide comparable results with some cells, unfortunately has been exploited little in recent years. This method is particularly useful for demonstrating, in retinas spread flat (whole mounts), very large ganglion cells (e.g. Schaper, 1899) or mammalian horizontal cells and their long axons (Kolmer, 1936). Some elements, such as the centrifugal fibers (Fig. 7), are demonstrated more readily with this than with the Golgi method (Dogiel, 1895; Cajal, 1896); others, such as horizontal cells in fishes — which can be impregnated well with the Golgi procedure — have not been stained with methylene blue (Dogiel, 1888). As pointed out by Pedler (1962a, 1965), apparent continuity of contiguous but separate processes may result from Golgi impregnation, particularly with the Cox variant. Artifactual continuity is a much more severe limitation, however, with methylene blue, the use of which led Dogiel and others to erroneous conclusions that neuronal interconnections were syncytial (Cajal, 1933).

In another category are the silver stains. In some silver procedures (as in the Golgi and Ehrlich methods) the entire cytoplasm is impregnated with a colored or dense material (Nauta method — Guillery and Ralston, 1964), while in others the silver is deposited predominantly on some subcellular material such as the neurofibrils (Glees method — Guillery, 1965) or membranes (Hortega method — Vogel and Kemper, 1962). For any particular conditions, therefore, (1) all cells or parts of cells containing the materials which react with silver will be stained, and (2) all cells or parts of cells lacking such materials will not be stained. Because of the first property, silver staining of some kinds permits studies of the size, shape, and regional distribution (Fig. 4, 5, 10, 11) of a population of cells of a given type (Gallego, 1953; Honrubia López, 1966). Because of the second property, however, sometimes only one of several subclasses, e.g. of horizontal cells (Testa, 1966; Parthe, 1967) or ganglion cells (Honrubia López, 1966), may be investigated usefully with silver staining. Because the finest dendrites are stained rarely with silver, the extent of dendritic fields shown by these methods is probably underestimated (Gallego, 1965). Finally, some silver stains may fail to demonstrate axons (Honrubia López, 1966; Parthe, 1967) of cells which appear to have axons in Golgi preparations (Cajal, 1892, 1896).

B. Photoreceptors

Heinrich Müller (1851, 1857) showed that vertebrate photoreceptors are of at least two types. In teleosts, mammals and nocturnal birds these are the classical rods and cones, easily distinguishable by many special features. A correlation noted long ago of the cones and rods with day and night vision, respectively, led to the various theories of the duplex retina (e.g. SCHULTZE, 1866). In lower fishes, amphibians, reptiles and diurnal birds (CAJAL, 1889, 1892) this distinction between rods and cones is frequently difficult to make; at the very least, their morphologic features overlap considerably.

In rather few species only a single class of photoreceptor cells has been observed. These include the apparently all-rod skate (DOWLING and RIPPS, in press) and some teleosts (e.g. ALI and HANYU, 1963), and the apparently all-cone ground squirrel (TANSLEY, 1961; DOWLING, 1964; HOLLENBERG and BERNSTEIN, 1966).

Nevertheless, as COHEN (1969 and this volume) has emphasized, most vertebrate retinas examined by suitable methods have been shown to have photoreceptors of at least two different classes. The role played by structural differences in the distal and proximal appendages of the photoreceptors in determining the functional distinctions between rods and cones is unknown. In any case, however, the differences in synaptic endings imply that patterns of synaptic connections are an important determinant of the retinal functions related to the existence of receptors of different morphological types. It is curious that the retinas of most of the teleosts examined by BLAXTER and JONES (1967) and BLAXTER and STAINES (1970) are pure-cone (all cones are not necessarily identical) in embryos prior to metamorphosis. Rods appear only at metamorphosis.

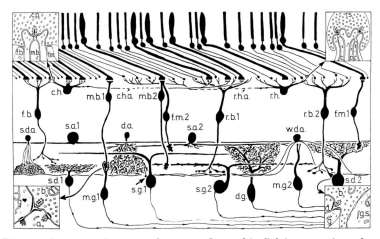

Fig. 2. Diagram showing major types of neurons observed in Golgi preparations of perifoveal human and rhesus monkey retinas. Horizontal cells: c.h., type A; and r.h., type B (both shown by electron microscopy to contact only cones); c.h.a., axon type A, and c.h.b., axon type B (both shown to contact only rods). Bipolar cells: r.b., for rods; f.b., flat; m.b., invaginating midget; f.m., flat midget. Amacrine cells: d.a., narrow-field diffuse; w.d.a., wide-field diffuse; s.d.a., stratified diffuse; s.a., unistratified. Ganglion cells: d.g., diffuse; s.d., stratified diffuse; s.g., unistratified; m.g., midget. Insets: diagrams of synaptic ultrastructure (cf. Figs. 17, 20, 24). Cone pedicle (c.p., upper left) is contacted by dendrites of flat (f.b.), flat midget (f.m.), and invaginating midget (m.b.) bipolar cells, as well as horizontal cell processes (unlabelled). Rod spherule (r.s., upper right) is contacted by dendrites of rod bipolar cell (c) flanked by horizontal cell processes (unlabelled). Synaptic endings of cone bipolar cells (b, lower left) make contact with processes of amacrine (a) and ganglion (g) cells. Synaptic endings of rod bipolar cells (b, lower right) make similar contacts and infrequently make close contact with a ganglion cell soma (g.s.) (reproduced from BOYCOTT and DOWLING (1969), by permission of the authors and The Royal Society of London)

The synaptic terminal of a typical rod, as in teleosts and mammals, is a tiny ovoid knob or *spherule* with a smooth surface, while the terminal of a typical cone cell is a large, pyramidal *pedicle* with a flattened base (Fig. 2). A number of fine filaments or *telodendria* extend horizontally from the rim of the pedicle. All the cone pedicles usually lie in the same horizontal plane, their bases delineating a border beyond which no receptor processes extend, while the rod spherules are found at many levels, in clusters between the distal parts of the cone terminals. This distinction is observed even in the "all rod" retinas of the guinea pig (SJÖSTRAND, 1961a), rat (DOWLING, 1967a), mouse (COHEN, 1967a) and fruit bat (PEDLER and TILLY, 1969), and in the periphery of the "all cone" tree shrew (SAMORAJSKI et al., 1966).

A few scattered receptors of a second type may be overlooked unless they are sought diligently. For example, MUNK (1965) was able to find typical rods in the retina of a polypterid fish, formerly thought to be all-cone. The Golgi and Ehrlich procedures may be particularly useful in differentiating receptor types by their synaptic terminals. TRETJAKOFF's demonstration of telodendria on the "long" receptors and their absence from the "short" receptors of lamprey lateral eye led WALLS (1935) to designate the latter rods and the former cones. In rapid-Golgi preparations of an American brook lamprey *(Entosphenus)*, I have found that while their levels of termination are consistent with this identification, endings of both types are large pedicles with telodendria (STELL, unpublished observations), and others have observed with the electron microscope that cells of both types are conelike in other respects (YAMADA and ISHIKAWA, 1967; DUNN, personal communication). Elasmobranch rods and cones, clearly distinguishable by many other criteria, both have telodendria (NEUMAYER, 1897). In Golgi and ultrastructural studies I (1972) have observed both rods and cones in *Squalus acanthias*, reported by WALLS (1942) to contain only rods; receptors of both types are provided with telodendria. In the labrid teleosts (ENGSTRÖM, 1963) some cones are more "rodlike" than others, their pedicles being less complex and more distally located than those of the more typical ones (STELL, unpublished observations). In amphibians, although rods and cones are clearly distinguishable by the structure of their outer and inner segments, synaptic endings of both are large and may bear numerous telodendria (CAJAL, 1892; DOWLING and WERBLIN, 1969). Terminals of receptors of all types in frog lie at about the same level (NILSSON, 1964b). The green rod is otherwise a conventional rod, its distal appendage occupying the most distal level and its perikaryon the most proximal of all the receptors. The perikarya of the red rods, however, are the most distally located (like those of teleostean or mammalian cones) and the position of the cone perikarya is intermediate. A similar intermingling of the properties of classical rods and cones is seen also in reptiles (WALLS, 1942; CRESCITELLI, 1965; DUNN, 1966; UNDERWOOD, 1968).

None of these observations answers the question of why a conelike outer segment should (or should not) be associated with a pedicle, or a rodlike one with a spherule. It is clear, though, that the synaptic connections' of a receptor are another important determinant of its function, and that the basic information analyzed by the more proximal layers of most vertebrate retinas is supplied by at least two functional classes of photoreceptors. The sobering fact that receptors of a second type have eventually been encountered in most retinas thought previously to be all-cone or all-rod encourages a diligent search for diversity in the remaining few. It seems possible that receptors in some pure-type retinas may be differentiated only according to their synaptic connectivities, rather than by the customary structural distinctions between the receptors themselves.

C. Intermediate Neurons

The principal elements of the intermediate neuronal (inner nuclear) layer are the bipolar, horizontal, and amacrine cells; it must be emphasized that this is *not* a "bipolar cell layer". The cell bodies of the bipolars occupy mainly the middle and the distal thirds of this layer and send processes into both synaptic layers. The horizontal cells occupy one or more strata near the first synaptic (outer plexiform) layer, into which they send their processes; the amacrine cells have an analogous relationship to the second synaptic (inner plexiform) layer. Bipolar cell bodies may be "displaced" to the layer of receptor cell bodies, and amacrine cell bodies to the inner plexiform or ganglion cell layers, especially in groups such as lower fishes and amphibians. The layer of intermediate neurons may likewise be supplemented by displaced ganglion cells, as in birds. Displacements of photoreceptor and horizontal cells have not been described. Still other cells (see p. 128, "Miscellaneous") may be observed among the intermediate neurons which are not easily assigned to any of the major neuronal categories noted above, and the cell bodies of the glial cells of MÜLLER are commonly found in this layer. The types of cells observed in the intermediate neuronal (inner nuclear) layer will now be considered in turn.

Bipolar Cells

SCHIEFFERDECKER (1886) noted that there are at least two distinct types of bipolar cells in the retinas of most vertebrates. CAJAL (e.g. 1892, 1893) emphasized the association of these two subtypes with the two classes of photoreceptors: the larger bipolar with rods, the smaller with cones. The larger variety generally had a broader and deeper dendritic tree, a larger cell body located closer to the receptors, and a stouter axon with a more proximal and enlarged terminal than that of the cone bipolar. Similar distinctions could be made even in groups, such as amphibians and reptiles, in which the receptors were not the classical rods and cones of mammals. NEUMAYER (1897) described the same two bipolar cell classes in selachian fishes, which had not been examined by CAJAL, and STELL and WITKOVSKY (unpublished) have observed at least five types of bipolars in the predominantly rod retina of *Mustelus*.

In recent years, BROWN and MAJOR (1966) described the same two varieties of bipolar cells in the cat — the larger with a more proximal and the smaller with a more distal terminal in the second synaptic layer. In Golgi preparations of the chick, SHEN et al. (1956) observed bipolars of three types, none corresponding exactly to CAJAL's classification: (1) cells with a wide dendritic tree, a large soma located nearest of the three to the receptors, and an axon terminating mainly in the first and fourth (of four) acetylcholinesterase-positive strata of the second synaptic layer; (2) cells with a small soma located in the middle of the layer of

interneurons and an axon terminating mainly in the second and third acetylcholinesterase-positive strata and also in the distalmost, acetylcholinesterase-
negative stratum of the second synaptic layer; and (3) cells with a small soma
located between the somata of the other two bipolar varieties, and an axon
terminating very close to the ganglion cell bodies. In Golgi preparations of several
species of marine teleosts, PARTHE (1972) has observed bipolar cells of
five types: two judged to be rod bipolars, having a deep field of dendrites ending in
small knobs at many levels, a large distally-located cell body, and an axon ending
in a large expansion near the ganglion cell bodies; and three judged to be cone
bipolars, having flat dendritic arborizations, smaller and more proximally located
cell bodies, and shorter and finer axons. The subtypes of rod and cone bipolars
appear to differ mainly in the horizontal extent of their dendritic tree (large or
small). In a number of teleosts, the terminals of the large bipolars appear to form
a square "grid" with the same spacing as units of the cone mosaic (GOODLAND,
1966; WITKOVSKY and DOWLING, 1969). In the teleost *Callionymus*(dragonet), the
bipolars in the "all-cone" dorsal half of the retina terminate at two levels to form
a pair of grids, while in the mixed ventral half of the retina the pattern is the same
as in these other teleosts (VRABEC, 1966).

CAJAL (1892) described in the fovea of chameleon and birds an unusually small
variety of bipolar which might contact only one cone. Similar bipolars were
observed subsequently in man and monkeys and described as "monosynaptic"
by POLYAK (1941). Such "midget" bipolars have been found also in the dog
(SHKOL'NIK-YARROS, 1968) but not in the rabbit (RAVIOLA and RAVIOLA, 1967;
SHKOL'NIK-YARROS, 1968). POLYAK (1941) further differentiated the diffuse bipolar
cells into "mop", "brush" and "flat-top" varieties, all of which he thought to
receive inputs from both rods and cones. BOYCOTT and DOWLING (1969), in their
definitive study (cf. Fig. 2), were unable to distinguish clearly a brush bipolar.
Their conclusion that Polyak's mop type is Cajal's rod bipolar and that his flat-
top is a diffuse cone bipolar is reasonable within the limits imposed by light microscopy, and consistent with the typical observation that in most vertebrates the
cone bipolars closely resemble Polyak's flat-tops (e.g. RAVIOLA and RAVIOLA,
1967). According to BOYCOTT and DOWLING, the diameter of the dendritic field
of rod bipolars in primates is roughly 15—20 μm, anywhere in the central area. It
appears, however, that the dendrites branch more profusely in the perifovea,
where rods are plentiful, than in the foveal slope, where they are scarce. The
diameter of the dendritic field of the diffuse cone bipolars (flat-tops) in primates
varies considerably with retinal region: in this case it appears that the number of
cones contacted is constant, so that the size of the dendritic tree varies with the
distance between adjacent cones. The diameter of the dendritic field of the midget
bipolars is about the same as the diameter of the cone synaptic endings. Studies of
monkey retina with optical microscopy and the Golgi-EM procedure (KOLB et al.,
1969) have revealed two kinds of midget bipolar cell. Each cone in the perifovea
appears to contact two midget bipolar cells — an "invaginating" midget and a
"flat" midget; cones in the periphery may contact also an additional invaginating
midget bipolar. POLYAK (1941) noted the presence of two varieties of midget
bipolars but did not observe the all-important difference in dendrites revealed by
KOLB et al.

It is worthy of note that other Golgi-EM studies (see below) have shown the unreliability of deducing connections only from the appearance of Golgi preparations in the light microscope. In goldfish, for example, Cajal's "bipolar for rods" (Fig. 25) sends dendrites also to cones (STELL, 1967), while the diffuse flat-topped bipolar, believed by POLYAK (1941) to go to both rods and cones, is shown by electron microscopy to be connected exclusively to cones (KOLB, 1970). Similarly, the diffuse mop bipolars of primates, thought by POLYAK to be connected to photoreceptors of both types, are in fact connected only to rods (KOLB, 1970; MISSOTTEN, 1968). Therefore, the connections of bipolars (and other cells) inferred from optical microscopy must be confirmed by electron microscopy.

LANDOLT (1871) described in amphibians a small appendage, different from nearby processes identifiable as receptor cells, extending between the first synaptic layer and the external junctional layer (external limiting membrane). These appendages were identified later as processes from a class of bipolar cells found commonly in the retinas of amphibians, reptiles and large diurnal birds (CAJAL, 1889, 1892), sharks and rays (RETZIUS, 1896; NEUMAYER, 1897; SCHAPER, 1899), and lampreys (STELL, unpublished observations). In dogfish shark, at least, Landolt's clubs are present on some bipolar cells but absent from others. In some animals both ordinary and displaced bipolars bear this appendage (CAJAL, 1892). Landolt's clubs have never been observed in teleosts, and although POLYAK (1941) reported having seen a few in chimpanzee, BOYCOTT and DOWLING (1969) never found any in their Golgi preparations of primate retinas. LIPETZ (in preparation) notes that it is difficult to correlate the presence of Landolt's clubs with any cellular or retinal feature which might indicate their function. Bipolar cells with Landolt's clubs are found in all vertebrate groups except teleosts, nocturnal birds and mammals; that is, they occur in groups in which the rod terminals are *not* the classical spherules. Perhaps the presence of these appendages will be associated with a demonstrable difference in the dendritic connections of bipolar cells which have or do not have them.

Displaced bipolar cells, having cell bodies in the most proximal portion of the layer of photoreceptor cell bodies, are found in birds, reptiles, amphibians, and lower fishes (CAJAL, 1892; NEUMAYER, 1897; MUNK, 1964; STELL, unpublished observations). Although no remarkable difference between these and the ordinary bipolars has been described, apart from the location of the cell body, appropriate studies may yet reveal some distinction according to types or patterns of synaptic contacts.

Horizontal Cells

The horizontal cells were first observed with conventional histological methods and designated the *Zwischenkörnerschicht* by HEINRICH MÜLLER (1851, 1857). These cells were described in many animals, primarily fishes, in the succeeding four decades (reviewed in STELL 1966, 1967). In lampreys (W. MÜLLER, 1874), sharks and rays (HAMA, personal communication; YAMADA and ISHIKAWA, 1965; STELL, 1965a, 1966), goldfish and carp (CAJAL, 1889, 1892; YAMADA and ISHIKAWA, 1965; STELL, 1966, 1967; WITKOVSKY and DOWLING, 1969) and some other teleosts (ANCTIL, 1969), the two strata of these unusually large cells form a distinct, con-

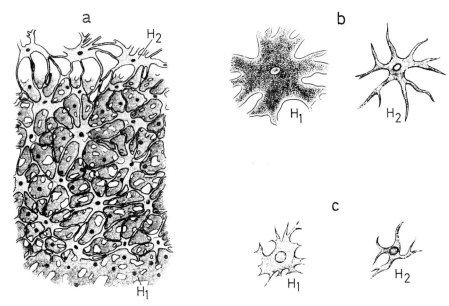

Fig. 3. Drawings of horizontal cells from teleostean retinas, seen from vitreal side; retinas fixed in chromic acid and macerated. H_1 first row (external); H_2 second row (probably intermediate). *a*, intact horizontal cell layers of a perch, *Acerina*. *b*, isolated cells from *Acerina* (H_1) and *Perca* (H_2). *c*, isolated cells from *Cyprinus carpio* (reproduced from H. Müller (1857) by permission of Akademische Verlagsgesellschaft)

tinuous but perforated layer just proximal to the first synaptic layer. As revealed by early methods (Fig. 3) they were polygonal to stellate in form and took most stains very lightly; since they appeared also to have no axon, they were considered to be supporting cells (Schiefferdecker, 1886). Later, Cajal (1892) demonstrated with the Golgi method that teleostean horizontal cells in the two distalmost strata had dendrites extending into the first synaptic layer and an axon running horizontally; he did not observe an axon terminal. Neumayer (1897) and Stell and Witkovsky (unpublished) were unable to demonstrate axons on selachian horizontal cells with the Golgi method. More recently, Stell (1967), O'Daly (1967b) and Parthe (1967) have also reported difficulty in locating horizontal cell axons in teleosts. The capriciousness of the Golgi method may account for this difficulty, however, and Kaneko (1970) has demonstrated axon-like appendages on some horizontal cells of goldfish injected intracellularly with the dye, Procion Yellow.

The illustrations of Cajal (1892) imply that the dendrites of horizontal cells of the first row (external or sclerad type) end at the level of the cone pedicles while those of the horizontal cells of the second row (intermediate or vitread type) end at the level of the rod spherules. Cajal's "internal horizontal cells," the "anucleate concentric cells" of Schiefferdecker and others, have never been shown to send processes into the synaptic layer. Since the presence of processes extending into the outer plexiform layer is a defining property of horizontal cells, I suggested (Stell, 1967) the use of an alternative term for these anomalous elements. Because so little is known except for their appearance in histologic sections, I shall designate them here the "cylindrical processes" of the intermediate neuronal layer, a term which simply recognizes their observed form. The various horizontal cells may be designated according to the number of layers, proceeding from distal to proximal (so that, for example, Cajal's intermediate horizontal cells are called "horizontal cells of the second row") — or, if known, the class of receptors with which they establish contact (the same cells in goldfish: "horizontal cells for rods").

As noted recently (e.g. ANCTIL, 1969), horizontal cells vary remarkably in size and variety in different species of teleosts. CAJAL (1892) observed that retinas having a high proportion of rods tend to have more rows and larger cells than retinas having a high proportion of cones. The cells in the distal row are small and polygonal, while those in successive proximal layers are progressively larger and more stellate (VILLEGAS and VILLEGAS, 1963; TESTA, 1966; PARTHE, 1967). In retinas in which the rods and cones are nearly equal in number, such as those of the teleostean family Labridae (ENGSTRÖM, 1963; ANCTIL, 1969; STELL, unpublished observations), the horizontal cells are reduced to a single thin but prominent layer of distal cells and an inconspicuous layer (seen in Golgi preparations) of proximal cells. In retinas strongly dominated by rods, such as those of the families Mugilidae, Gerridae, and Centropomidae, there are as many as four layers of true horizontal cells (PARTHE, 1972). The so-called "stellate amacrines" (TESTA, 1967; PARTHE, 1967) are really horizontal cells for rods (PARTHE, 1969, 1970, 1972; STELL and LAUFER, unpublished observations). Sincethe horizontal cells of teleosts are largest and most varied in animals with rod-dominated retinas, it is curious that the cells in three of the four layers appear to send dendritic processes only to the cone pedicles (PARTHE, 1972). Since a preponderance of rods is usually accompanied by a reduction in the number of bipolar cells, the larger horizontal cells are like glia in that they appear to fill in the spaces between the sparsely distributed bipolars. It should be noted that PARTHE (1972) has shown that the horizontal cells for rods occupy the fourth horizontal cell layer in the Mugilidae and Gerridae but the third (of four) in Centropomidae. STELL (1965a, 1966) observed this location in *Centropomus* but, like others, overlooked the fourth horizontal cell row, which is evident only in well impregnated Golgi specimens. It should be clear from the preceding discussion that the retinas of teleosts are so varied in structure that it is not meaningful to refer to "the" fish retina.

Much less is known about the variety of horizontal cells in other vertebrates, with the possible exception of a few mammals. CAJAL (1892) believed that there were three rows of these cells in all vertebrates, although it was difficult to demonstrate an element corresponding to the teleostean cylindrical processes. Only one horizontal cell type has been identified clearly in amphibians (DOWLING and WERBLIN, 1969) and some reptiles (YAMADA and ISHIKAWA, 1965), but LASANSKY (1971) has observed at least two types in a turtle. The "horizontal cells" described by PEDLER (1963) in the layer of photoreceptor cell bodies of reptilian retinas are in all probability the displaced bipolar cells illustrated in this group by CAJAL (1892). CAJAL (1892, 1896) described two rows of horizontal cells in birds and mammals; a distal row of small, flattened cells with long protoplasmic

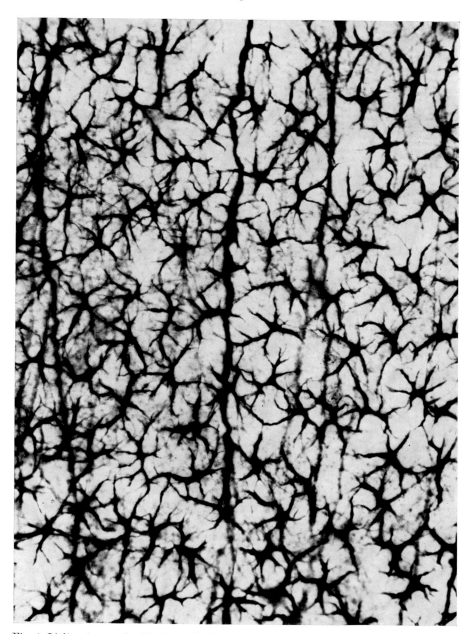

Fig. 4. Light micrograph of horizontal cells in flat-mounted cat retina stained with reduced silver, viewed from vitreal surface (reproduced from Gallego (1965), by permission of the author and Masson et Cie.)

dendrites, and a proximal row of cells with large cell bodies and long, thick processes. Polyak (1941, 1957) observed only one type of horizontal cell in man and other primates. Boycott and Dowling (1969) described two types of horizontal

cell in monkey retina by means of the Golgi method, provisionally designating one
a cone type ("A") and the other a rod type ("B") (Fig. 2). Further studies of these
cells with the electron microscope (KOLB, 1970) showed that "dendrites" of both
types contact only cones, however. Little evidence suports GALLEGO's (1964—1965,
1965) contention that there are at least two rows of horizontal cells in all mammals
and that other investigators have succeeded in demonstrating only the proximal,
axon-bearing cells in primates.

Using the Golgi method, CAJAL (1896) described two types of horizontal cell
in essentially adult form in ten-day postnatal cats and dogs: "external" (small
with a fine axon) and "internal" (larger, with a thick axon). PRINCE and McCON-
NELL (1964), in routine histological sections of rabbit retina, described giant cells,
in the distal intermediate neuronal layer which are probably typical large (proximal) horizontal cells. GALLEGO (1964—1965; 1965) described a plexus of horizontal
cells in silver-stained flat preparations of cat retina (Fig. 4). The cells are small
(6 μm by 9 μm in the central area, with a dendritic field about 76 μm by 89 μm)
and no axon can be seen, even in preparations in which the finest axons of some
other cells are visible. These are interpreted as the equivalent of Cajal's external
horizontal cells. GALLEGO believed that the internal horizontal cells of CAJAL
(which have axons) are present in this retina, but that they are not stained by this
procedure. DOWLING et al. (1966) described two types of horizontal cells in Golgi
preparations of cat and rabbit: large cells (300 μm to 500 μm dendritic field) with
no apparent axon, designated "external"; and small cells (100 μm to 200 μm
dendritic field), frequently with an axon, designated "internal". These identi-
fications, based on the apparent proximity of the cell body to the first synaptic
layer and the presence or absence of an axon, are inconsistent with Cajal's identification of the "external" horizontal cells as the smaller ones (CAJAL, 1892). In
canine retinas prepared by GALLEGO's (1953) technique, HONRUBIA LÓPEZ (1966)
observed horizontal cells with dendritic fields measuring 160 μm to 200 μm and
having no axon, which he designated "external". LEICESTER and STONE (1967)
described horizontal cells of a single type in the cat using a silver stain; like the
ones described by GALLEGO (1965), they are small (soma 15—20 μm, dendritic
field about 100 μm) and appear to lack an axon. These cells were not seen in preparations stained intravitally with methylene blue. HONRUBIA and ELLIOTT
(1969a) described in cat and rabbit, by silver staining, two types of horizontal
cells located in a single layer (Fig. 5). In rabbit, about half of the horizontal cells
observed are "symmetric" and half are "asymmetric" (i.e., one process is disproportionately longer and has fewer branches that the others), whereas in cat only
one cell in twenty-five is "asymmetric". Cells appear similar in both species except
that they are on the average larger in the rabbit than in the cat (dendritic fields
of 100—1000 μm compared to 80—350 μm). UYAMA (1951) also described horizontal
cells in cats.

BOYCOTT (personal communication) has observed symmetric and very occasionally asymmetric horizontal cells in the cat, using several different Golgi procedures. Asymmetric cells stained more commonly in the rabbit. He finds also two
subtypes of these cells: a large one contacting many cones (about 100 in cat,
possibly more in rabbit) and having no apparent axon; and a smaller one having
finer and more numerous dendrites and a distinct axon. The axons, which are very

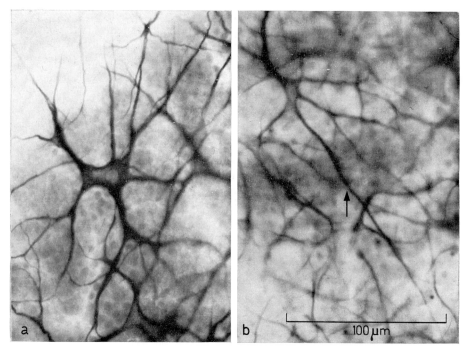

Fig. 5. Light micrographs of horizontal cells in flat-mounted cat retina stained with reduced silver, viewed from vitreal surface. *a*, "symmetrical" type from peripheral retina. *b*, "asymmetrical" type; one process (arrow) is much longer than any others. Same magnification for *a* and *b* (reproduced from HONRUBIA and ELLIOTT (1969), by permission of the authors and the American Medical Association)

fine (0.3—1.0 μm), sometimes can be followed from the cell body to thicker (4 μm in cat, more in rabbit) expansions, from which arise fine terminals which appear to contact rod spherules. These axons are some 200—500 μm in length but sometimes follow a tortuous path, so that the cell body and axonal ending may be quite close together. It is not yet clear whether Boycott's axonless cells are identical with the silver-stained cells of GALLEGO (above) or constitute a third class.

BOYCOTT is the only recent investigator who has succeeded in following a complete horizontal cell axon from cell body to terminal in any mammal. Complete horizontal cells have been seen only rarely, and CAJAL (1892, 1909—1911) and POLYAK (1941) observed mainly fragments which they interpreted as axons and axon terminals. CAJAL (1892, plate IV, Fig. 5) and POLYAK (1941, Fig. 77c) illustrated complete horizontal cells in chicken and chimpanzee, respectively, however. KOLMER (1936) noted that in flat preparations of horse retina stained intravitally with methylene blue he could follow some horizontal cell axons for many centimeters from the cell body without coming to a terminal arborization. Unfortunately this claim is not supported by any illustrations or confirmation. A long process which runs horizontally from a horizontal cell body may not be a true axon, even if longer than other processes of that cell. LASANSKY (in preparation) has analyzed some short axon-like processes of horizontal cells in Golgi preparations of turtle retina by electron microscopy; the fine-structural relationships of these "axons" are indistinguishable from those of the "dendrites". The asymmetrical processes of some horizontal cells in cat and rabbit (DOWLING et al., 1966; HONRUBIA and ELLIOTT, 1969a) and the very short "axons" in the fruit bat (PEDLER and TILLY, 1969) and teleosts (PARTHE, 1967) may be quasi-axonal processes of this sort.

KOLB (1970 and personal communication) has shown by Golgi-EM that the "axon terminals" of horizontal cells in monkey, cat and guinea pig make contact with rods but not cones, independent of the subtypes A and B described by BOYCOTT and DOWLING (1969). BOYCOTT's demonstration of continuity leaves little doubt that in the cat, at least, a single horizontal cell contacts both cones and rods, as implied by the physiological studies of STEINBERG (1969a, b). Probably the same is true of other mammals, although it remains to be confirmed. In this respect the horizontal cells of mammals differ significantly from those of teleosts, in which the separation of rod and cone horizontal cells has been demonstrated morphologically (STELL, 1967) and electrophysiologically (LAUFER and MILLÁN, 1970, but see LAUFER et al., 1971).

These observations raise numerous questions. Why do the mammalian horizontal cells provide for separation of specific rod and cone-contacting zones but allow interaction, or at least summation, by interconnecting them as parts of the same cell, while the teleostean rod-and cone-contacting zones are segregated completely by locating them on separate cells? How do the teleostean horizontal cell axons (if present) terminate, and how is horizontal cell function varied according to the presence or absence of axons? As will be discussed later, their interconnections with other cells show that the horizontal cells are true neurons, whether or not they have a true axon. It has been noted, however, that they differ from "typical neurons" — especially in fishes — in size, form, and cytoplasmic structure (YAMADA and ISHIKAWA, 1965), histochemistry (LESSELL and KUWABARA, 1963; MIZUNO, 1964; PARTHE, 1967) and electrophysiology (SVAETICHIN, 1967). Much of importance surely remains to be learned about the detailed morphology and diversity of the horizontal cells. The available evidence implies that a major role of these cells is to mediate local interactions between photoreceptor cells or between receptors and bipolar cells. The evidence of fine structure supports this conclusion (see below).

Amacrine Cells

With the exception of the displaced varieties, the amacrine cells are elements whose cell bodies lie in the most proximal part of the intermediate neuronal layer and whose processes extend into the second synaptic layer (Fig. 2). These cells are more diversified and numerous than the horizontal cells, and even less well known. CAJAL's (1892) word "amacrine" (no axon) is appropriate because for the most part the amacrine cells are characterized by having numerous identical processes and no obvious polarity. Their ultrastructural relationships identify them clearly as neurons, however. The cells have been classified according to their processes as "stratified" (the many branches of the main process ramifying in the same planar level of the second synaptic layer — "unistratified" if at a single level, "bistratified" or "multistratified" if at two or more levels) or "diffuse" (really "unstratified", with their many branches ending or synapsing without respect to the stratification of the synaptic layer). Cells vary further in the specific level(s) of stratification, caliber of processes, fineness or multiplicity of branching, size of field, symmetry of field, and so forth. Another type of cell, the "associational amacrine" found in birds by Cajal (1896), is not truly "amacrine". This cell (Fig. 6) has a single dendritic tuft in the second synaptic layer, near the cell body; from this an axon extends horizontally to terminate in an extensive arborization at the same level some distance away (up to 1 mm in chicken and pigeon). As noted by Cajal, these are true short-axon neurons, presumably comparable to some horizontal cells. Although of considerable interest because of their relation to the centrifugal nerve fibers (Cajal, 1896), they have not been investigated in this century.

Cajal (1892, 1893) emphasized that in the most highly differentiated retinas, those of small birds and reptiles, five to seven sublayers can be distinguished in the inner plexiform (second synaptic) layer. He observed (in 1893) the largest number of sublayers in the species in which the amacrine cells were most numerous and diversified. He also observed (in 1896) that during development of cat and dog retinas the synaptic sublayering becomes evident only when small ganglion cells and multistratified amacrine cells can be discerned. Each layer was thought to comprise the terminal arborizations of bipolar cells of one class, enclosed between the more distal arborizations of amacrines of the corresponding class and the more proximally located dendrites of ganglion cells of one class. This observation clearly relates the role of amacrine cells to the functioning of the bipolar-ganglion cell synapses.

A curious but not very useful generalization is that the retinas with the most highly differentiated amacrine cells and second synaptic layer — those of amphibians, reptiles, and birds — are similar also in having bipolar cells with Landolt's clubs, receptor cells which are less clearly differentiated than typical (mammalian) rods and cones at the synaptic level, and ganglion cell functions which are complex (to be discussed later). In lampreys and selachians also the differentiation of rods and cones is atypical and some bipolar cells have Landolt's clubs (see above), but the complexity of organization of their second synaptic layer and ganglion cell behavior is not yet known. Although activity, habitat and visual acuity can be correlated well with the degree of specialization of the retina (including the second synaptic layer) in some teleostean species, in most the correlation is poor (Ali et al., 1968). Furthermore, although the second synaptic layer may be stratified and highly developed in some teleosts (Anctil, 1969), teleostean bipolar cells lack Landolt's clubs and teleostean rods and cones are conventional (Cajal, 1892).

Vrabec (1966) has described amacrine cells in the highly differentiated ventral retina of the teleost Callionymus using intravital methylene blue staining. He observed large and small monostratified amacrines which ramify in the most proximal sublayers of the second synaptic layer, bistratified amacrines which ramify in the distal and intermediate level of that layer (near the terminations of the cone bipolar cells), and diffuse amacrines. Testa (1966) and Parthe (1967) have described in several Caribbean teleosts only some "piriform" amacrines which are unistratified cells ramifying in any of five sublayers in the second synaptic layer; and the "interstitial" or displaced amacrines, described below, which are unistratified cells whose soma is in the same layer as their processes.

In addition to the "associational" and displaced varieties, CAJAL (1896) distinguished amacrine cells of three types in birds: stellate; bi- and tri-stratified (very abundant in birds); and giant unistratified (unique to birds and reptiles). In rapid Golgi preparations of chick retina, SHEN et al. (1956) observed amacrine cells of three types: large stellate cells ramifying in the distalmost zone of the second synaptic layer; unistratified cells ending in one of four sublayers; and nonstratified (diffuse) cells whose oblique processes pass through all four sublayers. In methylene blue preparations of pigeon retina MATURANA and FRENK (1965) observed large, flattened stellate amacrines in the distalmost part of the second synaptic layer, and "parasol" (unistratified) amacrines with a 6—7 μm soma and a 100—200 μm field of processes ramifying in one synaptic sublayer. In Golgi-Cox preparations of 9—17 day chick embryos, CASTRO (1966) observed mainly amacrines of the unistratified variety, each sending out dozens of finely branched processes covered with spines. These cells ramified in one of three levels of the second synaptic layer; cells ramifying in the most distal level had the largest dendritic fields, and those ramifying in the most proximal level had the smallest fields. CASTRO observed other varieties of amacrine cells on occasion. Using a reduced silver method, BOYCOTT and DOWLING (1969) were able to demonstrate at least five sublayers of the second synaptic layer of the pigeon retina. They accounted for sublayers two, three and four mainly by the branchings of the unistratified amacrine cells.

In mammals, KOLMER (1936) observed three types of amacrine cells: stratified (mainly unistratified), diffuse, and horizontal (stellate). In Golgi preparations of rabbit, RAVIOLA and RAVIOLA (1967) observed both unistratified and diffuse types, both of which might end at any level of the inner plexiform layer.

SHKOL'NIK-YARROS (1968) observed mainly diffuse amacrines in rabbit with the Golgi method. In the cat, LEICESTER and STONE (1967) observed that most amacrine cells stained with silver or methylene blue were stellate with a small (8—20 μm) soma and four or five flattened processes which spread over a field of

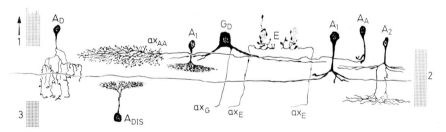

Fig. 6. Drawing of amacrine cells, displaced ganglion cells, and efferent (centrifugal) fibers as seen in radial sections of avian retinas prepared by methylene blue or Golgi method. *1* intermediate neuronal (inner nuclear) layer; *2* second synaptic (inner plexiform) layer; *3* ganglion cell layer. Amacrine cells: A_1 ordinary unistratified; A_{DIS} displaced unistratified; A_2 bistratified; A_D diffuse. AA associational "amacrine" cell with horizontally directed axon and terminal (ax_{AA}); here viewed somewhat obliquely, the terminal is flattened in the plane of the retina. G_D displaced ganglion cell (Dogiel's cell), with its axon (ax_G). Efferent axons (ax_E) terminate in arborizations (E) beneath and around unidentified amacrine cell bodies (redrawn with modifications from CAJAL (1892, 1896))

70–660 μm (in methylene blue) or 55–180 μm (in silver). They did not describe any ordinary unistratified, parasol-shaped, or diffuse cells, but after silver staining noted a few "globular amacrines" with a large cell body in the middle of the synaptic layer and processes reaching as far as the somata of both bipolar and ganglion cells. Boycott and Dowling (1969) (Fig. 2) observed "diffuse" amacrines of three types in Golgi preparations of monkey retina: "wide-field", ramifying in the most proximal (vitread) part of the synaptic layer; "narrow-field"; and "stratified", ending in all levels of the synaptic layer but branching maximally in only one of the three levels. They found stratified amacrines of three types: "unistratified", usually ramifying in the distal half of the synaptic layer with processes spreading over a field less than 500 μm in diameter and sometimes bearing spines; "large unistratified", with a field as broad as 1 mm; and "bistratified", with processes spreading over fields about 100 μm in diameter in the most distal and most proximal levels of the second synaptic layer. They did not observe any multistratified amacrines.

Miscellaneous

In addition to the horizontal, bipolar, and amacrine cells, several other varieties have been described in the intermediate neuronal layer. The cylindrical processes of this layer in teleosts (Cajal's internal horizontal cells), mentioned earlier, are variable in distribution but present in all teleostean species examined by electron microscopy (Yamada and Ishikawa, 1965; Stell, 1966; O'Daly, 1967a; Parthe and Laufer, personal communication). Cajal (1892) described also in teleostean retina "small stellate cells", whose somata lay among the bipolars and which sent out fine, undulant processes to both synaptic layers. Similar cells were described in selachian retinas by Neumayer (1897) and in one group of birds by Cajal (1896), both employing the rapid Golgi method. The "stellate amacrines" described by Testa (1966) do not correspond to these cells of Cajal but are horizontal cells for the rods, as indicated above. The "undulant amacrine cells" of O'Daly (1967a), described as similar in architecture to the small stellate cells, are actually bundles of small-diameter (a few hundred Ångstrom units) extracellular tubules (see below). Catecholamine-containing cells with processes similar in form and extent to those of Cajal's cells have been described through fluorescence microscopy in teleosts by Ehinger et al. (1969) (Fig. 26) and in primates by Ehinger and Falck (1969). Since the perikarya of these adrenergic cells appear to be confined to the most proximal part of the intermediate neuronal layer, however, they are not identical to the small stellate cells described before. Boycott (personal communication) has, however, observed in goldfish a Golgi-stained cell which matches the fluorescent cells of Ehinger et al. In Golgi preparations of the teleost *Eugerres*, Parthe, Laufer and Stell (unpublished observations) have observed a plexus of fine processes similar to those described by Ehinger et al. (1969) among the horizontal cells. Since we have not yet established the cell of origin or the fine structure of these processes, it remains possible that these are the clusters of small extracellular tubules. Polyak (1941, 1957) described certain cells in the intermediate neuronal layer of primates in which the usual dendritic/axonal polarity appeared to be reversed; he designated these "centrifugal bipolars". Missotten (1965a) found no evidence for such cells in human retina by serial sectioning and electron microscopy, and he suggested that they are "rare or absent" from adult human retina. A reasonable explanation for the discrepancy has been proposed by Boycott and Dowling (1969). Noting that Polyak's observations were made in young or immature animals, they suggest that the "centrifugal bipolars" are simply the bipolar developmental stage of diffuse amacrines, which lose the distal process ("axon") later during maturation. In the rhesus monkey, Cohen (personal communication) has observed a single case, believed to be aberrant, of a bipolar cell with a terminal bag in the first synaptic layer. The terminal contained a synaptic lamella and vesicles but made no apparent synaptic contacts.

Bipolar and amacrine cells may be "displaced" from the usual location of the major class of cells. Such displacement is not just an occasional anomaly but a regular structural feature of many retinas. A population of displaced bipolar cells is probably found in the most proximal (vitread) part of the photoreceptor cell layer in some members of all groups except teleosts and higher mammals (see above). Cell bodies of displaced ganglion cells — otherwise typical ganglion cells, with dendrites in the inner plexiform layer and axons which cross that layer to exit in the optic nerve — are found in the most proximal part of the intermediate neuronal layer among the amacrine cell bodies. They were observed by Cajal (1892) in all vertebrates except fishes. They are especially common in birds, where they were first described by Dogiel and are therefore called "Dogiel's cells" (Dogiel, 1895; Cajal, 1896; Shen et al., 1956; Maturana and Frenk, 1965). Giant displaced ganglion cells of a rather different variety, large multipolar cells with a large axon, are found frequently within the inner plexiform layer in *Mugil* (Parthe, 1968). Similar cells have been observed in this layer in rabbit retina by Honrubia and Grijalbo (1967). In lampreys, it is said, all the ganglion cells are displaced to the scleral side of the second synaptic layer, and the optic fiber layer is displaced to the scleral side of the ganglion cells (Walls, 1942; citing Tretja-koff, 1915). Displaced amacrine cells, large multipolar cells without axons, have been observed in the second synaptic layer of certain teleostean retinas (Parthe, 1967). They form two definite layers, one near the layer of interneurons (the "interstitial amacrines" of Testa 1966 or "external interstitial amacrines" of Parthe, 1967) and one near the layer of ganglion cells ("displaced amacrines" of Testa or "internal interstitial amacrines" of Parthe). Displaced amacrines were observed in various animals by Cajal (1892).

D. Efferent (Centrifugal) Fibers

Cajal (1889, 1892) observed a system of small nerve fibers which appear to leave the retinal nerve fiber layer, traverse the second synaptic layer and terminate on or near the amacrine cells (Fig. 6). This system was highly developed in birds; he observed it also in the dog, but was uncertain about its occurrence in other animals. Cajal (1933) believed later that he might have demonstrated it in amphibians. These fibers, which he termed centrifugal, were described in detail after intravital staining with methylene blue (Cajal, 1896). They varied in different avian species and were found to be especially well developed in the pigeon.

Three types of terminals were observed in the pigeon: the "pericellular nest", a cluster of several fusiform or ellipsoidal terminals clasping the proximal surface of the amacrine cells; the "inferior (or basal) branches", short preterminal collaterals ending freely on several nearby amacrine cells; and "ascending or long filaments", one to three per terminal, climbing between amacrine cells to terminate at the distal margin of the amacrine cell layer. He believed that the efferent fibers ended mainly on the soma and descending dendritic stalk of the "association amacrines", sending only collaterals to the ordinary amacrines. The associational cells, in turn, appeared to end on distant ordinary amacrines and perhaps bipolar-ganglion cell articulations. Thus it appeared that the system might provide for central control of information leaving the cone system, since it is the bipolar cells for cones which terminate in the distal sublayers of the second synaptic layer (Cajal, 1933). Cajal (1896) noted that the efferent fibers he observed corresponded to only one of the two classes described in birds by Dogiel (1895); he found no clear evidence for fibers of a second type.

Later studies of the avian centrifugal fiber system, cited in Cowan and Wenger (1968) and the comprehensive reviews of Ogden (1968) and Cowan (1970), will be summarized here briefly. Maturana and Frenk (1965) presented a description of methylene blue-stained fibers in pigeon retina which differs from Cajal's. They observed endings of two types: "convergent", forming a terminal nest as described by Cajal, and "divergent", branching in a fanlike flat terminal at the boundary between amacrine cell somata and the second synaptic layer (Fig. 7). The convergent type appeared mainly to form a pericellular nest around the "parasol" (ordinary unistratified) amacrines, the divergent type mainly to end on the base of dendrites of the flat (stellate) amacrines. Each type seemed also to end on the base of the primary dendrites of displaced ganglion cells, without forming a terminal nest (Fig. 8). Associational cells were not observed, but Cowan (1970) suggests that they were represented by the flat amacrines in these investigators' preparations. Maturana and Frenk estimated the number of terminals (fibers) as 10^5 per retina, or about one per 1500 μm^2. From counts of cells and axons in the isthmo-optic nucleus and tract, Cowan (1970) estimates that each axon branches in the optic nerve and retina to form a total of 8—10 terminals.

The central connections of these fibers were pursued by Cowan and his colleagues (Fig. 9). Cowan et al. (1961) followed degeneration of central fiber tracts

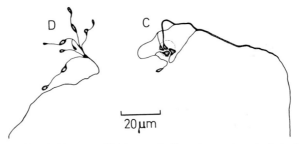

Fig. 7. Convergent (C) and divergent (D) forms of efferent axon terminals in the retina of the pigeon, stained intravitally with methylene blue, mounted flat, and viewed from the vitreal side (reproduced from Maturana and Frenk (1965) by permission of the authors and Science; copyright 1965 by the American Association for the Advancement of Science)

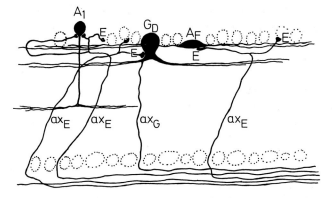

Fig. 8. Relationships of efferent axons (ax_E) and terminals (E) to cells in avian retina, represented schematically in radial section. Efferent axons terminate upon unistratified amacrine cells (A_1), flat amacrine cells (A_F) and displaced ganglion cells (G_D). Cf. Fig. 6 (reproduced from MATURANA and FRENK (1965) by permission of the authors and Science; copyright 1965 by the American Association for the Advancement of Science)

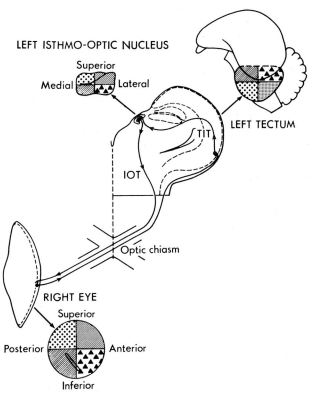

Fig. 9. Diagram of organization of visual centripetal-centrifugal (efferent) loop in birds. IOT, isthmo-optic tract; TIT, tecto-isthmal tract (reproduced from COWAN (1970) by permission of the author, the British Medical Bulletin, and the Journal of Anatomy)

in the pigeon with silver, Nauta, and Nissl staining after unilateral enucleation. Degeneration of the main pathways (tectum, thalamus) was demonstrable after about 12 days. About 26 days after enucleation, changes appeared also in the isthmo-optic nucleus and tract; the pattern of degeneration suggested that these later changes were retrograde — that is, that the isthmo-optic nucleus was the source of efferent fibers to the retina. This was confirmed by Cowan and Powell (1963), who followed antegrade degeneration into the contralateral optic nerve and retina with the Nauta method after placing unilateral lesions in the isthmo-optic nucleus and tract. They estimated the total number of efferent fibers per eye as 10^4 (one per 100 centripetal fibers in the optic nerve). More fibers went to the thicker temporal retina, which includes the region subserving binocular vision, than to other areas. The terminations in the retina could not be seen well with the Nauta method but were demonstrated in the electron microscope by Dowling and Cowan (1966). They observed synaptic terminals of efferent fibers on free processes (including proximal processes of amacrine cells) in the synaptic layer and on the base of cell bodies (apparently of ordinary amacrine cells); they observed no terminations on displaced ganglion cells and few terminations having the distribution of pericellular nests. McGill et al. (1966a, b), using both antegrade and retrograde degeneration, showed that the system is retinotopic (the chain of neurons, retina — tectum — isthmo-optic — retina, returns to the same retinal region from which it originated). Although the precision of this representation is greater than the resolution of the methods employed, it is limited at least by the ganglion cell: efferent fiber convergence ratio of 100 : 1. This implies also that each fiber covers a rather large area. Holden (1966a, 1968a, b) has confirmed the existence of the tecto — isthmo-optic — retinal pathway electrophysiologically, and has shown (J. Physiol., 1970) that many isthmo-optic neurons have large (10—20°), complex visual receptive fields. In an important contribution, Miles (1970) has shown that electrical stimulation of avian isthmo-optic axons alters the excitability of ganglion cells, usually by decreasing inhibition.

Regarding the evidence for retinal efferent systems in other vertebrate groups, reviewed by Brindley (1960) and Ogden (1968), I must agree with Ogden that "the bird is the only vertebrate for which conclusive evidence of an efferent retinal projection is available". Nevertheless, since inconclusive evidence often leads to conclusive experiments, I shall review some of the data.

In teleosts, Aichel (1896) has reported seeing centrifugal fibers with classical methods. He described endings of two types: a flat arborization close to the amacrine cells, and a diffuse spray spanning the second synaptic layer. Catois (1902) observed that fibers descend from diencephalic nuclei (including the lateral geniculate nucleus) to end in free arborizations in the "deep layers" of the retina in teleosts. Witkovsky and Dowling (1969) and Witkovsky (1971) have observed, in carp and dogfish shark, that the loosely myelinated axons which form a dispersed layer immediately proximal to the amacrine cells, make synaptic contacts with amacrine cells and processes of cone bipolar cells (in carp) in the nearby second synaptic layer. The similarity of these elements to the efferent fibers in birds strongly implies that there is an efferent fiber system in fish. The identity of these axons as centrifugal, rather than axons of displaced ganglion cells, axons of ordinary ganglion cells following an anomalous course to the optic nerve (cf. Munk, 1964), or axons of intraretinal associative neurons (Cajal, 1896; Gallego and Cruz, 1965), has not yet been confirmed experimentally.

Maturana (1958) has offered anatomical, and Branston and Fleming (1968) physiological, indirect evidence for centrifugal fibers to the amphibian retina. Lázár (1969), having destroyed the "optic centre" of the frog on one side, followed degenerating fibers through the contralateral optic nerve to the retina.

Since birds are so closely related to reptiles, it is remarkable that a well developed efferent system has not been demonstrated in reptilian retina. Armstrong (1950) concluded that there are no centrifugal fibers in the lizard when he failed to find any normal fibers in the central stump of the optic nerve, with silver staining, eleven weeks after unilateral enucleation. This experiment is not conclusive.

Among mammals, only the dog was believed by Cajal (1892) to have centrifugal fibers. In confirmation, Honrubia López (1966) demonstrated plentiful centrifugal fibers terminating in the most proximal part of the intermediate neuronal layer by silver staining of retinal whole-mounts. Ventura and Mathieu (1959), on the other hand, were able to demonstrate only a few such fibers per retina (each ramifying over an area of 1 mm² or more) in the retina of the dog, rabbit, cat, or man, using a similar technique. Cragg (1962) examined the retina and optic nerve of the rabbit distal to a nerve crush. He was unable to identify degenerating efferent fibers with certainty but believed he could follow a few as far as the retinal nerve fiber layer with the Nauta method. With the Glees stain, however, he observed degenerating boutons in the *first* synaptic (outer plexiform) layer. He concluded that although these "cannot be regarded as convincing evidence of a centrifugal projection", nevertheless there is such a projection to the retina, although it may be small. In the cat, Brindley and Hamasaki (1961, 1962 a, b) were unable to establish the presence of efferent fibers by either anatomical or physiological experiments. Brooke et al. (1965) claim to have demonstrated degenerating centrifugal fibers in cat, monkey, and pigeon by electron microscopy. It is not clear, however, that the quality of their preparations allowed them to distinguish degenerating fibers from other processes. In methylene blue preparations of human retina, Kolmer (1936) observed some processes suggestive of centrifugal fibers. Polyak (1941) saw apparently centrifugal fibers in some Golgi preparations of chimpanzee but not of other primates, and Boycott and Dowling (1969) found no positive evidence of them in Golgi preparations of monkey retina. Wolter's (1961, 1965) demonstration of centrifugally-directed "degeneration bulbs" in human pathological material is not evidence for the existence of efferent fibers. Honrubia et al. (1967) and Honrubia and Elliott (1968), using silver methods and whole mounts, have provided the best evidence for what may be efferent fibers in the retina of monkey and man. They find that some fibers in the retinal nerve fiber layer, more intensely argyrophilic than the optic (centripetal) fibers, follow an oblique course into the second synaptic layer. There they branch and diminish in caliber as they go and finally terminate mainly in the most proximal part of the layer of interneurons; a few fibers, however, appear to cross that layer to terminate in the first synaptic layer. Here it must be recalled that when vascularization is intraretinal, as in mammals, one must distinguish between the innervation of retinal neurons and vessels. Since this distinction cannot be made readily by light microscopy, especially of whole mounts, it may be possible to demonstrate in some animals a true centrifugal fiber system which is completely different in function from the classical one of birds. The centrifugal systems of all non-avian retinas will require much further study before their presence can be accepted and their significance appreciated.

E. Ganglion Cells

In recent years special attention has been drawn to the ganglion cells of frog retina by LETTVIN and his collaborators, because "it is the capacity of the ganglion cells to combine the information impinging on them into an operation", and "each morphological cell type has a different pattern of connectivity [and operation] entirely determined by its shape" (MATURANA et al., 1960). In methylene blue preparations of unspecified retinal regions they observed two classes of ganglion cells according to size: small, with a $7-10\,\mu$m soma and $50-350\,\mu$m dendritic field, and large, with a $20\,\mu$m soma and a dendritic field as large as $600\,\mu$m. Taking into account shape as well as size they distinguished cells of five morphological classes, which they called: "one-level restricted field", "one-level broad field", "many-level H-distribution", "many-level E-distribution", and "diffuse". According to size, complexity and level (probable bipolar connections) of the dendritic field and the size of the perikaryon, they assigned to each cell type one of the five complex ganglion cell functions which they had discovered. The possibility that amacrine cells play some role in generating complex ganglion cell behavior was not mentioned by these authors. Further support for a role of dendritic tree architecture in coding comes from the matching development of cells and response types in frog tadpoles (POMERANZ and CHUNG, 1970).

The Lettvin scheme would be more plausible if the complexity of frog ganglion cells were reproduced in other animals in which ganglion cell behavior is as complex as in the frog. Unfortunately, detailed information on the form of ganglion cell dendrites is scarce. Recent research on ganglion cell complexity, furthermore, has taken a completely different approach (DOWLING, 1968; DUBIN, 1969, 1970; and below).

It was shown first in cat and other mammals by GALLEGO (1954) that the dendritic fields of ganglion cells, as measured in silver preparations, are much smaller than the functional receptive fields (e.g. KUFFLER, 1952). GALLEGO proposed therefore that the direct bipolar-ganglion cell dendritic contacts might determine only the properties of the center of the receptive field, the periphery effect being mediated by some other neurons, probably amacrine cells. Further studies lend credence to this view. GALLEGO's (1965) more complete studies on flat preparations of cat retina stained with silver showed two classes of ganglion cells: large, with dendritic fields as broad as $160\,\mu$m in the retinal center, and small,

---

Fig. 10. Determination of minimum dendritic field boundaries in flat-mounted cat retina stained with reduced silver and examined from vitreal side. a, drawing of all stained cells in a portion of temporal retina including area centralis. b, dendritic fields represented by joining tips of visible dendrites with straight lines (reproduced from HONRUBIA LÓPEZ (1966) by permission of the author and the Archivos de la Sociedad Oftalmológica Hispano-Americana)

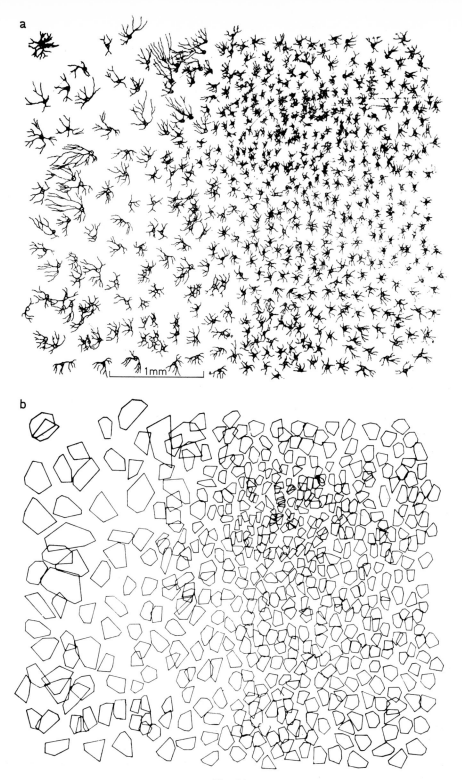

a

1mm

b

Fig. 10

with dendritic fields everywhere in the retina almost as small as the cell body, some 20—25 μm in the retinal center. He suggested that published electrophysiological studies concerned only the activity of the large cells. STONE (1965) noted in similar preparations of cat retina that ganglion cells were smallest and most densely packed in the central area; "arms" of relatively high cell density extend horizontally from this center into both nasal and temporal regions of the retina (cf. Fig. 10). In Golgi and methylene blue preparations of cat, BROWN and MAJOR (1966) also observed ganglion cells with either large (400—700 μm) or small (70—200 μm) dendritic fields. They saw no evidence for stratification of either the second synaptic layer or the ganglion cell dendrites. From flat preparations of cat retina stained with silver or methylene blue, LEICESTER and STONE (1967) classified 242 peripheral cells into four categories (Fig. 11): (1) "deep multidendrite", 161 cells with 3—5 dendrites crossing the synaptic layer obliquely to form a 70—710 μm wide dendritic field close to the interneuronal layer; (2) "loose single dendrite", 26 cells with one dendrite ramifying in an 18—135 μm wide field close to the intermediate neuronal layer; (3) "dense single dendrite", 22 cells, similar to the "loose single" type but with more and finer dendrites forming a 30—110 μm wide field; and (4) "shallow multidendrite", 29 cells, like "deep multidendrite" but with a smaller field (38—165 μm) and soma, and ending in the middle third of the second synaptic layer. In the central area, 20 % of the cells resembled the smallest of the peripheral deep multidendrite cells, while 80 % resembled the smallest of the

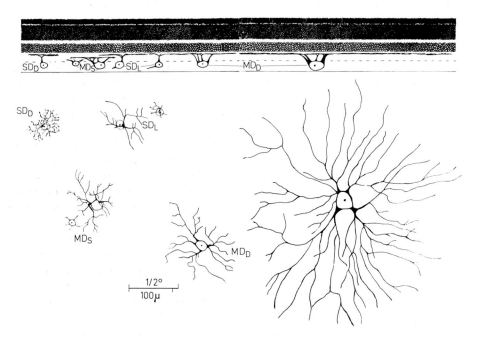

Fig. 11. Camera lucida drawings (below) of ganglion cells from peripheral retina of the cat, stained with methylene blue and viewed from vitreal side; same cells are projected on radial section in corresponding position above. SD_L loose single dendrite; SD_D dense single dendrite; MD_S shallow multidendrite; MD_D deep multidendrite (reproduced, slightly modified, from LEICESTER and STONE (1967) by permission of the authors and Vision Research)

peripheral loose single dendrite cells. The distribution observed in this study was unimodal for size within each class of cells; the bimodal distributions observed by GALLEGO (1965) and BROWN and MAJOR (1966) probably reflect their practice of categorizing cells by size of field alone, overlooking the differences in shape between the three classes of smaller cells. LEICESTER and STONE indicate that in the retinal center the physiological field centers are so much larger than the small cells that it is difficult to correlate them straightforwardly. They find no multi-stratified or diffuse ganglion cells in cat. HONRUBIA and ELLIOTT (1969 b, 1970) describe in silver-stained flat preparations of cat retina a giant unistratified multi-polar ganglion cell which is compatible with the description of the deep multi-dendrite cells above. SHIBKOVA and VLADIMIRSKII (1969) also describe size and numbers of ganglion cells in the cat. The account of LEICESTER and STONE appears to be the most comprehensive description of the varieties of retinal ganglion cells in any non-primate mammal.

In silver-stained dog retina, HONRUBIA LÓPEZ (1966) observed only "giant" ganglion cells, with fields measuring $30-50 \mu m$ centrally and up to 1 mm in the periphery. In silver-stained rabbit retina, only the "giant" (dendritic field $1-5$ mm) and smaller ordinary ganglion cells have been described, along with a displaced variety (HONRUBIA and GRIJALBO, 1967). In rat retina stained with methylene blue, BROWN (1965) observed two types of ganglion cells: "tight" with an arborization averaging $282 \mu m$ in diameter, with many branches in the distal second synaptic layer; and "loose" having a larger (average $397 \mu m$) arborization of few dendritic branches in the proximal second synaptic layer. He correlated these cells, respectively, with electrophysiological units lacking or having a center-surround organization. Differences in size of the anatomical and physiological field centers – as much as $200-300 \mu m$ – were accounted for by light-scattering, anatomical convergence (receptor-bipolar-ganglion cell), and failure of staining methods to demonstrate the full width of the dendritic tree. The same factors might explain equally well the discrepancies noted in cat by LEICESTER and STONE (above).

BOYCOTT and DOWLING (1969) have updated the classic description of primate ganglion cells by POLYAK (1941, 1957). They can distinguish five varieties (none of them multistratified): "midget" (also designated "monosynaptic", although each midget bipolar-ganglion cell complex appears to make lateral interconnections through other cells), found in all regions of the retina, with a dendritic field of 10 μm or less; "diffuse" – also as described by POLYAK – found in all retinal regions, with a dendritic field of $30-75 \mu m$, larger in the periphery of the retina than in the center; "stratified diffuse", very commonly observed cells which ramify in one of three levels of the second synaptic layer, most often distal and least often proximal; "unistratified", similar to the small diffuse or shrub types of POLYAK, with dendritic trees variable in size but generally around $200 \mu m$; and "displaced". These authors never observed the giant ganglion cells observed by POLYAK but note that he also did not see them in the perifovea. They point out a correlation between the "midget" system and the tiny receptive field centers observed physiologically in primate fovea.

Little is known of any direct interconnections between ganglion cells. Although KOLMER (1936) observed axon collaterals ending in fine branchlets on the perikarya of neighboring ganglion cells, such recurrent collaterals have been observed so infrequently that probably they should be disregarded. Gallego (1965b) and Gallego and Cruz (1964–1965), for example, point out that in most mammals they have seen no evidence for them. These authors describe instead a new type of "associational nerve cells" in the ganglion cell layer. Found in dog and man, mainly in the periphery, the silver-stained cells have axons as long as 5 mm which appear to terminate around the base of the ganglion cell dendrites. The observation of these cells is reported also by Honrubia López (1966) and Honrubia et al. (1967). It may seem curious that such cells have never been observed before by the numerous other investigators who have examined canine and human retinas. Certainly one should be cautious. However, the flat preparations used by Gallego and Honrubia are much more favorable than sectioned Golgi material for observing such horizontally extensive cells, and it is possible that they are stained more easily with silver than with methylene blue.

III. The Ultrastructure of Cell Contacts and Patterns of Connections
A. Methods

The electron microscope, which permits both the resolution of very fine processes and the identification and characterization of many neurons, has revolutionized the study of retinal organization. It has not yet been possible to avoid significant loss of information about the living state due to chemical reaction, solvent extraction, and molecular rearrangement. But major technical innovations of the past decade have greatly increased the potential for retaining ultrastructural information; these include aldehyde fixation, epoxy embedding, and staining with heavy metals in addition to vast improvements in electron optics. Very few investigators of the retina have taken full advantage of the best techniques and instruments now available. The result has been considerable inefficiency in retrieval of information and difficulty in interpretation of flawed and incomplete data.

One consequence of recent technical innovations has been a constantly changing view of most ultrastructural details, for example of synapses (cf. Gray, 1959; Uchizono, 1967) and other membrane junctions (cf. Pappas and Bennett, 1966; Brightman and Reese, 1969); in any case few correlations can be made between the fine structure and function of retinal synapses. While many retinal junctions appear similar to synapses in the brain in which transmission is thought to be chemical, such similarities are insufficient to identify the transmitter or to indicate whether the synapse is excitatory or inhibitory.

The potential of the electron microscope for ultra-fine dissection of synaptic relationships was realized early by Sjöstrand (1953, 1954), De Robertis and Franchi (1956), and Ladman (1958), who reconstructed the synaptic contacts of photoreceptor cells largely by analyzing many randomly oriented sections. Most recently, Dowling and his associates (e.g. Dowling, 1968; Dowling and Boycott, 1965a, b, 1966; Dubin, 1969) have made extensive use of information from single sections and fragmentary reconstructions from short series of sections. Even a complete morphological analysis of cells and their connections would be of little utility unless something is known of their electrophysiology or chemistry, in which case one might be able to choose between alternative schemes of interconnections suggested by even fragmentary morphological data. Dowling and Werblin (1969) and Werblin and Dowling (1969) have demonstrated well the productivity of this approach.

Limitations of single or random sectioning have been stated forcefully by Sjöstrand (1958, 1959, 1961, 1965, 1969) and his associates (Sjöstrand and Nilsson, 1964; Allen, 1969), who have advocated complete three-dimensional reconstruction from serial ultrathin sections. Applying this method under ideal conditions, one should be able to specify all the cellular relationships (or contacts) within a given small piece of tissue. In practice, problems of interpretation are encountered even with long series of uniformly ultrathin sections. When the dimensions of processes are close to the thickness of the sections, or when their surface membranes are sectioned obliquely, the continuity or separateness of processes can no longer be distinguished reliably (PEDLER and TILLY, 1965, 1966b). As indicated by MISSOTTEN (1965a), probably these factors led SJÖSTRAND (1958) to the supposition, later amended (SJÖSTRAND, 1965), that the so-called "synaptic vacuoles" of rod synaptic complexes were not connected to any other cellular structure.

According to SJÖSTRAND's (1969) "principle of detailed circuitry analysis of nervous centers", the Golgi studies of CAJAL and POLYAK and the ultrastructural findings of DOWLING and his associates (which achieve not a circuit diagram but rather a "composite illustration") suffer from a deficiency which can be corrected only by complete reconstruction from serial sections. ALLEN's (1965, 1969) analysis of a single completely reconstructed bipolar cell teledendron illustrates not only the interesting and no doubt important details of microcircuitry which may be observed only with this technique, but also the limitations of this method for exploring and reconstructing circuits on a slightly larger — but still very microscopic — scale. Because serial sectioning alone is ill suited for the reconstruction of large cells or groups of cells, ALLEN's purpose, to reveal the complete circuitry of a number of retinal cells, is not only unfulfilled but probably incapable of fulfillment by reasonable effort.

Many of the deficiencies of the straightforward serial sectioning approach are overcome by the "Golgi-EM" method, involving thin-sectioning of selected cells impregnated by the Golgi procedure (STELL, 1964, 1965a; BLACKSTAD, 1965 and in NAUTA and EBBESSON, 1970). Golgi impregnation results in the deposition in the cytoplasm of an electron-dense material containing silver and chromium (STELL, 1965a and unpublished observations), identified by BLACKSTAD (personal communication) as silver chromate. In favorable preparations, the material is confined within the plasma membrane of the cell and extends even into some processes as narrow as a few hundreds of Ångstrom units. The processes of the impregnated cell thus present a very distinctive appearance in the electron microscope. The principle of the Golgi-EM method is simply that the impregnating material which makes an entire cell observable in the light microscope is also a marker which identifies the processes of that cell in the electron microscope, so that they can be followed easily even if very fine or parallel-sectioned. Numerous studies, some using serial sectioning to obtain additional quantitative and topographical data, have confirmed the validity of the Golgi-EM approach (KOLB, 1969; KOLB et al., 1969; MISSOTTEN, 1968; LASANSKY, 1971). A particular advantage of Golgi-EM is that cells of a specific type may be preselected so that only restricted regions of greatest interest need be analysed. Sampling of large numbers of cells of a chosen variety or of the relationships of larger cells is accomplished much more quickly than with the straight serial sectioning-reconstruction method. By fixing the retina first in neutral-buffered aldehydes, both LASANSKY (1971) and MISSOTTEN (personal communication) have obtained preservation of fine structure in Golgi preparations superior to that observed in earlier studies (STELL, 1965a, 1967).

A new tool for the study of synaptic organization is the scanning electron microscope. The standard scanning instrument provides great depth of focus, and resolution intermediate between that of optical and transmission electron microscopes, but is restricted to the examination of structures on or very near the surface of the specimen. Although, as noted by HANSSON (1970b), it may help "to obtain a proper picture of the organization of cells without the laborious reconstructions from large series of thin sections", this is at present a hope rather than an attainment. In my opinion the high-voltage transmission microscope holds great promise for studies of synaptic architecture. The great penetrating power of high-voltage electrons permits examination of relatively thick sections (1 μm or more at 500 kV). Such sections are cut easily and rather few are required to include an entire cell. Three-dimensional information within the full thickness of the section may be recorded in original stereoscopic pairs of electron micrographs. Similar advantages may be afforded by high-resolution scanning transmission electron microscopy.

B. Direct Connections between Photoreceptor Cells

Studies with the optical microscope demonstrated at least two structural features of photoreceptors cells which might suggest a special interconnection between them: the close mutual association of "double" receptors (e.g. MÜLLER, 1857; WALLS, 1942) and the horizontally oriented terminal filaments, or teloden-dria, particularly of cone pedicles (CAJAL, 1892). A greater variety of contacts between receptor cells can be seen with the electron microscope.

Distal to the external junctional layer (limiting membrane), receptors may contact one another in various ways. In the goldfish I have observed rod inner segments lodged in indentations around the cone inner segments, near the junctional layer. The separation of membranes was regular but wide; sub-surface cisterns were found in the cones along the area of contact (STELL, 1965c). In the same location in human retina, UGA et al. (1970) described small rod-rod and rod-cone contacts in which the membranes of both inner segments were increased in thickness and density and separated by a narrow gap (total width 140 Å). The large area of close membrane contact between inner segments of teleostean double cones was first examined by SJÖSTRAND and ELFVIN, cited by SVAETICHIN (1956) who suggested that the apposition is a giant synapse. The contact is characterized by large areas of mutually apposed membranes, separated by a very regular but not particularly narrow cleft; each cone contains a single flattened sub-surface cistern, which lies beneath the entire area of apposition in each cone of the pair (ENG-STRÖM, 1963; STELL, 1965c; BOROVYAGIN, 1966; BERGER, 1967). Perhaps, as suggested by BERGER, the function of these specializations is to insulate from one another two cells which for some reason must be anatomically close but functionally separate. Other special but not obviously synaptic contacts have been shown between inner segments of double cones in a nocturnal gecko (PEDLER and TILLY, 1964). The principal and accessory members of double cones in pigeon (COHEN, 1963b), frog (NILSSON, 1964b) and chicken (MATSUSAKA, 1967) contact one another continuously from distal inner segment to synaptic ending. Receptors in the grouped main retina of the deep-sea teleost *Scopelarchus* are joined to one another along their inner segments by "regions of tight membrane junctions" of unspecified type (LOCKET, 1970a); they are not so joined in either the grouped accessory or the non-grouped main retina of this fish. Although it is difficult to imagine that these unusual relationships fail to serve some purpose, that purpose remains obscure. COHEN (1963b) suggested that such close pairing makes the cells an optical unit for sampling the same small field but handling the information in two different ways.

With the exception noted above, receptor cells are generally prevented by Müller's cell processes from contacting each other within the layer of receptor cell bodies. At the level of their synaptic terminals, however, the photoreceptors may contact one another in a variety of ways. COHEN (1969) described special contacts of human foveal cones, but DOWLING (1965) observed no contacts between cone pedicles in monkey fovea (perifovea). Simple apposition, without evidence of any junctional specialization, is observed occasionally between adjacent rod spherules in adult teleosts (LOCKET, 1969; STELL unpublished); single cone synaptic endings in herring larva (BLAXTER and JONES, 1967); cone pedicles in ground squirrel

(HOLLENBERG and BERNSTEIN, 1966), although DOWLING (1964) described these contacts as synaptic; rod and cone terminals in amphibians (NILSSON, 1964b; BOROVYAGIN, 1966; DOWLING and WERBLIN, 1969); and pedicles of double cone members in chicken (GOVARDOVSKII and KHARKEIEVITCH, 1967), for example. Infrequently some increase in electron-density of such casually apposed membranes may be observed (LOCKET, 1969; LASANSKY, 1971). COHEN (1969) noted that pedicles of human foveal cones make numerous contacts with one another at many levels, characterized by dense membranes on both sides of a 200 Å gap. He observed similar but fewer contacts in monkey fovea. Usually, however, such densities are found only where at least one of the contacting members is an elongated process of a photoreceptor terminal, i.e. a terminal filament or telodendron.

Contacts between receptors of guinea pig by means of long telodendria were first studied in the electron microscope by SJÖSTRAND (1958). He observed that each β-receptor terminal (pedicle) sent out about four processes and that each α-receptor terminal (spherule) was contacted at its proximal pole (where the bipolar dendrites exit) by three or four such processes. The junctional region resembled a synapse in that the membranes were closely and regularly apposed and increased in density, and small vesicles were found in the terminal of the telodendron (although not clustered at the membrane). There were at first "clear-cut indications that there is a definite fundamental pattern" for these inter-receptor contacts (SJÖSTRAND, 1959). Later, however, what was thought to be a square or triangular network was shown by extensive reconstruction from serial sections to be a nearly random array (Fig. 12), "although over a certain limited area the long processes from the β-type receptor cells appear to be preferentially oriented in one particular direction" (SJÖSTRAND and MOUNTFORD, cited in SJÖSTRAND, 1965). Each pedicle sends out both short processes, which terminate on adjacent receptors, and long ones, which bypass one to three rows of receptor terminals to terminate as far away as 10 μm. Either short or long processes may contact the surface of either pedicles or spherules (SJÖSTRAND and MOUNTFORD, ibid.). The ultrastructure of the telodendron-spherule contacts and the disorder-liness of the pattern are similar in the rabbit, except that here the pedicles are mutually interconnected in pairs and each spherule receives one telodendron at the side as well as several at the proximal pole (SJÖSTRAND, 1969). Telodendronal interconnections of widely separated pedicles in rabbit retina were recognized also by RADNÓT and LOVAS (1968).

Similar interreceptor contacts were described in primate retinas by MISSOTTEN et al. (1963), COHEN (1964, 1965a), and MISSOTTEN (1965a, b). In the perifovea, each cone pedicle emits several telodendria, and most (but not all) rod spherules are contacted by one telodendron (MISSOTTEN et al., op. cit.; COHEN, 1965a). With osmium fixation, the electron density of the contacting membranes was increased to give a "desmosomal" appearance (COHEN, 1965a; MISSOTTEN, 1965a); with aldehyde fixation the contact appeared to be very close (UGA et al., 1970). Similar contacts were shown in grey squirrel (COHEN, 1964). The apparent cytoplasmic continuities or "bridges" between receptors observed in some preparations of pigeon (COHEN, 1963b) and primates (COHEN, 1965a) are, as stated, unquestionably artifacts. Specialized cone contacts with densification of membranes have been observed in human fovea by COHEN (1969).

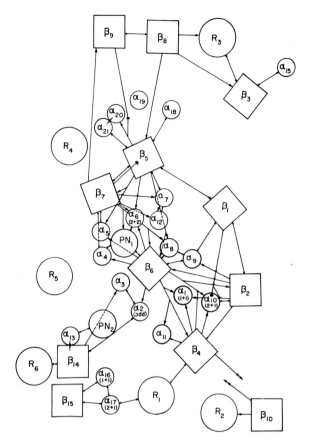

Fig. 12. Diagram of photoreceptoral interconnections in guinea pig retina, reconstructed from electron micrographs of 98 serial ultra-thin sections. The synaptic endings are placed accurately in the horizontal plane and all connections are diagrammed; but the endings are shown smaller than their actual size and the true length and course of the interconnecting processes is not shown. Receptors are α (rod) or β (cone); PN is paranuclear (as opposed to spherule distant from nucleus) variety of α. R indicates receptors at the edges of the reconstructed area which could not be analyzed completely enough for classification. Large arrows show contacts of longer telodendria; small double-ended arrows show direct contacts of adjacent endings; lines with rounded ends show processes invaginated into α_{10} and β_5. Double-headed arrows (lower right) show processes which were cut at the end of the series and could not be followed. Not all α-spherules were invaginated; those that were are identified as to pattern (see text): α_1 and α_{16}, $1 + 1$; α_{10} and α_{17}, $2 + 1$; α_6, $2 + 2$; and α_2, odd (a single large bulbous invaginating process). β_{15} had a synaptic ribbon but no invagination (reproduced from Sjöstrand (1965, 1969) by permission of the author. Originally published by Little, Brown and Co. and the University of California Press; reprinted by permission of J. and A. Churchill and the Regents of the University of California)

By means of conventional and Golgi-EM procedures, Lasansky (1971) has revealed that the "basal processes" (telodendria) of turtle photoreceptors make contacts of three kinds: (1) symmetrical junctions, with specializations of functional membranes but no clusters of vesicles, between two telodendria near their pedicles of origin; (2) asymmetrical junctions between a telodendron and an unidentified process away from the pedicles, with junctional specializations only in the telodendron; and (3) junctions in the recessed cone synaptic complexes. The third type is most interesting, since it may provide a pathway for action of cones at a distance (LASANSKY observed some telodendria as long as 40 μm in Golgi preparations). Since these telodendria make no specialized junctions with *distant* pedicles, or telodendria, any interaction between widely separated receptors might be mediated by the small horizontal cell terminal, which makes specialized contacts with both pedicle and telodendron. In the mudpuppy, receptor telodendria make contact with second-order cells (DOWLING and WERBLIN, personal communication).

Although the broad areas of membrane apposition between receptor terminals in frog appeared undifferentiated, unlike the punctate and markedly differentiated contacts in mammals and turtle, analysis of serial sections showed characteristic patterns of contact in the frog (NILSSON, 1964b). Each terminal of a red rod is in contact with three other terminals, those of another red rod, a single cone, and the principal member of a double cone. Each terminal of a single cone, and each terminal of the principal member of a double cone, is in contact with the endings of three red rods. The synaptic endings of green rods, and the accessory elements of double cones, make no direct contacts with terminals of other receptors. This work illustrates an important application of reconstruction from serial sections. Such characteristic patterns might play an important role in the processing of photoreceptor information. NILSSON follows SJÖSTRAND (1958, 1969) in suggesting that this system enhances contrast by lateral inhibition or perhaps plays some role in the rapid phase of adaptation. Although conceding that such arrangements might allow for "coordination" of the influence of simultaneously stimulated groups of receptors, MISSOTTEN et al. (1963) favored the simpler function of mechanical stabilization of synaptic elements. Admittedly the latter view is consistent with the lack of evidence for interactions between receptors (TOMITA, 1965; WERBLIN and DOWLING, 1969). SJÖSTRAND's, NILSSON's, and LASANSKY's findings of well-developed patterns of interconnections and specialized junctions between receptors suggest strongly, however, that synaptic connections will be found if appropriately sought. Although BAYLOR and FUORTES (1970) observed that the receptive field of a cone in the turtle was about equal to the diameter of the cone, more recently BAYLOR et al. (1971) have obtained firm evidence for presynaptic interaction of neighboring turtle cones.

C. Interconnections of Photoreceptor, Bipolar and Horizontal Cells

Features and Variety of Contacts in the First Synaptic Layer

The essential features of the synaptic junctions of photoreceptor cells were described first by SJÖSTRAND (1953). He noted that the spherules of α-type (rod) receptors in guinea pig were penetrated by dendrites at their distal pole; therefore he called the synapse "intracellular". In the spherule he observed "minute rodlets, which stain intensely with osmium, and minute granules which represent the major part of the synaptic cytoplasm". By comparing the appearance of hundreds of spherules sectioned in various orientations, SJÖSTRAND (1954) was able to begin reconstructing their synaptic configuration. The synapse included moderately recessed bipolar dendrites and even more deeply enclosed "bläschenförmige Gebilde" (later called "synaptic vacuoles"), and the osmiophilic rodlets of his earlier description were seen to be cross-sections of a long ribbon which lay in a groove or ridge projecting between a pair of the large vacuoles. DE ROBERTIS and FRANCHI (1956) observed similar structures in rabbit retina. Their meritorious designation of the synapse as "invaginating" rather than "intracellular" has been generally adopted; but no connection has ever been shown between the plasma membrane of the receptor cell and the synaptic ribbon, which they described as a "process of the synaptic membrane". DE ROBERTIS and FRANCHI described Sjöstrand's granules as "synaptic vesicles" and showed that they were clustered along the presynaptic membrane and ribbon; smaller vesicles were shown in postsynaptic processes. Since presynaptic vesicles were found to vary with the state of adaptation, the authors inferred a role in chemical synaptic transmission. The distribution of vesicles observed by these authors is rarely seen, however, in better preserved specimens, in which they are spread rather uniformly throughout the presynaptic cytoplasm (see e.g. LASANSKY, 1969, 1971), and the uptake of peroxidase (LASANSKY, 1967) suggests that some vesicles may take up rather than release material. In a well controlled experiment, MOUNTFORD (1964) was unable to find a statistically significant difference between populations of vesicles in a given type of receptor cell in light and dark. CRAGG (1969) has confirmed MOUNTFORD's findings under more severe conditions. Fetal and neonatal rats reared in total darkness were exposed to daylight. The marked effect of brief (3 min or less) first exposure was a decrease in the average width of the synaptic terminals and a corresponding increase in concentration of synaptic vesicles; the cytoplasmic density of invaginated postsynaptic processes was also increased. Thus there is no firm evidence for a role of the small vesicles in synaptic transmission in the first synaptic layer.

It is likely that the synaptic ribbon or "synaptic lamella" (LADMAN, 1958) is important for synaptic transmission. It is found close to the presynaptic membrane in the synaptic terminals of rods, cones and bipolar cells (MISSOTTEN, 1960; KIDD, 1962; FINE, 1962), as well as in the acousticolateralis system (e.g. SMITH and RASMUSSEN, 1965). In the latter, and apparently in some photoreceptor terminals in the median eye of the frog (KELLY and SMITH, 1964), this organelle is located specifically at the point of contact with the dendrite of an afferent neuron. It may not function directly in the final events of transmission, however. A more remote role may be suggested by the occurrence of similar but non-synaptic organelles in mammalian pinealocytes (WOLFE, 1965), the inner segments of photoreceptor cells in lamprey (YAMADA and ISHIKAWA, 1967), bipolar cell dendrites (LASANSKY, 1971; STELL, 1972), and cell processes and perikarya within the intermediate neuronal layer (ALLEN, 1969). According to NILSSON and CRESCITELLI (1969) the appearance of the b-wave of the electroretinogram in bullfrog is correlated closely with the appearance of vesicles and synaptic ribbons in photoreceptor cells. The same correlation was found in the more slowly developing green frog, suggesting that "the process of transmission is required ... to generate a *b*-wave and this process is dependent upon the vesicles" (NILSSON and CRESCITELLI, 1970). In the mouse, however, processes in both synaptic layers already contained numerous synaptic lamellae and vesicles even before the earliest appearance of the electroretinogram (OLNEY, 1968). In the rat, too, the photoreceptor terminals were well developed several days before the earliest signs of a *b*-wave (WEIDMAN and KUWABARA, 1967, 1968, 1969). Likewise in the chicken the synaptic layers are differentiated ultrastructurally (SHIRAGAMI, 1969) and histochemically (SHEN et al., 1956) at 15—16 days, while electrical activity is not detected until several days later (WITKOVSKY, 1963). The appearance of the full electroretinogram (including *b*-wave) only after both synaptic layers are well developed may be due not to the maturation of the synapses of the second layer, as suggested by WEIDMAN and KUWABARA (1968), but to the simultaneous development of the photoreceptor response mechanism. A causal relationship between synaptic structure and electrical activity cannot be determined from the experiments cited above. To date, however, it appears that the *b*-wave of the electroretinogram cannot precede morphologic differentiation of the receptor terminals; thus all evidence is consistent with the view that the *b*-wave is dependent upon some function of the receptor synaptic endings and that the synaptic lamella and vesicles are instrumental in that function.

The spatial relationships of the presynaptic organelles and postsynaptic processes were investigated further by LADMAN (1958), who described the invagination of a complex "bipolar dendrite" into the rod spherules of the rat on the basis of individual sections. His data could be interpreted by a model similar to SJÖSTRAND's incorporating more than one invaginated postsynaptic element. LADMAN described a newly found structure, the "arciform density", between the edge of the synaptic lamella and the plasma membrane; similar structures have been observed subsequently in many vertebrates. SJÖSTRAND (1958) presented a more detailed analysis of rod spherules and their relationships in guinea pigs, based on serial sections cut parallel to the axis of the receptor. The spherule analyzed in detail was in contact with five invaginated processes. The two processes identified as bipolar cell dendrites (but not traced to their cell body) did not invaginate deeply. The other three processes, called "synaptic vacuoles", penetrated the spherule more deeply but could not be traced to processes outside the invagination. I believe, however, that careful examination of SJÖSTRAND's published micrographs reveals very slender processes, nearly parallel to the plane of sectioning, with which the "vacuoles" may be continuous; indeed, subsequent studies by SJÖSTRAND (1969) confirm this interpretation. The ridge of presynaptic cytoplasm containing the synaptic ribbon protrudes between two "vacuoles" which are paired; this configuration is characteristic of both rod and cone synaptic complexes in most vertebrates (Fig. 13).

MISSOTTEN and his colleagues (MISSOTTEN et al., 1963; MISSOTTEN et al., 1964; MISSOTTEN, 1965a, b) produced evidence raising serious doubts about SJÖSTRAND's interpretation. Serially sectioning rod and cone synaptic regions in human retina perpendicular as well as parallel to the photoreceptor axis, they were able to follow elements corresponding to SJÖSTRAND's "synaptic vacuoles" out of the rod and cone invaginations. The pair of these more deeply-invaginated processes was bisected by the presynaptic ridge and ribbon. In the cone complexes they were accompanied regularly by a less deeply invaginated central process to form a characteristic postsynaptic "triad", but in the rod complexes there were as many as five less deeply invaginated processes and the relationships, as in guinea pig rods, were less regular than in cone synaptic complexes. Lateral elements were positively identified as terminals of horizontal cells in goldfish by STELL (1964), using the Golgi-EM method. Except for their close association with the pre-synaptic lamella, both lateral and central processes had few features suggesting synaptic contacts. The postsynaptic membrane thickening or "fuzz" observed easily in synapses in the second synaptic layer was seen only rarely in processes invaginated into the photoreceptor cells (DOWLING and BOYCOTT, 1966).

LASANSKY (1971), however, has observed specialized membrane contacts (junctions) in this region in turtle (Fig. 14, 15). In "proximal (i.e. close to

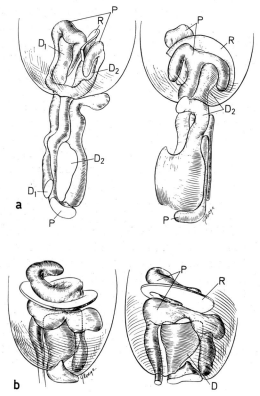

Fig. 13. Relationships of α-receptor (rod) spherules and postsynaptic processes in guinea pig retina, reconstructed from electron micrographs of serial ultrathin sections (see text). *a*, spherule complex of 1 + 2 type shown in orthogonal projections. "Proximal vacuoles" (*P* probably horizontal cell endings) are branches of a single process within the invagination; two "distal vacuoles" (*D* probably bipolar cell dendrites) enter independently and penetrate somewhat less deeply. *b*, spherule complex of 1 + 1 type seen at two different angles from the horizontal. Deep process (*P*) spirals around ridge containing synaptic ribbon (*R*), while slightly invaginating process (*D*) ends bluntly. Receptor cell membranes in invagination are not shown (reproduced from SJÖSTRAND (1969) by permission of the author. Originally published by the University of California Press; reprinted by permission of the Regents of the University of California)

the receptor) junctions" at the side of the presynaptic ridge, an opaque layer coats
the cytoplasmic side of the membrane of the lateral processes. Opposite some ribbons the lateral processes are joined together along a narrow (50—60 Å) "medial
gap" in a postsynaptic *dyad*. Opposite other ribbons the lateral processes are
separated by a 500—800 Å space containing a moderately dense amorphous

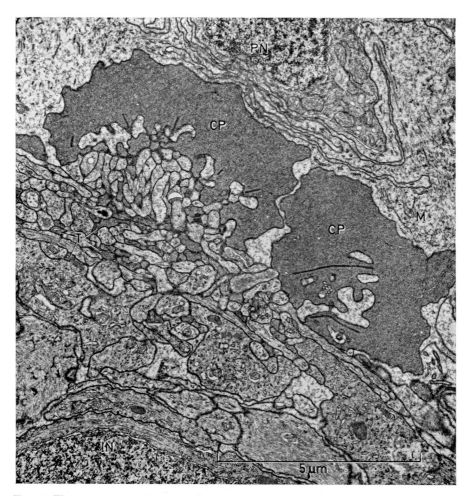

Fig. 14. Electron micrograph of cone pedicles and related processes in the first synaptic (outer
plexiform) layer of the turtle. Retina was fixed with glutaraldehyde followed by osmic acid,
stained *en bloc* with uranyl acetate, sectioned radially, and the sections stained with uranyl
acetate and lead. *PN* photoreceptor nucleus. *CP* cone pedicle. *T* photoreceptor cell telodendria.
IN intermediate (inner nuclear) layer neuron. *M* Müller's (glial) cell (LASANSKY, unpublished)

material and are accompanied by a third element to form a *triad*. Because of the
dense material and the special location of the central process, the latter may be
said to form an "apical junction" in this region. Lateral processes in dyads also
make less deeply invaginated "distal junctions", showing variable specializations

of membranes and the intercellular gap, with small visual cell processes. The apical junctions at turtle cones are somewhat similar to the structures in human rod triads described recently by UGA et al. (1970).

In general, small vesicles are present in the lateral invaginated processes which contact the rods (DE ROBERTIS and FRANCHI, 1956; SJÖSTRAND, 1958) but absent from those which contact cones (MISSOTTEN, 1965a, b; COHEN, 1965, 1967; STELL, 1967). Subsurface cisterns were observed opposite the lateral process in rod spherules, and dense submembranal plaques were seen in the lateral processes contacting the cones, in goldfish (STELL, 1964, 1967) and *Centropomus* (STELL, 1966). The observation that the subsurface cisterns and submembranal plaques disappear in osmium-fixed goldfish retina after prolonged dark-adaptation (STELL, 1964) was based on only a few specimens. The problem should be re-examined with several different schedules of light- and dark-adaptation, and, because different fixatives exert different effects on labile membrane structures (e.g. KABUTA et al., 1968), with both osmium and aldehyde fixation.

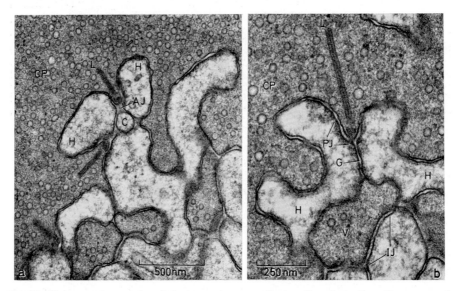

Fig. 15. Electron micrographs of photoreceptor synaptic junctions in turtle retina, prepared as in Fig. 14. *a*, triadic complex in which two horizontal cell processes (*H*) and central process (*C*) form an apical junction (*AJ*) with the cone pedicle (*CP*) at the synaptic lamella (*L*). *b*, dyadic complex in which two horizontal cell processes, separated only by a narrow medial gap (*G*), form a proximal junction (enclosed by *PJ*) with the cone pedicle. Farther from the pedicle a visual cell process (*V*) makes an invaginating junction (*IJ*) with a slightly penetrating process (LASANSKY, unpublished)

Complex terminals, including cone pedicles, make special contacts not only with invaginating elements but also with processes which contact the proximal face of the terminal superficially. Densely-staining surface contacts with extra-foveal cones were observed in rhesus monkey by COHEN (1961), who considered that some might be desmosomal and others synaptic. Similar densities shown by

FINE and ZIMMERMAN (1962) in human retina were interpreted as attachment plates. MISSOTTEN et al. (1963) estimated the number of such superficial contacts with human perifoveal cones at about 500, and MISSOTTEN et al. (1964) designated the processes making these contacts as dendrites of bipolar cells. Simple attachment, therefore, is an unlikely function. Surface contacts showing "regularly periodic striations in the intersynaptic space" were described in alligator by KALBERER and PEDLER (1963). Dense surface contacts, usually showing some periodicity, have been described or illustrated also in primates by PEDLER and TILLY (1965), DOWLING (1965) and DOWLING and BOYCOTT (1966); in various

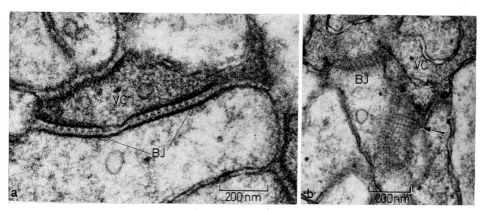

Fig. 16. Electron micrographs of basal junctions (*BJ*) in turtle retina prepared as for Fig. 14. In *a*, electron-opaque cross-bars are seen in the gap between a visual cell process (*VC*) and a postsynaptic dendrite. In *b*, a lattice pattern is seen clearly in a tangentially sectioned basal junction (arrow) as well as in an obliquely sectioned one (*BJ*). (reproduced from LASANSKY (1969) by permission of the author and The Rockefeller University Press)

other mammals, birds, reptiles and amphibians by EVANS (1966); in chicken by MATSUSAKA (1967); in frog by DOWLING (1968) and in mudpuppy by DOWLING and WERBLIN (1969); and in the spiny dogfish *(Squalus)* by STELL (1972). Wherever the classical rod-cone distinction is clear, these superficial contacts appear to be made only with cone pedicles. LASANSKY (1969) described the basal junctions (superficial contacts without ribbons) on cone pedicles in turtle retina (Fig. 16). In cross-sections, the gap between the junctional membranes appears to be bridged partially or completely by transverse filaments or cross-bars; in sections parallel to the plane of the junction, a regular square lattice appears. Because of superposition effects and Moiré patterns, one must be cautious in interpreting images such as these. According to LASANSKY (1971) the basal junctions, which number hundreds per pedicle in the turtle, are of two types according to location: "superficial", in which a caplike junctional area involves only the tip of the postsynaptic process contacting the receptor terminal; and "invaginated", in which a sleevelike junctional area involves an invaginating process near its tip, and both narrow- and wide-gap junctions are made with lateral processes in dyads and triads.

Identification of Cells and Patterns of Interconnection (I)

Conventional ultrastructural studies including serial sectioning revealed partially the patterns of cellular interconnections in a number of species, especially
monkey and man (MISSOTTEN and colleagues, 1963 et seq., and below). A more
complete description of interconnections in goldfish was obtained with Golgi-EM
by STELL (1964, 1965a, 1967), who analyzed limited numbers of the two types
each of bipolar and true horizontal cells described by CAJAL (1892) in cyprinid
retinas. STELL found that the lateral elements in the triads are processes from
horizontal cells while the central elements are dendrites of bipolar cells, and that
in general the interconnected cells form two separate or nearly separate systems:
one of cones with bipolars and distal (external or "sclerad") horizontal cells, and
one of rods with large bipolars and proximal ("intermediate", "vitread") horizontal
cells. STELL noted that the large bipolars contact not only rods but also a small
proportion of cones, and LAUFER (personal communication) has observed similar
mixed rod-cone input to several cells of the "rod bipolar" type in the teleost
Eugerres. Although this point cannot be considered established without more
careful and extensive studies, it is significant that the large bipolar cells of goldfish
retina identified by intracellular injection of Procion Yellow (KANEKO, 1970)
probably respond maximally to red light (KANEKO and HASHIMOTO, 1969). The
implication that large bipolar cells carry signals from red-sensitive cones as well as
rods is consistent with the observation that the ganglion cell receptive fields
dependent upon the activation of the rod and red-cone systems are coincident in
cyprinids (RAYNAULD, 1969). STELL did not observe basal junctions in goldfish and
he did not consider any role for interreceptor processes, horizontal cell axons, or
other elements than the above in the complex contacts of the first synaptic layer.

Cursory studies of the teleost *Centropomus* (STELL, 1965a, 1966) with the
Golgi-EM procedure suggested that processes of horizontal cells of the first and
second rows contact only cone pedicles, while those of the third row contact only
rods. I overlooked a fourth row of horizontal cells, observed in Golgi preparations
by PARTHE (1970), which appear to contact cones. Previous studies on *Centropomus*
and other Caribbean teleosts (VILLEGAS, 1960, 1961; VILLEGAS and VILLEGAS,
1963) produced no evidence for receptor-horizontal cell contacts, but the methods
used were inappropriate for demonstrating interconnections. YAMADA and
ISHIKAWA (1965), using conventional methods for ultrastructure, reported that
processes from horizontal cells of both distal and proximal rows in carp make contact with cone pedicles. Since they and WITKOVSKY and DOWLING (1969) reported
only two horizontal cell layers in carp, it is likely that they have overlooked an inconspicuous third layer containing rod horizontal cells. MELLER and ESCHNER (1965)
also observed contacts between horizontal cells and photoreceptor terminals in
goldfish with conventional methods. They gave no details but illustrated a contact
between a cone pedicle and a first-row horizontal cell process identical to that
described by STELL (1964, 1967). Partly on the basis of serial sections, BOROVYAGIN
(1966) reported that the processes of horizontal cells of the first row in the pike
(Esox) branch to terminate on either side of the presynaptic lamella of the cone
pedicles; fifteen to twenty cones are contacted in this way by each first-row
horizontal cell. Dendrites of bipolar cells penetrate the pedicles less deeply than

horizontal cell processes, as in other animals, but their terminals appear unrelated to the lamella-horizontal cell cluster. Processes from horizontal cells of the second row were followed as far as the first synaptic layer but their relationships with receptor terminals could not be established. Basal junctions, processes making superficial contacts, and interreceptor contacts were not described in any of the above studies of teleosts.

In 1964, HAMA (personal communication) observed contacts between photo-receptor cells and processes from horizontal cells of the first row in the retina of the Japanese red ray. YAMADA and ISHIKAWA (1965) described similar contacts in a number of sharks and rays and observed that the horizontal cell process may bifurcate outside or within the invaginated synaptic complex. They noted that processes from horizontal cells of the second row "may" reach the rod ending to form a synapsis; since the distinction between rods and cones may not be clear in the animals they examined, this statement is uninformative. In the spiny dogfish, *Squalus*, more than 95 % of the photoreceptor cells are rodlike except with respect to the synaptic terminal, which is more complex than teleostean rod spherules and extends many telodendria (STELL, 1972). Since the processes of horizontal cells of the first row are far more numerous than the scarce and widely scattered pedicles of the typical cones, many if not all of these processes must terminate upon the rods. The connections of horizontal cells of the second row, and of bipolar cells of any type, are unknown in *Squalus*. All the above studies have shown that the horizontal cell processes in sharks and rays terminate in the lateral position by the presynaptic ridge.

Patterns of contacts in the first synaptic layer of amphibians have not been well established. BOROVYAGIN (1966) examined the frog but the relationships between dendrites and photoreceptors were "not quite clear". According to EVANS (1966), in the frog two rod and two cone endings "alternate diagonally around a central portion which contains the bipolar invagination distributing connections to all four neighboring cells". This is at variance with the careful account of NILS-SON (1964b) on interreceptor contacts in another species of frog and with the report by DOWLING (1968) that processes of both bipolar and horizontal cells invaginate into photoreceptor terminals in the usual way. Superficial contacts were observed by both EVANS and DOWLING, but the identity of the cells making these or invaginated contacts has not been established. While it may be true that in the frog "the outer plexiform layer is not apparently greatly different from the outer plexiform layer of other vertebrates" (DOWLING, 1968), such statements unfortunately emphasize similarities which are well known rather than encourage investigation of the extent of universality or diversity. Some differences have been indicated by DOWLING and WERBLIN (1969) in their description of the mudpuppy, *Necturus*, in which the triadic arrangement of the unidentified processes which invaginate the cone synaptic endings is irregular if present at all. Superficial contacts, at least some of which form dense basal junctions, are found on the proximal surface of both rod and cone terminals; synaptic ribbons may or may not be present at these contacts. All the contacts made by postsynaptic processes with rod termi-nals are "superficial" in the mudpuppy.

Relationships of photoreceptor synaptic endings in various reptiles were described by PEDLER (1963). His suggestion that lateral inhibition between photoreceptor cells is mediated by processes from cells of other types is supported only by his failure to observe direct interreceptor contacts and his assumption that photoreceptors are mutually inhibitory. Pedler and Tansley (1963) described contacts of cone pedicles with processes of "horizontal", bipolar, and Müller's cells and other cone cells in diurnal geckoes, but were uncertain of their relationships. The "horizontal cells" described by these authors are surely the displaced bipolar cells described by Cajal (1892) in lizard retinas. The vesicle-filled "cisternae" which they observed in "cone" synaptic endings are surrounded by two cell membranes; they are, therefore, likely to be processes of another (probably photo-receptor or horizontal) cell. Similar structures, identity unknown, are seen in the cone pedicles of some teleost fishes ("a structure resembling a big vacuole", Villegas, 1960; and Stell, unpublished observations). They may be comparable to the unusually deeply-inserted "vacuole", designated $V3$, in the guinea pig rod spherule reconstructed by Sjöstrand (1958). Kalberer and Pedler (1963) describe rudimentary clusters of "simple" (rodlike) and "complex" (conelike) receptor terminals in alligator, and some basal junctions on the complex pedicles, but there is no evidence on the identity of the interconnected processes.

In the tortoise, Yamada and Ishikawa (1965) observed branches of horizontal cell processes which extend toward the cone terminals; sometimes these processes could be followed to a "typical synaptic contact" (presumably lateral element in dyad or triad) with a pedicle. Similar relationships were reported in *Emys* by Borovyagin (1966). Patterns of interconnections in *Pseudemys* have been deter-mined by Lasansky (1971) using the Golgi-EM procedure. He observed that terminals of horizontal cell processes, whether long and axon-like or short, are the lateral elements in dyads and triads. Dendrites of bipolar cells of the two varieties observed, one possessing and the other lacking a Landolt's club, all make superficial or invaginated basal junctions. Cone telodendria terminate upon horizontal cell processes within the invagination, probably at the "distal junction of the lateral processes" (see above) although these cannot be identified clearly in Golgi-EM preparations. The central processes of triads in turtle resemble the dendrites of midget bipolars in monkey by their position and by occasionally con-taining a ribbon, and their distal junctions resemble basal junctions. The cells providing central processes were not impregnated in these Golgi studies but accord-ing to these features are probably another variety of bipolar. This failure of all cells of one class to reveal themselves should be taken into account when inter-preting Golgi preparations with the light microscope.

The plan of the first synaptic layer in avian retinas has been neglected except by Pedler. Tilly and Pedler (1966) reconstructed the connections of central parafoveal receptors in pigeon from serial sections and described the endings of horizontal cells as a "terminal ring containing protrusions" or a "claw-shaped array with end-bulbs". The general plan of the connections between the horizontal cells and the photoreceptors "has been described", in conference but not in print.

Regarding mammals, Dowling (1964) stated that processes of both horizontal and bipolar cells enter the pedicles in the thirteen-lined ground squirrel but did not describe the relationship in detail. He estimated that the horizontal cells each contact 50—75 receptors, the bipolars 3—9. Stell (1965b) observed with the Golgi-EM method that the dendrites of a horizontal cell in monkey terminated as the lateral processes in cone triads. Nevertheless, some investigators (e.g. EVANS, 1966; ORDY and SAMORAJSKI, 1968) have persisted in designating all invaginated postsynaptic processes as "bipolar dendrites". In the guinea pig, further studies by SJÖSTRAND and MOUNTFORD (cited in SJÖSTRAND, 1965, 1969) showed four patterns of processes in relation to the spherules (cf. Fig. 13): (1) two deeply-inserted processes (unfortunately called "proximal vacuoles", meaning "closer to the receptor", as in LASANSKY, 1971) and two slightly-inserted ("distal vacuoles"); (2) one deeply-inserted, branching to two within the invagination, and two slightly-inserted; (3) one deeply inserted (curled or branched) and one slightly-inserted (simple, unbranched); and (4) deeply-inserted processes not traceable out of the invagination (as in SJÖSTRAND, 1958). The slightly-inserted processes were usually identified as bipolar cell dendrites (criteria not stated); in one case, however, such a process was traced to a β-receptor pedicle. The source of the deeply-inserted processes was obscure except in one case in which it was identified as a Müller's cell. The β-type pedicles contained 3—5 synaptic ribbons, each related to a pair of deeply-inserted terminals; all appeared to be terminals of a single branching process of a horizontal cell (SJÖSTRAND, 1969). In the rabbit (SJÖSTRAND and NILSSON, 1964; SJÖSTRAND, 1969) the rod spherules and their interconnections appeared identical to those of the guinea pig. The cone pedicles in rabbit were found to contain at least ten synaptic ribbons, each associated with a pair of deeply-inserted terminals derived from horizontal cells; in this case, the two terminals of each pair usually arise from different processes.

MISSOTTEN and his colleagues made significant progress in serial-sectioning studies of the human perifovea. MISSOTTEN et al. (1963) observed 2—7 processes (deeply- plus slightly-inserted) in each rod spherule invagination. Each cone was penetrated by about 25 triads (75 processes) and contacted superficially by some 500 processes. It was assumed that the three members of each triad arose from three different bipolar cells, but they had not been traced to their origins. All processes penetrating rod or cone terminals, however, could be followed outside the invagination. MISSOTTEN et al. (1964) observed that some bipolar cells, probably the flat-top variety of POLYAK (1941), made only superficial contacts (did not participate in triads). Although POLYAK predicted that such cells should contact more than one cone pedicle, these authors were unable to establish multiple contacts in their moderately long series of sections. The central elements of at least 5—7 of 25 triads were shown to be dendrites from the same cell, identified as a midget bipolar. Other central elements in the triads, it was believed, were derived from horizontal cell dendrites. While they could not be traced to their origin, the lateral (deeply-inserted) processes in both rod and cone synaptic endings were interpreted as terminals of horizontal cell axons; in one case, however, a lateral element was traced from a cone triad to a cell body (unidentified) nearby. The slightly-inserted processes in rod spherules were supplied by dendrites from a large bipolar cell, believed to be identical to the "mop" variety of POLYAK; these dendrites made no contact of any kind with cones *en route* to their termination. MISSOTTEN's monograph (1965a) is an excellent compilation of these studies.

YAMADA and ISHIKAWA (1965) observed in serial sections of human paramacular retina that processes of the horizontal cells "enter into the rod spherule"; horizontal cell processes could not be followed to cones. The results of other investigations do not support their interpretation. PEDLER and TILLY (1965) reconstructed a cone pedicle located 1—2 mm from fovea of a rhesus monkey. The 77 invaginated postsynaptic processes were found to enter along an annular region halfway between the center and the periphery of the proximal surface of the pedicle. Their identification of the remainder of the processes contacting this surface as Müller's cells is without justification, as shown by the studies of MISSOTTEN (1965a, 1969) and KOLB (1969). They were unable to further specify the identity, number, and arrangements of the penetrating processes. DOWLING (1965) estimated from their appearance in individual thin sections that cone pedicles in the fovea (perifovea) of the monkey are penetrated by about twelve triads, or a total of 36 processes. Why, he wondered, should it be necessary to make so many contacts with a single postsynaptic (midget bipolar) cell? It should be emphasized, however, that POLYAK (1941) stated that even in the fovea each cone contacts several post-synaptic cells of different types (BOYCOTT and DOWLING, 1969). DOWLING and BOYCOTT (1966) accept that the lateral processes in triads are processes of horizontal cells and that the central are dendrites of midget ("monosynaptic") bipolars; thus the horizontal cells are "in a very strategic position" for regulating receptor-bipolar synaptic transmission.

The most complete and interesting information about synaptic patterns in the first synaptic layer of primates comes from ultrastructural analysis of serial sections of Golgi preparations. KOLB et al. (1969) were first to realize that while some midget bipolar cells send all their dendritic terminals to the center of the cone triads (and the central processes in all triads of a given cone are derived from the same bipolar cell), other bipolars send all their dendrites to pairs of superficial contacts located on either side of the central invaginating terminals. It was reasonable to call the former the "invaginating" and the latter the "flat" midget bipolar cell, and to conclude that at least some cone pedicles are connected to a second (flat) midget bipolar as well as to the conventional invaginating one. Once the flat midget bipolar had been discovered by electron microscopy, it was possible to distinguish both varieties of midgets by light microscopy according to the level of their cell bodies and axon terminals as well as the form of their dendritic arborizations (Fig. 2). Further details were reported by KOLB (1970) in her outstanding paper on rhesus monkey retina (Fig. 17). Each cone pedicle in the near peri-fovea contacts about 25 triads. The 25 central processes of the triads comprise all of the dendritic terminals of a single invaginating midget bipolar cell. These terminals are flanked by the 50 superficial dendritic terminals (25 pairs) from a single flat midget bipolar cell. The 50 lateral processes of the triads (25 pairs) are all terminals of horizontal cell dendrites; usually no more than one member of each pair of lateral processes is derived from the same horizontal cell. Each horizontal cell makes contact with all the cone pedicles within its dendritic field; more terminals go to the cone(s) near the center of that field than to the peripheral ones. Each cone is contacted by processes from a number of horizontal cells, probably about equal to the numbers of cones contacted by each horizontal cell. The dendrites of the small horizontal cells (identified as type "A" by BOYCOTT and DOWLING, 1969) make contact with the 6 or 7 cone pedicles in their dendritic field (Fig. 18); those of the large horizontals (type "B" of BOYCOTT and DOWLING), with 10—12 pedicles. The cone pedicles are also contacted at about 25 superficial junctions (one per triad) by dendrites from each diffuse (flat-top) cone bipolar cell; since there are hundreds of superficial junctions on a pedicle, each cone therefore is connected to dendrites of a number of diffuse cone bipolars. Each of these bipolars contacts superficially all the cone pedicles (about 6) within its dendritic field. Processes identified in the light microscope as horizontal cell axon terminals

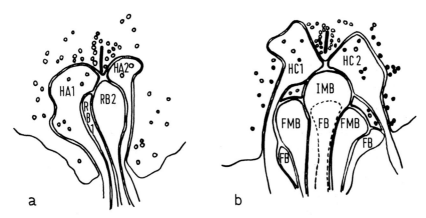

Fig. 17. Summary diagram of connections made by rod spherule (a) and cone pedicle (b) in monkey retina, as revealed by electron micrography of serial sections of Golgi-impregnated bipolar and horizontal cells. Around the presynaptic ribbon of the rod are lateral axon terminals from two different horizontal cells (HA 1, HA 2) and central dendrites from different rod bipolar cells (RB 1, RB 2). Around the ribbon of a cone pedicle is a triad comprising the central dendrite of an invaginating midget bipolar cell (IMB) and two horizontal cell dendrites, usually from different cells (HC 1, HC 2). Astride the dendrite of the invaginating midget bipolar is a pair of processes from a flat midget bipolar cell, which intrudes slightly into the invagination and contacts the pedicle. Scattered around and among the invaginating processes are numerous dendrites of flat bipolar cells, which make superficial contacts with the cone pedicle (reproduced from KOLB (1970) by permission of the author and the Royal Society of London)

of either type "A" or "B" always terminated in one of the pair of lateral (deeply-inserted) processes in many rod spherules; therefore each rod is connected to two of these processes. Dendrites of rod (mop-type) bipolar cells terminated only as the small central (slightly-inserted) processes postsynaptic to rod spherules; only one of these 1—4 central processes was provided by a given rod bipolar cell. One rod bipolar sectioned serially contacted at least 33 spherules but made no apparent contact with the many other spherules in its dendritic field. MISSOTTEN (1968) has confirmed most of these observations in human retina also by serial sectioning of Golgi preparations. Of six midget bipolars selected randomly, four made invaginated and two made superficial contacts, as described by KOLB. One parafoveal diffuse flat bipolar made only superficial contacts with cone pedicles. Each of four near-peripheral horizontal cells which he examined sent dendrites to 7—9 cone pedicles (i.e. probably to all the pedicles in its dendritic field), generally to only one of two lateral elements in the triads. Rod bipolar cells from the near periphery contributed only one dendritic terminal each as slightly-inserted processes to about

50 rod spherules. Missotten did not follow horizontal cell axons. He has not observed rod horizontal cells but does not reject the possibility that they exist.

What may be concluded about synaptic function from the ultrastructure and patterns of interconnections in the first synaptic layer ? The synaptic lamella is probably functionally presynaptic (see above), but whether to the bipolar or horizontal cell, or both, is unclear. Furthermore, as noted by Lasansky (1971) even the finest observed details of structure have not yet established whether the central process of a triad in an apical junction is postsynaptic to the photoreceptor, the horizontal cell, or both. Even the role of the basal junctions, which appear by their simple and direct relationship to receptor terminals to mediate transmission from receptor to bipolar cell, is open to question. In view of the pluripotentiality of the processes of retinal horizontal and amacrine cells, are there compelling reasons to suppose that bipolar cell dendrites are incapable of bidirectional action ? I believe that the possibility of bipolar-receptor transmission is not excluded by the available evidence, although the substantial differences in latency of receptor and bipolar cell responses to light in *Necturus* (Miller and Dowling, 1970) imply that basal junctions are at least chemical synapses of receptors upon bipolars. The considerable differences in geometry, ultrastructure, and patterns of interconnection at the basal and apical (superficial and invaginated) receptor-bipolar junctions probably signify differences in their synaptic functions. It is noteworthy that the same receptor which makes exclusively basal junctions with bipolar cells of one type also makes exclusively apical junctions with bipolars of another type. Thus complex receptor pedicles and their contacts comprise a *divergent* system, in which information from one receptor is distributed through many synapses to a number of bipolar cells of different types. It is to be expected, therefore, that a variety of functional interconnections will be detected upon recording intracellularly from a variety of bipolar cells in retinas having complex receptor terminals.

Interpretation of the synaptic relationships of horizontal cells poses special problems in all animals. At least some processes of horizontal cells must be postsynaptic to receptor terminals. From the amplification of receptor voltages (see e.g. Naka and Rushton, 1967; Baylor et al., 1971) as well as the long latency of response (Miller and Dowling, 1970) it follows that transmission to these processes is mediated chemically. Most uncertain is the pathway by which horizontal cells affect the activity of other cells.

In groups above fishes, in which at least some horizontal cells appear to have axons, the simplest view is that these cells are horizontally oriented bipolar neurons, postsynaptic to cones and presynaptic to rods. On physiological grounds, however, this interpretation must be discarded, for Steinberg (1969a, b) demonstrated clearly that rod as well as cone signals contribute to the activity of horizontal cells in the cat. The horizontal cells of mammals, at least, may therefore be functionally polarized in the sense that distinct parts of the cell, concerned with interconnecting exclusively cones or rods, are separated by a long, slender cytoplasmic bridge. This bridge may connect as well as separate; but although axonal in appearance, it conducts passively in either direction, and the fine processes at either end may be pre- as well post-synaptic to the photoreceptor cells. In fishes, in which horizontal cell axons have been observed only rarely, the problem of functional pathways is even more starkly evident.

Conventional synapses between processes of horizontal cells and nearby elements may provide in some instances a pathway for chemical transmission from horizontal to bipolar cell. Junctions having some features of presumed chemical synapses (clusters of vesicles and dense, widely separated membranes) were described in the cat and rabbit by DOWLING et al. (1966). The apparently presynaptic element in these junctions was a horizontal cell; the postsynaptic element, a bipolar cell dendrite or perikaryon. These investigators observed similar junctions of horizontal cells with small processes (unidentified) in the pigeon and rat but not monkey or man. OLNEY (1968) found similar horizontal-bipolar cell contacts in the mouse. LASANSKY (1971) has observed such conventional synapses of horizontal cells only rarely in the turtle. Some conventional synapses between processes making contacts with photoreceptors, interpreted as horizontal-to-bipolar cell synapses, were noted in *Necturus* by DOWLING and WERBLIN (1969). WITKOVSKY and DOWLING (1969) observed conventional synaptic contacts between horizontal cell perikarya or processes and other elements, which could be either pre- or post-synaptic, in carp. HAMA (personal communication) demonstrated synaptic boutons, of unknown cellular origin, on the surface of the horizontal cell somata in the Japanese red ray. Conventional contacts have been observed so infrequently, however, that perhaps one should seek a more general route for the output of horizontal cells. Since unambiguous ultrastructural evidence of synapsis has not usually been observed, horizontal cells might be pre- or post-synaptic to either receptor or bipolar cells.

In the absence of direct demonstration by the use of microelectrodes, synaptic interaction must be inferred largely from the geometry or patterns of interconnections of cells. DOWLING and BOYCOTT (1966) emphasized that the horizontal cell processes, with their lateral position in the triads, are "in a very strategic location to regulate transmission from receptor to bipolar dendrite" and suggested, by analogy to the amacrine cells in the second synaptic layer, that the horizontal cells might mediate reciprocal lateral presynaptic interaction between receptors. Because horizontal cell processes in goldfish make contact with receptors over a wide area away from the synaptic lamella, STELL (1966, 1967) inferred that receptor and horizontal cells interact bidirectionally. DOWLING and WERBLIN (1969) suggested, however, that the invagination of postsynaptic processes in *Necturus* permits interaction between horizontal and bipolar cell processes, and WERBLIN and DOWLING (1969) proposed that the horizontal cells mediate the surround effect by acting postsynaptically upon bipolar cells, rather than presynaptically upon receptor terminals. In support of this they found that an annular stimulus which markedly altered the response of bipolar cells to a centered spot had no effect on the response of receptor cells. BAYLOR et al. (1971), however, have demonstrated that in the turtle the membrane potential of cone cells is influenced by changes in membrane potentials of nearby cones and horizontal cells. This observation clearly implies the existence of synaptic transmission in both directions between receptor and horizontal cells in the turtle. The discrepancy between the findings of BAYLOR et al. and WERBLIN and DOWLING may be explained most simply by species-specific differences in functional organization. It is noteworthy that the conventional horizontal-bipolar cell synapses observed in the mudpuppy were seen only rarely in the turtle (LASANSKY, 1971). Alternatively it might be supposed that any presynaptic effects upon *Necturus* photoreceptors are so localized to the region of contact that they cannot be recorded by electrodes inserted distal to the pedicle, or that such effects are demonstrable only with very small or brief stimuli as in the turtle. These alternatives could be tested by intracellular recording and stimulation of cell pairs in *Necturus*.

BYZOV (1967) and BYZOV and TRIFONOV (1968) altered the local electroretinogram of turtle, and MAKSIMOVA (1969) and NAKA (1971) have influenced ganglion cell discharge in teleosts, by polarizing horizontal cells with extrinsic current. BYZOV's suggestion that current flowing electrotonically from horizontal to receptor cell regulates receptor-bipolar transmission remains to be proven, however.

The patterns of contacts between cone pedicles and horizontal and bipolar cell processes observed in primates by MISSOTTEN (1968) and KOLB (1970) give further evidence for bidirectional interaction between receptors and horizontal cells. I reason as follows. The surface by which cone pedicles and midget bipolar cells are joined is subdivided into small patches, distributed widely over the pedicles and

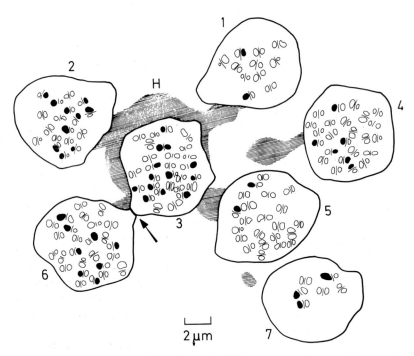

Fig. 18. Connections of cones with dendrites of a Golgi-impregnated type A horizontal cell in monkey perifovea. This horizontal projection from electron micrographs of serial vertical sections shows the location of nearly all triads, represented by the presynaptic ribbon and a pair of lateral postsynaptic processes. The perikaryon (*H*) and major dendrites of the horizontal cell are shaded. Each lateral process contributed by the Golgi-impregnated cell is black. Pedicles 3 and 6 contact one another by a stubby process (arrow) (reproduced from KOLB (1970) by permission of the author and the Royal Society of London)

dendritic trees. This pattern apparently is not designed to increase the area of mutual contact, which remains small. Rather, it assures that numerous patches of cone-bipolar cell contact, which otherwise would be equivalent, are associated with different combinations of processes from several nearby horizontal cells. Such an arrangement would be pointless if the horizontal cell processes were only post-synaptic; but if some were presynaptic, they could provide a complex array of pre- or post-synaptic modulatory influences upon cone-bipolar synaptic transmission. This argument does not apply to some other mammals, for example the guinea pig, in which all the lateral processes contacting a given cone pedicle are derived from the same horizontal cell (SJÖSTRAND, 1969). Even in the guinea pig, however, the horizontal cell processes may function presynaptically. Since each process reaches all the triads of a given pedicle, it might be supposed to act as a gate controlling transmission from that cone to numerous bipolar cells. In man and monkeys, in contrast, different horizontal cells in various combinations might cooperate in subtle and complex modulation of transmission from one cone to the same midget bipolar cell(s) over many parallel pathways.

Finally the difference between mammals and bony fishes must be emphasized. While it cannot be said that teleostean horizontal cells lack axons (see above), it does appear that the function of interconnecting groups of rods (or rods and rod bipolars), which is accomplished in mammals by the so-called "axon terminal" of the horizontal cell, is given to an entirely separate cell in teleosts. An important implication of this distinction is that in mammals some admixture of rod and cone information may be expected even in bipolar cells which are connected exclusively to rods or cones. In teleosts, on the other hand, the categories of receptor information carried by bipolar cells should depend entirely upon the patterns of direct receptor-bipolar cell contact, since all known indirect connections between receptor and bipolar cells (via horizontal cells) retain specificity for the same receptor system.

The connections revealed by KOLB and MISSOTTEN have other important implications for information processing in man. Since the diffuse cone bipolars and cone horizontal cells receive inputs from all of the six or more cones in their receptive field, and since it is unlikely that this many closely clustered cones contain the same photolabile pigment, the diffuse cells are color-coded only if some of their apparently similar dendritic contacts differ functionally. Midget bipolar cells of both types appear to carry monochromatic information since they receive from only one cone (POLYAK, 1941; BOYCOTT and DOWLING, 1969; KOLB, 1969). Of course, if cone-bipolar transmission is subject to modification as suggested above, these chromatic properties also might be modified. The speculations of MATURANA et al. (1968) concerning the mechanisms of color-coding in man are interesting but are based upon obsolete structural information.

The notion of *anatomical convergence* is commonly used to indicate the complexity of functional organization (cf. ORDY and SAMORAJSKI, 1968). The variety of retinal cells and their interconnections greatly limits the value of this approach, as a brief reconsideration of contacts in the first synaptic layer will show. "Convergence" means a progressive reduction in the number of information channels at successive (distal to proximal) levels. The rod system in teleosts and mammals is highly convergent, mainly because of the large number of input channels per bipolar cell and also because of the small number of output channels per receptor. If the number of receptors far exceeds the number of bipolar cells (as in the rod-dominated retinas of the mouse and rat), then the *convergence ratio* (receptors: bipolars) is a useful indicator of the *average minimum* number of receptors to which a bipolar cell sends dendrites. The average minimum figure is meaningful, however, only when the populations of receptor and bipolar cells approach being uniform, e.g., all rods and all rod bipolars with similar connectivities. The failure of predictability in a heterogeneous system may be illustrated by a simple example. Assume a retina having two bipolar cells and one hundred receptors, each receptor contacting only one bipolar cell. In one case let both bipolar cells contact fifty receptors; in another case let one contact ninety-nine receptors and the other only one receptor. Unless connections are examined directly, convergence appears to be the same in both cases; but the functional consequences of the difference are enormous. A concrete illustration is the first synaptic layer of the monkey para-fovea (KOLB, 1970). Since each diffuse (flat) cone bipolar cell contacts about six cones and each cone is contacted by about six flat bipolars, the overall ratio of cones to flat bipolars is about 1 : 1. Since each cone is contacted also by one invaginated and one flat midget bipolar cell, the overall ratio of cones to cone bipolar cells is about 1 : 3 — a number which completely obscures the important fact that some bipolars receive from only one cone, others from six. Thus were PEDLER and TILLY (1965) led by the 1 : 3 to 1 : 4 ratio of nuclei in monkey para-fovea to the untenable (and untrue) conclusion that there is no midget cone-bipolar (1 : 1) system.

A final comment concerns the relation between cells and operations. Ganglion cells, for example, are commonly described according to their functions as "motion detectors", "contrast detectors", "opponent-color cells", and so forth. Such operations are functions of multicellular assemblies, not properties of individual units. It is erroneous, therefore, to designate rod cells as "simple light-recording cells" and cones as "contrast-enhancing elements" or "color-discriminating cells" (cf. SJÖSTRAND, 1969). Each differentiating operation requires at least two circuit elements. Given what is known about functional properties of vertebrate photo-receptors, this means that each operation requires two or more cells.

Landolt's Clubs

Some bipolar cells in certain vertebrate groups (see above) send this clublike appendage from their dendritic tree to the external junctional layer (limiting membrane). COHEN (1963b) identified as Landolt's clubs some small, rather featureless processes in horizontal sections at the level of this layer in the pigeon retina. The clubs were shown to contain many mitochondria and a pair of ciliary basal bodies — thereby resembling rudimentary receptor inner segments, bearing no outer segment — in frog (LIPETZ, 1965; RICHARDSON and LIPETZ, in preparation), newt (HENDRICKSON, 1964, 1966), chicken (GOVARDOVSKII and KHARKEIEVITCH, 1967) turtle (RICHARDSON and LIPETZ, in preparation), and lungfish (LOCKET, 1970b). Similar elements, not identified as Landolt's clubs, were illustrated also in turtle by WALD and DE ROBERTIS (1961, Fig. 8), in frog by NILSSON (1964, Fig. 2, 5—7), and in toad by LASANSKY (1965, Fig. 11). Usually the club is separated from the photoreceptors by processes of Müller's cells (GOVARDOVSKII and KHARKEIEVITCH, 1967), but sometimes direct contact is made, without signs of synapsis (HENDRICK-SON, 1964, 1966; RICHARDSON and LIPETZ, in preparation). YAMADA and ISHIKAWA (1967) observed "synaptic" ribbons surrounded by small vesicles in the inner segments of lamprey photoreceptors at the level of the ciliary appendages of the Landolt's clubs. Although the ribbons were near the surface of the receptor, the receptor and the club were not especially close together and their membranes were not modified in this region. Also it should be recalled that similar ribbons, with associated vesicles, have been observed in a number of places where no synapsis can be identified (e.g. WOLFE, 1965; ALLEN, 1969; LASANSKY, 1970). Therefore it does not seem likely that, as YAMADA and ISHIKAWA propose, "the ribbon in the inner segment has some functional relationship to the Landolt's club".

For further understanding of the function of Landolt's clubs, therefore, we must look either to the internal structure of the clubs or to the relationships and functions of the cells which bear clubs. HENDRICKSON (1964, 1966), COHEN (1969) and LOCKET (1970b) have emphasized the striking resemblance of bipolar cells, especially those bearing Landolt's clubs, to photoreceptor cells. Since many receptors bear "9 + 0" cilia, and the photoreceptor outer segments are themselves derived from such cilia, the 9 + 0 ciliated club may well be a receptor of some kind. LIPETZ (1965, and in preparation, and RICHARDSON and LIPETZ, in preparation) suggests that the large number of mitochondria in the club provides an energy source essential for the functioning of the club. LOCKET (1970b), noting that Landolt's clubs in the lungfish extend to the pigmented epithelium, proposes that they provide a pathway for metabolites which circumvents the photoreceptors and Müller's cells. These are only speculations, however; the function of the Landolt's clubs is totally obscure.

Specialized Junctions between Horizontal Cells

Close appositions of horizontal cells have been observed in fishes. VILLEGAS (1960) observed that the processes of horizontal cells (apparently of the second row) in the marine teleost *Centropomus* make "membrane to membrane connections . . . with the processes of the other horizontal cells", forming "a structural unit, like a net". VILLEGAS and VILLEGAS (1963) described 60 Å appositions between adjacent horizontal cells of the first, second, or third row in *Centropomus*. HAMA (personal communication) observed a "broad close contact between the neighboring outer horizontal cells" in the Japanese red ray. A similar "fused membrane structure" was seen by YAMADA and ISHIKAWA (1965) between adjacent horizontal cells of a given row in sharks and rays, in which they comprised about 50 % of the area of mutual contact, and in carp. The authors inferred from the presence of these membrane structures that the horizontal cells of fishes form an electrical syncytium. In the tortoise, in contrast, the large horizontal cells were joined by occasional "attachment plaques" but not membrane fusions. "Tight junctions" were observed between adjacent horizontal cells in three teleostean species by BOROVYAGIN (1966), who suggested that they "point to the possible existence of a functional interaction within the layers of these cells". Similar junctions were observed in the dogfish shark, *Squalus* (STELL, 1966), the teleosts

Mugil (O'Daly, 1967a, b) and carp (Witkovsky and Dowling, 1969), and certain deep-sea teleosts (Locket, 1970a). In *Mugil*, at least, the cylindrical processes (see above) also were closely apposed. In all cases the junctions have been found only between cells of the same row — i.e. first row to first row, but not first to second, etc. Such junctions between horizontal cells, which to date have been observed only in fishes, are of interest because of the summation and interaction of the responses of horizontal cells (certain S-potentials) over distances which far exceed the dimensions of any retinal cell — virtually the entire retina in carp, for example (Norton et al., 1968), compared to $50-200\,\mu$m in the frog and 400 to $600\,\mu$m in the turtle (Byzov, 1968b). Since close membrane appositions, or "tight junctions", were described between cells known to be electrically coupled (e.g. Pappas and Bennett, 1966), it seemed reasonable to look for evidence of electrical coupling between horizontal cells in fishes. Early attempts, however, failed to demonstrate any such coupling in teleosts (Negishi et al., 1968). As O'Daly (1967a) indicated, the tight junctions which he and others have described in fishes differ from the usual electrical junctions by their absence of any internal (e.g. pentalaminar or gap) structure. Furthermore, in the electron microscope O'Daly (1967b) demonstrated non-transport adenosine triphosphatase and Brzin and Drujan (1969) showed acetylcholinesterase activity, between or along the surfaces of the closely apposed membranes of teleostean horizontal cells. These observations were taken to support the idea that the propagation of S-potentials from one horizontal cell to another is mediated by a metabolic junction ("transferapse") rather than a conventional electrical or chemical synapse (Negishi et al., 1968). Recently, however, Kaneko (1971) has demonstrated electrical coupling of horizontal cells of the first row in the retina of the dogfish shark, *Mustelus*, confirming the location of his micropipettes by injection of Procion dyes. The same cells were not coupled to bipolars or to horizontal cells of the other rows. Brightman and Reese (1969) have demonstrated that the junctional structure most regularly associated with electrical coupling is a close apposition of membranes between which is a $20-30$ Å gap in the form of a hexagonal lattice. Such a "gap junction" is demonstrable after staining tissues *en bloc* with uranyl acetate at pH 5 before dehydration, or by infiltrating the hexagonal gap with lanthanum hydroxide during fixation. I have observed a patchy "gap"-like structure in a specimen of guitarfish (an elasmobranch) stained *en bloc* with uranyl acetate (Fig. 19). It seems probable, therefore, that the horizontal cells of fishes are interconnected through low-resistance gap junctions. It is not known whether the enzyme activities found here are found in other gap junctions or are uniquely retinal.

Interconnections of yet another kind have been claimed to exist between horizontal cells in fishes. Villegas (1961) observed "strips of fibrous and dense cytoplasm", composed of closely packed small tubules, which "line" the horizontal cells of the second row in *Centropomus* and thus separate them from adjacent cells of the first and third rows. Villegas and Villegas (1963) interpreted these "strips of fibrilose cytoplasm" as intercellular bridges, making the horizontal cell network a true syncytium. Borovyagin (1966) described similar structures in the horizontal cell layers of pike and perch retinas, as did Locket (1970a) in deep-sea teleosts, and both drew the same conclusion about their function. None of these

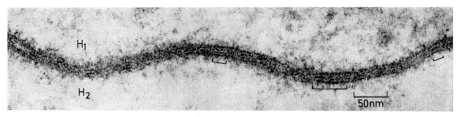

Fig. 19. Electron micrograph of transversely sectioned junction between two horizontal cells (H_1, H_2) of the first row in the adult guitarfish, *Rhinobatus lentiginosus*; fixed by perfusion with aldehydes, postfixed with osmic acid, stained *en bloc* with uranyl acetate,and section stained with lead. Because of the thickness of the section and the tortuosity of the membranes, junctional details are not observed easily, but in some places (brackets) a narrow gap or pentalaminar structure can be discerned. Its structure, the overall dimensions, and the presence of flocculent material on the cytoplasmic side of the membranes suggest that this is a gap junction (STELL, unpublished)

investigators, however, has presented unequivocal evidence for true cytoplasmic bridges which either contain or consist of small tubules. STELL (1966) observed similar tubules in the extracellular space near horizontal cells in the dogfish *(Squalus)*, goldfish, and *Centropomus*. In dogfish, rows of parallel tubules formed a nearly continuous layer over the distal surface of the horizontal cells of the first row. In goldfish, small extracellular tubules were seen in scattered clusters in the first synaptic layer near the surface of the horizontal cells, occupying irregular spaces measuring up to $1-2\,\mu$m in the largest dimension. In *Centropomus*, large masses of parallel tubules occupied spaces between horizontal cells of the first, second and third rows; these tubule-filled spaces could be as large as $5\,\mu$m vertically and several times that horizontally. The tubules were never enclosed in cell membranes. WITKOVSKY and DOWLING (1969) have observed tubules in the first synaptic layer of carp retina; their description and interpretation are in agreement with those of STELL (1966) for goldfish. The regions containing the extracellular tubules show a strong diastase-resistant periodic acid-Schiff reaction (PARTHE, personal communication). Because the rather cylindrical extracellular compartments containing tubules follow a sinuous course between horizontal and bipolar cells in *Mugil*, these compartments were designated processes of "undulant amacrine cells" by O'DALY (1967a). Careful examination of his micrographs, however, reveals that these tubular masses are not enclosed by membranes but are extracellular. Small clusters of extracellular tubules were observed also in the first synaptic layer of the mudpuppy, *Necturus*, by DOWLING and WERBLIN (1969). I suggest that these slender tubules, which appear to be fairly rigid structures, serve to maintain a rather large constant colume of extracellular fluid in this region. By virtue of their macromolecular properties they may help to maintain or control the ionic composition of the extracellular fluid around the horizontal cells.

Other Interconnections

Little is known about the kind, number and distribution of apparently unspecialized contacts between cells of the intermediate neuronal layer. Such contacts may, for instance, mediate some of the still unknown mechanisms underlying the horizontal cell responses which show different polarity at different wavelengths (C-responses; see e.g. Svaetichin et al., 1965) or the nonlinear addition of certain horizontal cell responses observed by Negishi (1971). Villegas (1961), among others, has recorded some observations on the direct contacts (not "tight", but devoid of intervening glia) between bipolar and horizontal cells at different levels; no special pattern emerges. I noticed that at least some rod horizontal cells are separated from cone bipolar cells in goldfish by leaflets of Müller's cells (this was seen clearly in the original prints for Fig. 5, Stell, 1965a). Since such glial "insulation" maintains the isolation of rod and cone subsystems established by the separate synaptic connections of the receptors, it is important to investigate such relationships much more carefully. This is particularly true of the cylindrical processes (Cajal's internal horizontal cells) of carp and goldfish, which – although joined to one another by close appositions – are not distinguished by special contacts of any sort with other cells (Yamada and Ishikawa, 1965; Stell, 1966; O'Daly, 1967a) and yet are said to generate or conduct S-potentials (Kaneko, 1970).

D. Interconnections of Bipolar, Ganglion, and Amacrine Cells

Knowledge of interconnections in the second synaptic layer has advanced slowly because of their forbidding complexity. Recent structural and electrophysiological findings, however, have provided new insights and reawakened interest in the detailed analysis of its complex structure.

Features and Variety of Contacts in the Second Synaptic Layer

Missotten (1960) noted that the synapses between bipolar and ganglion cells in human retina are characterized by a «crête synaptique» (synaptic ribbon) in the bipolar cell terminal; thus they resemble the synapses between photoreceptors and bipolar cells. In man the ribbon contacts of bipolars are smaller than those of the receptors and are characterized by dense substance in the synaptic gap and an increase in density of the postsynaptic membrane; unlike the contacts in the first synaptic layer, they are not invaginated. Cohen (1961) made similar observations in the monkey and mouse. Fine (1962) noted in human retina not only the ribbons but also "interrupted densities" of membranes (having some features of conventional synapses) at contacts with unspecified processes.

The complexity of structures and relationships was indicated by Kidd (1961, 1962) in his description of cat and pigeon (cf. Fig. 20). Ribbon synapses were observed equally often in both animals; frequently the ribbon appeared in a presynaptic ridge ("corner") extending between two processes which appeared different in ultrastructure. "Conventional" synapses, with small vesicles clustered at one side of a large, irregular gap between two membranes of increased density, were 4–5 times as common as ribbon synapses in the cat and 10 times as common as ribbon synapses in the pigeon. "Spine" synapses, like those observed in Purkyně cells of the cerebellum, were observed rarely in either animal. Finally, "serial" synapses of two types were described. In the first type, a terminal which was postsynaptic to a process containing a ribbon formed a conventional synapse back

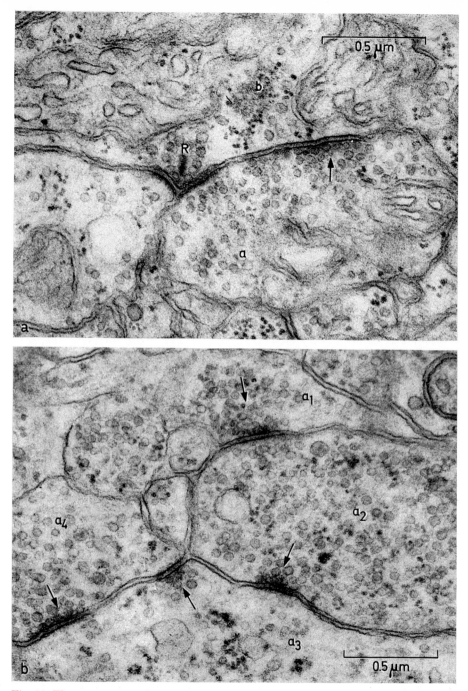

Fig. 20. Electron micrographs of synaptic contacts in the second synaptic (inner plexiform) layer of the frog retina, fixed in osmic acid. a, synapse of a bipolar cell axon terminal (b) with a dyad, one member of which contains vesicles and is probably an amacrine cell process (a). A ribbon synapse (R) and a conventional synapse (arrow) are reciprocally related. b, serial arrangement of conventional synapses ($a_1 \rightarrow a_2 \rightarrow a_3 \rightarrow a_4$), and reciprocal conventional synapses ($a_3 \leftrightarrows a_4$) (reproduced, slightly modified, from DOWLING (1968), by permission of the author and the Royal Society of London)

upon that process; this type occurred with equal frequency in both cat and pigeon. In the second type, a number of terminals were joined by conventional synapses, one terminal appearing to be postsynaptic to another but presynaptic to a third, and so on; serial synapses of this type were rare in the cat but not uncommon in the pigeon, although even in the latter they occurred only about one-tenth as frequently as ribbon synapses. KIDD indicated the important relationship between the relative frequency of simple and serial conventional synapses and the number of amacrine cells, in pigeon versus cat.

In human retina, VILLEGAS (1964) noted the presence of synaptic lamellae in some presynaptic processes of the second synaptic layer and described individual and serial conventional synapses. MISSOTTEN et al. (1964) reported that the synaptic ending of human perifoveal rod bipolar cells contains a synaptic ribbon opposite a pair of postsynaptic processes, one of which usually contains small vesicles and one of which does not, and suggested that this implies some special organization. MISSOTTEN (1965a) also described serial conventional synapses in the human perifovea.

Identification of Cells and Patterns of Interconnection (II)

The nature of the interconnected processes and the significance of their relationships was outlined in the classic papers of DOWLING and BOYCOTT (1965a, b, 1966, 1969) on the second synaptic layer of man and rhesus monkey. They defined the constellation of a bipolar cell terminal ridge (containing a synaptic ribbon) with a pair of postsynaptic processes as a "dyad". In the present account, however, I shall use "dyad" in the restricted sense of only the two processes which contact the bipolar terminal at the presynaptic ridge, since this is more consistent with MISSOTTEN's usage of "triad" in describing the first synaptic layer. DOWLING and BOYCOTT observed occasional conventional synapses from one process of a dyad back to the bipolar cell, like those described by KIDD (1962) as the first type of serial synapse, and called them "reciprocal synapses". Since it is really the *pair* of synapses, ribbon and conventional, which are reciprocal, I shall call this a "feedback" synapse. After tracing processes in short series of sections, DOWLING and BOYCOTT were able to distinguish amacrine cell processes from ganglion cell dendrites in single sections by their characteristic differences in ultrastructure. Usually in these primates one element of each dyad is contributed by a ganglion cell, the other by an amacrine cell. They found also an occlusive "tight junction" (five-layered apposition, narrower than a gap junction; cf. section above on horizontal cell junctions) between a bipolar cell terminal and a ganglion cell body. Since this was observed only rarely and has not been confirmed by other investigators, it should not yet be taken as a regular feature of cellular interconnections in primate retinas.

In addition to the ribbon (bipolar-dyad) and feedback (amacrine-bipolar at a ribbon) synapses, DOWLING and BOYCOTT observed a variety of other contacts including amacrine-bipolar (no ribbon), amacrine-amacrine (some serial), and amacrine-ganglion cell. They proposed that, as suggested earlier by GALLEGO (1954), the amacrine cells mediate lateral interactions which are responsible for the opponent surround of the functional receptive field of the ganglion cell. The major evidence for this point of view is that, as noted earlier, the diameter of receptive

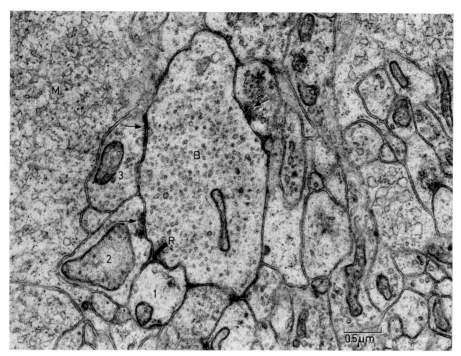

Fig. 21. Electron micrograph of second synaptic (inner plexiform) layer of the rabbit retina, fixed with osmic acid, near the ganglion cell layer. The synaptic ending of a rod bipolar cell (*B*), with its synaptic ribbon (*R*), makes apparently synaptic contact with four processes (1—4). The thickness of the membranes of dyadic processes 1 and 2 is increased and extra-cellular material of moderate density is present in the gap at the ribbon synapse. Processes 2 and 4, and possibly 3, make conventional synapses (arrows) upon the bipolar cell terminal; the synapse from process 2 is of the feedback type, reciprocal with a ribbon synapse on the same process. *M* Müller's cell (reproduced from RAVIOLA and RAVIOLA (1967), by permission of the authors and the Wistar Press)

field "surrounds" exceeds the diameter of ganglion cell dendritic fields (DAW, 1968, showed for example that virtually the entire retina constitutes the surround for ganglion cells in the goldfish); that the transient phenomena associated with surrounds appear to be properties of amacrine and ganglion cells only (cf. WERBLIN and DOWLING, 1969); and that center and surround signals reach the ganglion cell over independent channels in the cat (ENROTH-CUGELL and PINTO, 1970). In fish, at least, the horizontal cells must be considered as an alternative pathway for the influence of the surround. As observed by NORTON et al. (1968), the receptive fields of S-potentials (probably horizontal cell responses) cover virtually the entire retina in cyprinid fish. That the horizontal cells of catfish mediate some ganglion cell surround functions is implied directly by both suprathreshold receptive field pro-perties (NAKA and NYE, 1970) and the effects of passing current into horizontal cells (NAKA, 1971). Although there can be little doubt that some properties of surrounds are contributed by amacrine cells, the possibility of a role for other horizontal elements — even in higher vertebrates — deserves further consideration.

In the enthusiasm which has greeted the work of Dowling and Boycott, several other excellent publications have attracted less attention than they deserve. Raviola and Raviola (1967) independently observed in the rabbit (Fig. 2) the same basic structures as were noted in primates: bipolar cell terminals with ridges containing synaptic ribbons, contacted by a pair of processes — one an amacrine and the other a ganglion cell; conventional synapses of amacrine cell processes on bipolar cell terminals, both feedback (from a dyadic process) and separate; very rare contacts of bipolar terminals with ganglion cell somata; and series of conventional synapses. These authors observed some close appositions of processes within the inner plexiform layer, but they observed no bipolar-ganglion (axosomatic) or bipolar-bipolar synapses. They considered the bipolar-ganglion and bipolar-amacrine synapses to be excitatory, and the amacrine-bipolar synapses to be inhibitory, by virtue of their similarities to synaptic complexes studied by Walberg (1965) and Rall et al. (1966). Raviola and Raviola (personal communication) have observed also some bipolar-amacrine axosomatic ribbon synapses in which one element of the postsynaptic dyad is the amacrine cell perikaryon, in a macaque monkey and rabbit. In the mouse, Cohen (1967a) observed dyads and feedback synapses; the two dyadic processes often appeared slightly different although usually both contained vesicles; he could not establish their identity, however. Some bipolar cell terminals made contact with ganglion cell somata, but without special signs of synapsis.

In goldfish, Meller and Eschner (1965) illustrated large end clubs of bipolar cells containing many mitochondria and small vesicles but few synaptic lamellae, and surrounded almost completely by contacts with small bouton-like processes. Goodland (1966) described similar contacts in the teleost *Cottus* and identified the large end clubs as terminals of the large (diffuse or rod) bipolar cells. The same arrangement has been noted in carp by Witkovsky and Dowling (1969) and in goldfish by Stell (unpublished). Although serial sectioning was not applied in any of these cases, the number of vesicle-containing processes contacting the terminal appeared to exceed the number of ribbon contacts, and the large bipolar cells were not observed to make direct contact with ganglion cell bodies as claimed by Cajal (1892). In the dogfish shark, *Mustelus*, Witkovsky and Stell (1971) have observed typical gap junctions between synaptic endings of bipolar cells, implying that these endings are coupled electrically.

In the frog, Dowling (1968) observed about the same number of ribbon synapses per unit volume of inner plexiform layer as in man, but about four times as many serial conventional synapses as in man. Since there were many amacrine-amacrine and amacrine-ganglion cell synapses, and since both elements of most dyads appeared to be amacrine cells, it was apparent that most transmission from bipolar to ganglion cells was indirect, i.e. mediated by amacrine cell processes. Thus, "in the retina of the frog, the complexity of response behaviour of the ganglion cells is best explained by the wealth and variety of interactions among the processes of the amacrine cells" (Dowling, 1968). The greater structural and functional complexity of the retina in primates and cat in contrast to frog and pigeon suggested also the principle that retinal complexity is augmented where the visual cortex is absent or diminished in size and importance. In large part this principle has been confirmed by further comparative studies.

In the mudpuppy, *Necturus*, DOWLING and WERBLIN (1969) observed that serial or feedback conventional contacts, as well as individual conventional contacts on amacrine, bipolar or ganglion cells, were about seven times as frequent as ribbon contacts. In about half of the ribbon contacts both members of the dyad were amacrine cells; in the other half, one was an amacrine and the other a ganglion cell. In the relative numbers and frequencies of synapses of different kinds and in the mixture of direct bipolar-ganglion cell contacts with indirect bipolar-amacrine-ganglion cell contacts, the mudpuppy is therefore intermediate between primate and frog.

Continuing such observations, DUBIN (1969, 1970) surveyed the numbers of synaptic contacts of various kinds in the second synaptic layer of eight vertebrates in which the behavior of ganglion cells is known to show varying degrees of complexity. Counts of all apparent synaptic contacts were made in large montages of electron micrographs covering the full thickness of the second synaptic layer, and the data were treated statistically. In general, the prevalence of ribbon synapses was about the same in all eight species. The ratio of amacrine to bipolar synaptic terminals correlated well with the functional complexity of ganglion cells, increasing in order from primate and rat, through rabbit and ground squirrel, to frog and pigeon (Table 1). As pointed out previously, this increase is ascribable largely to the absolute increase in prevalence of amacrine cells and their terminals. The overwhelming importance of the second synaptic layer, stressed also by DOWLING and BOYCOTT (1969), is stated most explicitly by DUBIN: "The bipolar cell may be thought of as the neural element which brings information of the visual world to the inner plexiform layer . . . (and) thus we may expect the major processing of the visual image within the retina to occur after the message reaches (that) layer, especially in the more complex animals". Center-surround interaction

Table 1. Summary of quantitative data on the size and relative numbers of conventional and ribbon synapses in eight vertebrate species, arranged in order of increasing ganglion cell complexity as well as conventional: ribbon synaptic ratio. (Reproduced from DUBIN 1969 and 1970 by permission of the author and the Wistar Press)

| | Corrected size in Å | | Ratio of Conventional: Ribbon synapses | |
	Ribbon synapses	Conven. synapses	Before correction	After correction
Human parafovea	1905	2868	1.7	1.3
Monkey fovea	1976	3069	2.6	2.0
Monkey parafovea	2127	2699	1.9	1.7
Monkey periphery	1717	2877	3.0	2.2
Cat no. 1	1926	2808	2.6	2.5
Cat no. 2	2140	3064	3.0	2.7
Rat	1649	2608	3.3	2.8
Rabbit no. 1	1603	3343	5.1	3.7
Rabbit no. 2	1442	3174	5.3	3.8
Ground squirrel	1730	2860	6.7	5.5
Frog *R. cates.*	984	2469	8.8	6.3
Frog *R. pip.* no. 1	796	2488	9.1	6.4
Frog *R. pip.* no. 2	800	2301	10.9	9.5
Pigeon	1987	2342	10.8	10.8

and selective responsiveness to motion or direction of motion are considered by Dubin likely to be functions mediated by the amacrine cell system.

The organization of the second synaptic layer in carp retina was discussed by Witkovsky and Dowling (1969). Both elements of the dyad could be identified in 34 ribbon synapses; half of the dyads were amacrine-amacrine and half were amacrine-ganglion cell pairs. Their conclusion was that one of every four processes which contact the surface of a bipolar cell terminal is therefore a ganglion cell dendrite. This is doubtful, however, because many of these processes, especially those at the surface of the large (mixed) bipolar terminals, appear not to be involved in ribbon contacts. In fact, terminals of the large bipolars in teleosts appear to be covered almost entirely with small synaptic boutons (see above) and contain few ribbons, while ribbon synapses appear to be a more constant feature of the terminals of the small bipolar cells. This apparent distinction between the synaptic relationships of the axonal terminals of mixed and cone bipolar cells is not mentioned by Witkovsky and Dowling, viz. in their determination of numbers of conventional and ribbon synapses. According to these numbers the carp retina is intermediate in complexity — like that of the rabbit and ground squirrel — but if the distribution of the synaptic contacts of bipolar cells is inhomogeneous in these fishes, total numbers of junctions in the entire synaptic layer are of dubious value. For example, the density of conventional synaptic junctions may be significantly higher at the surface of the large bipolar terminals than elsewhere in the synaptic layer.

As stated by Dubin (1969), "one desires to know how individual cells are interconnected", and "a quantitative picture ... begins to indicate such information". An important limitation of this approach, acknowledged by Dubin, is that the individuality of different synaptic junctions — the inhomogeneity of the synaptic layer — may be easily overlooked. While the differences in synaptic relationships of bipolar cells of different types might be specified easily in carp, in which their synaptic terminals are distinguished without difficulty, such inhomogeneity is much less obvious in other animals. Dubin was able to find evidence for the segregation of synapses into two sublayers in electron micrographs of the second synaptic layer of the pigeon retina (Fig. 22); yet *numerous* sublayers (at least four) can be demonstrated in this layer by various light-microscopic methods (e.g. Cajal, 1892; Boycott and Dowling, 1969), and some of these sublayers differ from one another chemically (see below: e.g. Shen et al., 1956; Ehinger, 1967). If this stratification has not been demonstrated by electron microscopy, then either the data obtained or the way they have been analyzed must be inappropriate. It remains likely, then, that other subtle but no less interesting features of synaptic organization have not been made apparent by the method of quantitative electron microscopic analysis as it has been employed.

A further limitation of studies such as those of Dowling and Dubin is that their analyses — of dyads, for example — have been restricted largely to two-dimensional relationships, even though both Dowling and Boycott and Dubin have done some serial sectioning. The complete description of three-dimensional relationships by analysis of serial ultrathin sections is the major accomplishment of Allen (1965, 1969), who has described in detail all of the contacts with a single bipolar cell terminal in human peripheral retina (Fig. 23). This cell can be identified probably as a rod bipolar by virtue of its form (including level of perikaryon and shape and level of synaptic terminal), location, and ultrastructure. The twenty-three ribbon contacts are distributed more or less randomly over the terminal (although the ribbons tend to be confined to axonal swellings or "boutons"), while

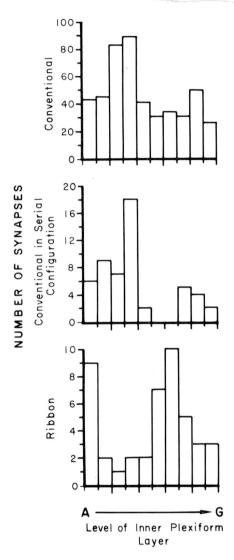

Fig. 22. Graphic representation of the distribution of conventional (top), serial (middle), and ribbon synapses (bottom) in the second synaptic (inner plexiform) layer of pigeon retina. In montages covering large areas, this layer was divided into ten equally thick horizontal bins (strata). Statistically corrected numbers of synaptic contacts in each bin are given. Although conventional synapses are everywhere more numerous than ribbon synapses, the ratio of conventional to ribbon synapses is very much reduced in the proximal (inner) half of the synaptic layer (reproduced from DUBIN (1969 and 1970) by permission of the author and the Wistar Press)

the conventional ("simple") synapses are localized mainly to a few patches of its surface. As many as four processes are clustered in a "synaptic complex" at each ribbon; the most frequent groupings are: two amacrines ("axons"); three amacrines; one amacrine + one ganglion cell ("dendrite"); and two amacrines + one ganglion cell. Some individual processes make contact at as many as four ribbons. Most but not all of the dendrites (8 of 11) contact the bipolar cell terminals

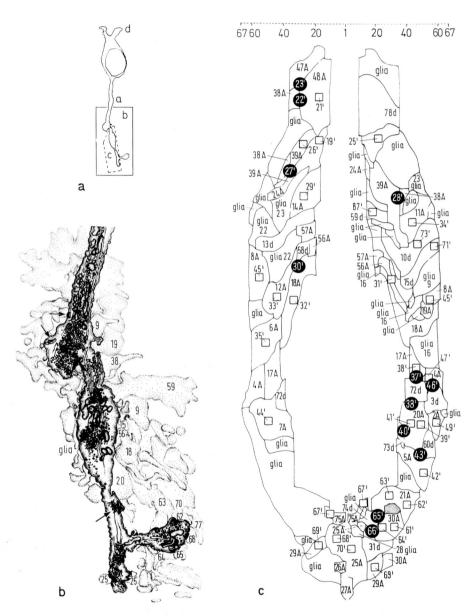

Fig. 23. (Legend see p. 173)

at a ribbon synapse, and several make contact at two or three different ribbons. Most of the amacrine cell terminals also make contact at ribbons, but more amacrine than ganglion cell contacts are not associated with any ribbon. The amacrine cell terminals in each ribbon complex are usually interconnected serially with one another and reciprocally with the bipolar cell. Particularly striking is the predominance of amacrine cell contacts (61 "axons" but only 11 "dendrites" contact this bipolar cell terminal) in this "simple" retina. Similar results were obtained with several other bipolar cells which were analyzed less completely.

ALLEN's special contribution is his demonstration of the complexity of the small groups of processes at the surface of the bipolar cell terminal. It is evident that in a two-dimensional analysis most of the ribbon synaptic clusters, including those having three or four postsynaptic elements, would appear to be either an amacrine-amacrine or an amacrine-ganglion cell dyad. ALLEN (1969) suggests that these small groups of processes "function as finer and still finer units, each at a different but not entirely separate integrative level". The difficulty with his analysis is that at best only this ultrafine local circuitry was revealed. In a very long series of sections, only one process was followed between two cells whose perikarya could be located, and none of the processes which contacted the bipolar terminal analyzed in detail made synaptic contacts with any other synaptic process. Furthermore, despite careful analysis "the precise relationship of the synaptic ribbon to the individual synapsing processes remains unclear".

The complex clusters of processes from cells of at least three different types in the second synaptic layer of the retina resemble in certain respects the "glomeruli" found elsewhere in the central nervous system (e.g. PETERS and PALAY, 1966; AKERT and STEIGER, 1967; SZENTÁGOTHAI, 1969). In some cases the evidence suggests that axo-axonic contacts of this sort mediate presynaptic inhibition (WALBERG, 1965). RALL et al. (1966) suggest that the interconnections of the retinal amacrine cell terminals, like the interconnections of the granule cell gemmules in the olfactory bulb, provide a pathway for both lateral and self-

---

Fig. 23. Reconstruction of relationships of axon terminal of a human peripheral bipolar cell (probably rod type) from 67 consecutive serial ultrathin sections (scale at top). a, outline drawing of entire reconstructed cell: d, base of dendrites (terminal arbor not reconstructed); a, axon; b, area covered by Fig. 23 b; c, area covered by Fig. 23 c. b, drawing representing three-dimensional appearance of axon terminal of bipolar cell shown in 23 a; the form and location of synaptic ribbons, vesicles, mitochondria, microtubules, and ribosomes in the axon are shown. Stippled processes are those which contact the axon in sections 1—5 of the series; numbered processes form synapses or are glia (Müller's cells) making extensive contact. c, topographic projection of surface of bipolar cell axon, region indicated by dotted lines in Fig. 23 a. Ribbon synapses are indicated by black circle with primed number; conventional contacts on axon, by open square with primed number; number + d, dendrite (presumably ganglion cell); number + A, axonal bouton (presumably amacrine cell); stippled area, attachment of laterally projecting bipolar terminal bouton. All glial contacts are indicated but not all are numbered (reproduced, with modifications, from ALLEN (1969), by permission of the author. Originally published by the University of California Press; reprinted by permission of the Regents of the University of California)

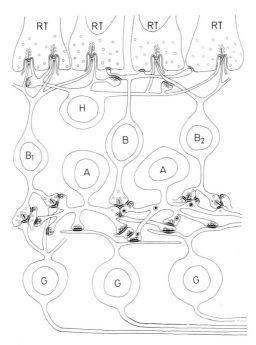

Fig. 24. Summary diagram of synaptic contacts in the frog's retina. *RT*, receptor terminals; *H*, horizontal cell; *B*, bipolar cells; *A*, amacrine cells; *G*, ganglion cells; *1*, first synaptic (outer plexiform) layer; *2*, second synaptic (inner plexiform) layer. In the first synaptic layer are invaginated receptor-bipolar and receptor-horizontal, superficial receptor-bipolar, and conventional horizontal-bipolar cell synapses. In the second synaptic layer are bipolar-amacrine and bipolar-ganglion ribbon, amacrine-bipolar (feedback) conventional, amacrine-amacrine (single and serial) and amacrine-ganglion cell synapses. The diagram suggests that some bipolar cells (B_1) make mainly direct synapses with ganglion cells, while others (B_2) make primarily or entirely indirect synapses through amacrine cells; the former is thought to be associated with "simple", the latter with "complex" ganglion cell organization (reproduced from Dowling (1968), by permission of the author and the Royal Society of London)

inhibition of the bipolar (as of the mitral) cell[1]. Meanwhile, the function of synapses in the second synaptic layer of the retina remains poorly known. The best data available suggest that many cells in both the amacrine and ganglion cell layers are cholinergic, and many are dopaminergic. Dense-cored vesicles, though to indicate the presence of catecholamines, have been described in cells of both layers, particularly in processes of amacrine cells (see section on histochemistry, below). Flattened vesicles, which may be related to inhibitory synaptic transmission (Uchizono, 1967), have not yet been observed in the retina.

---

[1] This seems to be a sound hypothesis. It appears, however, that the olfactory granule cells are not truly amacrine, since they probably receive centrifugal or collateral inputs to the proximal portion of a long axonal process from which project small spines (Cajal, 1905; Cragg, 1962). If this be so, then it is misleading to refer to the gemmules as "dendrites". A detailed analysis of synaptic relationships in the olfactory bulb by Price and Powell (1970a—e) should be consulted particularly by those interested in retinal amacrine cells and centrifugal fibers.

WERBLIN and DOWLING (1969) propose reasonably that the transience of the amacrine cell response may be explained by negative feedback inhibition of bipolar-to-amacrine cell synaptic transmission (at ribbon synapses) by means of the conventional amacrine-to-bipolar (reciprocal or feedback) synapses. They suggest also that in *Necturus* the two classes of ganglion cells, which respond with either transient or maintained responses, correspond respectively to the indirect bipolar-amacrine-ganglion cell and the direct bipolar-ganglion cell pathways suggested by the ultrastructural findings of DOWLING and WERBLIN (1969). DOWLING's schema of retinal interconnections is shown in Fig. 24. Confirmation of many of these ideas awaits direct investigation of functional connections between these neurons with multiple microelectrodes and analysis of the ultrastructure of contacts of known function between pairs of identified cells.

E. Second Synaptic Layer: Efferent (Centrifugal) Fibers

The ultrastructure of efferent fibers and details of their relationships are known with certainty only in birds. DOWLING and COWAN (1966) identified efferent fibers in electron micrographs of pigeon retinas fixed 3—7 days after lesions were placed in the isthmo-optic nucleus or tract. The myelinated fibers which constitute this system cross the second synaptic layer to terminate on or near the amacrine cells. Their terminals contain tight clusters of vesicles and the electron density of the pre- and post-synaptic membranes is enhanced. Usually these terminals contact free processes (not perikarya) of what appear to be amacrine cells; these are thought to be the "ordinary" amacrines of CAJAL (1896). Only a small minority of endings are seen more distally in the amacrine cell layer, where they are found on a pyriform perikaryon thought to be an "associational amacrine cell" of CAJAL. Cajal's "pericellular nests" are not observed, perhaps because so little tissue is surveyed in the electron microscope. DOWLING and COWAN believe they observed some displaced ganglion cells, none of which received terminals of efferent fibers. Their criteria for recognizing such a cell are not specified, however, and it should be recalled that MATURANA and FRENK (1965) believed, on the basis of their light microscope studies, that some efferent fibers terminate on the dendrites but not the soma of displaced ganglion cells. IRALDI and ETCHEVERRY (1967) reported finding granulated vesicles in these terminals in the pigeon and rat but did not present adequate data to support their identification as "efferent".

MELLER and ESCHNER (1965) observed axosomatic contacts on perikarya in the amacrine cell layer in birds, a reptile, and the rhesus monkey. MISSOTTEN (1965a) noted that amacrine cell bodies in man were often in contact with a process containing small vesicles, and frequently a "black and straight line" (apparently a flattened subsurface cistern) was found in the amacrine cell beneath the contact with this process. DOWLING and BOYCOTT (1966) observed in man and monkey contacts similar to those described by MISSOTTEN; they identified the contact as bipolar-to-amacrine but believed it not to be synaptic. RAVIOLA and RAVIOLA (1969) found that subsurface cisterns are rather common in amacrine cells of the rabbit retina; many are found opposite processes of Müller's cells, however, and only a few are associated with specialized intercellular contacts. WITKOVSKY (1971) has observed myelinated nerve fibers which terminate synaptically on cells or processes (amacrine or bipolar) at the distal margin of the second synaptic layer in carp and dogfish. Because of their course, location, and form of termination they appear to be homologous to the efferent fibers of birds; their central origin, however, has not yet been demonstrated. Some of these terminals contain dense-cored vesicles.

F. Glia-Neuronal Relationship and Extracellular Space

The retinal neurons are surrounded by a restricted extracellular fluid, the volume and composition of which may be influenced by the relationships with and function of nearby retinal cells. Some knowledge of the relationships of neurons, glial cells and the extracellular space is necessary for a fuller understanding of retinal function.

The ubiquitous glial cell of the vertebrate retina is the radial fiber of Müller, which extends from the vitreal surface of the retina to the level of the receptor inner segments. Many investigators (e.g. Fine and Zimmerman, 1962) have emphasized the marked difference in internal structure of the distal ("cytoplasmic") and proximal ("fibrous") parts of these Müller's cells. As indicated by Cohen (1963a), because of their peculiarities of structure the Müller's cells cannot be identified simply with one of the classical types of central nervous system gliocytes. They have some of the properties of ependymal, oligodendroglial and astroglial cells. As discussed elsewhere, they penetrate between — and more or less surround — neuronal perikarya of all kinds, and they send their processes among the neuronal processes in the synaptic layers. These features can be discerned readily in Müller's cells impregnated with the Golgi method (Cajal, 1892). Missotten's (1965a) electron microscopical observation that "here and there the cell bodies (of neurons) touch each other, and each dendrite is in contact with many other nerve fibers over most of its length" could be applied to most retinas.

The perikarya of the Müller's cells are normally found in the proximal half of the intermediate neuronal layer. Their end feet, which in at least some animals are arranged in orderly rows (Pedler, 1962; Lessell and Kuwabara, 1963), form a continuous layer between the ganglion cells (including their axons) and the basement membrane which bounds the vitreous body. The end feet are simply apposed to one another, without special junctions; the extracellular space between them appears patent. The distal (apical) extremities of the Müller's cells look more highly differentiated. The external junctional layer (limiting membrane) comprises mainly the special junctions of these cells with each other and with visual cells. The glia-glial appositions are close, perhaps gap junctions (Miller and Dowling, 1970) like those between ependymal cells in the brain (Brightman and Reese, 1969); these appositions are circumvented by only a few larger spaces. The glia-neuronal junctions, however, are all wider and of the adhering type (Lasansky, 1965; Cohen, 1965b).

Certain features of the Müller's cells near the external junctional layer suggest that they play an important role in the metabolic support of neurons. Plentiful mitochondria are found in the distal part of the cell, and in the toad Lasansky (1961) described an especially close relationship of these mitochondria to the plasma membranes apposed to processes of photoreceptor cells. Distal from the junctional layer, the apex of the Müller's cell in most species is elaborated into large numbers of long microvilli, which extend between the inner segments of the photoreceptor cells; intracellular vesicles, channels or infoldings project inward from the base of the microvilli. WALD and DE ROBERTIS (1961) and PEDLER (1963) suggested that these cellular structures facilitate the exchange of fluids or metabolites, particularly with the radial fins (see COHEN, this volume) of receptors in retinas which lack intraretinal vessels. This suggestion was supported by the findings of MARCHESI et al. (1964) that nucleoside phosphatase activity, prominent in many salt- and water-transporting tissues, is associated with the outer surfaces of glial membranes in the retina — particularly around the microvilli of the Müller's cells and between them and the photoreceptors. LASANSKY (1965) proposed that these cells mediate active transport of water and ions and facilitate removal of CO_2 from the neurons and of excess potassium ions from the extracellular space. Other features of the Müller's cells, such as their high content of glycogen and oxidative enzymes (LESSELL and KUWABARA, 1963), are consistent with an important metabolic role.

Alternatively it has been proposed that the Müller's cells are the functional equivalent of a true extracellular space. The extracellular volume of the retina was determined to be on the order of 30% of the total retinal volume by chemical methods (AMES and HASTINGS, 1956). Therefore the finding of an apparently very small extracellular volume in electron micrographs led a number of investigators to the conclusion that the glial cells are the true "extracellular" compartment for the neurons (SJÖSTRAND, 1958, 1961; FINE and ZIMMERMAN 1962; SJÖSTRAND and NILSSON, 1964). COHEN (1963a) acknowledged, however, that the small space between cell membranes could be filled with a hydrated, conductive gel, rather than the non-conductive lipid matrix proposed by SJÖSTRAND, and emphasized that "there is probably significant extracellular space" in the optic fiber layer and both plexiform layers. COHEN (personal communication) has found, furthermore, that the pigmented epithelium usually remains attached to the "isolated" rabbit retina. The extracellular volume measured by AMES and HASTINGS is therefore probably greatly overestimated because of the large volume of interreceptor (optic ventricle) space included in their measurements. HANSSON (1970b) reported finding "extracellular material" on the surface of rod spherules in freshly frozen and dried rat retina examined in the scanning electron microscope. Since the fracture plane commonly passed *through* cells in freshly frozen tissue, rather than *between* them as in tissue fixed chemically before fracturing (HANSSON, 1970a), this material may be only cellular debris.

Studies with electron-opaque tracers have supported at least the patency of the true extracellular space, if not the large volume indicated by chemical probes. LASANSKY and WALD (1962) injected ferrocyanide ions into the vitreous body of the toad and localized it in the electron microscope after conversion to Prussian blue. All of the dense end-product was detected in true intercellular channels. LASANSKY (1965), observing that intravitreous ferrocyanide diffuses exclusively by these channels — even through the partially occluded distal junctional layer — concluded that "ions and metabolites are not forced to traverse glial cytoplasm in order to reach the retinal neurons". Lanthanum hydroxide and ruthenium red, both very small tracers, penetrate between the end feet of Müller's cells in the dogfish (STELL and WITKOVSKY, unpublished observations). Even the small protein, horseradish peroxidase (mol. wt. *ca.* 40000), diffuses easily from the vitreous body to pigment epithelium only through extracellular channels in the toad (LASANSKY, 1967). In the cat, however, the larger particles of thorium dioxide (diameter $\geq$ 100 Å) diffuse through the extracellular spaces from the vitreous body but appear to be stopped by the external junctional layer (SMELSER et al., 1965).

Using a newly developed technique of freeze-substitution, VAN HARREVELD and KHATTAB (1968) showed a considerable extracellular space in the optic fiber layer and the most proximal part of the second synaptic layer of mouse retina, in which these spaces appear smaller after preparation by conventional methods. Unfortunately this technique cannot be applied to the study of relationships deeper than a few tens of micrometers from the surface of the tissue. These authors showed also that pretreatment of the tissue with 5 mM glutamate, a procedure known to reduce the volume of the extracellular space as determined chemically (AMES, 1956), results in swelling of processes in the second synaptic layer, as shown earlier by WALD and DE ROBERTIS (1961) using chemical fixation without freezing. This evidence supports the contention of VAN HARREVELD and KHATTAB that the very small extracellular space evident in most electron micrographs is an artifact due to changes in volume of cellular processes during conventional chemical fixation. In agreement, HANSSON (1970a) noted that "the intercellular space [in the optic fiber layer, examined with the scanning electron microscope] was lost if the tissue pieces were not frozen or fixed immediately" after excision.

Finally it should be recalled that the extracellular space in fish retinas contains tubules which may serve to maintain some constancy of magnitude or composition of the extracellular fluid there (see pp. 162 – 163). Although these observations seem therefore to establish that the true extracellular space in the retina is substantial, FABER (1969) estimated from measurements of intra- and extra-retinal resistances that the extracellular volume in rabbit retina is less than 1% of the total retinal volume.

These relationships of neurons to glial cells and extracellular compartments have important physiological implications. The spatial distribution of electrical current sources and sinks and extracellular resistances, which reflects these relationships, determines functional properties which are usually detected by extracellular recording. For example, most of the total transretinal resistance is localized near the retinal epithelium. The locus of this high resistance, called the "R-membrane", has been thought at various times to be one of such real structures as BRUCH's membrane, the membranes of the epithelial cells, or the distal junctional layer (see review in BRINDLEY, 1960). BYZOV (1968) confirmed the location of the resistance change in frog retina, with and without the pigmented epithelium, by marking with Niagara Sky Blue dye. The major component of the resistance change lies between the space around the receptor inner segments and the interior of the epithelial cells; a minor component is located between the interreceptor space and the interior of the layer of receptor cell bodies (the precise localization within this layer was uncertain). These observations are consistent with the suggestion of COHEN (1965b) that the continuum of occluding junctions between the apices of the epithelial cells — combined with the high resistance of the apical membrane of those cells — is the major component of the R-membrane (although, since many junctions formerly thought to be occluding have since been shown to have a gap structure, these epithelial junctions should be re-examined with newer methods). The component of transretinal resistance across the external junctional layer may be due either to the junctions in that layer which partly reduce the extracellular passage through it or to the plasma membranes of photo-receptors and Müller's cells, depending upon whether the location of the recording electrode — believed certain in only 25% of the cases — was in fact extra- or intra-cellular. If this interpretation of the R-membrane is correct, then one may predict that the extracellular resistance will be elevated wherever the space appears to be restricted by special junctions. Such an increase in resistance should be expected, for example, across the layers of horizontal cells in fishes, where broad areas of close apposition are found (see above).

MILLER and DOWLING (1970) have recorded responses in *Necturus* which have the properties of an intracellular local b-wave. By injecting dye from the recording pipette they localized these responses to Müller's cells. Although the responses came usually from the proximal end foot, it appeared that the polarity of the poten-tial change was the same everywhere in the cell. From measurements of the second derivative of response amplitude at different levels, FABER (1969) concluded that the generator of the b-wave in rabbit has a sink in the first synaptic layer and a diffuse source extending from this to the vitreous body. Since the only cell which could generate such a response is the Müller's cell, he proposes that the b-wave is the response of that cell to an increase in extracellular concentrations of potassium ion during activation of synapses in the first synaptic layer; a similar mechanism is proposed by MILLER and DOWLING. Such behavior, which has been observed in other glial cells (see most recently BAYLOR and NICHOLLS, 1969), requires some restriction of the extracellular space and some proximity of the Müller's cell processes to that space at the presumed site of generation; both of these require-ments are met (see above). MILLER and DOWLING (1970) propose reasonably that the lateral spread of the b-wave is mediated through electrical coupling of Müller's cell apices at gap junctions in the external junctional layer.

An additional function of Müller's cells is the ensheathing of axons in the optic fiber layer, at least in most reptiles (PEDLER, 1963), birds (LADMAN and SOPER, 1962), amphibians and mammals (COHEN, 1963a). In many cases the enclosure is incomplete, so that large bundles of naked axons are held together loosely by the Müller's cell (COHEN, 1963a; PEDLER, 1963; VILLEGAS, 1964; SJÖSTRAND and NILSSON, 1964; VAN HARREVELD and KHATTAB, 1968). Since moderate activity of such loosely sheathed unmyelinated nerve fibers has been shown to cause significant changes in the concentration of K^+ in the extra-cellular fluid in *Necturus* optic nerve (KUFFLER et al., 1966; ORKAND et al., 1966), the possibility of interaction of some sort between these naked axons — even if not in direct contact with one another — must be considered.

In teleosts, at least, all axons in the optic fiber layer of the retina are wrapped in individual sheaths of loose myelin. The cells responsible for these sheaths are not the Müller's cells but small oligodendrocytic elements scattered throughout the optic fiber layer (O'Daly, 1967a). The presence of glial cells of other varieties in the retina will not be considered here. As we have seen earlier, although the horizontal cells of fishes present certain features usually associated with neuroglial cells, they are unquestionably neuronal in many significant respects. Lessell and Kuwabara (1963) may be consulted for a general review and discussion of retinal neuroglia.

G. Simpler Vertebrate Visual Organs

Although the cellular structure of median eyes is not well known, it is clearly much simpler than that of the lateral eyes (see reviews by Oksche, 1965; and Van de Kamer, 1965). In the ammocoete (larval lamprey), for example, Owsiannikow (1888) observed only two cell layers — one of "rods" and their cell bodies and one of ganglion cells and their axons — joined by a layer of fine neural processes. Retzius (1895) applied the Golgi method to the ammocoete; he was not able to observe the layering noted by Owsiannikow and succeeded mainly in impregnating supporting cells and some bipolar neurons with centripetal axons. Tretjakoff (1915) studied the same organ by intravital staining with methylene blue. He demonstrated cells of three types: sensory cells with long or short axons, the latter ending in branched terminals; ganglion cells with one or several long dendrites, also branching terminally; and "pigmented" and "basal" supporting cells. Kelly and Van de Kamer (1960) described in the frog a plexus of fibers, thought to be efferent, which after silver staining appear to form end-loops on the receptors and supporting cells in the frontal eye.

The fine structure of synapses in the median eye of the larval lamprey was described by Collin and Meiniel (1968). The photoreceptor terminals contain a synaptic lamella, surrounded by clear vesicles, and some dense-cored vesicles. Contacts at the ribbon are only superficial (not invaginated), with but a single postsynaptic process which seems in all cases to have the same ultrastructure. This observation implies that the only neurons in the median eye of the lamprey larva are the photoreceptors and ganglion cells. This situation should be extremely favorable for studying the simple synaptic output of the photoreceptor cells. Similar organization was observed in the median eye of the turtle by Vivien-Roels (1969).

The epiphyseal (pineal) and frontal (parapineal) eyes of frogs have been studied by a number of investigators. Eakin et al. (1963) described synaptic endings of receptors containing ribbons and vesicles but said nothing about their relations to other processes. Oksche and von Harnack (1963) described receptor, supporting and ganglion cells and a plexiform zone comprising terminals with synaptic lamellae and vesicles, invaginated postsynaptic processes, and some appositions of membranes having increased density. Kelly and Smith (1964) observed three types of photoreceptor synaptic junctions: (1) a ribbon synapse in which the postsynaptic process appears to be dendritic; (2) a ribbon synapse in which a postsynaptic process appears to be axonal, containing many small vesicles; and (3) a non-ribbon synapse in which the photoreceptor contains a subsurface cistern opposite a process containing many vesicles. These authors reconstructed the contacts of a portion of one receptor terminal, revealing a very complex pattern of invaginating processes of different types (to some extent these relation-ships are reminiscent of those shown in the first synaptic layer in the lateral eye of *Necturus* by DOWLING and WERBLIN 1969). Although KELLY and SMITH hesitated to identify anything but receptor and supporting cells, it appears from the variety of synaptic structures that not only ganglion cells but probably also processes of at least another type — intrinsic interneuron, efferent axon or receptor telodendria — make contact with receptor bases.

In birds (COLLIN, 1968), rat (WOLFE, 1965; ARSTILA, 1967), cat and monkey (WARTENBERG, 1968), synaptic ribbons and associated vesicles are found within some pinealocytes. There is no evidence that the ribbons are associated with synaptic junctions in these animals, in which the pinealocytes appear not to be photoreceptive.

Most of what is known about the function of photoreceptive pineal organs is derived from studies of spike discharges in the afferent nerve. The functional organization is rather simple (see review of DODT, 1966). HANYU et al. (1969) have recorded intracellular responses resembling S-potentials from the median eye of three species of teleosts. The possibility of recording simultaneously the activity of interneurons and ganglion cells in such a system, which may comprise cells of only three types, should encourage more detailed investigations.

Another very simple system is the diffuse cutaneous sensitivity to light of the lamprey (YOUNG, 1935). Ammocoetes respond to photic stimulation of the skin by a delayed body twitch. The afferent limb of this reflex travels through the lateral line nerve. Although the photoreceptive cells have not been identified in this system, it is possible that they resemble the photoreceptors of the lateral and median eyes in having a pigmented distal appendage of modified ciliary origin and an afferent ribbon synapse. Electro- and mechanoreceptive cells in the verte-brate acousticolateralis system are very similar to photoreceptors except in their lack of the specialized outer segment (see review in STELL, 1967). It is possible, therefore, that the activity of vertebrate photoreceptors and their synapses, undistorted by interactions with other cells close by, can be studied in this simple system by recording from the lateral line nerve and receptors.

IV. Experimental Correlation of Structural and Functional Organization

Functional circuitry cannot be revealed by structural information alone. One can hope, however, to approach a circuit by understanding the electrical or chemical function of individual circuit elements or small groups of elements whose structure and relationships are known. In a normal, fully developed retina this has been accomplished electrically by recording from single units, marking and identifying the cells morphologically, and making correlations with the fine structure of similar cells as revealed in special morphological studies (e.g. Golgi-EM). Even more powerful is the technique of stimulating and recording from pairs (or larger assemblies) of coupled identified cells (e.g. BAYLOR et al., 1971, KANEKO, 1971). The aim here is to reduce the circuitry examined to a manageable size and complexity by examining only a restricted part of it. The same purpose might be accomplished instead by starting with a system which is less complex, either by virtue of naturally primitive development (phylogenetic, developmental, or a special organ system) or by means of natural or induced regression or deficiency (genetic, pharmacological, or toxic). Only the approach of identifying single units in normal adult retina has been applied extensively. The function of synapses might be approached also by histochemical (see review by Lolley, 1969) and physiological studies of the localization or effects of chemicals which might be instrumental in synaptic transmission, such as acetylcholine and cholinesterase, catecholamines, and amino acids. In this section I shall discuss the direct correlation of structural and functional organization by means of physiological localization, dye injection, and the morphological localization and functional analysis of substances plausibly implicated is synaptic processes. The reader is referred also to recent reviews of retinal physiology by Imbert (1970) and Tomita (1970), in addition to other chapters in this Handbook.

A. Morphological Identification of Functional Units

In a tissue as neatly stratified as the vertebrate retina, one might hope to identify functional units simply by comparing their depth, or sequence during electrode penetration, to the known stratification of retinal units. This approach was applied with some success by Brown and Wiesel (1959), who localized units "physiologically" in the cat in the order of their appearance during penetration, and by Negishi, Svaetichin, and Parthe (personal communication), who find that in certain teleostean retinas the different types of S-potential are recorded at characteristically different relative depths. Since the retina may be deformed or displaced by a penetrating electrode, however, stable and identifiable references for the conversion from relative to absolute depth may not be available, especially in the isolated retina. Such behavior may account for Svaetichin's (1953) localization of teleostean hyperpolarizing slow potentials to cones, and of impulse-discharges to horizontal cells, on the basis of depth indicated on the electrode micromanipulator[2].

It is predictable that units which can be obtained frequently with micro-electrodes are relatively numerous, and that those which can be penetrated easily, held for a long time, and obtained with larger microelectrodes are rather large cells. Tomita's (1957) finding that S-potentials could be recorded in carp with pipettes ar large as 5 μm at the tip allowed the interpretation that these potentials were either intracellular recordings from extremely large cells or massed extra-cellularly recorded responses. Gouras (1960) noted, however, that although he could occasionally record S-potentials in bream with electrodes having tips as

---

[2] Motokawa (cf. 1970, p. 8) introduced the term "S-potential" in the late 1950's to designate a *class* of slow, local, perhaps intracellular responses to light. Electrode localization experiments had shown that at least some of these responses arose not from cones, as was first supposed, but from more proximal cells. The identity of those cells could not be established at that time. Contrary to common usage, therefore, "S-potential" does not mean "horizontal cell response". Some S-potentials are responses of horizontal cells; others may be responses of receptor or bipolar cells, for example (e.g. Bortoff, 1964). Responses should be identified by cell of origin if known; "S-potential" should be applied only to local slow potentials of unknown origin, as was originally intended.

large as 20 μm in diameter, the largest potentials were obtained with finer pipettes. His observation implied that the S-potentials were generated by moderately large cells but could be recorded quasi-intracellularly with large-tipped microelectrodes. MITARAI (1958) surveyed eight teleostean species. He was able to record S-potentials easily only in the four species in which the bipolar cells were scarce but horizontal cells were large and numerous. TOMITA (1963) noted that the "ease of obtaining S-potentials in fish, turtle, and frog agrees with the order of size of horizontal cells in these animal forms". A correlation between the numerous classes of S-potentials and assorted large horizontally-oriented cells in certain marine teleosts has been emphasized also by SVAETICHIN (1967) and TESTA (1966).

Naturally the functional properties of single units and the conditions under which they can be observed experimentally reflect the morphology of those units. TOMITA's (1965) earliest evidence that he had succeeded in recording from single cones was that the responses to a 100 μm spot and to diffuse illumination were not significantly different. While not ruling out interactions over distances less than 100 μm, this evidence did serve to differentiate them from other units generating typical S-potentials, such as horizontal cells, with their much larger receptive fields. OIKAWA et al. (1959) recorded slow potentials of two kinds in carp — from units with a large area of summation, interpreted as horizontal cells, and from units summing over a small area, tentatively identified as cones (but perhaps bipolar cells, cf. KANEKO, 1970). The studies of KANEKO and HASHIMOTO (1969) and KANEKO (1970) confirm in large part the localizations implied by such correlations. In various teleosts, MITARAI et al. (1961) and MITARAI (1963, 1964) observed that the S-potentials encountered closest to the photoreceptors had the characteristics of a photopic (cone) system, while the properties of some S-potentials recorded farther proximally were mixed (rod and cone). While horizontal cells of the distal row (or rows) are good candidates for generators of the purely photopic S-potentials, since they appear to be connected exclusively to cones, no teleostean horizontal cell is known to be connected to both rods and cones (STELL, 1966, 1967; PARTHE, 1970, 1972; LAUFER and STELL, unpublished observations). The mixed units might therefore be horizontal cells which receive some indirect input, not yet discerned morphologically, from photoreceptors of a second class; cells which are known to receive both rod and cone inputs directly, like the large bipolar cells of the goldfish (STELL, 1967); or summed independent activity of cells of two different types recorded partially extracellularly. Finally, KANEKO and HASHIMOTO (1968) were able to identify spike-generating units in the intermediate neuronal layer in the frog by showing that they responded antidromically to stimulation of the optic nerve, and GOURAS (1969) has obtained additional information about classes of ganglion cells in the monkey by determining conduction velocities for antidromic spikes in single units.

The most direct approach to identification of single units is the deposition of some marking substance, usually a dye, at the tip of the recording micro-electrode. The electrode is then withdrawn, the tissue is fixed and sectioned, and the marked cells are located and identified. Earlier attempts included the injection of crystal violet followed by frozen-sectioning (MAC NICHOL et al., 1957;

Mac Nichol and Svaetichin, 1958); lithium carmine detected by light microscopy (Mitarai, 1958, 1960, 1963, 1964; Mitarai et al., 1961; Svaetichin et al., 1961) or electron microscopy (Yamada and Ishikawa, 1965); silver ions followed by formalin fixation (Oikawa et al., 1959); ferricyanide followed by fixative containing ferrous ion, converting it to Turnbull's Blue (Tomita et al., 1959; Gouras, 1960; Lipetz, 1963); and Trypan Blue (Bortoff, 1964) or the closely related Niagara Sky Blue (Kaneko and Hashimoto, 1967, 1968; Byzov, 1968; Werblin and Dowling, 1969; Miller and Dowling, 1970). With most of these methods the localizations were inconstant because of difficulty in expelling the dye or recovering the marked cells, or they were insufficient for localization to a single cell because of diffusion of the marker. All failed to meet the criteria for reliable localization discussed by Kaneko and Hashimoto (1967). An important criterion is that the response be recorded even after injection of the dye. Although Lasansky and Fuortes (1969) met this with the other criteria in studies on leech photoreceptors using Niagara Sky Blue, other investigators have applied this dye with less care. Kaneko and Stell (unpublished observations) observed that lithium carmine is an unreliable localizing dye because it may be taken up by cells from the extracellular space; therefore a cell containing lithium carmine need not have been penetrated by the dye-filled pipette. Werblin and Dowling (1969) obtained improved fixation and visualization of unmarked cells by employing dichromate rather than acetate in the fixative for Niagara Sky Blue marking. But because this dye does not diffuse into fine cell processes in concentration sufficient to make them visible, even a single, well-injected cell in a good histological preparation (Werblin and Dowling, 1969) often cannot be positively identified.

Clearly, then, what is desired is a "dye which spreads into fine branches of cells, survives fixation and routine histological procedures, and permits reconstruction of cell shapes" (Stretton and Kravitz, 1968). Such are the Procion (triazinyl halide-coupled) dyes introduced for marking large invertebrate nerve cells by Stretton and Kravitz, and employed successfully by Kaneko (1970, 1971) to mark smaller neurons in the vertebrate retina. This method yields images of cells comparable to those obtained with the Golgi methods (Fig. 25). The structural interconnections of Procion-stained cells of a given type may thus be established by Golgi-EM studies of cells of the same type. Next priority should be given the development of a technique permitting ultrastructural as well as optical study of a dye-marked cell, as in the Golgi-EM procedure, so that one may determine the connections of specific cells of known functional organization.

Important information has been derived from recent dye-marking experiments. Responses judged to be cone potentials by their superficial location and negligible area of summation (Tomita, 1965) were localized to cones by means of Niagara Sky Blue (Kaneko and Hashimoto, 1967). With the same dye other units were localized to the intermediate neuronal layer in the frog (Kaneko and Hashimoto, 1968) and carp (Kaneko and Hashimoto, 1969), but because of the variety of cells in this layer and the failure of the dye to reveal fine processes, the cells of origin could not be identified. Bortoff (1964) and especially Werblin and Dowling (1969) were more successful at identifying single units in Necturus with dyes of this type because of the simplicity of the retina and the great size of its cells. In

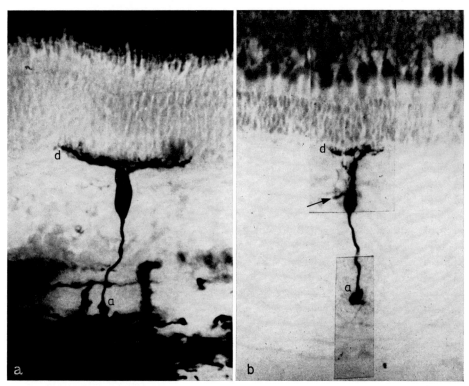

Fig. 25. Light micrographs of bipolar cells (mixed rod-cone type) from the retina of the goldfish. *d*, dendritic arbor in first synaptic (outer plexiform) layer; *a*, axon terminal near ganglion cell layer. *a*, entire cell impregnated by the rapid Golgi procedure, in a section 80 μm thick (reproduced from STELL (1966)). *b*, cell injected with Procion Yellow M—4RS by micro-iontophoresis from a recording micropipette; contrast reversed photographically to emphasize similarity to Golgi preparation. This is a montage of photographs of several sections through the same cell. Small cylindrical process (arrow) may have been stained by leakage of dye from the marked cell (reproduced from KANEKO (1970) by permission of the author and the Journal of Physiology)

particular, these authors were able to distinguish marked horizontal, bipolar and amacrine cells from one another by consistent differences in their shape and location. Hyperpolarizing responses of units with small receptive fields were localized to photoreceptors; of these units with large field, to horizontal cells. Graded slow potentials with opposite polarities for focal and annular stimulation were localized to bipolar cells. Units with only transient depolarization and spike discharge at *on* and *off* were found to be amacrine cells; responses localized to ganglion cells were usually transient but differed in the form and polarity of the slow potential changes and in the duration of the spike discharges. Subtypes of cells, e.g. rods vs. cones, were not distinguished from one another either morphologically or physiologically in these studies, however. Single units having physiological properties similar to those in *Necturus*, described in the retina of cyprinid fish by KANEKO and HASHIMOTO (1969), were localized to cells of the same classes as for *Necturus* by Kaneko (1970), using the fluorescent dye Procion Yellow M4RS. These elegant studies prove the feasibility of recording from and distinguishing morphologically different subtypes of cells having perikarya no more than 10 μm in diameter, although there is no evidence that the responses of either small (pure-cone) bipolars or rod horizontal cells have been recorded in cyprinids. It is worthy of note that although the spectral response functions of all the units studied by Kaneko and Hashimoto (1969) seemed to reflect cone activity only, Kaneko (personal communication) has observed more recently that the responses of large bipolar cells in goldfish are a function of the rod system in well dark-adapted retinas. Thus the mixed rod-cone input to these cells predicted from Golgi-EM studies (Stell, 1967) is confirmed.

It is clear that many S-potentials are the intracellularly recorded responses of horizontal cells. Kaneko (1971) has shown most clearly that this is the origin of L-type responses in the dogfish shark, *Mustelus*, using Procion dyes. Kaneko (1970), however, observed unexpectedly that S-potentials of either the L- or C-type may be localized either to horizontal cells of the first row or to cylindrical processes (Cajal's internal horizontal cells) in goldfish retina by electrophoretic injection of Procion Yellow. These localizations should be repeated in order to determine whether some as yet unknown connections of the cylindrical processes and horizontal cells, or artifacts of the marking method, are responsible for Kaneko's observations.

Svaetichin, Negishi, and Parthe (personal communication) have localized S-potentials with different spectral properties to horizontal cells of different rows in several species of Caribbean teleosts. These investigators have found injection under pressure to be superior to microelectrophoresis for localizing teleostean S-potentials with lithium carmine. This technique, however, is clearly applicable only to cells large enough to be penetrated and held with large micropipettes, and because lithium carmine fails to delineate the finer cell processes, it is useful only in the case of these gigantic, regularly stratified cells, which can easily be identified and correlated with their structure revealed in Golgi preparations.

B. Histochemical Localization and Pharmacology of Substances Possibly Related to Synaptic Transmission

Anatomic and physiologic data imply that transmission in most retinal synapses is chemical rather than electrical. Histochemical and pharmacologic methods have been applied in attempts to identify and localize substances which may play a role in synaptic transmission.

Acetylcholine and Acetylcholinesterase

Anfinsen (1944) analyzed fresh-frozen sections of bovine retina by microchemical methods and found acetylcholinesterase (AChE) in large amounts, primarily in the two synaptic layers. Koelle and Friedenwald (1950) localized AChE activity histologically in the vertebrate retina mainly to bipolar cells, but their method was not specific for AChE and was subject to diffusion artifacts. With an improved method, Koelle et al. (1952) demonstrated specific AChE activity only in the cell bodies and processes of amacrine cells in the cat retina. HEBB et al. (1953), studying the rabbit retina with KOELLE's methods, observed activity in cell bodies in the layers of intermediate neurons and ganglion cells, but not in either synaptic layer. FRANCIS (1953), using a modification of KOELLE's procedure, found no enzyme activity in the frog; activity in both synaptic layers, primarily the second, in minnow and sheep; and activity in the second synaptic layer only, in pigeon, chick, cat, guinea pig, rat, rabbit and pig. The stratification of reaction product in the pigeon and chick, and the diffuseness of it in the cat and guinea pig, paralleled the appearance of the second synaptic layer in ordinary histologic sections stained with silver. In sheep and rabbit, some cells of the proximal margin of the layer of intermediate neurons (amacrine cells ?) were positive for the enzyme. LEPLAT and GEREBTZOFF (1956), also using a modified KOELLE's method, demonstrated mainly diffuse positivity only in the second synaptic layer and optic nerve of cat and rabbit. EICHNER (1956) found activity mainly in the distal and proximal regions of the second synaptic layer, and inconstantly in the first synaptic layer, in the retina of man. Later, EICHNER (1958), using GEREBTZOFF's method, observed strong AChE activity in both synaptic layers and mild activity in the cellular layers in man. ERÄNKÖ et al. (1961), however, observed some stratification of end-product in the second synaptic layer of frog, rat, mouse and rabbit, and found some faint activity in the first synaptic layer after prolonged incubation in the substrate mixture. Some cells in the amacrine and ganglion cell layers also were stained. VIALE and APPONI (1961), using GEREBTZOFF's technique, observed strong activity in the ganglion cell somata and three broad regions of the second synaptic layer in man. They observed *two* layers of moderate activity in the first synaptic layer. ESILÄ (1963) demonstrated AChE activity in the second synaptic layer of several mammals, including man, attributed mainly to amacrine but also to bipolar cells. In the rat, LEWIS and SHUTE (1965) observed AChE activity in the nerve fiber layer and ganglion cells, throughout the second synaptic layer, and in all intermediate neurons. Because activity was only "small" in the ganglion cells, and because LEWIS and SHUTE had found AChE activity in the endoplasmic reticulum of all known cholinergic cells studied elsewhere in the nervous system they concluded that the enzyme is located in the centrifugal fibers. This conclusion is inconsistent with the number and distribution of centrifugal fibers, as discussed earlier.

NICHOLS and KOELLE (1969) studied AChE and nonspecific ChE in the retina of the pigeon and several mammals (Fig. 26). In the mainly cone retinas of the pigeon and ground squirrel they found two dense layers (sometimes 3—4 in pigeon) of AChE activity in the second synaptic layer; both ordinary and displaced amacrine cell bodies also were positive. In the primarily rod retinas of the rabbit, rat, and cat the amacrine cells were not clearly positive, and staining of the second synaptic layer was diffuse except for two dense bands at the extremes of the diffuse layer in the rabbit. No nonspecific ChE activity was found in the second synaptic layer in any species. In the first synaptic layer, on the contrary, all ChE activity was found to be nonspecific, and this was only present in the pigeon and ground squirrel. Moderate AChE positivity was detected in the isthmo-optic nucleus but could not be traced into the retina. The authors concluded that the specific AChE activity is localized mainly to amacrine cells and suggested that it mediates the release of a non-cholinergic inhibitory transmitter.

The localization of acetylcholinesterase mainly to amacrine and ganglion cells was suggested also by other experiments. The appearance of the enzyme was followed in developing frog retina by BOELL et al. (1955) and in chick by SHEN et al. (1956). Activity in both cases appears first in the cytoplasm of the ganglion cells, then in the second synaptic layer and the amacrine cells. The staining of the second synaptic layer is at first diffuse, later stratified. In the frog, at least, much of the

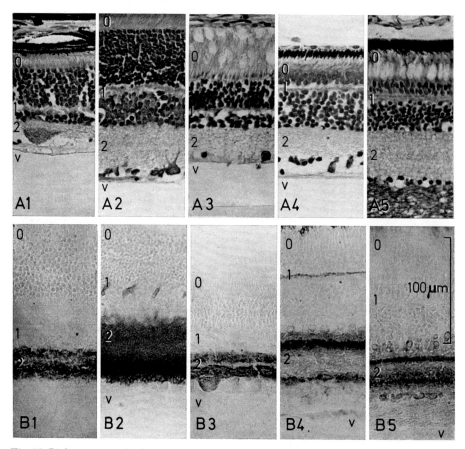

Fig. 26. Light micrographs demonstrating the presence of acetylcholinesterase histochemically in vertically sectioned retinas. *A*, normal histology in sections stained with cresyl violet; *B*, sections stained for cholinesterase activity by the gold thiocholine method. *A 1* and *B 1* cat; *A 2* and *B 2* rat; *A 3* and *B 3* rabbit; *A 4* and *B 4* ground squirrel; *A 5* and *B 5* pigeon. *0* layer of photoreceptor outer segments; *1* first synaptic (outer plexiform) layer; *2* second synaptic (inner plexiform) layer; *V* vitreous body. Magnification is the same for all figures, which are mounted with distal margin of second synaptic layer aligned. All activity in second synaptic layer and adjacent amacrine (in *B 4* and *B 5* only) and ganglion cells is specific for acetylcholinesterase. Activity in first synaptic layer, *B 4*, and in capillary endothelial cells in and near the first synaptic layer in *B 2*, is due to non-specific cholinesterase (reproduced, slightly modified, from NICHOLS and KOELLE (1969), by permission of the authors and the Wistar Press)

activity in the synaptic layer is probably derived from the ganglion cells, since it appears at a stage of development when little activity is observed in amacrine cell bodies. Some faint but definite activity was seen also in the first synaptic layer in the frog.

RAVIOLA and RAVIOLA (1962) observed that in the adult rabbit acetyl-cholinesterase activity is greatest in the ganglion cell layer and in two strata of the second synaptic layer. Activity is less intense in neonatal rabbits but similar to that of adults in distribution; however, the synaptic stratum showing higher activity is the more proximal one in neonates but the more distal one in adults. Activity increases most markedly during the sixth to eighth postnatal days and attains the adult level and distribution by the eleventh day. The authors note that the electroretionogram first appears in the rabbit around the eighth postnatal day (NOELL, 1958). In young rabbits, LEE et al. (1967) observed that AChE activity appeared first in ganglion cell perikarya. During development it became very strong in ganglion cells and moderately strong in some cell bodies in the layer of intermediate neurons and in the proximal part of the second synaptic layer. Small but definite AChE activity was seen also in the first synaptic layer. A relation to visual function was indicated by the effect of suturing the eyelids for various durations. During eight weeks of eyelid closure the enzyme activity was reduced gradually until barely detectable. After the eyelids were reopened, AChE activity recovered somewhat in about three days and appeared normal in as little as seven days. NICHOLS and KOELLE (1967) performed related experiments on rabbit retina, also showing that AChE is produced by both amacrine and ganglion cells. In normal adult rabbits, activity was localized mainly in two broad bands in the second synaptic layer and in ganglion cell perikarya; only a "suggestion" of positive reaction was seen in amacrine cell bodies. Three to six months after sectioning the optic nerve they observed exactly the same pattern except for the absence of staining among the ganglion cells. This implied "that the amacrine cells are the major source of the AChE" of the second synaptic layer. That ganglion cells as well as amacrines synthesize enzyme was confirmed by another experiment. Five to seven hours after intravenous injection of DFP, an irreversible inhibitor of AChE, enzyme activity was found only in the perikarya of amacrine and ganglion cells; it was found in the synaptic regions only later in recovery.

LIBERMAN (1962) determined the content of acetylcholinesterase quanti-tatively in newborn rats raised in the dark from the age of three days to 17 weeks. The total activity of true AChE was less in these than in light-reared controls, while non-specific cholinesterase and glycolytic activity were identical in the two groups. GLOW and ROSE (1964) followed *de novo* synthesis of AChE quanti-tatively in rats after injection of DFP. One eye of each animal was occluded and sutured shut while the other was illuminated continuously. Although the initial loss of activity was the same in both eyes, activity recovered more rapidly in the illuminated eye. The authors' conclusion that synthesis of enzyme was induced by acetylcholine released during light-dependent synaptic activity is reasonable although not the only one possible.

Pharmacophysiological information on the retinal cholinergic system is scarce. NOELL and LASANSKY (1959) reported that optic fiber discharges in intact rabbit and cat retina are qualitatively similar to Renshaw discharge in response to acetylcholine, nicotine, prostigmine or procaine applied microiontophoretically near the ganglion cell bodies. Val'tsev (1965) observed lasting depression of the b-wave and transient depression followed by enhancement of the d-wave of the electroretinogram in isolated perfused frog retina, with atropine but not hexamethonium. Straschill (1968) found that the ganglion cell discharge in the intact barbiturate anesthetized cat is evoked by eserine and excited or inhibited by acetylcholine, injected intra-arterially. Cholinolytic agents blocked these effects of acetylcholine but had no effect on light-induced activity of ganglion cells. These and the extensive observations of Ames and Pollen (1969) on ganglion cell discharge in isolated perfused rabbit retina imply that there are one or more kinds of cholinergic synapses along both excitatory and inhibitory pathways from photoreceptors to ganglion cells. The specific synapses having these properties have not been identified.

Catecholamines

Early attempts to localize catecholamines and related substances in the retina were not remarkably successful. For example, Leplat and Gerebtzoff (1956) were able to localize certain amines (possibly adrenaline or noradrenaline) only to the outer segments and nuclei of photoreceptor cells in cat and rabbit. Eränkö et al. (1961) localized activity of monoamine oxidase (MAO) to photoreceptors and both synaptic layers in the frog and several mammals but concluded that the localization was nonspecific because it was not altered by inhibitors. Kojima et al. (1961) found MAO activity highest in Müller's fibers at the level of the "inner layers and outer reticular layer" in human retina. Mitarai (1963) localized MAO to Müller's cells in fish retina and claimed that the distribution and amount of activity depend upon the state of adaptation, being concentrated in the proximal foot-pieces of these cells in dark-adapted, but distributed throughout the cells (total activity increased) in light-adapted retina. Neither of the latter two studies, however, demonstrated satisfactorily that the enzyme was specific for monoamines.

Catecholamine (CA) content in the retina (plus pigment epithelium) of toad, frog and rabbit was determined quantitatively by Drujan et al. (1965). Total CA content was greater in light- than in dark-adapted retinas of all three animals. Noradrenaline could not be detected in frog and toad and was present only in trace amounts in the rabbit. Adrenaline was present in significant amounts in the frog and toad but only in traces in the rabbit. Total content of adrenaline was greater in the dark-adapted than in the light-adapted toad. Substantial amounts of dopamine were present in all three animals, more in dark- than in light-adapted ones. Drujan and Díaz Bórges (1968) observed that continuous exposure of dark-adapted toads to bright light leads to depletion of 50% (maximum, implying that a fraction is bound) of the adrenaline in the retina within minutes. Repletion is more rapid in the dark than in the light.

Studies of catecholamines have been advanced considerably with the introduction of the formaldehyde-condensation, induced fluorescence method of Falck and Hillarp (Falck et al., 1962; see also Corrodi and Jonsson, 1967). It was first applied successfully to the retina by Malmfors (1963), who observed a plexus of fluorescent (CA-containing) fibers terminating around non-fluorescent cell bodies at the proximal margin of the intermediate neuronal layer in the rat. Some amacrine cell bodies and processes were also fluorescent. The fluorescence was unchanged 48 hours after opticotomy and bilateral cervical sympathectomy, and therefore probably a property of neurons having cell bodies within the retina. The effects of inhibitors suggested that the fluorescence was due to a primary amine, probably dopamine or norepinephrine. HÄGGENDAL and MALMFORS (1965) showed by a biochemical method that the dominant substance in rabbit, as in toad and frog (see DRUJAN et al., 1965), was dopamine. All fluorescence in rabbit was localized to one of three strata in the second synaptic layer: one adjacent to the layer of intermediate neurons — this was the layer most richly supplied with fibers and most intensely fluorescent — and two of lesser density and intensity in the middle and most proximal regions of the second synaptic layer. The histochemical controls used by MALMFORS were reapplied here, with the same results. In the rabbit, cat, and Cebus monkey, LATIES and JACOBOWITZ (1966) obtained results similar to those of HÄGGENDAL and MALMFORS. In cat and monkey the proximal two synaptic strata were much less marked than in the rabbit, and a "modest number" of fluorescent cell bodies were observed in the ganglionic and second synaptic layers. EHINGER (1966a, b) expanded and systematized such data after observing retinas of rat, mouse, cat, rabbit, guinea pig and Cynomolgus monkey with the Falck-Hillarp technique. He designated the distal, intermediate and proximal fluorescent plexuses, respectively as the "outer, middle, and inner adrenergic fiber layers" (AFL). The outer AFL, prominent in all six species, included processes forming basket-like enclosures around non-fluorescent perikarya in the amacrine cell layer, as previously described by MALMFORS. The middle AFL, seen well only on the rat, mouse, rabbit, and guinea pig, comprised processes from ordinary adrenergic amacrine cells as well as "adrenergic eremite (stellate) cells" confined to the middle AFL. The inner AFL, less marked than the outer AFL but more so than the middle AFL, was prominent only in the rabbit. In all species EHINGER observed "alloganglionic" cells — i.e. fluorescent cells which appeared otherwise similar to other, non-fluorescent cells in the ganglion cell layer.

EHINGER and FALCK (1969) observed marked differences between the adrenergic systems in retinas of Old World (Cynomolgus) and New World, or platyrhine (Aotes, Saimiri, Saguinas and Cebus) monkeys. CA-containing cells and fibers were scarce in Cynomolgus, and only the outermost AFL was seen. In Cebus, on the other hand, all three AFL's were prominent, and in the other three species both the outer and middle AFL were visible. In all of the platyrhine monkeys, fluorescent "adrenergic junctional cells" constituted as much as 10% of all cells in the same layer; alloganglion cells were far less common, and eremite cells were encountered only rarely. Most striking in the platyrhines was a plexus of adrenergic fibers the majority of which were confined to the proximal part of the intermediate neuronal layer. In Cebus and spider monkey some adrenergic fibers extended as far distally as the cone synaptic endings in the first synaptic layer. A "scant but definite adrenergic plexus" had been seen earlier in Saimiri (EHINGER et al., 1969) and Cebus (LATIES and JACOBOWITZ, 1966). Cebus was unique in having numerous "adrenergic pleomorph cells" — large neurons, variable in form and location in the interneuron layer, from which fine processes extended horizontally and to both synaptic layers. It is remarkable that the retinas of these New World monkeys, which are adapted structurally and functionally to a wide range of habitats, are nevertheless more similar to one another than to the retina of *Cynomolgus* in the development of the adrenergic system. The authors stress with justification the hazard of drawing conclusions about one species (of primates) from studies on another.

SANO et al. (1968) observed the usual three adrenergic fiber layers in the second synaptic layer of the dog retina, the most distal being most intense, and observed both fluorescent and non-fluorescent amacrine and ganglion cell bodies. After intravitreal injection of dopamine, a much larger number of amacrine cells ("almost all') were fluorescent. A relationship between dopamine and retinal function, first suggested by MALMFORS (1953) in relation to reserpine-induced photophobia, was implied more directly by the data of NICHOLS et al. (1967). In albino rabbit and guinea pig these investigators observed greater fluorescence of cells and fibers after one hour in the light than after 24 hours in total darkness; since the fluorescence was already very intense in dark-adapted rats, no increase could be detected in these animals after light-adaptation. In all three animals, however, biochemical measurements showed an increase in dopamine content after exposure to light. KRAMER (1971), in contrast, showed that tritiated dopamine is taken up by the cat's retina in the dark and released in the light. By autoradiography, KRAMER et al. (1971) localized dopamine mainly to processes near the junction of the intermediate neuronal and second synaptic layers, as well as to some perikarya in both amacrine and ganglion cell layers.

In the pigeon, chicken and duck, EHINGER (1967), using the method of FALCK and HILLARP, observed the same three adrenergic fiber layers (AFL) of the second synaptic layer as in mammals. Fibers of the outer AFL, the densest and broadest of the three, often formed baskets around non-fluorescent cell bodies in the amacrine cell layer but never penetrated farther than halfway into the intermediate neuronal layer. Fibers of the middle AFL were intermediate also in density and intensity; those of the inner AFL, very sparse. Both large and small fluorescent cell bodies were seen in the amacrine cell layer, and cells of both classes appeared to contribute fibers to all three AFL's. About five sublayers can be shown by means of phase contrast microscopy in the second synaptic layer of avian retinas treated by the Falck-Hillarp procedure. It is very difficult to correlate these five layers with the three fluorescent ones, even by careful comparison of the same section with the two methods: it appears most likely that the fluorescent layers correspond to the first, third and fifth dark layers shown in the positive phase contrast micrograph.

In a most interesting study with the technique of FALCK and HILLARP on the retina of teleosts, mainly perch and trout, EHINGER et al. (1969) also observed three adrenergic fiber layers in the second synaptic layer. The inner AFL was most intense and constant, the outer less so, and the middle difficult to discern regularly. They noted also some fluorescent cell bodies and occasional alloganglion cells and demonstrated a remarkable plexus of fluorescent fibers forming "baskets of terminals of synaptic character around the middle horizontal cells" (Fig. 27) which resembles somewhat the adrenergic plexus observed in this layer in New World monkeys by EHINGER and FALCK (1969). In perch and cichlid fish,

EHINGER et al. (1969) demonstrated fluorescent fibers interconnecting cells in the
amacrine cell layer with the plexus among the horizontal cells; the number of
fluorescent fibers was equal to the number of fluorescent amacrine cell perikarya
"in favourable preparations". These cells are comparable to the "small stellate
cells" of CAJAL (1892; see above), except that in his account the somata were not
located in the amacrine cell layer. Neither he nor we (PARTHE, LAUFER, and STELL,
unpublished observations) ever observed a direct connection between the cell
bodies in the amacrine cell layer and the fiber plexus among the horizontal cells,

a

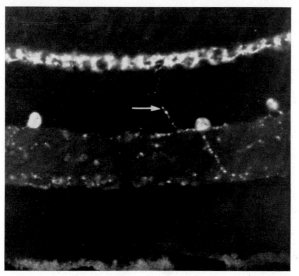

b

Fig. 27. Light micrographs of teleostean retinas prepared by the Falck-Hillarp procedure to
demonstrate catecholamines by fluorescence. *a,* survey micrograph of perch retina showing
frequency of fibers crossing intermediate neuronal layer (arrows); fluorescence enhanced by
pretreatment with β-methyl-noradrenaline. $\times$ 280. *b,* cell in amacrine cell layer of cichlid fish
retina sends a process (arrow) distally toward adrenergic plexus surrounding horizontal cells.
$\times$ 160 (reproduced from EHINGER et al. (1969), by permission of the authors. Springer-Verlag)

in Golgi preparations of a variety of teleostean retinas, although Boycott (personal communication) has seen such a cell in goldfish. The conclusion that this new adrenergic plexus "lays the morphological groundwork for interaction between cells and fibers of the inner and outer plexiform layers" seems to be premature. The authors demonstrated by microspectrofluorometry that the CA in the fluorescent condensate is dopamine.

Amphibians and reptiles have been examined with this technique by Laties (personal communication). He has observed that in the frog nearly all the catecholamine is contained in a few cells scattered in the amacrine layer; a weakly fluorescent outer AFL is seen near this layer. In *Necturus*, CA-containing processes ramify very widely in the second synaptic layer, and some fluorescent processes ascend to the first synaptic layer as in fish and platyrhines (see above). These ascending processes have been observed only rarely to date, however, in *Necturus*. In the reptile, *Gecko gecko*, Laties has demonstrated a population of amacrine cells rich in CA-containing elements (estimated 1/15 of all amacrines), and occasional CA-containing cells in the ganglion cell layer. He has not yet observed any CA-containing processes to the first synaptic layer of *Gecko*. He has observed catecholamine-containing amacrine cells also in the retinas of several snakes.

An interesting correlation has been established between catecholamine content, as determined by fluorescence or chemical methods, and the appearance

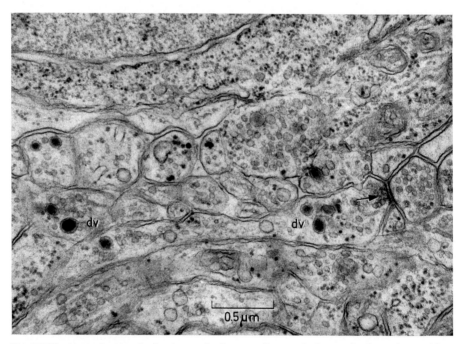

Fig. 28. Electron micrograph of processes in the second synaptic (inner plexiform) layer of frog retina containing large, dense-cored vesicles (*d v*). One process containing them appears to make a conventional synaptic contact (arrow) upon another process; the vesicles close to this contact are typically clear and small (reproduced from Dowling (1968), by permission of the author and the Royal Society of London)

in the electron microscope of large vesicles containing dense centers, called "dense-cored", "granular" or "nucleated" vesicles (see e.g. IRALDI and DE ROBERTIS, 1964 for review). They were first described in the retina by MALMFORS (1963), who observed them in some cells and processes in the amacrine and second synaptic layers of the rat. IRALDI and ETCHEVERRY (1967) described dense-cored vesicles of different sizes in structures identified as amacrine cell processes and efferent terminals in the rat, and noted some also in the perikarya in the amacrine and ganglion cell layers. They also cited their unpublished studies showing such vesicles in the processes in pigeon retina identified as centrifugal fibers by DOWLING and COWAN (1966). OLNEY (1968b) found dense-cored vesicles in cell bodies in the ganglion and amacrine cell layers and in processes in the second synaptic layer, but only "rarely ... close to the synaptic cleft". Similarly, dense-cored vesicles were seen often in the carp in amacrine cell bodies but only rarely in their synaptic terminals (DOWLING and WITKOVSKY, 1969). They were noted also in some processes in the second synaptic layer of the marine teleost *Mugil* by O'DALY (1967a). DOWLING (1968) found a few large dense-cored vesicles in some processes of amacrine cells in the frog (Fig. 28), and DUBIN (1969) observed them in some amacrine cell processes in each of the eight vertebrates which he surveyed.

Dense-cored vesicles have not yet been described in a distribution corresponding to the peri-horizontal cell network of EHINGER et al. (1969), although WITKOVSKY and DOWLING (1969) reported such vesicles in horizontal cell perikarya and processes in carp. According to O'DALY's (1967a) description of *Mugil*, the only structural elements having suitable architecture and relationships to be the pericellular network are the masses of extracellular tubules which he designated "undulant amacrine cells". OLNEY (1968) observed occasional dense-cored vesicles in the perikarya of horizontal cells in the mouse. The identification of certain electron-dense organelles in human horizontal cells as "dense-cored vesicles" by DIETERICH (1969) is not justified by his published electron micrographs. SOBRINO and GALLEGO (1968—1969), however, have observed dense-cored vesicles in horizontal cell processes as well as "bipolar cell" perikarya in the cat, rabbit, and monkey.

Electrophysiological evidence also suggests that catecholamines act as transmitters in retinal synapses but fails to indicate which synapses are involved. STRASCHILL (1968) found that intra-arterial injection of noradrenaline into the intact barbiturate-anesthetized cat excited, while serotonin and ergometrine inhibited and D-amphetamine either excited or inhibited, ganglion cell discharge. Adrenolytic agents blocked the effect of injected noradrenaline and D-amphetamine, but neither they nor reserpine had any effect on light-induced ganglion activity. In isolated perfused rabbit retina, AMES and POLLEN (1969) observed that noradrenaline and dopamine generally enhanced off-responses and inhibited on-responses of ganglion cells, and that the α-adrenergic blocking agent phentolamine had the opposite effect. Although these data suggest that an adrenergic synapse is involved in horizontal circuits such as those which generate on-off and center-surround firing patterns — consistent with the localization of dopamine to amacrine cells — the authors emphasize that the effects of propanolol and cocaine are not explained completely in terms of this system.

Amino Acids

KURIYAMA et al. (1968) noted that γ-aminobutyric acid (GABA) is present in the retina of dog and ox and that its absence from the optic nerve implies that it is a product of intrinsic retinal neurons. They found GABA, GABA transaminase and glutamic acid transaminase in whole detached rabbit retina, although in only a fraction of their activity in whole brain. Analyzing frozen sections quantitatively they determined GABA and glutamic transaminase activity to be highest in the second synaptic (and ganglion cell) layer. Assuming that the photoreceptor and bipolar cells are excitatory, they suggested the associational cells of GALLEGO and CRUZ (1965) and horizontal and amacrine cells as likely GABA neurons. According to GRAHAM et al. (1968), retinal GABA content falls with illumination.

KISHIDA and NAKA (1967) concluded from their experiments on isolated perfused bullfrog retina that amino acids only influence excitability at bipolar-ganglion cell synapses and probably do not act as transmitters. STRASCHILL (1968), nevertheless, found intra-arterial strychnine and L-glutamate to excite, and GABA to inhibit, ganglion cell discharge in intact barbiturate-anesthetized cats. Strychnine, however, failed to abolish light-induced inhibition. In isolated perfused rabbit retina, glycine and GABA inhibited both spontaneous and evoked ganglion cell activity, while glutamate increased transiently only the spontaneous activity (AMES and POLLEN, 1969). Since very high concentrations were required, the role of these amino acids in synaptic transmission is open to question. Although picrotoxin and strychnine enhanced both spontaneous and evoked activity at low concentrations, uncertainty about their mode of action prohibits a clear interpretation. SILLMAN et al. (1969) observed that isotonic sodium aspartate or sodium glutamate applied to the receptor surface of isolated frog retina eliminates all of the electroretinogram except component P III (distal), thought to reflect the activity of receptor cells alone. This effect resembles closely the effect of an excess of chemical transmitter on post-synaptic cells, and MURAKAMI (personal communication) has found that sodium aspartate depolarizes horizontal cells in carp. While these observations are consistent with amino acid-mediated synaptic transmission from photoreceptor to bipolar cell, they might be explained equally well by a less specific effect on synaptic (chemical *or* electrical) membranes to that proposed by KISHIDA and NAKA.

V. Closing Remarks

The study of the interconnections of retinal cells has advanced remarkably in the past 5 or 10 years. The factors responsible for this have been especially: (1) greater use of classical neurohistological methods for examining the architecture of entire cells; (2) increasing correlation of the fine structure of specific cellular junctions with the architecture of cells forming them; (3) improvement of microelectrode techniques so that it may be possible to record from any retinal cell; (4) development of superior methods for morphological identification of cells which have been characterized functionally in microelectrode studies. These advances are such that more complete understanding of retinal interconnections will be most limited by the availability of better molecular-chemical methods for analyzing specific cells or parts of cells, especially their synaptic membranes, and chemical transmitter substances.

Numerous problems, nevertheless, await solutions even with techniques currently available. Among these are: the function of double receptors and the nature of their interconnections; the function of interreceptoral (telodendronal) contacts; the function of the synaptic ribbon; the role for invagination of post-synaptic processes at receptor synapses; the pre- and post-synaptic interactions in complex ribbon-synaptic junctions, and in particular the site and mechanism of horizontal cell action in these complexes; the synaptic mechanism at basal junctions; the question of the function and ubiquity of horizontal cell axons; the nature of horizontal cell coupling at mutual close appositions — electrical or metabolic — and the role of acetylcholinesterase and nucleoside phosphatase at those appositions; the chemical synaptic transmitter at any synaptic junction in the retina, and the role of acetylcholine, dopamine, and amino acids in the retina; pathways responsible for nonlinear addition of S-potentials; the function of teleostean cylindrical processes, in particular their relation to S-potenials and its basis; identity and function of the network of dopaminergic fibers around teleostean horizontal cells; the role and origin of small extracellular tubules in teleostean retinas; the function of Landolt's clubs and the bipolar cells which bear them; the occurrence and function of efferent fiber systems; the role of stratification of processes and synapses in the second synaptic layer; differences in interconnections in the second synaptic layer (either individual junctions or multicellular patterns) of cells of different subtypes or cells in different sublayers; the prevalence and role of such unusual cells as the "associational amacrines" of CAJAL and the "associational ganglion cells" of GALLEGO and CRUZ; the identification of the structural elements and their connections which subserve adaptation or specific analytic functions such as color, brightness, contour, and movement; and so on.

It should be clear that it is no longer useful or interesting to the physiologist merely to describe the histology of the retina or to observe the ultrastructure of some cells or cellular junctions encountered at random. What must be known are the structural and functional relationships between identified pairs of cells or larger groups of cells. For the greater part, the methods needed for revealing these relationships are available and need only to be judiciously and diligently applied.

VI. References

AICHEL, O.: Zur Kenntnis des histologischen Baues einiger Teleostier. Inaug.-Diss. Erlangen (1896). Cited by FRANZ (1913).

AKERT, K., STEIGER, U.: Über Glomeruli im Zentralnervensystem von Vertebraten und Invertebraten. Schweiz. Arch. Neurol. Neurochir. Psychiat. **100**, 321—337 (1967).

ALI, M. A., ANCTIL, M.: Correlation entre la structure rétinienne et l'habitat chez *Stizostedion vitreum vitreum* et *S. canadense.* J. Fish. Res. Bd. Canada **25**, 2001—2003 (1968).

— — MOHIDEEN, H. M.: Structure rétinienne et la vascularisation intraoculaire chez quelques poissons marins de la région de Gaspé. Canad. J. Zool. **46**, 729—745 (1968).

— HANYU, I.: A comparative study of retinal structure in some fishes from moderately deep waters of the Western North Atlantic. Canad. J. Zool. **41**, 225—241 (1963).

ALLEN, R. A.: Isolated cilia in inner retinal neurons and in retinal pigment epithelium. J. Ultrastruct. Res. **12**, 730—747 (1965).

Allen, R. A.: The retinal bipolar cells and their synapses in the inner plexiform layer. In: Straatsma, B. R., Hall, M. O., Allen, R. A., Crescitelli, F. (Eds.): The retina: morphology, function and clinical characteristics. UCLA Forum in Medical Sciences No. 8, pp. 101—143. Berkeley-Los Angeles: University of California Press 1969.

Ames, A., III.: Studies in water and electrolytes in nervous tissue. II. Effect of glutamate and glutamine. J. Neurophysiol. 19, 213—223 (1956).

— Hastings, A. B.: Studies in water and electrolytes in nervous tissue. I. Rabbit retina: Methods and interpretation of data. J. Neurophysiol. 19, 201—212 (1956).

— Pollen, D. A.: Neurotransmission in central nervous tissue: a study of isolated rabbit retina. J. Neurophysiol. 32, 424—442 (1969).

Anctil, M.: Structure de la rétine chez quelques téléostéens marins du plateau continental. J. Fish. Res. Bd. Canada 26, 597—628 (1969).

Anfinsen, C. B.: The distribution of cholinesterase in the bovine retina. J. biol. Chem. 152, 267—278 (1944).

Armstrong, J. A.: An experimental study of the visual pathways in a reptile (Lacerta vivipara). J. Anat. 84, 146—167 (1950).

Arstila, A. U.: Electron microscopic studies on the structure and histochemistry of the pineal gland of the rat. Neuroendocrinology Suppl. 2, 101 pp. (1967).

Baylor, D. A., Fuortes, M. G. F.: Electrical responses of single cones in the retina of the turtle. J. Physiol. (Lond.) 207, 77—92 (1970).

— — O'Bryan, P. M.: Receptive fields of cones in the retina of the turtle. J. Physiol. 214, 265—294 (1971).

— Nicholls, J. G.: Changes in extracellular potassium concentration produced by neuronal activity in the central nervous system of the leech. J. Physiol. (Lond.) 203, 555—569 (1969).

Berger, E. R.: Subsurface membranes in paired cone photoreceptor inner segments of adult and neonatal Lebistes retinae. J. Ultrastruct. Res. 17, 220—232 (1967).

Blackstad, T. W.: Mapping of experimental axon degeneration by electron microscopy of Golgi preparations. Z. Zellforsch. 67, 819—834 (1965).

Blaxter, J. H. S., Jones, M. P.: The development of the retina and retinomotor responses in the herring. J. Mar. Biol. Ass. U.K. 47, 677—697 (1967).

— Staines, M.: Pure-cone retinae and retinomotor responses in larval teleosts. J. Mar. Biol. Ass. U.K. 50, 449—460 (1970).

Boell, E. J., Greenfield, P., Shen, S. C.: Development of cholinesterase in the optic lobes of the frog (Rana pipiens). J. exp. Zool. 129, 415—452 (1955).

Borovyagin, V. L.: Submikroskopicheskaya morfologiya i strukturnaya vzaimosvyaz' retseptornykh i gorizontal'nykh kletok setchatki ryada nizhnykh pozvonochnykh. Biofizika 11, 810—817 (1966). English translation: Submicroscopic morphology and structural connexion of the receptor and horizontal cells of the retina of a number of lower vertebrates. Biophysics 11, 930—940 (1966).

Bortoff, A.: Localization of slow potential responses in the Necturus retina. Vision Res. 4, 627—635 (1964).

Boycott, B. B., Dowling, J. E.: Organization of the primate retina: light microscopy. Phil. Trans. roy. Soc. Lond. B 255, 109—184 (1969).

Branston, N. M., Fleming, D. G.: Efferent fibers in the frog optic nerve. Exp. Neurol. 20, 611—623 (1968).

Brightman, M. W., Reese, T. S.: Junctions between intimately apposed cell membranes in the vertebrate brain. J. Cell Biol. 40, 648—677 (1969).

Brindley, G. S.: Physiology of the Retina and the Visual Pathway. London: Edward Arnold 1960.

— Hamasaki, D. I.: Absence of early degeneration of fibers in the orbital part of the cat's optic nerve after transection of the intracranial part. J. Physiol. (Lond.) 159, 88 P (1961).

— — Histological evidence against the view that the cat's optic nerve contains centrifugal fibres. J. Physiol. (Lond.) 163, 25 P—26 P (1962a).

— — Evidence that the cat's electroretinogram is not influenced by impulses passing to the eye along the optic nerve. J. Physiol. (Lond.) 163, 558—565 (1962b).

Brooke, R. N. L., Downer, I. de C., Powell, T. P. S.: Centrifugal fibres to the retina in the monkey and cat. Nature (Lond.) 207, 1365—1367 (1965).

Brown, J. E.: Dendritic fields of retinal ganglion cells of the rat. J. Neurophysiol. **28**, 1091—1110 (1965).
— Major, D.: Cat retinal ganglion cell dendritic fields. Exp. Neurol. **15**, 70—78 (1966).
Brown, K. T., Murakami, M.: Rapid effects of light and dark adaptation upon the receptive field organization of S-potentials and late receptor potentials. Vision Res. **8**, 1145—1171 (1968).
— Tasaki, K.: Localization of electrical activity in the cat retina by an electrode marking method. J. Physiol. (Lond.) **158**, 281—295 (1961).
— Wiesel, T. N.: Intraretinal recording with micro-pipette electrodes in the intact cat eye. J. Physiol. (Lond.) **149**, 537—562 (1959).
Brzin, M., Drujan, B. O.: Activity and histochemical and cytochemical localization of cholinesterases in fish retina. Proc. 2nd Intl. Mtng. Intl. Soc. Neurochem. p. 110 (1969).
Bullivant, S., Loewenstein, W. R.: Structure of coupled and uncoupled cell junctions. J. Cell Biol. **37**, 621—632 (1968).
Byzov, A. L.: Gorizontal'nye kletki setchatki — regyulyatory sinapticheskoi peredachi. Fiziol. Zhur. SSSR (Sechenov) **53**, 1115—1124 (1967). Engl. Transl.: Horizontal cells of the retina as the regulators of synaptic transmission. Neurosci. Transl. (F.A.S.E.B.) **3**, 268—276 (1968).
— Localization of the R-membrane in the frog eye by means of an electrode marking method. Vision Res. **8**, 679—700 (1968 a).
— The component analysis of electroretinogram in the retina of coldblooded vertebrates and the regulative function of horizontal cells. In: Advances in electrophysiology and pathology of the visual system. 6. ISCERG Symposium, Erfurt, 1967, pp. 217—230. Leipzig: G. Thieme 1968 b.
— O roli gorizontal'nykh kletok v mekhanizme adaptatsii setchatki. (On the role of horizontal cells in the mechanism of retinal adaptation. In Russian with English summary and figure legends). Neirofiziologiia **1**, 210—218 (1969).
— Trifonov, Yu. A.: Gipoteza ob elektricheskoi obraznoi svyazi v sinapticheskoi peredache fotoretseptory — neirony vtorogo poryadka setchatki pozvonochnykh. (Hypothesis on the electrical feed-back in synaptic transmission between the photoreceptors and second-order neurons in the vertebrate retina. In Russian with English summary). Sinapticheskie Protsessy v Afferentnykh Sistemakh. Trudy vtorogo simpoziuma po voprosam obshchei fiziologii (P. G. Kostyuk), Akad. Nauk Ukrain. S.S.R. pp. 231—248. Kiev: Naukova Dumka 1968.
Cajal, S. R. y: Sur la morphologie et les connexions des éléments de la rétine des oiseaux. Anat. Anz. **4**, 111—121 (1889).
— La rétine des vertébrés. Cellule **9**, 121—225 (1892).
— Neue Darstellung vom histologischen Bau des Centralnervensystems. Retina. Arch. Anat. Physiol., Anat. Abt. **1893**, 399—410.
— Die Retina der Wirbelthiere. Wiesbaden: Bergmann 1894.
— Nouvelles contributions a l'étude histologique de la rétine, et la question des anastomoses des prolongements protoplasmiques. J. Anat. Physiol. **32**, 481—543 (1896).
— Histologie du système nerveux (transl. L. Azoulay), II, 296—367. Paris: A. Maloine 1909—1911. Reprinted in Madrid: Instituto Ramón y Cajal 1955.
— Estudios sobre la degeneración y regeneración del sistema nerviosa. Madrid 1913—1914. Engl. transl.: Degeneration and regeneration of the nervous system (transl. R. M. May). London: Oxford Univ. Press 1928. Reprinted in New York: Hafner 1959.
— Neuronismo o reticularismo? Trav. Lab. Rech. Biol. **28** (1933). Engl. transl.: Neuron theory or reticular theory? (transl. M. Purkiss and C. A. Fox). Madrid: Instituto Ramón y Cajal 1954.
— Los problemas histofisiológicos de la retina. XIV. Concil. Ophtal. Madrid **1933**, 1—8.
Castro, G. de O.: Branching pattern of amacrine cell processes. Nature (Lond.) **212**, 832—834 (1966).
Catois, E. H.: Recherches sur l'histologie et l'anatomie microscopique de l'encéphale chez les poissons. Bull. Scient. France Belg. **36**, 1—166 (1902).
Cohen, A. I.: The fine structure of the extrafoveal receptors of the rhesus monkey. Exp. Eye Res. **1**, 128—136 (1961).

COHEN, A. I.: Vertebrate retinal cells and their organization. Biol. Rev. **38**, 427—459 (1963a).
— The fine structure of the visual receptors of the pigeon. Exp. Eye Res. **2**, 88—97 (1963b).
— Some observations on the fine structure of the retinal receptors of the American gray squirrel. Invest. Ophthal. **3**, 198—216 (1964).
— Some electron microscopic observations on inter-receptor contacts in the human and macaque retinae. J. Anat. **99**, 595—610 (1965a).
— A possible cytological basis for the "R" membrane in the vertebrate eye. Nature (Lond.) **205**, 1222—1223 (1965b).
— An electron microscopic study of the modification by monosodium glutamate of the retinas of normal and "rodless" mice. Amer. J. Anat. **120**, 319—356 (1967a).
— Ultrastructural aspects of the human optic nerve. Invest. Ophthal. **6**, 294—308 (1967b).
— Rods and cones and the problem of visual excitation. In: STRAATSMA, B. R., HALL, M. O., ALLEN, R. A., CRESCITELLI, F. (Eds.): The Retina: Morphology, Function and Clinical Characteristics. UCLA Forum in Medical Sciences No. 8, pp. 31—62. Berkeley-Los Angeles: University of California Press 1969.
— Rods and cones. (This volume).
COLLIN, J.-P.: Rubans circonscrits par des vésicules dans les photorécepteurs rudimentaires épiphysaires de l'Oiseau: *Vanellus vanellus* (L), et nouvelles considérations phylogénétiques relatives aux pinéalocytes (ou cellules principales) de Mammifères. C.R. Acad. Sci. Paris **267**, 758—761 (1968).
— MEINIEL, A.: Les synapses de l'organe pinéal de l'ammocète. C.R. Acad. Sci. Paris **266**, 1293—1295 (1968).
CORRODI, H., JONSSON, G.: The formaldehyde fluorescence method for the histochemical demonstration of biogenic monoamines. A review on the methodology. J. Histochem. Cytochem. **15**, 65—78 (1967).
COWAN, W. M.: Centrifugal fibres to the avian retina. Brit. med. Bull. **26**, 112—118 (1970).
— ADAMSON, L., POWELL, T. P. S.: An experimental study of the avian system. J. Anat. **95**, 545—562 (1961).
— POWELL, T. P. S.: Centrifugal fibres in the avian visual system. Proc. Roy. Soc. B **158**, 232—252 (1963).
— WENGER, E.: The development of the nucleus of origin of centrifugal fibers to the retina in the chick. J. comp. Neurol. **133**, 207—240 (1968).
CRAGG, B. G.: Centrifugal fibers to the retina and olfactory bulb and composition of the supraoptic commissures in the rabbit. Exp. Neurol. **5**, 406—427 (1962).
— Structural changes in naive retinal synapses detectable within minutes of first exposure to daylight. Brain Res. **15**, 79—96 (1969).
CRESCITELLI, F.: The spectral sensitivity and visual pigment content of the retina of *Gekko gekko*. In: Ciba Foundation Symposium on Colour Vision, pp. 301—322. Boston: Little, Brown and Company 1965.
DANILOVA, L. B., ROKHLENKO, K. D., BODRYAGINA, A. V.: Electron microscopic study on the structure of septate and comb desmosomes. Z. Zellforsch. **100**, 101—117 (1969).
DARTNALL, H. J. A., ARDEN, G. B., IKEDA, H., LUCK, C. P., ROSENBERG, M. E., PEDLER, C., TANSLEY, K.: Anatomical, electrophysiological and pigmentary aspects of vision in the bush baby: an interpretive study. Vision Res. **5**, 399—424 (1965).
DAW, N.: Colour-coded ganglion cells in the goldfish retina: extension of their receptive fields by means of new stimuli. J. Physiol. (Lond.) **197**, 567—592 (1968).
DE ROBERTIS, E., FRANCHI, C. M.: Electron microscope observations on synaptic vesicles in synapses of the retinal rods and cones. J. biophys. biochem. Cytol. **2**, 307—318 (1956).
DETWILER, S. R.: Vertebrate Photoreceptors. New York: MacMillan 1943.
DIETERICH, C. E.: Elektronenmikroskopische Untersuchungen über die synaptischen Formationen der Photoreceptoren einiger Mammalier. Virchows Anat. Gesellsch. **62**, 595—596 (1967).
— Elektronenmikroskopische Untersuchungen über die Photoreceptoren und Receptorensynapsen bei reinen Stäbchen und Zapfennetzhäuten. Albrecht v. Graefes Arch. klin. exp. Ophthal. **174**, 289—320 (1968).
— Feinstrukturelle Untersuchungen an den Horizontalzellen der menschlichen Netzhaut. Z. Zellforsch. **98**, 277—289 (1969).

DODT, E.: Vergleichende Physiologie der lichtempfindlichen Wirbeltier-Ephiphyse. Nova Acta Leopoldina N.F. **31**, 219—235 (1966).

DOGIEL, A. S.: Über das Verhalten der nervösen Elemente in der Retina der Ganoiden, Reptilien, Vögel und Säugethiere. Anat. Anz. **3**, 133—143 (1888).

— Die Retina der Vögel. Arch. mikr. Anat. **44**, 622—648 (1895).

DOWLING, J. E.: Structure and function in the all-cone retina of the ground squirrel. In: Symposium on the Physiological Basis for Form Discrimination, pp. 17—23. Providence, R. I.: Brown University 1964.

— Foveal receptors of the monkey retina: Fine structure. Science **147**, 57—59 (1965).

— The site of visual adaptation. Science **155**, 273—279 (1967a).

— Visual adaptation: its mechanism. Science **155**, 584—585 (1967b).

— Synaptic organization of the frog retina: an electron microscopic analysis comparing the retinas of frogs and primates. Proc. roy. Soc. B **170**, 205—228 (1968).

— Organization of vertebrate retinas. Invest. Ophthal. **9**, 655—680 (1970).

— BOYCOTT, B. B.: Neural connections of the retina: Fine structure of the inner plexiform layer. Cold Spr. Harb. Symp. quant. Biol. **30**, 393—402 (1965a).

— — Neural connections of the primate retina. In: ROHEN, J. (Ed.): The Structure of the Eye, II, pp. 55—68, Symposium, Wiesbaden 1965. Stuttgart: Schattauer 1965b.

— — Organization of the primate retina: electron microscopy. Proc. roy. Soc. Lond. B **116**, 80—111 (1966).

— — Retinal ganglion cells: A correlation of anatomical and physiological approaches. In: STRAATSMA, B. R., HALL, M. O., ALLEN, R. A., CRESCITELLI, F. (Eds.): The retina: Morphology, function and clinical characteristics. UCLA Forum in Medical Sciences No. 8, pp. 31—62. Berkeley-Los Angeles: University of California Press 1969.

— BROWN, J. E., MAJOR, D.: Synapses of horizontal cells in rabbit and cat retinas. Science **153**, 1639—1641 (1966).

— COWAN, W. M.: An electron microscope study of normal and degenerating centrifugal fiber terminals in the pigeon retina. Z. Zellforsch. **71**, 14—28 (1966).

— RIPPS, H.: Visual adaptation in the retina of the skate. J. gen. Physiol. **56**, 491—520 (1970).

— WERBLIN, F. S.: Organization of the retina of the mudpuppy, *Necturus maculosus.* I. Synaptic structure. J. Neurophysiol. **32**, 315—338 (1969).

DRUJAN, B. D., DIAZ BÓRGES, J. M.: Adrenaline depletion induced by light in the dark-adapted retina. Experientia (Basel) **24**, 676—677 (1968).

— — ALVAREZ, N.: Relationship between the contents of adrenaline, noradrenaline and dopamine in the retina and its adaptational state. Life Sci. **4**, 473—477 (1965).

DUBIN, M. W.: The inner plexiform layer of the retina: A quantitative and comparative electron microscopic analysis in several vertebrates. Ph. D. Dissertation, The Johns Hopkins University 1969.

— The inner plexiform layer of the vertebrate retina: A quantitative and comparative electron microscopic analysis. J. comp. Neurol. **140**, 479—506 (1970).

DUKE-ELDER, S.: The eye in evolution. System of ophthalmology, Vol. I. St. Louis: Mosby 1958.

DUNN, R. F.: Studies on the retina of the gecko *Coleonyx variegatus.* I. The visual cell classification. J. Ultrastruct. Res. **16**, 651—671 (1966).

EAKIN, R. M., QUAY, W. B., WESTFALL, J. A.: Cytological and cytochemical studies on the frontal and pineal organs of the tree frog, *Hyla regilla.* Z. Zellforsch. **59**, 663—683 (1963).

EHINGER, B.: Adrenergic neurons in the retina. Life Sci. **5**, 129—131 (1966a).

— Adrenergic retinal neurons. Z. Zellforsch. **71**, 146—152 (1966b).

— Adrenergic nerves in the avian eye and ciliary ganglion. Z. Zellforsch. **82**, 577—588 (1967).

— FALCK, B.: Adrenergic retinal neurons of some New World monkeys. Z. Zellforsch. **100**, 364—375 (1969).

— — LATIES, A. M.: Adrenergic neurons in teleost retina. Z. Zellforsch. **97**, 285—297 (1969).

EICHNER, D.: Zur Frage der Fermentlokalisation in der Netzhaut des Menschen. Z. Zellforsch. **44**, 339—344 (1956).

— Zur Histologie und Topochemie der Netzhaut des Menschen. Z. Zellforsch. **48**, 137—186 (1958).

ENGSTRÖM, K.: Structure, organization and ultrastructure of the visual cells in the teleost family *Labridae*. Acta Zool. **44**, 1—41 (1963).

ENROTH-CUGELL, C., PINTO, L.: Algebraic summation of centre and surround inputs to retinal ganglion cells of the cat. Nature (Lond.) **226**, 458—459 (1970).

ERÄNKÖ, O., NIEMI, M., MERENMIES, E.: Histochemical observations on esterases and oxidative enzymes of the retina. In: SMELSER, G. S. (Eds.): The structure of the eye, pp. 159—171. New York-London: Academic Press 1961.

ESILÄ, R.: Histochemical and electrophoretic properties of cholinesterases and non-spceific esterases in the retina of some mammals, including man. Acta Ophthal. **77** (Suppl.), 1—113 (1963).

EVANS, E. M.: On the ultrastructure of the synaptic region of visual receptors in certain vertebrates. Z. Zellforsch. **71**, 499—516 (1966).

FABER, D. S.: Analysis of the slow transretinal potentials in response to light. Ph. D. Dissertation, State University of New York at Buffalo 1969.

FALCK, B., HILLARP, N.-Å., THIEME, G., TORP, A.: Fluorescence of catecholamines and related compounds condensed with formaldehyde. J. Histochem. Cytochem. **10**, 348—354 (1962).

FINE, B. S.: Synaptic lamellas in the human retina: an electron microscopic study. J. Neuropath. exp. Neurol. **22**, 255—262 (1962).

— ZIMMERMAN, L. E.: Müller's cells and the "middle limiting membrane" of the human retina. Invest. Ophthal. **1**, 304—326 (1962).

FRANCIS, C. M.: Cholinesterase in the retina. J. Physiol. (Lond.) **120**, 435—439 (1953).

FRANZ, V.: Sehorgan. In: OPPEL, A. (Hrsg.): Lehrbuch der vergleichenden mikroskopischen Anatomie der Wirbelthiere, Bd. 7. Jena: G. Fischer 1913.

GALLEGO, A.: Sinapsis a nivel de la capa plexiforme externa de la retina. An. Inst. Farm. Esp. **1**, 145—180 (1952).

— Procedimiento de impregnación argentica de la retina entera. An. Inst. Farmacol. esp. **2**, 171—176 (1953).

— Conexiones transversales retinianas. An. Inst. Farmacol. esp. **3**, 31—39 (1954).

— Description d'une nouvelle couche cellulaire dans la rétine des mammifères et son rôle fonctionelle possible. An. Inst. Farmacol. esp. **13—14**, 175—180 (1964—1965).

— Connexions transversales au niveau des couches plexiformes de la rétine. Actualités Neurophysiol. **6**, 5—27 (1965). Reprinted in An. Inst. Farmacol. esp. **13—14**, 181—204 (1964—1965).

— CRUZ, J.: Células nerviosas de asociación en la capa de células ganglionares de la retina de los mamíferos. An. Inst. Farmacol. esp. **13—14**, 205—209 (1964—1965).

— — Mammalian retina: associational nerve cells in ganglion cell layer. Science **150**, 1313—1314 (1965).

GLOBUS, A., LUX, H. D., SCHUBERT, P.: Somadendritic spread of intracellularly injected tritiated glycine in cat spinal motoneurons. Brain Res. **11**, 440—445 (1968).

GLOW, P. H., ROSE, S.: Effects of light and dark on the acetylcholinesterase activity of the retina. Nature (Lond.) **202**, 422—423 (1964).

GOODLAND, H.: The ultrastructure of the inner plexiform layer of the retina of *Cottus bubalis*. Exp. Eye Res. **5**, 198—200 (1966).

GOURAS, P.: Graded potentials of bream retina. J. Physiol. (Lond.) **152**, 487—505 (1960).

— Antidromic responses of orthodromically identified ganglion cells in monkey retina. J. Physiol. (Lond.) **204**, 407—419 (1969).

— LINK, K.: Rod and cone interaction in dark-adapted monkey ganglion cells. J. Physiol. (Lond.) **184**, 499—510 (1966).

GOVARDOVSKII, V. I., KHARKEIEVITCH, T. A.: Elektronnomikroskopicheskoe issledovanie setchatki ptits (*Gallus bankiva domestica*). (Electronmicroscopic study of retina in birds (*Gallus bankiva domestica*). In Russian). Arkh. Anat. Gistol. Embriol. **52**, 53—61 (1967).

GRAHAM, L. T., LOLLEY, R. N., BAXTER, C. F.: Effect of illumination upon levels of γ-aminobutyric acid and glutamic acid in frog retina *in vivo*. Fed. Proc. **27**, 463 (1968).

GRAY, E. G.: Axo-somatic and axo-dendritic synapses of the cerebral cortex: an electron microscope study. J. Anat. **93**, 420—433 (1959).

GUILLERY, R. W.: Some electron microscopical observations of degenerative changes in central nervous synapses. In: SINGER, M., SCHADÉ, J. P. (Eds.): Degeneration Patterns in the Nervous System. Progr. Brain Res. 14, 57—76. New York-Amsterdam: Elsevier 1965.

— RALSTON, H. J.: Nerve fibers and terminals: Electron microscopy after Nauta staining. Science 143, 1331—1332 (1964).

HAFT, J. S., HARMAN, P. J.: Evidence for central inhibition of retinal function. Vision Res. 7, 499—501 (1967).

— Further remarks on evidence for central inhibition of retinal function. Vision Res. 8, 319—323 (1968).

HÄGGENDAL, J., MALMFORS, T.: Identification and cellular localization of the catecholamines in the retina and the choroid of the rabbit. Acta physiol. scand. 64, 58—66 (1965).

HANNOVER, A.: Über die Netzhaut und ihre Gehirnsubstanz bei Wirbelthieren, mit Ausnahme des Menschen. Arch. Anat. Physiol. wissen. Med. (J. MÜLLER), pp. 320—345 (1840).

— De la rétine et de sa substance cérébrale dans les animaux vertébrés l'homme excepté. In: Recherches microscopiques sur le système nerveux, pp. 37—56, 64—67. Copenhagen: P. G. Philipsen 1844.

HANSSON, H.-A.: Scanning electron microscopy of the rat retina. Z. Zellforsch. 107, 23—44 (1970a).

— Scanning electron microscopic studies on the synaptic bodies in the rat retina. Z. Zellforsch. 107, 45—53 (1970b).

HANYU, I., NIWA, H., TAMURA, T.: A slow potential from the *epiphysis cerebri* of fishes. Vision Res. 9, 621—623 (1969).

HEBB, C. O., SILVER, A., SWAN, A. A. B., WALSH, E. G.: A histochemical study of cholinesterase of rabbit retina and optic nerve. Quart. J. exp. Physiol. 38, 185—191 (1953).

HEIMER, L.: The tracing of pathways in the central nervous system. Proc. I.E.E.E. 56, 950—959 (1968).

HENDRICKSON, A. E.: Regeneration of the retina in the newt (*Diemictylus v. viridescens*): an electron microscope study. Ph. D. Dissertation, University of Washington (Seattle) 1964. Reproduced by University Microfilms, Inc., Ann Arbor, Mich. and University Microfilms Ltd., High Wycomb, England.

— Landolt's club in the amphibian retina: A Golgi and electron microscope study. Invest. Ophthal. 5, 484—496 (1966).

HILLARP, N.-Å., FUXE, K., DAHLSTROM, A.: Central monoamine neurons. In: VON EULER, U. S., ROSELL, S., UVNAS, B. (Eds.): Mechanisms on release of biogenic amines, pp. 31—37. New York-Oxford: Pergamon 1966.

HOLDEN, A. L.: An investigation of the centrifugal pathway to the pigeon retina. J. Physiol. (Lond.) 186, 133 P (1966a).

— Two possible visual functions for centrifugal fibres to the retina. Nature (Lond.) 212, 837—838 (1966b).

— Antidromic activation of the isthmo-optic nucleus. J. Physiol. (Lond.) 197, 183—198 (1968a).

— The centrifugal system running to the pigeon retina. J. Physiol. (Lond.) 197, 199—219 (1968b).

— Receptive properties of centrifugal cells projecting to the pigeon retina. J. Physiol. (Lond.) 210, 155 P (1970).

HOLLENBERG, M. J., BERNSTEIN, M. H.: Fine structure of the photoreceptor cells of the ground squirrel (*Citellus tridecemlineatus tridecemlineatus*). Amer. J. Anat. 118, 359—374 (1966).

HONRUBIA LÓPEZ, F. M.: Estudio de los campos anatómicos de las células ganglionares de la retina. Arch. Soc. oftal. hisp.-amer. 26, 693—720 (1966).

HONRUBIA, F. M., ELLIOTT, J. H.: Efferent innervation of the primate retina. Invest. Ophthal. 7, 618 (1968).

— — Horizontal cell of the mammal retina. Arch. Ophthal. 82, 98—104 (1969a).

— — Dendritic fields of the retinal ganglion cells in the cat. Invest. Ophthal. 8, 461 (1969b).

— — Dendritic fields of the retinal ganglion cells in the cat. Arch. Ophthal. 84, 221—226 (1970).

— GRIJALBO, M. P.: Estudio de las células ganglionares de la retina. Arch. Soc. oftal. hisp.-amer. 27, 796—804 (1967).

Honrubia, F. M., Grijalbo, M. P., Elliott, J. H.: Fibras centrífugas en la retina de los primates. Arch. Soc. oftal. hisp.-amer. **27**, 561—569 (1967).

Imbert, M.: Aspects récents de la physiologie des voies visuelles primaires chez les vertébrés. J. Physiol. (Paris) **62** (Suppl. 1), 3—59 (1970).

Iraldi, A. P., Robertis, E. de: Ultrastructure and function of catecholamine containing systems. In: Proc. 2nd Internat. Congr. Endocrinol., London. Excerpta Medica Internat. Congr. Ser. **83**, 355—363 (1964).

— Etcheverry, G. J.: Granulated vesicles in retinal synapses and neurons. Z. Zellforsch. **81**, 283—296 (1967).

Kabuta, H., Tominaga, Y., Kuwabara, M.: The rhabdomeric microvilli of several arthropod compound eyes kept in darkness. Z. Zellforsch. **85**, 78—88 (1968).

Kalberer, M., Pedler, C.: The visual cells of the alligator: An electron microscopic study. Vision Res. **3**, 323—329 (1963).

Kaneko, A.: Physiological and morphological identification of horizontal, bipolar and amacrine cells in goldfish retina. J. Physiol. (Lond.) **207**, 623—633 (1970).

— Electrical connections between horizontal cells in dogfish retina. J. Physiol. (Lond.) **213**, 95—105 (1971).

— Hashimoto, H.: Recording site of the single cone response determined by an electrode marking technique. Vision Res. **7**, 847—851 (1967).

— — Localization of spike-producing cells in the frog retina. Vision Res. **8**, 259—262 (1968).

— — Electrophysiological study of single neurons in the inner nuclear layer of the carp retina. Vision Res. **9**, 37—55 (1969).

Kelly, D. E., Smith, S. W.: Fine structure of the pineal organs of the adult frog, *Rana pipiens*. J. Cell Biol. **22**, 653—674 (1964).

— Kamer, J. C. van de: Cytological and histochemical investigations on the pineal organ of the adult frog (*Rana esculenta*). Z. Zellforsch. **52**, 618—639 (1960).

Kidd, M.: Electron microscopy of the inner plexiform layer of the retina. Proc. Anat. Soc.: "Cytology of nervous tissue", pp. 88—91. London: Taylor and Francis 1961.

— Electron microscopy of the inner plexiform layer of the retina in the cat and the pigeon. J. Anat. **96**, 179—187 (1962).

Kishida, K., Naka, K.-I.: Amino acids and spikes from the retinal ganglion cells. Science **156**, 648—650 (1967).

Koelle, G. B.: The elimination of enzymatic diffusion artefacts in the histochemical localization of cholinesterases and a survey of their cellular distributions. J. Pharmacol. **103**, 153—171 (1951).

— Friedenwald, J. S.: The histochemical localization of cholinesterase in ocular tissues. Amer. J. Ophthal. **33**, 253—256 (1950).

— — Allen, R. A., Wolfand, L.: Localization of specific cholinesterase in ocular tissues of the cat. Amer. J. Ophthal. **35**, 1580—1584 (1952).

Kojima, K., Iida, M., Majima, Y., Okada, S.: Histochemical studies on monoamine oxidase (MAO) of the human retina. Jap. J. Ophthal. **5**, 205—210 (1961).

Kolb, H.: Organization of the outer plexiform layer of the primate retina: Electron microscopy of Golgi-impregnated cells. Proc. roy. Soc. Lond. B **258**, 261—283 (1970).

— Boycott, B. B., Dowling, J. E.: A second type of midget bipolar cell in the primate retina. Appendix to Boycott, B. B. and Dowling, J. E.: Organization of the primate retina: light microscopy. Phil. Trans. roy. Soc. Lond. B **255**, 109—184 (1969).

Kolmer, W.: Die Netzhaut (Retina). In: von Mollendorff, W. (Hrsg.): Handbuch der mikroskopischen Anatomie des Menschen, 2. Bd., 3. Teil, S. 295—468. Berlin: Springer 1936.

Kramer, S. G.: Dopamine: A retinal neurotransmitter. I. Retinal uptake, storage, and light-stimulated release of H3-dopamine in vivo. Invest. Ophthal. **10**, 438—452 (1971).

— Potts, A. M., Mangnall, Y.: Dopamine: A retinal neurotransmitter. II. Autoradiographic localization of H3-dopamine in the retina. Invest. Ophthal. **10**, 617—624 (1971).

Kuffler, S. W.: Neurons in the retina: Organization, inhibition and excitation problems. Cold Spr. Harb. Symp. quant. Biol. **17**, 281—292 (1952).

— Nicholls, J. G., Orkand, R. K.: Physiological properties of glial cells in the central nervous system of amphibia. J. Neurophysiol. **29**, 768—787 (1966).

KURIYAMA, K., SISKEN, B., HABER, B., ROBERTS, E.: The γ-aminobutyric acid system in rabbit retina. Brain Res. **9**, 161—164 (1968).

LADMAN, A. J.: The fine structure of the rod-bipolar cell synapse in the retina of the albino rat. J. biophys. biochem. Cytol. **4**, 459—466 (1958).

— HUTCHINGS, S.: Electron microscopic localization of cholinesterase activity in normal rat retina. Preliminary observations. Anat. Rec. **151**, 375 (1965).

— SOPER, E. H.: Preliminary observations on the fine structure of Müller's cells of the avian retina. Proc. Fifth Internat. Congr. Electr. Micr. **2**, R-6 (1962).

LANDOLT, E.: Beitrag zur Anatomie der Retina vom Frosch, Salamander und Triton. Roux' Arch. mikr. Anat. **7**, 81—100 (1871).

LASANSKY, A.: Morphological bases for a nursing role of glia in the toad retina. Electron microscope observations. J. biophys. biochem. Cytol. **11**, 237—243 (1961).

— Functional implications of structural findings in retinal glial cells. In: Biology of neuroglia (Eds. DE ROBERTIS, E. D. P. and CARREA, R.). Amsterdam: Elsevier 1965. Progr. Brain Res. **15**, 48—72 (1965).

— The pathway between hyaloid blood and retinal neurons in the toad. Structural observations and permeability to tracer substances. J. Cell. Biol. **34**, 617—626 (1967).

— Basal junctions at synaptic endings of turtle visual cells. J. Cell Biol. **40**, 577—581 (1969).

— Synaptic organization of cone cells in the turtle retina. Phil. Trans. Roy. Soc. Lond. B, **262**, 365—381 (1971).

— FUORTES, M. G. F.: The site of origin of electrical responses in visual cells of the leech, *Hirudo medicinalis*. J. Cell Biol. **42**, 241—252 (1969).

— WALD, F.: The extracellular space in the toad retina as defined by the distribution of ferrocyanide. A light and electron microscope study. J. Cell Biol. **15**, 463—479 (1962).

LATIES, A. M., JACOBOWITZ, D.: A comparative study of the autonomic innervation of the eye in monkey, cat, and rabbit. Anat. Rec. **156**, 383—396 (1966).

LAUFER, M., MILLÁN, E.: Spectral analysis of L-type S-potentials and their relation to photopigment absorption in a fish (*Eugerres plumieri*) retina. Vision Res. **10**, 237—251 (1970).

— — VANEGAS, H.: Retinal adaptation and S-potentials. Proc. Int. Union Physiol. Sci., 25th Congr. Munich 1971. **IX**, 1000 (1971).

LÁZÁR, GY.: Efferent paths of the frog's optic centre. Acta morph. Acad. Sci. hung. **17**, 341 (1969).

LEE, S. H., PAK, S. Y., CHOI, K. D.: A histochemical study of cholinesterase activity in rabbit's retinae. Yonsei med. J. **8**, 1—7 (1967).

LEICESTER, J., STONE, J.: Ganglion, amacrine and horizontal cells of the cat's retina. Vision Res. **7**, 695—705 (1967).

LEPLAT, G., GEREBTZOFF, M. A.: Localisation de l'acetylcholinesterase et des médiateurs diphénoliques dans la rétine. Ann. Oculist. **189**, 121—128 (1956).

LESSELL, S., KUWABARA, T.: Retinal neuroglia. Arch. Ophthal. **70**, 671—678 (1963).

LETTVIN, J. Y., MATURANA, H. R., McCULLOCH, W. S., PITTS, W. H.: What the frog's eye tells the frog's brain. Proc. I.R.E. **47**, 1940—1951 (1959).

— — PITTS, W. H., McCULLOCH, W. S.: Two remarks on the visual system of the frog. In: ROSENBLITH, W. A. (Ed.): Sensory communication, pp. 757—776. Cambridge-New York: Technology Press and Wiley 1961.

LEWIS, P. R., SHUTE, C. C. D.: Fine localization of acetylcholinesterase in the optic nerve and retina of the rat. J. Physiol. (Lond.) **180**, 8 P—10 P (1965).

LIBERMAN, R.: Retinal cholinesterase and glycolysis in rats raised in darkness. Science **135**, 372—373 (1962).

LIPETZ, L. E.: Glial control of neuronal activity. I.E.E.E. Trans. Milit. Electr. MIL-**7**, 144—155 (1963).

— Information processing in the frog's retina. Document No. AD 614249, Clearinghouse for Federal Scientific and Technical Information, U.S. Dept. of Commerce. 75 pp. (1965).

— The Landolt Club: I. Historical resume. In preparation.

LOCKET, N. A.: The retina of *Poromitra nigrofulvus* (Garman): an optical and electron microscope study. Exp. Eye Res. **8**, 265—275 (1969).

— Deep-sea fish retinas. Brit. med. Bull. **26**, 107—111 (1970a).

Locket, N. A.: Landolt's club in the retina of the African lungfish, *Protopterus aethiopicus* Heckel. Vision Res. **10**, 299—306 (1970 b).

Lolley, R. N.: Metabolic and anatomical specialization within the retina. In: Lajtha, A. (Ed.): Handbook of Neurochemistry, Vol. 2, Chap. 20, pp. 473—504. New York: Plenum Press 1969.

Macnichol, E. F., Jr., Macpherson, L., Svaetichin, G.: Studies on spectral response curves from the fish retina. Paper No. 39, Natl. Physical Lab., Teddington 1957.

— Svaetichin, G.: Electric responses from the isolated retinas of fishes. Amer. J. Ophthal. **46** (3, part II), 26—46 (1958).

Maksimova, E. M.: Vliyanie vnutrikletochnoi polyarizatsii gorizontalnykh kletok na aktivnost' ganglioznykh kletok setchatki ryb. Biofizika **14**, 537—544 (1969). English transl.: Effect of intracellular polarization of horizontal cells on ganglion cell activity in the fish retina. Neurosci. Transl. (F.A.S.E.B.) **11**, 114—120 (1970).

Malmfors, T.: Evidence of adrenergic neurons with synaptic terminals in the retina of rats demonstrated with fluorescence and electron microscopy. Acta. physiol. scand. **58**, 99—100 (1963).

Marchesi, V. T., Sears, M. L., Barrnett, R. J.: Electron microscopic studies of nucleoside phosphatase activity in blood vessels and glia of the retina. Invest. Ophthal. **3**, 1—21 (1964).

Marks, W. B.: Difference spectra of the visual pigments in single goldfish cones. Ph. D. Dissertation, The Johns Hopkins University 1963.

— Visual pigments of single goldfish cones. J. Physiol. (Lond.) **178**, 14—32 (1965 a).

— Visual pigments of single cones. In: Ciba Foundation Symposium on Colour Vision, pp. 208—213. Boston: Little, Brown and Comp. 1965 b.

Matsusaka, T.: Lamellar bodies in the synaptic cytoplasm of the accessory cone from the chick retina as revealed by electron microscopy. J. Ultrastruct. Res. **18**, 55—70 (1967).

Maturana, H. R.: Efferent fibres in the optic nerve of the toad (*Bufo bufo*). J. Anat. **92**, 21—27 (1958).

— Frenk, S.: Synaptic connections of the centrifugal fibres in the pigeon retina. Science **150**, 359—361 (1965).

— Lettvin, J. Y., McCulloch, W. S., Pitts, W. H.: Anatomy and physiology of vision of the frog (*Rana pipiens*). J. gen. Physiol. **43** (6 part 2), 129—175 (1960).

— Uribe, G., Frenk, S.: A biological theory of relativistic colour coding in the primate retina. Arch. biol. med. exp. Suppl. **1**, 30 pp. (1968).

McGill, J. I., Powell, T. P. S., Cowan, W. M.: The retinal representation upon the optic tectum and isthmo-optic nucleus in the pigeon. J. Anat. **100**, 5—33 (1966 a).

— — — The organization of the projection of the centrifugal fibres to the retina in the pigeon. J. Anat. **100**, 35—49 (1966 b).

Meller, K., Eschner, J.: Vergleichende Untersuchungen über die Feinstruktur der Bipolarzellschicht der Vertebratenretina. Z. Zellforsch. **68**, 550—567 (1965).

Miles, F. A.: Centrifugal effects in the avian retina. Science **170**, 992—995 (1970).

Miller, R. F., Dowling, J. E.: Intracellular responses of the Müller (glial) cells of the mudpuppy retina: their relation to b-wave of the electroretinogram. J. Neurophysiol. **33**, 323—341 (1970).

Missotten, L.: Étude des synapses de la rétine humaine au microscope électronique. Proc. Eur. Reg. Conf. Electr. Micr., Delft 1960 **2**, 818—821 (1960).

— The Ultrastructure of the Human Retina. Brüssel: Arscia Uitgaven 1965 a.

— The synapses in the human retina. In: Rohen, J. (Ed.): The Structure of the Eye, II. Symposium, Wiesbaden 1965, pp. 17—28. Stuttgart: Schattauer 1965 b.

— The synaptic relations of visual cells and neurons in the human retina, studied by electron microscopy, after silver impregnation. Presented to Ass. Res. Ophthal. meeting, Chicago, Ill., Oct. 26, 1968.

— Appelmans, M., Michiels, J.: L'ultrastructure des synapses des cellules visuelles de la rétine humaine. Bull. Mém. Soc. franç. Ophthal. **76**, 59—82 (1963).

Mitarai, G.: The origin of the so-called cone potential. Proc. Jap. Acad. **34**, 299—304 (1958).

— Determination of ultramicroelectrode tip position in the retina in relation to S-potential. J. gen. Physiol. **43** (6, part 2), 95—99 (1960).

— Function of glia (in Japanese). Seitai Kagaku **14**, 36—48 (1963).

MITARAI, G.: Further identification of the site of origin and the spectral response curve of S-potential (in Japanese). Seitai Kagaku 15, 38—46 (1964).
— Glia-neuron interaction in carp retina, glia potentials revealed by microelectrode with lithium carmine. In: SENO, S., COWDRY, E. V. (Eds.): Intracellular Membraneous Structure. Sympos. Soc. Cell Chem. Suppl. 14, 549—558 (1965).
— SVAETICHIN, G., VALLECALLE, E., FATEHCHAND, R., VILLEGAS, J., LAUFER, M.: Glia-neuron interaction and adaptational mechanisms of the retina. In: JUNG, R., KORNHUBER, H. (Hrsg.): Neurophysiologie und Psychophysik des visuellen Systems, Symposium Freiburg 1960, S. 463—481. Berlin-Göttingen-Heidelberg: Springer 1961.
MIZUNO, K.: Histochemical studies on the glial cell in the retina and optic nerve. Acta Soc. Ophthal. Jap. 68, 1567—1573 (1964).
MOTOKAWA, K.: Physiology of Color and Pattern Vision. Berlin-Heidelberg-New York: Springer 1970.
MOUNTFORD, S.: Effects of light and dark adaptation on the vesicle populations of receptor-bipolar synapses. J. Ultrastruct. Res. 9, 403—418 (1963).
MÜLLER, H.: Zur Histologie der Netzhaut. Z. wiss. Zool. 3, 234—237 (1851).
MÜLLER, H.: Anatomisch-physiologische Untersuchungen über die Retina bei Menschen und Wirbelthieren. Z. wiss. Zool. 8, 1—122 (1856).
— Bau und Wachstum der Netzhaut des Guppy (Lebistes reticulatus). Zool. Jahrb., Abt. allgem. Zool. Physiol. Tiere 63, 275—322 (1952).
MÜLLER, W.: Über die Stammesentwicklung des Sehorgans der Wirbelthiere. Leipzig: Vogel 1874.
MUNK, O.: The eye of Calamoichthys calabaricus Smith, 1965 (Polypteridae, Pisces). Vidensk. Medd. Dansk naturh. Foren. 127, 113—126 (1964).
— Omosudis lowei Gunther, 1887, a bathypelagic deep-sea fish with an almost pure-cone retina. Vidensk. Medd. Dansk naturh. Foren. 128, 341—355 (1965).
NAKA, K.-I.: Receptive field mechanism in the vertebrate retina. Science 171, 691—693 (1971).
— NYE, P. W.: Receptive field organization of the catfish retina: Are at least two lateral mechanisms involved ? J. Neurophysiol. 33, 625—642 (1970).
— RUSHTON, W. A. H.: The generation and spread of S-potentials in fish (Cyprinidae). J. Physiol. (Lond.) 192, 437—461 (1967).
NAUTA, W. J. H., EBBESSON, S. O. E. (Eds.): Contemporary research methods in neuroanatomy. Berlin-Heidelberg-New York: Springer 1970.
NEGISHI, K.: Reduction and enhancement of S-potential observed with two simultaneous light stimuli in the isolated fish retina. Vision Res., Suppl. 3, 65—76 (1971).
— LAUFER, M., SVAETICHIN, G.: Excitation spread along horizontal and amacrine cell layers in the teleost retina. Nature (Lond.) 218, 39—40 and 69 (1968).
NEUMAYER, L.: Der feinere Bau der Selachier-Retina. Arch. mikr. Anat. 48, 83—111 (1897).
NICHOLS, C. W., JACOBOWITZ, D., HOTTENSTEIN, M.: The influence of light and dark on the catecholamine content of the retina and choroid. Invest. Ophthal. 6, 642—646 (1967).
— KOELLE, G. B.: Acetylcholinesterase: method for demonstration in amacrine cells of rabbit retina. Science 155, 577—478 (1967).
— — Comparison of the localization of acetylcholinesterase and non-specific cholinesterase activities in mammalian and avian retinas. J. comp. Neurol. 133, 1—15 (1969).
NILSSON, S. E. G.: An electron microscopic classification of the retinal receptors of the leopard frog (Rana pipiens). J. Ultrastruct. Res. 10, 390—416 (1964a).
— Interreceptor contacts in the retina of the frog (Rana pipiens). J. Ultrastruct. Res. 11, 147—165 (1964b).
— CRESCITELLI, F.: Changes in ultrastructure and electroretinogram of bullfrog retina during development. J. Ultrastruct. Res. 27, 45—62 (1969).
— — A correlation of ultrastructure and function in the developing retina of the frog tadpole. J. Ultrastruct. Res. 30, 87—102 (1970).
NOELL, W. K., LASANSKY, A.: Effects of electrophoretically applied drugs and electrical currents on the ganglion cell of the retina. Fed. Proc. 18, 115 (1959).
NORTON, A. L., SPEKREIJSE, H., WOLBARSHT, M. L., WAGNER, H. G.: Receptive field organization of the S-potential. Science 160, 1021—1022 (1968).

O'Connell, C.P.: The structure of the eye of *Scardinops caerulea*, *Engraulis mordax*, and four other pelagic marine teleosts. J. Morph. **113**, 287—329 (1963).

O'Daly, J.A.: Ultraestructura y citoquímica de la retina de los teleósteos. Doctoral Thesis, Universidad Central de Venezuela (1967a).

— ATPase activity at the functional contacts between retinal cells which produce S-potential. Nature (Lond.) **216**, 1329—1331 (1967b).

Ofuchi, Y.: Electron microscopic histochemistry of nucleoside phosphatases of the retina. III. Fine structural localization of nucleoside phosphatases in the photoreceptor synapses. (In Japanese with English Summary). Acta Soc. Ophthal. jap. **72**, 515—522 (1968).

Ogden, T.E.: On the function of efferent retinal fibres. In: Structure and function of inhibitory neuronal mechanisms. Proc. 4th Internat. Meeting Neurobiol., Stockholm, 1966, pp. 89—109. Oxford-New York: Pergamon Press 1968.

Oikawa, T., Ogawa, T., Motokawa, K.: Origin of so-called cone action potential. J. Neurophysiol. **22**, 102—111 (1959).

Oksche, A.: Survey of the development and comparative morphology of the pineal organ. Progr. Brain Res. **10**, 3—29 (1965).

— Harnack, M. von: Elektronenmikroskopische Untersuchungen am Stirnorgan von Anuren (zur Frage der Lichtrezeptoren). Z. Zellforsch. **59**, 239—288 (1963).

Olney, J.W.: Centripetal sequence of appearance of receptor-bipolar synaptic structures in developing mouse retina. Nature (Lond.) **218**, 281—282 (1968a).

— An electron microscopic study of synapse formation, receptor outer segment development, and other aspects of developing mouse retina. Invest. Ophthal. **7**, 250—268 (1968b).

Ordy, J.M., Samorajski, T.: Visual acuity and ERG-CFF in relation to the morphologic organization of the retina among diurnal and nocturnal primates. Vision Res. 8, 1205—1225 (1968).

Orkand, R.K., Nicholls, J.G., Kuffler, S.W.: Effect of nerve impulses on the membrane potential of glial cells in the central nervous system of amphibia. J. Neurophysiol. **29**, 788—806 (1966).

Owsiannikow, Ph.: Über das dritte Auge bei *Petromyzon fluviatilis*. Mem. Acad. Imp. Sci. St. Petersbourg **36**, (9), 1—26 (1888).

Pappas, G.D., Bennett, M.V.L.: Specialized junctions involved in electrical transmission between neurons. Ann. N.Y. Acad. Sci. **137**, 495—508 (1966).

Parthe, V.: Células horizontales y amacrinas de la retina. Acta Cient. Venez. Supl. **3**, 240—249 (1967).

— Células ganglionares dislocadas de la retina de los teleósteos. Acta Cient. Venez. **19**, 13 (1968).

— Conexiones de las células horizontales y estrelladas de la retina de los teleósteos. Acta Cient. Venez. **20**, 84 (1969).

— Clasificación morfológica de las células horizontales de la retina. Acta Cient. Venez. **21**, (Supl. 1), 19 (1970).

— Horizontal, bipolar and oligopolar cells in the teleost retina. Internat. Sympos. Visual Processes in Vertebrates, Santiago 1970. Vision Res. **12**: 395—406 (1972).

Pedler, C.: The radial fibres of the retina. Doc. Ophthalmol. **16**, 208—220 (1962a).

— Some observations on the fine structure of the visual-cell synapse. XIX. Concil. Ophthal. 1962. I, 645—651 (1962b).

— The fine structure of the radial fibres in the reptile retina. Exp. Eye Res. **2**, 296—303 (1963).

— Rods and cones: a fresh approach. In: Ciba Foundation Symposium on Colour Vision, pp. 52—83. Boston: Little, Brown and Comp. 1965.

— Tansley, K.: The fine structure of the cone of a diurnal gecko (*Phelsuma inunguis*). Exp. Eye Res. **2**, 39—47 (1963).

— Tilly, R.: The nature of the gecko visual cell. A light and electron microscopic study. Vision Res. **4**, 499—510 (1964).

— Tilly, R.: The serial reconstruction of a complex receptor synapse. In: Rohen, J. (Ed.): The structure of the eye, II. Symposium, Wiesbaden 1965, S. 29—53. Stuttgart: Schattauer 1965.

— — The reconstruction of the outer plexiform layer in the retina. Proc. Sixth Internat. Congr. Electr. Micr. Kyoto, 1966, pp. 497—498. Tokyo: Maruzen 1966a.

PEDLER, C., TILLY, R.: A new method of serial reconstruction from electron micrographs J. roy. micr. Soc. **86**, 189—197 (1966b).

— — The retina of a fruit bat (*Pteropus giganteus* Brunnich). Vision Res. **9**, 909—922 (1969).

PETERS, A., PALAY, S. L.: The morphology of laminae A and A_1 of the dorsal nucleus of the lateral geniculate body of the cat. J. Anat. **100**, 451—486 (1966).

POLYAK, S. L.: The retina. Chicago: University of Chicago Press 1941.

— The vertebrate visual system (Ed. KLÜVER, H.), pp. 207—287. Chicago: University of Chicago Press 1957.

POMERANZ, B., CHUNG, S. H.: Dendritic-tree anatomy codes form-vision physiology in tadpole retina. Science **170**, 983—984 (1970).

PRINCE, J. L., POWELL, T. P. S.: An experimental study of the origin and the course of the centrifugal fibres to the olfactory bulb in the rat. J. Anat. **107**, 215—237 (1970a).

— — The afferent connexions of the nucleus of the horizontal limb of the diagonal band. J. Anat. **107**, 239—256 (1970b).

— — The morphology of the granule cells of the olfactory bulb. J. Cell Sci. **7**, 91—124 (1970c).

— — The synaptology of the granule cells of the olfactory bulb. J. Cell Sci. **7**, 125—156 (1970d).

— — An electron-microscopic study of the termination of the afferent fibres to the olfactory bulb from the cerebral hemisphere. J. Cell Sci. **7**, 157—185 (1970e).

PRINCE, J. H., McCONNELL, D. G.: Retina and optic nerve. In: PRINCE, J. H. (Ed.): The rabbit in eye research, Chapt. 13, pp. 385—448. Springfield, Illinois: C. C. Thomas 1964.

RADNÓT, M., LOVÁS, B.: Beitrag zur Feinstruktur der Kaninchennetzhaut. Klin. Mbl. Augenheilk. **152**, 242—246 (1968).

RALL, W., SHEPHERD, G. M., REESE, T. S., BRIGHTMAN, M. W.: Dendrodendritic synaptic pathway for inhibition in the olfactory bulb. Exp. Neurol. **14**, 44—56 (1966).

RAVIOLA, E., RAVIOLA, G.: Ricerche istochimiche sulla retina di coniglio nel corso dello sviluppo postnatale. Z. Zellforsch. **56**, 552—572 (1962).

— — Subsurface cisterns in the amacrine cells of the rabbit retina. J. Submicr. Cytol. **1**, 35—42 (1969).

RAVIOLA, G., RAVIOLA, E.: Light and electron microscopic observations on the inner plexiform layer of the rabbit retina. Amer. J. Anat. **120**, 403—426 (1967).

— — TENCONI, M. T.: Sulla organizzazione dello strato granulare esterno e della membrana limitante esterna nella retina di coniglio. Z. Zellforsch. **70**, 532—553 (1966).

RAYNAULD, J.-P.: Rod and cone responses of ganglion cells in goldfish retina: A microelectrode study. Ph. D. Dissertation. The Johns Hopkins University 1969.

RETZIUS, G.: Über den Bau des sogenannten Parietalauges von Ammocoetes. Biol. Untersuch. N.F. **7**, 22—25 (1895).

— Zur Kenntnis der Retina der Selachier. Zoologiska Studier, Festskr. W. Lilljeborg, pp. 19—28. Uppsala: Almqvist-Wiksells 1896.

RICHARDSON, T. M., LIPETZ, L. E.: The Landolt club: II. Its fine structure in the retinas of frog and turtle. In preparation.

ROHEN, J. W.: Das Auge und seine Hilfsorgane. In: Handbuch der mikroskopischen Anatomie des Menschen (v. MÖLLENDORFF), III/4. Berlin-Göttingen-New York-Heidelberg: Springer 1964.

SAMORAJSKI, T., ORDY, J. M., KEEFE, J. R.: Structural organization of the retina in the tree shrew (*Tupaia glis*). J. Cell Biol. **28**, 489—504 (1966).

SANO, Y., YOSHIKAWA, H., KONISHI, M.: Fluorescence microscopic observations on the dog retina. Arch. histol. jap. **30**, 75—81 (1968).

SCHAPER, A.: Die nervösen Elemente der Selachier-Retina in Methylenblaupräparaten. Festschr. 70 Geburts. C. Kupffer, pp. 1—10. Jena: G. Fischer 1899.

SCHIEFFERDECKER, P.: Studien zur vergleichenden Histologie der Retina. Arch. mikr. Anat. **28**, 305—396 (1886).

SCHULTZE, M.: Zur Anatomie und Physiologie der Retina. Arch. mikr. Anat. **2**, 165—174, 175—286 (1866).

SHEN, S.-C., GREENFIELD, P., BOELL, E. J.: Localization of acetylcholinesterase in chick retina during histogenesis. J. comp. Neurol. **106**, 433—461 (1956).

Shibkova, S. A., Vladimirskii, B. M.: Strukturnaya organizatsiya ganglioznykh kletok temporal'noi oblasti setchatki koshki. (The structural organization of ganglion cells of the temporal region of the retina in the cat. In Russian). Dokl. Akad. Nauk S.S.S.R. **186**, 461—464 (1969).

Shiragami, M.: Electron microscopic study of synapses of visual cells. II. The morphogenesis of synapses of visual cells in the chick embryo (In Japanese). Acta Soc. Ophthal. jap. **72**, 1060—1073 (1968).

Shkol'nik-Yarros, E. G.: K morfologii bipolyarnykh kletok setchatki. Arhk. Anat. Gistol. Embriol. **54**, 30—37 (1968). Engl. Transl.: Morphology of the bipolar cells of the retina. Neurosci. Transl. (F.A.S.E.B.) **7**, 778—784 (1969).

Sillman, A. J., Ito, H., Tomita, T.: Studies on the mass receptor potential of the isolated frog retina. I. General properties of the response. Vision Res. **9**, 1435—1442 (1969).

Sjöstrand, F. S.: The ultrastructure of the retinal rod synapses of the guinea pig eye. J. appl. Phys. **24**, 1422 (1953).

— Die routinemäßige Herstellung von ultradünnen (ca. 200 Å) Gewebschnitten für elektronenmikroskopische Untersuchungen der Gewebszellen bei hoher Auflösung. Z. wiss. Mikr. **62**, 65—86 (1954).

— Ultrastructure of retinal rod synapses of the guinea pig eye as revealed by three-dimensional reconstructions from serial sections. J. Ultrastruct. Res. **2**, 122—170 (1958).

— The ultrastructure of the retinal receptors of the vertebrate eye. Ergeb. Biol. **21**, 128—160 (1959).

— Electron microscopy of the retina. In: Smelser, G. K. (Ed.): The structure of the eye, pp. 1—28. New York-London: Academic Press 1961 a.

— Topographic relationship between neurons, synapses and glia cells. In: Jung, R., Korhuber, H. (Eds.): Neurophysiologie und Psychophysik des visuellen Systems, Symposium Freiburg 1960, S. 13—22. Berlin-Göttingen-Heidelberg: Springer 1961 b.

— The synaptology of the retina. In: Ciba Foundation Symposium on Colour Vision, pp. 110—144. Boston: Little, Brown and Company 1965.

— The outer plexiform layer and the neural organization of the retina. In: Straatsma, B. R., Hall, M. O., Allen, R. A., Crescitelli, F. (Eds.): The retina: Morphology, function and clinical characteristics. UCLA Forum in Medical Sciences No. 8, pp. 63—100. Berkeley-Los Angeles: University of California Press 1969.

— Nilsson, S. E. G.: The Structure of the rabbit retina as revealed by electron microscopy. In: Prince, J. H. (Ed.): The rabbit in eye research, Chapt. 14, pp. 449—513. Springfield, Ill.: C. C. Thomas 1964.

Smelser, G. K., Ishikawa, T., Pei, Y. F.: Electron microscopic studies of intra-retinal spaces. Diffusion of particulate materials. In: Rohen, J. W. (Ed.): The structure of the eye, II. Symposium, Wiesbaden 1965, S. 109—120. Stuttgart: Schattauer 1965.

Smith, C. A., Rasmussen, G.: Degeneration in the efferent nerve endings in the cochlea after axonal section. J. Cell Biol. **26**, 63—77 (1965).

Sobrino, J. A., Gallego, A.: Granulated vesicles in the mammalian retina. An. Inst. Farm. Esp. **17—18**, 343—347 (1968—1969).

Steinberg, R. H.: Rod and cone contributions to S-potentials from the cat retina. Vision Res. **9**, 1319—1329 (1969a).

— Rod-cone interaction in S-potentials from the cat retina. Vision Res. **9**, 1331—1344 (1969b).

Stell, W. K.: Correlated light and electron microscope observations on Golgi preparations of goldfish retina. J. Cell Biol. **23**, 89 A (1964).

— Correlation of retinal cytoarchitecture and ultrastructure in Golgi preparations. Anat. Rec. **153**, 389—397 (1965a).

— Discussion: Dendritic contacts of horizontal cell in monkey retina. In: Rohen, J. (Ed.): The structure of the eye, II. Symposium, Wiesbaden 1965, S. 27—28. Stuttgart: Schattauer 1965b.

— Observations on some ultrastructural characteristics of goldfish cones. Amer. Zool. **5**, 435 (1965c).

— The structure of horizontal cells and synaptic relations in the outer plexiform layer of the goldfish retina, as revealed by the Golgi method and electron microscopy. Ph. D. Dissertation, University of Chicago 1966.

STELL,W.K.: The structure and relationships of horizontal cells and photoreceptor-bipolar synaptic complexes in goldfish retina. Amer. J. Anat. **121**, 401—424 (1967).
— The structure and morphologic relations of rods and cones in the retina of the spiny dog-fish, *Squalus*. Comp. Bioch. Physiol. **42** (2 A), 141—151(1972).
STONE,J.: A quantitative analysis of the distribution of ganglion cells in the cat's retina. J. comp. Neurol. **124**, 337—352 (1965).
— Structure of the cat's retina after occlusion of the retinal circulation. Vision Res. **9**, 351—356 (1969).
STRASCHILL,M.: Actions of drugs on single neurons in the cat's retina. Vision Res. **8**, 35—47 (1968).
STRETTON,A.O.W., KRAVITZ,E.A.: Neuronal geometry: Determination with a technique of intracellular dye injection. Science **162**, 132—134 (1968).
SVAETICHIN,G.: The cone action potential. Acta physiol. scand. **29** (Suppl. 106), 565—600 (1953).
— Spectral response curves from single cones. Acta physiol. scand. **39** (Suppl. 134), 17—46 (1956).
— Células horizontales y amacrinas de la retina: propiedades y mecanismos de control sobre las bipolares y ganglionares. Acta Cient. Venezolana, Suppl. **3**, 254—276 (1967).
— LAUFER,M., MITARAI,G., FATEHCHAND,R., VALLECALLE,E., VILLEGAS,J.: Glial control of neuronal networks and receptors. In: JUNG,R., KORNHUBER,H. (Eds.): Neurophysiologie und Psychophysik des visuellen Systems, Symposium Freiburg 1960, S. 445—456. Berlin-Göttingen-Heidelberg: Springer 1961.
— NEGISHI,K., FATEHCHAND,R.: Cellular mechanisms of a YOUNG-HERING Visual System. In: Ciba Foundation Symposium on Colour Vision, pp. 178—203. Boston: Little, Brown and Comp. 1965.
SZENTÁGOTHAI,J.: The anatomical substrates of nervous inhibitory functions. Acta Morph. Acad. Sci. hung. **17**, 325—327 (1969).
TESTA,A.S. DE: Morphological studies on the horizontal and amacrine cells of the teleost retina. Vision Res. **6**, 51—59 (1966).
TOMITA,T.: A study on the origin of intraretinal action potential of cyprinid fish by means of pencil-type microelectrode. Jap. J. Physiol. **7**, 80—85 (1957).
— Electrical activity in the vertebrate retina. J. Opt. Soc. Amer. **53**, 49—57 (1963).
— Electrophysiological study of the mechanisms subserving color coding in the fish retina. Cold Spr. Harb. Sympos. quant Biol. **30**, 559—566 (1965).
— Electrical activity of vertebrate photoreceptors. Quart. Rev. Biophys. **3**, 179—222 (1970).
— MURAKAMI,M., SATO,Y., HASHIMOTO,Y.: Further study on the origin of the so-called cone action potential (S-potential). Its histological determination. Jap. J. Physiol. **9**, 63—68 (1959).
TOYODA,J.-I., HASHIMOTO,H., OHTSU,K., TOMITA,T.: Receptor-bipolar transmission in the retina. Personal communication. (Presented at Neurosciences Research Program Work Session on the Retina, March 23, 1970).
TRETJAKOFF,D.: Die Parietalorgane von *Petromyzon fluviatilis*. Z. wiss.Zool. **113**, 1—112 (1915).
UCHIZONO,K.: Synaptic organization of the Purkinje cells in the cerebellum of the cat. Exp. Brain Res. **4**, 97—113 (1967).
UGA,S., NAKAO,F., MIMURA,M., IKUI,H.: Some new findings on the fine structure of the human photoreceptor cells. J. Electr. Micr. **19**, 71—84 (1970).
UNDERWOOD,G.: Some suggestions concerning vertebrate visual cells. Vision Res. **8**, 483—488 (1968).
UYAMA,Y.: Die Retina des Säugetieres (II). Med. J. Osaka Univ. **2**, 629—673 (1951).
VAL'TSEV,V.B.: Role of cholinergic structures in outer plexiform layer in the electrical activity of frog retina (In Russian). Zh. Vysshei Nervnoi Deyatel'nosti (Pavlova) **15**, 934 (1965). Engl. Transl. Fed. Proc. **25** (Transl. Suppl.), T 765—T 766 (1966).
VAN DE KAMER,J.C.: Histological structure and cytology of the pineal complex in fishes, amphibians and reptiles. Progr. Brain Res. **10**, 30—48 (1965).
VAN HARREVELD,A., KHATTAB,F.I.: Electron microscopy of the mouse retina prepared by freeze-substitution. Anat. Rec. **161**, 125—140 (1968).

Ventura, J., Mathieu, M.: Exogenous fibres of the retina. Canad. Ophthal. Soc. Trans. 21, 184—196 (1959).

Viale, G., Apponi, G.: Histochemische Untersuchungen über die Cholinesterasen in der menschlichen Netzhaut. Z. Zellforsch. 55, 673—678 (1961).

Villegas, G. M.: Electron microscopic study of the vertebrate retina. J. gen. Physiol. 43 (6, part 2), 15—43 (1960).

— Comparative ultrastructure of the retina in fish, monkey and man. In: Jung, R., Kornhuber, H. (Hrsg.): Neurophysiologie und Psychophysik des visuellen Systems, Symposium Freiburg 1960, S. 3—13. Berlin-Göttingen-Heidelberg: Springer 1961.

— Ultrastructure of the human retina. J. Anat. 98, 501—513 (1964).

— Villegas, R.: Neuron-glia relationship in the bipolar cell layer of the fish retina. J. Ultrastruct. Res. 8, 89—106 (1963).

Vivien-Roels, B.: Étude structurale et ultrastructurale de l'épiphyse d'un Reptile: *Pseudemys scripta elegans*. Z. Zellforsch. 94, 352—390 (1969).

Vogel, F. S., Kemper, L.: A modification of Hortega's silver impregnation method to assist in the identification of astrocytes with electron microscopy. J. Neuropath. exp. Neurol. 21, 147—154 (1962).

Vrabec, Fr.: A new finding in the retina of a marine teleost, *Callionymus lyra* L. Folia Morphol. 14, 143—147 (1966).

— Obenberger, J., Bolkova, A.: Effect of intravitreous vincristine sulfate on the rabbit retina. Amer. J. Ophthal. 66, 199—204 (1968).

Walberg, F.: Axoaxonic contacts in the cuneate nucleus, probable basis for presynaptic depolarization. Exp. Neurol. 13, 218—231 (1965).

Wald, F., Robertis, E. de: The action of glutamate and the problem of the "extracellular space" in the retina. Z. Zellforsch. 55, 649—661 (1961).

Walls, G. L.: The visual cells of lampreys. Brit. J. Ophthalmol. 19, 129—148 (1935).

— The Vertebrate Eye and Its Adaptive Radiation. Bloomfield Hills, Michigan: Cranbrook Institute of Science 1942. Reprinted. New York-London: Hafner 1963.

Wartenberg, H.: The mammalian pineal organ: Electron microscopic studies on the fine structure of pinealocytes, glial cells and on the perivascular compartment. Z. Zellforsch. 86, 74—97 (1968).

Weidman, T. A., Kuwabara, T.: Correlation of morphogenesis and physiology of the developing retina. Invest. Ophthal. 6, 453 (1967).

— — Postnatal development of the rat retina. Arch. Ophthal. 79, 470—484 (1968).

— — Development of the rat retina. Invest. Ophthal. 8, 60—69 (1969).

Werblin, F. S., Dowling, J. E.: Organization of the retina of the mudpuppy, *Necturus maculosus*. II. Intracellular recording. J. Neurophysiol. 32, 339—355 (1969).

Witkovsky, P.: An ontogenetic study of retinal function in the chick. Vision Res. 3, 341—355 (1963).

— Synapses made by myelinated fibers running to teleost and elasmobranch retinas. J. Comp. Neurol. 142, 205—222 (1971).

— Dowling, J. E.: Synaptic relationships in the plexiform layers of carp retina. Z. Zellforsch. 100, 60—82 (1969).

— Stell, W. K.: Gross morphology and synaptic relationships of bipolar cells in the retina of the smooth dogfish, *Mustelus canis*. Anat. Rec. 169, 456—457 (1971).

Wolfe, D. E.: The epiphyseal cell: an electron-microscopic study of its intercellular relationships and intracellular morphology in the pineal body of the albino rat. Progr. Brain Res. 10, 332—376 (1965).

Wolter, J. R., Liss, L.: Zentrifugale (antidrome) Nervenfasern im menschlichen Sehnerven. Albrecht v. Graefes Arch. Ophthal. 158, 1—7 (1956).

— Silver carbonate techniques for the demonstration of ocular histology. In: Smelser, G. K. (Ed.): The structure of the eye, pp. 117—138. New York-London: Academic Press 1961.

— The reactions of the centrifugal nerves of the human eye: After photocoagulation, occlusion of the central retinal artery and bilateral enucleation. In: Rohen, J. W. (Ed.): The structure of the eye, II. Symposium, Wiesbaden 1965, S. 85—95. Stuttgart: Schattauer 1965.

YAMADA, E.: Some observations on the fine structure of the human retina. Fukuoka Acta Med. **57**, 163—182 (1966).

— ISHIKAWA, T.: The fine structure of the horizontal cells in some vertebrate retinae. Cold Spr. Harb. Sympos. quant. Biol. **30**, 383—392 (1965).

— — The so-called "synaptic ribbon" in the inner segment of the lamprey retina. Arch. histol. jap. **28**, 411—417 (1967).

— — HATAE, T.: Some observations on the retinal fine structure of the snake *Elaphe climacophora*. In: UYEDA, R. (Ed.): Electron Microscopy 1966. Proc. 6th Intl. Congr. Electr. Micr., Kyoto 1966, **2**, 495—496. Tokyo: Maruzen Co. Ltd. 1966.

YASUZUMI, G., TEZUKA, O., IKEDA, T.: The submicroscopic structure of the inner segments of the rods and cones in the retina of *Uroloncha striata* var. *domestica* Flower. J. Ultrastruct. Res. **1**, 295—306 (1958).

YOUNG, J. Z.: The photoreceptors of lampreys. I. Light-sensitive fibres in the lateral line nerves. J. exp. Biol. **12**, 229—238 (1935).

Chapter 4

Interactions between Light and Matter

By

David Mauzerall, New York, New York (USA)

With 6 Figures

Contents

The past twenty years have seen an enormous growth of research in photo-chemistry. Aside from the simple increase in population, in general, and of science in particular, this growth can be assigned several causes. One is the realization that whole chemistries lay waiting to be discovered, since the reactions of mole-cules in excited electronic states can be distinctly different from those in their ground states. Whole "Beilsteins" of singlet and triplet state organic chemistries can be imagined! A second cause is the availability of the required instrumentation: automatic recording spectrophotometers, analytical methods such as chromato-graphy and mass spectrometry, and of course nuclear magnetic resonance for structural work. Last but not the least of these technical advances has been the introduction of lasers. They have not only transformed the closed field of physical optics into an energetic discipline, but have vastly extended the time scale of photochemistry. Recent measurements show that processes occurring in 10^{-12} s can now be studied directly. Finally, the two fundamental photoprocesses of biology — the energy conversion of photosynthesis and the information conversion of vision have provided a challenge worthy of any photochemist. They have spurred a great deal of experimental and theoretical work aimed at explaining their basic photochemistry.

In this introduction to photochemistry I will deal largely with reactions in the condensed phase, since these are more relevant to biological problems. Molecular energy states, which introduce the common language of Molecular Orbital Theory will be described first. This will be followed by a description of the interaction of radiation and matter, first from the classical and then very briefly from the quantum mechanical viewpoints. The life history of an excited state will then be described. Photochemical reactions will appear as one of a large number of alternatives. Finally, three classes of photochemical reactions which may be of some consequence in photobiological systems will be considered.

Molecular Energy States

In the absence of collisions, isolated atoms and molecules exist in definite, discreet energy states. They can only change their energy in discreet steps by the absorption or emission of photons. In complex molecules, there may be many ways to distribute this energy among rotational, vibrational and even electronic motions. The lowest possible such state of a molecule is called the ground state. The translational motion of molecules can always be treated classically and assigned the $1/2\,kT$ energy per degree of freedom where k is the Boltzman constant and T the absolute temperature. The other molecular motions: rotational, vibrational and electronic must be treated by quantum mechanical means; i.e., the appropriate Schroedinger equation must be solved for the energy states. The interelectronic repulsion of multi-electron atoms and molecules makes the problem very difficult. Moreover, the many degrees of freedom can interact with one another making the general problem quite unsolvable. However, the energies, or frequencies, of such motions are reasonably well separated: rotations, $\sim 10\ \mathrm{cm}^{-1}$ or 3×10^{11} cps; vibrations, $\sim 10^3\ \mathrm{cm}^{-1}$ or 3×10^{13} cps; electronic, $3 \times 10^4\ \mathrm{cm}^{-1}$ or 10^{15} cps. Thus they are treated as independent at the first level of approximation.

The calculations of the detailed energy levels from basic constants and quantum mechanical theory are straightforward for the hydrogen atom, have proceeded with great difficulty for heavier atoms, require many approximations for diatomic molecules, and are reduced to a bare scaffolding of semi-emperical calculation for molecules of interest to biologists. The simplest model of the energy levels in large conjugated molecules is that of free electrons in a box. The single bonds of the molecular skeleton are assumed to form the box and the (π) electrons in the conjugated double bonds move freely in this box. For a one dimensional box with walls of infinite potential, the Schroedinger equation is directly solved for the energy levels: $E = h^2 n^2/8\,m\,a^2$, (Murrell, 1963; Kauzmann, 1957), where a is the length of the box, h is Planck's constant, m is the mass of the electron, and n is an integer, i.e., a quantum number. This simple model is able to predict the energy of the first absorption band of simple polymethine dyes:

$$[R_2N = CH\,(-CH = CH)_n - NR_2]^+,\ (\text{Fig. 1 a}).$$

With each additional double bond (i.e., as n increases), the maximum of the absorption band shifts to the red by about 100 nm. This is in good agreement with the calculated red shift of about 130 nm. By way of contrast it should be pointed out that the spectra of a series of simple polyenes, $CH_3(-CH = CH)_n - CH_3$ (Fig. 1 b), or of polyenes with a heteroatom at only one end, e.g. retinal, do not follow such a simple relation. Their first absorption band tends to a low energy or long wavelength limit as n increases. Comparison of these two kinds of molecules show that all the bonds are equivalent for the polymethines (Fig. 1 a), while this is not the case for the polyenes (Fig. 1 b). In the first case, the equivalence of the bonds means their lengths, and also the potential seen by the "free" electron is constant, as was assumed in the calculation quoted above. In the case of the polyenes this potential is periodic with a period equal to the single plus double bond length, because the electron has a different potential in a double bond (positive nuclei closer together) than in a single bond. The energy levels calculated from a model with periodic potentials (no longer so easy) fit experimental observations fairly well. In particular, they predict the long wavelength limit to be 550 nm. For $n = 15$, decapreno-β-carotene, 510 nm is observed (MURRELL, 1963). Carotenes are the best example of such pigments in nature. Their maximum depth of color is reddish unless they are

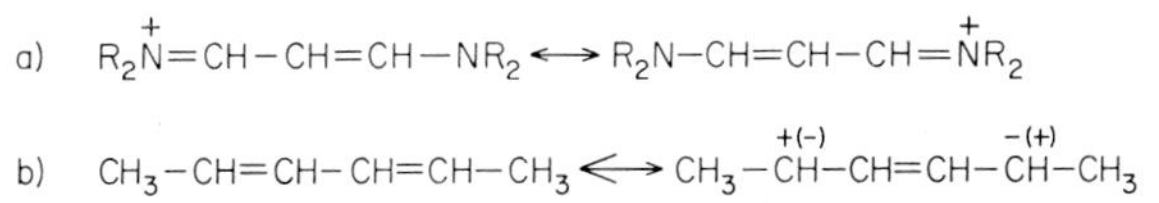

a) $R_2\overset{+}{N}=CH-CH=CH-NR_2 \longleftrightarrow R_2N-CH=CH-CH=\overset{+}{N}R_2$

b) $CH_3-CH=CH-CH=CH-CH_3 \overset{+(-)}{\underset{}{\longleftrightarrow}} CH_3-\overset{+(-)}{CH}-CH=CH-\overset{-(+)}{CH}-CH_3$

Fig. 1. Equivalence of the bonds in a polymethine (a) and non-equivalence of bonds in a polyene (b)

complexed to a particular protein as in the lobster shell, or in rare cases have conjugated heteroatoms at both ends of the molecules. The absorption bands are also predicted to be fully allowed (see below) i.e., to have an extinction coefficient of $\sim 10^5\,M^{-1}\,cm^{-1}$, as is observed. The free electron model is readily extended to a closed ring. In this case, all the energy levels except the lowest are degenerate: two different states of angular momenta exist, but are of equal energy. A classical picture is that the electrons can rotate clockwise or counter-clockwise about the ring. Transitions between adjacent levels can take place between states of angular momenta of similar or opposite sign. In the first case, the change in angular momentum is 1 unit (an "allowed" transition). In the second case, the change is $2n + 1$ units, where n is the quantum number of the lower level. Such transitions are "forbidden" and therefore weak. These weak transitions are also expected to be at longer wavelengths because of Hund's rule, to be explained later.

The description just presented explains the observed spectra of cyclic aromatic and heterocyclic molecules, including the porphyrins, quite well. The latter molecules are characterized by a very strong band near 400 nm — the Soret band — and the relatively weak bands in the visible. The latter are responsible for the characteristic red colors of the porphyrins.

I have spent some time on the free electron model because I believe it gives a simple, direct picture of electronic energy levels. However, the very simplicity of this model is also its limitation. By far the most work on molecular spectroscopy is accomplished with the molecular orbital (MO) approach (MURRELL, 1963). It is especially well suited to conjugated systems which show long wavelength absorptions and are thus of interest in biology. An orbital is the space about a nucleus which is occupied by the electron. It represents the probability of finding the electron in a certain volume of space. This probability is proportional to the square of the wave function, found by solving the Schroedinger equation. For the hydrogen atom these functions are analytic and are quoted in all text books (KAUZMANN, 1957). In multi-electron atoms, the inter-electronic interactions are assumed to change the energy levels of the hydrogen-like orbitals, but not their basic character. Molecular orbitals are constructed by adding mathematically the orbitals of the individual atoms of the molecule. In principle all of the electrons in a molecule could — and should — be included in the molecular orbital. However, it is expedient to neglect the core electrons and, usually, even the sigma (σ) electrons. For atoms in the first row of the periodic table, the core electrons are the 1 s electrons localized at each atom. The electrons in σ orbitals are responsible for the single bonds holding atoms together; they are made of $2s$ and of $2p$ orbitals which point toward each atom of the bond. They are characterized by cylindrical symmetry about the internuclear axis. The p orbitals perpendicular to the single bond can overlap, or bond, with adjacent similarly situated p obitals to form π bonds. A diagram of these orbitals is given in Fig. 2. Thus, for large molecules (number of atoms > 2) it is often assumed that the framework or geometry of the molecule is determined by the σ electron bonds and only the π electron energy levels, largely responsible for near UV and visible spectra, are calculated. With the advent of fast computers, it has become possible to explore both the effect of the core and σ electrons on the π levels and the effect of geometry on these energy levels. Both effects are quite important, the latter especially for the problem of cis-trans isomerization which sooner or later has to be faced. The simpler theory approximates a wave function with a linear combination of atomic orbitals (LCAO). The coefficients of these atomic orbitals are optimized by variational calculus to minimize the total energy. The simultaneous multi-electron coulomb repulsions make the problem intractable, so these interactions are usually approximated by an "effective" potential. Certain procedures have been developed to attempt to correct these insufficiencies. These are the self consistent field (SCF) and configuration interaction (CI) methods. In the former the electrostatic field obtained after the first solution of the problem is used as the start for a second calculation and this, is reiterated until the field changes by less than a prescribed amount. In the latter those configurations or energy states defined by a first calculation that have the correct properties such as symmetry are allowed to interact to produce new energy states which are combinations of the previous ones. The energy levels resulting from such calculations are best classified by their symmetry properties, but a simpler and more qualitative classification has come into wide use.

A very useful classification of molecular spectroscopic absorptions or transitions is that developed by KASHA (1960). The classification designates the orbitals chiefly involved in the transitions. As we have seen orbitals are designated σ if they have cylindrical symmetry along an internuclear axis (Fig. 2a). They are called π if they have a nodal plane in the internuclear axis. If this axis is called z, σ molecular orbitals are made up of s and p_z or two p_z atomic orbitals, and π molecular orbitals are made up of two p_x or two p_y atomic orbitals. On atoms such as nitrogen or oxygen singly or doubly bonded (meaning σ or $\sigma + \pi$ bonded) to carbon, one of the p orbitals of the heteroatom is left in the lurch, so to speak, and it is called an n orbital (nonbonding). Because of differing symmetries, the nonbonding orbital on a conjugated nitrogen such as pyridine or Schiff bases is sometimes called l.

Some comment on the use of the word symmetry is required. Since we are flailing about in a sea of approximations, it would be nice if we had a reliable beacon guiding us towards our goal. Symmetry is just such a guide. Any physical property of a molecule, such as absorption of light, must be invariant to symmetry operations such as rotations about an axis or reflection in a plane which leave the molecule in a state physically indistinguishable from the previous state. In forming molecular orbitals, for example, their symmetry must reflect the symmetry of the molecule. Thus s and p_z orbitals combine because they both have cylindrical symmetry about the internuclear axis, and the resulting σ orbital has the same symmetry. Likewise, s and p_x orbitals will not combine because p_x has a different symmetry (rotation of $180°$ with change of sign of the wave function) referred to the internuclear axis than has the s orbital (spherical, i.e. rotation by any angle with no change of sign).

An illustration of these concepts is shown for formaldehyde (Fig. 2). An s and two p (p_z and p_y) orbitals have already been mixed on the carbon atom (sp^2 hybridization) to give two sp hybrid orbitals bonding to hydrogen and an sp_z hybrid along the axis to oxygen. This orbital and the corresponding one from oxygen combine to form two molecular orbitals, σ a bonding orbital of lower energy than its constituents and σ^* a nonbonding orbital of higher energy (Fig. 2a, b). These are in the plane of the molecule. The second type of molecular orbitals, π and π^* are similarly formed by combination of the two p_x orbitals. These are situated perpendicular to the molecular plane and the wave functions change sign on reflection in the plane. The n orbital is the "left over" p_y orbital on oxygen. The lower energy of the bonding orbitals accounts for the stability of the molecule when these are filled with electrons. This is caused by the electrons being concentrated between the nuclei and thus in a region of low potential. Contrariwise, the antibonding orbitals keep the electrons away from this cozy region and pay the price of lesser binding energy. Note that the bonding and the antibonding orbitals have the same basic symmetry.

An electronic state of a molecule is defined by the distribution of electrons among the various orbitals. The lowest or ground state is that in which the lowest orbitals are filled with two electrons of opposite spin (more of which later). An excited state is that in which one or more of these electrons occupy higher energy orbitals. The electronic transitions from one state to another are usually accomplished by the absorption or emmission of a photon of light. The $\pi \rightarrow \pi^*$ transitions (Fig. 2b) are allowed, but the $n \rightarrow \pi^*$ are partially forbidden because of poor spatial overlap. As can be seen in Fig. 2, the π and π^* orbitals occupy a good part of the same region of space and have the same symmetries. The n and π^* orbitals occupy quite different regions of space and have different symmetries. Since weak transitions are often allowed on certain vibrational distortions, the $n \rightarrow \pi^*$ bands often show fine vibrational structure. Their absorption of polarized light also differ — the electronic displacement in the $\pi \rightarrow \pi^*$ transition is seen to be in the plane of the CH_2O molecule (Fig. 2a) while that of the $n \rightarrow \pi^*$ transition will be perpendicular to this plane. One also expects a particular solvent effect on $n \rightarrow \pi^*$ transitions. Proton donors will hydrogen bond to the n electrons (the classical "unshared pair"). This will stabilize these electrons, i.e. lower the n level, and thus shift the $n \rightarrow \pi^*$ transition to shorter wavelengths, often burying it under the far

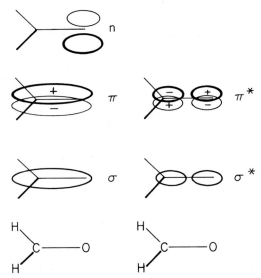

Fig. 2a. Classification of molecular orbitals in formaldehyde: positive and negative signs show phase of wave function; only the same phases will overlap. The CH_2 orbitals are omitted

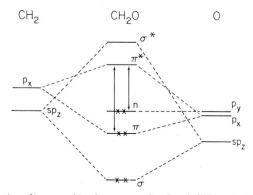

Fig. 2b. The correlation diagram for the energy levels of CH_2 and O with formaldehyde. Individual levels of the C—H bonds are omitted. The $n - \pi^*$ and $\pi - \pi^*$ transitions are shown by arrows, crosses denote electron occupation in the ground state of formaldehyde

stronger $\pi \rightarrow \pi^*$ transition. Thus protonic solvents change the spectra of such molecules in a predictable way. The $\pi \rightarrow \pi^*$ transitions usually show the fundamental shift to longer wavelength on increasing the refractive index of the solvent (Basu, 1964). These characteristics: transition probabilities, polarization and solvent dependence are used as operational criteria to label bands as $\pi \rightarrow \pi^*$ or $n \rightarrow \pi^*$ in the observed spectra of molecules.

In all of the above discussion, we have assumed that the electron spins are paired in each orbital and remain so on transition to a new level. The electron spin is a purely quantum mechanical concept characterizing the magnetic interactions of the electron. It is classically caricatured as a finite sized electron spinning on a fixed axis, thus giving rise to a magnetic moment. The component of this magnetic moment along the external magnetic field axis is quantized, and can only have the value $\pm 1/2$ in appropriate units. The Pauli exclusion principle states that no two electrons in an atom or molecule can have all the same quantum numbers. In our simplified pictures, this means that two electrons can occupy a single orbital only if their spins are opposite. Since ordinary organic molecules have an even number of electrons and their energy levels are well separated, their ground states have all electrons paired. They are said to be in the singlet state since in a magnetic field no splitting of these states is possible. For those few molecules having two orbitals of identical energy (said to be "degenerate") with only two electrons present, the electrons need not be paired. Whether or not the electrons pair depends on the inter-electron interactions. If they are unpaired (i.e. one in each orbital with parallel spins) the state is said to be a triplet. In a magnetic field this state will split into three levels. Oxygen is the well known example of such a ground state triplet, and the prediction of this state is a triumph of simple molecular orbital theory. Methylene or carbene and its derivatives are recently discovered examples of divalent carbon which have such triplet ground states (TROZZOLO, 1968; KIRMSE, 1964).

Whereas the unpaired electrons in the ground state triplet must occupy iso-energetic (degenerate) orbitals, those in the electronically excited triplet state occupy orbitals of different energy. Consequently, excited triplet states are far more common. Since, according to the rules of quantum mechanics, spin reversal is forbidden on electric dipole excitation, direct ground state singlet to excited triplet state transitions are not allowed. Experimentally they are found to be about a million times weaker than allowed transitions for molecules composed of light, i.e. first row, elements. The angular momentum of the spin change must be supplied by some other interaction, such as that of spin-orbit coupling. This is very weak for first row atoms.

The spin of the electron is neglected in simple molecular orbital theory, so singlets and triplets have the same energy. A more elaborate theory corrects this neglect, but only at the cost of more complex calculations. In general, the lowest lying triplet state almost always lies below the first excited singlet in energy. This is usually explained with a reference to Hund's rule from atomic spectroscopy: the state of maximum multiplicity is lowest in energy. The difference in energy is caused by electron correlation effects. The electrons tend to be closer together in a singlet state and the Coulombic repulsion raises its energy over that of the triplet state.

Physical Properties of Excited States

In general not only the chemistries, but the physical properties of excited states of molecules differ from those of the familiar ground state. The geometrical structure can be strikingly different (BRAND and WILLIAMSON, 1963), e.g. the linear acetylene molecule is skewed into a trans-bent form in the lowest excited singlet state, with a $C-C-H$ angle of $120°$ and a $C-C$ bond length of 1.4 A (1.2 A in the ground state). Planar ground state formaldehyde becomes slightly pyramidal [H_2-C-O angle is $20°$] in the $n-\pi^*$ singlet state. Aromatic systems, and presumably polyenes, are less changed in the excited state. We shall return to this point when discussing cis-trans isomerizations. The dipole moment of the molecule usually changes on excitation. Again formaldehyde is an excellent example. The ground state has a dipole moment of 2.3 Debyes, the more electronegative oxygen being negative. From the simple MO picture of the $n - \pi^*$ state (Fig. 2) one might expect the dipole moment to decrease or even to change sign. This can be measured via the Stark effect, which is the effect of an electric field on the absorption band. In fact, the dipole moment of the first excited singlet state is reduced to 1.5 Debyes, but with the same sign (Freeman and Klemperer, 1964). As is often the case with the simple MO pictures, the ionic effects are overstated.

One might anticipate that the acid-base properties of such molecules as formaldehyde would also be changed in the excited states. One of the first examples of this kind was 1-naphthol, whose hydroxyl group is 10^7 times more acidic in the lowest excited singlet state (pK: 2.0) than in the ground state (pK: 9.2). This was found by noting a large difference between absorption and fluorescence changes as a function of pH. Not only have many such examples been studied (Weller, 1961), but this is a method of studying ultra-fast reactions. The singlet excited state lives for only a few nanoseconds, so proton transfers near diffusion limited rates ($10^{10} M^{-1} s^{-1}$) are readily measured. This kind of pK change in the excited state may be of importance in rhodopsin, since it is believed that the retinal is bound as a protonated Schiff base to the lipo-protein. It would make an ideal primary photochemical event: fast and efficient.

Excitation

Excited states of molecules are most readily populated by absorption of photons of the appropriate frequencies. However, excitation by collisions with electrons in gases or solids, or by ionizing radiation, and even by chemical reactions are possible. Only a bare outline of the interaction of radiation and matter can be given. I will closely follow the excellent description of Kauzman (1957), to which I refer the reader for a more extensive treatment. (References to this source will be given by K, followed by the page number.) The volumes of Feynman et al. (1963) also contain an excellent treatment of this subject. A qualitative discussion can be found in Reid (1957), and more quantitative discussions in books by Heitler (1954) or Hameka (1965).

The discussion will be simplified by neglecting magnetic interactions from the onset. We thus sacrifice optical activity and magnetic dipole interactions such as nuclear and electron magnetic resonance. We start with the classical description of the electric field. Lines of force radiate from a fixed charge or one with constant velocity. Only an accelerated charge emits a propagated field: since the signal from the accelerating charge travels at the speed of light a "kink" develops at the place in the field prior in time to this acceleration. This kink is maximal at right angles to the motion of the charge and zero in line with it. An oscillating dipole has a rather complex field distribution in the near field region (distance comparable to separation of charges), but simplifies further out. This is the fundamental model of an emitter or antenna. The elementary dipole is associated with the electron bound to a positive nucleus. The classical equation of motion for a single bound electron states that the sum of the forces should be zero.

inertial	restoring	radiation damping	fields
$- m\ddot{x}$	$- kx$	$- \mu\dot{x}$	$+ eE = 0 .$ (1)

Wherein m and e are the mass and charge of the electron, E is the electric field, k and μ are constants, x is a displacement and dot refers to differentiation with respect to time. The solution of the equation with only the first two terms is that of simple harmonic motion:

$$x = x_0 e^{i\omega_0 t} \qquad \omega_0^2 = k/m. \tag{2}$$

This represents an infinite wave train (ω is the circular frequency, $= 2\pi\nu$, $\nu =$ linear frequency) and to rescue the first law of thermodynamics, it was necessary to include the classical radiation damping term. The solution to the equation with the first three terms is:

$$x = x_0 e^{i\omega_0 t} e^{-\omega' t} \qquad \omega' = \mu/2m \qquad \omega_0^2 = k/m - (\mu/2m)^2. \tag{3}$$

This represents a damped harmonic oscillation, and thus a finite amount of energy is emitted by the excited dipole. Because of the finite lifetime of the source $(Y = 1/\omega')$, and therefore the finite wave train emitted, one can determine its frequency only to an accuracy given by about one cycle out of the total cycles in the time Y. Thus $\Delta\omega_0 = 2\pi\Delta\nu \cong 2\pi/Y$. For radiation at 600 nm this corresponds to a linewidth of about 4×10^{-5} nm and a lifetime of about 3×10^{-8} sec. This result hints of the uncertainty principle of quantum mechanics which gives a similar answer: $\Delta E \Delta t = h/2\pi$, where ΔE is the uncertainty in energy, related to frequency by another fundamental result of quantum mechanics $E = h\nu$. The linewidth is experimentally determined by dispersing the light with a spectrograph. This instrument essentially disperses the component frequencies of the electric field (light) in space so that the amplitude as a function of frequency may be resolved. Conversely, the electric field at a point in space as a function of time is the sum of the amplitudes of the component frequencies. By taking the Fourier transform of the damped harmonic motion of our pure dipole emitter one obtains the amplitudes as a function of frequency. The intensity, or flow of energy in the various components is proportional to the square of the amplitudes, and the result $(K, p\ 566)$ is,

$$I(k) = \frac{E_0^2}{16\,\pi^2} \frac{1}{\omega'^2 + (\omega - \omega_0)^2} \qquad \omega = \text{variable circular frequency.} \tag{4}$$

This is the equation for a Lorentzian line shape having a width at half intensity of ω'/π. The intensity is kept finite at the resonance frequency $\omega = \omega_0$ by the finite lifetime, related to ω'.

Many other effects cause the emission to be further broadened. In a gas the motion of the molecules in the direction of the emitted radiation leads to a frequency shift because of the Doppler effect. The change in wavelength will be proportional to the ratio of velocity of the emitter to the frequency of the radiation. For a molecule of mass 20, emitting at 600 nm, the shift would be about 10^{-4} nm. The gas will usually have a Maxwellian distribution of velocities, and thus one must integrate over this distribution. The resulting line shape, as a function of frequency, is Gaussian $(K, p\ 568)$. At any high pressure of the gas, collisions will cause a large broadening of the emitted radiation. A complete treatment is very difficult, but if one assumes the wave train is simply interrupted at each collision, then one can consider a sort of lifetime broadening given by the mean collision time:

$$\Delta\omega_0 = \sqrt{2}\,v n \sigma \qquad n = \text{molecules cm}^{-3}, \ \sigma = \text{cross section,} \ v = \text{velocity.} \tag{5}$$

For visible radiation 600 nm, this comes to about 3×10^{-4} nm for a gas of molecular weight 50 at one atmosphere and $300°$ K. Simple extrapolation to the condensed phase ($n \simeq 10^{23}$ molecules cm^{-3}) yields a pressure broadening of about 3 nm. This is not too dissimilar to what is actually observed in nonpolar solvents. However, as we approach the case of perpetual collisions, a more realistic picture of a collision is required. This means considering the effect of the electrical fields of the neighboring molecules on the emitter, i.e. one must include the fourth, time dependent, term in Eq. (1). The results strongly depend on the exact form of this term. Experimentally it is observed that on changing from a nonpolar to a polar solvent, emissions and absorptions broaden and fine structure is blurred out (Jaffe and Orchin, 1962).

This brings us to the interesting case of the external field being electromagnetic radiation, i.e. what is the response of a bound electron to light ? The fourth term in Eq. (2) is replaced by $e|E_0| \sin \omega t$, and the resulting inhomogeneous differential equation has the solution of the homogeneous Eq. (4) and a particular solution:

$$X_1 = A e^{i(\omega t - \phi)} \qquad \text{where } \tan \phi = \frac{2 \omega \omega'}{\omega_0^2 - \omega} . \qquad (6)$$

There is thus a phase lag in the displacement: ~ 0 at $\omega \ll \omega_0$, $\pi/2$ at $\omega = \omega_0$ and π for $\omega \gg \omega_0$. A very useful parameter is the polarizability, α, defined as the ratio of induced moment to driving force. The in phase or real component, α, is responsible for the index of refraction of a substance, and the out of phase, or imaginary, component, α', is responsible for the light absorption by the substance. The usefulness of the polarizability is extended by the fact that a simple redefinition in quantum mechanical terms allows all classical results to be conserved (K, p 569).

$$\alpha = \frac{e X_0}{|E_0|} = \frac{e^2}{m} \frac{\omega_0^2 - \omega^2}{(\omega_0^2 - \omega^2)^2 + 4\omega'^2\omega^2} , \qquad (7)$$

$$\alpha' = \frac{e X_0 \tan \phi}{|E_0|} = \frac{2 e^2}{m} \frac{\omega \omega'}{(\omega_0^2 - \omega^2)^2 + 4\omega'^2\omega^2} . \qquad (8)$$

At very low frequencies, Eq. (7) gives the contribution to the static polarizability of the resonance of ω_0. It also predicts the zig-zag shape of the polarizability, unfortunately named anomalous dispersion, through the absorption region. The absorption line, of Lorentzian shape, is given by Eq. (8).

We have assumed an isotropic oscillator. For a fixed linear oscillator only that component of the electromagnetic field along the dipole is effective. For random orientation of the dipoles, the components of the induced moments perpendicular to the field cancel and the average over $\cos^2 \theta$ is 1/3. The observed polarizability is one third of the maximum observable.

Operationally, the tendency to absorb light is defined by the molar absorptivity, a, which we define by the Beer-Lambert law:

$$I = I_0 10^{-a M l} \qquad a = -\frac{1}{M} \frac{d \log_{10} I}{d \ell} = -\frac{1}{2.3 M I} \frac{d I}{d \ell} \qquad (9)$$

where I_0 is the incident light intensity of a collimated beam, M is the molarity of the solution and ℓ is the path length of light in the solution. But the gradient of the light intensity is the product of the density of molecules, n molecules per cc, and the rate of energy dissipation per molecule, W. This rate is given by the energy density times the fraction dissipated per cycle times the frequency:

$$W = \frac{|E_0|^2}{2} \alpha' \omega \ . \tag{10}$$

The intensity of a light beam is defined as the radiant energy passing a unit area per second (see FEYNMAN 1-31-10, 1963):

$$I = \frac{c}{8\pi} |E_0|^2 \ , \tag{11}$$

by combining these relations, $a = \dfrac{4\pi N}{2303\,c}\, \omega\alpha'$ $(K, p579)$. \hfill (12)

We note that pure radiation theory says that the absorption coefficient increases with frequency, and it is well known that analytical spectrophotometry in the ultraviolet is more sensitive than in the infrared.

For molecules in solution the discrete vibrational and rotational structure of the absorption by a complex molecule is smeared into a broad band by the fluctuating forces from both intra- and intermolecular interactions. Failing a detailed understanding of these things, a useful parameter remains; the area under the absorption band. In fact, a suitably normalized integral of the absorption as a function of frequency yields the number of oscillators. This suggests a useful concept — the oscillator strength, defined, for convenience, by the experimental absorptivity:

$$f = \frac{2303\,mc}{\pi\,Ne^2} \int a\,dv \ . \tag{13}$$

Classically, f is thought of as the number of electrons absorbing the light. Experimentally, f is often found to be less than unity, and such bands are said to be forbidden. For example, the practically colorless rare earths have $f \sim 10^{-7}$ in the visible; the weak color of the transition metal ions is caused by an f of $\sim 10^{-4}$; the intense color of permanganate ion has an f of 0.03, and many dyes have an f of ~ 1. The Soret band of many porphyrins has an f of ~ 2, because of a degenerate excited state.

The above picture of the interaction of radiation and matter is inadequate. It founders on Planck's law, and is replaced by quantum electrodynamics. This theory has been highly developed and at present accounts for all experimental data on simple systems. Unfortunately the erudite mathematical brilliance which surrounds the modern theory makes it accessable only to the initiates of advanced theoretical physics. I will restrict myself to a verbal description of a simplified view of the quantum theory of radiation, and refer the reader to the references for a more serious discussion (HEITLER, 1954; FEYNMAN et al., 1963, Vol. III; HAMEKA, 1965).

For simple systems the time independent Schroedinger equations can be completely solved for all cases of interest, and the resulting eigenvalues, or stationary energy states, of such systems can be determined. In the presence of a variable field or interaction, however, the full time-dependent equation must be used. A simplification is possible if the perturbation is slow, meaning the rate of its change is less than the energy difference between the states divided by h. The system is thought of as being always in a definite eigenstate for all values of the perturbation. The Born-Oppenheimer approximation to separate nuclear and electronic motions in molecules is a good example of such a slow change. The energy levels of the system can be solved for any value of the parameters, e.g. the equilibrium internuclear distance. If the perturbation is rapid, meaning the rate of its change is equal to or greater than the energy difference over h, the system will oscillate between states at a frequency $\nu_{k0} = E_k - E_0/h$. The final result depends on the frequency of the radiation causing the perturbation. If it is equal to an eigen frequency of the system, that state of energy E_k is populated with ever increasing probability with a corresponding decrease of energy in the field. If in fact the original state is the excited state of energy E_k, the state will be depopulated and the ground state population increased as the energy of the field increases, i.e. what is known as stimulated emission occurs. If the frequency does not correspond to any natural frequency, only transients of frequency ν_{k0} occur, corresponding to the classical "in phase" polarizability. These transients can lead to observable effects, namely, Rayleigh, Brouillin and Raman scattering (Fabelinskii, 1968). Since the system can exist only in distinct energy states, the absorption from (or emission to) the field can occur only in bundles of energy or quanta, thus leading to the famous equation

$$h\nu_{k0} = E_k - E_0. \tag{14}$$

It must be noted that the above statements are true for ordinary light intensities. For intensities now available from giant pulse lasers, nonlinear effects or multiphoton processes become observable, and are deduced from a complete treatment (Bloembergen, 1965).

In the quantum mechanical theory the induced moment is given by the transition moment times the field. The transition moment is:

$$m_{k0} = \int \psi_k^*(\mathrm{ex})\,\psi_0\,ds \tag{15}$$

where e is the charge and x the coordinate of the electrons, ψ_k and ψ_0 are the wave functions for the two states, and the integral is over all space. This transition moment is readily related to the classical oscillator strength or absorption coefficient (K, $p644$):

$$|m_{k0}|^2 = \frac{3\,h\,e^2\,f}{4\pi\,\omega_{k0}\,m} = \frac{6909\,h\,c}{4\,\pi^2\,N\,\omega_{k0}} \int a\,d\nu\,. \tag{16}$$

Similarly one can define a polarizability [Eq. (16), (12), and (13)] and thus take over the classical results directly, simply redefining the oscillator strength. A very basic quantity is the transition probability per unit time per unit of radiation energy density, also known as the Einstein Coefficient:

$$B_{0k} = \frac{(2\,\pi)^3}{3\,h^2}\,|m_{k0}|^2 = B_{k0}. \tag{17}$$

The B_{0k} refer to absorption of a photon, and B_{k0} to stimulated emission. The equality follows from an examination of the properties of the definition of m_{k0}, Eq. (15).

What if there is no radiation field present? The above discussion implies that the gound state remains in the ground state and this satisfies thermodynamics. However, it also says the excited state remains excited and this is contrary to experience—fluorescence is a fairly common phenomena. The error is in quantizing the energy levels of matter and ignoring the quantization of the radiation field.

There must always be present the black body and zero point fields. Dirac showed that quantum field theory leads to both spontaneous and stimulated emission. We will calculate this probability by following Einstein. The results of quantum mechanics must agree with Planck's law, which is empirical. The energy per unit volume $\varrho(\omega)$ at frequency ω in a black body at temperature T is

$$\varrho(\omega) = \frac{h\,\omega^3}{\pi^2\,c^3\,(e^{h\omega/2\pi kT}-1)} \tag{18}$$

(In fact, Planck's law can now be derived from quantum mechanical radiation theory.) For a system of N_0 atoms in a ground state and N_k atoms in an excited state, the distribution of atoms at thermal equilibrium is given by Boltzmann's equation:

$$N_k/N_0 = e^{-(E_k-E_0)/kT} = e^{-h\omega_{k0}/2\pi kT} \, . \tag{19}$$

At thermal equilibrium the rate of up transitions must equal that of down transitions:

$$N_k[A_{k0} + \varrho(\omega)\,B_{k0}] = N_0\,B_{0k}\varrho(\omega) \, , \tag{20}$$

where A_{k0} is the spontaneous transition probability. So

$$\varrho(\omega) = \frac{A_{k0}}{B_{k0}} \left[\frac{N_0}{N_k} - 1\right]^{-1} = \frac{A_{k0}}{B_{k0}} / (e^{h\,\omega/2\pi kT} - 1) \, . \tag{21}$$

and thus by Planck's law,

$$\frac{A_{k0}}{B_{k0}} = \frac{h\,\omega_{k0}^3}{\pi^2\,c^3} \text{ or } A_{k0} = \frac{8\,\pi\,\omega_{k0}^3}{3\,h\,c^3}\,|m_{k0}|^2 \, . \tag{22}$$

EINSTEIN pointed out that if there are no spontaneous transitions, one could never reach thermodynamic equilibrium between the states. Simple, but very revealing. Spontaneous transitions play the same role as collisions in kinetic gas theory. We also see that a very common observation, fluorescence, can only be explained by a purely quantum mechanical description of both the radiation field and the energy levels of the material.

The reciprocal of A_{k0} is the mean lifetime of the system. This is how the problem of classical radiation damping is side stepped. For a fully allowed transition (classical $f = 1$) at 6000 A, this lifetime is $\sim 5 \times 10^{-9}$ s. The dependence of this probability on the third power of the frequency explains the observation that many more molecules fluoresce in the ultraviolet than in the infrared. The long radiation lifetime in the latter case leads to quenching by collisions, energy transfer or radiationless transitions.

Since the only variable left in the above equations is the transition moment, its value determines the strength or forbiddenness of a transition. Precise selection rules can be enumerated for atomic transitions in the gas phase, but they are largely irrelevant for large, asymmetric molecules in condensed phases. For highly symmetrical molecules, such as benzene or metalloporphyrins, the electric dipole transition to the lowest excited state is essentially zero and becomes allowed largely through vibrational mixing with electronic states. For such cases Group theory can be used as an elegant means to decide the allowedness of a transition without elaborate quantum mechanical calculation (COTTON, 1963). The selection rule which remains true from the atomic case is the forbiddenness of transitions where the number of unpaired spins changes — that is from a singlet to a triplet state for molecules with light atoms. The oscillator strength, hence the absorption coefficient, of such a transition is typically 10^{-6} of an allowed transition. It is important to note that the radiative lifetime of such a forbidden state is correspondingly lengthened, since these variables are determined by the same parameter, namely the transition moment.

An important rule in molecular spectroscopy is the Franck Condon principle. It states that the coordinates of the nuclei will not change during an allowed electronic transition. The forces acting on the nuclei may, and often do, change so that vibrations are of different energies in excited states. The rule follows from the separation of electronic and nuclear motion in setting-up the Schroedinger equation. A simple consequence of this rule is the impossibility of observing a direct radiation transition between the cis and trans states of a molecule.

De-Excitation

We have just seen that if a molecule in an excited singlet state is left alone for a few nanoseconds, it will return to the ground state by emission of a photon. This process is called fluorescence (Parker, 1968). For real molecules in condensed media, life is rarely so uneventful. The excited molecule may make radiationless transitions from higher excited singlets to the lowest state. This is called internal conversion. The excess energy is dissipated as heat. The excited molecule may make a transition from an excited singlet to a triplet state. This is called inter-system crossing. The triplet state may emit a quantum and drop to the ground state. This is called phosphorescence. It has a lifetime of milliseconds or more because the transition is "forbidden". It may be thermally kicked back from the triplet into an excited singlet state, which in turn may fluoresce, leading to delayed fluorescence. It may also drop by radiationless transition to the ground state, dribbling the energy out as heat. Both excited singlet and triplet states may be quenched by energy transfer, either to identical molecules or to molecules with lower lying singlet or triplet states respectively. If the excited molecule avoids these pitfalls, or occasionally in cooperation with these changes, it may undergo the long awaited photochemical reaction.

There has been a rather abrupt change in assigning a-priori probabilities of radiationless transfers. The change is a good example of subjective estimate of probabilities versus estimates based on measurement. Whereas a few molecules such as fluorescein, are observed to have a quantum yield of fluorescence near unity even at room temperatures, most simple molecules have far smaller yields. It was previously thought that such molecules were simply undergoing very rapid internal conversion to the ground state. Intersystem crossing to the triplet state was dismissed as "forbidden." Measurements have shown that for most fairly rigid molecules, exactly the opposite is true: the sum of the quantum yields of fluorescence and of intersystem crossing to the triplet state is nearly unity (Lamola and Hammond, 1965; Medinger and Wilkinson, 1965). A good example of such a molecule is chlorophyll a in dilute solution (Bowers and Porter, 1967). The a-posteriori explanation of these results is that these molecules have a large gap between their ground and first excited singlet states. The radiationless transition requires crossing to high vibrational levels of the ground state, with more or less coupling to the surrounding fluid or lattice. The theories of this transition differ on this latter point (HENRY and KASHA, 1969; PHILLIPS et al., 1968; TING, 1969; JORTNER, RICE, and HOCHSTRASSER, 1969). They also differ on how they remove the Born-Oppenheimer approximation, i.e. that electronic and nuclear motions are separable. It is just the breakdown of this separability that usually allows the radiationless transitions between electronic energy levels. For molecules whose chromophores are fairly rigid, the small number and larger spacing of vibrational levels slows this energy transfer. Vice versa, "loose" molecules have a large number of such levels, and radiationless transitions will predominate. Such molecules, e.g. carotenes and retinal are notable for their lack of fluorescence. It is still believed that the radiationless transition between higher excited singlets and the first excited singlet is very rapid. The high density of overlapping states in these regions promote conversion. With a few extremely rare exceptions — azulene is the clearest exception — fluorescence is observed only from the lowest excited singlet, no matter how much higher the molecule is excited. Even if it is excited by ionizing radiation, the same unique fluorescence is seen. The discovery of picosecond $(10^{-12}$ s) laser pulses has suddenly opened the exciting possibility of measuring these previously unobservable processes (DE MARIA et al., 1969).

Intersystem crossing is the particular case of radiationless transfer in which the electron spin quantum number changes. The two mechanisms by which this occurs are spin-orbit coupling and electron overlap with paramagnetic species (MURRELL, 1964). In the former, electrons in orbits which come very close to the nucleus generate time varying magnetic fields which interact with the electron spin. In the latter, singlet and triplet states are mixed either directly by interacting magnetic fields of the paramagnetic species or indirectly by interacting with a common charge transfer state between the excited molecule and the paramagnetic species. The spin-orbit interaction is greater the larger the atomic number of the nucleus, and the smaller the separation of the states. The high efficiency of intersystem crossing of carbonyl compounds when they are in the longer lived $n - \pi^*$ state is most likely explainable by this mechanism.

Measurements of the intersystem crossing can be made by a variety of methods: out phosphorescence, flash photolysis and chemical reactions. The most direct, namely emission from the triplet state, is observable only in rigid glasses, usually at very low temperatures.

The relatively long lifetime ($> 10^{-3}$ s) of the triplet state renders it highly suspectible to quenching by impurities, especially oxygen, with which it comes into contact by diffusion in non-rigid solution. Flash photolysis measures the absorption, and thus the concentration, of the triplet state directly. However, a quantum yield requires a monochromatic exciting flash and thus a difficult, but possible, measurement (BOWERS and PORTER, 1967). The chemical methods make use of a triplet scavanger molecule that is known to undergo a definite reaction of known yield on interaction with a triplet molecule. The *cis* → *trans* isomerization of piperylene is often used (WAGNER and HAMMOND, 1968).

Photochemistry

I will limit this discussion to three classes of photochemical reactions, the first two being relevant to the two largest areas of photobiology. The three classes of photochemical reactions we will mention are redox reactions, isomerizations, and bond making and breaking processes. A large number of recent books cover various aspects of photochemistry, particularly organic photochemistry: Calvert and Pitts (1966) and Kan (1966). The series "Progress in Reaction Kinetics", edited by Porter and "Advances in Photochemistry," edited by Noyes, Hammond, and Pitts are very useful. Mention must also be made of some general books on photophysiology: Giese (1964), Thomas (1965), and Kamen (1963).

(1) Redox reactions would seem to be the most natural photochemical reactions. In fact, with quanta of sufficiently high energy direct photoionization occurs through absorption in the continua. For most molecules this occurs only in the vacuum ultraviolet. The more usual mechanism is that the electron in the high energy orbital in the excited state can jump or tunnel to a neighboring acceptor, thus leading to (photo) oxidation of the pigment and reduction of the acceptor.

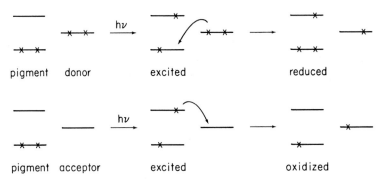

Fig. 3. Electron transfer into and out of excited states to appropriate donors and acceptors

Conversely, an electron from a donor can drop into the hole left in the previously highest filled orbital of the excited state, leading to (photo) reduction of the pigment and oxidation of the donor (Fig. 3). Such over-simplified "solid state" diagrams would give the impression that photoreduction or oxidation would occur at will simply by changing the environment of the excited state. This is true to a considerable extent, particularly for molecules such as benzenoid aromatics or the porphyrins. But for many molecules, chemistry enters more decisively. Because of the peculiarities of chemical bonding, the energy levels shift considerably on reduction or oxidation, and thus one process may be highly favored over the other. For example, ketones, particularly benzophenone, are very readily photoreduced in a solvent such as isopropyl alcohol or aliphatic amines (Cohen and Chao, 1968), whereas phenylenediamines are just as readily photooxidized, often by impurities in the solvent (Linschitz et al., 1967). The primary reactions usually generate reactive free radicals. These may be stabilized by reaction with proton donors or acceptors, and they in turn again usually dimerize or disproportionate to give the final stable products; for example, the ketyl radicals first formed on reduction of photoexcited benzophenone dimerize to benzpinacol. The delicacies of both the individual chemical bonding and the ensuing chemical reactions contribute to making organic photochemistry still a bit of an art.

The fact that the photo redox reaction in free solution requires that the excited state find an acceptor or donor has an important consequence. Since the singlet excited state lives only a few nanoseconds, the concentration of donor or acceptor must be of the order of 0.1 M for it to become a neighbor to the excited state during this short lifetime. Thus the triplet state with a much longer lifetime is highly favored for these reactions, particularly if the donor or acceptor is at low concentration. Much experimental evidence tends to support this contention (WAGNER and HAMMOND, 1968). On the other hand, for a photoreaction occurring in a complex of pigment and donor or acceptor, no such disfavoring of the singlet state occurs. This is the case in the primary photoreaction in photosynthetic bacteria. An electron is transferred from a special excited bacteriochlorophyll to an unknown acceptor in a time estimated to be 10^{-11} s (ZANKEL et al., 1968). This reaction occurs even at the temperature of superfluid helium (MCELROY et al., 1969). At these low temperatures, the electron returns by a temperature independent process requiring about 30 msec. Thus a mechanism involving quantum mechanical tunneling of the electron through a considerable barrier is indicated. These data are leading to an understanding of the trapping process in the energy conversion step in photosynthesis, and so to an understanding of the efficiency of these processes, so fundamental to all life on earth.

Photosensitized oxidations are an important class of reactions carried out by irradiating pigments in the presence of the donor and air. They may be particularly relevant to "photodynamic action" or cellular damage and death caused in light by many pigments both synthetic, such as acridine orange, and natural, such as the porphyrins. Study of the specificity of many such reactions precluded general free radical chain reactions involving oxygen and had led to the hypothesis of some kind of active "photoperoxide" formed from the photoexcited pigment and oxygen. It now seems that many such reactions can be explained by the interaction of the photoexcited dye in the triplet state and oxygen. The oxygen is left in one of its two excited singlet states; $^1\Sigma g^+$ at 13000 cm^{-1} (770 nm) and $^1\Delta_g$ at 7700 cm^{-1} (1300 nm) (FOOTE, 1968; WAYNE, 1969). The former is too short lived to be of much interest, but the latter is long lived — the transition being forbidden by both symmetry and spin pairing rules. This excited singlet oxygen has been detected in photosensitized reactions and when generated by chemical means it duplicates photosensitized oxidations.

Isomerization Reactions

Photochemical isomerization reactions range from simple cis-trans isomerizations around double bonds to the unbelievably complex isomerizations observed in rigid polycyclic carbon ring systems. A good example of the latter is the isomerization of santonin to lumisantonin (KAN, 1966). No less than three bonds change partners during this rearrangement. We will restrict ourselves to the case of cis-trans isomerizations where sufficient complications can already be found. The problem may be summarized: for very simple molecules, such as ethylenes, theory makes definite predictions, but experiments have been very difficult; for complex molecules, such as stilbenes, experiments are easy, and a vast literature exists, but theory becomes very complex.

The expectation that different excited states of molecules may have different geometries is borne out in the case of ethylenes. In this simplified discussion, one may assume that the hydrogens of ethylene itself are so labelled that cis and trans forms can be differentiated. The important parameter is the degree of rotation about the carbon — carbon double bond. The energies required for rotation in the ground state and in various excited states of ethylene have been calculated by elaborate molecular orbital theories (Murrell, 1963) and are shown in Fig. 4. The ground state requires about 63 kcal/mole (22 000 cm^{-1}) to rotate to the perpendicular orientation and this can be looked upon as the energy of binding of the two electrons in the π orbital. Both the lowest excited singlet state and the triplet

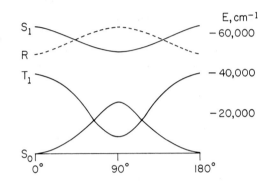

Fig. 4. Potential energy curves for twisting ethylene

state have minimum energy in the perpendicular configuration. These states correspond roughly to having one of the π electrons transferred to an antibonding π^* orbital and the minimum overlapping of this antibonding orbital occurs at 90°. The R bands are singlet and triplet states involving promotion of an electron about equally to a 3 s (Rydberg) and combination of antibonding π^* orbitals. Thus on optical excitation to the lowest singlet state, ethylene would easily relax to the perpendicular geometry. Radiationless transitions could return it to the cis or trans ground state. Alternatively, the lowest excited singlet state may, by intersystem crossing, pass to the triplet state possibly via a higher level triplet state. Molecules in the nearly perpendicular lowest triplet state could then intersystem cross to the distorted cis or trans ground states. These would then thermally relax to their normal vibrational levels. The equality of the triplet and ground energy levels near 90° ± 20° geometry would favor their interaction. The thermal isomerization could follow a similar path. It is understood that a large number of other vibrational, and librational levels exist for such a molecule in a condensed state. This allows small amounts of energy to be rapidly dissipated.

The trivial or chemical mechanism for isomerization of double bonds must also be mentioned. In this a free radical, either thermally or photochemically generated, adds to one end of the double bond. The resulting single bonded radical is free to rotate about this bond, whereupon the original radical addend leaves. Obviously, a little free radical can go a long way towards ruining an experiment. This mechanism tends to produce the thermodynamic equilibrium mixture of cis and trans isomers. It is the most likely explanation of the well known isomerization of carotenoids by iodine.

Unfortunately, experimental data on simple ethylenes is very meager. The sum of the quantum yields for trans → cis and cis → trans isomerization of 1.2 dichloro-ethylene when irradiated directly in the singlet-triplet band (~ 30000 cm^{-1}), induced by high pressure O_2, is 1 ± 0.1 (CALVERT and PITTS, 1966). This suggests a common triplet state for both cis and trans forms, as shown in Fig. 4. A similar conclusion was drawn from the isomerization of 2-butene sensitized by the triplet state of benzene (PHILLIPS et al., 1968). The fact that no emission can be detected from olefin triplet states also tends to support the diagram of Fig. 4. Radiative transitions between states with such different geometries is very unlikely because of the Franck-Condon restrictions. This impediment is of course added to the already severe restriction of spin pairing.

The data on the photoisomerization of stilbenes are far more voluminous than on simple ethylenes. However, here the complexity is in the molecular spectra and energy levels. A detailed diagram such as Fig. 4 for stilbenes is not yet available. It is believed that the delocalized orbitals present in stilbene (and retinal also) will lead to potential minima in many excited states more similar to that of the ground state than to the perpendicular states of ethylene. One reason for the difficulty was pointed out by DYCK and McCLURE (1962). Many excited states contribute to the upper levels, and a good part of the excitation energy is in the benzene rings. Thus one has exchanged experimental simplicity for theoretical complexity. To further complicate matters, it has been found by MOORE et al. (1963) that cis-stilbene isomerizes to 2.2' dihydrophenanthrene. This yellow substance slowly reforms cis-stilbene or is oxidized by oxygen to phenanthrene. Arguments have been advanced that the singlet (SALTIEL, 1968) or the triplet state (MUSZKAT, 1967) are inter-mediates in the photoisomerization of stilbenes. The detailed mechanism of even so simple a photochemical reaction is still much a matter of debate. The weak, but real, fluorescence of trans stilbene indicates that the singlet trans excited state has a definite lifetime before isomerizing. The mechanism of radiationless transition to a vibrationally excited (twisted) ground state is very interesting. It has been much discussed as a general mechanism of radiationless transition to the ground state (PHILLIPS, 1968). It is amusing that some years ago, such "isomerized" or distorted molecules were invoked to explain the occurrence of long lived excited states (PRINGSHEIM, 1949) now largely ascribable to triplet states.

The question of the temperature dependence of cis-trans photoisomerization has been studied by FISCHER and his collaborators GEGIOU et al. (1968). By varying temperature and viscosity independently, their effects on the quantum yields of the cis and trans interconversions can be evaluated. It is found that the quantum yield of trans to cis conversion decreases with decreasing temperature and increasing viscosity. The cis to trans conversion is far less sensitive to these para-meters. Thus the phototransition to the usually more sterically hindered and thus thermodynamically less stable isomer requires an energy and a volume of acti-vation, whereas the reverse transition does not.

The extreme of an increasingly viscous solvent is the lattice of a crystal of the photoreactive molecule. The work of SCHMIDT and his collaborators on crystals of substituted cinnamic acids has turned up a variety of extremely relevant information (Schmidt, 1967) on this subject. It was known from very early work that certain of these acids photodimerized in the crystal to cyclobutane carboxylic acids. To poorly summarize much excellent work the trans acids occur in photoactive α and β crystal forms and light stable γ forms. The latter differ from the others by having a nearest neighbor parallel molecule distance of 5 A as compared to 3.5 to 4 A distance in the α and β forms. Thus one angstrom can determine the photochemical fate of a molecule. Presumably the molecule cannot obtain the energy necessary for the extra deformation during its limited lifetime. The α and β forms photodimerize specifically to α-truxillic acid and β-truxinic acid respectively. These are both related to trans-cinnamic acids, and their symmetry accurately reflects the symmetry of the crystals. Crystals of the cis isomers in general isomerize to the light stable trans crystals or dimerize to the same isomer from the corresponding photoactive trans forms. Thus it seems photoisomerization to the trans forms and recrystallization into the various crystal structures occurs, followed by photodimerization of the photoactive trans forms. The β-methyl cis cinnamic acid crystal is actually light stable, and the trans form both dimerizes and isomerizes to the photostable, but thermodynamically less stable, cis form.

These intriguing results show what can be accomplished by careful experimentation and a combination of the techniques now available.

Bond Making and Breaking Processes

We will only mention two classes of such reactions: The Norrish type II process and the Woodward-Hoffman cycloaddition rules. The former is a good example of the difficulties involved in predicting the behavior of a more complex system from the known behavior of a simpler system. In this case the increase of complexity is disarmingly simple: the addition of one further CH_2 group in a homologous series.

The photochemistry of acetone and the ethyl ketones had been investigated and the overall reaction was found to be (Cundall and Davis, 1967):

$$CH_3COCH_3 \xrightarrow{h\nu} CO + CH_3CH_3 + CH_4 + \cdots .$$

The mechanism was shown to be α bond cleavage followed by both dimerization of the methyl radicals and their attack on acetone. However, on making a simple minded extension to methyl propyl ketone a dramatically new and much simpler reaction occurred:

$$CH_3COCH_2CH_2CH_3 \xrightarrow{h\nu} CH_3COCH_3 + CH_2 = CH_2 .$$

This reaction is intra molecular, involves no "free" radicals and clearly requires the increased molecular complexity to exist. The enol form of acetone is very likely the first product. The possibility of a bi-radical intermediate is suggested by the finding of 1-methyl cyclobutanol as an additional product.

The Woodward-Hoffmann rules on cyclo additions are the brilliant result of combining seemingly odd and irrelevant experimental observations with modern bonding theory (Woodward and Hoffmann, 1965). In the closing (or opening) of a conjugated molecule to a cycle it was observed that the rotation of the groups to be newly bonded (or broken) was sterospecific. Moreover the specificity depended on ring size and on whether the reaction was thermal or photochemical (Fig. 5). This can be explained by considering the sign of the wave function of the particular molecular orbital involved (HOFFMANN and WOODWARD, 1968). Alternatively one can construct a level correlation diagram (LONGUET-HIGGINS and ABRAHAMSON, 1965). One classifies the level (Fig. 6) by the symmetry of the operation (plane, m, for disrotatory; two fold rotation axis, C_2, for conrotatory). Only orbitals of the same symmetry may mix, and so this diagram shows that disrotatory process on

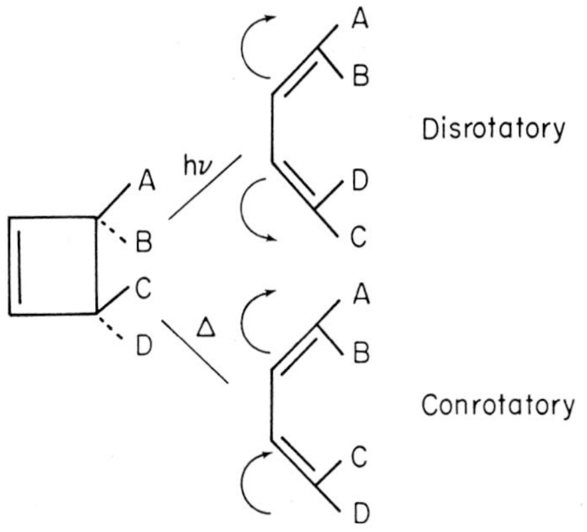

Fig. 5. Definition of electrocyclic reactions; hν: photochemical; Δ: thermal

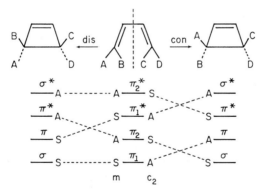

Fig. 6. Symmetries and energy levels of the orbitals involved in an electrocyclic reaction

ring closure is thermally forbidden. The occupied orbital π_2 goes uphill to an excited state of the cyclobutene. The conrotatory process is thermally allowed — there is no net difference in energy in the bonding, to the present level of approximation. Conversely for the excited state of the butadiene, the opposite holds true since one electron is in the π^* orbital and it is the disrotatory state that is allowed. Moreover for the hexatriene, the rules are reversed and so on. The detailed theoretical treatment of the excited states unfortunately complicates the simple picture presented here. These principles have found wide application to a great variety of reactions, and have been in turn supported by experimental evidence. They apply, of course, to strictly concerted reactions.

With these few examples from the expanse of photochemistry we have reached the end of our exposition of the interaction of radiation and matter. More has been omitted, unfortunately, than has been added, and I can only urge the reader to refer to the more general works and reviews for further knowledge.

References

Basu, S.: Theory of solvents effects on molecular electronic spectra. Advanc. Quant. Chem. **1**, 145—169 (1964).
Bennett, R. G., Kellogg, R. E.: Mechanisms and rates of radiationless energy transfer. In: Porter, G. (Ed.): Prog. reaction kinetics, Vol. 4, pp. 215—238. Oxford: Pergamon Press 1967.
Bloembergen, N.: Nonlinear optics. New York: W. A. Benjamin 1965.
Bowers, P. G., Porter, G.: Quantum yields of triplet formation in solutions of chlorophyll. Proc. roy. Soc. A **296**, 435—441 (1967).
Brand, J. C. D., Williamson, D. G.: The structure of electronically excited organic molecules. Advanc. Phys. Org. Chem. **1**, 365—423 (1963).
Calvert, J. G., Pitts, J. M.: Photochemistry. New York: John Wiley 1966.
Cohen, S. G., Chao, H. M.: Photoreduction of aromatic ketones by amines. Studies of quantum yields and mechanism. J. Amer. chem. Soc. **90**, 165—173 (1968).
Cotton, F. A.: Chemical applications of group theory. New York: Interscience 1963.
Cundall, R. B., Davis, A. S.: Primary processes in the gas phase photochemistry of carbonyl compounds. In: Porter, G. (Ed.): Prog. reaction kinetics, Vol. 4, pp. 149—213. Oxford: Pergamon Press 1967.
DeMaria, A. J., Glenn, W. H. Jr., Brienza, M. J., Mack, M. E.: Picosecond laser pulses. Proc. IEEE **57**, 2—25 (1969).
Dyck, R. H., McClure, D. S.: Ultraviolet spectra of stilbene, ϱ-monohalogen stilbenes, and azobenzene and the trans to cis photoisomerization process. J. chem. Phys. **36**, 2326—2345 (1962).
Fabelinskii, I. L.: Molecular scattering of light. New York: Plenum Press 1968.
Feynman, R. P., Leighton, R. B., Sands, M.: The Feynman lectures on physics, Vols. I and II. Reading: Addison-Wesley 1963.
Foote, C. S.: Photosensitized oxygenations and the role of singlet oxygen. Accts. Chem. Res. **1**, 104—110 (1968).
Freeman, D. E., Klemperer, W.: Dipole moments of excited electronic states of molecules: The $'A_2$ state of formaldehyde. J. chem. Phys. **40**, 604—605 (1964).
Gegiou, D., Muszkat, K. A., Fischer, E.: Temperature dependence of photoisomerization. VI. The viscosity effect. J. Amer. chem. Soc. **90**, 12—18 (1968).
Giese, A. C.: Photophysiology, Vol. I and II. New York: Academic Press 1964.
Hameka, H. F.: Advanced quantum chemistry. Reading: Addison-Wesley 1965.
Heitler, W.: The quantum theory of radiation, third ed. Oxford: Clarendon Press 1954.
Henry, B. R., Kasha, M.: Radiationless molecular electronic transitions. Ann. Rev. Phys. Chem. **19**, 161—192 (1968). Ann. Reviews, Palo Alto.
Hoffmann, R., Woodward, R. B.: The conservation of orbital symmetry Accts. Chem. Res. **1**, 17—22 (1968).
Jaffé, H. H., Orchin, M.: Theory and application of ultraviolet spectroscopy, Chap. 9. New York: J. Wiley 1962.
Jortner, J., Rice, S. A., Hochstrasser, R. M.: Radiationless transitions in photochemistry. In: Pitts, J. N. Jr., Hammond, G. S., Noyes, W. A., Jr. (Eds.): Advances in photochemistry, Vol. 7, pp. 149—310, New York: Interscience, 1969.
Kamen, M. D.: Primary processes in photosynthesis. New York: Academic Press 1963.
Kan, R. O.: Organic photochemistry. New York: McGraw-Hill 1966.
Kasha, M.: Paths of molecular excitation. In: Augenstine, L. G. (Ed.): Radiation research. Suppl. 2, pp. 243—275. New York: Academic Press 1960.
Kauzmann, W.: Quantum chemistry, Chap. 14—16. New York: Academic Press 1957.

KIRMSE, W.: Carbene chemistry. New York: Academic Press 1964.

LAMOLA, A. A., HAMMOND, G. S.: Mechanisms of photochemical reactions in solutions XXXIII. Intersystem crossing efficiencies. J. chem. Phys. **43**, 2129—2135 (1965).

LINSCHITZ, H., OTTOLENGHI, M., BENSASSON, R.: The one electron oxidation of triplet diphenyl-p-phenylenediamine by the Diimine. J. Amer. chem. Soc. **89**, 4592—4599 (1967).

LONGUET-HIGGINS, H. C., ABRAHAMSON, E. W.: The electronic mechanism of electrocyclic reactions. J. Amer. chem. Soc. **87**, 2045—2046 (1965).

McELROY, J. D., FEHER, G., MAUZERALL, D. C.: On the nature of the free radical formed during the primary process of bacterial photosynthesis. Biochim. biophys. Acta (Amst.) **172**, 180—183 (1969).

MEDINGER, T., WILKINSON, F.: Mechanism of fluorescence quenching in solution. Trans. Faraday Soc. **61**, 620—630 (1965).

MOORE, W. M., MORGAN, D. D., STERMITZ, F. R.: The photochemical conversion of stilbene to phenanthrene. The nature of the intermediate. J. Amer. chem. Soc. **85**, 829—830 (1963).

MURRELL, J. N.: The theory of the electronic spectra of organic molecules, Chap. 5. New York: John Wiley 1963.

MUSZKAT, K. A., GEGIOU, D., FISCHER, E.: Temperature dependence of photoisomerization. IV. Evidence for the involvement of triplet states in the direct photoisomerization of stilbenes. J. Amer. chem. Soc. **89**, 4814—4815 (1967).

PARKER, C. A.: Photoluminescence of solutions. New York: Elsevier 1968.

PHILLIPS, D., LEMAIRE, J., BURTON, C. S., NOYES, W. A., JR.: Isomerization as a route for radiationless transitions. In: NOYES, W. A., HAMMOND, G. S., PITTS, J. N. (Eds.): Advances in photochemistry, Vol. 5, pp. 329—363. New York: Interscience 1968.

PRINGSHEIM, P.: Fluorescence and phosphorescence. New York: Interscience 1949.

REID, C.: Excited states in chemistry and biology. London: Butterworths 1957.

SALTIEL, J.: Perdeuteriostilbene. The triplet and singlet paths for stilbene photoisomerization. J. Amer. chem. Soc. **90**, 6394—6400 (1968).

SCHMIDT, G. M. F.: The photochemistry of the solid state. In: Reactivity of the photoexcited organic molecule. Thirteenth international conference on chemistry, pp. 227—284. New York: Interscience 1967.

THOMAS, J. B.: Primary photoprocesses in biology. Amsterdam: North-Holland 1965.

TING, C. H.: Theory on the radiationless transitions in large polyatomic molecules. Photochem. Photobiol. **9**, 17—31 (1969).

TROZZOLO, A. M.: Electronic spectroscopy of arylmethylenes. Accts. Chem. Res. **1**, 329—335 (1968).

WAGNER, P. J., HAMMOND, G. S.: Properties and reactions of organic molecules in their triplet states. In: NOYES, W. A., HAMMOND, G. S., PITTS, J. N. (Eds.): Advances in photochemistry, Vol. 5, pp. 21—156. New York: Interscience 1968.

WAYNE, R. P.: Singlet molecular oxygen. In: PITTS, J. N., JR., HAMMOND, G. S., NOYES, W. A., JR. (Eds.): Advances in photochemistry, Vol. 7, pp. 311—372, New York: Interscience 1969.

WELLER, A.: Fast reactions of excited molecules. In: PORTER, G. (Ed.): Progress in reaction kinetics, Vol. 1, pp. 187—214. New York: Pergamon Press 1961.

WOODWARD, R. B., HOFFMANN, R.: Sterochemistry of electrocyclic reactions. J. Amer. chem. Soc. **87**, 395—397 (1965).

ZANKEL, K. L., REED, D. W., CLAYTON, R. K.: Fluorescence and photochemical quenching in photosynthetic reaction centers. Proc. nat. Acad. Sci. (Wash.) **61**, 1243—1249 (1968).

Chapter 5

The Structure and Reactions of Visual Pigments

By

Allen Kropf, Amherst, Massachusets (USA)

With 11 Figures

Contents

I. Introduction

The complementary themes of form and function are used in our analysis of a diversity of objects ranging from buildings to protein molecules. We often try to infer function from form, and though more often than not we cannot make the inference, we continue to trust in the existence of a connection. So it is with the structures and reactions of the visual pigments of the retina where the search has led us to expect that we can discover the secret of the visual receptor, its ability to translate the message of photons into the language of the nervous system, by studying the structure and dynamics of visual pigment molecules.

It has been known for about a century that light causes a bleaching of the retina whereupon the pink or violet color of this tissue fades through orange and yellow intermediate hues to a pale white translucence. This bleaching reaction can be carried out in solutions of visual pigments as well, and we now are also able to assign molecular structures as well as names to the chemical compounds giving rise to the colors we see in solution and in the retina.

In addition to what we know of the photochemistry of the visual process, there is the precise layering of the rod and cone outer segments of the visual cells, the structures which contain and are themselves partially constructed of the visual pigments. It is not only the photochemical bleaching that we try and ally to visual excitation, but also the ordered environment in which it naturally occurs. Though this review sets forth a view of our current understanding of the structure of the visual pigments and their reactions in solution and in the visual receptors, it is written with the hope that the reader will either be stimulated to utilize the information to pursue the question of the molecular aspects of visual cells' functioning or else be in a better position to appreciate the advances which will almost certainly be made in the near future.

II. Structure of Visual Pigments

1. The Chromophore

The property of absorbing visible and ultraviolet light in the proportions shown in Fig. 1 derives, from a small portion of the visual pigment molecule, its retinylidene chromophore (WALD, 1935). This structure, shown in Fig. 2, results from the combination of the 11-*cis* isomer of either retinaldehyde (HUBBARD and WALD, 1952) or 3,4 dehydroretinaldehyde (WALD et al., 1955) with an amino group associated with opsin (BOWNDS, 1967; AKHTAR et al., 1968). The aldehyde, in turn, is formed from the corresponding vitamin A and the interrelationships between these compounds and the visual pigments have been the subject of many investigations which have recently been summarized with a personal perspective by WALD (1968).

Though the chemical structure of vitamin A had been known for many years prior to 1944, it was not until then that MORTON and his coworkers (MORTON and GOODWIN, 1944; BALL et al., 1948) recognized that retinaldehyde, or retinene as it was called then, was related to vitamin A simply as an aldehyde is related to an alcohol. Since that discovery, several geometrical isomeric forms of retinaldehyde have been synthesized from simple compounds of known structure (ROBESON et al., 1955). As a result of this synthetic work the primary structure, which describes the covalent linkages betweens atoms, has been determined, but the exact conformations and thus the resulting shapes of the geometrical isomers of retinaldehyde have not been established. The most straightforward technique for accomplishing this, X-ray diffraction analysis of crystalline solids, has not been reported for retinaldehyde or retinol (vitamin A). There have been reports of X-ray diffraction studies of some compounds which are structurally related to retinaldehyde and the results of these investigations are probably relevant to its three dimensional structure.

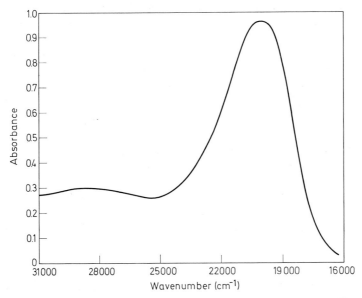

Fig. 1. Absorption spectrum of frog rhodopsin in 2% digitonin solution, pH = 6.5. Since absorbance is plotted versus wavenumber (i.e. wavelength-l), the spectral curve represents the shape common to all rhodopsin solutions (DARTNALL, 1953). Frog rhodopsin differs from other rhodopsins only in its distinctive value for λ_{max} (502 nm = 19900 cm⁻¹)

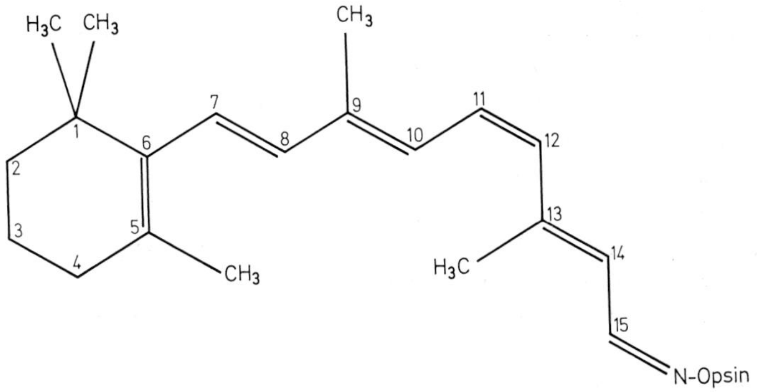

Fig. 2. Skeletal outline of the 11-*cis* retinylidene chromophore with conventional numbering of the carbon atoms in the structure. The nitrogen atom may be protonated in visual pigments. The attachment to opsin is to a primary amino group of either lysine (BOWNDS, 1967; AKHTAR et al., 1968) or phosphatidyl ethanolamine (POINCELOT et al., 1969). The chromophore shown derives from retinaldehyde. When there is a double bond between carbon atoms 3 and 4, the visual pigments are called porphyropsins and cyanopsins (WALD, 1937; WALD et al., 1953)

a) **Conformation.** The nearest chemical relative of retinaldehyde whose 3-dimensional structure has been studied is vitamin A acid (STAM and MACGILLAVRY, 1963). Fig. 3 is a representation of this compound showing some of the bond angles and bond lengths they found. Several aspects of this structure are noteworthy, especially when we consider the stereochemistry of the retinylidene chromophore, since we can reasonably assume that the same factors determine the bond lengths, bond angles and interatomic distances in the chromophore as in vitamin A acid.

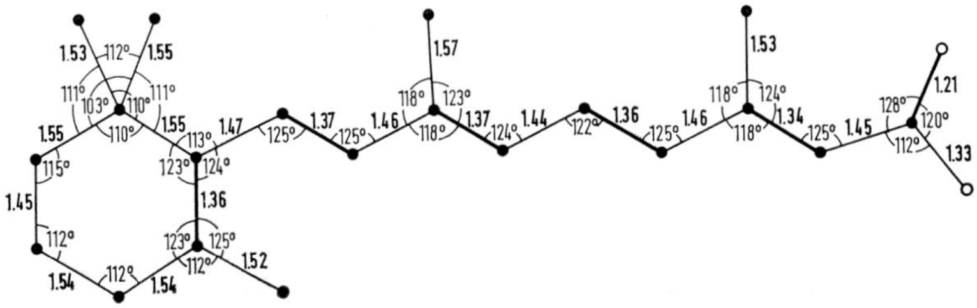

Fig. 3. Bond angles and bond distances in retinoic acid (Vitamin A acid) as determined by X-ray analysis (STAM and MACGILLAVRY, 1963). Bond lengths are in Ångstrom units (10^{-8} cm) and have a mean deviation of between 0.02 and 0.03 Å. Bond angles, which are determined to one-tenth of a degree, have been rounded off to the nearest degree. The dihedral angle between the plane of the six-membered substituted cyclohexene ring and the polyene side chain is 35° in this compound. Open circles (○) are the oxygen atoms while the closed circles (●) represent the carbon atoms. Hydrogen atoms were not resolved in this study

Initially we may note the strong alternation of bond lengths in the polyene side chain, insuring the integrity of the concept of single and double bonds with the consequent possibilities for geometrical isomerism about the double bonds. In addition there is an unexpected geometry associated with the single bond between atoms 6 and 7, referred to as an s-*cis* conformation. This structure, since it is found in β-carotene (STERLING, 1964), 15,15-dehydro-β-carotene (SLY, 1964), canthaxanthin (BART and MACGILLAVRY, 1968a), 15,15-dehydrocanthaxanthin (BART and MACGILLAVRY, 1968b) and all-*trans* vitamin A acid (STAM and MAC-GILLAVRY, 1963), although not in all-*trans*-β-ionylidene crotonic acid (EICHHORN and MACGILLAVRY, 1959), can be assumed to be present in retinaldehyde itself, although the factors responsible for an s-*cis* rather than an s-*trans* conformation about the 6—7 single bond are not evident. The consequence of this geometry in a planar molecule is an overcrowding due to the interference of the methyl group on C-5, with the hydrogen atom on C-8. This is depicted in Fig. 4a where the arrangement of atoms, as they would exist if atoms 5—8 were in all the same plane, is shown. But the X-ray diffraction results just mentioned show a dihedral angle of about 35° between the planes determined by atoms 5—6—7 and 6—7—8. The significance of this result for the structure of the chromophore in visual pigments is the following. In the 11-*cis* configuration, the only configuration known to occur naturally in visual pigments (HUBBARD and WALD, 1952), the steric hindrance between the methyl group on C-13 and the hydrogen atom on C-10 is somewhat more severe, as seen in Fig. 4b, than it is for the arrangement shown in Fig. 4a. Since this latter crowding leads to a twist of some 35°, we expect the twist in the polyene side chain, when it is in the 11-*cis* isomeric form, to be somewhat greater than 35°. Thus the retinylidene chromophore is severely twisted in two places and as a result it is impossible to have the light absorbing portion of the visual pigment molecule confined to a single plane.

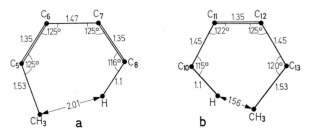

Fig. 4. Bond angles, bond lengths and interatomic distances in: a) the *cis* configuration of the C_6—C_7 single bond in vitamin A acid; b) the *cis* configuration of the C_{11}—C_{12} double bond in vitamin A acid. In each instance the atoms involved are assumed to be planar. In fact, the dihedral angle in a) is know to be approximately 35°. Bond lengths and bond angles are taken from STAM and MACGILLAVRY (1963)

Though there have been no direct experimental studies to confirm this last remark, a recent NMR study by PATEL (1969) lends it strong support. This NMR study, which was carried out on the stereoisomers of retinaldehyde, indicated a variation in proton-proton interactions indicated by small shifts in the resonance frequencies of absorption of 220 MHz radiation. The NMR spectra were interpreted as indicating a skew geometry at the C_{12}—C_{13} single bond in 11-*cis* and 11,13-*dicis*-retinaldehyde. Until X-ray diffraction measurements are made on the stereoisomers of retinaldehyde or chemically related species, however, we will have to content ourselves with the present structural evidence from the X-ray and NMR studies referred to, which indicates that the structure present in all known visual pigments, the 11-*cis* retinylidene chromophore, is markedly nonplanar.

b) Electronic Structure. In addition to the spatial arrangement of atoms in the chromophore, it is important to know the distributions of charge which are predicted by quantum mechanical theory. Knowledge of the electronic structure leads to a better understanding of the spectral and electrochemical properties of the system and enables us to make predictions concerning these properties. At present, though, sufficiently refined models and calculating procedures have not been developed to allow completely accurate and reliable answers to be given to questions involving electronic structure. Some calculations have been reported, though, which do point the way to future developments in this area.

The first calculations of the electronic structure of the isomers of retinaldehyde were performed by BERTHOD and PULLMAN (1960). Their calculations, based upon the method of PARISER and PARR (1953), led to the curious conclusion that the isomers of retinaldehyde should fall into two electronically similar groupings. One group was all-*trans*, 11-*cis* and 9-*cis* and the second group was 13-*cis*, 11,13-*dicis* and 9,13-*dicis*. A year later PULLMAN and PULLMAN (1961) extended the earlier calculations to include the effect of the interruption of conjugation on the isomerization behavior of the isomers. Unfortunately, they dealt with the ground state rather than the electronically excited states. Since most is known about the light promoted isomerization of retinaldehyde, conclusions drawn from ground state potential energy curves have no relevance to the photochemical behavior of the chromophore.

More recently there have been several reports of quantum mechanical calculations based on the methods of PARISER and PARR (1953) and of POPLE (1955) (PPP methods) (WIESENFELD and ABRAHAMSON, 1968; INUZUKA and BECKER, 1968; FRATEV et al., 1968; LANGLET et al., 1969; NASH, 1969). These calculations, which have taken advantage of newer procedures, provide more detailed proposals for the photochemical behavior of the all-*trans* and 11-*cis* isomers of retinaldehyde. Judging by INUZUKA and BECKER's results it seems clear that calculated energy differences between the ground states of all-*trans* and 11-*cis* retinaldehyde are subject to a large error, since their calculated value of 55 kcal/mole is much greater than the experimental value of about 0.15 kcal/mole (HUBBARD, 1966). WIESENFELD and ABRAHAMSON adjusted their calculated results by inserting the experimental value into their calculational scheme. Both calculations produce reasonable values for the wavelength of maximum absorption of the main peak in retinaldehyde, 380 nm and 378 nm, and WIESENFELD and ABRAHAMSON's calculations also account for retinaldehyde's weaker spectral peaks at around 290 nm and 254 nm. The PPP method, though, at its present stage of refinement, is unable to account quantitatively for the different intensities of absorption and the small differences in λ_{max} values among the different geometrical isomers for both the main absorption band and the smaller peaks, the so-called "*cis* peaks", which occur on the short wavelength side of the main absorption band.

NASH (1969) has coupled a PPP calculation with one involving the interaction of the non-bonded atoms in order to theoretically estimate the conformation of 11-*cis* retinaldehyde relative to the planar all-*trans* form. He too finds that the energy of the π-electrons is lower in the planar all-*trans* conformation, but the relief of steric hindrance as a result of twisting about the $10-11$, $11-12$, $12-13$ and 13-Me bonds by around $40°$, $10°$, $40°$ and $70°$ respectively in 11-*cis* retinaldehyde can more than compensate for the energy increase due to poorer π-electron conjugation. Consequently the total energy of 11-*cis* retinaldehyde can be calculated to be slightly less than that of the planar all-*trans* form, thus rationalizing the experimental result found earlier by HUBBARD (1966).

Though the PPP method is currently the method favored by those making quantum-mechanical calculations on conjugated systems, there are other approaches which are easier to grasp qualitatively and are more closely related to classical concepts of chemical bonding and electromagnetic behavior of charged particles. An almost purely classical model of partial oscillators (ZECHMEISTER et al., 1943) was used to rationalize the appearance and intensities of polyene spectra, particularly the relative height of the "*cis* peaks". Though this model was useful in understanding why, of all the geometrical isomers, the all-*trans* isomer absorbed most strongly and why a bend in the conjugated system was necessary to generate both a "*cis* peak" and a lower specific absorption in the main absorption band, it did not allow for quantitative calculations of intensities and charge distributions.

Furthermore, the relationship to basic quantum-mechanical principles was not clear.

Another approach, which can be called the "molecules-in-molecules" or resonance force method (SIMPSON, 1955), has been used successfully by MERZ et al. (1965) to rationalize the spectra of a large number of polyenes. In this method the observed electronic properties of relatively simple molecules, such as ethylene, are used as parameters in calculating the properties of the composite system. Each property of the larger conjugated system is assumed to result from the electronic interactions of the smaller units making it up. An idea similar to this had previously been used by FARRAR et al. (1952) to calculate the spectral peaks of vitamin A_2 precursors and derivatives, but that *ad hoc* method was not based on any theoretical foundation. The "molecules in molecules" approach leads to a relatively simple procedure for obtaining numerical results for λ_{max} 's of all the peaks in the spectra of polyenes and, along with certain modifications, such as inclusion of charge-transfer states, provides results which agree closely with experiment (MERZ et al., 1965). The method has been seriously challenged, though, in a study of the polarization of the lowest energy allowed transition of β-ionylidene crotonic acid (PARKHURST and ANEX, 1966). The direction of the transition moment which they determined lay outside of the range predicted by the resonance force method, even when charge-transfer corrections were included.

We must conclude this discussion by noting the progress that has been made in calculating the electronic structure of retinaldehyde and related polyenes but observing that there still do not exist any theoretical results which would allow us to interpret and predict with confidence the physical and chemical behavior of retinaldehyde and its derivatives, especially the visual pigments.

2. Structure of Rhodopsin

Unlike its 11-*cis* chromophore, whose primary structure is firmly established and where the spatial arrangement of atoms can be estimated with a certain amount of assurance, the protein portion of visual pigments, opsin, is still largely "terra incognita". Recently there have been a number of publications dealing with (1) the nature of the primary linkage joining the chromophore to the protein in rhodopsin and (2) the composition of pure rhodopsin. Each of these topics will be reviewed in turn.

a) **The Protein-Chromophore Linkage in Rhodopsin.** MORTON and his coworkers (BALL et al., 1949) were the first to provide experimental evidence which suggested an aldimine or Schiff base linkage between retinaldehyde and opsin in visual pigments. This type of bonding, resulting from the reaction of retinaldehyde with a primary amine:

$$C_{19}H_{27}CHO + RNH_2 \rightarrow C_{19}H_{27}CH = N - R + H_2O$$

retinaldehyde + primary amine → retinylidine R amine + water

was suggested on the basis of experiments whereby retinaldehyde condensed with simple primary amines, such as methyl amine, to form amphoteric aldimines. The behavior of the resulting retinylidene amino compound was reminiscent of certain of the intermediates and artifacts which are produced when rhodopsin is bleached. The Schiff base hypothesis and the evidence supporting it was thoroughly reviewed by Morton and Pitt (1957).

The Schiff base hypothesis, though widely accepted, was not directly verified until quite recently. It was then elegantly demonstrated by two groups working independently and using slightly different variations of the same basic procedure (Bownds, 1967; Akhtar et al., 1967). Their technique was to fix the chromophore to its site in metarhodopsin and then to chemically dissect this derivative and identify that fragment containing the chromophore. This was accomplished by the following series of reactions:

$$\text{rhodopsin} \xrightarrow{\text{light}} \text{metarhodopsin} \xrightarrow{\text{NaBH}_4} \text{N-retinylopsin} \xrightarrow[\substack{\text{hydrolysis} \\ \text{(NaOH)}}]{\text{basic}}$$

N-retinyl lysine + amino acids.

The resulting fragment, N-retinyl lysine, was identified chromatographically. Bownds (1967) compared its mobility in two different solvent systems to that of synthetic N-retinyl lysine (NRL) and in the other instance (Akhtar et al., 1968), where tritiated 11-*cis* retinaldehyde had been used to synthesize rhodopsin, the resulting NRL was co-chromatographed with synthetic, non-radioactive NRL. 50% of the radioactive counts were found precisely at the same location on the chromatogram as NRL, while the remaining radioactivity was spread diffusely over the chromatographic plate.

These two carefully executed experiments have firmly established that the retinyl chromophore is attached to the ε-NH$_2$ group of lysine in metarhodopsin and, by strong inference, the same attachment site binds the chromophore to opsin in rhodopsin as well. Presumably the attachment site in rhodopsin is inaccessible to NaBH$_4$ (Bownds and Wald, 1965), as it seems to be to NH$_2$OH, H$^+$ and OH$^-$ as well. On the other hand, metarhodopsin, which is a pH indicator (Matthews et al., 1963) as well as being liable to decomposition by NH$_2$OH, has its aldimine bond exposed and thus is able to be reduced by NaBH$_4$.

This line of reasoning and the conclusion that the 11-*cis* retinyl ε-amino lysine (11-*cis* R-εAL) structure exists in native rhodopsin has recently been challenged by Poincelot et al. (1969) whose experiments led them to conclude: "(1) The 11-*cis* retinylidene chromophore is bound by a Schiff base linkage to the lipid phosphatidyl ethanolamine rather than to a protein in agreement with a suggestion of Krinsky. (2) The chromophore after photoisomerization remains bound to the lipid at the stage of metarhodopsin$_{478}$I but migrates to an ε-amino group of a lysine unit in the backbone protein in the intermediate reaction metarhodopsin$_{478}$I $\rightarrow$ metarhodopsin$_{380}$II, a process known to occur at physiological temperatures *in vitro* and *in vivo* at times of the order of 100 μsec." This conclusion has been reinforced by the experiments of Akhtar et al. (1969) and more fully documented by Poincelot et al. (1970). Though the role of the protein portion of opsin is still important to their scheme, phospholipid, known to be present in large amounts in rod outer segments (Collins et al., 1952b) and some rhodopsin solutions (Krinsky, 1958), is accorded an equally important role.

The evidence presented by Poincelot et al. (1969, 1970) for retinylidene phosphatidylethanolamine (N-RPE) as the chromophoric structure in native rhodopsin rests upon the isolation from both lyophylized rod outer segments

(ROS) and lyophylized cetyl trimethylammonium bromide (CTAB) rhodopsin solutions of a compound behaving both chromatographically and spectroscopically as N-RPE. This compound was obtained by treating lyophylized ROS and lyophylized CTAB extracts, which had been isolated and purified by a procedure developed by ERHARDT et al. (1965), with acidified methanol which was $2.5(10)^{-4}$M in HCl. The lyophylized material which was extracted with acidified methanol had either been dark adapted, and thus presumed to contain rhodopsin exclusively, or else had been exposed to light under conditions which favored the formation of metarhodopsin I. POINCELOT et al. (1970) assert that their procedures removed more than 90% of the retinaldehyde from these lyophylized powders although careful documentation for that crucial aspect of their experiments is lacking. AKHTAR et al. (1969), who directly reduce trichloroacetic acid denatured rhodopsin with $NaBH_4$, recover about 40% of the tritiated retinaldehyde in the form of N-RPE. When an identical rhodopsin solution was not acid treated but instead transformed to metarhodopsin II and then reduced, less than 10% of the original radioactivity was then soluble in methanol, presumably as N-RPE.

Another approach to this problem was recently made by HALL and BACHRACH (1970) who injected frogs with sodium ^{32}P orthophosphate and then three days and six days later isolated each group of outer segments and analyzed an extract from them. Upon chromatographic purification they found insufficient radioactivity in the rhodopsin peak to account for even one mole of PE per mole of rhodopsin. They conclude ". . . that purified visual pigment contains no phospholipid and that retinal cannot be bound to PE in native visual pigment."

In addition to the experiment just cited, there are several other reasons for exercising caution before accepting the suggestion of a primary linkage between retinaldehyde and phosphatidyl ethanolamine in rhodopsin which, following exposure to light, becomes a retinylidene ε-lysyl linkage (POINCELOT et al., 1969; KIMBEL et al., 1970; AKHTAR et al., 1970). The first objection derives from the experiments of AKHTAR et al. (1968). They took the precaution of having 0.05 M hydroxylamine, as well as $NaBH_4$, present when their rhodopsin containing a tritium labelled chromophore was exposed to light. If the chromophore behaved in the following way:

$$\text{retinylidene-PE} + H_2O \rightarrow \text{retinaldehyde} + \text{phosphatidyl ethanolamine}$$

$$\xrightarrow{\text{opsin}} \varepsilon\text{-N-retinylidene-lysyl-opsin}$$

which would be expected according to the scheme of binding suggested by POINCELOT et al. (1969), then retinaldehyde would have been trapped when it appeared, even briefly, by NH_2OH. Since the recovery of the chromophore as ε-N-retinyllysine was about one-half the amount expected on the basis of control experiments with synthetic ε-N-retinyl-lysine, it is clear that the chromophore did not appear in a form which was subject to reaction with NH_2OH.

BOWNDS and WALD (1965) also searched for vitamin A in a rhodopsin solution which was irradiated in the presence of $NaBH_4$. They found that $NaBH_4$ reduced the chromophore when the pigment was at the stage of metarhodopsin II, and not before this stage had been reached. Had retinaldehyde appeared at any time in the interval between rhodopsin and metarhodopsin II, it would have been reduced to the alcohol, i.e. vitamin A. But since vitamin A was not detected, it seems unlikely that the retinylidene chromophore passed through the aldehyde stage if it transferred from PE to opsin. Though the direct reaction of a Schiff base with a primary amine (trans-imination) is known (Jencks, 1969), it is questionable in the case of rhodopsin whether the stereochemistry around the chromophoric site and the availability of appropriate amino groups would conspire to make the metarhodopsin I $\rightleftharpoons$ metarhodopsin II equilibrium a trans-imination reaction. To this author, then, the available evidence favors an ε-N-retinylidene-lysyl-opsin attachment in native rhodopsin.

b) The Structure and Composition of Opsin. The composition of opsin, the protein which constitutes the major portion by weight of the rhodopsin molecule, has only come under intensive study in the last five years. Judging by the number of papers which have appeared, though, it seems as though we may know its primary structure in the near future. The importance in knowing the detailed structure of opsin is that it may enable us to assess rhodopsin's role in (1) influencing the microscopic structure of the ROS, (2) aligning the chromophore in the ROS, (3) mediating between the light absorbing event and the initiation of a neural excitation.

Since rhodopsin constitutes about 15—35% of the dry weight of the outer segment (Hubbard, 1954), we must assign it an important role when accounting for the regular arrangement of molecules in the ROS. The regular arrangement manifests itself in the microscopic appearance of the ROS (Sjöstrand, 1953). In addition, the studies of Dowling and coworkers (see for example Dowling, 1967) have demonstrated the deleterious effects that rhodopsin deprivation has on

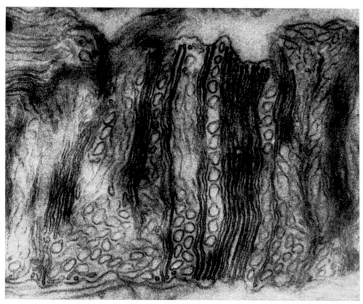

Fig. 5. A longitudinal section of a rod outer segment showing an early stage of degeneration. Stacks of intact disks alternate with distended disks that have often pinched off to form large vesicles and tubules. At this stage the rod maintains much of its normal cylindrical shape. Magnification: ×20000. [From Dowling and Gibbons (1961). Used with permission.]

the structure of the ROS (DOWLING and GIBBONS, 1961), Fig. 5. Here one sees directly that the regular structure becomes disrupted without rhodopsin.

Though the role of rhodopsin and opsin as structural components of the ROS is established, the structure of opsin, in the same sense as we used that term when speaking of the structure of its chromophore, is not yet known. Not only is the 3-dimensional placement of each atom in opsin not known, but even the sequence of the amino acids and other units which constitute the primary structure of this protein has not been determined. This is due largely to the difficulties associated with obtaining sizeable quantities of homogeneous material of known purity. Detergents are necessary to bring rhodopsin and opsin into solution and the detergents used are not so selective as to solubilize visual proteins alone. In addition to their lack of specificity, it is generally thought that the detergent molecules surround the protein, in the form of a micelle, and thus help to defeat purification procedures. Purification usually depends upon differential interactions between proteins and other types of substances, such as chromatographic absorbents and sieves, and among proteins themselves in electrophoresis. Such difficulties appear to have been largely overcome in recent work where apparently monodisperse chromoprotein, as judged by its behavior in gel electrophoresis, has been obtained by means of column chromatography on a variety of adsorbents (HELLER, 1968a; SHIELDS et al., 1967; SHICHI et al., 1969).

The purified proteins have been analyzed by the standard methods now available for determining the amino acid composition (SPACKMAN et al., 1958) and, in addition, HELLER (1968a) has examined rhodopsin for other components as well. The amino acid analyses, as well as some results obtained by other workers by other methods, are included in Table 1. Each of the rhodopsin analyses was done on proteins which had been purified in somewhat different ways, including the use of different ionic and non-ionic detergents. The rhodopsin solutions were chromatographed on columns of different materials, some of which acted mainly by absorption and some by "molecular sieving". It is encouraging, therefore, to see how similar the results are. One begins to feel confident that a pure, unique rhodopsin species can be prepared. As more is done to solve this problem we should know how important the reported differences in composition are and how they originate.

One of the more surprising findings has been the small, but apparently stoichiometric quantity of hexosamine covalently linked to rhodopsin (HELLER, 1968a). The linkage of a heptasaccharide through a peptide bond between the β-carboxyl group of aspartic acid and the amino group of glucosamine has recently been found in egg albumin (GOTTSCHALK, 1966) and in bovine pancreatic DNAse (CATLEY et al., 1969), a protein having a similar amino acid composition and molecular weight to rhodopsin, but varying significantly in other ways. HELLER and LAWRENCE (1970) have shown that the hexasaccharide residue is attached to rhodopsin in exactly this same manner.

The amino acid compositions presented in Table 1 are consistent with the findings that rhodopsin is an acidic protein (BRODA and VICTOR, 1940), although it is not yet known how many of the Asp and Glu residues are present as amides. The number of methionine (Met) residues is constant for all analyses and agrees well with the results of the CNBr digestion of rhodopsin derivatives (MILLAR et al.,

Table 1. Amino acid composition of cattle rhodopsin

Amino Acid	No. of residues						
	(1)	(2)	(3)	(4)	(5)[a]	(6)	(7)
Asp	18	15	19	15	—	—	—
Glu	21	21	25	27	—	—	—
Thr	20	17	25	17	1	—	—
Ser	17	12	15	5	—	—	—
Lys	13	10	8	5	1	—	—
Arg	10	6	6	5	—	—	—
His	6	4	4	3	—	—	—
Cys (as Cys SO$_3$H)	6	5	6	3	—	—	—
Met	8	8	8	8	—	—	—
Pro	19	13	16	14	1	—	—
Gly	19	16	21	17	—	—	—
Ala	22	20	25	20	3	—	—
Val	20	20	15	17	—	—	—
Ileu	14	13	10	12	1	—	—
Leu	22	20	20	20	—	—	—
Tyr	11	11	10	11	—	25	11
Phe	21	19	22	20	—	—	—
Try	—	5	—	6.5	3	15	5

[a] Analysis of a selected portion of the rhodopsin molecule.

(1) Shields et al. (1967) — purified rhodopsin in digitonin solution. 8 methionine residues per rhodopsin molecule were assumed and the molecular weight was calculated to be 28 600.

(2) Heller (1968) — purified rhodopsin in cetyl trimethylammonium bromide solution. The molecular weight was calculated to be 27 700 and in addition to the amino acids listed, 1 mole of rhodopsin was found to contain 3 moles of glucosamine, 2 moles of mannose and 1 mole of galactose.

(3) Shichi et al. (1969) — purified rhodopsin in Emulphogene BC-720 (Non-ionic poly-ethoxyethanol detergent) solution. 8 methionine residues per rhodopsin molecule were assumed and the molecular weight for opsin was calculated to be 27 700. In addition to the amino acids listed one mole of rhodopsin was found to contain 8 moles of hexosamine.

(4) Azuma and Kito (1967) — digitonin solution of rhodopsin. If we assume there are 8 methionine residues per molecule of rhodopsin we calculate a molecular weight of 25 000. Tryptophan content was estimated spectrophotometrically.

(5) Bownds (1967) — moles of amino acid/mole of N-retinyl-lysine in purified peptide resulting from a pronase digest of N-retinyl-opsin. These amino acids are adjacent to that lysine residue which is linked to retinaldehyde in metarhodopsin.

(6) Hubbard (1969) — spectrophotometric determination on native and denatured rhodopsin. A molar extinction of 40 600 for rhodopsin at 498 nm was utilized in the calculation.

(7) Collins et al. (1952a) — spectrophotometric determination. We have calculated the values listed by assuming a molecular weight of 30 000 for rhodopsin.

1969). However, these workers found 19 peptides from the tryptic digestion of carboxymethylated-opsin. These peptides were presumed to arise from peptide bond cleavage adjacent to both Arg and Lys residues. The results of HELLER (1968a) and SHICHI et al. (1969) would lead to a maximum of 17 and 15 peptides respectively from trypsin digestion and it is not now clear how these inconsistencies will be resolved.

We are still unable to describe opsin in meaningful structural terms and are restricted to approximating it as a globular structure, about 40 Å in diameter, with a largely hydrophobic exterior, and with a cleft capable of accepting and binding certain conjugated aldehydes closely related to 11-*cis* retinaldehyde.

3. Non-Covalent Interaction between Chromophore and Protein

The extensive differences in properties, particularly spectral, between N-retinylidene opsin and rhodopsin (MORTON and PITT, 1955) have prompted most investigators to propose non-covalent interactions in addition to the covalent aldimine bond between the chromophore and protein in native visual pigments (DARTNALL, 1957; HUBBARD, 1958a; KROPF and HUBBARD, 1958; BLATZ, 1965; IRVING et al., 1969; KITO et al., 1968). These interactions, to be enumerated below, were meant to account for:

(1) the large and variable shift in absorption spectrum from around 440 nm in N-retinylidene amines in aqueous or ethanolic solution to the longer wavelengths of λ_{max} in rhodopsin, iodopsin, various unstable intermediates containing the all-*trans* retinylidene chromophore and the many species of visual pigments with spectral properties ranging from 433 nm in the green rod pigment of the frog (DARTNALL, 1967) to 562 nm in chicken iodopsin (WALD et al., 1955). We can also include the vitamin A_2 based pigments, called by WALD the porphyropsins and cyanopsins, which range from 510 nm Brown Wrasse porphyropsin (BROWN and BROWN in WALD, 1960) to 620 nm in tadpole and other species' cyanopsin (LIEBMAN and ENTINE, 1967).

(2) the stereospecificity of reaction between opsin and the geometric isomers of retinaldehyde and other related molecules (HUBBARD and WALD, 1952; NELSON et al., 1969) best illustrated by the monotonous occurrence of the 11-*cis* isomer as chromophore. Its geometry alone has been the one found in all naturally occurring visual pigments which have been examined (see, for example, HUBBARD, 1958b).

(3) the thermodynamic stability of visual pigments. Schiff bases, which serve as models for rhodopsin, hydrolyze spontaneously over pH intervals where visual pigments are exceedingly stable (MORTON and PITT, 1955).

(4) the asymmetry in the absorption of circularly polarized light by the chromophores in rhodopsin (CRESCITELLI and SHAW, 1964; TAKEZAKI and KITO, 1967; CRESCITELLI et al., 1966; KITO et al., 1968), porphyropsin (CRESCITELLI and SHAW, 1964) and isorhodopsin (TAKEZAKI and KITO, 1967). This asymmetry is abolished when the pigment solution is bleached by light or heat (TAKEZAKI and KITO, 1967; CRESCITELLI et al., 1966), the chromophore itself exhibiting no optical activity.

a) **The Spectral Properties of Visual Pigments.** The carbon and hydrogen atoms constituting the chromophore of a visual pigment molecule are joined by covalent bonds which have very little ionic character. The hydrocarbon chromophore is thus not particularly reactive toward the side chain groups present in proteins like opsin or toward phospholipid molecules which might be present as part of a visual pigment entity. An additional covalent interaction, in addition to the Schiff base linkage, is thus not considered likely nor has any been found. Ionic-covalent interactions have been suggested by one author (Blatz, 1965), but he has since withdrawn his proposal (Blatz et al., 1968) because evidence he gathered to test this hypothesis seemed to him to be in conflict with its predictions.

Other theories have been concerned with various aspects of the group of van der Waals interactions (Kropf and Hubbard, 1958; Irving et al., 1969), concentrating mainly on the interactions between the protein and chromophore. The protein is considered as a conforming bed containing charged and/or polarizable groups, and the chromophore, both in its ground and excited states, is represented by a conjugated system of electrons which is assumed to be polarizable. As a group, these theories all involve a net attractive interaction between either charged groups on the protein, groups such as carboxylate (COO^-), histidyl

$$CH{=\!\!=\!\!=}C{-}CH_2{-} \atop {}^{+}NH \qquad NH \atop {\searrow}CH{\diagup}$$

,

arginyl $(-(CH_2)_3-NH-\underset{\underset{NH}{\|}}{C}-NH_3{}^+)$ or lysyl $(-(CH_2)_4-NH_3{}^+)$ or polarizable groups such as thiol $(-SH)$, phenyl $\left(\bigcirc\right)$, indol $\left(\text{indol structure } C{-}CH_2{-}, \underset{H}{N}, CH\right)$, alkyl $(-R)$

or others and the pi-electron cloud of the excited state retinylidene chromophore. The attractive interactions are proposed to account for the narrowing of the gap between the ground and excited state by lowering the energy of the excited state relative to the isolated chromophore and thus bringing it nearer to the ground state energy level. The range in λ_{max} for visual pigments from different sources is supposed to result from the fact that different visual pigments, though containing exactly the same chromophore, differ in their opsins. The electronic environment provided by the different opsins will presumably vary sufficiently by virtue of a changed amino acid arrangement and/or differing conformation of the protein in the neighborhood of the chromophore to produce the shifts in absorption spectra which are observed. The same arguments are involved when trying to account for the variations in λ_{max} for prelumi, lumi, meta- and isopigments (Yoshizawa and Wald, 1963; Hubbard et al., 1959). With the characterization of chicken pre-lumiiodopsin, which was found to have λ_{max} at ~ 640 mμ (Yoshizawa and Wald, 1967), the range of spectral maxima for pigments based on retinaldehyde, which must be accounted for theoretically, now extends from 430 nm to 640 nm.

Recently another suggestion has been made by Shichi et al. (1969). It is that the spectrum of rhodopsin results from a charge transfer interaction of the kind proposed by Mulliken (1952) for benzene-iodine complexes. They base their suggestion on the fact that the oscillator strength, which is a measure of the absorption intensity of a spectral band, increases by about a factor of two when going from 11-cis retinaldehyde to rhodopsin. This is taken to mean that a change in character of the spectral transition has occurred. Had they compared, however, the oscillator strengths for rhodopsin and Schiff base analogues of rhodopsin, which was done by ERICKSON and BLATZ (1968) and IRVING et al. (1969), they would have found the character and intensity of the absorption peaks to be similar. HELLER (1968b) also suggested a charge transfer mechanism in order to account for the spectral properties of rhodopsin. He proposed a specific chemical structure, involving covalent bonding between the retinylidene chromophore and lysine and cysteine, to buttress his proposal. The structure he suggested had already been synthesized by PESKIN and LOVE (1963) and found by them to be a poor model for rhodopsin because its spectral shift was to shorter wavelengths compared to free retinaldehyde and hence in the opposite direction to that which was needed for a rhodopsin model compound.

The best model, then, which accounts for the spectral properties of visual pigments seems to be the somewhat unspecific one of a protonated, retinylidene-amino, Schiff base whose van der Waals interaction with opsin gives rise to the characteristic absorption peaks shown in Fig. 1. Opsins from different species having different amino acid sequences and perhaps with different configurations give rise to the varying spectral maxima and molar absorptions which have been found (DARTNALL, 1965). The value of the lack of specificity in this model is that it allows one to encompass all of the data on visual pigments. As more is known of the structures of opsins, more specific proposals to account for the spectral properties should be forthcoming.

b) Circular Dichroism of Visual Pigments. One manifestation of an electronic transition occurring in an asymmetric environment is the phenomenon of circular dichroism, CD, and its complement, optical rotatory dispersion, ORD (MASON, 1963). The magnitude of circular dichroic absorption, the difference in absorption of left handed and right handed circularly polarized light, is a measure of the mirror asymmetry of the chromophore or, in the case of a chromophore which is itself symmetric, of the environment. A spectral transition will exhibit circular dichroism when it arises in a system which is asymmetric, such as a preferentially right or left twisted conjugated system, or when it is due to an electronic transition in a symmetrical chromophore which occurs in the immediate neighborhood of a perturbing, asymmetric group of atoms (MASON, 1963).

CD has been observed in both rhodopsin, with a twisted chromophore, and isorhodopsin, where the chromophore is presumed to be planar, and we agree with KITO et al.'s (1968) "tentative speculation" that the asymmetry induced in the chromophore absorption is due to the asymmetric environment provided by opsin and is not due to an induced right or left handed twist in the retinylidene chromophore, as had been earlier suggested by CRESCITELLI et al. (1966) and TAKEZAKI and KITO (1967). The values vary as follows for the relative molar CD absorbances, $A_{CD} = \dfrac{\varepsilon_L - \varepsilon_R}{\varepsilon_{\max}}$ where ε_L and ε_R are the molar absorbances for left handed and right handed circularly polarized light and where $\varepsilon_{\max}$ is for monochromatic, unpolarized light at $\lambda_{\max}$.

Cattle rhodopsin: $A_{CD} = 2 \quad (10)^{-4}$ at $\lambda = 500$ nm (TAKEZAKI and KITO, 1967),

$$A_{CD} = 2.5\,(10)^{-4} \text{ at } \lambda = 340 \text{ nm}$$

Frog rhodopsin: $A_{CD} = 5$ $(10)^{-4}$ at $\lambda = 490$ nm (Crescitelli et al., 1966), and
$\qquad\qquad\quad A_{CD} = 6$ $(10)^{-4}$ at $\lambda = 340$ nm

Squid rhodopsin: $A_{CD} = 5$ $(10)^{-4}$ at $\lambda = 495$ nm (Kito et al., 1968).
$\qquad\qquad\quad A_{CD} = 3.3\,(10)^{-4}$ at $\lambda = 350$ nm

Since these CD spectra are due to the same 11-*cis* retinylidene chromophore, we take this as support for the interpretation that the differences in the magnitude of the CD spectra are due to the diversity in amino acid composition (Heller, 1969) and consequent variation in protein conformation which attend these changes. In addition, a twist in the conjugated chromophore might be expected to produce fine structure in the CD spectral peak (Charney, 1965) and the observed peaks are all symmetrical and without fine structure.

If we pursue the interpretation that circular dichroic absorption is a measure of the strength of interaction between asymmetric groups of opsin and a symmetrical chromophore, we can see in the magnitude of the CD absorption in rhodopsins from different species a reflection of the details of the fit between chromophore and opsin. In addition to the variation among rhodopsins, we note that the molar CD absorbance is less in isorhodopsin than in rhodopsin even though the molar absorbances, ε_{max}, vary in the reverse way. This result, along with the interpretation of CD favored here, supports the suggestion of Kropf and Hubbard (1958) that the 9-*cis* configuration of the chromophore in isorhodopsin fits opsin at the same site as the 11-*cis* chromophore, but less perfectly.

Few complete studies of CD and ORD have been made on visual pigments, and intermediates in the photolytic breakdown of these pigments have not been characterized by this technique, probably because of the experimental difficulties which are involved. We think such studies will be particularly helpful in enhancing our understanding of the conformational changes accompanying the bleaching process.

III. Photolysis of Visual Pigment Solutions

The effect of light on rhodopsin, as well as on all other visual pigments which have been studied to date, is to initiate a cascade of reactions leading to an all-*trans* derivative of retinaldehyde (Hubbard et al., 1965). The sequence of reactions and names of the intermediate compounds are summarized in Fig. 6. The values of λ_{max} are for solutions of cattle rhodopsin where the thermal sequence proceeds only in the direction leading to metarhodopsin. In vertebrates there is the eventual appearance of all-*trans* retinaldehyde and opsin, but in solutions of visual pigments extracted from invertebrates, metarhodopsin is often the end product of the bleaching sequence (Hubbard and St. George, 1958; Brown and Brown, 1958).

1. Pre-Lumi Pigments

Light initially transforms rhodopsin into pre-lumirhodopsin, a change originally observed by Yoshizawa and Kito (1958). Cattle pre-lumirhodopsin is stable below about $-140°$ and has been studied most thoroughly at "liquid nitrogen" temperatures, around $-196°$ (Yoshizawa and Wald, 1963). However, it has also been detected as a transient intermediate in flash photolytic experiments at $-25°$ C

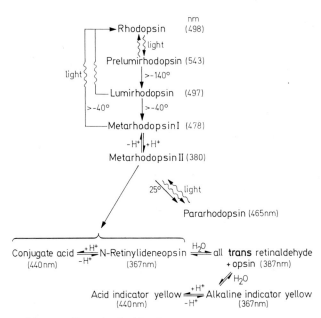

Fig. 6. Sequence of intermediates in the bleaching of cattle rhodopsin. Values in parentheses are λ_{max}'s for the compounds at that temperature where they are stable. All the known interconversions are shown. [Modified from BRIDGES (1967).]

(GRELLMAN et al.,1962). Since pre-lumirhodopsin is formed when either rhodopsin or isorhodopsin is irradiated at $-195°$ C, and since it reverts to these two stable pigments when it is irradiated, the system: rhodopsin $\underset{}{\overset{light}{\rightleftharpoons}}$ prelumirhodopsin $\overset{light}{\rightleftharpoons}$ isorhodopsin is reminiscent of the rhodopsin, metarhodopsin I, isorhodopsin system (HUBBARD and KROPF, 1958), where it was concluded that the transformations were photoisomerizations. Thus pre-lumirhodopsin, which in the case of the cattle pigment has $\lambda_{max} = 543$ nm and $\varepsilon_{543} = 72000$, was assumed to consist of an all-*trans* retinylidene chromophore attached to opsin where the protein still had essentially the same secondary and tertiary structure as rhodopsin (YOSHIZAWA and WALD, 1963). The supposition that the chromophores of all pre-lumi pigments were all-*trans* was consistent with the experimental data for cattle and squid rhodopsins (YOSHIZAWA and WALD, 1964), but difficulties of interpretation arose when iodopsin was irradiated at $-195°$ C and then warmed.

The chromophore of iodopsin has an 11-*cis* conformation (WALD et al., 1955), and when it is irradiated at $-195°$, prelumi-iodopsin is formed (YOSHIZAWA and WALD, 1967). But unlike pre-lumirhodopsin, prelumi-iodopsin returns to iodopsin by a thermal, dark reaction at $-140°$. Spontaneous, *trans* → *cis* isomerization in polyenes at $-140°$ is unprecedented (ZECHMEISTER, 1962), and special explanations are necessary if one is to assume an all-*trans* conformation for the chromophore in prelumi-iodopsin. YOSHIZAWA and WALD have tried to incorporate prelumi-iodopsin into their structural scheme by proposing that the chromophore of prelumi pigments is constrained by the "frozen" protein from assuming its all-*trans* planar configuration. They account for the different behavior on warming of prelumi-rhodopsin and prelumi-iodopsin by suggesting that in the former case the protein relaxes rapidly enough to accommodate the all-*trans* planar chromophore before it has a chance to return to a *cis*-configuration, but that the chromophore of prelumi-iodopsin relaxes to its 11-cis configuration before the protein "unfreezes".

This description of the pre-lumi pigments as having twisted all-*trans* chromophores may account for some aspects of their spectra, such as their large red shifts relative to rhodopsin and iodopsin (Yoshizawa and Wald, 1963, 1967), but other details are inconsistent with this structural hypothesis. Chief among these is the behavior of isoiodopsin, which also seems to form prelumi-iodopsin when irradiated at $-195°$. Prelumi-iodopsin formed from isoiodopsin also reverts to iodopsin when warmed to $-140°$. It is then difficult to understand how a 9-*cis* chromophore and an 11-*cis* chromophore can pass to the same twisted all-*trans* chromophore without one of them proceeding through a planar all-*trans* form. If a common all-*trans* form is involved, then one must explain the nature of the forces which would displace the chromophore from its all-*trans* conformation once it had reached that state. This argument, along with other less crucial arguments (Kropf, 1969), lead this author to conclude that we do not yet understand the structural changes attendant upon the initial photochemical reaction, rhodopsin (or iodopsin) $\xrightarrow{\text{light}}$ pre-lumirhodopsin (or prelumi-iodopsin).

2. Lumirhodopsin

The intermediate which follows pre-lumirhodopsin as it thermally decays to more stable products is lumirhodopsin. This transformation from pre-lumi to lumi pigment is characterized by a decrease in absorbance at long wavelengths and the appearance of a new spectral species. In the case of cattle lumirhodopsin λ_{max} lies at 497 nm, but even though specific lumirhodopsins seem to have similar ranges of thermal stability (Hubbard et al., 1959) λ_{max} varies from species to species. Table 2 lists the properties of lumirhodopsins from a number of vertebrate species. The variation in the spectral properties of these pigments is similar to that encountered in the rhodopsins themselves (Dartnall, 1965). Presumably the opsins from different species contain different sequences of amino acids and hence provide environments consisting of different polar and non-polar groups near which the chromophore resides. It is tacitly assumed here that spectral variation is associated with protein variation.

Table 2. Spectral properties of lumirhodopsins from five animal species (see Hubbard et al., 1959) Lumirhodopsins (at —65° C)

Species	λ_{max} (nm)	ε_{max} (l/mole-cm)	Half-band width (cm^{-1})
Monkey	500	56000	3500
Cattle	497	54000	3760
Chicken	513	49000	3630
Bullfrog	515	43500	4100
Cusk	519	54000	3630

Cattle lumirhodopsin exists in solution at 25° for only a very short time after a rhodopsin solution is flashed with light (ERHARDT et al., 1966; PRATT et al., 1964) and because it is difficult to carry out chemical tests at temperatures below $-40°$, where lumirhodopsin is stable almost indefinitely, we know very little about the chemical reactivity of this intermediate. Measurements of the kinetics of formation of lumirhodopsin do yield results which can be used to infer certain limited information about its structure. The values found for the enthalpy of activation for the reaction pre-lumirhodopsin $\rightarrow$ lumirhodopsin are all around 11 kcal/mole and the corresponding entropy of activation has been found to range from 0 to $+20$ cal/°K-mole (GRELLMAN et al., 1962; PRATT et al., 1964). These values indicate that the activated complex, with a configuration intermediate between pre-lumi and lumi-rhodopsin, does not differ greatly from pre-lumirhodopsin. If we further assume that pre-lumirhodopsin does not differ significantly from rhodopsin in the conformation of opsin, then we are led to conclude that there is only a very small change in protein conformation between rhodopsin and lumirhodopsin.

The most significant change occurs in the chromophore. There seems little question that the chromophore has adopted a normal, all-*trans* conformation at the stage of lumirhodopsin. At $-65°$ as well as at $-195°$ lumirhodopsin is readily interconvertible by light into *cis* pigments, such as rhodopsin and isorhodopsin (HUBBARD et al., 1959; YOSHIZAWA and WALD, 1963). In addition, its spectrum is similar to other *trans* chromoproteins, like metarhodopsin I, into which it readily converts at around $-40°$ C (HUBBARD et al., 1959). On the other hand, at $-65°$ C metarhodopsin I cannot be photochemically transformed into rhodopsin or isorhodopsin (HUBBARD et al., 1959) even though *trans* $\rightarrow$ *cis* photoisomerization is possible at this temperature. We take this to mean that in the lumi pigments opsin retains the configuration it had in rhodopsin or isorhodopsin when it was accommodating the 11-*cis* or 9-*cis* chromophore, while in metarhodopsin the protein conformation is altered. It is therefore almost solely the chromophore which changes in the transformation of rhodopsin into lumirhodopsin.

3. Metarhodopsin

At temperatures around $-40°$ and higher lumirhodopsin spontaneously changes into metarhodopsin (WALD et al., 1950). We will begin our discussion of vertebrate metarhodopsin by noting that two forms of the compound, metarhodopsin I and metarhodopsin II, have been distinguished (MATTHEWS et al., 1963). This finding now brings the vertebrate system into closer register with the visual pigment system of the molluscs, where acid and alkaline metarhodopsins have been known for some time (HUBBARD and ST. GEORGE, 1958; BROWN and BROWN, 1958). The two forms of the cattle pigment exist in a reversible equilibrium which can be represented as:

$$\text{metarhodopsin I} + \text{H}^+ \rightleftharpoons \text{metarhodopsin II}$$

$$\lambda_{max} = 478 \text{ nm} \qquad \text{pK} = 6.4 \qquad \lambda_{max} = 380 \text{ nm.}$$

In the case of the squid pigment the equilibrium is as follows:

$$\text{acid metarhodopsin} \rightleftharpoons \text{H}^+ + \text{alkaline metarhodopsin}$$

$$\lambda_{max} = 500 \text{ nm} \qquad \text{pK} = 7.7 \qquad \lambda_{max} = 380 \text{ nm.}$$

In the squid system Hubbard and St. George interpreted the metarhodopsin equilibrium as due to the reversible binding of a proton to the nitrogen atom of the $>C=N-$ bond. Studies of model compounds (Pitt et al., 1955) show that the protonated Schiff base of retinaldehyde has its λ_{max} at longer wavelengths than the non-protonated form. To consistently hold to the hypothesis that the Schiff base nitrogen atom is being deprotonated in the vertebrate system when metarhodopsin I forms metarhodopsin II, opsin must take up two protons in the reaction. In favor of such an interpretation is the measured entropy change for the transformation of cattle metarhodopsin I $\rightarrow$ metarhodopsin II, where a value of 46.5 cal/°-mole was found by Matthews et al. (1963) while Ostroy et al. (1966) obtained $\Delta S = 34$ cal/°-mole. Since the standard entropy change for the corresponding reaction in the squid pigment system is -8 cal/°-mole (Hubbard and St. George, 1958), one could readily postulate a sizeable conformational change in cattle opsin which might result from a change in its total charge when changing from one metarhodopsin species to the other.

In addition to the equilibrium thermodynamic data for the acid-base metarhodopsin equilibrium just cited, the rate of formation of metarhodopsin I from lumirhodopsin has been studied and energies and entropies of activation measured (Hubbard et al., 1965; Erhardt et al., 1966). These two studies differed substantially, with Hubbard et al. reporting a value of $\Delta H^{\pm}$ of 60 kcal/mole and $\Delta S^{\pm}$ of 160 cal/°-mole whereas Erhardt et al. reported three parallel first order processes with $\Delta H^{\pm}$'s equal to 16.3, 3.6 and 5.3 kcal/mole and $\Delta S^{\pm}$ values of about $-2.2, -52$ and -50 cal/°-mole respectively.

A more recent analysis by Rapp et al. (1970) of the lumirhodopsin $\rightarrow$ metarhodopsin decay has considerably altered this controversy. If Rapp et al.'s data are analyzed in terms of a single first order decay, then at temperatures between $-13.8°$ and $-20.5°$, $\Delta H^{\pm}$ is found to be 70.5 kcal/mole and $\Delta S^{\pm}$ is around 205 cal/°-mole. These values are much more line with Hubbard et al.'s results and are consistent with their interpretation that in the transition from lumi- to metarhodopsin there is a major loosening or unfolding of the protein structure. Rapp et al. find that they can analyze their data as well in terms of the second order reaction scheme: lumirhodopsin $+ E \xrightarrow{k_{LE}}$ metarhodopsin I, where E represents a species other than lumirhodopsin. In this case they find that $\Delta H^{\pm} = 22.4$ kcal/mole and $\Delta S^{\pm} = 18$ cal/°-mole. These latter values certainly imply a different molecular interpretation than the first order analysis provides. It seems clear that other than kinetic methods of analysis are needed in order to elucidate the structural changes occurring when lumirhodopsin changes into metarhodopsin I.

4. Pararhodopsin and N-Retinylidene-Opsin

Pararhodopsin (Wald, 1968), MRH_{465} (Ostroy et al., 1966), P_{470} (Williams, 1968), transient orange (Lythgoe, 1937; Lythgoe and Quilliam, 1938) are all names for the same species. It was known for some time that pararhodopsin appeared in a rhodopsin solution after it had been exposed to light, but not until the experiments of Matthews et al. (1963) could one identify a sequential pathway leading to its formation. They showed that pararhodopsin could be formed both photochemically and by a dark reaction from metarhodopsin II. Pararhodopsin would not reform metarhodopsin II by a thermal pathway, but the conversion could be carried out by light (i.e. metarhodopsin II $\underset{\text{light}}{\overset{-25°\rightarrow}{\rightleftharpoons}}$ pararhodopsin).

WILLIAMS (1968) studied the photochemical transformation in greater detail and suggested that pararhodopsin might well be a 13-*cis* isomeric chromoprotein, the same proposal which MATTHEWS et al. (1963) had made. This hypothesis is not accepted by OSTROY et al. (1966) who interpret their own experiments as indicating that pararhodopsin is an all-*trans* chromoprotein whose protein conformation differs from metarhodopsins I and II. This conformation is characterized by a λ_{max} at around 465 nm. Since pararhodopsin has been identified *in vivo* in the living rat retina (CONE and BROWN, 1969; CONE and COBBS, 1969) it is clearly important that its structure and properties be determind in order to help complete our understanding of the photochemistry of rhodopsin.

N-retinylidene opsin (NRO), the Schiff base union of all-*trans* retinaldehyde with an amino group on opsin, was defined by MORTON and PITT (1955) as a derivative of rhodopsin where no migration of the retinylidene chromophore from its original binding site had taken place. The derivatives which have been called NRO, acid and alkaline indicator yellow, are distinguished by their spectral maxima which are at 440 nm and 365 nm respectively. They are found under conditions where retinaldehyde could have randomly joined with amino groups on opsin, and it is thus not surprising that there are a variety of views as to whether NRO is an intermediate occurring before or after the appearance of free retinaldehyde (MATTHEWS et al., 1963; OSTROY et al., 1966; MORTON and PITT, 1969). This places NRO in just the opposite position of pararhodopsin, for with NRO the structure is known but its position in the photolytic scheme is in dispute, while pararhodopsin's place in the photolytic scheme seems well established but there are divergent views as to its structure.

Summarizing our knowledge of the intermediates which occur during the photolysis of rhodopsin, we are led to conclude that though the sequence, interconvertability and stabilities of the intermediates are known in most cases, we are still at an elementary level when we try and speak of their structures. Perhaps the newer and more sophisticated techniques developed for studying the conformational structures of proteins, such as circular dichroism and nuclear magnetic resonance, can be applied to the study of these compounds as well.

IV. Physico-Chemical Properties of Rod Outer Segments (ROSs)

1. Structure of the ROS

Before discussing the chemistry of visual pigments *in situ*, a very brief introduction will be given to the structure of the rod outer segment in order to set the stage, so to speak, for the scenery and action which follow. Electron microscopic studies, beginning with SJÖSTRAND'S (1949) pioneering work and continuing through the present (NILSSON, 1969) have shown that photoreceptors are highly organized structures. Though electron microscopy has not yet shown us how to locate the pigment molecules or any other class of substance in the photoreceptors, the regular structure which electron microscopy has revealed can accommodate the visual pigment in only a few ways. As Fig. 7 so clearly shows, there are only two different kinds of layers in the ROS which can accomodate the number of pigment molecules known to be present there. Though most workers favor the interpretation that rhodopsin is present in the darker staining layers rather than the alternate lighter layers, it suffices for the discussion of the photochemistry in the outer segments that pigment be situated in one set of layers or the other.

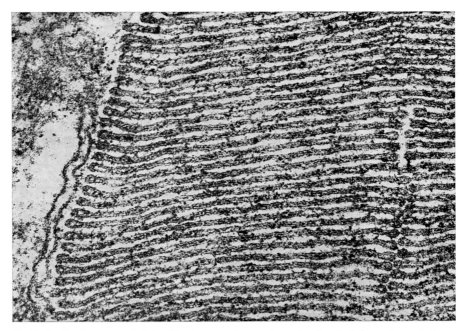

Fig. 7. Rod outer segment disks. The two membrane elements of each disk are in most places closely packed ×131000. [From NILSSON (1969). Used with permission.]

The outer segment has often been likened to a cylindrical stack of coins, although, because of the convoluted membranes enclosing each disk, a more accurate analogy would transform the coins into clover leaves (FERNANDEZ-MORAN, 1959). Electron microscopic studies show that there are approximately 1000 of these disks in ROSs from a variety of animals. Furthermore, estimates based upon the absorption of light by the ROSs and the optical properties of rhodopsin place the number of pigment molecules in the outer segment at around 40×10^6 (CONE, 1963). Assuming that each disk is like any other leads to an estimate of 40000 visual pigment molecules per disk. With this information we may then ask what the arrangement of the rhodopsin molecules in a disk is. The available evidence which can help us to deal with this question does not come from electron microscopy, since electron micrographs do not show the location of the visual pigment in the outer segments. Rather we must first examine those experiments in which the outer segments have been studied with polarized light.

2. Dichroic Properties of ROSs

W. J. SCHMIDT (1938) discovered the remarkable dichroic properties of ROSs which he described in a series of papers beginning in 1935. He demonstrated that when ROSs were viewed by white light directed down their long axes, they appeared pink. The absorption and color were independent of the plane of polarization of the viewing light. When viewed with polarized light propagating perpendicularly to the long axis, though, a striking difference was seen. The same absorption of green light and resulting pink color was noted for light polarized with its electric vector perpendicular to the long axis of the rod outer segment, but for polarized white light whose electric vector was parallel to the long axis, the outer segment was essentially transparent. SCHMIDT correctly explained these results as being a consequence of the orientation of the visual purple molecules in the ROS.

For the present discussion this result of SCHMIDT's is the most significant, although he was more concerned with the birefringent rather than the dichroic properties of ROSs. His analysis of the form birefringence of ROSs, performed long before the advent of the electron microscope, led to the proposal that there

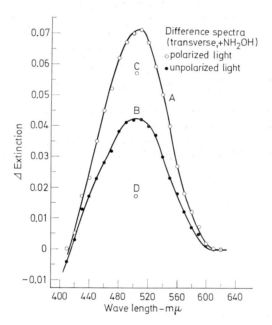

Fig. 8. Orientation of rhodopsin in a frog rod. These measurements were made on a single rod outer segment lying on its side in 30% methyl cellulose solution in the presence of 0.1 M hydroxylamine. Four types of absorption spectrum were recorded: (1) Dark-adapted outer segment in plane-polarized light, the electric vector of which was oriented at right angles to the rod axis; (2) the same, but with the plane of polarization turned through 90° so as to be parallel with the rod axis; (3) the same, in unpolarized light; and (4) after wholly bleaching the rhodopsin. The curves and points shown in the figure are: A: (1) - (4); B: (3) - (4); C: (1) - (2) — only the maximum extinction is shown; and D: the same (2) - (4). Were the rhodopsin chromophores oriented ideally in the transverse plane, D would be zero, and the ratio of maximum extinctions A/B would be 2.0; here it is 1.8. About 80% of the rhodopsin extinction is concentrated in the transverse plane in this instance. [From WALD et al. (1963). Used with permission.]

were transverse alternating layers of submicroscopic thickness consisting of lipid and protein, a deduction spectacularly confirmed by the studies of Sjöstrand (1953). Schmidt's other suggestion, that the lipid molecules are oriented parallel to the length of the rods while the protein molecules are transversely arranged, has not yet been thoroughly tested, although it is not unreasonable in view of current ideas about membrane structure (Branton and Park, 1968). The conclusion that protein molecules are transversely arranged and the finding that the chromophores are mainly arranged in the same way are coincidental, since they arise from different observations. From what we know about the structure of proteins containing prosthetic groups, there is no necessary correlation between the axes of the protein and its covalently bound prosthesis.

Subsequent studies with a variety of animal species (Denton, 1959; Liebman, 1962; Wald et al., 1963) have confirmed Schmidt's finding that the visual pigment chromophore is oriented chiefly in the transverse plane of the ROS. The more recent microspectrophotometric studies (Liebman, 1962; Wald et al., 1963) using frog ROSs have shown that while chromophore absorption is mainly confined to the transverse plane, there is still measurable optical density parallel to the rod axis which is due to rhodopsin. Referring to Fig. 8, Wald, Brown and Gibbons find that $E_\perp/E_{||} \approx 4.5$ while Liebman, as shown in Fig. 9, finds $E_\perp/E_{||} = 5.3$, where $E_\perp$ and $E_{||}$ are measures of the absorption of monochromatic light polarized in perpendicular and parallel directions to the ROS long axis. Liebman calculates an increase in absorption by a factor of 1.4 due to orientation, while stating that perfect orientation would lead to a factor of 1.5. On the other hand, Wald et al.,

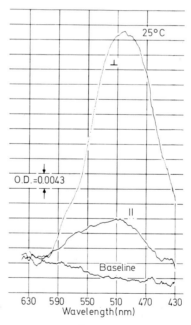

Fig. 9. Spectra recorded in polarized light. $||$ denotes recording made with light electric vector parallel to rod long axis. $\perp$ was recorded after $||$, with light electric vector perpendicular to the rod long axis. The ratio $\epsilon_\perp/\epsilon_{||}$ is 5.3. [From Liebman (1962). Used with permission.]

by comparing $E_\perp$ with $E_{\text{unpolarized}}$ find the ratio $E_\perp/E_{\text{unpolarized}} = 1.8$ and conclude that "about 80% of the rhodopsin extinction is concentrated in the transverse plane in this instance". One would like to be able to use these figures to arrive at a value for either the degree of orientation of the chromophores or the fraction of chromophores lying completely in the transverse plane of the ROS. But, recalling the conclusions reached earlier, that — (1) the chromophore of the visual pigment molecule is incapable of assuming a planar structure, (2) the deviation from planarity, while reasonably large, is not accurately known, and (3) the quantitative aspects of light absorption by the chromophore, particularly the dependence upon non-planarity or the absorption by "partial chromophores", is not reliably known at present — we are forced to take the view, already expressed by WALD et al. (1963), that the dichroism of the ROS can only strongly suggest that the chromophores are principally oriented in the transverse plane. It seems impossible at present to make any quantitative judgements about the degree of orientation or the fraction of pigment molecules oriented in a particular way.

A further aspect of the question of orientation of rhodopsin in the ROS is the possibility of orientation or organization within the transverse planes of the disks. One can conceive of many possible forms of organization which are all consistent with the dichroic properties already described for the ROS. These range from complete 2-dimensional randomness of the chromophore within a disk to crystalline-like order of these structures, with, perhaps, the direction of chromophore orientation varying from disk to disk and thus presenting the aspect of 2-dimensional disorder when viewed down the rod axis.

3. Energy Transfer

The question of chromophore organization within the outer segment has been addressed in an indirect way by asking whether the transfer of electronic excitation energy from one molecule of rhodopsin to another was possible in rod outer segments. HAGINS and JENNINGS (1959) were first in an attempt to explicitly answer this question with their study of photodichroism of rhodopsin in the ROSs of rabbits and frogs and in the excised retina of the rabbit. They initially showed that irradiation with plane polarized light of dark adapted ROSs which had been fixed in agar produced a marked dichroism, due, presumably, to the preferential photolysis of rhodopsin molecules whose chromophores were principally oriented in the same direction as the electric vector of the irradiating light beam. They suggested that this result was to be understood as due to preferential bleaching of those outer segments which were aligned at right angles to the polarization axis of the light. Those ROSs whose long axes were parallel to the polarization axis would not absorb light, and the probability of bleaching a ROS would increase as the angle made with the polarization axis increased from °0 to 90°. Though the ROSs were presumed to be arranged randomly in three dimensions in the agar and so did not necessarily all have their long axes in a plane perpendicular to the direction of propagation of the light, the dichroism found by HAGINS and JENNINGS could be understood on the basis of this qualitative description and dichroic properties of the ROSs which have just been described.

On the other hand, when plane polarized light was directed onto a rabbit's retina and the photodichroism measured, none was found. Fig. 10 shows the results from the retina and from the agar suspensions of ROSs. In the case where a retina was used, the direction of propagation of the polarized light was along the axis of the ROS and thus perpendiular to the planes of the disks in which the chromophores are presumed to lie. The results of SCHMIDT, DENTON and the other workers already mentioned showed that there was no preferred alignment of chromophores in the plane of the disks. HAGINS and JENNINGS' experimental findings, then, lead us to the conclusion that rhodopsin molecules, whose chromophore absorption axes are not parallel and are even perpendicular to the direction of the electric vector of the polarized light beam, are photolyzed as readily as their more favorably aligned neighbors.

The same lack of photodichroism was found by PAK and HELMRICH (1968) in retinas even when they were frozen at − 10° C or fixed in 2.5% glutaraldehyde for

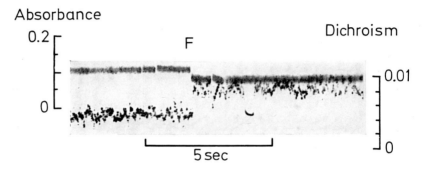

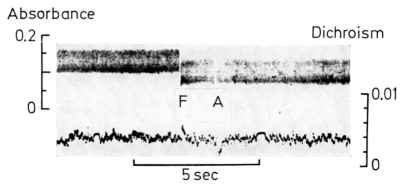

Fig. 10. a Effect of a plane-polarized bleaching flash F on the absorbance and dichroism of a gel containing rhodopsin in randomly oriented rods. Measuring wavelengths, 486 nm, temperature 3° C. b Effect of a plane-polarized bleaching flash F on the absorbance and dichroism of the rhodopsin in oriented, axially illuminated rods of an excised rabbit eye. Measuring wavelength, 485 nm; temperature, 3° C; deflection at A produced by mirror used to introduce bleaching flashes. [From HAGINS and JENNINGS (1959). Used with permission.]

an hour before being exposed to a polarized bleaching light. These workers used the early receptor potential response, rather than direct spectrophotometry to assay the rhodopsin absorption before and after the retina was bleached with polarized light. Though their technique did not preclude bleaching by de-polarized light which may have been back-scattered by the filter paper on which the retinas were placed, their results, taken together with those of HAGINS and JENNINGS, make it clear that plane polarized light propagated parallel to the long axis of the ROS affects equally those molecules whose chromophore axes are principally oriented parallel to the electric vector of the light beam as well as those oriented perpendicularly to the direction of polarization.

Perhaps the simplest explanation of these repeated failures to produce photo-dichroism is the following. Polarized light quanta, though initially absorbed only by rhodopsin molecules whose chromophore axes are principally oriented along the direction of the electric vector of the light beam, can be transferred to neigh-boring molecules whose axes are less favourably inclined. This radiationless transfer of electronic excitation energy would occur before the molecule which originally absorbed the photon had a chance to undergo *cis* to *trans* isomerization, and subsequently bleach.

The theory of intermolecular energy transfer as formulated by FÖRSTER (1951) showed that molecules could pass energy from one to another by an efficient, radiationless process over distances approaching 100 Å. The efficiency of this energy transfer, as measured by R_0, a defined average transfer distance over which transfer occurred with a probability of 1/2, was shown to depend upon the ab-sorption and fluorescence properties of the donor and acceptor species as well as on their mutual orientation. The explicit relationship for energy transfer for the process represented by A^* (excited) $+ A'$ (unexcited) $\rightarrow A$ (unexcited) $+ A'^*$ (excited) is:

$$R_0^6 = \frac{9000 \ln 10 \; \chi^2 \; \eta_0}{128 \; \pi^5 \; \bar{n}^4 \; N} \int_0^\infty \varepsilon_{A'} (\bar{v}) \, f_{A*} (\bar{v}) \, \frac{d \bar{v}}{\bar{v}^4} \tag{1}$$

where η_0 — fluorescence quantum yield of the donor A^*.

    $\bar{n}$ — refractive index of the solvent medium.

    N — Avogadro's number.

    χ — a dimensionless, geometric factor, varying between 0 and 1, deter-mined by the orientations in space of the transition dipole moments of A and A'. For randomly oriented molecules $\chi = 2/3$.

  $\varepsilon_{A'} (\bar{v})$ — the molar extinction coefficient of A' at the wavenumber $\bar{v}$.

  $f_{A*}(\bar{v})$ — the normalized fluorescence intensity of A^* at $\bar{v}$.

HAGINS and JENNINGS (1959) considered energy transfer as an explanation for their inability to produce dichroism in the retina. Though they were handi-capped by lack of data pertaining to the fluorescence of rhodopsin and its quantum yield, they made some simplifying assumptions about these parameters and cal-culated a maximum average transfer distance, R_0, of 19 Å. Since that time the fluorescence spectrum and fluorescence quantum yield of rhodopsin have been measured by Guzzo and Pool (1968) and so a more reliable value of R_0 can now be calculated, containing fewer crucial assumptions about the light absorbing and fluorescent properties of rhodopsin. Using the measured fluorescence and absorption spectra shown in Fig. 11, and the reported fluorescence quantum efficiency of 0.005, we have calculated R_0 using Eq. (1) and found it to be equal to 19 Å, coincidentally exactly the same value as calculated by Hagins and Jennings! (It should be noted that the equation for R_0 shows, by its 1/6th power dependence, a certain insensitivity to variations in the parameters. However, the fortuitous agreement in this case arises from a cancellation of large errors. Hagins and Jennings assumed a value of 1 for η_0, when in fact η_0 is equal to 0.005. Offsetting this was their assumption that the fluorescence maximum of rhodopsin was at 750 nm. Thus the overlap integral, $\int_0^\infty \varepsilon_{A'}(\bar\nu)\, f_{A*}(\bar\nu)\, \dfrac{d\bar\nu}{\bar\nu^4}$, between the absorption and fluorescence spectra of rhodopsin which they calculated was several orders of magnitude less than that calculated here.)

The experimental and theoretical evidence thus seem to favor the hypothesis that short range energy transfer in retinal rods is possible. This transfer would presumably occur within the confines of a single disk, as the distance between disks ranges from 200—400 Å (Sjöstrand, 1953) and so is much larger than the 19 Å we have calculated to be physically reasonable. It should also be noted that energy transfer over distances of several thousand Å is ruled out by the calculations and the direct light-microscopic observations of Hagins and Jennings (1959) and Liebman (1962), who found no detectable spread of rhodopsin bleaching when a carefully defined area of a ROS was illuminated through a slit and observed under the light microscope.

Though calculations from reasonably well established physical theory show that short range energy transfer is possible and experiments designed to look for the

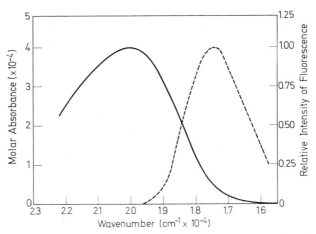

Fig. 11. Absorption spectrum (————) and fluorescence spectrum (— — — —) of cattle rhodopsin in solution. Molar absorbance is plotted vs. wavenumber for a 2% digitonin solution at 25° C. Arbitrary fluorescence emission intensity vs. wavenumber is for a digitonin solution at 3° C excited by light of approximately 490 nm (Guzzo and Pool, 1968)

consequences of short range energy transfer give results which are most easily interpreted, as PAK and HELMRICH (1968) have said, to mean that "short range ($\leqq 100$ Å) radiationless migration of energy is a possibility that does not appear to be excluded", the following considerations concerning the size and placement of rhodopsin molecules leads us to question this conclusion.

As mentioned earlier, there are around 40 million rhodopsin molecules distributed among approximately 1000 disks. If the 40000 molecules in one disk are spread over the outer surface of this disk only, and we approximate the retinal disk by a cylinder with a diameter of $1.5(10)^{-4}$ cm and negligible height, than there is a surface area of $3.5(10)^{-8}$ cm² available for the rhodopsin molecules. As for the area taken up by the rhodopsin molecules, this can be estimated by assuming each molecule to be a sphere of molecular weight 28000 daltons (SHIELDS et al., 1967; HELLER, 1968a; SHICHI et al., 1969) and a partial specific volume of 0.725 cc/g, a value typical of many proteins (SOBER, 1968). The resultant idealized rhodopsin molecule has a radius of 20 Å or a projected area of $1.25(10)^{-13}$ cm². The total area subtended by 40000 of these spherical rhodopsin molecules would be $6(10)^{-9}$ cm² or about 1/6th of the disk's surface area. The simplest assumption about the distribution of these molecules, namely that they are randomly distributed, thus leads to an average separation between rhodopsin centers of over 50 Å. At 50 Å, moreover, the probability of radiationless transfer between rhodopsin molecules is equal to about one in five hundred (0.002), a negligibly small value.

One possibility which may be considered, in view of the spatially linked chromoproteins found in mitochondria (LEHNINGER, 1967), is that rhodopsin molecules are arranged in clusters of four to ten with their chromophores facing each other, and thus close enough to transfer energy from one chromophoric group to another one oriented differently. This possibility has already been discussed and dismissed by HAGINS and JENNINGS (1959) on the grounds that "there is no clear evidence of large clusters (of rhodopsin) in electron micrographs of rods". Aside from one report on the ROS of Necturus (BROWN et al., 1963) subsequent electron microscopic work and low angle X-ray diffraction studies (BLASIE et al., 1965; BLASIE et al., 1969) all bear out HAGINS and JENNINGS' conclusions.

We are left to conclude that though the experimental observations on photodichroism can best be interpreted as giving evidence of intermolecular transfer of electronic excitation energy, the relevant structural considerations suggest that energy transfer of the kind suggested by FÖRSTER is not physically reasonable in retinal rods. Whether a new kind of energy transfer is operative in this case or other explanations will be found to apply, such as 2-dimensional rotational diffusion as suggested by HAGINS and JENNINGS (1959), remains to be determined by future experiments.

4. Photolysis of Visual Pigments in situ

The discussion just concluded indicates the pronounced effects which the organization of pigment molecules have on the light absorbing properties of the ROS and one might next ask whether there are also differences in the photochemical processes occurring in the visual pigments as a result of this organization. Though the amount of data relating to this question is small, there are experimental fundings which do indicate that photochemistry is fundamentally the same in the retina and in solution.

a) Light Absorption in situ. Photochemistry begins with the absorption of a photon (Stark, 1908; Einstein, 1912), in this case by a single molecule of visual pigment. If the rhodopsin molecules in the ROSs were organized in a regular array, as in a crystal, and if they interacted strongly enough so that the excitation was delocalized, we might expect a shift in the absorption peak and a change in shape of the absorption band from their values in solution (Kasha, 1959). Since the absorption properties of visual pigments *in situ* are essentially identical with their spectra in solution (Dartnall, 1961; Wald et al., 1963; Liebman and Entine, 1968), we may reasonably assume that the electronic state of the excited rhodopsin molecule is the same in the ROS as it is in solution.

Since the excitation process appears to be the same both in solution and in the rod, what physico-chemical processes other than transformation to prelumirhodopsin, and consequent *cis* to *trans* isomerization, might occur following light absorption? A proposal of Rosenberg (1966) is that the ROS can behave as a photoconductor and conduct charges generated by excited rhodopsin molecules. He proposes that *cis* to *trans* isomerization is a process competing with photoconduction, photoisomerization serving simply as an adaptive process allowing vision to operate over a wide dynamic range. Many of Rosenberg's hypotheses relating photoconduction and visual excitation derive from his experiments with β-carotene (Rosenberg, 1959, 1961) and retinaldehyde (Rosenberg and Harder, 1967), where he has measured action spectra, activation energies and other parameters characteristic of the photoconductive process. Recently Cherry (1968) has critically reviewed his own and Rosenberg's measurements of photoconductivity in β-carotene and has also discussed the relationship between the solid state measurements and the corresponding processes in ROSs, where quite different molecular arrangements and concentrations prevail. Because of the apparent low efficiency of photoconduction in the model systems (Cherry, 1968) as compared with the high quantum efficiency of visual excitation (Hecht et al., 1942) and photoisomerization of rhodopsin (Dartnall et al., 1936) and the lack of evidence that the action spectrum for photoconduction is the same as the absorption spectrum in either β-carotene or other model pigments, it would seem that the photoconduction theory of the receptor process (Rosenberg, 1966) has a number of fundamental difficulties to contend with before it can be considered as a valid alternative to the photochemical theory (Wald, 1968) of visual excitation.

b) Pre-Lumirhodopsin and Lumirhodopsin. Pursuing the photochemical sequence in the ROS, we find that the earliest intermediate found in rhodopsin solutions, pre-lumirhodopsin, has also been detected by Pratt et al. (1964) in rod particle suspensions. The absorption spectrum of pre-lumirhodopsin appears to be the same in cattle ROSs and in glycerol-water solutions. Transformation of pre-lumirhodopsin to lumirhodopsin was found to proceed both in solution and in the ROS by three parallel first order reactions, each with the same activation energy. Within experimental error the activation energy for the transformation in the ROS and in solution was 11 kcal/mole. In view of Rapp et al.'s (1970) reevaluation of a similar kinetic analysis of lumirhodopsin decay in solution, it would be worthwhile reexamining the kinetics of the decay of pre-lumirhodopsin as well. There has been no study to date of the orientation of pre-lumirhodopsin in the ROS.

PRATT et al. (1964) probably also produced lumirhodopsin in cattle ROSs since they detected a shift in λ_{max} from about 520 nm to about 502 nm by warming a suspension containing pre-lumirhodopsin from $-195°$ C to $-110°$ C. They did not report either a difference spectrum or the spectrum of the bleached products from which a difference spectrum could be calculated, and thus one can only infer the presence of lumirhodopsin in their ROS suspension. Likewise, GUZZO and POOL (1969) in reporting on the fluorescence spectra at $-196°$ C of the intermediates of rhodopsin bleaching in ROSs present suggestive evidence that lumirhodopsin is present and responsible for an emission spectrum with a peak at 600 nm. Since they are exciting a mixture of pigments in the ROS, and since the fluorescence spectrum of rhodopsin and that presumed for lumirhodopsin occur at the same value of λ_{max}, we must treat their suggestion with caution. They also report that λ_{max} for the fluorescence peak is at 600 nm when ROSs are examined, but at 580 nm for a digitonin extract. It is difficult to assess the significance of this result because of the uncertainty of the chemical species giving rise to the fluorescence emission. A difference in fluorescence spectrum of this magnitude, if it is confirmed, could indicate a significant difference in the conformation of lumirhodopsin and/or a difference in its interactions with neighboring molecules in the ROS as compared with the digitonin micelle.

c) **Metarhodopsin and Pararhodopsin.** Though evidence for the presence of pre-lumi and lumi-rhodopsin is meager, there are a variety of studies which converge on the point that the various metarhodopsins occur in ROS suspensions and living retinas. Though the terms metarhodopsin I and II have only been used explicitly in recent *in vivo* studies (BAUMANN, 1966; DONNER and REUTER, 1969), these intermediates most likely account for the spectral transients absorbing at around 480 nm and 380 nm seen in earlier photochemical studies of frog ROSs (BRIDGES, 1962; LIEBMAN, 1962). In addition to *in situ* spectrophotometric recordings, rapid electrical recordings of retinal events have demonstrated the formation and decay of these intermediates in rat (CONE, 1967; ARDEN et al., 1966), squid (HAGINS and McGAUGHY, 1967) and other animals. EBREY (1968) has also studied the formation and decay of the metarhodpsins and has compared their kinetic parameters *in situ* with the corresponding energies and entropies of activation in solution. EBREY finds that the kinetic parameters for the decay of meta I and meta II do not greatly differ between rhodopsin solutions and rhodopsin in ROSs or the intact retina. The available data are usually from different species, such as cattle, rabbit and rat, and it is not clear whether the agreement in kinetic properties is fortuitous or whether metarhodopsins I and II have structures which are almost invariant among vertebrates and which are transformed at almost the same rate regardless of the environment. More careful experiments will need to be done to examine these questions, but it seems abundantly clear that metarhodopsin appears as an intermediate in rhodopsin photochemistry at about the same time and with very nearly the same properties both in solution and in the retina.

ERP studies (Cone, 1967; Ebrey, 1968) have also shown that pararhodopsin appears in the retina following the formation of metarhodopsin II, confirming an earlier finding of Hagins (1957) in his study of the rabbit retina. In addition to the electrical signaling of its presence, pararhodopsin has been detected spectro-photometrically in the living rat retina (Frank and Dowling, 1968; Cone and Brown, 1969) and the latter authors, together with a parallel study by Cone and Cobbs (1969), have placed pararhodopsin in a bleaching-regeneration cycle which may point the way to an understanding of the mechanism or regeneration of rhodopsin. Finally we should take note of the absence of pararhodopsin in the flash illuminated squid retina (Hagins and Mc Gaughy, 1967), a result consistent with the solution photochemistry of squid rhodopsin (Hubbard and St. George, 1958).

d) Orientation of Rhodopsin and Vitamin A in situ. In contrast to some other photobiological processes, such as photosynthetic phosphorylation, which do not involve light induced movement of atoms or molecules, the first photochemical effect in vision is a photoisomerization wherein a long, carotenoid segment of the visual pigment molecule substantially changes its shape (Hubbard and Kropf, 1959). Since this change takes place in the ROS, which is itself highly ordered, it would seem eminently reasonable to try to follow the stereochemical course of visual pigment photochemistry *in situ* in order to enhance our knowledge of the stereochemical path followed in passing from rhodopsin through pre-lumi, lumi, and metarhodopsin to Vitamin A and incidentally learn something about the constraints placed on this process by the organization of the ROS itself. Cone and Brown (1967) have indicated one way of studying the orientation of rhodopsin by noting that the early receptor potential is evoked from ROSs only when the rhodopsin is oriented and the amplitude of the ERP varies exactly in parallel with the loss in orientation of the pigment. It is very difficult to observe the orientations of pre-lumi-, lumi- and the metarhodopsins in ROSs largely due to the fact that they are very transient species. In addition their absorption spectra overlap considerably and they are almost never present alone. Consequently there have been very few studies of the orientation of these particular rhodopsin photoproducts *in situ* and the orientation studies that have been carried out have largely been confined to polarized light analyses of vitamin A, a bleaching product many steps removed from rhodopsin.

Denton (1959) studied the fluorescence changes following illumination of frog, salamander and conger eel retinae and noted the slow appearance of a yellow-green emission which was largely polarized parallel to the long axis of the ROS. He interpreted this to mean that vitamin A, which emits a characteristic greenish yellow fluorescence (Moore, 1957), oriented parallel to the rod's long axis. Wald et al. (1963) also studied the orientation of rhodopsin's photoproducts in the frog retina by measuring the spectrum before and after bleaching with unpolarized light travelling down the rod axes. They found the intensities of the spectral changes to be identical in the ROS and in solution. Since they already knew that rhodopsin was oriented perpendicular to the rod axis, they concluded that vitamin A was largely oriented in the same manner, otherwise the intensity relations would have differed. Wald et al.'s method is more direct than Denton's, since they measure the spectrum directly. Thus their interpretation, which places the vitamin A orthogonal to its orientation in DENTON's model, seems more firmly grounded. But we are still left with DENTON's observation of a fluorescent product, appearing after the retina has been exposed to light, which emits most strongly in a direction parallel to the long axis of the ROS. WALD et al.'s and DENTON's studies have uncovered an intriguing aspect of visual photochemistry. It is clear that more thorough studies along these lines are needed in order to uncover the details of the molecular anatomy and physiology of the visual excitation process.

V. Summary

Reporting in 1970 on the extent of our knowledge of the structure and reactions of visual pigments leaves one in awe of the prodigious effort that has been expended to advance our understanding to its present state and a little impatient that we are not getting on more quickly with the job of answering the interesting questions which have become apparent as a result of these advances. Thus we know many of the details of the molecular structure of rhodopsin's 11-*cis* retinylidene chromophore, but we don't know its exact shape, though we can say with some assurance that the polyene chain is severely twisted. We know the bonding scheme in the chromophore and can assign bond lengths and bond angles in many cases, but our uncertainty of the exact conformation, together with certain inadequacies of quantum-mechanical calculational schemes, have prevented theoreticians from making reliable estimates of such quantities as intensities and wavelengths of absorption, differences in ground state energies, and pathways of photochemical reactions in molecules like the geometric isomers of vitamin A, retinaldehyde, retinylidene Schiff bases and ultimately the visual pigments.

A question which has been extensively studied experimentally and is presently at the center of an important controversy is that of the protein-chromophore linkage in rhodopsin, and presumably other visual pigments as well. Just when it seemed clear that the resolution of this question had been achieved and a Schiff base linkage to the ε-amino group of lysine identified as the group to which retinaldehyde was attached in rhodopsin, evidence was presented contradicting this conclusion. The new evidence seemed to point to a Schiff base linkage between 11-*cis* retinaldehyde and the amino group of phosphatidylethanolamine, a lipid found in rod outer segment extracts. The evidence for these two models has been reviewed and my reasons for favoring the Schiff base lysine linkage presented.

Purification and analysis of the amino acids and sugars making up the glycoprotein portion of rhodopsin have been almost completely achieved and it seems reasonable to expect that rhodopsin may soon have its amino acid sequence determined. Apart from the somewhat unusual circumstance of finding a hexasaccharide covalently attached to the β-carboxyl group of an aspartic acid residue, and a higher than average proportion of hydrophobic amino acid residues, there is nothing in the amino acid composition of opsin which suggests an explanation for its unique behavior. Three aspects of this behavior, the spectral shift in the chromophore when native visual pigment is formed, the wide variation in wavelength over which the spectral maxima of visual pigments and their photoproducts occur and the circular dichroism exhibited by several rhodopsins in the visible spectral region with the corresponding lack of optical activity of the chromophoric group, retinaldehyde, have been discussed. The phenomena can be accounted for in qualitative terms but the model proposed is only one of several possibilities and since quantitative predictions cannot be made from it, its validity cannot be critically tested.

Our understanding of the nature of the intermediates in the bleaching reaction of visual pigments, pre-lumi-, lumi-, meta-, para-rhodopsin and N-retinylidene opsin, has not progressed sufficiently to allow us to ascribe complete chemical structures to all these species. It seems clear that a very early step in the photolysis sequence is a *cis* to *trans* isomerization of the chromophore, but we have questioned whether this occurs at the pre-lumi stage or not. The thermal reformation of chicken iodopsin from pre-lumi iodopsin at around − 180° C raises doubt about the all-*trans* configuration of the chromophore in pre-lumi iodopsin and in other pre-lumi pigments as well. However, there seems little doubt that lumi pigments have all-*trans* chromophores. It also seems to be agreed that the conformation of opsin in the visual pigment, where it occurs combined with the 11-*cis* chromophore, differs from free opsin which results from the bleaching reaction. What is at issue is the nature and extent of the conformational change in going from lumi- to metarhodopsin and from metarhodopsin to later products. The nature of pararhodopsin is very largely undecided; the structure of N-retinylidene opsin is not in doubt, but its position and role in the bleaching sequence is.

The reactions of visual pigments in solution should properly serve as a prelude to the main theme of visual excitation, in this case the molecular transformations which initiate the neurophysiological events in the rod or cone cell. The fact that the pigment containing organelle, the outer segment, is not simply a sac filled with randomly oriented visual pigment molecules in solution, but is rather a neatly organized repeating structure of dichroic membranous discs consisting partially of organized light absorbers, leads one to expect that this structure is important for the proper functioning of the visual receptor. Knowing this much, we still do not understand the mechanism of transduction wherein the photochemical transformation of a single visual pigment molecule initiates an electrochemical event powerful enough to trigger the cell's response. Other physicochemical consequences of chromophore organization such as intermolecular electronic energy transfer have been explored. The experiments which have been carried out to detect the consequences of short-range intermolecular energy transfer lead to the interpretation that the process occurs. Yet simple, well-grounded calculations suggest the opposite conclusion. Not only are we uncertain about the biological utility of electronic energy migration over distances of the order of 100 Å, but we are still unclear as to whether we should interpret the experiments showing a lack of induced dichroism in ROSs to mean that energy migration in fact occurs, since physical considerations make it seem unreasonable.

Following the absorption of light and assuming that the physical fate of the photon energy does not influence visual excitation, we have outlined the photochemical steps known to occur in the intact rod outer segment. Intermediates similar to, if not identical with, those in solution have been detected. The evidence for pre-lumi and lumirhodopsin *in situ* is not as convincing as the solution evidence and their characteristic behavior in the rod structure have not been fully determined. The presence and properties of meta- and pararhodopsin are more fully documented and they have been positioned in a visual cycle. The organized matrix provided by the rod outer segment has been exploited only a few times for studies on the orientational changes of the pigment molecules as they undergo their stepwise transformations. In addition to studies of the orientation of rhodopsin, the positioning of vitamin A *in situ* has been observed, although different methods of detection have led to different suggestions for its orientation. A study of the *in situ* stereochemical changes, together with the added information we now have and are continuing to acquire about the structure and properties of the visual pigments, may soon lead to a more detailed understanding of the still mysterious process whereby the relatively enormous structure of the rod or cone is coupled to the universe of light by its smallest messenger, a single photon.

Acknowledgements

I would like to express my gratitude to Prof. ERNST FISCHER and other members of the Weizmann Institute for their hospitality when much of this review was being written. During this period I held an NIH Special Fellowship, and I have also been, and continue to be, supported by NIH grant EY 00201. I also wish to acknowledge this support with special thanks.

References

AKHTAR, M., BLOSSE, P. T., DEWHURST, P. B.: The active site of the visual protein, rhodopsin. Chem. Commun. 631—632 (1967).
— — — Studies on vision: The nature of the retinal-opsin linkage. Biochem. J. 110, 693—702 (1968).
— HIRTENSTEIN, M. D.: Chemistry of the active site of rhodopsin. Biochem. J. 115, 607—608 (1969).
ARDEN, G. B., IKEDA, H., SIEGEL, I. M.: New components of the mammalian receptor potential and their relation to visual photochemistry. Vision Res. 6, 373—384 (1966).
AZUMA, M., KITO, Y.: Studies on optical rotation, circular dichroism and amino acid composition of rhodopsin. Ann. Rep. Biol. Works Fac. Sci. Osaka Univ. 15, 59—69 (1967).
BALL, S., GOODWIN, T. W., MORTON, R. A.: Studies on vitamin A. 5. The preparation of retinene -vitamin A aldehyde. Biochem. J. 42, 516—523 (1948).
— COLLINS, F. D., DALVI, P. D., MORTON, R. A.: Studies in vitamin A. 11. Reactions of retinene with amino compounds. Biochem. J. 45, 304—307 (1949).
BART, J. C. J., MAC GILLAVRY, C. H.: The crystal and molecular structure of canthaxanthin. Acta Cryst. B 24, 1587—1606 (1968a).
— — The crystal structure of 15,15′-dehydrocanthaxanthin. Acta Cryst. B 24, 1569—1587 (1968b).
BAUMANN, C.: Der Einfluß von Metarhodopsin auf die Sehpurpurbleichung in der Isolierten Netzhaut. Vision Res. 6, 5—13 (1966).
BERTHOD, H., PULLMAN, A.: Aspects de la structure electronique du rétinène et de ses isomères d'intérêt biologique. Compt. Rend. Acad. Sci. (Paris) 251, 808—810 (1960).
BLASIE, J. K., DEWEY, M. M., BLAUROCK, A. E., WORTHINGTON, C. R.: Electron microscope and low-angle X-ray diffraction studies on outer segment membranes from the retina of the frog. J. Mol. Biol. 14, 143—152 (1965).
— WORTHINGTON, C. R., DEWEY, M. M.: Molecular localization of frog retinal receptor photopigment by electron microscopy and low-angle X-ray diffraction. J. Mol. Biol. 39, 407—416 (1969).
BLATZ, P. E.: The role of carbonium ions in color reception. J. gen. Physiol. 48, 753—760 (1965).

Blatz, P. E., Pippert, D. L., Balasubramaniyan, V.: Absorption maxima of cations related to retinal and their implication to mechanism for bathochromic shift in visual pigment. Photochem. Photobiol. 8, 309—315 (1968).

Bownds, D., Wald, G.: Reaction of the rhodopsin chromophore with sodium borohydride. Nature (Lond.) 205, 254—257 (1965).

— Site of attachment of retinal in rhodopsin. Nature 216, 1178—1181 (1967).

Branton, D., Park, R. B. (Ed.): Papers on Biological Membrane Structure. Boston: Little, Brown and Co. 1968.

Bridges, C. D. B.: Studies on the flash-photolysis of visual pigments. IV. Vision Res. 2, 215—232 (1962).

— Biochemistry of visual processes. Comprehensive Biochem. 27, 31—78 (1967).

Broda, E. E., Victor, E.: The cataphoretic mobility of visual purple. Biochem. J. 34, 1501—1506 (1940).

Brown, P. K., Brown, P. S.: Visual pigments of the octopus and cuttlefish. Nature (Lond.) 182, 1288—1290 (1958).

— Gibbons, I. R., Wald, G.: The visual cells and visual pigment of the mudpuppy, Necturus. J. Cell Biol. 19, 79—106 (1963).

Catley, J., Moore, S., Stein, W. H.: The carbohydrate moiety of bovine pancreatic deoxy-ribonuclease. J. biol. Chem. 244, 933—936 (1969).

Charney, E.: Optical activities of non-planar conjugated dienes. III. HMO calculation of the skew angle dependence. Tetrahedron 21, 3127—3139 (1965).

Cherry, R. J.: Semiconduction and photoconduction of biological pigments. Quart. Rev. 22, 160—178 (1968).

Collins, F. D., Love, R. M., Morton, R. A.: Studies in rhodopsin. 4. Preparation of rhodopsin. Biochem. J. 51, 292—298 (1952a).

— — — Studies in rhodopsin. 5. Chemical analysis of retinal material. Biochem. J. 51, 669—673 (1952b).

Cone, R. A.: Quantum relations of the rat electroretinogram. J. gen. Physiol. 46, 1267—1286 (1963).

— Early receptor potential: Photoreversible charge displacement in rhodopsin. Science 155, 1128—1131 (1967).

— Brown, P. K.: Dependence of the early receptor potential on the orientation of rhodopsin. Science 156, 536 (1967).

— — Spontaneous regeneration of rhodopsin in the isolated rat retina. Nature (Lond.) 221, 818—820 (1969).

— Cobbs, W. H.: Rhodopsin cycle in the living eye of the rat. Nature (Lond.) 221, 820—822 (1969).

Crescitelli, R., Shaw, T. I.: The circular dichroism of some visual pigments. J. Physiol. (Lond.) 175, 43—45P (1964).

— Mommaerts, W. F. H. M., Shaw, T. I.: Circular dichroism of visual pigments in the visible and ultraviolet spectral regions. Proc. nat. Acad. Sci. (Wash.) 56, 1729—1734 (1966).

Dartnall, H. J. A., Goodeve, C. F., Lythgoe, R. J.: The quantitative analysis of the photo-chemical bleaching of visual purple solutions in monochromatic light. Proc. roy. Soc. A 156, 158—170 (1936).

— The interpretation of spectral sensitivity curves. Brit. med. Bull. 9, 24—30 (1953).

— The Visual Pigments. New York: John Wiley & Sons, Inc. 1957.

— Visual pigments before and after extraction from visual cells. Proc. roy. Soc. 154B, 250—266 (1961).

— The spectral clustering of visual pigments. Vision Res. 5, 81—100 (1965).

— The visual pigment of the green rods. Vision Res. 7, 1—16 (1967).

Denton, E. J.: The contributions of the orientated photosensitive and other molecules to the absorption of whole retina. Proc. roy. Soc. B150, 78—94 (1959).

Donner, K. O., Reuter, T.: The photoproducts of rhodopsin in the isolated retina of the frog. Vision Res. 9, 815—847 (1969).

Dowling, J. E.: The organization of vertebrate visual receptors. In: Allen, J. M. (Ed.): Molecular Organization and Biological Function. New York: Harper & Row 1967.

DOWLING, J. E., GIBBONS, I. R.: The effect of vitamin A deficiency on the fine structure of the retina. In: SMELSER, C. K. (Ed.): The Structure of the Eye, pp. 85—99. New York: Academic Press Inc. 1961.

EBREY, T. G.: The thermal decay of the intermediates of rhodopsin *in situ*. Vision Res. 8, 965—982 (1968).

EICHORN, E. L., MAC GILLAVRY, C. H.: The crystal structure of the *trans* isomer of β-ionylidene crotonic acid. Acta Cryst, 12, 872—883 (1959).

EINSTEIN, A.: Thermodynamische Begründung des photochemischen Äquivalentgesetzes. Ann. Phys. 37, 832—838 (1912).

ERHARDT, F., OSTROY, S. E., ABRAHAMSON, E. W.: Protein configuration changes in the photolysis of rhodopsin. I. The thermal decay of cattle lumirhodopsin *in vitro*. Biochem. biophys. Acta 112, 256—264 (1966).

ERICKSON, J. O., BLATZ, P. E.: N-retinylidene-1-amino-2-propanol: A Schiff base analog for rhodopsin. Vision Res. 8, 1367—1375 (1968).

FARRAR, K. R., HAMLET, J. C., HENBEST, H. B., JONES, E. R. H.: Studies in the polyene series. Part XLIII. The structure and synthesis of vitamin A_2 and related compounds. J. chem. Soc. 2657—2668 (1952).

FERNÁNDEZ-MORÁN, H.: Fine structure of biological lamellar systems. Rev. Mod. Phys. 31, 319—330 (1959).

FÖRSTER, T.: Fluoreszenz organischer Verbindungen. Göttingen: Vandenhoeck & Ruprecht 1951.

FRANK, R. N., DOWLING, J. E.: Rhodopsin photoproducts: Effects on electroretinogram sensitivity in isolated perfused rat retina. Science 161, 487—489 (1968).

FRATEV, P., ATANSOV, B., PETKOV, D.: Quantum-chemical approach to the conversions of retinal during the rhodopsin cycle at the initial act of visual reception. Izv. Otd. Khim. Nauki Bulg. Acad. Nauk 1, 125—142 (1968).

GOTTSCHALK, A. (Ed.): Glycoproteins. New York: Elsevier Publ. Co. 1966.

GRELLMAN, K.-H., LIVINGSTON, R., PRATT, D. C.: A flash-photolytic investigation of rhodopsin at low temperature. Nature (Lond.) 193, 1258—1260 (1962).

GUZZO, A. V., POOL, G. L.: Visual pigment fluorescence. Science 159, 312—314 (1968).

— — Fluorescence spectra of the intermediates of rhodopsin bleaching. Photochem. Photobiol. 9, 565—570 (1969).

HAGINS, W. A.: Rhodopsin in a mammalian retina. Ph. D. thesis, Cambridge Univ., 1957.

— Mc GAUGHY, R. E.: Molecular and thermal origins of fast photoelectric effects in the squid retina. Science 157, 813—816 (1967).

— JENNINGS, W. H.: Radiationless migration of electronic excitation in retinal rods. Disc. Faraday Soc. 27, 180—190 (1959).

HALL, M., BACHRACH, A. D. E.: Linkage of retinal to opsin and absence of phospholipids in purified frog visual pigment 500. Nature (Lond.) 225, 637—638 (1970).

HECHT, S., SHLEAR, S., PIRENNE, M. H.: Energy, quanta and vision. J. gen. Physiol. 25, 819—840 (1942).

HELLER, J.: Structure of visual pigments. I. Purification, molecular weight and composition of bovine visual pigment 500. Biochem. 7, 2906—2913 (1968a).

— Structure of visual pigments. II. Binding of retinal and conformational changes on light exposure in bovine visual pigment 500. Biochemistry 7, 2914—2920 (1968b).

— Comparative study of a membrane protein. Characterization of bovine, rat, and frog visual pigments 500. Biochemistry 8, 675—679 (1969).

— LAWRENCE, M. A.: Structure of the glycopeptide from bovine visual pigment 500. Biochemistry 9, 864—869 (1970).

HUBBARD, R.: The molecular weight of rhodopsin and the nature of the rhodopsin-digitonin complex. J. gen. Physiol. 37, 381—399 (1954).

— On the chromophores of the visual pigments. Natl. Phys. Lab. Symp., No. 8, H. M. S. O. London, 153—169 (1958a).

— The thermal stability of rhodopsin and opsin. J. gen. Physiol. 42, 259—280 (1958b).

— The stereoisomerization of 11-*cis* retinal. J. biol. Chem. 241, 1814—1818 (1966).

— Absorption spectrum of rhodopsin: 280 nm absorption band. Nature (Lond.) 221, 435—437 (1969).

HUBBARD, R., BOWNDS, D., YOSHIZAWA, T.: The chemistry of visual photoreception. Cold Spr. Harb. Symp. quant. Biol. **30**, 301—315 (1965).
— BROWN, P. K., KROPF, A.: Vertebrate lumi- and meta-rhodopsins. Nature (Lond.) **183**, 442—450 (1959).
— KROPF, A.: Molecular aspects of visual excitation. Ann. N. Y. Acad. Sci. **81**, 388—398 (1959).
— ST. GEORGE, R. C. C.: The rhodopsin system of the squid. J. gen. Physiol. **41**, 501—528 (1958).
— WALD, G.: *cis-trans* isomers of Vitamin A and retinene in the rhodopsin system. J. gen. Physiol. **36**, 269—315 (1952).
INUZUKA, K., BECKER, R. S.: Mechanism of photoisomerization of the retinals and implications in rhodopsin. Nature (Lond.) **219**, 383—385 (1968).
IRVING, C. S., BYERS, G. W., LEERMAKERS, P. A.: Effect of solvent polarizability on the absorption spectrum of all-*trans*-retinylpyrrolidinium perchlorate. J. Amer. chem. Soc. **91**, 2141—2143 (1969).
JENCKS, W. P.: Catalysis in chemistry and enzymology. New York: McGraw-Hill Book Co. 1969.
KASHA, M.: Relation between exciton bands and conduction bands in molecular lamellar systems. Rev. mod. Phys. **31**, 162—169 (1959).
KIMBEL, R. L., Jr., POINCELOT, R. P., ABRAHAMSON, E. W.: Chromophore transfer from lipid to protein in bovine rhodopsin. Biochemistry **9**, 1817—1820 (1970).
KITO, Y., AZUMA, M., MAEDA, Y.: Circular dichroism of squid rhodopsin. Biochem. biophys. Acta **154**, 352—359 (1968).
KRINSKY, N. I.: The lipoprotein nature of rhodopsin. Arch. Ophthal. Part II, **60**, 688—694 (1958).
KROPF, A., HUBBARD, R.: The mechanism of bleaching rhodopsin. Ann. N. Y. Acad. Sci. **74**, 266—280 (1958).
— Photochemistry of visual pigments. In: Proceedings of the International School of Physics "Enrico Fermi" course XLIII. New York: Acad. Press 1969.
LANGLET, J., BERTHOD, H., PULLMAN, B.: Electron structure of retinal and its *cis* isomers. I. Semiempirical self-consistent molecular field method. J. Chim. Phys. Physicochim. Biol. **66**, 566—574 (1969).
LEHNINGER, A. L.: Molecular basis of mitochondrial structure and function. In: ALLEN, J. M. (Ed.): Molecular Organization and Biological Function. New York: Harper & Row 1967.
LIEBMAN, P. A.: *In situ* microspectrophotometric studies on the pigments of single retinal rods. Biophys. J. **2**, 161—178 (1962).
— ENTINE, G.: Cyanopsin, a visual pigment of retinal origin. Nature (Lond.) **216**, 501—503 (1967).
— — Visual pigments of frog and tadpole (*Rana pipiens*). Vision Res. 8, 761—775 (1968).
LYTHGOE, R. J.: The absorption spectrum of visual purple and of indicator yellow. J. Physiol. (Lond.) **89**, 331—358 (1937).
— QUILLIAM, J. P.: The relation of transient orange to visual purple and indicator yellow. J. Physiol. (Lond.) **94**, 399 —410 (1938).
MASON, S. F.: Optical rotatory power. Quart. Rev. **17**, 20—66 (1963).
MATTHEWS, R. G., HUBBARD, R., BROWN, P. K., WALD, G.: Tautomeric forms of metarhodopsin. J. gen. Physiol. **47**, 215—240 (1963).
MERZ, J. H., STRAUB, P. A., HEILBRONNER, E.: Berechnung der Absorptionsspektren von all-*trans* Polyenen mittels eines "Molecules in Molecules" Modelles. Chimia **19**, 302—314 (1965).
MILLAR, P. G., SHIELDS, J. E., HENRIKSEN, R. A., KIMBEL, Jr., R. L.: Peptide pattern studies on bovine rhodopsin. Biochem. biophys. Acta **175**, 345—354 (1969).
MOORE, T.: Vitamin A. Amsterdam: Elsevier Publ. Co. 1957.
MORTON, R. A., GOODWIN, T. W.: Preparation of retinene *in vitro*. Nature (Lond.) **153**, 405—406 (1944).
— PITT, G. A. J.: Studies on rhodopsin. 9. pH and the hydrolysis of indicator yellow. Biochem. J. **59**, 128—134 (1955).
— — Visual pigments. Fortschr. Chem. Org. Naturst. **14**, 244—316 (1957).
— — Aspects of visual pigment research. Advanc. Enzymol. **32**, 97—171 (1969).

MULLIKEN, R. S.: Molecular compounds and their spectra. II. J. Amer. chem. Soc. **14**, 811—824 (1952).

NASH, H. A.: The stereoisomers of retinal — a theoretical study of energy differences. J. theoret. Biol. **22**, 314—324 (1969).

NELSON, R., DE RIEL, J. K., KROPF, A.: 13-desmethyl rhodopsin and 13-desmethyl isorhodopsin: Visual pigment analogues. Proc. nat. Acad. Sci. (Wash.) **66**, 531—538 (1970).

NILSSON, S. E. G.: The ultrastructure of photoreceptor cells. In: Proceedings of the International School of Physics "Enrico Fermi", Course XLIII. New York: Academic Press 1969.

OSTROY, S. E., ERHARDT, F., ABRAHAMSON, E. W.: Protein configuration changes in the photolysis of rhodopsin. II. The sequence of intermediates in thermal decay of cattle metarhodopsin *in vitro*. Biochem. biophys. Acta **112**, 265—277 (1966).

PAK, W. L., HELMRICH, H. G.: Absence of photodichroism in the retinal receptors. Vision Res. **8**, 585—589 (1968).

PARISER, R., PARR, R. G.: A semi-empirical theory of the electronic spectra and electronic structure of complex unsaturated molecules. J. chem. Phys. **21**, 466—471 (1953).

PARKHURST, L. J., ANEX, B. G.: Polarization of the lowest energy allowed transition of β-ionylidine crotonic acid and the electronic structure of the polyenes. J. chem. Phys. **45**, 862—873 (1966).

PATEL, D. J.: 220 MHZ proton nuclear magnetic resonance spectra of retinals. Nature (Lond.) **221**, 825—828 (1969).

PESKIN, J., LOVE, B. B.: The reaction of l-cysteine with all-*trans* retinene. Biochem. biophys. Acta **78**, 751—753 (1963).

PITT, G. A. J., COLLINS, F. D., MORTON, R. A., STOK, P.: Studies on rhodopsin. 8. Retinylidene-methylamine, an indicator yellow analogue. Biochem. J. **59**, 122—128 (1955).

POINCELOT, R. P., MILLAR, P. G., KIMBEL, R. L., Jr., ABRAHAMSON, E. W.: Lipid to protein chromophore transfer in the photolysis of visual pigments. Nature (Lond.) **221**, 256—257 (1969).

— — — — Determination of the chromophoric binding site in native bovine rhodopsin. Biochemistry **9**, 1809—1816 (1970).

POPLE, J. A.: The electronic spectra of aromatic molecules. II. A theoretical treatment of excited states of alternant hydrocarbon molecules based on self consistent molecular orbitals. Proc. Phys. Soc. **68** A, 81—89 (1955).

PRATT, D. C., LIVINGSTON, R., GRELLMAN, K.-H.: Flash photolysis of rod particle suspensions. Photochem. Photobiol. **3**, 121—127 (1964).

PULLMAN, A., PULLMAN, B.: The *cis-trans* isomerization of conjugated polyenes and the occurrence of a hindered *cis* isomer of retinene in the rhodopsin system. Proc. nat. Acad. Sci. (Wash.) **47**, 7—14 (1961).

RAPP, J., WIESENFELD, J. R., ABRAHAMSON, E. W.: The kinetics of intermediate processes in the photolysis of bovine rhodopsin. 1. A re-examination of the decay of bovine lumirhodopsin 497. Biochem. biophys. Acta **201**, 119—130 (1970).

ROBESON, C. D., CAWLEY, J. D., WEISLER, L., STERN, M. H., EDDINGER, C. C., CHECHAK, A. J.: Chemistry of Vitamin A. XXIV. The synthesis of geometric isomers of Vitamin A *via* methyl β-methylglutaconate. J. Amer. chem. Soc. **77**, 4111—4119 (1955).

ROSENBERG, B.: Photoconduction and *cis-trans* isomerism in β-carotene. J. Chem. Phys. **31**, 238—246 (1959).

— Photoconduction activation energies in *cis-trans* isomers of β-carotene. J. chem. Phys. **34**, 63—66 (1961).

— A physical approach to the visual receptor process. Advanc. Rad. Biol. **2**, 193—241 (1966).

— HARDER, H. C.: Semiconduction and photoconduction activation energies of the retinals. Photochem. Photobiol. **6**, 629—641 (1967).

SCHMIDT, W. J.: Polarisationsoptische Analyse eines Eiweiß-Lipoid-Systems, erläutert am Außenglied der Sehzellen. Kolloid-Z. **85**, 137—148 (1938).

SHICHI, H., LEWIS, M. S., IRREVERRE, F., STONE, A. L.: Biochemistry of visual pigments. I. Purification and properties of bovine rhodopsin. J. biol. Chem. **244**, 529—536 (1969).

SHIELDS, J. E., DINOVO, E. C., HENRIKSEN, R. A., KIMBEL, R. L., Jr., MILLAR, P. G.: The purification and amino acid composition of bovine rhodopsin. Biochem. biophys. Acta **147**, 238—251 (1967).

SIMPSON, W. T.: Resonance force theory of carotenoid pigments. J. Amer. chem. Soc. **77**, 6164—6168 (1955).

SJÖSTRAND, F. S.: An electron microscope study of the retinal rods of the guinea pig eye. J. cell comp. Physiol. **33**, 383—403 (1949).

— The ultrastructure of the outer segments of rods and cones of the eye as revealed by the electron microscope. J. cell comp. Physiol. **42**, 15—44 (1953).

SLY, W. G.: The crystal structure of 15, 15'-dehydro-β-carotene. Acta Cryst. **17**, 511—528 (1964).

SOBER, H. A. (Ed.): Handbook of Biochemistry. Cleveland: Chemical Rubber Co. 1968.

SPACKMAN, D. B., STEIN, W. H., MOORE, S.: Automatic recording apparatus for use in the chromatography of amino acids. Anal. Chem. **30**, 1190—1206 (1958).

STAM, C. H., MacGILLAVRY, C. H.: The crystal structure of the triclinic modification of Vitamin A acid. Acta Cryst. **16**, 62—68 (1963).

STARK, J.: Weitere Bemerkungen über die thermische und chemische Absorption im Bandspektrum. Physik. Z. **9**, 889—894 (1908).

STERLING, C.: Crystal structure analysis of β-carotene. Acta Cryst. **7**, 1224—1228 (1964).

TAKEZAKI, M., KITO, Y.: Circular dichroism of rhodopsin and isorhodopsin. Nature (Lond.) **215**, 1197—1199 (1967).

WALD, G.: Carotenoids and the visual cycle. J. gen. Physiol. **19**, 351—371 (1935).

— Visual purple system in fresh-water fishes. Nature (Lond.) **139**, 1017—1018 (1937).

— The distribution and evolution of visual systems. In: FLORKIN, M., MASON, H. S. (Eds.): Comparative Biochemistry, Vol. 1. New York: Academic Press 1960.

— The molecular basis of visual excitation. Nature (Lond.) **219**, 800—807 (1968).

— BROWN, P. K., GIBBONS, I. R.: The problem of visual excitation. J. Opt. Soc. Amer. **53**, 20—35 (1963).

— — Smith, P. H.: Cyanopsin, a new pigment of cone vision. Science **118**, 505—508 (1953).

— — — Iodopsin. J. gen. Physiol. **38**, 623—681 (1955).

WALD, G., DURELL, J., ST. GEORGE, R. C. C.: The light reaction in the bleaching of rhodopsin. Science **111**, 179—181 (1950).

WIESENFELD, J. R., ABRAHAMSON, E. W.: Visual pigments: their spectra and isomerizations. Photochem. Photobiol. **8**, 487—493 (1968).

WILLIAMS, T. P.: Photolysis of metarhodopsin II: Rates of production of P470 and rhodopsin. Vision Res. **8**, 1457—1466 (1968).

YOSHIZAWA, T., KITO, Y.: Chemistry of the rhodopsin cycle. Nature (Lond.) **182**, 1604—1605 (1958).

— WALD, G.: Pre-lumirhodopsin and the bleaching of visual pigments. Nature (Lond.) **197**, 1279—1286 (1963).

— — Transformations of squid rhodopsin at low temperatures. Nature (Lond.) **201**, 340—345 (1964).

— — Photochemistry of iodopsin. Nature (Lond.) **214**, 566—571 (1967).

ZECHMEISTER, L.: *cis-trans* isomeric carotenoids, vitamins A and arylpolyenes. New York: Academic Press 1962.

— LE ROSEN, A. L., SCHROEDER, W. A., POLGÁR, A., PAULING, L.: Spectral characteristics and configuration of some stereoisomeric carotenoids including prolycopene and pro-γ-carotene. J. Amer. chem. Soc. **65**, 1940—1951 (1943).

Chapter 6

Generator Potentials in Invertebrate Photoreceptors

By

M. G. F. Fuortes and Paul M. O'Bryan, Bethesda, Maryland (USA)

With 28 Figures

Contents

The purpose of the present chapter is to describe the responses evoked by light in photoreceptor cells of invertebrates, and to discuss various views on the mechanisms leading to the photic responses. Since the structure and organization of invertebrate photoreceptors are described in detail elsewhere in this volume, a few general remarks on this topic will be sufficient in this chapter.

Organization of Photoreceptor Cells

The function of photoreceptors is to relay to the central nervous system information on the state of the light in the surrounds. Their ability to reproduce in their signals the properties of the surrounding light depends both on features of individual cells and on the manner in which the cells are organized: the messages produced by a single photoreceptor cell can signal only the intensity and the duration of the light impinging upon it; aggregates of receptor cells in association with one or more lenses can monitor in addition wavelength and location of the light and thus can recover information on colors and shapes.

The organization of photoreceptor cells is quite uniform in all vertebrates. Here millions of receptor cells are tightly packed in the outer layer of the retina, where an image is formed by a single lens. In invertebrates instead, the organization of photoreceptors varies greatly.

Two different types of invertebrate eyes have a structure sufficiently elaborate for the resolution of fine patterns and, in some cases, of colors. The first type is the *compound eye* of arthropods, which consists of hundreds or thousands of lenses, each of which is associated with a small group of photoreceptor cells arranged in radial symmetry and called *ommatidium*. The second type is the *chamber eye* of cephalopods such as squid and octopus which resembles in its optical properties the eyes of vertebrates. This eye possesses a single lens and an underlying regular array of many thousands of receptor cells.

Examples of moderate complexity are the eyes of the molluscs *Pecten* and *Aplysia*. *Pecten* (the common scallop) has about 80 small eyes, each with a single lens, a reflector and a total of perhaps 2000 photoreceptor cells arranged in two orderly layers. *Aplysia* has two small eyes consisting each of an almost spherical lens and an underlying layer of about 7000 receptors.

More primitive photoreceptor organs are the ocelli of arthropods, in which a single lens is placed above a small group of sensory cells. In some cases they resemble a single ommatidium of the compound eye. Other simple eyes which have recently been the object of physiological investigations are the eye of the annelid *Hirudo* which consists of a rudimentary lens and an irregular, multi-layered group of about 50 sensory cells; the eye of the nudibranch mollusc *Hermissenda* with only five photoreceptors and the eye of the crustacean *Balanus* (the barnacle) which contains only three receptor cells.

Finally, there are examples of photoreceptors which are not associated with a lens but are located under the translucent external tissues of the animal (KENNEDY, 1960; MILLECCHIA et al., 1966; SMITH et al., 1968a). In these cases, the light must suffer considerable attenuation and scattering before reaching the photosensitive cells.

Microvilli — Visual Pigment

A characteristic feature of photoreceptor cells is the presence of structures with large membrane areas, which are found in different animals in the form of villi, tubules, lamellae or discs (EAKIN, 1965). The morphology of the rod discs and cone lamellae of vertebrates is described by COHEN (1972) in this volume. It seems fairly certain that while the lamellae of the cones are infoldings of the cell membrane exposed to the outside, the discs of the rod are intracellular formations, disconnected from the external membrane (NILSSON, 1969).

In the large majority of cases, the photoreceptor cells of invertebrates possess microvilli. These are submicroscopic cylindrical extensions of the cell membrane, with a diameter of about 0.1 μ and a length of 1 μ to 2 μ. Among the examples mentioned in the previous section, only the photoreceptors in the distal layer of the eye of *Pecten* do not form microvilli but have in their place tubules wound in the form of an irregular spiral which are derived from cilia. Similar tubules of ciliary origin are found in the photoreceptors of chaetognaths according to EAKIN (1965).

In many cases the microvilli are concentrated in an elongated structure called *rhabdome*[1], which runs parallel to the long axis of the cell. Frequently but not always, neighboring photoreceptor cells form close contacts in this region. The microvilli greatly increase the area of the cell membrane. For instance in

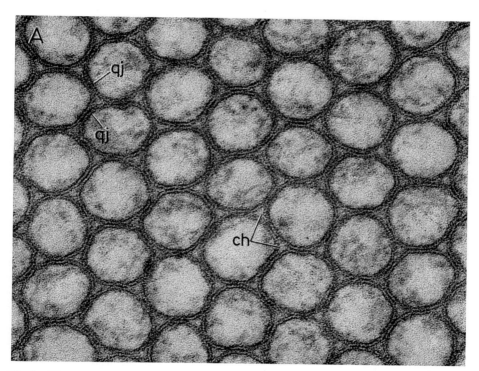

Fig. 1a. Microvilli. Electron micrograph of the cross-section of microvilli in the eye of *Limulus*. In this area, the microvilli are arranged in hexagonal symmetry, each making contact with six neighbors (in other areas, the arrangement is less regular). At the points of contact (*qj*) no separation is seen between the outer leaflets of the two adjacent membranes. These areas have been called "quintuple-layered junctions" by LASANSKY (1967). At the points where three microvilli meet (*ch*) the outer leaflets of the membranes separate from one another leaving a small channel of extracellular space. X 210,000 (from A. LASANSKY, unpublished)

---

[1] From the Greek *rhabdos*, meaning rod.

Limulus, where they are tightly packed, the area of the membrane occupied by microvilli is about 20 times larger than it would be if it were smooth.

The function of the microvilli is not known. It is not likely that their purpose is to increase the exchange between the cytoplasm and the external fluids, because in many cases a large fraction of their membrane is not exposed freely to the outside. As shown in Fig. 1, about one half of the surface of the microvilli of *Limulus* is directly in contact with other membranes and the other half faces narrow channels (about 10^{-12} cm^2 in cross section and 10^{-4} cm long) which are probably unsuitable for rapid exchange with the bulk of the extracellular fluid. It seems more probable then that the membrane expansion provided by the microvilli has the purpose of accommodating the visual pigment required for the function of the cell. Evidence that the visual pigment of invertebrate photoreceptors is located in the rhabdome was obtained by LANGER and THORELL (1966) by means of microspectrophotometric measurements, while electrophysiological investigations by SMITH and BROWN (1966) and HAGINS and McGAUGHY (1968) suggest that in *Limulus* and in squid the visual pigment is a component of the membrane of photoreceptor cells. It is reasonable to conclude from this that, in microvillar photoreceptors, the visual pigment is contained in

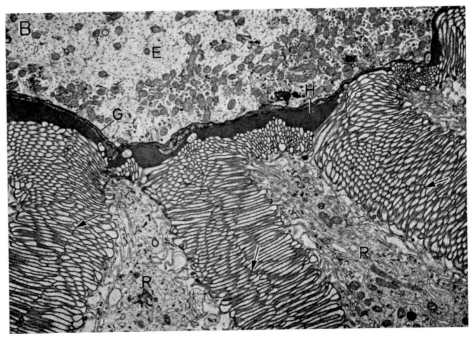

Fig. 1b. Electron micrograph of a section of *Limulus* ommatidium at the base of the eccentric cell dendrite. One sees parts of two retinular cells (*R*) and of the eccentric cell *E*. At this level the two types of cells are separated by a space occupied by hemolimph (*H*) and glia cell processes (*G*). Each retinula cell forms microvilli, which are seen partly in longitudinal section and partly in oblique or cross-sections. The microvilli of two adjacent cells meet at their tips (arrows) and from there quintuple-layer junctions (see Fig. 9 in LASANSKY, 1967) similar to those shown in Fig. 1a. X 17,000 (A. LASANSKY, unpublished)

the membrane of the microvilli, and to suggest in general that the significance of the membrane expansions characteristic of photoreceptors is to provide a framework ensuring the orderly arrangement of a large number of visual pigment molecules (EAKIN, 1965).

The Lateral Eye of Limulus

HARTLINE and his associates (see HARTLINE, 1933, 1934; HARTLINE and GRAHAM, 1932a, 1934) first used the eye of *Limulus* to study the relation between stimulus and response in a photoreceptor. This ancient animal has two lateral eyes each containing about 600 ommatidia. Each ommatidium contains about 12 cells, called *retinular cells*, which are arranged in radial symmetry like the sections of an orange, and usually a single cell of different shape, called the *eccentric cell*. On occasions the eccentric cell may be missing, or one ommatidium may contain two eccentric cells (Fig. 2).

The eccentric cell has a long distal process or dendrite. Around this and extending for some distance toward the periphery of the ommatidium, the cellular membranes form the rhabdome which appears in cross sections as a set of spokes radiating from a central hub, and consists of a dense array of microvilli (MILLER, 1957).

Both eccentric and retinular cells possess axons, the axon of eccentric cells being considerably larger than that of retinular cells (see WATERMAN and

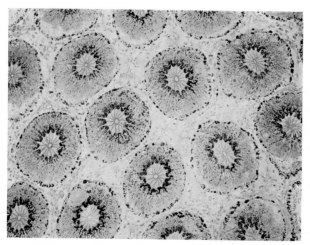

Fig. 2a. Ommatidia of *Limulus*. Cross-section of several ommatidia from the ventral eye of *Limulus*. Each ommatidium is formed by about 14 wedge-shaped retinular cells arranged in radial symmetry like the sections of an orange. At the centre of the ommatidium one sees at this level the dendrite of the eccentric cell. Around the dendrite is the central ring of the rhabdome which is formed by microvilli originating from both retinular and eccentric cells (see Fig. 12). The "spokes" departing from the central ring are made up of the retinular cell microvilli illustrated in Fig. 1b. The rhabdomeric spokes extend only about one third of the way towards the outside. Where they terminate, an accumulation of melanine pigment granules is present in the retinular cells. The retinular cell membrane peripheral to the spokes of the rhabdome is smooth and the membranes of adjacent cells are separated by glial processes and extracellular space. X 120 (A. LASANSKY, unpublished)

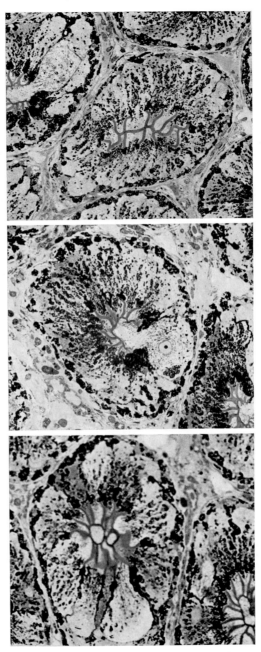

Fig. 2b. Ommatidia of *Limulus*. Cross-sections of three ommatidia to illustrate the not in-frequent absence of eccentric cell (left), the normal occurrence of one eccentric cell (centre) and the occasional finding of two eccentric cell dendrites in one ommatidium (right). These sections are at a lower level than in Fig. 2a. Eccentric cell bodies can be seen in two ommatidia. X 300 (A. Lasansky, unpublished)

WIERSMA, 1954; NUNNEMACHER and DAVIS, 1968). Thus, about 12 axons originate from each ommatidium and the individual bundles merge to form the long optic nerve which connects the eye with the optic lobe of the brain.

Hartline's Studies on Limulus. Input-Output Relation

HARTLINE's early research was performed by recording from small bundles of optic nerve fibers (Fig. 3). When he recorded from a filament originating from a single ommatidium, he found that the nerve impulses produced by illumination had the characteristics of single unit discharges (HARTLINE, 1935). This is a surprising observation, since the filament contained about 12 nerve fibers; it suggests that only one fiber from each ommatidium carries nerve impulses, or that the firing is perfectly synchronized in all fibers. In either case, the functional significance of this arrangement is unclear (see below). The presence of unitary discharges in a bundle of fibers offers, however, a favorable opportunity for the analysis of the relations between the stimulus and the response of a sensory unit. The records of Fig. 3 show several characteristics of these relations: it is seen in these records that an increase of light intensity by a factor of 1000 brings about an increase of the steady-state frequency of firing by a factor of only 3. Clearly,

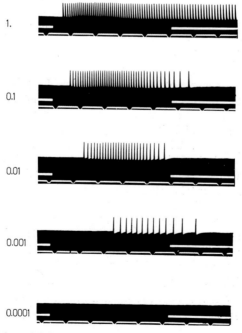

Fig. 3. Responses to lights of different intensity. Nerve impulses recorded from a single fibre of the nerve from the lateral eye of *Limulus*. Relative values of light intensity are given at left. Bottom line in each record gives 200 msec time marks. Illumination is signaled by interruption of the white line above the time marks. Raising light intensity decreases the latency, increases the frequency and prolongs the duration of the discharge (from HARTLINE, 1932)

"intensity information is considerably compressed in being translated into discharge frequency of the nerve fiber" (HARTLINE, 1967, p. 243).

Another property of these responses is the long delay between the onset of the stimulus and the beginning of the discharge. In the experiment of Fig. 3, the latency was about 500 msec when the light was dim (0.001) and decreased to about 100 msec with the brightest stimulus. Apparently the response is not a direct consequence of the absorption of light, but rather it is brought about by a process which develops slowly following illumination.

Finally, it is seen in the records that frequency of firing is higher at the beginning of the discharge than in the steady state. This illustrates the familiar property of sensory adaptation, common to many receptors. Due to this property, changes of stimulus intensity produce stronger responses than sustained stimuli of constant intensity.

It may be concluded from these results that the responses of visual cells "are not a faithful representation of the light stimuli, which were simple steps of intensity. To some extent, the receptor mechanism distorted the sensory information. This illustrates the broad principle established by the earliest studies of single sensory endings: receptors, by virtue of their inherent properties operate upon the information they collect from their surroundings to favor certain features of it. The processing of sensory data begins in the receptors" (HARTLINE, 1967, p. 244–245).

Generator Potentials in Eccentric Cells

HARTLINE's early results on *Limulus* give a valuable description of the relations between the stimulus and the firing of ommatidial cells but are insufficient for identifying the processes intervening between the primary photochemical reaction and the discharge of nerve impulses.

Important insight into this question was gained when HARTLINE, WAGNER and MACNICHOL (1952) succeeded in recording visual cell responses by means of intracellular micropipettes. They demonstrated with this method that the eccentric cell is depolarized by illumination and that nerve impulses are superposed on this depolarization. It appears from these results that nerve impulses originate in eccentric cells, as in other nerve cells, as a consequence of depolarization of the cell's membrane. The depolarization evoked by sensory stimuli is often called "generator potential" (BERNHARD and GRANIT, 1946).

Later studies on generator potentials and impulse discharge of eccentric cells[2] (MACNICHOL, 1956; FUORTES, 1958, 1959) showed that, in the steady state, the amplitude of the generator potential is approximately a linear function of the logarithm of the light intensity (Fig. 4a and 5a) and that the frequency of firing

---

[2] In the early experiments performed with microelectrodes (FUORTES, 1958), ommatidial cells of *Limulus* were classified in two groups according to the size of the spikes recorded in the different instances. In one type of cell, spike size ranged continuously between zero and 15 mV. In the other type peak-to-peak spike size was over 40 mV. These two types of cells were then tentatively identified with retinular and eccentric cells respectively. This classification was later confirmed by BEHRENS and WULFF (1965 and 1967) who passed a stain through the intracellular electrode used for recording, and so were able to identify the penetrated cell histologically.

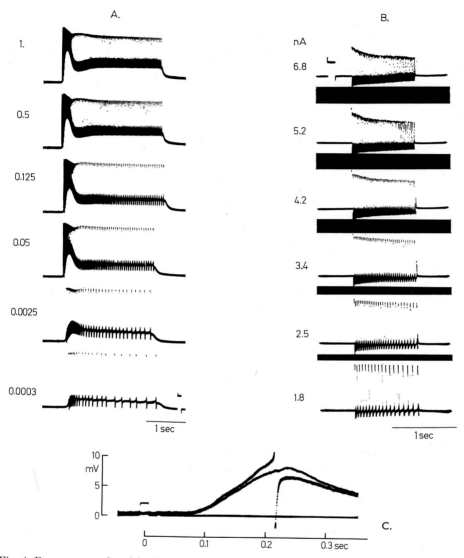

Fig. 4. Responses produced by light or by currents. a Responses to steps of light of different intensities, recorded intracellularly from an eccentric cell of *Limulus*. Small artifacts at the beginning and end of each record show the time at which the shutter was opened and closed. The square wave at the bottom is a 20 mV calibrating pulse. b Responses to depolarizing currents. Records from the same cell of Fig. 4a. Currents were passed through the intracellular electrode and a bridge was used to cancel the potential drops evoked by the currents across electrode and membrane resistance. Current intensity in nA (1 nA = 10^{-9} A) is indicated by the figures at the left. Square wave at start of top record is a 20 mV calibration pulse. c Latency of spike discharge. Flashes of light, indicated in lower trace were just threshold for eliciting firing of an eccentric cell. Records were taken by means of an intracellular electrode. Two responses are superimposed. The generator potentials develop slowly and an impulse is produced when the depolarization reaches a level of about 8 mV (modified from FUORTES, 1958)

is a linear function of the amplitude of the generator potential (Fig. 5b). Sustained depolarizing currents artificially passed through the membrane of an ommatidial cell were shown to produce regular trains of spikes (Fig. 4b) and the frequency of firing was observed to be linearly related to current intensity (Fig. 5c).

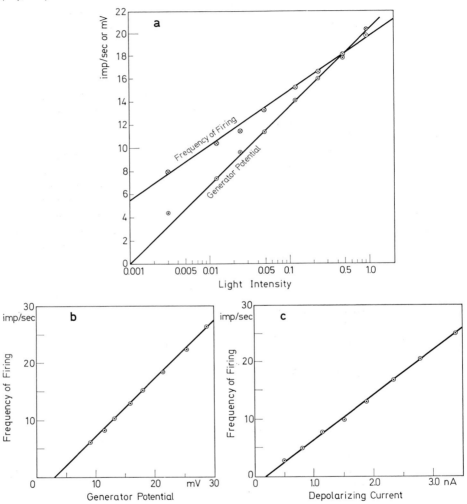

Fig. 5. a Plot of generator potential amplitude (dots) and frequency of firing (crosses) as a function of light intensity. The measurements were taken between 15 and 20 sec after the onset of a steady light. Both amplitude of the generator potential and frequency of firing are an approximately linear function of the logarithm of the light intensity. b Plot of frequency of firing as a function of amplitude of generator potential. Same data as in Fig. 5a. Frequency is a linear function of generator potential amplitude. The slope of this relation in eccentric cells of *Limulus* is usually between 0.75 and 2.5 imp/sec/mV. c Relation between frequency of firing and current intensity. Measurements were taken between 15 and 20 sec after the onset of a steady depolarizing current. Frequency is a linear function of current intensity. The slope of the relation ranges between 5 and 15 imp/sec/nA in most units (from FUORTES, 1958)

It was also noted in these studies that the long delay between the onset of a dim light and the firing of the first spike, which HARTLINE had already observed in his records from the optic nerve, is largely due to the slow development of the generator potential (Fig. 4c). This observation was later utilized as a basis for a phenomenological interpretation of the generator potential (FUORTES and HODGKIN, 1964). Furthermore, the generator potential showed adaptation, being larger at the beginning of a step of light than later on. This can account to a large extent for the adaptation of the impulse discharge.

Origin of Nerve Impulses in Eccentric Cells

It has been known for a long time (COLE and CURTIS, 1939) that the membrane resistance of nerve fibers is greatly reduced during spike activity. If the generator potential and the action potential were produced by the same area of membrane, the slow potential would be decreased by the impulses, because the generator current would evoke only a small potential drop across the low membrane resistance present during the spikes (cf. the analysis of end-plate potentials and spikes by FATT and KATZ, 1951). Intracellular records from visual cells of *Limulus* show instead that the generator potential persists during impulse firing, so that the impulses appear to "ride" on the slow depolarization. These findings are readily explained by assuming that the generator potential originates at the dendrite or soma of the eccentric cell while the spike originates at the axon and does not invade the soma.

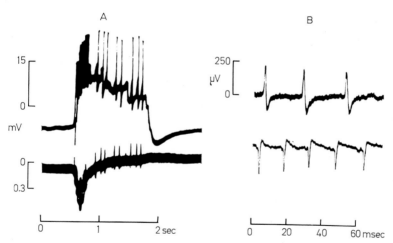

Fig. 6. Impulses recorded intra- and extracellularly from ommatidial cells. a One electrode was inserted in an eccentric cell of *Limulus*, and another electrode was just outside the same cell. Following a flash of light (signalled by a small artifact) the intracellular electrode recorded a depolarizing generator potential and positive-going spikes. The extracellular electrode recorded a negative-going generator potential with positive-going spikes. b Top record shows positive-going spikes recorded by an electrode located just outside the soma of an eccentric cell of *Limulus*. The bottom record was led off by an electrode placed on a nerve bundle emerging from the ommatidium. In this location, the nerve impulses give rise to negative-going spikes (modified from TOMITA, 1957)

Evidence supporting this interpretation was obtained by Tomita (1956, 1957). He recorded responses to light with an electrode inside the soma of an eccentric cell and another electrode just outside, and found that the generator potential had an opposite polarity in the two records. The spikes, however, gave, positive deflections in both cases. Negative-going spikes were recorded, instead, by an external electrode placed in the vicinity of the initial segment of the axon. (Fig. 6). Thus, the current associated with the generator potential is inward-directed at the soma and outward-directied at the axon, while the current accompanying nerve impulses has an outward direction at the soma and an inward direction at the axon. These results on impulses in eccentric cells were confirmed by Fuortes (1959) and by Purple and Dodge (1965).

Function of the Small Fibers in the Optic Nerve of Limulus

The function of the eccentric cell axons is clear: they generate nerve impulses in the vicinity of the cell soma and conduct them to their terminals in the optic ganglia. By contrast, the function of the axons of the retinular cells has not yet been ascertained.

It has been demonstrated that the small fibers of the optic nerve of *Limulus* conduct impulses if the nerve is stimulated electrically (see Fig. 7) (Borsellino et al., 1965)[3] but in spite of many efforts, nerve impulses have never been recorded from these fibers following illumination (see Waterman and Wiersma, 1954; Borsellino et al., 1965; Hartline, 1967). Since these axons are several centimeters long, they can transmit signals to the central nervous system only by means of nerve impulses. If they do not produce nerve impulses following illumination, then the only sensory signal reaching the nerve centers from each ommatidium will be the discharge of the eccentric cell, and no message at all will be transmitted by the ommatidia which do not possess an eccentric cell. The same problem arises with respect to the cells of the ventral eye, which do not seem to produce impulses and yet are connected to the central ganglia through thin and long axons (Millecchia, Bradbury and Mauro, 1966; Smith et al., 1968a, b). It may be suggested that these axons do not subserve any function, perhaps due to some oddity of evolution. It is possible, however, that impulses are produced by illumination in the living animal but are not recorded in the conventional preparations. A difficulty of this sort was encountered in the squid: nerve impulses could not be recorded for some time from the optic nerve of this animal following illumination, but were finally detected when appropriate care was taken to maintain the preparation in favorable conditions (MacNichol and Love, 1961). It is conceivable, therefore, that future work on *Limulus* may yet reveal impulse discharges in the small optic fibers of the lateral eye and of the ventral eye photoreceptors.[4]

---

[3] Ratliff et al (1966) quote unpublished results by H. Gasser and W. Miller and by C. Stevens and D. Lange showing that electrical stimulation of the optic nerve evokes two spikes: one travelling at about 2 m/sec. and the other at about 1 m/sec. E. A. Schwartz has recently confirmed these findings. In addition he has shown that if an electric shock is applied to the optic nerve shortly after delivering a flash of light to the eye, the first electrically evoked spike is decreased in amplitude but the slower spike is unchanged. This suggests that only the large optic nerve fibres discharge impulses following illumination of the eye (unpublished).

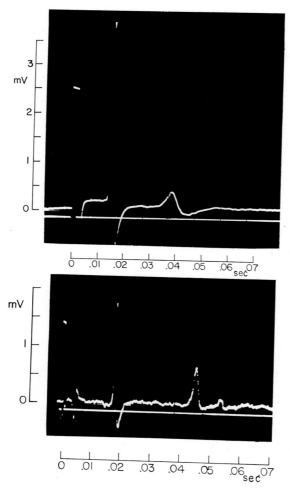

Fig. 7. Spikes produced by electrical stimulation of the optic nerve of *Limulus*. a Records taken from the whole optic nerve at a distance of about 3.5 cm from the point of stimulation. A large spike produced by the nerve impulses travelling along the large eccentric cell fibres is followed by a smaller and slower spike, produced by impulses in the smaller axons of retinular cells. b Records from dissected filament of optic nerve. The first recording electrode was at a distance of about 4 cm from point of stimulation. Both the first and the second spike were all-or-none, each produced by the activity of a single fibre (BORSELLINO, FUORTES, and SMITH, 1965)

---

[4] There are reasons to believe that the spike-producing mechanism may be damaged by experimental procedures which do not appreciably affect the generator potential. It is known from HARTLINE'S (1938) studies that cells in the eye of *Pecten* produce impulses; yet impulses were recorded only on very rare occasions by GORMAN and McREYNOLDS (1969) and McREYNOLDS and GORMAN (1970a, b) in an extensive investigation in which hundreds of cells were successfully impaled. When spikes were present, they disappeared soon after penetration although resting and generator potentials as well as membrane resistance did not change. Similar difficulties are encountered following penetration of visual cells in the Leech (LASANSKY and FUORTES, 1969).

Conductance Changes Following Illumination

Analysis of eccentric cell responses to light and to electric currents showed in addition that the resistance of the cell's membrane is decreased by illumination (this had already been suggested by TOMITA, 1956). This resistance is not altered, however, when the cell is depolarized by applied currents. It was inferred from this that the increase of conductance produced by light is not a consequence of the change of voltage, and the results were explained by assuming that the generator potential originated as a consequence of an increased permeability of the eccentric cell's membrane to ions (see Figs. 8 and 9).

The generator potential of this cell appeared similar to the end-plate potential and it was suggested that the generator potential was produced by a chemical transmitter liberated by the rhabdome of the retinular cells during illumination (FUORTES, 1959; RUSHTON, 1959).

The processes leading to generation of impulses following illumination in the eye of *Limulus* were then supposed to be as follows: 1) Light is absorbed in the rhabdome of retinular cells; 2) a chemical substance is liberated by these cells and diffuses to the adjacent eccentric cell dendrite; 3) this substance increases the permeability of the membrane of the dendrite and produces its depolarizatoin; 4) an outward current is produced at the proximal portions of the eccentric cell axon and impulses are generated there as a consequence of this current.

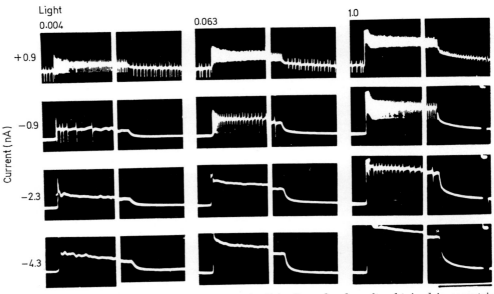

Fig. 8. Action of currents on generator potentials. Records of results obtained in eccentric cells of *Limulus*. Hyperpolarizing (−) or depolarizing (+) currents were applied through the intracellular microelectrode. Current intensity in nA is indicated by the figures at left. Responses were evoked by steps of light at different intensities (0.004, 0.063, 1.0). Each pair of records shows beginning and end of a response to illumination lasting 15 sec. (Only a small portion of the spikes is visible in the records.) The amplitude of the generator potential potential increases with increasing polarization of the membrane. Square waves at the right are 20 mV calibration pulses (FUORTES, 1959)

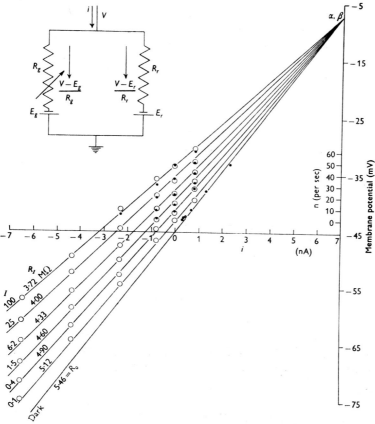

Fig. 9. Plots of the data illustrated in Fig. 8. Abscissa measures the intensity of the current through the microelectrode and ordinate measures the membrane potential of the cell. Membrane resistance in darkness was estimated to be $5.46 \, \mathrm{M}\Omega$ ($1 \, \mathrm{M}\Omega = 10^6 \, \Omega$). Thus, currents in darkness changed membrane potential as indicated by the lowest line (R_0). The vertical distance between this line and the open circles measures the amplitude of the generator potentials evoked by lights of different intensities, in the presence of currents. Light intensity I is given by the figures at left. For each light intensity, the relation between current and membrane voltage is linear. The slope of the lines through the experimental points gives the resistance of the cell membrane for the corresponding light intensity. It is seen that resistance decreases as light intensity is increased. The diagram shows the equivalent circuit proposed to explain the results. In darkness the resistance R_g is high and the voltage V across the membrane approaches the voltage of the battery E_r (about -70 mv). Illumination decreases the resistance R_g. Consequently the voltage V decreases to a value approaching the smaller voltage of the battery E_g (about -5 mv). The dots measure frequency of firing. Since frequency is linearly related to membrane potential (see Figs. 5b and c), these measurements plotted with the appropriate scale, fit the same straight lines which measure membrane potentials.
(RUSHTON, 1959)

Rectification and Electrical Coupling in Ommatidial Cells

Alternative interpretations of the generator potentials of Limulus have, however, been proposed. H. K. HARTLINE has pointed out (see FUORTES and

POGGIO, 1963, p. 449; BORSELLINO et al., 1965, p. 443) that the tight packing of the membranes at the rhabdome could be responsible for electrical coupling between retinular and eccentric cells, a suggestion which had already been advanced by TOMITA (1956) (see also TOMITA, KIKUCHI and TANAKA, 1960). If this were the case, then the experimental observations could be explained assuming that light evokes depolarization and increase of conductance of the retinular cells and that the changes measured from the eccentric cell simply reflect the alterations of the retinular cell membranes.

It became important therefore to study the properties of retinular cell responses and to investigate the possibility of electrical coupling between ommatidial cells.

The responses of retinular cells to light were recorded by different authors (MACNICHOL, 1956; FUORTES, 1958; BENOLKEN, 1961; KIKUCHI et al., 1965; SMITH, 1966) and were found to be essentially similar to those of eccentric cells: light produces depolarization and increases membrane conductance. Electrical coupling was investigated by means of experiments involving impalement of two cells in the same ommatidium (SMITH, et al., 1965; BORSELLINO, et al., 1965). Both electrodes were used for recording potentials and for passing currents in darkness and during illumination. It was found that the spikes and the irregular potential changes which occur spontaneously in dark-adapted preparations always appeared synchronously in the two impaled cells (Fig. 10). Moreover, when current was passed through one electrode, a potential drop was recorded from both impaled cells (Fig. 11). These observations indicate that all ommatidial cells are electrically coupled, as suggested by HARTLINE and by TOMITA. Similar interactions were later demonstrated by SHAW (1967a and b, 1968, 1969) in the Bee and in the Locust, and by BROWN et al. (1970) in the Barnacle.

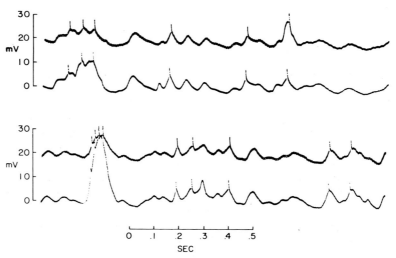

Fig. 10. Synchrony of potentials recorded from different ommatidial cells. Records from two retinular cells in the same ommatidium. The preparation was dark-adapted and produced irregular potential fluctuations (discrete waves, see Chapter 7 in this volume) in darkness. The fluctuations and the superimposed small spikes are synchronous in the two cells. Many waves have about the same size in the two records; other are larger in one or in the other

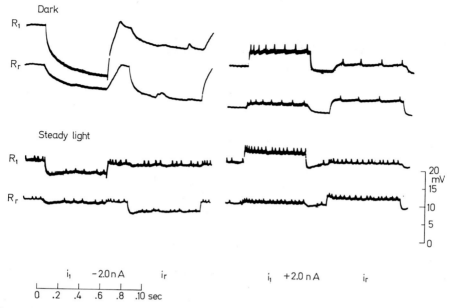

Fig. 11. Electrical coupling between ommatidial cells. Experiment as in Fig. 10, in a different ommatidium. A bridge circuit (Fig. 1 in FUORTES, 1959) was used to balance the potential drops evoked by the applied currents through the microelectrode. In each sweep a current of 2 nA was applied first to one (R_1) and then to the other (R_r) cell. Left hand records: hyperpolarizing currents; right hand records: depolarizing currents. The currents evoked potential drops in both impaled cells, indicating that the cells of an ommatidium are electrically coupled. Potential drops are larger and coupling is stronger in darkness (upper records) than during illumination (lower records). In the upper left-hand record, small potential fluctuation (see Fig. 10) appear synchronously in both cells. In the other records, the small spikes are also synchronous in the two cells (BORSELLINO et al., 1965)

The studies on Limulus showed in addition that current-voltage relations are quite different in retinular and in eccentric cells: the relation is approximately linear in eccentric cells (as had already been noted by FUORTES, 1959) but is strongly non-linear in retinular cells (Fig. 14). Similar rectification was demonstrated by BAUMANN (1968) in photoreceptors of the Honeybee. With hyperpolarizing currents, the input resistance of retinular cells increases; on occasions the increase occurs with some delay, giving rise to "hyperpolarizing" responses similar to those observed in other tissues (see GRUNDFEST, 1961). With depolarizing currents resistance decreases. In retinular cells then, resistance is decreased both when their membrane is depolarized by light and when it is depolarized by currents. Therefore the argument that the increase of conductance is not a consequence of the voltage change, does not apply to the retinular cells.

Microvilli in Eccentric Cells. Quintuple Junctions

The anatomical organization responsible for the electrical coupling was studied by LASANSKY (1967). In a detailed electronmicroscopical study of the

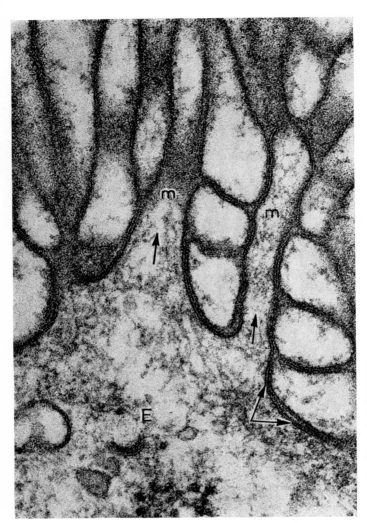

Fig. 12. Eccentric cell microvilli. Electron micrograph of a portion of the central ring of *Limulus* ommatidium. The arrows point to two microvilli *m* originating from the eccentric cell *E*. They interdigitate with other microvilli which presumably belong to a retinular cells. The double arrow indicates a quintuple-layered junction dividing in two unit membranes X 141000 (A. LASANSKY, unpublished)

ommatidia of Limulus he confirmed MILLER's (1957) finding that the rhabdome is made up of tightly-packed microvilli and observed in addition that, although the majority of microvilli are extensions of the retinular cells membranes, an appreciable number originates from the dendrite of the eccentric cell (Fig. 12). Where the membranes of adjacent microvilli come in contact, they form "five-layered structures", which result from direct apposition of the external electron-dense leaflets of two membranes. These structures are not unlike the *tight junctions* found in a variety of tissues in which neighboring cells are electrically coupled (KARRER, 1960; BENNETT et al., 1963; ROBERTSON, 1963; ROBERTSON et al., 1963; DEWEY and BARR, 1964)[5]. It is probable therefore that the tight

[5] These junctions have now been shown to be "gap junctions" and differ from the quintuple junctions of *Limulus* because the two outer leaflets of the membranes are separated by about 20 Å.

junctions joining the microvilli of *Limulus* are responsible for the electrical interactions between ommatidial cells. Since tight junctions exist between the microvilli of eccentric cells and those of retinular cells, the electrical coupling between these two types of cells can be readily explained.

LASANSKY (1967) and BASS and MOORE (1969) noted also that small extracellular spaces exist between each group of these microvilli. These provide a continuous network of extracellular channels through which the extracellular fluids can reach the rhabdomal membranes and permit electric currents to flow from the outside to the inside of the microvilli.

Action of Light on the Ommatidial Resistances

It was seen in *Limulus* that the degree of the electrical interaction (defined as the ratio between the potential drop in the cell to which current is applied and the potential drop in a second cell) changes both with currents and with light. In darkness, only minor changes occur with currents through the eccentric cell or with depolarizing currents through a retinular cell. When strong hyperpolarizing currents are applied to the retinular cell, however, its interaction with the eccentric cell or with other retinular cells is almost abolished. Significant decrease of the ommatidial interactions also occurs during illumination.

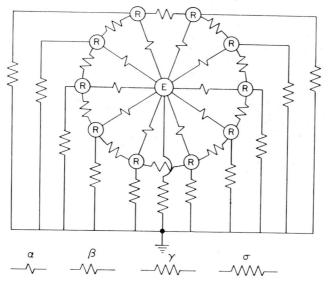

Fig. 13. Electric diagram of an ommatidium. The ommatidium is supposed to consist of one eccentric and ten retinular cells. α is the conductance between eccentric and retinular cells, β is the conductance between two adjacent retinular cells, γ and δ are the conductances from retinular and from eccentric cells to the outside. In order to determine the values of these four unknown conductances, it is necessary to use four independent measurements. These can be obtained combining the results of two types of experiments: one with microelectrodes in two retinular cells (see Fig. 11), the other with one electrode in a retinular and the other in the eccentric cell (SMITH and BAUMANN, 1969)

Analysis of the results based on the simplified linear network of Fig. 13 led to the conclusion that light increases most appreciably the conductance γ, of the membranes between retinular cells and outside: this could be the undifferentiated peripheral membrane or the portions of the microvillar membrane which are exposed to the outside (see Fig. 1a). The measurements were not sufficiently accurate to allow any safe conclusions on the changes produced by light in the other resistances: in particular, it was not possible to establish whether the conductance δ between eccentric cell and outside is altered by illumination.

The functional significance of the electrical interactions between the ommatidial cells of *Limulus* and of insects remains obscure.

Interpretation of Results on the Lateral Eye of Limulus

The anatomical and physiological studies just described brought to light several important properties of ommatidial cells but failed to clarify a number of pressing problems on their nature and functions.

The electrical investigations showed that retinular cells have non-linear current-voltage properties and demonstrated that all ommatidial cells are electrically coupled. It appears from these results that the approximate linearity of the current-voltage relations measured from eccentric cells results from a superposition and cancellation of the non-linearities of different electrically-coupled membranes. The demonstration of electrical interaction between retinular and eccentric cells is consistent with the view that light acts only on the retinular cells altering their conductance and potentials, and that these changes are transmitted to the eccentric cell through the tight junctions of the rhabdome. This notion is supported by Smith and Baumann (1969) who conclude that "the eccentric cell is not a primary photoreceptor but the second-order neuron in the visual pathway and is apparently connected to the retinular cells only by the electrotonic junctions" (p. 344, see also Kennedy, 1964, p. 103).

It should be noted, however, that the results on electrical interaction do not rule out the alternative possibility that the eccentric cell itself is a photoreceptor and in fact the anatomical observation that eccentric cells form microvilli (Lasansky, 1967) favors strongly this interpretation. According to Lasansky's unpublished counts, the eccentric cell gives to the central ring of the rhabdome at least as many microvilli as one retinular cell (the retinular cells form however additional microvilli along the spokes of the rhabdome). Since microvilli are typical of photoreceptors but not of second-order neurons, these findings suggest that the eccentric cell is itself a photoreceptor (Lasansky, 1967).

In order to prove directly the photoreceptive nature of eccentric cells it would be necessary to show that they develop a current following illumination. Since this has not been demonstrated so far, the question must be regarded as unresolved.

Analysis of Generator Potentials in Other Preparations

Electrical interactions and non-linearities make it very difficult to analyze and interpret the electrical responses of ommatidial cells. For this reason, the interpretation of generator potentials was attempted in simpler preparations.

MILLECCHIA et al. (1966) established the photoreceptor nature of the large cells of the "rudimentary ventral eye" of *Limulus* (DEMOLL, 1914) and determined that their responses to light are similar to those of the retinular cells of the lateral eye. The ventral eye photoreceptors can be very large (their major axis can be 200μ) and can be penetrated with microelectrodes under direct visual observation. They are often widely separated from one another (see CLARK et al., 1969); in these cases electrical coupling can be ruled out.

SMITH et al. (1968a) measured current-voltage relations in these cells and found that they are strongly non-linear. As for the retinular cells of the lateral eye, input resistance decreases if the cell is depolarized (Fig. 14). Moreover they found that resistance is approximately the same when the cell is depolarized by currents as when it is depolarized by light. From this they conclude that "the receptor potential in *Limulus* photoreceptor appears to be the consequence not of permeability changes in the cell membrane but of alteration in a light-sensitive constant-current generator" (p. 454). According to this view, light generates a current which depolarizes the membrane and membrane conductance increases as a consequence of the voltage displacement, due to the non-linear properties of the membrane.

SMITH et al. (1968b) found also that removal of sodium or of potassium from the external medium decreases the resting potential of ventral eye photoreceptors and transiently abolishes the responses to light. Decrease of membrane potential and abolition of the response were obtained also with 1 mM ouabain or by cooling the preparation below 2° C. Since absence of sodium or potassium, ouabain and low temperature are all known to inactivate the enzymes which drive the metabolic transport of sodium, the authors suggest that the resting potential of photo-

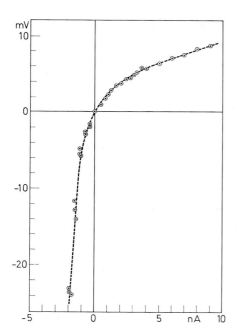

Fig. 14. Current-voltage relations in retinular cells. A microelectrode inserted in a retinular cell of *Limulus* lateral eye was used both for passing currents and for recording potential changes. A Wheatstone bridge was used to cancel the potential drop evoked by the current across the microelectrode itself. In this way, the changes of membrane potential could be measured. The slope of the current-voltage relation increases with hyperpolarizing and decreases with depolarizing currents

receptors in the ventral eye is partly maintained by active extrusion of sodium, not compensated by an equal influx of potassium (a mechanism referred to as "electrogenic sodium pump"). Generator potentials are interpreted assuming that light decreases or abolishes the electrogenic activity of the pump and in this way brings about a decrease of membrane potential.

This interpretation leads to two major difficulties. Firstly, the results obtained by SMITH et al. (1968 a and b) show that the agents which interfere with the sodium pump decrease membrane potential by no more than 20 mV. If light had the same effect it could therefore produce generator potentials of 20 mV at the most. Since the generator potential may exceed 50 mV of amplitude at the onset, the authors suggest that their interpretation applies only to the later phases of the response to steps of light which usually do not exceed 20 mV. This implies different mechanisms for production of the transient and steady-state phases of the response. Secondly, the authors observed that the generator potential is increased by hyperpolarizing currents and is decreased and eventually reversed by depolarizing currents. In order to explain this apparent contradiction, the authors assume that the electrogenic action of the pump is a function not only of light but also of membrane potential and that it reverses when the membrane is strongly depolarized.

One of the crucial arguments brought in support of the hypothesis that light produces depolarizing generator potential by inactivating an electrogenic sodium pump is that agents which are known to inhibit the activity of the pump reduce membrane potential and reduce or abolish the response to light. It has already been stated that SMITH et al. (1968a and b) found that membrane and generator potentials were decreased in the absence of external sodium or potassium as well as in the presence of 1 mM ouabain.

MILLECCHIA et al. (1966) and MILLECCHIA and MAURO (1969a) failed to observe any reduction in membrane potential when external sodium was absent or when potassium was decreased to 1 mM concentration.

FULPIUS and BAUMANN (1969) investigated the effects of ions on the photoreceptors of the Honeybee and found that resting potential is not decreased when either sodium or potassium are removed from the bathing medium. The response to light remains for many hours in potassium-free solutions. In the absence of sodium the response is decreased but not abolished (see below, p. 305).

The action of ouabain was investigated by BROWN et al. (1970, 1971) and by KOIKE et al. (1970). They found that the generator potential produced in the Barnacle photoreceptor by intense illumination is followed by long-lasting hyperpolarization (the same phenomenon had been observed previously by BENOLKEN, 1961 in photoreceptors of *Limulus*), and gave evidence showing that this after-hyperpolarization is due to the activity of an electrogenic sodium pump. It was seen in their experiments that 0.01 mM ouabain (one hundredth of the concentration used by SMITH et al., 1968) abolishes the hyperpolarization but has negligible effect on membrane potential and on the generator potential (Fig. 15).

These results seem to contradict the evidence offered by SMITH et al. (1968) and support the view that activity of an electrogenic sodium pump is not required

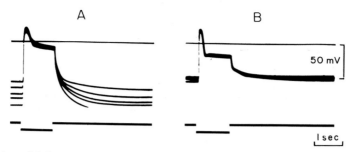

Fig. 15. Action of Oubain on generator potential. a Superimposed records of membrane potential changes produced by five consecutive steps of light of fixed intensity in a moderately
dark-adapted photoreceptor of barnacle. Flashes were applied at intervals of 10 sec. The
generator potential is followed by longlasting hyperpolarization: this accumulates with the
successive responses, giving rise to progressive increase of membrane potential. (See similar
results in *Limulus* in BENOLKEN, 1962.) b Same experiment 15 minutes after application of
0.01 mM oubain. Resting potential and generator potentials have changed only little but the
slow hyperpolarization has disappeared (BROWN et al., 1970)

for maintenance of the resting potential and for production of the responses to light
in photoreceptors.

Voltage-Clamp Studies on Photoreceptors

The ventral photoreceptors of *Limulus* are sufficiently large to be
penetrated by two microelectrodes without obvious damage. Taking advantage
of this property, MILLECCHIA and MAURO (1969b) investigated membrane
characteristics of these cells with the voltage-clamp method. One of the two
microelectrodes was connected to a unity gain amplifier ($X1$) and was used to
monitor the membrane potential of the cell. The output of $X1$ was led to a
feedback amplifier (A_1) which was connected to the second electrode and supplied
the currents required to clamp the membrane potential at any desired level
(Fig. 16b). Using this method they confirmed that current-voltage relations
in darkness are non-linear and time dependent.

With the membrane potential clamped at a fixed level, a step of bright light
produced an additional current, the light-induced current: its time course was
similar to that of the generator potential, showing a sharp transient followed by a
lower steady state. The light-induced currents depends non-linearly upon the
voltage at which the membrane is clamped and reverses at a membrane
potential of about 10 mV, which is invariant with respect to time and light
intensity. Current-voltage relations at the peak and in the steady state are
shown in Fig. 16a, b, while Fig. 16a, c illustrates that the amplitude of the light-
induced current increases with light intensity but the reversal point does not
change. These results are best explained assuming that light increases a
conductance with non-linear characteristics, associated with a constant
electromotive force of about +10 mV.

BROWN et al. (1968, 1969, 1970 and 1971) performed an extensive study
of the membrane characteristics of the photoreceptors of the Barnacle. As in the

experiments by MILLECCHIA and MAURO (1969b) they found that the current-voltage relations measured with voltage-clamp techniques were non-linear and time dependent in darkness.

The light-initiated current was also found to be a function of voltage, time and light intensity. It reversed at a voltage of +27 mV (average of 33 measurements) which was independent of light intensity and of time (Fig. 17). BROWN et al. (1970) measured also the instantaneous current-voltage curves by recording the current required to maintain membrane voltage at the desired level immediately

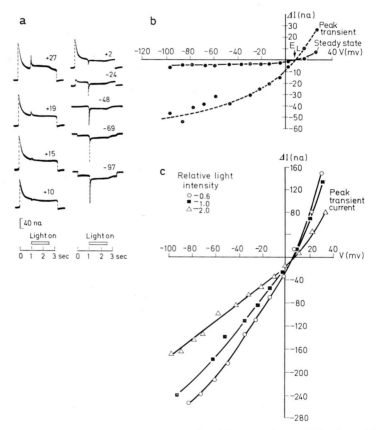

Fig. 16a. Currents produced by illumination in cells of the ventral eye of *Limulus*. *a* Membrane voltage was clamped at the values given by the figures near each record (mV). One second after application of the clamp, light of moderate intensity was flashed on the cell for 1.5 sec. With clamping voltages between −97 and +10 mV inward current was produced during the illumination. When the clamping voltage was between +15 and +27 mV the light-induced current was outward. As seen in the records, the current produced by illumination includes a sharp transient followed by a lower steady-state level. The values of these peaks and steady-state currents are plotted in *b* as a function of clamping voltage. The reversal potential E_L is the same for these two values of the light-induced current. *c* is a plot of the peak value of the light-induced current as a function of clamping voltage for different light intensities (relative light intensities expressed in logarithmic units). The light-induced currents are larger when the light is brighter, but the reversal potential does not change (MILLECCHIA and MAURO, 1969b)

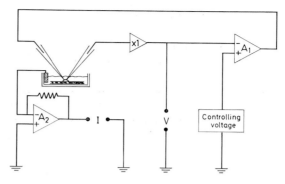

Fig. 16b, Voltage clamp in photoreceptors. Schematic diagram of
voltage clamp circuit. A unity-gain
amplifier of high input impedance
($X1$) measures the membrane potential (V) of the impaled cell. The
high-gain differential amplifier A_1
produces a current (I) proportional
to the difference between the voltage V and any chosen controlling
voltage. This current is fed back
into the cell through a second microelectrode and keeps the membrane potential of the cell
equal to the controlling voltage. The current I is measured by another differential amplifier A_2.
The output voltage of A_2 is equal to the current I times the feedback resistance (500 KΩ)
(MILLECCHIA and MAURO, 1969 b)

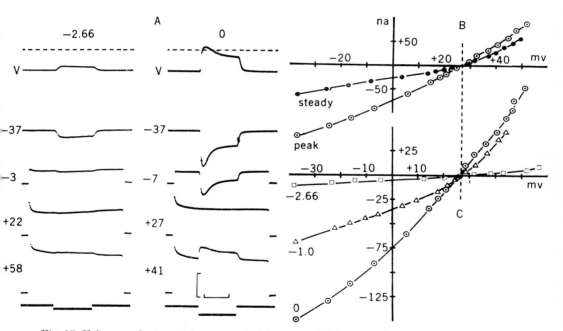

Fig. 17. Voltage and current changes evoked by steps of light in the Barnacle photoreceptor.
a Top row shows the generator potentials produced by dim or bright lights (the figures are the
logarithm of the light intensity in arbitrary units). The dashed line indicates zero potential.
The lower records show the currents developed during illumination at the same two intensities
when the membrane was clamped at the voltages given by the figures near each record. Downward deflection indicates inward membrane current. Vertical calibration line: 40 mV for
records in top row; 200 nA for other records. Horizontal line: 400 msec. b Current-voltage
relations of responses to bright light under voltage clamp, measured at the peak of the
response (open circles) and in the steady state (dots). c Current-voltage relations measured
at the peak of the responses for three different light intensities, given in logarithmic units by
figures at left (BROWN et al., 1970)

(2 msec) after the onset of the voltage pulse. It was seen in this way that the instantaneous light-induced current is linearly related to voltage (Fig. 18). This means that the voltage dependence of the light-induced current develops slowly, being inappreciable 2 msec after the beginning of the voltage pulse.

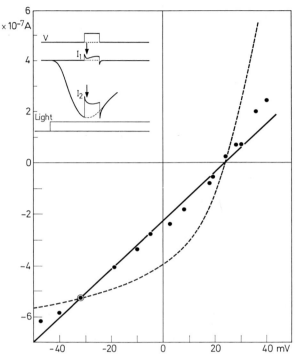

Fig. 18. Instantaneous current-voltage relation in Barnacle photoreceptors. Membrane potential was clamped at its resting value in darkness. Brief voltage steps were added first in darkness, giving a current I_1 and then at the peak of a response to light, giving a current I_2. The light-induced current $I_2 - I_1$ was measured 2 msec after the voltage pulse (arrow). These "instantaneous" values were plotted as a function of membrane potential. They fall on an approximately straight line. If the voltage step is applied before delivering the light, peak current varies with membrane potential as shown by the dashed line (Brown et al., 1970)

These results obtained with voltage-clamp method in photoreceptors of *Limulus* and Barnacle leave little doubt that the generator potential arises as a consequence of an increase of membrane conductance, as originally suggested by Fuortes (1959) for the eccentric cell of *Limulus*.

To explain the failure to detect a conductance change in the results by Smith et al. (1969a), Brown et al. (1970) pointed out that experiments with constant currents necessarily measure "steady-state" rather than "instantaneous" current-voltage relations. In these conditions the light-sensitive channels add only little to the membrane conductance, already increased by the depolarization, and therefore their contribution may remain undetected.

Action of Sodium Ions on the Generator Potential

It has been reported that the responses of several receptors are decreased but not completely abolished in the absence of external sodium. This has been observed to occur in the Pacinian corpuscle, (DIAMOND et al., 1958), in the muscle spindle of the frog (OTTOSON, 1964; CALMA, 1965) and in the crayfish stretch receptor (EDWARDS et al., 1963; OBARA, 1968; OBARA and GRUNDFEST, 1968).

KIKUCHI et al. (1962) found that the generator potential of cells in the lateral eye of *Limulus* is reduced in the absence of sodium and suggest that a small response may persist because some sodium ions remain trapped in the extracellular spaces surrounding the cell.

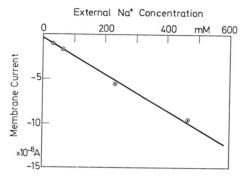

Fig. 19. Relation between membrane current and Sodium concentration. Membrane potential of a Barnacle photoreceptor was clamped at −30 mV (Resting potential is about −35 mV in these cells). The additional current produced by illumination was measured at four different concentrations of external sodium, Tris Chloride being used as a substitute. The light-induced inward current is linearly related to external sodium concentration (from BROWN et al., 1970)

BROWN et al. (1970) investigated the photoreceptors of the Barnacle replacing the external sodium chloride with Tris, sucrose, choline or urea. They found that the inward current initiated by light under voltage clamp is linearly related to the external sodium concentration but an appreciable current remains when sodium is completely removed from the external medium (Fig. 19). Correspondingly, the responses recorded without voltage clamp were reduced but not abolished in the absence of external sodium. The interpretation of these results is that illumination increases the permeability of the membrane to sodium ions and also to some other ion species. The same conclusion was reached by FULPIUS and BAUMANN (1969) is a study of the ionic requirements of the photoreceptors of the Honeybee.

In the photoreceptors of the *Limulus* ventral eye, removal of external sodium has a more complicated effect. The response to light is almost completely abolished a few minutes after exposing the preparation to the sodium-free solution but later on the cell partially recovers (SMITH et al., 1968b; MILLECCHIA and MAURO, 1969a). In order to explain these observations, MILLECCHIA and MAURO (1969b) suggest that, in normal conditions the response to light is a "pure sodium response, but when the external sodium is removed the cell becomes modified in such a way that it can now use other ions to produce the light response. This modification occurs slowly enough that initially a blockage phase occurs and then gradually the recovery phase develops" (p. 327).

Generator Potentials in the Eye of Scallop

It has been mentioned at page 280 that the eyes of the Scallop, *Pecten irradians,* contain two layers of photoreceptor cells. Cells of the proximal layer possess microvilli while the cells in the distal layer possess long, irregularly wound tubules derived from cilia (MILLER, 1958, 1960; BARBER, EVANS and LAND, 1967; GRAZIADEI and METCALF, 1970). HARTLINE (1938) found that fibers from the proximal layer discharge impulses during illumination of the eye; the fibers originating in the distal layer, instead "respond only to cessation of illumination or to reduction of its intensity (Fig. 20). No impulses at all (as a rule) are discharged during illumination but the "off" response is nevertheless strictly dependent for its excitation upon the preceding period of illumination" (HARTLINE, 1938, p. 477).

TOYODA and SHAPLEY (1967) impaled cells in the eye of Scallop with microelectrodes and found that illumination produced hyperpolarization. These cells were probably the cells of the distal layer which produce "off" responses. Depolarizing responses to light were not recorded in this study.

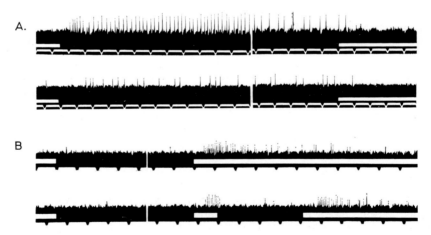

Fig. 20. Action potentials in the optic nerve of *Pecten.* a Records from a nerve fiber originating in the proximal cell layer. Illumination (signaled by interruption of the white band above the time marks) evokes sustained firing of these cells. Light intensity in the upper record was 100-fold brighter than in the lower. The records show beginning and end of responses lasting 30 sec. Time marks: 200 msec. b Responses recorded from a nerve fiber from the distal cell layer. The upper record shows that the preparation did not discharge impulses in darkness and remained silent during a 20 sec period of illumination (only the beginning and end of the illumination are shown). A prolonged discharge of impulses occurred when the light was interrupted. After cessation of this discharge, the experiment was repeated (lower record) and the eye was re-illuminated while the cell was discharging impulses. Illumination abruptly stopped the firing (HARTLINE, 1938)

GORMAN and McREYNOLDS (1969) and McREYNOLDS and GORMAN (1970 a, b) succeeded in recording intracellularly from cells in both layers of the Scallop retina and established (as predictable) that cells of the proximal layer are depolarized with light while those of the distal layer are hyperpolarized (Fig. 21). It may be thought that the cells in the distal layer are second-order neurons: if this were the case their hyperpolarizing response should be regarded not as a direct consequence of illumination but rather as an inhibitory synaptic potential. This interpretation, however, is unlikely because latency was found to be clearly shorter and the threshold to be much lower for the hyperpolarizing than for the depolarizing responses. It must be concluded therefore "that two independent types of photoreceptors that give opposite responses to light are present in the scallop retina" (GORMAN and McREYNOLDS, 1969, p. 310).

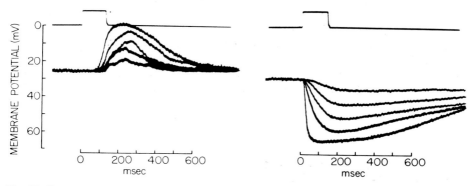

Fig. 21. Responses from photoreceptors of *Pecten*. Intracellular records of responses produced by pulses of light of different intensities. Cells in the proximal layer produce depolarizing responses whereas cells of distal layer are hyperpolarized by illumination. Both types of cells normally produce spikes (see Fig. 20) but their ability to discharge impulses is almost always impaired when they are punctured by the microelectrode (McREYNOLDS and GORMAN, 1970a)

It is mentioned in Chapter 12, of this volume that vertebrate cones also respond to light with hyperpolarization. The mechanisms underlying the response of cones, however, are different from those operating in the distal cells of the scallop because the generator potential of cones is associated with decrease of membrane conductance (TOYODA et al., 1969; BAYLOR and FUORTES, 1970; see also Chapter 12 in this volume) whereas the hyperpolarizing response of the distal cells of the Scallop is accompanied by a conductance increase (TOYODA and SHAPLEY, 1967; McREYNOLDS and GORMAN, 1970b).

The photoreceptors in the distal layer of the eye of Scallop are the only invertebrate receptors known to produce hyperpolarizing responses to light. They are also the only invertebrate cells studied in which the photoreceptive apparatus derives from cilia. It will be interesting to see if other invertebrate photoreceptors of ciliary origin also produce hyperpolarizing responses to light.

Generator Potentials in Other Structures

Visual responses have been studied in single photoreceptor cells of a variety of animals including molluscs, annelids, insect and crustaceans. In all cases, the cells had microvilli and produced depolarizing receptor potentials not unlike those already described with reference to *Limulus*. Examples of generator potentials of invertebrates are illustrated in Fig. 22. In the mollusc *Aplysia* (Fig. 22a) the response to a step of light includes a transient phase which may overshoot membrane potential, followed by a rapidly-adapting depolarization (JACKLETT, 1969). *Hermissenda*, another (nudibranch) mollusc, has primitive eyes containing only five synaptically interconnected, photoreceptor cells. Each cell responds to light with a depolarizing generator potential which may give rise to nerve impulses (Fig. 22h). Simultaneous recording from two photoreceptors revealed that impulses in one cell produce hyperpolarizing synaptic potentials and inhibit firing in the other cell.

Generator potentials in the Leech were studied by WALTHER (1965) and by LASANSKY and FUORTES (1969). They will be described briefly in the next section.

The first recordings of generator potentials from photoreceptors of insects were obtained by NAKA (1961) in the Honeybee. The same preparation was

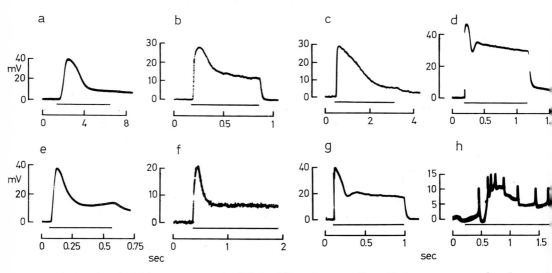

Fig. 22. Depolarizing generator potentials in different preparations. Responses were produced in all cases by steps of light (indicated by horizontal bars under the records) and were recorded by means of intracellular electrodes. a Photoreceptor of *Aplysia* (JACKLETT, 1969); b Locust (SHAW, 1967a); c cell from the proximal layer in the eye of Scallop. As in *Aplysia* (a) the depolarization decays rapidly to a low level. Damage produced by the microelectrode abolished the spikes which are normally produced by these cells (McREYNOLDS, unpublished); d Dragonfly (FUORTES, 1963); e Crayfish (EGUCHI, 1965); f Photoreceptor in the epistellar body of *Eledone moschata*. Repetitive nerve impulses are occasionally recorded from these cells (MAURO and BAUMANN, 1968); g Photoreceptor in the compound eye of the Honeybee Drone (NAKA and EGUCHI, 1962); h Photoreceptor of *Hermissenda* (DENNIS, 1967)

investigated later by NAKA and EGUCHI (1962) and by BAUMANN (1968). These cells respond to light with the usual depolarizing wave and usually no nerve impulses are produced. This is what one would expect since the cells are only about 0.5 cm long and the generator potential spreads effectively to the synaptic endings (BAUMANN, 1968). On rare occasion, however, trains of nerve impulses have been recorded superposed to the generator potentials of the bee and it is not yet known if they are normally produced in the living animal (as suggested by NAKA, 1961) or if they appear only in damaged preparations, as suggested by BAUMANN (1968).

Generator potentials were studied also in a number of other insects including the Dragonfly (FUORTES, 1963), the Locust (SCHOLES, 1965; SHAW, 1967), the Housefly (KIRSCHFELD, 1965), the Blowfly (WASHIZU, 1964) and the Grasshopper (WINTER, 1967). In the Dragonfly, Blowfly and Grasshopper it was demonstrated that the generator potential is associated with increased membrane conductance.

In general, comparative studies show that photoreceptors with a long axon produce nerve impulses (as in the case for the eccentric cell of *Limulus*, the photoreceptors of Leech, Scallop, *Hermissenda*, etc.) while receptors with a short axon generate slow potentials but not nerve impulses [as in the Dragonfly (Fig. 22d), Cockroach, Locust (Fig. 22b), Barnacle, Crayfish (Fig. 22e) etc.]. There are, however, some apparent exceptions since conducted impulses have never been demonstrated to occur in the retinular cells of the lateral eye or in the photoreceptors of the ventral eye of *Limulus*, which have a long axon (see page 290) and by contrast, nerve impulses have occasionally been recorded from the short photoreceptors of the Bee (see above).

Site of Origin of Visual Responses

It has been mentioned that in invertebrate photoreceptors the visual pigment is probably located in the membrane of the microvilli. Recent results have convincingly shown that the generator current is produced at or very near the site where light is absorbed. The results leading to this important conclusion were first obtained by HAGINS et al. (1962) and HAGINS (1965) in the eye of the Squid. The eye was sectioned to expose the whole length of the photoreceptors and light was applied over a short section of their length. It was then seen that inward current developed over the illuminated area while the other parts of the cell supplied outward current. The inward current associated with the generator potential of these cells is, therefore, localized to an area very close to the site where light is absorbed (Fig. 23). The same results were later obtained by F. BAUMANN (unpublished) in the Bee. It was not possible to establish from these experiments whether the inward current flows through the microvilli or through the adjacent, unspecialized regions of the membrane. Evidence that the inward current flows through the microvilli was obtained by LASANSKY and FUORTES (1969) in the photoreceptors of the Leech. These cells have a peculiar structure, illustrated in Fig. 24. The cell soma is roughly ovoid and its smooth external membrane is freely exposed to the outside. The microvillar membrane lines an internal cavity which is connected to the external fluids by narrow channels. When an electrode is inserted in the cytoplasm the usual resting potential is recorded in darkness and a

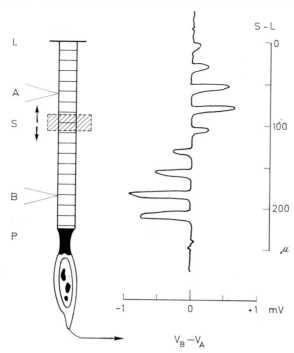

Fig. 23. Potential drop around the outer segments of Squid photoreceptors. Two micropipettes (A and B) were placed near the outer segments and a localized light (S) was applied over different regions of the outer segments. The ordinate in the plot measures the distance between the stimulus and the internal limiting membrane (L). The abscissa measures the difference of potential occurring between the two microelectrodes following illumination of different regions. The potential changes indicated that light produces a current which enters the outer segment in the area where the light is applied (HAGINS et al., 1962)

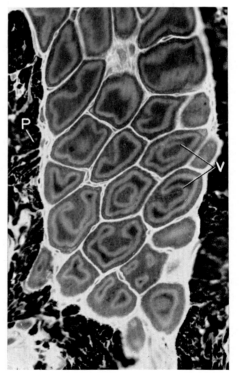

Fig. 24. Leech visual cells. Light micrograph showing part of an eye of the Leech (Hirudo medicinalis). The visual cells contain a large vacuole V which is different sections appears as an oval, an annulus or a U. The light ribbon outlining the vacuole is the "brush border" made up of microvilli. The vacuole communicates with the outside by means of narrow channels. The eye is surrounded by pigment cells P. X 500
(LASANSKY and FUORTES, 1969)

positive response is produced by light. If the electrode penetrates instead the microvillar cavity, no resting potential is recorded in darkness and light produces a negative response (Fig. 25). No response is detected if the electrode simply touches the smooth external membrane of the cell. These results indicate that light

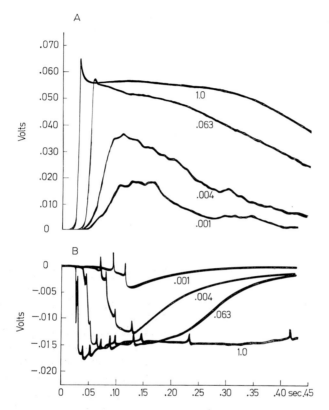

Fig. 25. Responses of visual cells of Leech. a A microelectrode was inserted in the cytoplasm of a cell in the eye of the Leech (see histological controls in LASANSKY and FUORTES, 1969). Brief flashes were applied at time zero. Relative light intensity is indicated near each trace. Visual cells of Leech produce repetitive spikes following illumination but these are easily lost upon impalement of the cell with the microelectrode (see p. 291). b Same experiment but with the electrode inserted in the vacuole, as shown by histological controls. Positive-going spikes superposed on a negative (extracellular) generator potential appears with decreasing latency as light intensity is increased (FUORTES and LASANSKY, 1969)

produces an inward current across the microvillar membrane while other parts of the cell supply a diffuse outward current.

It is reasonable to conclude from these results that the processes intervening between absorption of light and production of a membrane current remain localized to the areas where the light is absorbed.

Kinetics of Visual Responses

The nature of these processes remains unknown but some insight on their properties has been obtained by means of analysis of the kinetics of the responses of visual cells. HARTLINE (1934) had already observed that a long delay elapses between the start of a dim light and the firing of the first nerve impulse. A similar delay is also observed when generator potentials are recorded from the cell bodies of eccentric or retinular cells of *Limulus* (HARTLINE et al., 1952) as well as from other vertebrate and invertebrate photoreceptors. Thus, if brief flashes of light are delivered, no measurable response is recorded during the illumination but rather the generator potential develops well after the cell has returned to darkness.

Studies performed by DELANGE (1958) in man and by DEVOE (1962, 1963) in the Spider *Lycosa*, suggest an explanation of the properties of this delay. DELANGE (1958) noticed that the rapid attenuation of human perception of flicker with increasing frequency is similar to the attenuation of the output of an electrical filter consisting of a sequence of several R–C elements. Later DEVOE (1963) observed that a filter of this type reproduces well the response of *Lycosa* to sinusoidal lights and to changes of illumination in the form of steps. Some non-linear responses could also be duplicated by incorporating a non-linear stage in the filter (DEVOL, 1967).

The same type of filter (Fig. 26) is also suitable for duplicating responses elicited by dim lights in retinular or eccentric cells of *Limulus* (FUORTES and HODGKIN, 1964). If the filter is made up of ten stages ($n = 10$) a brief impulse at the input will produce a slowly-developing wave at the output, resembling the generator potential produced by visual cells following a flash of light. Since the filter of Fig. 26 has linear properties, the amplitude of its response will increase in proportion to the strength of the stimulus. This is not the case for visual cells, where response amplitude is approximately proportional to the logarithm of the light intensity (see Fig. 4) except for the small responses evoked by very dim lights. For these reasons, the linear model should not be expected to reproduce visual

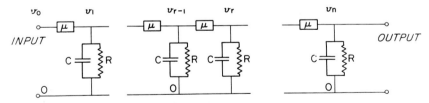

Fig. 26. Electrical network reproducing shape of visual responses. The network consists of n identical stages including two time constants each: $\tau_1 = RC$ and $\tau_2 = C/\mu$. The elements μ are unidirectional current sources characterized by the response $i_r = \mu v_{r-1}$ to the input v_{r-1}. In this form, the model is linear and therefore its responses are proportional to the strength of the stimulus (see Fig. 28a). The changes of visual responses brought about by adaptation (Fig. 27) can be simulated in the model by appropriate changes in the values of the resistances R. The rapid non-linearities which become prominent in visual responses when light intensity is increased, can be reproduced in the model by making the conductances $1/R$ linearly dependent upon the voltage of one the of last stages (see Fig. 28b) (Modified from FUORTES and HODGKIN, 1964)

responses over a wide range of intensities. It can be shown, however, that the multi-stage filter has properties which are suitable for reproducing the non-linearities of the responses of photoreceptors.

Relation between Sensitivity and Speed of Response

The starting point for this study is derived from the observation that light adaptation decreases sensitivity while it increases temporal resolution. It has been

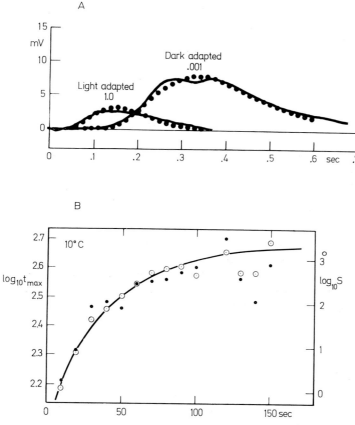

Fig. 27. Responses evoked in dark and light adaptation. a The larger response was evoked by a flash of intensity 0.001 in a dark-adapted preparation. The other response was elicited by a flash of intensity 1.0 applied a few seconds after a period of strong illumination. It is seen in results of this type that light adaptation decreased sensitivity and shortens the delay of the response. Similar changes of sensitivity and time scale are obtained in the model of Fig. 26 by changing the value of R. The dotted curves in the figure show the calculated output of the model with $\tau_1 = 42.6$ msec for the dark-adapted response and $\tau_1 = 18.8$ msec for the smaller, light-adapted response. All other parameters were the same in the two cases (from FUORTES and HODGKIN, 1964). b The dots ($\bullet$) measure time to peak (t_{max}) and the circles ($\circ$) measure sensitivity ($S = v_{max}/\varrho$) of responses evoked by flashes after exposing the preparation to a bright light. It is seen that S and t_{max} (a measure of τ_1) recover from light adaptation following approximately the same time course (from FUORTES and HODGKIN, unpublished)

known for some time that in the presence of a background light or after a period of intense illumination, a flash becomes less visible but flicker fusion frequency is increased. These results suggest that the response to a flash becomes smaller and faster with light adaptation and in fact these changes can be observed directly when responses of single visual cells are recorded (Fig. 27a). It is seen in experiments of this type that the changes of sensitivity (in the case of responses to flashes, sensitivity, S, is defined as peak voltage of the response, v_{max}, divided by the quantity of light Q, in the flash — FUORTES and HODGKIN, 1964) is much greater than the shortening of the time scale. In *Limulus*, a thousand-fold change in sensitivity is associated with a change in time scale by a factor of 2 to 2.5. As dark adaptation progresses, both sensitivity and time scale change, maintaining a constant relation: in any state of adaptation, sensitivity is proportional to a high power of the time scale, about the eight power in *Limulus* (Fig. 27b). Correlated changes of time scale and sensitivity similar to those observed in visual cells can be obtained in the electrical analogue by changing the value of the leak resistances R,

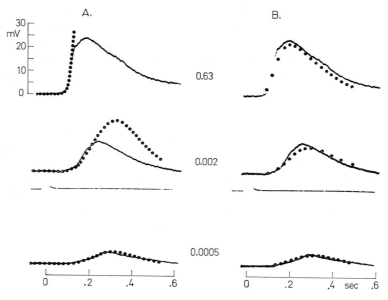

Fig. 28. Comparison of experimental and theoretical responses. Visual responses were produced by brief flashes of different intensities and were recorded from an eccentric cell of *Limulus* by means of an intracellular microelectrode. The theoretical curves (filled circles) are the output of the model of Fig. 26 for inputs in the form of an instantaneous impulse. In *A*, all components have fixed values (no feedback loop) so that the model had linear properties. In these conditions, the responses of the model grow in proportion with the strength of the input and, therefore, they increase much more than the experimental responses. When the conductances $1/R$ are made dependent upon voltages using the relation

$$\frac{1}{R} = \frac{1}{R_0}\left(1 + \frac{v_{n-1}}{w}\right)$$

(where R_0 and w are constants), the responses of the model change as shown in *B*, becoming similar to the responses of the cell (modified from FUORTES and HODGKIN, 1964 and MARIMONT, 1965) since the time scale of the output is proportional to R, while sensitivity (or gain) is proportional to R^{n-1}, where n is the number of stages of the filter. Thus, the changes produced by light and dark adaptation can be simulated in the model by altering the time constants τ_1 of its stages ($\tau_1 = RC$).

This general interpretation can now be used to explain the non-linearities of responses to light. Responses to flashes of increasing intensity are illustrated in Fig. 28. In part A of this figure, the experimental responses are compared to the outputs of the linear model for inputs of correspondingly increasing strength: the response to the dimmest flash is reproduced fairly well by the model, but the other two responses deviate sharply from the theoretical curves. It is noted, however, that experimental and theoretical curves coincide at early times, indicating that the non-linearities of visual responses become appreciable only some time after the start of the response.

These features can be reproduced in the model, introducing a feedback loop through which the voltage of one of the last stages controls the value of all the leakage resistances R. With this control loop, as the output increases, the gain is reduced and the time constants are shortened. With appropriate parameters, a good fit can be obtained between calculated and observed data, as shown in Fig. 28B. (See PINTER, 1966 for a description of some shortcomings of the model.) The non-linear model has been applied so far only to the study of responses in *Limulus*. The linear model, however, has been found adequate for reproducing the responses to dim lights in the bee (BAUMANN, 1968) and in the Squid (HAGINS, unpublished).

It would be desirable to determine at this point the physical nature of the processes in the visual cell which correspond to the components of the electrical model. One possibility is that each stage of the network corresponds to a chemical transformation (see WALD, 1965; BORSELLINO et al., 1965) so that the model represents a sequence of chemical reactions. The voltage v_n of the last stage would then represent the concentration of a substance which can react with the membrane, thus leading to a change of its conductance. This general interpretations seems to be reasonable but it must be recognized that alternative assumptions could be equally satisfactory.

References

BARBER, V. C., EVANS, E. M., LAND, M. F.: The fine structure of the eye of the mollusc *Pecten maximus*. Z. Zellforsch. **76**, 295—312 (1967).

BASS, L., MOORE, W. J.: An electrochemical model for depolarization of a retinula cell of *Limulus* by a single photon. Biophys. J. **10**, 1—19 (1970).

BAUMANN, F.: Slow and spike potentials recorded from retinula cells of the Honeybee drone in response to light. J. gen. Physiol. **52**, 855—875 (1968).

BAYLOR, D. A., FUORTES, M. G. F.: Electrical responses of single cones in the retina of the turtle. J. Physiol. (Lond.) **207**, 77—92 (1970).

BEHRENS, M. E., WULFF, V. J.: Light-initiated responses of retinular and eccentric cells in the *Limulus* lateral eye. J. gen. Physiol. **48**, 1081—1093 (1965).

— — Functional autonomy in the lateral eye of the horseshoe crab, *Limulus polyphemus*. Vision Res. **7**, 191—196 (1967).

BENNETT, M. V. L., ALJURE, E., NAKAJIMA, Y., PAPPAS, G. D.: Electrotonic junctions between teleost spinal neurons: Electrophysiology and ultrastructure. Science **141**, 262—264 (1963).

BENOLKEN, R. M.: Reversal of photoreceptor polarity recorded during graded receptor potential response to light in the eye of *Limulus*. Biophys. J. **1**, 551—564 (1961).

BERNHARD, C. G., GRANIT, R.: Nerve as a model temperature end organ. J. gen. Physiol. **29**, 257—265 (1946).

BORSELLINO, A., FUORTES, M. G. F., SMITH, T. G.: Visual responses in *Limulus*. Cold Spr. Harb. Symp. quant. Biol. **30**, 429—443 (1965).

BROWN, H. M., HAGIWARA, S., KOIKE, H., MEECH, R. W.: Membrane properties of a Barnacle photoreceptor examined by the voltage clamp technique. J. Physiol. (Lond.) **208**, 385—413 (1970).

— — — — Electrical characteristics of a barnacle photoreceptor. Fed. Proc. **30**, 69—78 (1971).

— MEECH, R. W., KOIKE, H., HAGIWARA, S.: Current-voltage relations during illumination: Photoreceptor membrane of a barnacle. Science **166**, 240—243 (1969).

— — SAKATA, H., HAGIWARA, S.: Voltage clamp of light receptor cells in the barnacle lateral eye. Proc. Internat. Union Physiol. Sci. **7**, 63 (1968).

CALMA, I.: Ions and the receptor potential in the muscle spindle of the frog. J. Physiol. (Lond.) **177**, 31—41 (1965).

CLARK, A. W., MILLECCHIA, R., MAURO, A.: The ventral photoreceptor cells of *Limulus*. I. The microanatomy. J. gen. Physiol. **54**, 289—309 (1969).

COHEN, A. I.: Rods and cones. Handbook of Sensory Physiology. This volume.

COLE, K. S., CURTIS, H. J.: Electric inpedance of the squid giant axon during activity. J. gen. Physiol. **22**, 649—670 (1939).

DeLANGE, H.: Research into the dynamic nature of the human fovea-cortex system with intermittent and modulated light. J. opt. Soc. Amer. **48**, 777—789 (1958).

DEMOLL, R.: Die Augen von *Limulus*. Zool. Jb. Abt. Anat. **38**, 443—464 (1914).

DENNIS, M. J.: Electrophysiology of the visual system of a nudibranch mollusc. J. Neurophysiol. **30**, 1439—1465 (1967).

DeVOE, R. D.: Linear superposition of retinal action potentials to predict electrical flicker responses from the eye of the wolf spider, *Lycosa Baltimoriana* (Keyserling). J. gen. Physiol. **46**, 75—96 (1962).

— Linear relations between stimulus amplitudes and amplitudes of retinal action potentials from the eye of the wolf spider. J. gen. Physiol. **47**, 13—32 (1963).

— A nonlinear model for transient responses from light-adapted Wolf Spider eyes. J. gen. Physiol. **50**, 1993—2030 (1967).

DEWEY, M. M., BARR, L.: A study of the structure and distribution of the nexus. J. Cell Biol. **23**, 553—585 (1964).

DIAMOND, J., GRAY, J. A. B., INMAN, D. R.: The relations between receptor potentials and the concentration of sodium ions. J. Physiol. (Lond.) **142**, 382—394 (1958).

EAKIN, R. M.: Evolution of photoreceptors. Cold Spr. Harb. Symp. quant. Biol. **30**, 363—370 (1965).

EDWARDS, C., TERZUOLO, C. A., WASHIZU, Y.: The effect of changes of the ionic environment upon an isolated crustacean sensory neuron. J. Neurophysiol. **26**, 948—957 (1963).

EGUCHI, E.: Rhabdom structure and receptor potentials in single crayfish retinular cells. J. cell. comp. Physiol. **66**, 411—429 (1965).

FATT, P., KATZ, B.: An analysis of the end-plate potential recorded with an intracellular electrode. J. Physiol. (Lond.) **115**, 320—370 (1951).

FULPIUS, B., BAUMANN, F.: Effects of sodium, potassium and calcium ions on slow and spike potentials in single photoreceptor cells. J. gen. Physiol. **53**, 541—561 (1969).

FUORTES, M. G. F.: Electrical activity of cells in the eye of *Limulus*. Amer. J. Ophth. **46**, 210—223 (1958).

— Initiation of impulses in visual cells of Limulus. J. Physiol. (Lond.) **148**, 14—28 (1959).

— Visual responses in the eye of the Dragonfly. Science **142**, 69—70 (1963).

— HODGKIN, A. L.: Changes in time scale and sensitivity in the ommatidia of *Limulus*. J. Physiol. (Lond.) **172**, 239—263 (1964).

— POGGIO, G. F.: Transient responses to sudden illumination in cells in the eye of *Limulus*. J. gen. Physiol. **46**, 435—452 (1963).

GORMAN, A. L. F., McREYNOLDS, J. S.: Hyperpolarizing and depolarizing receptor potentials in the scallop eye. Science **165**, 309—310 (1969).

GRAZIADEI, P. P. C., METCALF, J. F.: Ultrastructure of the retina in the scallop's eye. Fed. Proc. **29**, 393 (1970).

GRUNDFEST, H.: Ionic mechanisms in electrogenesis. Ann. N. Y. Acad. Sci. **94**, 405—457 (1961).

HAGINS, W. A.: Electrical signs of information flow in photoreceptors. Cold Spr. Harb. Symp. quant. Biol. **30**, 403—418 (1965).

— McGAUGHY, R. E.: Membrane origin of the fast photovoltage of squid retina. Science **159**, 213—215 (1968).

— ZONANA, H. V., ADAMS, R. G.: Local membrane current in the outer segments of squid photoreceptors. Nature (Lond.) **194**, 844—846 (1962).

HARTLINE, H. K.: The discharge of impulses in the optic nerve in response to flashes of light of short duration. Amer. J. Physiol. **105**, 45—46 (1933).

— Intensity and duration in the excitation of single photoreceptor units. J. cell. comp. Physiol. **5**, 229—247 (1934).

— The discharge of nerve impulses from the single visual sense cell. Cold Spr. Harb. Symp. quant. Biol. **3**, 245—249 (1935).

— The discharge of impulses in the optic nerve of *Pecten* in response to illumination of the eye. J. cell. comp. Physiol. **11**, 465—477 (1938).

— Visual receptors and retinal interaction (Nobel Lecture). Le Prix Nobel, 242—259 (1967).

— GRAHAM, C. H.: Nerve impulses from single receptors in the eye. J. cell. comp. Physiol. **1**, 277—295 (1932).

— — The spectral sensitivity of single visual cells. Amer. J. Physiol. **109**, 49—50 (1934).

— WAGNER H. G., MacNICHOL, E. F., JR.: The peripheral origin of nervous activity in the visual system. Cold Spr. Harb. Symp. quant. Biol. **17**, 125—141 (1952).

JACKLETT, J. W.: Electrophysiological organization of the eye of *Aplysia*. J. gen. Physiol. **53**, 21—42 (1969).

KARRER, H. E.: The striated musculature of blood vessels. II. Cell interconnections and cell surface. J. Biophy. Biochem. Cytol. **8**, 135—150 (1960).

KENNEDY, D.: Neural photoreception in a Lamellibranch mollusc. J. gen. Physiol. **44**, 277—299 (1960).

— The photoreceptor process in lower animals. In: A. C. GIESE (Ed.): Photophysiology, pp. 79—121. New York: Academic Press 1964.

KIKUCHI, R., IHNUMA, M., TACHI, S.: Different cellular components in ommatidia of horseshoe crab, *Tachypleus tridentatus*. Naturwissenschaften **52**, 265 (1965).

— NAITO, K., TANAKA, I.: Effect of sodium and potassium ions on the electrical activity of single cells in the lateral eye of the horseshoe crab. J. Physiol. (Lond.) **161**, 319—343 (1962).

KIRSCHFELD, K.: Discrete and graded receptor potentials in the compound eye of the fly (Musca). In: BERNHARD, C. G. (Ed.): The Functional Organization of the Compound Eye. Oxford: Pergamon 1965.

KOIKE, H., BROWN, H. M., HAGIWARA, S.: Post-illumination hyperpolarization of a barnacle photoreceptor cell. Fed. Proc. **29**, 393 (1970).

LANGER, H., THORELL, B.: Microspectrophotometric assay of visual pigments in single rhabdomes of the insect eye. In: BERNHARD, C. G. (Ed.): The Functional Organization of the Compound Eye, pp. 145—149. Oxford: Pergamon 1966.

LASANSKY, A.: Cell junctions in ommatidia of *Limulus*. J. Cell Biol. **33**, 365—383 (1967).

— FUORTES, M. G. F.: The site of origin of electrical responses in visual cells of the Leech *Hirudo Medicinalis*. J. Cell Biol. **42**, 241—252 (1969).

MacNICHOL, E. F., JR.: Visual receptors as biological transducers. In: GREWELL, R. G., MULLINS, L. J. (Eds.): Molecular Structure and Functional Activity of Nerve Cells, pp. 34—52. American Institute of Biological Sciences, Publ. No. 1, 34, Washington, D.C., 1956.

— LOVE, W. E.: Impulse discharges from the retinal nerve and optic ganglion of the squid. In: JUNG, R., KORNHUBER, H. (Eds.): The Visual System: Neurophysiology and Psychophysics, pp. 97—103. Berlin-Göttingen-Heidelberg: Springer 1961.

McREYNOLDS, J. S., GORMAN, A. L. F.: Membrane conductances and spectral sensitivities of *Pecten* photoreceptors. J. gen. Physiol. **56**, 376—391 (1970 a).

— — Photoreceptor potentials of opposite polarity in the eye of the scallop, *Pecten irradians.* J. gen. Physiol. **56**, 392—406 (1970 b).

MARIMONT, R.: Numerical studies of the Fuortes-Hodgkin *Limulus* model. J. Physiol. (London) **179**, 489—497 (1965).

MAURO, A., BAUMANN, F.: Electrophysiological evidence of photoreceptors in the epistellar body of *Eledone Moschata*. Nature (Lond.) **220**, 1332—1334 (1968).

MILLECCHIA, R., BRADBURY, J., MAURO, A.: Single photoreceptor cells in *Limulus polyphemus*. Science **154**, 1199—1201 (1966).

— MAURO, A.: The ventral photoreceptor cells of *Limulus* II. The basic photoresponse. J. gen. Physiol. **54**, 310—330 (1969 a).

— — The ventral photoreceptor cells of *Limulus* III. A voltage-clamp study. J. gen. Physiol. **54**, 331—351 (1969 b).

MILLER, W. H.: Morphology of the ommatidia of the compound eye of *Limulus*. J. biophys. biochem. Cytol. **3**, 241—248 (1957).

— Derivatives of cilia in the distal sense cells of the retina of *Pecten*. J. biophys. biochem. Cytol. **4**, 227—228 (1958).

— Visual photoreceptor structures. In: BRACHET, J., MIRSKY, A. E. (Eds.): The Cell. London: Academic Press 1960.

NAKA, K.-I.: Recording of retinal action potentials from single cells of the insect compound eye. J. gen. Physiol. **44**, 571—584 (1961).

— EGUCHI, E.: Spike potentials recorded from the insect photoreceptor. J. gen. Physiol. **45**, 663—680 (1962).

NILSSON, S. E. G.: The ultrastructure of photoreceptor cells. Proceedings of the International School of Physics "Enrico Fermi". Vol. 43. Processing of Optical Data by Organisms and by Machines, pp. 69—115. New York: Academic Press 1969.

NUNNEMACHER, R. F., DAVIS, P. P.: The fine structure of the *Limulus* optic nerve. J. Morph. **125**, 61—70 (1968).

OBARA, S.: Effects of some organic cations on generator potential of crayfish stretch receptor. J. gen. Physiol. **52**, 363—386 (1968).

— GRUNDFEST, H.: Effect of lithium on different membrane components of crayfish stretch receptor neurons. J. gen. Physiol. **51**, 635—654 (1968).

OTTOSON, D.: The effect of sodium deficiency on the response of the isolated muscle spindle. J. Physiol. (Lond.) **171**, 109—118 (1964).

PINTER, R. B.: Sinusoidal and delta function responses of visual cells in the *Limulus* eye. J. gen. Physiol. **49**, 565—593 (1966).

PURPLE, R. L., DODGE, F. A.: Interaction of excitation and inhibition in the eccentric cell in the eye of *Limulus*. Cold Spr. Harb. Symp. quant. Biol. **30**, 529—537 (1965).

RATLIFF, F., HARTLINE, H. K., LANGE, D.: The dynamics of lateral inhibition in the compound eye of *Limulus* I. In: BERNHARD, C. G. (Ed.): The Functional Organization of the Compound Eye, pp. 399—424. Oxford: Pergamon 1966.

ROBERTSON, J. D.: The occurrence of a subunit pattern in the unit membranes of club endings in Mauthner cell synapses in goldfish brains. J. Cell Biol. **19**, 201—221 (1963).

— BODENHEIMER, T. S., STAGE, D. E.: The ultrastructure of Mauthner cell synapses and nodes in goldfish brains. J. cell. Biol. **19**, 159—199 (1963).

RUSHTON, W. A. H.: A theoretical treatment of Fuortes' observation upon eccentric cell activity in *Limulus*. J. Physiol. (Lond.) **148**, 29—38 (1959).

SCHOLES, J.: Discontinuity in the excitation process in Locust visual cells. Cold Spr. Harb. Symp. quant. Biol. **30**, 517—527 (1965).

SHAW, S. R.: Simultaneous recording from two cells in the locust retina. Z. vergl. Physiol. **55**, 183—194 (1967 a).

— Coupling between receptors in the eye of the drone honeybee. J. gen. Physiol. **50**, 2480—2481 (1967 b).

— Organization of the locust retina. Symp. Zool. Soc. Lond. **23**, 135—163 (1968).

SHAW, S. R.: Interreceptor coupling in ommatidia of drone honeybee and locust compound eyes. Vision Res. **9**, 999—1029 (1969).

SMITH, T. G.: Receptor potentials in retinular cells in *Limulus*. Res. Lab. Elec. Quant. Prog. Rept., Mass. Inst. Techn. **81**, 242—248 (1966).

— BAUMANN, F.: The functional organization within the ommatidium of the lateral eye of *Limulus*. Progress Brain Res. **31**, 313—349 (1969).

— — FUORTES, M. G. F.: Electrical connections between visual cells in the ommatidium of *Limulus*. Science **147**, 1446—1448 (1965).

— BROWN, J. E.: A photoelectric potential in invertebrate cells. Nature (Lond.) **212**, 1217—1219 (1966).

— STELL, W. K., BROWN, J. E.: Conductance changes associated with receptor potentials in *Limulus* photoreceptors. Science **162**, 454—456 (1968a).

— — — FREEMAN, J. C., MURRAY, G. C.: A role for the sodium pump in photoreception. Science **162**, 456—458 (1968b).

TOMITA, T.: The nature of action potentials in the lateral eye of the horseshoe crab as revealed by simultaneous intra- and extracellular recordings. Jap. J. Physiol. **6**, 327—340 (1956).

— Peripheral mechanism of nervous activity in eye of *Limulus*. J. Neurophysiol. **21**, 245—254, (1957).

— KIKUCHI, R., TANAKA, I.: Excitation and inhibition on lateral eye of horseshoe crab. In: KATSUKI, Y. (Ed.): Electrical Activity of Single Cells. Tokyo: Igakushiin 1960.

TOYODA, J. I., NOSAKI, H., TOMITA, T.: Light-induced resistance changes in single photoreceptors of *Necturus* and *Gekko*. Vision Res. **9**, 453—463 (1969).

— SHAPLEY, R. M.: The intracellularly recorded response in the scallop eye. Biol. Bull. **133**, 490 (1967).

WALD, G.: Visual excitation and blood clotting. Science **150**, 1028—1030 (1965).

WALTHER, J. B.: Single cell responses from the primitive eyes of an annelid. In: BERNHARD, C. G. (Ed.): Functional organization of the compound Eye. New York: Pergamon Press 1965.

WASHIZU, Y.: Electrical activity of single retinular cells in the compound eye of the blowfly *Calliphora erythrocephala. Meig*. Comp. Biochem. Physiol. **12**, 369—387 (1964).

WATERMAN, T. H., WIERSMA, C. A. G.: The functional relation between retinular cell and optic nerve in *Limulus*. J. Exp. Zool. **126**, 59—85 (1954).

WINTER, D. L.: Intracellular responses from the grasshopper eye. Nature (Lond.) **213**, 607—608 (1967).

Chapter 7

Responses to Single Photons

By

M. G. F. Fuortes and Paul M. O'Bryan, Bethesda, Maryland (USA)

With 14 Figures

Contents

Responses of photoreceptor cells are usually brought about by the action of many photons. Therefore, they may reasonably be regarded as composite responses resulting from combination of numerous small responses, evoked each by the absorption of one photon.

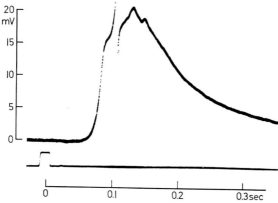

Fig. 1. Response of an eccentric cell to a flash of light. The response was recorded by means of an intracellular micropipette. A spike (only its undershoot is visible in the record) is discharged when the depolarization reaches about 15 mV. Note that the response develops after the flash (bottom trace) when the preparation has returned to darkness. The flash delivered about 10^7 absorbable photons. Responses of this type may be regarded as the combination of many smaller responses produced each by absorption of one photon

Study of the properties of the composite response is not sufficient for determining the features of responses to single photons: for instance, a composite response such as that illustrated in Fig. 1 could result from summation of single photon responses having the same shape as the composite wave and no dispersion in time, or of shorter responses arising with different latencies. In order to resolve this problem, it is necessary to record directly responses produced by absorption of single or of few photons, and therefore it is necessary to study the changes produced by very dim lights.

Energy in a Just-Visible Flash

Investigations of this type were first performed using psychophysical techniques, and the results of these studies showed not only that absorption of a single photon produces a response in a photoreceptor but also that this response propagates to other cells, giving perception of the light.

Langley (1889) measured the energy of a just-visible flash of light of 550 mμ and found that perception occurred when the energy at the cornea was about 3×10^{-9} erg. This is the energy carried by approximately 750 photons. It should be pointed out, however, that not all the photons reaching the cornea are absorbed by the visual pigments.

Later measurements (von Kries and Eyster, 1907; Barnes and Czerny, 1932) led to estimates which were about ten-fold lower. Hecht, Shlaer and, Pirenne (1942) in a classical study of visual thresholds in man, confirmed these lower values, establishing that a flash of green-blue light can be perceived six out of ten times if the energy at the cornea is $2.2-5.7 \times 10^{-10}$ ergs, corresponding to 54 to 148 photons.

Hecht et al. (1942) estimated that 4 percent of the light is reflected at the cornea; about 50 percent of the remaining light is absorbed before reaching the retina and perhaps 80 percent goes through the retina without being caught by the visual pigments. These corrections lead to the conclusion that the flash was seen when 5 to 14 photons were absorbed by the retinal photoreceptors. The stimuli used in these studies covered about 500 rods. "It is therefore unlikely that any one rod will take up more than one quantum out of the small number absorbed. In fact, the probability that 2 quanta will be taken up by a single rod is only about 4 percent. We may, therefore, conclude that in order to see it is necessary for only one quantum of light to be absorbed by each of 5 to 14 retinal rods." (Hecht, 1942, p. 47.)

Relation between Probability of Seeing and Flash Intensity

These figures of 5 to 14 absorbed photons per flash are average values. The actual number of photons absorbed in the retina following individual flashes will fluctuate around these averages: on some occasions they will be more than the expected value, on other occasions they will be fewer. These fluctuations are described by the Poisson distribution. If the mean number of photons absorbed per flash is Q, the probability $P(x)$ that *exactly* x photons are absorbed following a flash is:

$$P(x) = Q^x \frac{e^{-Q}}{x!} . \tag{1}$$

If x is the minimum number of photons which must be absorbed to produce perception, flashes delivering more than x photons will also be perceived. Thus the
probability of seeing a flash will be the same as the probability $D(x)$ that x or more
photons are absorbed. This is:

$$D(x) = \sum_{r=x}^{\infty} Q^r \frac{e^{-Q}}{r!} \tag{2}$$

Fig. 2 shows that the dependence of $D(x)$ upon Q changes for different values of x.
It can be deduced from this figure that if vision is evoked by a small number (x)
of photons, then the probability (D) to see the flash will increase gradually as
flash intensity (Q) is increased. But if the number of required photons is large, then
the probability D will increase rapidly with Q.

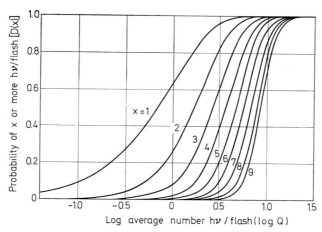

Fig. 2. Cumulative Poisson distribution. The abscissa measures in logarithmic scale the
average number of photons in a flash [Q in Eq. (2)]. The ordinate measures the probability
[$D(x)$ in Eq. (2)] that the actual number of photons in the flash is x or greater than x. The
curves give this probability for $x = 1, 2, \ldots, \ldots, 9$. (From HECHT, 1942)

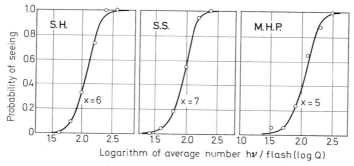

Fig. 3. Probability of seeing as a function of number of photons. Three experiments in different observers. The subjects were asked to report whether or not they saw the flashes at
the different intensities. Each point gives the results of 50 presentations. The curves are the
Poisson distributions of Fig. 2 for $x = 6, 7$ and 5 (from HECHT, 1942)

Hecht et al. (1942) explored this statistical problem and found that the probability of seeing grows with flash intensity as the curve $D(x)$ for x approximately equal to six (Fig. 3). In this way they confirmed by an independent method the estimates obtained by measurement of threshold energy.

Bouman and van der Velde (1948) performed similar experiments and confirmed the conclusion that absorption of one photon in a rod leads to the production of a transmitted signal but found that coincidence of the signals of only two rods (rather than of five or more) is sufficient to produce vision.

Statistical Studies on Limulus

Hartline, Milne, and Wagman (1947) investigated the fluctuations of threshold by recording from single fibers of the optic nerve of *Limulus* the impulses produced by dim flashes of light. For light intensities within an appropriate range a fraction of the flashes produces one or more nerve impulses while other flashes fail to evoke a response. "In most dark adapted preparations the intensity range within which frequency of responses is greater than zero and less than 100 % covers approximately one logarithmic unit . . . Light adaptation raises the threshold of the sense cell; at the same time the range of uncertainty is narrowed on a

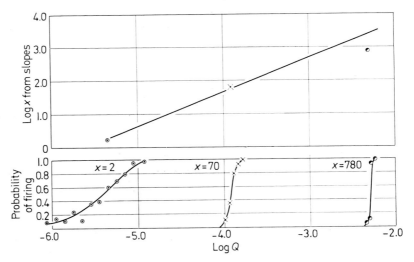

Fig. 4. Probability of firing in different states of adaptation. Responses were recorded from bundles of *Limulus* optic nerve, giving single unit activity. The response was the discharge of one or more spikes following flashes of different intensities. At the bottom, the left-hand plot shows the measurements taken when the preparation was deeply dark-adapted: probability of firing increases as $D(x)$ for $x = 2$ [see Eq. (2) and Fig. 2]. With increasing light (middle and right-hand plots) the measurements fit the same theoretical distributions with $x = 70$ and $x = 780$. In the upper graph, the abscissa measures the intensity of the flashes required to produce firing in 50 percent of the presentations (threshold intensity). Ordinate measures the value of x as determined by the fits at the bottom. On logarithmic coordinates, these measurements fall on a straight line with a slope of one, indicating a linear relation between average number of photons required for firing and threshold intensity of the flash, in different states of adaptation (from Hartline, unpublished)

logarithmic scale of intensities. This effect is reversed by dark adaptation."
(HARTLINE et al., 1947, p. 124.) WAGMAN, HARTLINE and MILNE (1949) and
HARTLINE (1959) studied in greater detail the fit of the results with the Poisson
distribution of relation (2). They established in this way that the probability that
an eccentric cell produces impulses following a flash increases with light intensity,
as would be expected if absorption of two photons were needed to produce firing,
when the cell is deeply dark adapted. The number of photons required for firing
increase with light adaptation, becoming very large when the preparation is
strongly light-adapted (Fig. 4).

These results can be explained assuming that firing occurs when the cell is
depolarized by a certain amount which does not change in dark or light adaptation.
In dark adaptation the response to one absorbed photon is large and summation of
two responses is sufficient to reach threshold. In these conditions there will be a
wide range of stimuli which may or may not produce a response: as seen in Fig. 2
and Fig. 4, an increase from 10 percent to 90 percent of the probability $D(x)$ that at
least two photons are absorbed requires an 8-fold increase of the average stimulus
strength (Q). With strong light adaptation, each absorbed photon gives a much
smaller response, so that many more than two responses must sum in order to
evoke firing. If 70 responses are needed (as in the middle plot of Fig. 4) probability
of firing increases from 10 to 90 percent with an increase of stimulus strength by a
factor of only 1.3. Thus, HARTLINE's results can be interpreted simply by assuming
that the size of the response of a visual cell to one absorbed photon decreases with
increasing light adaptation.

Subliminal Responses in Visual Cells of Limulus

As electrophysiological methods advanced it became possible to record not
only nerve impulses but also the smaller potential changes corresponding to

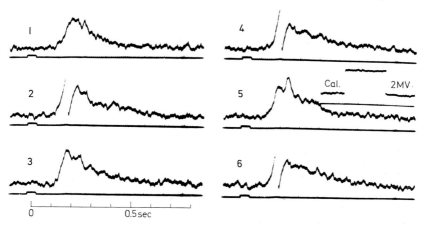

Fig. 5. Responses to dim flashes. Responses were recorded by a micropipette inserted in a
visual cell of *Limulus*. Flash intensity was adjusted to give a spike in 50 percent of the presentations. (Only the bottom portions of the spikes are seen in records 2, 4 and 6.) The responses of records 1 and 5 are "noisy" suggesting that they may be made up of few discrete
components (MACNICHOL, 1958)

subliminal activities. Changes of this type had already been detected in the eccentric cell of *Limulus* by laborious studies of excitability (Wagner et al., 1951) but could be recorded directly only some time later by MacNichol (1958). Intracellular records taken from dark-adapted preparations showed that dim flashes which do not evoke impulse firing produce a clearly-detectable wave of depolarization. Sharp, irregular potential changes were often found to be superposed to this slow wave (Fig. 5). Since the results of Fig. 4 indicate that absorption of two photons may be sufficient to evoke firing, it is reasonable to suppose that the subliminal responses recorded by MacNichol (1958) were brought about by only few photons. This suggests that absorption of a single photon may produce a detectable electrical wave.

Yeandle's Discrete Waves

Working in MacNichol's laboratory, Yeandle (1957 and 1958) found that irregular discrete waves can often be recorded from dark-adapted cells of the Lateral eye of *Limulus* (Fig. 6). Similar potential changes were later recorded from photoreceptors of the locust (Scholes, 1964 and 1965), of the ventral eye of *Limulus* (Millecchia et al., 1966) and of the Leech (Walther, 1966). If the preparation is fully dark-adapted, discrete waves are usually present even in complete darkness. Dim illumination increases the probability of their occurrence. Applying

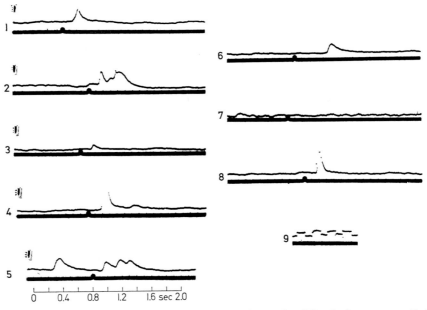

Fig. 6. Discrete waves in dark-adapted visual cells of *Limulus*. Dim flashes were applied as indicated by the signals under each record. The flashes are often followed by one or more potential transients (called "bumps" by Yeandle, 1957 and 1958) which may have different size and shape in individual cases. These transients often occur spontaneously, as in record 5, but their probability of occurrence increases following or during illumination. The square wave in record 9 is a 1 mV calibration (Yeandle, 1958)

the statistical methods which had already been used by HECHT et al. (1942) and by WAGMAN et al. (1949), YEANDLE found that the probability that a flash evokes at least one discrete wave increases with light intensity following relation (2) with $x = 1$ (Fig. 7). As already mentioned, this result is consistent with the notion that absorption of a single photon is sufficient for generation of a discrete wave.

It was noted later (FUORTES and YEANDLE, 1964) that fit of results of this type with the statistical curves given by relation (2) does not necessarily give a measurement of the number of photons required to produce a response, because the parameter x may represent some physical quantity other than the absorbed photons. For example, it could be supposed that a) light liberates a transmitter substance in the form of droplets whose number is proportional to the number of absorbed photons; b) the arrival of droplets at the membrane follows the Poisson distribution; c) each droplet, reacting with the membrane produces a discrete wave. With a sequence of processes of this type, the probability of occurrence of discrete waves would follow relation (2) with $x = 1$ even if many photons were required to produce a response. More generally, it should be kept in mind that statistical analysis can be useful for establishing whether a response is produced by one or by more photons only if there are reasons to believe that the statistical variability of the response is due to the fluctuations in the number of absorbed

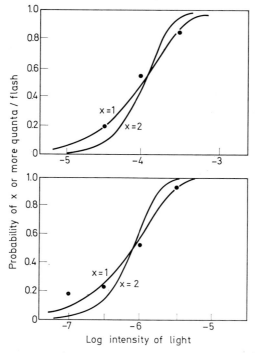

Fig. 7. Probability of occurrence of discrete waves. The dots measure the probability that one or more discrete waves occur following flashes of different intensity, as indicated in the abscissa (arbitrary logarithmic units). The two plots show the data obtained in two different cells. The solid lines measure the probability $D(x)$ for $x = 1$ or $x = 2$ (from YEANDLE, 1958)

photons. It is important, therefore, to supplement the statistical studies on the responses with estimates of the number of photons in the stimulus since these estimates will give at least a range of expected fluctuations in the number of absorbed photons.

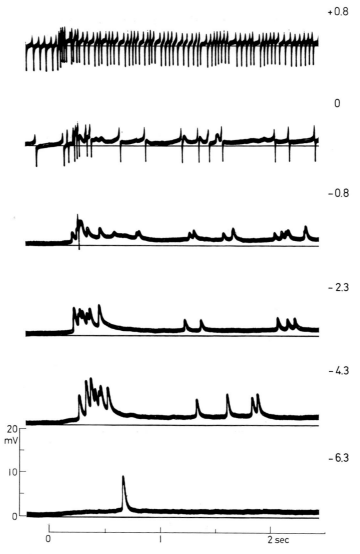

Fig. 8. Effects of currents on the responses produced by light. Depolarizing (+) and hyperpolarizing currents were passed through the impaling microelectrode and undesired potential drops were balanced by means of a Wheatstone bridge. Current intensity ($A \times 10^{-9}$) is indicated by numbers at right. Steps of dim light (same intensity in all records) were applied at time 0. Latency and size of discrete waves increases with hyperpolarizing currents. In the lower records (-4.3 and -6.3) a small, sustained depolarization is seen to occur following the illumination, starting before the discrete waves (Fuortes, 1959b)

Dual Nature of Responses to Dim Lights

The properties of the discrete waves of *Limulus* were later re-examined (FUORTES, 1959b). It was found that their average size is increased by hyper-

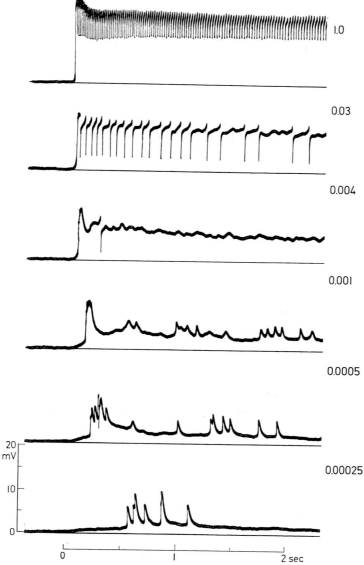

Fig. 9. Responses evoked by light of different intensities. Steps of light were applied at time 0. Light intensity is given in relative units by the figures at right. With dim lights, a slow, sustained depolarization precedes and underlies the discrete waves. Discrete waves are more numerous at the beginning of the stimulation, apparently corresponding to a peak in the sustained depolarization. As light intensity is increased, the discrete waves become smaller and more numerous (FUORTES, 1959b)

polarizing and decreased by depolarizing currents (Fig. 8). The discrete waves appear, therefore, to be associated with increased membrane conductance (see Adolph, 1964) as is the case for the generator potentials (Fuortes, 1959a).

When discrete waves are not already very numerous in darkness, dim illumination may produce only few isolated waves. As light intensity is increased the discrete waves become smaller and more numerous and eventually appear to fuse in a smooth generator potential (Fig. 9). Observations of this type suggest that generator potentials evoked by brighter light originate from summation of many discrete waves such as those recorded in isolation when the light is dim.

It was noted, however, that in certain conditions two components can be detected in the responses to dim illumination: when the preparation produces only few discrete waves in darkness, it can sometimes be shown that a step of dim light not only increases the frequency of the discrete waves but produces a smoothly-sustained depolarization with superimposed discrete waves. Responses of this type are illustrated in Fig. 10.

It is clear in this figure that the small, sustained depolarization cannot arise from fusion of the much larger discrete waves seen in the records. Rather, sustained depolarization and discrete waves must be the consequence of two different processes. This conclusion is supported by the observation that a sustained depolarization can sometimes be evoked in isolation by light dimmer than those required to produce discrete waves (Fuortes, 1959b).

These results suggest that generator potentials are not simply the result of summation of a single type of discrete wave, but include a second component

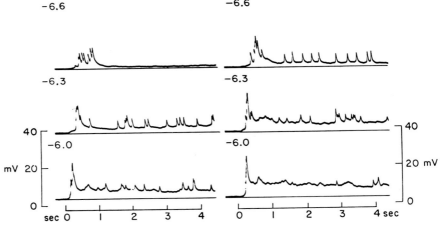

Fig. 10. Responses to dim lights in different states of adaptation. Steps of light of intensity indicated in arbitrary logarithmic units were applied at time 0. Left-hand records were taken when the cell was moderately dark-adapted; right-hand records when it was deeply dark-adapted. In all cases illumination evoked sustained depolarization with superposed discrete waves. With the dimmest lights, a transient burst of waves appeared at the beginning of the illumination. In this experiment light intensity −6.0 corresponded to about 6000 absorbable photons per sec through the rhabdome of the cell. It was estimated that less than 10% of the light through the rhabdome could be absorbed by the visual pigment (Borsellino and Fuortes, 1968a)

which gives rise to a smooth and sustained depolarization when the light is in the form of a step.

A possible interpretation of these results is suggested by the observations made by ADOLPH (1964). He noted that discrete waves can often be separated in two classes: some large, fast, others are small and slow (Fig. 11). ADOLPH's (1964) observations were later confirmed by DOWLING (1968) and by BORSELLINO and FUORTES (1968a). The fast waves were often found to be superimposed on the

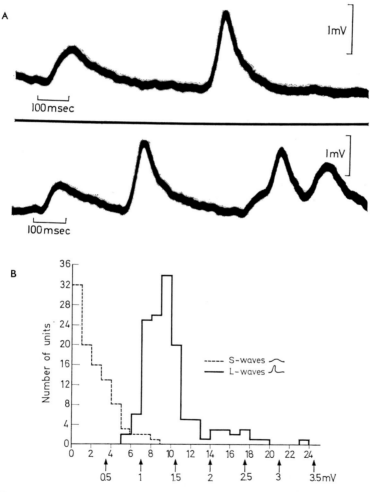

Fig. 11. Two types of discrete waves. a Potential changes recorded from a dark-adapted visual cell of *Limulus* with an intracellular micropipette. Some waves are small and slow (*S*-waves) others are larger and faster (*L*-waves). Some waves appear to be a combination of the two types: they start slowly and then proceed more rapidly giving rise to a sharp inflection as in the bottom record (from ADOLPH, 1964). b Amplitude histogram of discrete waves. The plot shows that discrete waves can sometimes be divided in two groups with respect to their peak amplitude (ADOLPH, 1964)

slower discrete waves and were regarded to be analogous to the sharp transients which had been studied in different experimental conditions by Fuortes and Poggio (1963) and by Benolken (1965).

If light can evoke two types of discrete waves, one slow and small and the other large and fast, then the two components of the responses to steps illustrated in Fig. 10 can be interpreted assuming that the sustained depolarization is produced by summation of numerous small slow waves; the fast waves appear as sharp transients superposed on this depolarization, just as they are often superposed on individual slow discrete waves.

Latency Distribution of Discrete Waves

The finding that light evokes two different types of electrical responses complicates the problem of determining the features of responses to single photons.

The results illustrated in Figs. 8, 9, and 10 suggest that dim steps of light evoke numerous small slow waves which fuse in a sustained generator potential; the fast waves may be missing altogether or they may be few, arising well after the sustained depolarization. It seems reasonable to infer from these observations that the slow waves are the more direct response to photon absorption, while the fast waves are an accessory process. If this inference is correct, it follows that statistics based on counts of the fast waves (which are more prominent in the records) may be unsuitable for determining the number of absorbed photons.

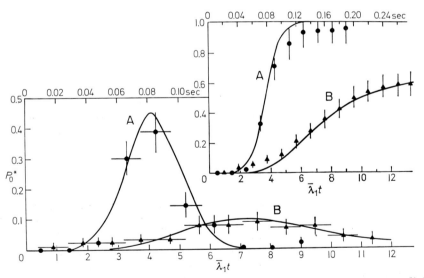

Fig. 12. Latency distribution of discrete waves evoked by flashes. Discrete waves were divided in two groups (S- and L-waves) as described in Fig. 11. Measurements of S-waves are indicated by circles and of L-waves by triangles. Probability of occurrence of waves of each group at given times after the flash was plotted after subtracting the probability that the same type of wave occurred in darkness. The dispersion of latencies is greater for the L-waves than for the S-waves. Inset: Same data plotted as cumulative distribution of the two types of waves. Solid lines are the integrals of the corresponding theoretical curves in the main figure (Borsellino and Fuortes, 1968a)

Starting from these considerations, BORSELLI NOand FUORTES (1968a) studied again the properties of the discrete waves of *Limulus*. They confirmed ADOLPH'S (1964) finding that discrete waves divide in two populations: in some cases the two types can be separated by means of amplitude histograms as in Fig. 11; in other instances the amplitude distribution is not clearly bimodal, but then the two populations can still be distinguished by plotting peak amplitude as a function of time of rise (see Fig. 5 in Chapter 8 of this volume). It was observed in addition that, when discrete waves are evoked by flashes, the two types differ not only in amplitude and time course, but also in their latency distribution: the fast waves suffer large latency fluctuations while the smaller, slow waves arise at an approximately constant time after the flash (Fig. 12).

In order to determine the properties of the responses to single photons, the usual statistical methods were applied in combination with estimates of the number of photons in the stimulus. In these experiments, the light traversed the rhabdome sideways rather than along its major axis. In these conditions, it was estimated that only about 1 percent of the photons impinging on the rhabdome was absorbed in the average by the visual pigments. With this assumption, the probability of

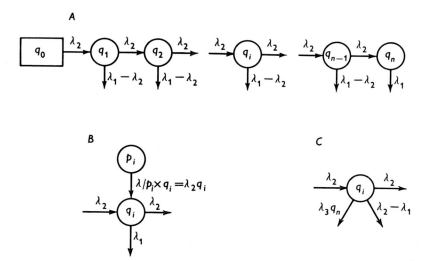

Fig. 13. Compartmental model reproducing visual responses. This model is analogous to the multi-stage filter already discussed in Chapter 6 of this volume. A compartment is defined as the state or location of particles (SOLOMON, 1961; BERMAN et al., 1962; BERMAN, 1963). The number of particles in compartment i is called q_i. λ_1 is the fractional rate of decay of the particles and λ_2 is their rate of transition from one compartment to the other. It is assumed that the number of particles can increase in this system, because the outflow $\lambda_2 q_i$ from compartment i into compartment $i+1$ is compensated by an equal influx $(-\lambda_2 q_i)$. This is accomplished by subtracting λ_2 from the decay rate λ_1. One obtains then:

$$\frac{dq_i}{dt} = -(\lambda_1 - \lambda_2)q_i - \lambda_2 q_i + \lambda_2 q_{i-1}) = -\lambda_1 q_i + \lambda_2 q_{i-1}.$$

B represents a chemical system which satisfies the relation above (see BORSELLINO et al. 1965) A typical compartment of the non-linear model is shown in C. Note that λ_3 in this diagram corresponds to the quantity $\bar{\lambda}_1/w$ (see equation 3) which introduces non-linear properties in the model (BORSELLINO and FUORTES, 1968a)

absorption of one or more photons for flashes of different strengths was found to fit the probability that the flash evoked one or more slow waves [see Eq. (2)]. The smallest recorded slow waves could then be reasonably ascribed to absorption of single photons. The fast waves were found to be more erratic, appearing in the average only in 10 percent of the instances in which the stimulus evoked a slow wave. They were considered, therefore, not as a primary response to light but as accessory events.

Interpretation of the properties of the two types of discrete waves was attempted making use of the model introduced by Fuortes and Hodgkin (1964) to explain the properties of generator potentials (Fig. 13) (see Chapters 6 and 8 in this volume). This model was supposed to represent a sequence of chemical transformations. More specifically, it was assumed that when a molecule of visual pigment is activated by a photon, it reacts with a substrate to form a new product E_1. This in turn reacts with another substrate to produce a substance E_2, and so on for n steps, until a substance E_n is generated which acts on the cell membrane producing an electrical response. Therefore, the latency of the response measures the time required for generation of the first molecule of the substance E_n.

It can be deduced from these assumptions that the dispersion in the latencies of responses to flashes decreases as the number of molecules E_n produced by the flash increases.

The experimental results show that the latency distribution of the smallest slow wave (ascribed to absorption of single photons) resembles the dispersion which would be expected if they were produced by at least 25 particles of the substance E_n. It follows that the processes which intervene between absorption of light and production of electrical responses involve amplification: from one excited molecule of visual pigment to 25 molecules of a substance capable of changing the properties of the membrane. This multiplication of particles could be brought about by a sequence of enzymatic reactions, as suggested by Wald (1956 and 1965), Borsellino et al. (1965) and Borsellino and Fuortes (1968a and b). Based on these conclusions, the following scheme was proposed: when a photon is absorbed by a molecule of visual pigment, a series of chemical reactions is initiated which leads to the production of about 25 molecules of a membrane-active substance. Each of these molecules produces a small "elementary response" and summation of the twenty-five elementary responses produced by absorption of one photon makes up a slow discrete wave. This would occur if each molecule (E_n) opened a "channel" permeable to ions. In order to explain the large fast waves, it is assumed that each molecule of the end-product E_n has a small probability of giving origin not to small elementary response but to a large effect, resulting in a large fast wave. This scheme would account well for the large dispersion in the latency of these waves.

Bass and Moore (1970) have proposed an alternative model to explain the generation of discrete waves. Contrary to the views of Borsellino and Fuortes (1968a), they assume that each photon activates only one channel. They conclude that this cannot explain the large potential change produced by a single photon. However, they show that the opening of a single, highly conductive channel in the membrane of the microvilli would lead to a sizable local depolarization and that, if the microvillar membrane were electrically excitable (as the axonal membrane) the depolarization could spread over a sufficiently large area to depolarize the whole cell by several millivolts.

It should be noted that electrical excitability would not be required if one accepted the conclusion that numerous membrane active particles are produced by a single photon. In addition, the multiplicative process in the production of membrane active particles could account for the remarkable stability of the responses to single photons, a property which cannot be readily explained assuming that "gain" is due to electrical excitability.

Summation of Discrete Waves

Some properties of visual responses can be usefully investigated by analyzing average rather than individual responses. To this purpose, a large number of flashes of fixed intensity is applied and all resulting responses are summed (BORSELLINO and FUORTES, 1968a). Dividing the voltage of this summed response by the number of flashes one obtains the average response to one flash at the given intensity, while dividing it by the total number of photons absorbed one obtains the average response to one photon. Some uncertainty attaches to this latter procedure when the number of absorbed photons cannot be measured directly.

Since light evokes two types of waves, the properties of average responses will be determined by the properties of both waves: their average shapes, probabilities of occurrence and latency distributions.

Average responses evoked by flashes of different intensities are shown in Fig. 14. As long as the light intensity is very dim, it is improbable that more than one photon per flash is absorbed by the visual pigment of a cell: in these conditions, the response to each absorbed photon remains independent of light intensity as one would expect. With brighter lights, however, the majority of flashes leads to absorption of two or more photons. The response to each photon becomes then smaller and faster; the initial part of the rising phase, however, coincides for all responses, showing that these changes do not occur immediately.

Delayed changes of this type can be reproduced by the chemical model proposed above (Fig. 13) assuming that the rates of decay λ_1 of all the intermediary products $E_1 \cdots E_n$ are increased by the concentration q_n of the membrane-active substance E_n:

$$\lambda_1 = \bar{\lambda}_1 \left(1 + \frac{q_n}{w}\right) \tag{3}$$

where $\bar{\lambda}_1$ and w are constants. The non-linearities of summation of discrete waves can, therefore, be ascribed to the same mechanisms which had been proposed to explain the non-linear features of the generator potentials evoked by bright flashes (see Chapters 6 and 8 in this volume).

The assumptions incorporated in relation (3) predict that if many photons are absorbed from a flash, each photon will still generate a slow discrete wave but each wave will be smaller because fewer particles of the substance E_n contribute to its production.

This interpretation based on study of average responses does not reveal what happens to the large waves when light intensity is increased. Study of individual responses to steps of light shows, however, that they certainly become smaller and perhaps disappear altogether, when bright lights are applied.

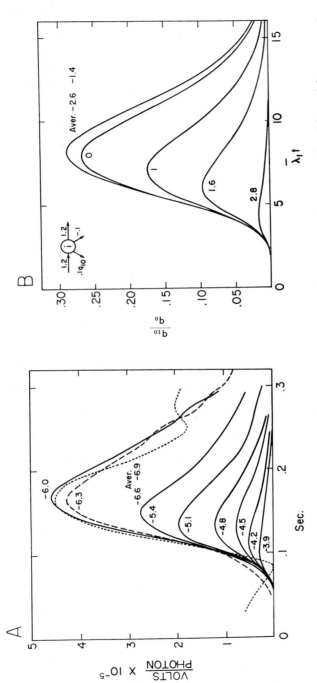

Fig. 14. a Summed responses to constant total number of photons and different flash intensities. Two thousand and forty eight responses to flashes of intensity − 6.9 (corresponding to about 7.5 absorbable photons per flash) were summed electronically, using a Mnemotron computer. The successive curves were obtained by doubling the quantity of light in the flash while halving the number of stimuli. Thus, 1024 responses were summed when flash intensity was − 6.6 (15 photons per flash), 512 responses for intensity − 6.3, etc. In this way, the total number of photons delivered was the same (15360) for all curves. The summed responses have a common rising phase but time to peak and peak amplitude decrease as light intensity is increased. b Responses of the model of Fig. 13 to inputs of different intensities. A chain of ten compartments is considered. The input is an impulse such as the sudden injection at time 0 of one or more particles in compartment 0. The unit in abscissa is the quantity $\bar{\lambda}_1 t$; the ordinate gives the number of particles in the tenth compartment (q_{10}) divided by the input (q_0). The rate of decay of particles increases as particles accumulate in the last compartment (see Fig. 13c and Legend). With this arrangement the (non-linear) responses of the model are similar to those observed experimentally (Borsellino and Fuortes, 1968a)

Responses to steps of light of different intensities were analyzed by DODGE et al. (1968) and by DODGE (1969). The results could be satisfactorily interpreted assuming 1) that the response is made up of discrete waves, each produced by absorption of one photon; 2) that the amplitude of these discrete waves decreases as light intensity is increased, the decrease occurring not immediately but with some delay. Dodge and his coworkers found that in order to fit the experimental responses obtained with different light intensities, it is necessary to assume in addition that the time course of the discrete waves is accellerated when higher intensities of illumination are used. Their conclusions are therefore in good agreement with the interpretations proposed by BORSELLINO and FUORTES, (1968 a and b).

References

ADOLPH, A.: Spontaneous slow potential fluctuations in the *Limulus* photoreceptors. J. gen. Physiol. **48**, 297—322 (1964).

BARNES, R. B., CZERNY, M.: Läßt sich ein Schroteffekt der Photonen mit dem Auge beobachten? Z. Physik. **79**, 436—439 (1932).

BASS, L., MOORE, W. J.: An electrochemical model for depolarization of a retinula cell of *Limulus* by a single photon. Biophis. J. **10**, 1—19 (1970).

BENOLKEN, R. M.: Regenerative transducing properties of a graded visual response. Cold Spr. Harb. Symp. quant. Biol. **30**, 445—450 (1965).

BERMAN, M.: The formulation and testing of models. Ann. N. Y. Acad. Sci. **108**, 182—194 (1963).

— SHAHN, E., WEISS, M. F.: The routine fitting of kinetic data to models; mathematical formalism for digital computers. Biophys. J. **2**, 275—287 (1962).

BOUMAN, M. A., VAN DER VELDEN, H. A.: The two quanta hypothesis as a general explanation for the behavior of threshold values and visual acuity for the several receptors of the human eye. J. opt. Soc. Amer. **38**, 570—581 (1948).

BORSELLINO, A., FUORTES, M. G. F.: Responses to single photons in visual cells of *Limulus*. J. Physiol. (Lond.) **196**, 507—539 (1968a).

— — Interpretation of responses of visual cells of *Limulus*. Proc. IEEE **56**, 1024—1032 (1968b).

— — SMITH, T. G.: Visual responses in *Limulus*. Cold Spr. Harb. Symp. quant. Biol. **30**, 429—443 (1965).

DODGE, F. A. JR.: Inhibition and excitation in the *Limulus* eye. Proc. internat. School of Physics "Enrico Fermi" **43**, 341—365 (1969).

— KNIGHT, B. W., TOYODA, J.: Voltage noise in *Limulus* visual cells. Science **160**, 88—90 (1968).

DOWLING, J. E.: Discrete potentials in the dark-adapted eye of the crab *Limulus*. Nature (Lond.) **217**, 28—31 (1968).

FUORTES, M. G. F.: Initiation of impulses in visual cells of *Limulus*. J. Physiol. (Lond.) **148**, 14—28 (1959a).

— Discontinuous potentials evoked by sustained illumination in the eye of *Limulus*. Arch. ital. Biol. **97**, 243—250 (1959b).

— HODGKIN, A. L.: Changes in time scale and sensitivity in the ommatidia of *Limulus*. J. Physiol. (Lond.) **172**, 239—263 (1964).

— POGGIO, G. F.: Transient responses to sudden illumination in cells in the eye of *Limulus*. J. gen. Physiol. **46**, 435—452 (1963).

— YEANDLE, S.: Probability of occurrence of discrete potential waves in the eye of *Limulus*. J. gen. Physiol. **47**, 443—463 (1964).

HARTLINE, H. K.: Light quanta and the excitation of single receptors in the eye of *Limulus*. Proc. 2nd internat. Congress Photobiol. Torino, (Italy) (1959).

— MILNE, L. J., WAGMAN, I. H.: Fluctuation of responses of single visual cells. Fed. Proc. **6**, 124 (1947).

Hecht, S.: The quantum relations of vision. J. opt. Soc. Amer. **32**, 42—49 (1942).
— Shlaer, S., Pirenne, M. H.: Energy quanta and vision. J. gen. Physiol. **45**, 819—840 (1942).
Kries, J. v., Eyster, J. A. E.: Über die zur Erregung des Sehorgans erforderlichen Energiemengen. Z. Sinnesphysiol. **41**, 373—394 (1907).
Langley, S. P.: Energy and vision. Phil. Mag. **27**, 1—23 (1889).
MacNichol, E. F.: Subthreshold excitatory processes in the eye of *Limulus*. Exp. Cell Res. Suppl. **5**, 411—425 (1958).
Millecchia, R., Bradbury, J., Mauro, A.: Single photoreceptor cells in *Limulus polyphemus*. Science **154**, 1199—1201 (1966).
Scholes, J.: Discrete subthreshold potential from the dimly lit insect eye. Nature (Lond.) **202**, 572—573 (1964).
— Discontinuity in the excitation process in Locust visual cells. Cold Spr. Harb. Symp. quant. Biol. **30**, 517—527 (1965).
Solomon, A. K.: Compartmental methods of kinetic analysis. In: Comar, C. L., Bomer, F. (Eds.): Mineral Metabolism, Vol. 1, Part A. New York: Academic Press 1961.
Wagman, I. H., Hartline, H. K., Milne, L. J.: Excitability changes of single visual receptor cells following flashes of light of intensity near threshold. Fed. Proc. **8**, 159 (1949).
Wagner, H. G., Hartline, H. K., Wagman, I. H.: Subliminal excitation and refractoriness in single photoreceptor elements. Fed. Proc. **10**, 141 (1951).
Wald, G.: Visual excitation and blood clotting. Science **150**, 1028—1030 (1965).
— The biochemistry of visual excitation. In: Gaebler, O. (Ed.): Enzymes: Units of Biological Structure and Function. New York: Academic Press 1956.
Walther, J. B.: Single cell responses from the primitive eyes of an annelid. In: Bernhard, C. G. (Ed.): Functional Organization of the Compound Eye. New York: Pergamon Press 1965.
Yeandle, S.: Studies on the slow potential and the effects of cations on the electrical responses of the *Limulus* ommatidium Ph. D. Thesis. The Johns Hopkins Univ. (1957).
— Electrophysiology of the visual system. Discussion. Amer. J. Ophthalmol. **46**, 82—87 (1958).

Chapter 8

Interpretation of Generator Potentials

By

John Z. Levinson, Murray Hill, New Jersey (USA)

With 9 Figures

Contents

Introduction

The light-evoked changes in the potential difference between the inside and outside of cells in *Limulus* ommatidia are being intensively investigated. Several different models have been proposed for the time-course of the voltage in eccentric cells[1] in response to a weak flash. This is therefore an interim report. "Interpretation", in the title, is a goal, not an achievement.

A thoroughgoing history of the study of the lateral eye of *Limulus* has been provided by Wolbarsht and Yeandle (1967), and brought up to date by Smith and Baumann (1969). The electrical model of the eccentric cell proposed by Fuortes (1959a) (Fig. 1) still stands — at least it is the one in terms of which people think. There exists an exciting controversy at the moment as to which of the four elements change directly in response to light and which of them change only as a result of other changes (Smith et al., 1968a, b. See Chapter 6 for a

---

[1] It may be that the process for which the models have been proposed takes place entirely in the retinular cells around the eccentric cell (Dowling, 1968; Smith and Baumann, 1969). No matter, whichever cell the process takes place in, it is worth trying to understand it; the models are not yet specific enough for this distinction, and experimentally the generator potentials in eccentric cells are too similar to the receptor potentials in retinular cells (until the former "go into action" potentials) for useful concern about the distinction.

discussion). This chapter deals with the process intervening between light absorption and generator potential — the results of the process will be visualized as a change in R_g (Fig. 1) for concreteness, but they are readily understood in other terms as well. Adaptation is taken to affect R_r. PURPLE and DODGE (1965) have elaborated R_r extensively, to include lateral and self-inhibition. Most of our attention will focus on the process whereby *excitation* ultimately affects voltage measurements made with a microelectrode with its tip inserted inside the cell.

The voltage swings from its resting value, E_r, towards the smallest value measured (during an action potential), E_g. If R_g is taken as infinite in the dark and has a minimum value of zero, these voltages may be equated with the emf's of the two cells in Fig. 1. Experiments with light and with injected currents led FUORTES (1959a) to estimate the value of the "dark resistance" as 4.7 meg Ω, while RUSHTON (1959) interpreted FUORTES' data to estimate 5.5 meg Ω. We shall

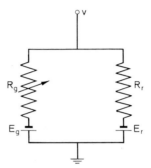

Fig. 1. Fuortes's model for eccentric cell generator potential

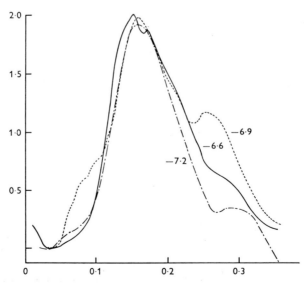

Fig. 2. Average generator potential response per photon

use a rounded-off value of 5 meg Ω as representative, and attribute it to R_r, in the dark-adapted state.

A weak light flash leads to voltage changes (the "generator potential") like those shown in Fig. 2, on the average. "Weak" means as little as a single photon. The curves shown are estimated to have originated in about 1/3 of the flashes — i.e. a photon was absorbed from only one flash in three. Response reaches a peak 150 msec or more after the 5 msec flash. Although the photochemical processes following a flash are well understood (WALD, 1961a, b, 1963) their time-course *in situ* has not been related to that of the response.[2] Fortunately, the response wave-form is a clue to several general possibilities.

Models of the Generator Potential

FUORTES and HODGKIN (1964) first showed that responses of the type of Fig. 2 are well approximated by an expression of the form

$$v(t) = v_0 (\lambda t)^n \exp(-\lambda t),\qquad(1)$$

with n, λ, constants. This form is met in electrical network theory and in statistics. In the former such an equation represents the impulse response of a cascade of n low-pass filters. In the latter, it is the Poisson distribution. Each of these, as we shall see, can be interpreted in terms relevant to photoreception, but the postulated processes differ widely despite their common output waveform. Further observations have led to a rather convincing resolution, as will be described below.

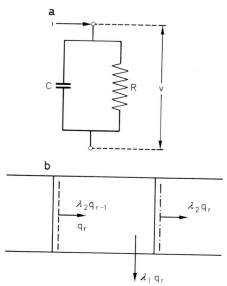

Fig. 3. *a* Network model of one of several stages. *b* Compartment equivalent of network model. If concentration inside the compartment is q_r, it "leaks" at a rate $\lambda_1 q_r$. The input rate is proportional to the concentration in the previous compartment, $\lambda_2 q_{r-1}$, but does not consist of an actual flow from it

---

[2] R. A. CONE and W. L. PAK, Chapter 12 Vol. 1 of this Handbook, discuss the relation between pigment changes and electrical events, in particular the "early receptor potential".

To understand this latest model, the earlier ones may be used to supply an elementary foundation. The network model is obvious (see Fig. 3a): An RC parallel network fed by a current source has the generic equation

$$i = C \frac{d\boldsymbol{v}}{dt} + \frac{\boldsymbol{v}}{R} \tag{2}$$

where $\boldsymbol{i}$ is the current into the network, $\boldsymbol{v}$ is the voltage across either the capacitor or the resistor, and C and R are their respective capacitance and resistance. An input current lasting the brief time Δt produces a voltage $\boldsymbol{v}(\Delta t) = \frac{i\Delta t}{C}$. In the absence of subsequent input, $\boldsymbol{v}$ will decay according to the law

$$\boldsymbol{v}(t) = \frac{i\Delta t}{C} e^{-t/CR} \tag{3}$$

This is the "impulse response" of a single stage RC lowpass-filter.

A cascade of such filters may be arranged by having the voltage of one stage, $\boldsymbol{v}_{r-1}$, control the current into the next stage proportionately, $\boldsymbol{i}_r = \mu \boldsymbol{v}_{r-1}$. Given the impulse $i_1 \Delta t$ into the first stage, the input to the second stage becomes $i_2(t) = \mu \boldsymbol{v}_1(t) = \frac{\mu i_1 \Delta t}{C} e^{-t/CR}$. Substituting into Eq. (2), and solving,

$$\boldsymbol{v}_2(t) = \frac{i_1 \Delta t}{C} \frac{\mu t}{C} e^{-t/CR}.$$

More generally, for the r th such stage,

$$\boldsymbol{v}_r(t) = \frac{i_1 \Delta t}{C} \left(\frac{\mu t}{C}\right)^{r-1} \frac{e^{-t/CR}}{(r-1)!}$$

or,

$$\boldsymbol{v}_r(t) = i_1 \Delta t \, \lambda R^r (\mu \lambda t)^{r-1} e^{-\lambda t}/(r-1) \,! \tag{4}$$

writing $\lambda = 1/CR$. Eqs. (1) and (4) have the same form.

But the generic Eq. (2), of the network model, applies in endless situations where growth and decay are involved. For instance, $\boldsymbol{v}_r$ might represent the concentration of an enzyme catalyzed by its precursor, of concentration $\boldsymbol{v}_{r-1}$. In this case, $\boldsymbol{i}_r$ would represent the rate of input of $\boldsymbol{v}_r$ due to $\boldsymbol{v}_{r-1}$, with μ as the conversion coefficient of enzyme $(r-1)$ into r. (More on this below.) The final output in response to a flash would be a smooth function of time, of a form like Fig. 2. Other n-stage processes may be postulated[3], all leading to Eq. (4).

The same equation is found on postulating a statistical process (LEVINSON 1966). The process assumed in the reference now seems inadequate to account for all of the most recent observations, but will serve as an elementary introduction to a more elaborate model to be discussed later. Also, for expository purposes, a concrete, specific, process will be suggested, rather than the general one given in the reference (with no intention of urging its likelihood).

Suppose absorption of a photon leads to the formation of an active site on the receptor membrane, such that sodium ions accumulate on it up to a certain number, n.

When this requisite number is finally present, the cluster momentarily behaves like a short-circuit through the membrane, thus raising its average conductance.

---

[3] Results would differ significantly if $\boldsymbol{v}_{r-1}$ did not "control" $\boldsymbol{i}_r$, but if instead $\boldsymbol{i}_r$ constituted a leak of actual charge from C_{r-1}. In that case, Eq. 2 would not apply in the first place (BORSELLINO et al., 1965).

I visualize it as exchanging Na^+ ions with the ambient solution on either side very readily, so that it acts as a "hole" in the membrane. (Unfortunately, I have no idea of the quantities involved here. As will be seen below, each cluster will have to assist in the passing of perhaps over a million electronic charges. It may be that this is too many to consider passing as Na^+ ions proper.) This problem is attacked in detail by BASS and MOORE (1970), but they give no theory for the temporal course of the "bumps". The depolarization voltage is taken to be proportional to the current constituted by the passage of these charges through the membrane.

If, on the average, Na^+ ions hit the site λ times per second, then t seconds after a site is activated it may be expected to have accumulated λt ions. However, this is only the expectation value. The probability that an active site will have accumulated some other number of ions, r, is given by the Poisson distribution, (e.g. WOODWARD, 1953)

$$P_r(\lambda t) = (\lambda t)^r e^{-\lambda t}/r!$$

If Q active sites are generated by a given flash, the number of sites ready to receive their r th ion at t seconds after the flash will be given by the number with $r - 1$ already accumulated: $Q P_{r-1}(\lambda t)$. Since these sites are each hit by Na^+ ions λ times per second, they will acquire their r th ion at this rate, and the clusters of r ions will be formed at this rate:

$$R = \lambda Q P_{r-1}(\lambda t) \quad \text{per sec}$$
$$= \lambda Q (\lambda t)^{r-1} e^{-\lambda t}/(r - 1)! \quad \text{per sec} \tag{5}$$

Eq. (5) is formally equivalent to the network-derived Eq. (4). The two models are therefore indistinguishable in simple stimulus-response experiments.

This apparent indistinguishability is a tempting challenge to the experimenter. Even though both models may be inadequate they provide a point of departure for further investigation.

On the multi-stage model, as presented, the generator potential is a continuous function of time. On the statistical model the potential is the result of discrete, quantal, events — only their distribution is continuous. A search for such events, and for their statistics, is indicated.

Generator Potential "Bumps"

"Discrete events" in the otherwise slow generator potential have been known for some time, (YEANDLE, 1957; FUORTES, 1959b; FUORTES and YEANDLE, 1964; ADOLPH, 1964), and more felicitously, called "bumps" among those who see them in their laboratories. However, no reliable inferences could be made at first as to the relationship between the slow potential and the bumps; in particular, it was not clear that the slow potential was simply the sum of a relatively large number of bumps (RUSHTON, 1961). It seems there are two kinds of bumps (ADOLPH, 1964; DOWLING, 1968), one, larger, quick-rising, and with longer latency than the other. Many of the large bumps were greater than the sustained slow potential itself (FUORTES, 1959b). A specific investigation of the response to dim flashes was undertaken to clear up these difficulties, by BORSELLINO and FUORTES (1968).

Again, the double distribution of large, quick, but long-delayed responses and of small, slow, extended and relatively immediate ones was found, Figs. 4, 5.

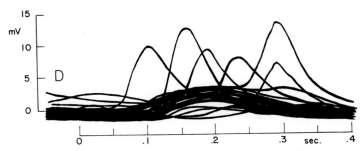

Fig. 4. Superimposed photographic records of generator potentials following flashes. Several small and slow waves are superimposed, approximately, while the larger and faster rising waves are more dispersed in time

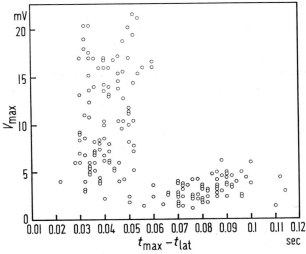

Fig. 5. Relation between peak amplitude and rise time of discrete waves. Two different populations are apparent

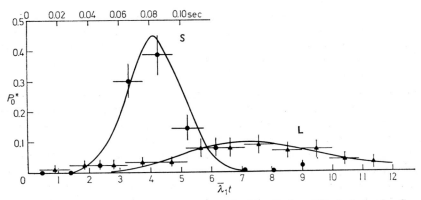

Fig. 6. Latency distribution of discrete waves evoked by flashes. Filled circles for S-waves, triangles for L-waves

Further, the large responses were relatively unpredictable in latency, while the small responses came on much more uniformly, Figs. 4, 6. These differences suggested that a response of the first type consisted of a single event and one of the second type was the sum of a large number of events (though smaller, the latter was more reproducible).

New Model of a "Bumpy" Generator Potential

BORSELLINO and FUORTES (1968) therefore investigated with great care the statistics of both kinds of responses. First, they considered how a multiple-event response might occur. For this they returned to an n-stage model, or, to use their term, an n-compartment model. On this model, different "particles" are produced in each compartment of a sequence. Photons produce the first particles. (The possibility of several particles per photon, rather than just one, is included.) Each of these, in turn, leads to the production of several particles in the next compartment, and so on, Fig. 3b. Finally, in the n th compartment there appears some number of particles per photon absorbed. ("Compartments" need not be taken literally — a perfectly valid model is that the particles are successively of different species, perhaps molecules of different enzymes each catalyzing the formation of several of the next species. WALD, 1961 b; FUORTES and HODGKIN, 1964; DE VOE, 1967; BORSELLINO et al., 1967.) Thus, Eq. (2) applies directly, and v_r may now represent the average number of particles in the r th compartment, q_r, and the current $i_r = \mu v_{r-1}$ may be replaced by μq_{r-1}. Transliterating Eq. (2) in this way into the Borsellino-Fuortes notation for the r th compartment:

$$\mu \, \mathbf{q}_{r-1} = C \frac{d\mathbf{q}_r}{dt} + \frac{1}{R} \, \mathbf{q}_r$$

or,

$$\frac{d\mathbf{q}_r}{dt} = \frac{\mu}{C} \, \mathbf{q}_{r-1} - \frac{1}{RC} \, \mathbf{q}_r = \lambda_2 \mathbf{q}_{r-1} - \lambda_1 \mathbf{q}_r \, (r = 1, \ldots, n) \tag{6}$$

where the λ's are rate constants. Decay rate (No. of particles vanishing per second per particle present) is symbolized by λ_1; accretion rate (No. of particles appearing per second per particle in the previous compartment) is represented as λ_2.

Obviously it is quite unrigorous to "replace" the continuous variable v_r by the discrete q_r. BORSELLINO and FUORTES allowed for discreteness, and took the statistics of the particles into account. For their rigorous treatment the original paper should be consulted. I shall indicate here more verbally than mathematically how their results may be understood.

Picture the photons of a flash of infinite area falling on an infinite array of receptors. On the average, M photons fall on a receptor, and instantly produce an average of $q_1(0)$ particles. Subsequently, two things happen: The new particles disappear at random at the average rate λ_1 per second, but, while they last, each one of them leads to the creation of a second type of particle at the rate λ_2 per second. (One may imagine that a "particle" acts like the "active site" of the statistical model mentioned above. A "rain" of two types of metabolically available other particles strikes them — one type leads to a "kill", the other to a "birth".) This process is repeated for each new particle, leading ultimately to a final (n th) type. Eq. (6) applies to the creation of the r th type of particle from the $(r-1)$th, in the average receptor. The solution for the ultimate particles is

$$\boldsymbol{q}_n(t) = K M (\lambda_1 t)^{n-1} e^{-\lambda_1 t}/(n-1)!, \tag{7}$$

for the average receptor (K is a photon conversion efficiency factor coined by this writer. BORSELLINO and FUORTES devote considerable attention to it).

Eq. (7), with generator potential understood for $\boldsymbol{q}_n(t)$, is thus of the same form as 1, 4 and 5, and provides a third model of the response, indistinguishable, by relatively direct means, from the n-stage model and the one-stage stochastic model, while combining aspects of both.

However, due to the random way in which the particles are created, the number of particles present in some other receptor may differ from the expected value, $\boldsymbol{q}_n(t)$. BORSELLINO and FUORTES show that the probability of finding x of the n th particles when the average is $\boldsymbol{q}_n(t)$ is also given by the Poisson distribution,

$$P_{n,x}(t) = \{\boldsymbol{q}_n(t)\}^x e^{-\boldsymbol{q}_n(t)}/x! \tag{8}$$

This distribution is a function of λ_2 and λ_1, by Eq. (7). Thus, in principle, repeated measurements should make it possible to discover, through the statistics, just how many of the ultimate particles are created per photon. $\lambda_2 = \lambda_1$, is a limiting case, the analog of the one treated by LEVINSON, while $\lambda_2 > \lambda_1$ implies increase by multiplication of particles each time new ones are formed, since accretion rate is greater than decay rate. In the former case, one photon leads to one "site" which passes current for a short time — "short" relative to the spread, in time, of site activation following the flash. This brief current creates one "bump". In terms of the compartment model, only one particle per photon reaches the nth compartment, and then generates its "bump". In the latter case, one photon leads to a shower of particles in the nth compartment, the "minibumps" due to the individual particles being lost within the overall bump constituting the generator potential, described by Eq. (7). Certainly no "minibumps" can be discerned within the slower, smaller responses of Fig. 4 — the ones assumed to consist of multiple-event responses. Of course, when many responses are averaged any underlying uncorrelated substructure is bound to be wiped out anyway. But it would seem to be unlikely that it would ever show up, even without averaging, given the difficulties of extracting the generator potentials from the noise.

Experimental Tests of the New Model

It is implicit in BORSELLINO's and FUORTES' formulation that each of the nth particles contributes a voltage "minibump", so that Eq. (7) for $\boldsymbol{q}_n(t)$ is taken proportional to the generator potential. One test of the model consists of the time to first occurrence of the minibumps, or, since they are not easy to discriminate when many overlap, the time to onset of a bump in response to a flash. If many minibumps constitute a bump, its onset will usually begin earlier than if only one minibump constitutes the whole bump. BORSELLINO and FUORTES calculated, as a function of time, the "probability of arrival of the first particle in compartment n" when responses were taken to consist of 1, 2, 5, 10, 20, 50, and 100 particles, Fig. 7. Measured latencies fell into two groups, as already indicated, Fig. 6. Data L, for the "large, fast-rising" responses indicate that they occur late and with wide scatter, while

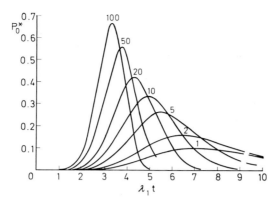

Fig. 7. Theoretical probability of arrival of first particle in compartment n as a function of time. These curves are calculated for $n = 10$ and $\lambda_1 = 1$. The parameter is the total number of particles arriving at the last compartment, taking into account multiplication of particles in each compartment

data S, for the "small, slow-rising" responses show more reproducible, shorter latencies. Solid curve L is that calculated on the assumption of one output particle per input photon, while curve S is for 25.

It seems as though we must indeed accept the existence of a complex response, comprising at least two types. Another test, not yet performed, would consist of finding the distribution of bump amplitudes, which must be a function of the number of minibumps per bump and of their timing. Unfortunately, this distribution is bound to be masked by the distribution due to different possible locations of photon absorption (BASS and MOORE, 1970) which would yield different voltages in accord with local differences in conductance as well as differences in distance from the recording microelectrode. Yet the relevant distribution lurks within the amplitude data and it would seem to be wasteful to take averages only.

However there is one worthwhile bit of information that should come out of the amplitude data readily. It does not bear on the minibump distribution, but on the way the bumps themselves interact. BORSELLINO and FUORTES (1968) have shown that linear summation of response ceases to hold if the number of absorbed photons from one flash exceeds even one or, at most, four. This is surprising, unless one assumes that every photon is absorbed at nearly the same site in the retinular cell. Latencies (Figs. 6 and 7) spread far less than bump width. Therefore when a particular flash elicits several bumps, they will be more or less simultaneous. These bumps should not affect each other's amplitude, if the gain control postulated in several studies applies (FUORTES and HODGKIN, 1964; MARIMONT, 1965; SPERLING and SONDHI, 1968). In their models gain is not reduced (adaptation) until late in the flash response, since the control signal is fed back from a late stage in the response process. SPERLING and SONDHI (1968) however, also suggest a feedforward control signal which might serve to reduce the response early on. Thus, what is suggested is that bump amplitude be studied as a function of intra-bump interval. The expectation is that the second of any two closely-spaced bumps should be smaller by an amount which is a function of their interval. This function could be indicative of the nature of the gain control. (Bumps with various spacings are best recorded not with flashes but with steady, weak illumination.)

Discussion

Several questions for future research arise: How is it that two different kinds of response to the same stimulus can occur in the same cell? If L-waves represent single-particle responses and S-waves 25-particle responses, why are there no (or undetectably few) 2-particle, 3-particle, etc. responses? Why are the L-waves more energetic than the S-waves, when the former seem to be due to single particles, while the latter are generated by about 25? In volt-seconds, L-waves appear to be about twice as intense. A single particle producing an L-wave is therefore about 50 times as effective as one participating collectively in producing an S-wave. In terms of energy (given by $\int (\text{volts})^2 \, dt/\text{resistance}$), if we assume resistance relatively constant during these small generator potential changes, L-wave particles are 200 times as effective as the others. It has been suggested (Adolph, 1964; Fuortes and Poggio, 1963; Dowling, 1968) that L-waves are regenerative. This would explain their relatively rapid rise time, but not their latency distribution, at first glance. Fig. 7 shows that L-waves may begin at any time during an S-wave, but it is difficult to understand how S-waves can give rise to L-waves on the downward slopes of the former. Assuming that regeneration is the result of the crossing of some threshold value by an S-wave, such a crossing occurs first on the rising phase. It becomes necessary to assume that the threshold value itself fluctuates, on a time-scale short compared with an S-wave, and with a comparable amplitude of fluctuation. More detailed study of the effect of hyperpolarization (Adolph, 1959) would possibly be fruitful.

Another question all of our models have yet to face is the problem of gain (Wald, Brown, and Gibbons, 1963). Sooner or later we will have to discover the metabolic process whereby the energy in a photon triggers the much greater energy output of the generator potential response. Even so there is a question as to whether it really is the energy gain which is the parameter of interest. Borsellino and Fuortes address themselves to the number of particles generated per photon, an attractive and useful gain criterion when it appears that the number is considerably greater than unity. But there are other "gains" in the transduction process. The ratio between any quantities of the same dimensionality may be a legitimate measure of gain. Once we discover the particular parameter which is *used* by the visual system, of course, that will become the quantity of particular interest. Meanwhile we are limited only to those quantities which are measured, explicitly or implicitly, in experiments.

Explicitly, we have for the stimulus, its light flux and wavelength, and for the response, its voltage and duration. From the former, the number of photons per flash or the energy per flash, may be derived. From the latter much more is inferrable (although in both cases really only imagination sets limits on inference). Voltage itself (or its integral in volt-seconds) is of possible interest, since it seems logarithmically related to stimulus luminance and this suggests that it may be related to sensation.

Fig. 4 contains the information needed for an estimate of the average energy of generator response to a single photon. If we take this energy to be expended as a current flow through the cell membrane resistance, R_r, the energy in the response may be written as

$$\mathscr{E} = \int_0^\infty \frac{v^2(t)\,dt}{R_r}. \tag{9}$$

where $v(t)$ is the generator potential. An explicit integration can be performed, since $v(t)$ is well approximated by an equation of the form of 7. Three constants need to be found and they may be found from Figs. 2 and 4, and Eq. (7): The maximum response voltage, v_{max}, λ_1 and n. Differentiation of 7 shows that the maximum is reached at a time for which $\lambda_1 t = (n-1)$. Substituting this into Eq. (7) we obtain the maximum value, q_{max}. Dividing this into 7, and taking v proportional to q we then obtain

$$v(t) = v_{max} (\lambda_1 t)^{n-1} e^{-\lambda_1 t} / (n-1)^{n-1} e^{-(n-1)} \tag{10}$$

as the analytical description of Fig. 2. v_{max} is about 3.3 mV for S-waves, from Fig. 4. There remain λ_1 and n. The latter is best taken as 10, following BORSELLINO and FUORTES. I have estimated λ_1 to fit Fig. 2 as about 54 per sec., while from BORSELLINO and FUORTES' Fig. 9 it would appear that they found a value of 52.5 gave the best fit — a check which is better than the precision of my estimate. We shall use $\lambda_1 = 53$ per sec.

Integrating 9, with 10 substituted for $v(t)$, and using STIRLING's approximation,

$$\{2(n-1)\}! \approx \{2(n-1)\}^{2(n-1)}\, e^{-2(n-1)} \sqrt{2\pi 2(n-1)}$$

we obtain for the energy $\dfrac{v_{max}^2 \sqrt{\pi(n-1)}}{R\lambda_1}$ joules $= 2.2 \times 10^{-13}$ joules. The energy in a

photon of 6000 A is 3.3×10^{-19} joules. The energy gain is therefore a factor of nearly 10^6. This estimate concerns an S-potential, and the energy gain for the single particle postulated for the L-potential is about 8 times as great. Both of these energy gains must ultimately be understood.

It is not enough to know that there is ample energy available in the resting potential difference between the inside and outside of the eccentric cell. Even a single action potential, with 10 times the voltage, though lasting only 1/10th as long, consumes an order of magnitude more energy; and action potentials may occur at 100 per sec at onset of illumination.[4] The gain in a vacuum tube or transistor is not to be understood in terms of the size of the battery which supplies the operating power. Power gains of 10^6 in single transistor stages are easily obtained, but from the results considered here there is no way of telling whether this analogy is even remotely applicable.

It may be noted in passing that if the gain of 10^6 is assumed to be distributed uniformly among 10 successive stages (or compartments, or enzymes) each stage

---

[4] In these energy calculations changes in membrane resistance have been ignored. In the case of the generator potential this would lead to negligible error, since even if the potential were entirely due to a proportional resistance drop this would only require a drop of 10%. For an action potential this argument is weaker by a factor of 2 — but the order of magnitude is enough for the argument.

need have a gain of only 4. For comparison, the gain per stage in particles is about 1.4 for a last-stage yield of 25.

Whether gain should be thought of in terms of particles or energy is a question, but not a resolvable one — until more explicit features of the model are substantiated, both points of view will have to be borne in mind for their possible heuristic value.

Another interesting definition of gain is possible, in terms of the number of electronic charges transferred through the cell membrane per photon absorbed. This may be found by integrating the current $v(t)/R_r$ for the response (10), to obtain

$$v_{max} \sqrt{2\pi(n-1)}/R_r\lambda_1 \text{ coulombs}$$

$$\approx 10^{-10} \text{ coulombs}$$

At 6.24×10^{18} electrons per coulomb, this amounts to about 6×10^8 electrons per absorbed photon (cf. HAGINS, 1965).

There is therefore no possibility that the quantity of charge transferred is generated by direct one-for-one photoionization or even ionization by the particles postulated in the models discussed. The quantity of charge is so large that the relevant process can only be a conductivity change or some equivalent thereof.

Other Waveform Tests of the Generator Response Model

Eq. (7) [or (10)] represents the *impulse response* of the eccentric cell to a very weak flash. An impulse response also implies the *frequency characteristic* of a linear system, i.e. given the impulse response, the response to sinusoidal luminance modulation (as a function of the modulation frequency) may be found by means of a Fourier transformation (STUART, 1961, p. 85). Linearity is crucial, and, as already noted, has been confirmed for flashes of up to 40 photons per ommatidium (or up to 4 photons absorbed, BORSELLINO and FUORTES, 1968), i.e. the average

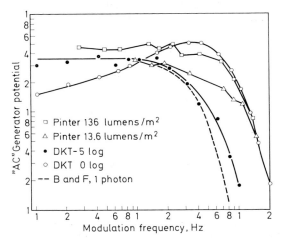

Fig. 8. Generator potential frequency characteristics as measured by two groups and as computed from BORSELLINO and FUORTES's results (as explained in text)

response amplitude is proportional to the number of photons absorbed, up to 4, then decreases (with progressively more of the decrease occurring in the later portions of the response)[5].

Given sufficient linearity, BORSELLINO's and FUORTES' results may be compared with frequency characteristics, which have been obtained for both retinular cells (PINTER, 1969) and eccentric cells (DODGE, KNIGHT, and TOYODA, 1968), Fig. 8. These constitute measurements of the sinusoidal amplitude fluctuations of the generator potential when the stimulating luminance itself is modulated sinusoidally around a fixed average luminance, at each of a number of modulation frequencies. Because of the need of at least a small steady luminance for modulation to be possible, the receptor in these experiments cannot be completely dark-adapted, but the condition can be approached. The "−5 log" curve of DODGE, KNIGHT, and TOYODA and the 13.6 lumens/m² curve of PINTER are for the lowest luminances they used. For comparison, one curve each is given for a higher luminance. As plotted, the generator potentials have been "normalized" by arbitrary factors to bring the low-frequency ends of the low-luminance curves approximately to the same level. The frequency scales have been preserved as published.

All the curves show steep high-frequency cut-offs approaching roughly equal slopes. Other comparisons must be qualitative − not only are these results for different preparations, they are for different cells. Nevertheless one is tempted to speculate that the apparently higher frequency cut-offs of the Pinter data are due to the absence of an intervening synapse. Surely someone must do a comparison study of eccentric and retinular cell frequency characteristics in the *same* ommatidium. Also, the relatively flat low-frequency part of the retinular data (duplicated at still higher luminances) suggests that the effect of inhibition in the eccentric cell does not extend back to the retinular cell. But without comparative data that is just speculation.

To compare the BORSELLINO and FUORTES results with these, we note that Eq. (10) gives, by a Fourier transformation (or directly from the n-stage model) the following relation for generator potential amplitude produced by constant-amplitude luminance modulation:

$$v_{\max} = v(0)\,(1 + \omega^2/\lambda_1^2)^{-n/2} \tag{11}$$

where $\omega = 2\pi f$ with f = modulation frequency in Hertz, and $v(0)$ is the response amplitude at frequency zero. Note that the value of λ_1 is that taken from the impulse response (53 per sec). This is the relation plotted by the dashed line in Fig. 8, with $v(0)$ adjusted to match the others.

It will be seen that the cut-off is at a little lower frequency than the DODGE, KNIGHT and TOYODA curve: whether this is due to use of data from different animals or due to more dark-adapted animals is moot, but the difference *is* consistent with the latter possibility, as the BORSELLINO and FUORTES data came from isolated weak flash stimuli rather than continuous illumination. The differ-

---

[5] Unfortunately, one can speak only of the *average* response — it has not been possible to collate the responses to one photon separately from those to two, or to three, etc. Had that been possible, one should have expected to find integral amplitude ratios, a fascinating outcome — if only it could be observed!

ence suggests that λ_1 was about 25 % higher for the bumps observed by Dodge, Knight and Toyoda than those observed by Borsellino and Fuortes.

Dodge, Knight and Toyoda did not confine themselves to the frequency characteristics of the eccentric cell response, but investigated their relation to bump shape also. However, they did not do this by recording the wave-shapes of individual responses to very weak flashes, rather, they measured the autocorrelation of relatively long records of the "noisy" generator potential response to sustained dim light, then showed that the frequency characteristics lead to correlation functions that match the measured ones. In effect, then, the bump shape was implicit in their measurements. *But the correspondence shown in Fig. 8 implies that these implicit shapes and the explicit ones of Borsellino and Fuortes are rather similar.*

Pinter (1966) also applied another test to the Fuortes-Hodgkin (1964) multi-stage, model: he measured the phase-frequency characteristic of the generator response and compared it with that predicted by the model, and found acceptable agreement. This is an excellent independent test of a model, but offers the same frustrating difficulty as does the gain-frequency characteristic, namely, that the amplitude of the response drops so rapidly in the critical high-frequency region that measurements become uncertain.[6] Phase measurements would otherwise be very promising: the phase-shifts are large, and, at high frequencies (beyond $\lambda_1/2\pi$ Hz) they differ strongly depending on the model one uses. For n-stages there is a maximum possible shift of $n \times 90°$ (i.e. 90° per stage) at the highest frequency. On another possible model (Kelly, 1969), viz. that the photo-product is delayed in producing the generator potential by having to diffuse towards the locale of the "generator", phase lag should increase as the square root of frequency, *without limit*. Since there is evidence (Veringa, 1964; Veringa and Roelofs, 1966) that in the human eye phase lag in the retina follows this law, it would be interesting to see if it holds in *Limulus*. No doubt the difficulties of measurement will yet be overcome.

Comparing human data with those of *Limulus* is, of course, questionable; all the more so in view of recent evidence that vertebrate receptors respond to light with hyperpolarization rather than with depolarization (Tomita, 1968). Despite this *caveat* the comparison cannot be avoided. The main spur is provided by the qualitative similarity in the frequency *vs.* attenuation characteristics of man and *Limulus* (*cf.* Figs. 5 and 6 in de Lange, 1958, and Fig. 2 in Pinter, 1966, and Fig. 1b in Dodge, Knight and Toyoda, 1968), especially in their high-frequency slopes, though they do differ by a factor of about two in frequencies of comparable attenuation, man being the faster. Whether such comparisons are valid will surely soon become evident, when the new techniques (Tomita, 1968; Kaneko and Hashimoto, 1967; Werblin and Dowling, 1969) are applied to the measurement of the dynamics of the hyperpolarizing generator potentials.

Nonlinear Generator Potentials

Linear responses, being simple, offer promise of easy interpretation, and hope of discovery of underlying processes. But it does not follow that nonlinear response

---

[6] High frequency measurements would also be highly desirable because they are less contaminated by inhibitory effects.

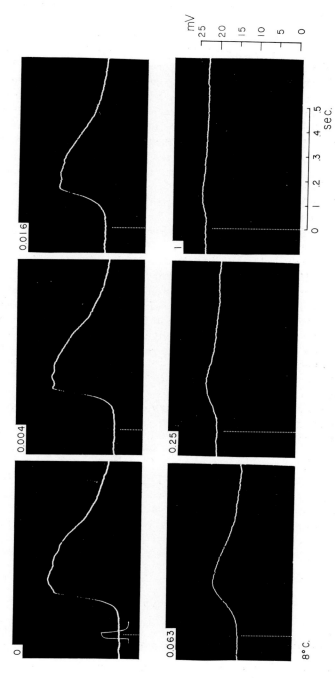

Fig. 9. Adaptation of the generator potential. The number in the left-hand corner of each record gives the intensity (in arbitrary units) of the steady adapting light. Equal test flashes were applied about 20 sec after turning on the adapting light

is completely intractable and may not even be of help on the same problems. A brief, but I hope useful, summary follows on existing interpretations of the non-linearity of response. When the ommatidium being studied is steadily illuminated, the generator potential response to a superposed weak flash is reduced and speeded up (Fig. 9). This result can be accounted for with some precision by hypothesising, that λ_1 increases as a linear function of generator potential (Fuortes and Hodgkin 1964; Levinson, 1966). Even closer correspondence between experimental and predicted response waveform is obtained by assuming λ_1 to depend on the penultimate ($n - 1$th rather than nth) stage response (Marimont, 1965).

From Eq. (10) it can be seen that the time scale of the response depends on λ_1, as time appears multiplied by λ_1. Increasing λ_1 speeds response. Further, absorbed in v_{max} is a factor of $\lambda_1^{-(n-1)}$, which implies that v_{max} drops strongly with increasing λ_1, i.e. with increasing steady illumination. Indeed, for high luminances the response approaches a logarithmic function of luminance, a result in accord with the Weber-Fechner Law. However, this success of the Fuortes-Hodgkin model is not conclusive as to the number of stages, n. The gain of the system can be made to depend quite strongly on luminance even though only one or two stages are controlled (Sperling and Sondhi, 1968).

Regardless of how gain is controlled, it seems to be a function of the system's output − i.e. a function of a process subsequent to at least a part of the luminance-to-potential transduction process. All writers on this subject have simply assumed (after Fuortes and Hodgkin, 1964) that one of the products of the process speeds up the rate of decay of the particles active in transduction. This is as far as we have gone. Since we do not yet know what any of the particles, or compartments, or stages are, we do not know how the decay rate can be affected. But it may be useful to know whether it is approximately a linear function of response level.

Summary

The temporal behavior of the generator potential response of eccentric (or retinular) cells has been described. It is attributed to a process having approximately ten successive stages, in each of which the relevant material, referred to as "particles", is somewhat greater than in the previous stage, until a multiplication factor of 25 particles per photon is attained.

Several experimental measures supporting this interpretation were adduced, but chemical and thermal experiments were omitted.

The problem of explanation of the high energy gain (not 25, but more like 10^6) between photon and generator potential bump was emphasized, and left for the reader.

References

Adolph, A. R.: Spontaneous slow potential fluctuations in the *Limulus* photoreceptor. J. gen. Physiol. **48**, 297—322 (1964).
— Discrete slow potentials and threshold-level spikes in the *Limulus* ommatidium. Vision Res. **11**, 371—376 (1971a).
— Recording of optic nerve spikes underwater from freelymoving horseshoe crab. Vision Res. **11**, 979—984 (1971b).

BARLOW, R. B., Jr., KAPLAN, E.: *Limulus* lateral eye: properties of receptor units in the unexcised eye. Science **174**, 1027—1028 (1971).

BASS, L., MOORE, W. J.: An electrochemical model for depolarization of a retinula cell of *Limulus* by a single photon. Biophys. J. **10**, 1—19 (1970).

BORSELLINO, A., FUORTES, M. G. F., SMITH, T. A.: Visual responses in *Limulus*. Cold Spr. Harb. Symp. quant. Biol. **30**, 429—443 (1965).

— — Responses to single photons in visual cells of *Limulus*. J. Physiol. (Lond.) **196**, 507—539 (1968).

DODGE, F. A., Jr., KNIGHT, B. W., TOYODA, J.: Voltage noise in *Limulus* visual cells. Science **160**, 88—90 (1968).

DOWLING, J. E.: Discrete potentials in the dark-adapted eye of *Limulus*. Nature (Lond.) **217**, 28—31 (1968).

FUORTES, M. G. F.: Initiation of impulses in visual cells of *Limulus*. J. Physiol. (Lond.) **148**, 14—28 (1959a).

— Discontinuous potentials evoked by sustained illumination in the eye of *Limulus*. Arch. ital. Biol. **97**, 243—250 (1959b).

— HODGKIN, A. L.: Changes in the time scale and sensitivity in the ommatidia of *Limulus*. J. Physiol. (Lond.) **172**, 239—263 (1964).

— POGGIO, G. F.: Transient responses to sudden illumination in cells in the eye of *Limulus*. J. gen. Physiol. **46**, 435—452 (1963).

— YEANDLE, S.: Probability of occurrence of discrete potential waves in the eye of *Limulus*. J. gen. Physiol. **47**, 443—463 (1964).

HAGINS, W. A.: Electrical signs of information flow in photoreceptors. Cold Spr. Harb. Symp. quant. Biol. **30**, 403—418 (1965).

KANEKO, A., HASHIMOTO, H.: Recording site of the single cone response determined by an electrode marking technique. Vision Res. **7**, 847—851 (1967).

KELLY, D. H.: Diffusion model of linear flicker responses. J. opt. Soc. Amer. **59**, 1665—1670 (1969).

DE LANGE, H., Dzn.: Research into the dynamic nature of the human fovea-cortex system. J. opt. Soc. Amer. **48**, 777—784 (1958).

LEVINSON, J.: One-stage model for visual temporal integration. J. opt. Soc. Amer. **56**, 95—97 (1966).

MARIMONT, R.: Numerical studies of the Fuortes-Hodgkin *Limulus* model. J. Physiol. (Lond.) **179**, 489—497 (1965).

PINTER, R. B.: Sinusoidal and delta-function responses of visual cells of the *Limulus* eye. J. gen. Physiol. **49**, 565—593 (1966).

PURPLE, R. L., DODGE, F. A.: Interaction of excitation and inhibition in the eccentric cell in the eye of *Limulus*. Cold Spr. Harb. Symp. quant. Biol. **30**, 529—537 (1965).

RUSHTON, W. A. H.: A theoretical treatment of Fuortes' observations upon eccentric cell activity in *Limulus*. J. Physiol. (Lond.) **148**, 29—38 (1959).

— In: McELROY, W. D., GLASS, H. B. (Eds.): Light and Life. Baltimore: Johns Hopkins Press 1961.

SMITH, T. G., STELL, W. K., BROWN, J. E.: Conductance changes associated with receptor potentials in *Limulus* photoreceptors. Science **162**, 454—456 (1968).

— — — FREEDMAN, J. A., MURRAY, G. C.: A role for the sodium pump in photoreception in *Limulus*. Science **162**, 456—458 (1968).

— BAUMANN, F.: The functional organization within the ommatidium of the lateral eye of *Limulus*. In: AKERT, K., WASER, P. G. (Eds.): Progress in Brain Research, Vol. 31. Amsterdam: Elsevier 1969.

SPERLING, G., SONDHI, M.M.: Model for visual luminance discrimination and flicker detection. J. opt. Soc. Amer. **58**, 1133—1145 (1968).

SREBRO, R., YEANDLE, S.: J. gen. Physiol. **56**, 751—767 (1970b).

STUART, R. D.: An Introduction to Fourier Analysis. New York: John Wiley & Sons 1961.

TOMITA, J.: Electrical response of single photoreceptors. Proc. IEEE **56**, 1015—1023 (1968).

VERINGA, F. J.: Phase shifts in the human retina. Nature (Lond.) **197**, 998—999 (1963).

— Electro-optical stimulation of the human retina as a research technique. Doc. Ophthalm. **18**, 72—82 (1964).

Veringa,F.J., Roelofs,J.: Electro-optical interaction in the retina. Nature (Lond.) **211**, 321—322 (1966).

DE Voe,R.D.: A nonlinear model for transient responses from light-adapted wolf spider eyes. J. gen. Physiol. **50**, 1993—2030 (1967).

Wald,G.: General discussion of retinal structure and visual function. In: Smelser,G.K. (Ed.): The Structure of the Eye. New York: Academic Press 1961a.

— The molecular organization of the visual system. In: McElroy,W.D., Glass,B. (Eds.): Light and Life. Baltimore: Johns Hopkins Press 1961b.

— Brown,P.K., Gibbons,J.R.: The problem of visual excitation. J. opt. Soc. Amer. **53**, 20—35 (1963).

Werblin,F.S., Dowling,J.E.: Organization of the retina of the mudpuppy, *Necturus maculosus*. II. Intracellular recording. J. Neurophysiol. **32**, 339—355 (1969).

Wolbarsht,M.L., Yeandle,S.J.: Visual processes in the *Limulus* eye. Amer. Rev. Physiol. **29**, 513—542 (1967).

Woodward,P.M.: Probability and Information Theory, p. 11. London: Pergamon Press 1953.

Yeandle,S.: Studies on the Slow Potential and the Effects of Cations on the Electrical Responses of the *Limulus* Ommatidium (with an appendix on the quantal nature of the slow potential), Doctoral Dissertation, The Johns Hopkins University, 1957.

— Srebro,R.: Latency fluctuations of discrete waves in the *Limulus* photoreceptor. J. opt. Soc. Amer. **60**, 398—401 (1970a).

Chapter 9

Optical Properties of the Compound Eye

By

C. G. BERNHARD, G. GEMNE, Stockholm (Sweden)

and

G. SEITZ, Erlangen (Germany)

With 11 Figures

Contents

The interest in the dioptric apparatus of the arthropod compound eye goes back to the microscopic studies made about 300 years ago by HODIERNA (1644), HOOKE (1665), SWAMMERDAM (1644–1673, see 1737–1738) and VAN LEEUWENHOEK (1698, see 1800 and 1939). One and a half centuries later, MÜLLER (1826) presented his mosaic theory according to which an erect image is formed in the eye due to the alleged circumstance that each ommatidium only monitors light coming from that part of the object directly facing it. On the basis of the optical characteristics, EXNER (1891) differentiated between three main types of compound eyes, one catoptric, and the two dioptric (apposition and superposition) types. In the catoptric type, the light is transmitted to the photoreceptors by total reflection from the walls of the proximal corneal prolongations. This uncommon type of eye (in e.g. the crab, *Phronima*) will not be treated in the present review.

The structural differences between apposition and superposition eyes are in the spatial relations between the lens system and the rhabdom, and in the arrangement of pigment surrounding the ommatidium. In the apposition eye (e.g. of flies, bees and diurnal butterflies), the rhabdom extends to the proximal part of the optical system, and pigment surrounds the ommatidium without gross migratory movements. In the superposition eye (e.g. of fireflies and moths), the rhabdom is located at a distance from the proximal part of the lens system. The narrow light-transmitting structure between the lens system and the rhabdom is surrounded by a sleeve of wedge-shaped pigment cells. The position of the pigment granules is dependent on the prevailing illumination.

Evidence that the honeybee can discriminate colours was presented by von Frisch (1914) on the basis of behavioural experiments. That the bee can orient with the aid of structural characteristics of the surroundings was shown by Hertz (1929—1931). Somewhat later, von Frisch (1949) made the important discovery that the honeybee also uses for its orientation the plane of linearly polarized light of the blue sky. Many morphological and physiological studies have later been made on the visual mechanisms underlying the performances demonstrated in the early behavioural experiments. During the last decades, increasing attention has been paid to the dioptric apparatus. The problems concerning the light transmission in the ommatidial optical components, with their rather limited dimensions, have been attacked by various experimental methods, model experiments and mathematical analyses.

In this review, we describe first the basic optical characteristics of the apposition eye (Chapter I) with special reference to that of the fly (Diptera), which has served as object for many studies. In Chapter II we discuss the structures which subserve the analysis of polarized light in flies (Diptera), crayfish and crabs (Crustacea). Light transmission in the superposition eye of fireflies (Coleoptera) has been treated separately (Chapter III), since the superposition theory was originally based on observations in these insects. A discussion of wave guide transmission and longitudinal pupil in the superposition eye (Chapter IV) deals mainly with recent studies on night moths (nocturnal Lepidoptera). Finally, selective transmission due to the influence of various microcomponents is treated in Chapter V with reference to observations made on butterflies and moths (diurnal and nocturnal Lepidoptera), and on horseflies (Diptera).

I. Light Transmission in the Eye of Flies

The Ommatidium of the Fly — An Aligned Optical System

Among apposition eyes, those of Diptera have been the subject of thorough analysis. Our discussion will deal mainly with *Musca* and *Calliphora*, which offer good opportunities for obtaining data on the optical properties with the application of a wide variety of techniques.

The corneal lens of these insects is composed of thin, negatively birefringent lamellae, the optical axes of which are normal to the lamellar surfaces (Stock-hammer, 1956). Because of the relatively large radius of the surface curvatures in the blowfly *(Calliphora erythrocephala)*, the whole corneal lens may be regarded as a negatively birefringent crystal. The optical axis is, however, parallel to the ommatidial axis; therefore, a double image does not arise in the ommatidium (Seitz, 1969a).

The optical density varies between the different corneal lamellae. This is because of a variation in the arrangement of the constituent molecules. In *Calliphora* (SEITZ, 1968a and b), the refractive index attains a maximum about $1\,\mu$ beneath the outer surface, the lowest value being found near the inner corneal surface (see Fig. 1). It was demonstrated that the refractive index of the whole cornea does not vary in the direction transverse to the ommatidial axis. Therefore, the refractive power of the lens depends only on the curvatures of the distal and proximal surfaces, the refractive index of the lens and of the outer and inner media. In Diptera, like in Hymenoptera, the crystalline cone is isotropic (STOCKHAMMER, 1956; SEITZ, 1968a, 1969a). The crystalline cone of the fly (soft pseudocone, see WEBER, 1954) is an extracellular structure without granular inclusions (PEDLER and GOODLAND, 1965; TRUJILLO-CENÓZ and MELAMED, 1966; SCHNEIDER and LANGER, 1966) and is optically homogeneous as shown in the interference microscope (SEITZ, 1968a). At a wavelength of 546 nm, the refractive index is 1.337

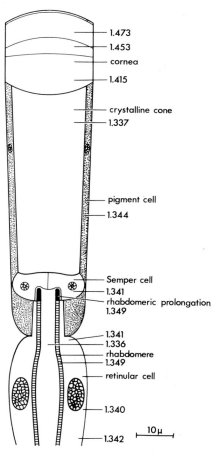

Fig. 1. Schematical drawing of an ommatidium in the center of the *Calliphora* eye. The total refractive index of the corneal lens in $n_e = 1.444$ (from SEITZ, 1968a)

($n_e = 1.337$), i.e. of the same order of magnitude as that of the human eye vitreous.

Such an optical system has to fulfill two conditions in order to yield a one-to-one relationship between an object point and its image point. The refracting surfaces have to be spherical, and their centers must be located on a straight line — the optical axis of the system. These conditions are met in the ommatidia in the central part of the blowfly eye, whereas the axes of the peripheral ommatidia are curved (Washizu, Burkhardt, and Streck, 1964). A central ommatidium can thus be regarded as an aligned optical system with two refracting spherical surfaces, the characteristics of which can be calculated on the basis of geometrical optics. There is a slight variation between ommatidia in the refractive index and magnitude of the surface curvatures of the corneal lenses as well as in the length of the crystalline cone. Since the cornea is a thick lens, located between two media with different optical densities, the focal distances must be calculated with relation to the corresponding principal planes. In the blowfly (Seitz, 1968 b), it was found that these are located in front of the vertex of the outer corneal surface (see Fig. 2).

According to calculations, the focal plane on the image side is located at the tips of the rhabdomeric prolongations. In the focal plane, the radius of the perimeter

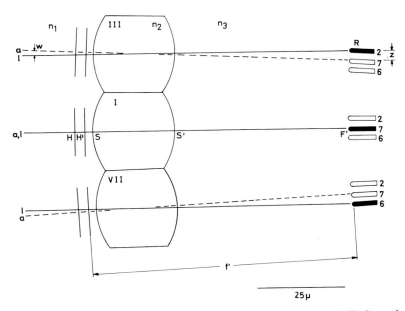

Fig. 2. Schematical drawing of a section through the compound eye of *Calliphora* along the broken line in Fig. 4. The divergence angle between the axes (a) of neighbouring ommatidia is $w = 2.5°$. The cornea is a thick lens with refractive index n_2, located between two media with different refractive indices (n_1 and n_3). The focal distance (f') is measured from the principal point H'. The image focal point (F') lies at the tip of the rhabdomeric prolongation (R). The rhabdomeres Nos. 2, 7, and 6, respectively, are reached by parallel rays (1) coming from a point light source; S and S' the vertices of the corneal lens (from Seitz, 1968 b)

circumscribing them is about 3 μ. Parallel rays farther away from the ommatidial axis will not reach the rhabdomeric prolongations. Using this value of the radius, the opening angle of a single ommatidium − as well as that of a rhabdomere − can be calculated with the aid of the Gauss equation. The calculated opening angle for an ommatidium in the central part of the eye is 6−7°, which is in accordance with that obtained on the basis of measurements of the pseudopupil (BURKHARDT, et al., 1966). For the *Calliphora* rhabdomere − the diameter of which averages 1.2 μ − the opening angle was found to be about 1.2° (SEITZ, 1968a).

"Visual Fields" in Flies and Other Arthropods; the Ommatidium as Optical Unit?

Table 1

Animal	Divergence angle	Opening angle of the ommatidium	Opening angle of the single sense cell	Author	Method
Calliphora	2.5°	7.6°	—	AUTRUM and WIEDEMANN (1962)	ES
	3.2° H 2.7° V	—	3.5° H 3.0° V	WASHIZU et al. (1964)	S, E
	2.3°	7.0°	—	BURKHARDT et al. (1965)	PP
	—	7.0°	1.2°	SEITZ (1968a)	C
Musca	3.9° H 2.3° V	—	7.7°	KIRSCHFELD (1965)	E
	3.9° H	—	2.5—3.0°	VOWLES (1966)	E
	—	—	3.0°	SCHOLES and REICHARDT (1969)	E
Apis	2.8° H 1.4° V	7.2° H 6.7° V	—	AUTRUM and WIEDEMANN (1962)	ES
	—	6.8°	—	KUIPER (1962)	ES
Aeschna	—	2.4°	—	WIEDEMANN (1965)	ES
Libellula	—	—	1.2—1.8°	HORRIDGE (1969a)	E
Dixippus	7.5°	9.8°	—	AUTRUM and WIEDEMANN (1962)	ES
Epargyrus	1.3°	—	2.1°	DØVING and MILLER (1969)	E
Phausis	—	8°	—	SEITZ (1969b)	C
Locusta	2.4° H 1.1° V	4.0° H 3.9° V	—	AUTRUM and WIEDEMANN (1962)	ES
	—	—	6.6° DA 3.4° LA	TUNSTALL and HORRIDGE (1967)	E
Limulus	4—15°	16°	—	WATERMAN (1954)	E
	6.7°	12°	—	KIRSCHFELD and REICHARDT (1964)	E

Abbreviations: C, calculations; DA, dark-adapted; E, electrophysiology; ES, eye-scalps; H, horizontal; LA, light-adapted; PP, pseudo-pupils; S, sections; V, vertical.

Table 1 is a comparison of the values of the divergence angle between the axes of neighbouring ommatidia (interommatidial angle), their opening angle (acceptance angle) and the opening angle of the single sense cell. A comparison between these three values can be made in flies and locusts, where they are all available. Since the opening angle of the ommatidia is larger than the divergence angle, the visual fields of neighbouring ommatidia overlap: this is not in accord with the mosaic theory of Müller (1826). According to this theory, the ommatidium receives light only within the morphological opening angle (the divergence angle). Since the physiological opening angle (acceptance angle) in both flies and locusts is greater than the morphological angle, a one-to-one optical correspondence is not possible in the compound eye of these animals between an object point and its image point.

A comparison of the opening angle of the ommatidium and that of the sense cell in the locust and the dragonfly, which both have fused rhabdoms, indicates that in these species the ommatidium can be regarded as an optical unit, since the angles mentioned are of the same order of magnitude. Calculations based on the theory of diffraction lead to the same conclusion in Hymenoptera (Barlow, 1952), where the diffraction disc covers the whole rhabdom. Diffraction studies on Limulus (Reichardt, 1961) and Musca (Kirschfeld, 1965) indicate that in these species, however, the ommatidia are not to be regarded as units in the optical sense. Nor is that the case in Calliphora; the small diameter of the main maximum of the Fraunhofer diffraction allows a further resolution in the ommatidium (Seitz, 1968a). The reader is referred to Goldsmith (1964), Smith et al. (1965), Borsellino et al. (1965), Ratliff (1966), Shaw (1969) for data on the functional coupling of different sense cells in one ommatidium.

In cap preparations of the dipteran eye, it has been shown that the seven rhabdomeres in one ommatidium are never lit up simultaneously when the eye is illuminated from a point light source (Autrum and Wiedemann, 1962; Wiedemann, 1965; Kirschfeld, 1967). These observations also show that, in the dipteran compound eye, the diffraction disc covers only that rhabdomere which is lit up. It should be pointed out that the value of the electrophysiologically obtained opening angle of the retinular cell (2.5—3.5°) is larger than that shown by application of geometrical optics on the rhabdomere (1.2°) as seen in Table 1. The explanation is that electrophysiological responses to illumination from a point light source are obtained not only when light with the highest intensity of the diffraction disc hits the rhabdomere. Since the light represented by the slopes of the main maximum of the Airy curve is also efficient, a larger opening angle value is obtained.

Optical Characteristics of the Neuro-Ommatidium

In the dipteran eye, the so-called neuro-ommatidium acts as a functional unit (Kirschfeld, 1967; see further Trujillo-Cenóz, this volume). In these insects — as described by Dietrich (1909) and Fernandez-Morán (1958) — each ommatidium has eight sensory cells, the rhabdomeres of which do not fuse (open rhabdom). Only seven rhabdomeres can be seen in each cross-section, irrespectively of the level at which the section is made, since rhabdomere No. 7 lies distally to

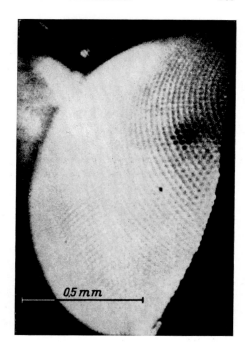

Fig. 3a. Pseudopupil in the eye of the white *Musca* mutant illuminated from inside. When observed in the direction of their axes, rhabdomeres with the same directional sensitivity appear as dark spots, since they absorb more light than the surrounding tissue (from KIRSCHFELD, 1967)

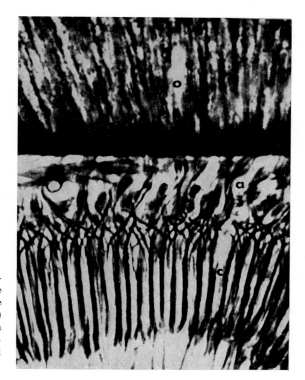

Fig. 3b. Light micrograph of silver-stained section through the eye of *Musca domestica* showing the course of photoreceptor axons (*a*) to the region of cartridges (*c*) in the *lamina ganglionaris*; *o*, ommatidia (from EICHENBAUM and GOLDSMITH, 1968)

rhabdomere No. 8 along the same axis (Trujillo-Cenóz, 1965; Trujillo-Cenóz and Melamed, 1966; Langer, 1966). The rhabdomeric microvilli are oriented perpendicularly to the axial surfaces of each retinular cell. In addition, the microvilli of the eighth rhabdomere are normal to those of the seventh (Langer and Schneider, 1969).

When a bundle of parallel rays from a point light source enters the central part of the eye, seven ommatidia appear dark ("pseudopupil") when observed in the direction of light incidence (see Fig. 3a). In these units, rhabdomeres receive light, although the interommatidial angle is 2—3° (Seitz, 1968a and b). If an eye scalp is illuminated from a point source, seven rhabdomeres in seven neighbouring ommatidia appear bright when observed microscopically *from behind* (Wiedemann, 1965; Kirschfeld, 1967). An explanation of this observation can be offered by geometrical optics as shown in Fig. 4 (see Seitz, 1968b). As a result of the optical properties of the eye, one object point has seven rhabdomeric image points with a specific arrangement. The functional coupling between these rhabdomeres — which "look" at the same point, i.e. have the same directional sensitivity — is of great significance for the further channelling of information (Cajal and Sanchez, 1915; Meyer, 1951; Trujillo-Cenóz and Melamed, 1966; Braitenberg, 1966 and 1967). Thus, the above-mentioned pattern of projection is matched by the neuronal organization in such a way that there is a possibility for the information about the image points to be "superimposed" centrally. For this mechanism, the term "neural superposition" was proposed by Kirschfeld (1967).

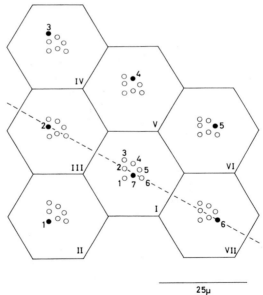

25μ

Fig. 4. Schematical drawing of seven ommatidia (I—VII) in the central part of the blowfly eye. Since rhabdomeres Nos. 1—7 (filled circles) look in the same direction, they are lit up simultaneously by parallel rays from a point light source. The illuminated rhabdomeres have the same mutual arrangement as their corresponding ommatidia. Broken line indicates the plane of section illustrated in Fig. 2 (from Seitz, 1968b)

Regulation of Rhabdomal Light Flux

In *Calliphora*, the optical density of the rhabdomeres is higher than that of the surrounding structure; the rhabdomeres and their processes may therefore serve as light guides. If variations in the prevailing illumination are followed by structural changes in the tissues at the borders of the rhabdomeres (e.g. granular movement), changes in the refractive indices may occur (cf. pigment migration in Lepidoptera, described below). It has been claimed that such a dynamic adaptation mechanism exists in the *Musca* eye (KIRSCHFELD and FRANCESCHINI, 1969). In the dark-adapted *Calliphora* eye, small pinocytotic vesicles — which are very few in the illuminated eye — seem to arise at the bases of the rhabdomeric microvilli (GEMNE and SEITZ, 1970). These vesicles have a lower refractive index than the rest of the sense cell. Thus, in darkness, light transmission in the rhabdomeres — which act as wave guides — may be increased by this change of refractive index of the sense cell medium adjacent to the rhabdomeres Nos. 1—6 (SEITZ, 1970; GEMNE and SEITZ, 1970). The cited results indicate the existence in the Dipteran eye of two subsystems as first pointed out by KIRSCHFELD and FRANCESCHINI (1969) for the *Musca* eye. According to their assumption, one system — represented by the sense cells Nos. 1—6 — works in darkness, with higher sensitivity and lower spatial resolution. The second system — composed of the cells Nos. 7 and 8 — works in daylight, with higher spatial resolution and lower absolute sensitivity.

Similar changes in the optical system of other arthropods have been discussed (HORRIDGE and BARNARD, 1965; HORRIDGE, 1966; FAHRENBACH, 1969; VARELA and PORTER, 1969). Pupil reactions have also been described following the circadian rhythm (in *Tenebrio molitor*, WADA and SCHNEIDER, 1967, 1968) and induced by light (in *Dytiscus marginalis* and *Lethocerus americanus*, WALCOTT, 1969). In the last-mentioned three species, the iris pigment cells form a variable diaphragm. Thus, among apposition eyes there is a variety in the outfit of light-attenuating mechanisms. The functional range of these pupils, however, seems to be rather limited (see e.g. KIRSCHFELD and FRANCESCHINI, 1969; SEITZ, 1970) in comparison with the pupil of the superposition eye (see Chapter IV).

II. The Compound Eye as Analyzer of Polarized Light

The ability of arthropods to distinguish between polarized and non-polarized light was indicated by studies on phototaxis in *Daphnia* and *Drosophila* (VERKHOVSKAYA, 1940). However, VON FRISCH (1949) was the first to present clear evidence that the arthropod compound eye has the ability of analyzing polarized light. In classical behaviour investigations, he showed that the honeybee is able to orient according to the plane of linearly polarized light of the blue sky (for detailed references, see VON FRISCH, 1964 and 1968). It is now well known that a great number of arthropods react to polarized light (see MAZOKHIN-PORSCHNYAKOV, 1969, p. 136).

Reception of polarized light has been studied with electrophysiological methods in investigations on the electroretinogram (AUTRUM and STUMPF, 1950; DE VRIES et al., 1953; LÜDTKE, 1957; AUTRUM and VON ZWEHL, 1962; GIULIO, 1963) and optic lobe impulses (SELETSKAYA, 1956) in insects as well as in studies on the optic nerve fiber activity in *Limulus* (Waterman, 1950). A more direct approach was made by recording intracellular responses in flies (Kuwabara and Naka on *Lucilia*, 1959; Burkhardt and Wendler on *Calliphora*, 1960). When the plane of linearly polarized light was turned 90° from the position of maximal response, the response was reduced to a value corresponding to a change in the intensity by 50 %. Similar results have been obtained later in experiments on the crayfish (Shaw, 1966, 1969).

Anisotropism of the insect cornea and rhabdomeres was observed in experiments on the mechanism of reception of polarized light (Stockhammer, 1956). The data on the dioptric apparatus, however, do not explain the analyzing ability, since the crystal optical axis of the corneal lens is parallel to that of the ommatidium. In the rhabdomeres, however, the direction of lower refractive index (the crystal optical axis) is normal to that of the anatomical axis of the ommatidium, resulting in maximal birefringence of incident light. The problem of polarized light detection has been studied also by Fernandez-Moran (1956 and 1958) who demonstrated a tubular structure of the rhabdomeres and assumed a relationship between the direction of the tubuli and the ability of the arthropod eye to analyze polarized light. However, the reason for the dependence of the receptor response on the plane of polarization remained an open question. If the rhabdomeres are dichroic due to an ordered structural arrangement of the photopigment, the degree of light absorption in the photopigment would depend on the plane of the polarized light. It is known that the frog rod outer segments are dichroic (Schmidt, 1934, 1935, and 1938). The dichroism is due to the orderly arrangement of rhodopsin in the outer segment lamellae (Denton, 1959; Liebman, 1962, see also Wald et al., 1962 and 1963). The role of the insect photopigment in the rhabdoms (rhodopsin, see Goldsmith, 1958) in analyzing polarized light was demonstrated in microspectro-photometric studies on single rhabdomeres in *Calliphora* (Langer, 1965 and 1966; Langer and Thorell, 1966a and b). These experiments showed that there is dichroic absorption between 450 and 540 nm — with a maximum between 500 and 520 nm — which disappears after bleaching (see Fig. 5). The direction of maximal

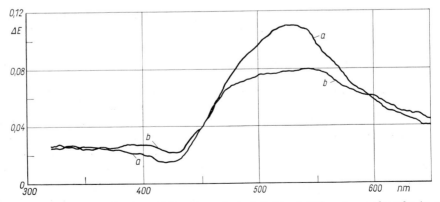

Fig. 5. Variation in extinction of linearly polarized light of different wavelengths in one rhabdomere of the *Calliphora* compound eye. Curve *a*, polarization plane with maximal absorption; *b*, polarization plane perpendicular to that in *a*. Dichroic absorption is demonstrated within the wavelength region of about 450—540 nm (from Langer, 1965)

light absorption (direction of dichroic absorption maximum) and the direction of the lower refractive index (crystal optical axis of the rhabdomere) coincide with the microvillar axis (STOCKHAMMER, 1956; LANGER, 1966; SEITZ, 1969a). The ability of rhabdomeres No. 1—6 in the *Calliphora* eye to analyze polarized light is based on the fact that these three directions coincide.

The rhabdom in the compound eye of decapods and of some other crustaceans consists of an axial series of interleaved layers (PARKER, 1895; RUTHERFORD and HORRIDGE, 1965; EGUCHI, 1965; EGUCHI and WATERMAN, 1966, 1968; HORRIDGE, 1967; KUNZE, 1968). Each layer is made up of closely packed, parallel microvilli which are orthogonal to one another in alternate layers. Such interlocking rhabdoms have no dichroism when the incident light is parallel to the axes of the rhabdoms. Microspectrophotometric measurements of isolated crayfish rhabdoms (*Orconectes virilis* and *O. immunis*) illuminated transversely demonstrate that the major absorption axis matches the microvillar axis (WATERMAN et al., 1969) and that the dichroism shows the same dependence on the wavelength as in the blowfly (LANGER, 1965). This dichroism is also lost when the photosensitive pigments are bleached. WATERMAN et al. (1969) came to the conclusion that, in a zone close to the surface, the absorbing dipoles of the chromophores lie parallel to the microvillar membrane but are otherwise randomly oriented. The functional dichroism of the rhabdom thus arises from its specific structural geometry (see also WATERMAN, 1966, 1968). Because all the microvilli of any one cell have the same orientation, the layers of microvilli constitute two sets of orthogonal polarization analyzers when illuminated along the visual axis of the ommatidium.

III. Light Transmission in the Eye of Fireflies

The superposition theory of EXNER (1891) was based on considerations concerning the structural characteristics of the eye of the firefly *(Lampyris = Phausis splendidula)*. His classical studies were performed on eye scalp preparations where the pigment and cone cells were replaced by a homogeneous medium. Since he obtained only one image proximally to the tips of the pseudocones, he concluded that light rays coming from one object point in front of the eye enter the facets of several neighbouring ommatidia and converge to one image point at the level of the rhabdomeres. Similar, although less clear-cut, observations on eyes from crustaceans and night moths led him to the definition of the term "superposition eye".

In this connection, recent data on the optical properties of the *Phausis* eye are of special interest. According to EXNER, the crystalline cone is built up of concentric layers with different optical densities, the axial part having the highest refractive index. Although these differences have been questioned on the basis of phase contrast microscopy (KUIPER, 1962), recent interference microscopic studies (SEITZ, 1969 b, see Fig. 6) show that the refractive index of the corneal lens and the hard pseudocone is highest in the axial parts ($n_e = 1.520$), and decreases toward the periphery ($n_e = 1.456$). The lens can be regarded as a homogeneous spherical lens, the pseudocone functioning as a lens cylinder. According to constructions on the basis of the laws of refraction, parallel rays are focussed to a point within the pseudocone, from which they emerge as approximately parallel rays. Total reflection occurs in the pseudocone when the angle of incidence exceeds 4°. For larger angles the light is refracted into the surrounding pigment cells where it is absorbed, or it may again escape through the facet after multiple total reflection in the dioptric apparatus. Thus, the opening angle is about 8°. Attached to the lens system are the four cone cells, whose proximal prolongations form a crystalline tract approximately 150 μ long. The tract has a higher optical density ($n_e = 1.349$) than the surrounding pigment cells ($n_e = 1.340$) and may therefore serve as a light pipe to the rhabdomeres. Similar recent data on the optical characteristics of the light-transmitting system in other insects, as well as observations on the premises for the formation of a superposition image, have made EXNER's theory doubtful. Actually, EXNER himself suggested that in certain crustaceans the cone and tract may act as a wave guide. This idea was also expressed by KIRCHHOFFER (1908). Measurements of the refractive indices of the crystalline cone and tract on the one hand and the surrounding media on the other supported this proposition (DE BRUIN and CRISP, 1957). In later studies on crustaceans, superposition images were not obtained (NUNNEMACHER, 1959 and 1960; KUIPER, 1962) and in scalp preparations from frozen eyes of *Phausis*, the image was found to be located behind the rhabdomeres (NUNNEMACHER, 1959 and 1960).

When EXNER's experiments on *Phausis* scalp preparations – deprived of all soft tissue – were repeated, his results could be confirmed (KUIPER, 1962; DØVING and MILLER, 1969), whereas the image seen in scalp preparations with *intact* tracts

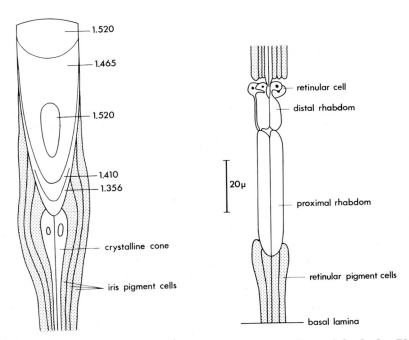

Fig. 6. Schematical drawing of an ommatidium in the compound eye of the firefly, *Phausis*. (The drawing to the right is a continuation of the left hand drawing.) The dioptric apparatus (corneal lens and hard pseudocone) consist of layers with different optical densities. The cornea acts as a homogeneous, spherical lens, the pseudocone as a lens cylinder (from SEITZ, 1969b)

was taken to represent the pattern of illumination radiating from the cut ends of the tracts (Døving and Miller, 1969). Studies on the American firefly, *Photuris versicolor*, have also led to the conclusion that a superposition image can only be obtained when the cellular tissue is replaced by a homogeneous medium (Horridge, 1968 and 1969 b). In this species, thread-like structures — assumed to serve as light guides — lead from the crystalline cone to the vicinity of the basal lamina (basement membrane). At all levels, these "crystalline threads" seem to be separated from the rhabdoms. According to Exner, the *Phausis* iris pigment is mainly located proximal to the dioptric system in the light-adapted state. However, in the dark-adapted state the pigment surrounds the pseudocone, the superposition effect thus being due to the elimination of the optical insulation of the ommatidia. In *Photuris versicolor*, on the other hand, it was found that all pigment does not migrate distally during dark-adaptation (Horridge, 1968 and 1969 b), so that partial optical insulation of the ommatidia remains in the dark-adapted state. Thus, in an eye of this type the "crystalline threads" may serve as wave guides irrespective of the pigment position.

IV. Light Guide Transmission and Longitudinal Pupil in Lepidoptera

Among Lepidoptera there are interesting variations in the arrangement of the different components of the light-transmitting system (see Fig. 7). In diurnal butterflies, the rhabdom extends to the crystalline cone, the proximal part of which may form a short tract. In nocturnal moths, as well as in the skipper butterflies (Hesperidae), the crystalline cone is separated from the rhabdomeres. The distal, filamentous extensions of the retinular cells constitute a long, narrow "crystalline" tract (diameter about $5\,\mu$), surrounded by a tube-shaped sleeve formed by the elongated "iris cells". In the moths, these cells contain migrating pigment granules, whereas in the skippers they are transparent and devoid of pigment. Reference has already been made to studies which suggest that in the firefly — like in crustaceans — the tract may function as a wave guide. Calculations of the light path in the ommatidium of the tobacco hornworm moth *(Manduca sexta)*, based on refractometric measurements (Allen, 1968), show that about 80 % of incident light (within 3°) is transmitted in the tract, the refractive index of which ($n_e = 1.523$) is higher than that of the surrounding "iris cells" ($n_e = 1.371$). On the basis of these data, as well as of data obtained in recent optical studies and electrophysiological experiments on dark-adapted moth eyes (Døving and Miller, 1969), it was concluded that the tract functions as a wave guide and that no superposition image exists in the retinula of the dark-adapted moth eye. It has also been demonstrated (Kunze, 1969, 1970) that projection of light into one single facet caused a complete glow in the dark-adapted eye of the moth, *Ephestia kühniella* and in mutant eyes not containing shielding pigment. Position and size of the glow are independent of the position of the illuminated facet within the glow area. Further, projection of a moving stripe pattern into one single facet elicited optomotor responses of the moth. The results confirm the superposition theory which therefore can not be entirely disposed with. This controversial issne has been further discussed in recent papers by Horridge (1971) and Kunze (1972).

The position of the screening pigment depends on the brightness of the surrounding light (Bernhard and Ottoson, 1964). The migration of pigment granules from a distal position in darkness (Fig. 8a) to a proximal position in bright light (Fig. 8b) causes attenuation of the light reaching the rhabdomeres by as much as 2—3 log units (Bernhard and Ottoson, 1964; Bernhard et al., 1963; Höglund, 1966). As a result of this attenuation, the photoreceptors become more dark adapted and increase their sensitivity (Höglund, 1963 and 1966; Post and Goldsmith, 1965). This light-attenuation mechanism serves to increase the range of intensity discrimination by 2—3 log units. These results, obtained in gross recordings, have been confirmed by mathematical calculations (Allen and Bernard, 1967) and are in agreement with measurements from recent intracellular recordings (Höglund and Struwe, 1971). Also, they are compatible with re-

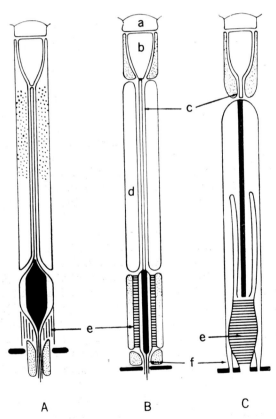

Fig. 7. Schematical drawing of three types of Lepidopteran compound eyes. A, moth; B, skipper (a Hesperid butterfly); C, typical diurnal butterfly. Stippled areas show the location of pigment granules. Black areas represent rhabdom. Abbreviations: *a*, cornea; *b*, crystalline cone; *c*, crystalline tract; *d*, "iris" cell; *e*, tracheolar tapetal structures; *f*, nerve fibers (from Miller et al., 1968)

cent microelectrode studies on the crayfish (SHAW, 1969). In electrophysiological experiments on *Manduca*, it has been found that the attenuation is almost uniform between 350 nm and 650 nm with a slightly lower screening effect above 600 nm (HÖGLUND and STRUWE, 1970). Measurements of the spectral absorption of the pigment granules are consistent with these data (HÖGLUND et al., 1970).

It has been pointed out that, because of the small dimensions of the tract, the energy is transmitted in discrete modes. The number of modes is dependent on the wavelength of the incident light, on the diameter and refractive index of the tract, and on the indices of refraction and absorption in the surrounding medium (MILLER et al., 1968). It was concluded that in both light- and dark-adapted moths *(Manduca, Elpenor, Cecropia, Polyphemus)* the tracts propagate light (ALLEN, 1968; MILLER et al., 1968), and that the tract and surrounding migrating pigment function as a longitudinal pupil (see KUIPER, 1962); in the

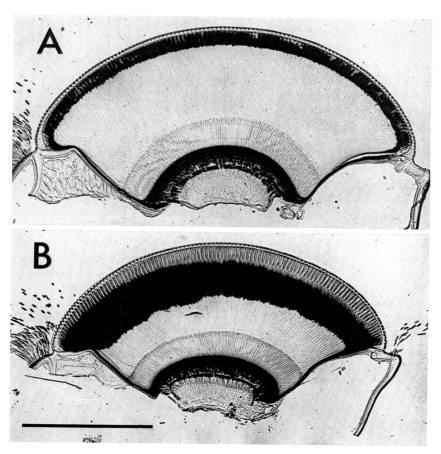

Fig. 8. Light micrographs of longitudinal sections through the compound eye of the night-moth, *Celerio euphorbiae*, showing iris pigment position in (a) dark-adaptation (with maximal "glow"), (b) light-adaptation (minimal "glow"). Horizontal bar, 1 mm (from HÖGLUND, 1966)

light-adapted state, when the tract is surrounded by the pigment sleeve, the high-order modes are attenuated more than the lower-order ones. The light guide principle also applies to the skipper's eye which, however, lacks the longitudinal pupil function, since the "iris cells" are devoid of pigment (Miller et al., 1968).

V. Optical Effects of Ommatidial Ultrastructures
Corneal Mechanisms for Refraction Matching and Selective Filtering

Electron microscopy has revealed the existence of ultrastructures in the optical system of butterflies which result in selective transmission of possible biological significance. When light of normal incidence hits a smooth corneal lens surface, about 5 % is reflected and can be observed as specular reflection from the surface of the facet, e.g. in the bee eye (see Bernhard et al., 1965). In many Lepidoptera, however, the surface layer of the corneal lens forms a hexagonal array of 200—250 mμ high, cone-shaped protuberances, first observed by Bernhard and Miller (1962), who termed them corneal nipples (Fig. 9). In an attempt to elucidate the phylogeny of this surface structure, a survey of several insects was made (Bernhard et al., 1970a). The results indicate a definite anagenetic trend in the appearance of nippled corneas and show that such "full-sized" nipples occur only in Trichoptera and Lepidoptera with predominance among nocturnal Lepidopteran species. Ontogenetically, the morphological characteristics of the array are determined by the spatial relationship between the corneagenous cell microvilli and an epicorneal lamina distal to the microvilli (Gemne, 1966a and b, 1970, 1971).

Model experiments, spectrophotometric studies and mathematical analysis showed that the nipple array serves as an impedance transformer, causing a gradual transition between the refractive indices of air and cornea. This transition results in a several hundred-fold decrease of reflection in the middle range of the visible spectrum and a consequent increase in the transmission of light by about 5 % (Bernhard et al., 1963 and 1965; Miller et al., 1966). This small increase may play a role for night-active species at intensities near absolute threshold. The possible role of decreased internal reflection in vision as well as the role of decreased external reflection in camouflage, have been discussed (Bernhard et al., 1965; see also Bernhard, 1967; Bernhard et al., 1968; Miller et al., 1968).

In many Diptera, more specifically in Tabanidae, various spectacular, coloured reflection patterns can be seen when the eye is illuminated with white light. Electron microscopy, refractometric investigations and theoretical calculations have shown that the patterned, selective reflection from the cornea of the horsefly, *Hybomitra lasiophthalma*, is caused by a superficial layering in the corneal material (Fig. 10). The system of layers acts as a transmission quarter-wavelength inter-ference filter, the selectivity of which depends on the optical path (Bernard and Miller, 1968a and b). Such interference filters may serve to enhance colour-contrast in the optical system (see also Miller et al., 1968).

In many arthropod eyes, there is a brilliant reflection from the deep region of the eye. Thus, in dark-adapted *nocturnal moths* (with the iris pigment in "dark-

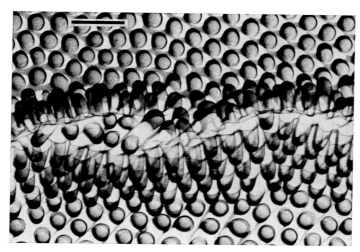

Fig. 9. Electron micrograph of facet surface replica of the night-moth, *Cerapteryx graminis* L., showing corneal nipples about 250 mμ high. In a fold of the replica, the actual form of the nipples is revealed. Horizontal bar, 0.5 μ (from BERNHARD et al., 1970)

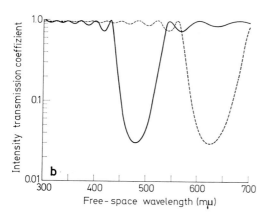

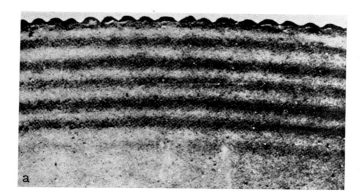

Fig. 10. a Electron micrograph showing alternating layers of varying electron density near the corneal surface cf the horsefly, *Hybomitra lasiophthalma*. These layers — with different optical densities — function as transmission quarter-wavelength filters giving rise to a corneal reflection pattern of coloured stripes. The dense layers are about 890 Å thick (from BERNARD and MILLER, 1968a). b Theoretical characteristics of transmission filters for typical 12-layer, blue (————) and orange (- - - - -) filters in the horsefly cornea (from BERNARD and MILLER, 1968b)

position", see Fig. 8a) light reflected from many ommatidia is seen as a bright spot ("glow") of unsaturated light in the part of the eye facing the observer (Leydig, 1864). This reflection occurs in the tracheolar layer proximal to the rhabdoms (Exner, 1891) and causes a repeated passage of the light through the receptors. When the eye is light adapted, the iris pigment migrates distally into "light-position" (see Fig. 8b), more incident light is absorbed by the pigment and the glow disappears. There is thus a relation between the pigment position and the size of the glow (see Höglund, 1966).

Selective Reflection from Tracheolar Structures

In *diurnal butterflies*, a discrete, coloured glow can also be seen (Exner, 1891) with the aid of the ophthalmoscope. Recent electron microscopic investigations have shown the existence of regularly spaced platelets (see Fig. 11), formed by the taenidial ridges of the tracheoles at the proximal end of the rhabdomeres (Miller and Bernard, 1968). Taking into account the dimensions and refractive indices, Miller et al. (1968) arrived to the conclusion that this multilayered structure is to be regarded as a quarter-wavelength interference filter which selectively reflects light back through the photoreceptors and the optical system giving rise to the coloured glow. This selective interference filter system in diurnal butterflies was concluded to act as a tapetum that enhances colour-contrast.

In this context, reference should be made to recent studies on the diurnal butterfly *Heliconius erato*, the courting behaviour of which is selectively released by light of the same colour as that of the insect's red forewing band (Crane, 1955). There is a selective reflection of light — seen as a red glow — from tracheolar structures which are similar in principle to those described above. This reflected light, which is approximately the colour of the forewing band, causes a significant selective enhancement of the photoreceptor response between 610 and 640 nm as shown in electrophysiological experiments (Bernhard et al., 1970b). These data indicate that the reception of a colour cue, serving as a releaser of an innate behavioural pattern, can be selectively enhanced by a specialized mechanism within the optical system. Further evidence has been obtained in an extended study by Struwe (1972), who also discusses certain objections made by Swihart and Gordon (1971).

Fig. 11. Electron micrograph of a longitudinal section through the compound eye of the monarch butterfly, *Danaus plexippus*, showing the selectively reflecting tracheolar platelet system proximal to the rhabdom (*rh*). Bar, 2 μ (from Miller and Bernard, 1968)

References

ALLEN, J. L.: The optical functioning of the superposition eye of a nocturnal moth. Thesis, Massachusetts Institute of Technology 1968.

— BERNARD, G. D.: Superposition optics, a new theory, Massachusetts Institute of Technology, Research Laboratory of Electronics. Quart. Progr. Rep. No. 86, 113—122 (1967).

AUTRUM, H., STUMPF, H.: Das Bienenauge als Analysator für polarisiertes Licht. Z. Naturforsch. **5 B**, 116—122 (1950).

— WIEDEMANN, I.: Versuche über den Strahlengang im Insektenauge (Appositionsauge). Z. Naturforsch. **17**, 480—482 (1962).

— ZWEHL, V. VON: Zur spektralen Empfindlichkeit einzelner Sehzellen der Drohne *(Apis mellifica)*. Z. vergl. Physiol. **46**, 8—12 (1962).

BARLOW, H. B.: The size of ommatidia in apposition eyes. J. exp. Biol. **29**, 667—674 (1952).

BERNARD, G. D., MILLER, W. H.: Interference filters in the corneas of Diptera, Investigative Ophthalmology **7** (4), 416—434 (1968a).

— — Physical optics of invertebrate eyes, Massachusetts Institute of Technology, Research Laboratory of Electronics. Quart. Progr. Rep. No. 88, 106—111 (1968b).

BERNHARD, C. G.: Structural and functional adaptation in a visual system. Endeavour **26**, 79—84 (1967).

— BOËTHIUS, J., GEMNE, G., STRUWE, G.: Eye ultrastructure, colour reception and behaviour. Nature (Lond.) **226**, 865—866 (1970b).

— GEMNE, G., MØLLER, A. R.: Modification of specular reflexion and light transmission by biological surface structures. Quart. Rev. Biophys. **1**, 89—105 (1968).

— — SÄLLSTRÖM, J.: Comparative ultrastructure of corneal surface topography in insects with aspects on phylogenesis and function. Z. vergl. Physiol. **67**, 1-25 (1970a).

— HÖGLUND, G., OTTOSON, D.: On the relation between pigment position and light sensitivity of the compound eye in different nocturnal insects. J. Insect Physiol. **9**, 573—586 (1963).

— MILLER, W. H.: A corneal nipple pattern in insect compound eyes. Acta physiol. scand. **56**, 385—386 (1962).

— — MØLLER, A. R.: Function of the corneal nipples in the compound eyes of insects. Acta physiol. scand. **58**, 381—382 (1963).

— — — The insect corneal nipple array. Acta physiol. scand. **63**, Suppl. 243 (1965).

— OTTOSON, D.: Quantitative studies on pigment migration and light sensitivity in the compound eye at different light intensities. J. gen. Physiol. **47**, 465—478 (1964).

BORSELLINO, A., FUORTES, M. G. F., SMITH, T. G.: Visual response in *Limulus*. Cold Spr. Harb. Symp. quant. Biol. **30**, 429—443 (1965).

BRAITENBERG, V.: Unsymmetrische Projektion der Retinulazellen auf die Lamina ganglionaris bei der Fliege *Musca domestica*. Z. vergl. Physiol. **50**, 212—214 (1966).

— Patterns of projections in the visual system of the fly. 1. Retina-Lamina projections. Exp. Brain Res. **3**, 271—298 (1967).

DE BRUIN, G. H. P., CRISP, D. T.: The influence of pigment migration on the vision of higher Crustacea. J. exp. Biol. **34**, 447 (1957).

BURKHARDT, D., DE LA MOTTE, I., SEITZ, G.: Physiological optics of the compound eye of the blow fly. In: BERNHARD, C. G. (Ed.): The Functional Organization of the Compound Eye. Wenner-Gren Center International Symposium series, Vol. 7, pp. 51—62. Oxford: Pergamon Press 1966.

— WENDLER, L.: Ein direkter Beweis für die Fähigkeit einzelner Sehzellen des Insektenauges, die Schwingungsrichtung polarisierten Lichtes zu analysieren. Z. vergl. Physiol. **43**, 687—692 (1960).

CAJAL, S. R., SÁNCHEZ, D.: Contribución al conocimiento de los insectos. Trab. Lab. Invest. Biol. Madrid **13**, 1—168 (1915).

CRANE, J.: Imaginal behaviour of a Trinidad butterfly, *Heliconius erato hydara* Hewitron, with special reference to the social use of color. Zoologica **40**, 167—195 (1955).

DENTON, E. J.: The contributions of the orientated photosensitive and other molecules to the absorption of whole retina. Proc. roy. Soc. B **150**, 78 (1959).

DIETRICH, W.: Die Facettenaugen der Dipteren. Z. wiss. Zool. **92**, 465—539 (1909).

Døving, K. B., Miller, W. H.: Function of insect compound eyes containing crystalline tracts. J. gen. Physiol. **54**, 250—267 (1969).

Eguchi, E.: Rhabdom structure and receptor potentials in single crayfish retinular cells. J. cell. comp. Physiol. **66**, 411 (1965).

— Waterman, T. H.: Fine structure patterns in crustacean rhabdoms. In: Bernhard, C. G. (Ed.): The Functional Organization of the Compound Eye. Wenner-Gren Center International Symposium Series, Vol. 7, p. 105. Oxford: Pergamon Press 1966.

— — Cellular basis for polarized light perception in the spider crab, *Libinia*. Z. Zellforsch. **84**, 87 (1968).

Eichenbaum, D. M., Goldsmith, T. H.: Properties of intact photoreceptor cells lacking synapses. J. exp. Zool. **169**, 15—32 (1968).

Exner, S.: Die Physiologie der facettirten Augen von Krebsen und Insekten. Leipzig-Wien: F. Deuticke 1891.

Fahrenbach, W. H.: The morphology of the eyes of *Limulus*. II. Ommatidia of the compound eye. Z. Zellforsch. **93**, 451—483 (1969).

Fernandez-Morán, H.: Fine structure of the insect retinula as revealed by electron microscopy. Nature (Lond.) **177**, 742—743 (1956).

— Fine structure of the light receptors in the compound eyes of insects. Exp. Cell Res. Suppl. **5**, 586—644 (1958).

Frisch, K. von: Demonstration von Versuchen zum Nachweis des Farbensinnes bei angeblich total farbenblinden Tieren. Verhandl. Dtsch. Zool. Ges. in Freiburg. Berlin 1914.

— Die Polarisation des Himmelslichtes als orientierender Faktor bei den Tänzen der Bienen. Experientia (Basel) **5**, 142—148 (1949).

— Tanzsprache und Orientierung der Bienen. Berlin-Göttingen-Heidelberg: Springer 1965.

— The dance language and orientation of bees. London: Oxford University Press 1968.

Gemne, G.: Ultrastructural ontogenesis of cornea and corneal nipples in the compound eye of insects. Acta physiol. scand. **66**, 511—512 (1966a).

— Fine structure of the insect cornea and corneal nipples during ontogenesis. In: "Electron Microscopy 1966". Proc. Sixth Internat. Congr. El. Micr., Kyoto, Vol. II, pp. 511—512. Tokyo: Maruzen Co. 1966b.

— Ultrastructure of epicorneal topography and morphogenesis in insects with aspects on phylogenesis and function. Thesis, Karolinska Institutet (1970).

— Ontogenesis of corneal surface ultrastructure in nocturnal Lepidoptera. Phil. Trans. roy. Soc. (Lond.) B **262** (843), 343—363 (1971).

— Seitz, G.: Electron and light microscopic evidence for a light-induced pupil reaction in the apposition eye of the blowfly. Acta physiol. scand. **79**, 30A (1970).

Giulio, L.: Elektroretinographische Beweisführung dichroitischer Eigenschaften des Komplexauges bei Zweiflüglern. Z. vergl. Physiol. **46**, 491—495 (1963).

Goldsmith, T. H.: The visual system of the honeybee. Proc. nat. Acad. Sci. (Wash.) **44**, 123—126 (1958).

— Fine structure of the retinulae in the compound eye of the honeybee. J. Cell Biol. **14**, 489—494 (1962).

— The visual system of insects. In: Rockstein, M. (Ed.): The Physiology of Insecta, Vol. 1 pp. 397—462. New York: Academic Press 1964.

Hertz, M.: Die Organisation des optischen Feldes bei der Biene. I. Z. vergl. Physiol. **8**, 693—748 (1929).

— Die Organisation des optischen Feldes bei der Biene. II. Z. vergl. Physiol. **11**, 107—145 (1930).

— Die Organisation des optischen Feldes bei der Biene. III. Z. vergl. Physiol. **14**, 629—674 (1931).

Hodierna, G. B.: L'occhio della mosca. Discorso fisico. Palermo: Per Decio Cirillo 1644.

Höglund, G.: Glow, sensitivity changes and pigment migration in the compound eye of nocturnal Lepidoptera. Life Sci. **1963**, 275—280.

— Pigment migration, light screening and receptor sensitivity in the compound eye of nocturnal Lepidoptera. Acta physiol. scand. **69**, Suppl. 282 (1966).

— Langer, H., Struwe, G., Thorell, B.: Spectral absorption by screening pigment granules in the compound eyes of a moth and a wasp. Z. vergl. Physiol. **67**, 238—242 (1970).

Höglund, G., Struwe, G.: Pigment migration and spectral sensitivity in the compound eye of moths. Z. vergl. Physiol. **67**, 229—237 (1970).

— — Pigment migration and illumination of single photoreceptors in a moth. Z. vergl. Physiol. **74**, 336—339 (1971).

Hooke, R.: Micrographia Obs. XXXIX: Of the eyes and head of a grey drone-fly, and of several other creatures. 175—180 (1665).

Horridge, G. A.: Perception of edges versus areas by the crab, *Carcinus*. J. exp. Biol. **44**, 247—254 (1966).

— Perception of polarization plane, colour and movement in two dimensions by the crab, *Carcinus*. Z. vergl. Physiol. **55**, 207 (1967).

— Pigment movement and the crystalline threads of the firefly eye. Nature (Lond.) **218**, 778—779 (1968).

— Unit studies on the retina of dragonflies. Z. vergl. Physiol. **62**, 1—37 (1969a).

— The eye of the firefly, *Photuris*. Proc. roy. Soc. B **171**, 445—463 (1969b).

— Alternatives to superposition images in clear-zone compound eyes. Proc. roy. Soc. B **179**, 97—124 (1971).

— Barnard, P. B. T.: Movement of palisade in locust retinula cells when illuminated. Quart. J. Micr. Sci. **106**, 131—135 (1965).

Kirchhoffer, O.: Untersuchungen über die Augen pentamer Käfer. Arch. Biontol. (Berl.) **5**, 235—287 (1908).

Kirschfeld, K.: Das anatomische und das physiologische Sehfeld der Ommatidien im Komplexauge von *Musca*. Kybernetik **2**, 249 (1965).

— Die Projektion der optischen Umwelt auf das Raster der Rhabdomere im Komplexauge von *Musca*. Exp. Brain Res. **3**, 248 (1967).

— Reichardt, W.: Die Verarbeitung stationärer optischer Nachrichten im Komplexauge von Limulus. Kybernetik **2**, 43—61 (1964).

— Franceschini, N.: Optische Eigenschaften der Ommatidien im Komplexauge von *Musca*. Kybernetik **5**, 47—52 (1968).

— — Ein Mechanismus zur Steuerung des Lichtflusses in der Rhabdomeren des Komplexauges von *Musca*. Kybernetik **6**, 13—21 (1969).

Kuiper, J. W.: The optics of the compound eye. Symp. Soc. exp. Biol. **16**, 58 (1962).

Kunze, P.: Die Orientierung der Retinulazellen im Auge von Ocypode. Z. Zellforsch. **90**, 454 (1968).

— Eye glow in the moth and superposition theory. Nature (Lond.) **223**, 1172—1174 (1969).

— Verhaltensphysiologische und optische Experimente zur Superpositionstheorie der Bildentstehung in Komplexaugen. Verh. dtsch. Zool. Ges. (Köln) **64**, 234—238 (1970).

— Comparative studies of arthropod superposition eyes. Z. vergl. Physiol. **76**, 347—357 (1972).

Kuwabara, M., Naka, K.: Response of a single retinula cell to polarized light. Nature (Lond.) **184**, 455—456 (1959).

Langer, H.: Nachweis dichroitischer Absorption des Sehfarbstoffes in den Rhabdomeren des Insektenauges. Z. vergl. Physiol. **51**, 258—263 (1965).

— Grundlagen der Wahrnehmung von Wellenlänge und Schwingungsebene des Lichtes. Verh. dtsch. Zool. Ges. (Göttingen) 195—233 (1966).

— Schneider, L.: Zur Struktur und Funktion offener Rhabdome in Facettenaugen. Verh. dtsch. Zool. Ges. (Würzburg) 1969. Zool. Anz. Suppl. **33**, 494—503 (1969).

— Thorell, B.: Microspectrophotometric assay of visual pigments in single rhabdomeres of the insect eye. In: Bernhard, C. G. (Ed.): The Functional Organization of the Compound Eye. Wenner-Gren Center International Symposium Series, Vol. 7, p. 145. Oxford: Pergamon Press 1966a.

— — Microspectrophotometry of single rhabdomeres in the insect eye. Exp. Cell Res. **41**, 673—676 (1966b).

Leeuwenhoek, A. van: The select works of Antoni van Leeuwenhoek. Translated from the Dutch and Latin editions published by Samuel Hoole. London: G. Sidney 1800.

— Alle de brieven van Antoni van Leeuwenhoek (in Dutch and English). Deel I. Amsterdam: Swets and Zeitlinger 1939.

Leydig, F.: Das Auge der Gliederthiere. Tübingen: Laupp 1864.

Liebman, P.A.: In situ microspectrophotometric studies on the pigments of single retinal rods. Biophys. J. **2**, 161 (1962).

Lüdtke, H.: Beziehungen des Feinbaues im Rückenschwimmerauge zu seiner Fähigkeit, polarisiertes Licht zu analysieren. Z. vergl. Physiol. **40**, 329—344 (1957).

Mazokhin-Porschnyakov, G.A.: Insect vision. New York: Plenum Press 1969.

Meyer, G.F.: Versuch einer Darstellung von Neurofibrillen im Zentralnervensystem verschiedener Insekten. Zool. Jb., Abt. Anat. u. Ontog. **71**, 413—426 (1951).

Miller, W.H., Bernard, G.D.: Skipper glow. Massachusetts Institute of Technology, Research Laboratory of Electronics. Quart. Progr. Rep. No. 88, 114—119 (1968).

— — Allen, J.L.: The optics of insect compound eyes. Science **162**, 760 (1968).

— Møller, A.R., Bernhard, C.G.: The corneal nipple array. In: Bernhard, C.G. (Ed.): The Functional Organization of the Compound Eye. Wenner-Gren Center International Symposium Series, Vol. 7, pp. 51—62. Oxford: Pergamon Press 1966.

Müller, J.: Zur vergleichenden Physiologie des Gesichtsinnes des Menschen und der Thiere. Leipzig: C. Cnobloch 1826.

Nunnemacher, R.F.: The retinal image of arthropod eyes. Anat. Rec. **134**, 618 (1959).

— The structure and function of arthropod eyes. Proc. 3rd Intern. Congr. Photobiol. Copenhagen, 428—429 (1960).

Parker, G.H.: The retina and optic ganglia in decapods, especially in *Astacus*. Mitt. Zool. Station Neapel **12**, 1 (1895).

Pedler, C., Goodland, H.: The compound eye and first optic ganglion of the fly. J. roy. Micr. Soc. **84**, 161—179 (1965).

Post, C.T., Jr., Goldsmith, T.H.: Pigment migration and light-adaptation in the eye of the moth, *Galleria mellonella*. Biol. Bull. **128**, 473—487 (1965).

Ratliff, F.: Selective adaptation of local regions of the rhabdom in an ommatidium of the compound eye of *Limulus*. In: Bernhard, C.G. (Ed.): The Functional Organization of the Compound Eye. Wenner-Gren Center International Symposium Series, Vol. 7, pp. 187—191. Oxford: Pergamon Press 1966.

Reichardt, W.: Über das optische Auflösungsvermögen der Facettenaugen von *Limulus*. Kybernetik **1**, 57—69 (1961).

Rutherford, D.J., Horridge, G.A.: The rhabdom of the lobster eye. Quart. J. micr. Sci. **106**, 119 (1965).

Schmidt, W.J.: Dichroismus des Außengliedes der Stäbchenzellen der Froschnetzhaut verursacht durch den Sehpurpur. Naturwissenschaften **22**, 206 (1934).

— Doppelbrechung, Dichroismus und Feinbau des Außengliedes der Sehzellen vom Frosch. Z. Zellforsch. **22**, 485 (1935).

— Polarisationsoptische Analyse eines Eiweiß-Lipoid-Systems, erläutert am Außenglied der Sehzellen. Kolloid-Z. **85**, 137 (1938).

Schneider, L., Langer, H.: Die Feinstruktur des Überganges zwischen Kristallkegel und Rhabdomeren im Facettenauge von *Calliphora*. Z. Naturforsch. **21 B**, 196—197 (1966).

Scholes, J., Reichardt, W.: The quantal content of optomotor stimuli and the electrical responses of receptors in the compound eye of the fly *Musca*. Kybernetik **6**, 74—80 (1969).

Seitz, G.: Der Strahlengang im Appositionsauge von *Calliphora erythrocephala* (Meig.). Z. vergl. Physiol. **59**, 205—231 (1968a).

— Der dioptrische Apparat im Insektenauge. Verh. dtsch. Zool. Ges. (Innsbruck) 361—367 (1968b).

— Polarisationsoptische Untersuchungen am Auge von *Calliphora erythrocephala* (Meig.). Z. Zellforsch. **93**, 525—529 (1969a).

— Untersuchungen am dioptrischen Apparat des Leuchtkäferauges. Z. vergl. Physiol. **62**, 61—74 (1969b).

— Nachweis einer Pupillenreaktion im Auge der Schmeißfliege. Z. vergl. Physiol. **69**, 169—185 (1970).

Seletskaya, L.I.: Perception of polarized light by the compound eye, in bees. Biofizika **1**, 155—157 (1956).

Shaw, S.R.: Polarized light responses from crab retinula cells. Nature (Lond.) **211**, 92 (1966).

— — Interreceptor coupling in ommatidia of drone honeybee and locust compound eyes. Vision Res. **9**, 999—1029 (1969).

SMITH, T. G., BAUMANN, F., FUORTES, M. G. F.: Electrical connections between visual cells in the ommatidium of *Limulus*. Science **147**, 1446—1448 (1965).

STOCKHAMMER, K.: Zur Wahrnehmung der Schwingungsrichtung linear polarisierten Lichtes bei Insekten. Z. vergl. Physiol. **38**, 30—83 (1956).

SWAMMERDAM, J.: Bybel der Natuure, Vols. I—II. Published by H. Boerhaave. Leyden: Severinus, B. Vander and P. Vander 1737—1738.

SWIHART, S. L., GORDON, W. C.: Red photoreceptor in butterflies. Nature (Lond.) **231**, 126—127 (1971).

TRUJILLO-CENÓZ, O.: Some aspects of the structural organization of the arthropod eye. In: Cold Spr. Harb. Symp. quant. Biol. **30**, 371—381 (1965).

— MELAMED, J.: Electron microscope observations on the peripheral and intermediate retinas of dipterans. In: BERNHARD, C. G. (Ed.): The Functional Organization of the Compound Eye. Wenner-Gren Center International Symposium Series, Vol. 7, pp. 339—361. Oxford: Pergamon Press 1966.

TUNSTALL, J., HORRIDGE, G. A.: Electrophysiological investigation of the optics of the locust retina. Z. vergl. Physiol. **55**, 167—182 (1967).

TUURALA, O.: Histologische und physiologische Untersuchungen über die photomechanischen Erscheinungen in den Augen der Lepidopteren. Suomal. Tiedeakat. Toim. (Annls Acad. scient. fennicae) A 4 **24**, 1—69 (1954).

VARELA, F. G., PORTER, K. R.: Fine structure of the visual system of the honey-bee *(Apis mellifera)*. I. The retina. J. Ultrastruct. Res. **29**, 236—259 (1969).

VERKHOVSKAYA, I. S.: Effect of polarized light on phototaxis. Byull. Mosk. o-va Isp. Prip. Otd. Biol. **49**, 101—113 (1940).

VOWLES, D. M.: The receptive fields of cells in the retina of the housefly *(Musca domestica)*. Proc. roy. Soc. B **164**, 552—576 (1966).

DE VRIES, H., SPOOR, A., JELOF, R.: Properties of the eye with respect to polarized light. Physica **19**, 419—432 (1953).

WADA, S., SCHNEIDER, G.: Eine Pupillenreaktion im Ommatidium von *Tenebrio molitor*. Naturwissenschaften **54**, 542 (1967).

— — Circadianer Rhythmus der Pupillenweite im Ommatidium von *Tenebrio molitor*. Z. vergl. Physiol. **58**, 395—397 (1968).

WALCOTT, B.: Movement of retinula cells in insect eyes in light adaptation. Nature (Lond.) **223**, 971—972 (1969).

WALD, G., BROWN, P. K., GIBBONS, I. R.: Visual excitation: a chemo-anatomical study. In: BEAMENT, J. W. L. (Ed.): Biological Receptor Mechanisms. Symp. Soc. exp. Biol. **16**, 32 (1962).

— — — The problem of visual excitation. J. opt. Soc. Amer. **53**, 20 (1963).

WASHIZU, Y., BURKHARDT, D., STRECK, P.: Visual field of single retinula cells and inter-ommatidial inclination in the compound eye of the blowfly *Calliphora erythrocephala*. Z. vergl. Physiol. **48**, 413—428 (1964).

WATERMAN, T. H.: A light polarization analyzer in the compound eye of *Limulus*. Science **111**, 252—254 (1950).

— Directional sensitivity of single ommatidia in the compound eye of *Limulus*. Proc. nat. Acad. Sci. (Wash.) **40**, 252 (1954).

— Polarotaxis and primary photoreceptor events in Crustacea. In: BERNHARD, C. G. (Ed.): The Functional Organization of the Compound Eye. Wenner-Gren Center International Symposium Series, Vol. 7, p. 493. Oxford: Pergamon Press 1966.

— Systems theory and biology — view of a biologist. In: MESAROVIĆ, M. D. (Ed.): Systems Theory and Biology (Proceedings of the 3rd Systems Symposium, Case Institute of Technology), p. 1. New York: Springer-Verlag 1968.

— FERNÁNDEZ, H. R., GOLDSMITH, T. H.: Dichroism of photosensitive pigment in rhabdoms of the crayfish, *Orconectes*. J. gen. Physiol. **54**, 415—432 (1969).

WEBER, H.: Grundriß der Insektenkunde, 3. Aufl. Stuttgart: Fischer 1954.

WIEDEMANN, I.: Versuche über den Strahlengang im Insektenauge (Appositionsauge). Z. vergl. Physiol. **49**, 526—542 (1965).

Chapter 10

Inhibitory Interaction in the Retina of Limulus

By

Haldan K. Hartline and Floyd Ratliff, New York, New York (USA)

With 53 Figures

Contents

Introduction

The interplay of excitation and inhibition lies at the foundation of nervous integrative function. Modern neurophysiology builds on Sherrington's analysis of motor function, extending his concepts to all the sensory systems and to the infinite complexity of the higher nervous centers (cf. Granit, 1966). Antagonistic processes in vision recall Hering; the role of inhibition in vision was clearly recognized by Mach. Sherrington (1897) himself ventured into this field, but it was Granit's work that played an essential role in introducing Sherringtonian concepts in the study of retinal function. "The retina is a nervous center" writes Granit, quoting Cajal, and this he proceeds to confirm, exhibiting the interplay of excitation and inhibition in the retinal action potential and in the unitary discharges of retinal ganglion cells.

Complexity, in structure and function, is the most outstanding characteristic of the vertebrate retina. The opportunity to study retinal integrative processes in a much simpler retina is provided by an arachnoid, the horseshoe "crab", *Limulus polyphemus*, in whose compound eye a non-ganglionic synaptic plexus lies behind the mosaic of ommatidia, interconnecting them. The average light intensity in the small section of the visual field "seen" by each ommatidium primarily determines the neural discharge initiated by that receptor unit; secondarily, however, each unit influences and is influenced by the activity of its neighbors. As a result, the patterns of optic nerve activity are complexly modified representations of the spatial and temporal patterns of light and shade on the receptor mosaic; that is, a stage of retinal integrative action, mediated by the plexus, intervenes in the transmission of receptor information to the higher visual centers — even in this comparatively primitive eye of *Limulus*.

In *Limulus*, a discharge of impulses in any particular optic nerve fiber is elicited only by illumination of the particular ommatidium from which it arises; the influence exerted on it laterally by neighboring ommatidia is inhibitory. Each receptor unit thus has its own small central excitatory field (its own small section of the total visual field), with an inhibitory surround provided by its neighbors.

Each ommatidium is a neighbor of its neighbors; inhibitory influences are exerted mutually. The interplay of excitation generated by the light stimulus to each receptor unit with the inhibition exerted by all the neighbors in each one's field, each of them influencing the others mutually, determines the net outcome — the final pattern of activity in the optic nerve fibers. The interaction can be complex for complex patterns of light and shade, but the mechanism in principle is simple and quantitatively understandable. The lateral eye of *Limulus* is a good preparation in which to study this basic principle of retinal integration.

In this article we will survey the state of knowledge of inhibitory interaction in the retina of *Limulus*. We will begin with a brief summary of the relevant anatomical and histological features of the compound eye and its retina and give a brief treatment of its basic physiology and our experimental methods. Following a review of receptor properties, we will outline the basic properties of the lateral inhibition and then take up in detail the quantitative experimental and theoretical features of the interaction for steady conditions of retinal illumination. We will then review the work that has been done on cellular mechanisms operating in the *Limulus* retina, confining ourselves largely to the inhibitory processes. This will make more understandable the review that will follow of the dynamic properties of the inhibitory interaction. Finally, we will discuss some of the consequences of mutual inhibitory interaction in visual systems — the enhancement of contrast and the accenting of contours and fluctuations in retinal patterns of light and shade.

Anatomy

The lateral, compound eyes of *Limulus* (Fig. 1) are almond shaped bulges on the carapace. In a medium sized adult "crab" (25 cm broad), the eyes are approximately 15 mm long (anterio-posterior) by 10 mm wide (dorso-ventral). They are coarsely facetted — the individual ommatidia are readily seen without magnification. In clear eyes, large pseudo-pupils are readily discerned — the regions in which ommatidia point in the direction of the observer, who then looks into their depths. The pupils change in location, size and shape as the eye is viewed from

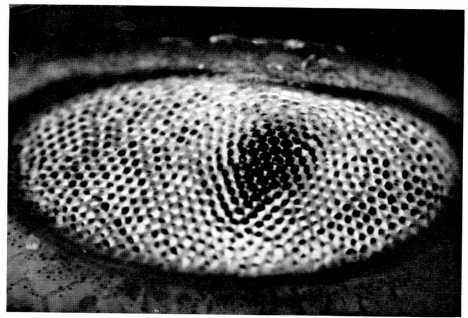

Fig. 1. Corneal surface of a compound lateral eye of *Limulus*. From RATLIFF (1961)

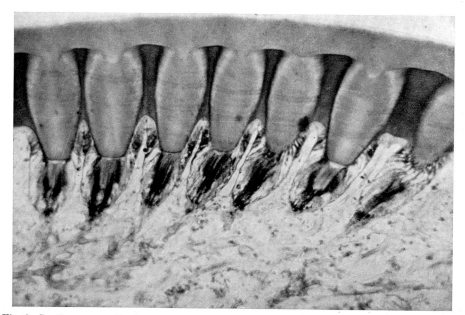

Fig. 2. Section perpendicular to cornea through a portion (approximately 1.5 mm) of the compound lateral eye of *Limulus*, showing 7 ommatidia. The cornea is above; the crystalline cones project downward to the sensory portions of the ommatidia, which have been partially bleached to reveal the retinulae. The fibers of the optic nerve and plexus show faintly below. Photomicrograph prepared by MILLER. From HARTLINE et al. (1952)

various directions. The facet of each ommatidium is approximately 0.1 mm in diameter in large animals; they are spaced, center to center about 0.2 mm apart. In young animals the facets are smaller and more closely packed. Each ommatidium

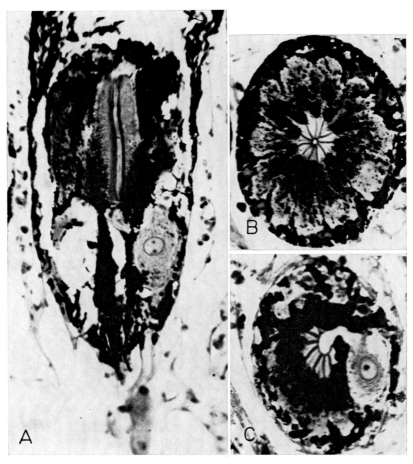

Fig. 3. Photomicrographs of the sensory portion of an ommatidium of the compound eye of *Limulus*. From Miller (1957). (a) Longitudinal section of an ommatidium made in an axial plane. The eccentric cell body with its nucleus is seen at the right. The axon arising from it may be seen distinctly a short distance below the cell body, but the site of its origin is partially hidden by pigment. The distal process of the eccentric cell is seen along the full length of the axial canal. The rhabdom is the relatively clear space centered on the distal process. The remainder of the retinular cells is shrouded by pigment. The axons of the retinular cells are not visible. (Height of Fig. approximately 0.2 mm.) (b) Transverse section between the eccentric cell body and the crystalline cone. In the center is the axial canal, occupied by the dendritic distal process of the eccentric cell. The wedgeshaped retinular cells are arranged radially around the axial canal; their boundaries are in axial planes which divide in two the densely stained spoke-like rays of the rhabdom. (c) Transverse section of an ommatidium at the level of the entrance of the distal process into the axial canal. The eccentric cell, with its nucleus, is seen on the right. The rhabdom is incomplete where the distal process enters the axial canal

accepts light over a solid angle with a half-width of about 5–10° (WATERMAN, 1954; KIRSCHFELD and REICHARDT, 1964). The optical axes of the ommatidia diverge markedly, so that the axes of the marginal ommatidia are oblique to the corneal surface. As a result, the overall visual field of each eye covers (in air) more than a complete hemisphere, the fields of two eyes overlapping except where the flaring carapace obstructs the downward lines of sight. These physical details are pertinent to any studies concerned with the pattern vision of *Limulus*.

Although light from a point in the visual field of a *Limulus* eye enters many ommatidia (within the pseudopupil as seen from that point), it is nevertheless possible for experimental purposes to insure the stimulation of just one, by focussing a small spot of light on its facet. Even better is the use of fiber optics — glass or plastic "light pipes" which are no larger than the corneal facet. The large size and wide separation of the facets in the eyes of adult *Limulus* make good "optical isolation" of individual ommatidia fairly easy, but perfect isolation requires care. Reflection and scatter within the rather thick (approximately 0.1 mm) cornea can

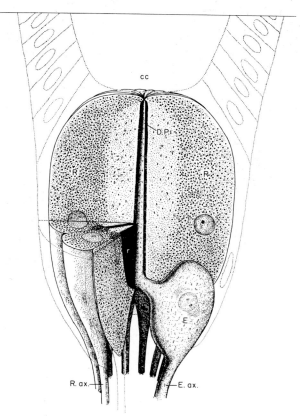

Fig. 4. Schematic drawing of an ommatidium of the lateral eye of *Limulus* as seen in longitudinal and transverse cut-away sections. *CC*, Crystalline cone; *r*, rhabdom; *R*, retinular cell; *D. P.*, distal process; *E. ax.*, eccentric cell axon; *R. ax.*, retinular cell axon. From RATLIFF et al. (1963)

spread appreciable amounts of light into neighboring ommatidia, enough to vitiate quantitative experiments on inhibitory interactions.

Fig. 2 shows a section of the lateral eye of *Limulus*, cut perpendicularly to the corneal surface. Each of the prominent crystalline cones, fused firmly to the cornea, is a lens which focuses a real image of a distant object at a distance of about 0.1 mm proximal to its tip, in the interior of the sensory portion of its ommatidium. The sensory structures are discernible in the figure, some cut open by the section, shrouded by a "hull" of dense melanin pigment (partially bleached in preparing this section). Almost no light escapes the dense pigment shroud, once it has been led into the ommatidium. For a study of the fine structure of the dioptric apparatus of the compound eye of *Limulus*, see Fahrenbach (1968).

The sensory structure of the ommatidium, as seen under the light microscope, is shown in one longitudinal and two cross sections in Fig. 3. The three-dimensional arrangement of the major parts is shown schematically in Fig. 4. A cluster of 10—15 wedge-shaped "retinular cells" are arranged around a central canal "like the segments of an orange" (Grenacher, 1879). The central canal is occupied by the distal process of a bipolar cell — the "eccentric cell". The retinular cells bear the rhabdom which presumably is the carrier of the light sensitive visual pigment. This pigment is a retinal rhodopsin, according to Hubbard and Wald (1960). In fresh preparations, and in conventionally prepared histologic sections, the rhabdom is clear and refractile. Prepared for electron microscopy, it stains with osmium and was shown by Miller (1957) to be composed of densely packed microvilli, giving

Fig. 5. Electron micrograph of the photosensitive part of a *Limulus* ommatidium in cross section. The distal process of the eccentric cell fills the axial canal. Centered on it, much like the hub and spokes of a wheel, is the rhabdom which is formed by microvillous projections from the inner margins of the eleven wedge-shaped retinular cells. From Ratliff et al. (1963)

it the "honey comb" appearance characteristic of the retinulae of arthropods (and some other invertebrate groups). The rhabdomere of each retinular cell is borne on the surface of the inner 1/3 of the thin edge of the cell; the junctional surfaces of the rhabdomeres of adjacent cells being almost indistinguishable, so that the whole rhabdom is a single fused structure. Thus in cross section (Fig. 5) the rhabdom resembles spokes of a wheel.

The retinular cells themselves contain a considerable number of spherical melanin granules. These granules, in dark-adapted eyes, are distributed in the outer portions of the cells, leaving a clear zone around the rhabdom (see Fig. 3). Light adaptation causes this pigment to migrate inwards, completely filling the spaces in the retinular cells between the "spokes" of the rhabdom (MILLER, 1958). Strong light, shining for 8—12 hours will produce heavy packing of the granules (Fig. 6), which disperse gradually during subsequent return to darkness. The exact time course of the pigment migration has not been measured. These pigment movements must have a profound effect on the sensitivity of the receptors — unfortunately quite unexplored. In fresh preparations the retinular cells appear pink, contrasting with the purplish-black of the heavily pigmented "hull" of the ommatidium — a tough capsule which surrounds the sensory structure.

The retinular cells fill the interior of the "hull" giving it a pronounced turgidity. Their distal tips are close to the proximal tip of the crystalline cone; from the proximal tip of each cell a nerve fiber arises to run in the optic nerve. The small bundles of these fibers from the proximal end of each ommatidium appear in Fig. 2 — they can be seen readily in fresh preparations.

The second type of cell in the sensory portion of the ommatidium is a bipolar neuron, the "eccentric cell". Its globular cell body (about $50\,\mu$ in diameter) is

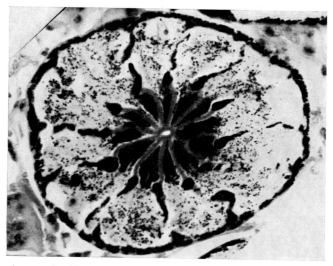

Fig. 6. Light micrograph of light-adapted *Limulus* ommatidium in transverse section. In this fully light adapted ommatidium, pigment granules have migrated into the areas between the rays of the rhabdom. Compare with the rhabdom of the dark adapted ommatidium shown in Fig. 3. Micrograph by MILLER

located eccentrically at the proximal end of the retinula. It sends a distal dendrite up the axial canal of the retinula — in the position that the pith of Grenacher's "orange" would occupy. A large axon emerges from the proximal pole of the eccentric cell, accompanying the small bundle of retinular cell axons into the optic nerve. Lasansky (1967) and Fahrenbach (1969) report the occurrence of a few microvilli arising from the dendritic process of the eccentric cell. This raises the question as to whether this cell may not itself be a photoreceptor.

In *Limulus* there is usually one eccentric cell in each ommatidium, but about 5% of the ommatidia contain two (Fig. 7 left), and in one instance three were observed. In the Japanese genus, *Tachypleus*, ommatidia with multiple eccentric cells occur more frequently (Tomita, 1957; Kikuchi and Ueki, 1965). Rarely, ommatidia are found containing no eccentric cells (Fig. 7 right) — their retinulae lack the regular radial organization of normal ones.

Back of the layer of ommatidia, the nerve fiber bundles thread their way through a loose mass of fibrous tissue (connective tissue, blood vessels, digestive gland, and in breeding season even sperm or eggs!). These bundles (Fig. 8) collect into larger bundles and eventually into the optic nerve, which runs rostrally forward and then curves caudally back and ventrally down to enter the optic lobes of the brain (circumoesophageal ganglion). In old adult animals the space back of the curved cornea is usually curtained off from the body cavity by a tough chitinous partition made up of irregular trabeculae. The optic nerve emerges through one or more fenestrations in this partition. Young adults lack this partition and are to be preferred in experiments involving sectioning the eye, for

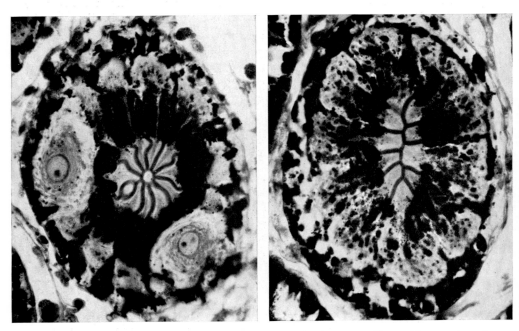

Fig. 7. Transverse sections through the rhabdom of a *Limulus* ommatidium with two eccentric cells (left) and the rhabdom of an ommatidium with none (right). Micrographs by Miller

although the partition can be cut away, doing so is likely to injure the neural tissue underlying it.

On leaving the proximal tips of the ommatidia and for one or two millimeters before collecting in the optic nerve, the nerve fibers from both the retinular and eccentric cells give off copious small branches (Fig. 9). These branches extend laterally in festoons (Fig. 8), forming a network back of the layer of ommatidia. This plexus, first described by GRENACHER (1879), furnishes the neural pathway over which the inhibitory interactions are exerted. It is indeed a "retina" — a little net. No nerve cell bodies have been detected in the plexus, but microscopy reveals numerous clumps of neuropile in which fine nerve fibers are densely packed in what seems to be a completely tangled mass. In electron micrographs MILLER has shown that the innermost nerve fibers in these clumps are packed with synaptic vesicles, and numerous points of contact between fibers have the structure of synaptic areas. The clumps of neuropile seem to be especially numerous around the axons of the eccentric cells, which send short branches into them. They are also present around the fibers from the retinular cells, and elsewhere throughout the plexus. Recent studies by MILLER (1965), by WHITEHEAD and PURPLE (1970) and by SCHWARTZ (1970) have revealed many additional features of the plexus,

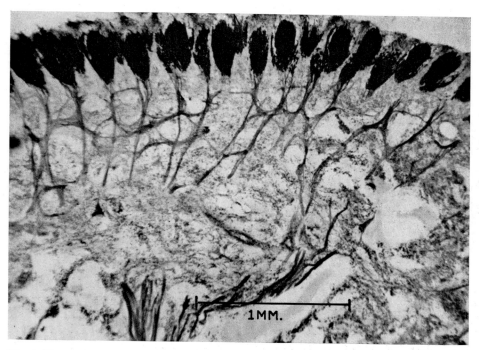

Fig. 8. Photomicrograph of a section, perpendicular to the cornea, through part of lateral eye of an adult *Limulus*. At the top of the figure are shown the heavily pigmented sensory portions of the ommatidia. Bundles of nerve fibers, silver stained, are shown emerging from the om-matidia, with the plexus of interconnecting fibers and a portion of the optic nerve below. The chitinous cornea and crystalline cones that appear in Fig. 3 were stripped away prior to fixation. Prepared by MILLER. From HARTLINE et al. (1956)

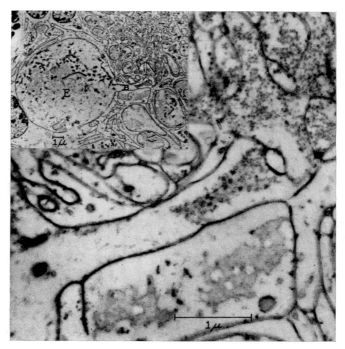

Fig. 9. Electron micrographs of eccentric cell axon *E*, branch *B*, and neuropile *N*. The inset (upper left corner) is at low magnification. The remainder of the figure shows the same branch, and neighboring portions of the eccentric cell and neuropile, at high magnification. Note the small circular outlines of what appear to be synaptic vesicles in the region of the neuropile. From Hartline et al. (1961)

neuropile and synaptic regions. But the pattern of branching and interconnection is still unknown — we have no "wiring diagram" of the *Limulus* retina.

Basic Physiology and Experimental Methods

Physiologic studies of the *Limulus* eye have been based, in large part, on the unitary analysis of the optic nerve activity. For this, the lateral eye is excised with 1—2 cm of optic nerve, mounted in the side of a moist chamber with the back of the eye and the optic nerve bathed in sea water, defibrinated blood or *Limulus* "Ringer's solution". It usually survives in good condition for eight hours or more. The nerve sheath is tenuous and easily stripped away, allowing small bundles of nerve fibres to be gently teased apart. These are slung over a pair of electrodes in the input circuit of a conventional electronic amplifier. On illumination of the eye, the character of the discharge of impulses in such bundles usually implies the activity of many nerve fibers. Single units can be isolated by splitting the bundles with sharp needles into finer and finer strands — after the original method of Adrian and Bronk (1928) — until the regular train of uniform action potentials is obtained which characterizes the activity of just one fiber (Hartline and

GRAHAM, 1932). Alternatively, optical isolation of any ommatidium represented in a fiber bundle can be almost equally effective in obtaining single unit activity. This latter method has the advantage that unwanted scatter of light, exciting receptors that neighbor the ommatidium singled out for study, can be monitored by noting extraneous impulses. For analyzing the interactions of several individual units, they are isolated and recorded simultaneously in several channels of amplification. In such studies it is especially important to avoid effects of light scatter from one receptor or group of receptors to another. Failure to do so can obscure weak inhibitory effects or vitiate quantitative studies of stronger ones.

A second physiological method employs intracellular microelectrodes. For this, the lateral eye is excised and sectioned by a sharp blade, cutting perpendicularly to the relatively soft chitinous cornea but taking care not to drag the sensory structures away from the crystalline cones. Ommatidia thus exposed, covered by a thin layer of bathing solution, can be probed by microelectrodes. The "husk" of the ommatidium is tough, and electrodes must be stiff and sharp. In some ommatidia the eccentric cell can be discerned faintly inside the husk of the ommatidium and penetrated directly. Usually, however, the probing is blind, guided by the nature of the electrical activity encountered. Intracellular electrical recording offers a powerful method for analyzing the cellular mechanisms of excitation and inhibition in the eye, as will be detailed later in this article. The drawbacks of the method are of course the injuries that are inevitable to the unit penetrated by the electrode and to its neighbors and its connections by the sectioning blade.

The discharge of impulses in a single fiber dissected from the optic nerve can be elicited by illumination of one and only one ommatidium — the one from which it arises. Dissection of a fiber in the optic nerve has been made right up to the tip of the ommatidium which on searching the eye had been found to elicit the fiber's discharge. Bundles from the optic nerve, containing several active fibers can have those fibers activated individually, one after another, by illuminating particular individual ommatidia successively. This is usually so, but exceptions occur: two discriminable fibers are sometimes found with partially synchronized discharges, excited by illumination of just one ommatidium. These instances occur with about the frequency with which ommatidia containing two eccentric cells are observed in histologic sections. (In one instance illumination of an ommatidium elicited a fairly well synchronized discharge of what were clearly three different sized spike action potentials; that ommatidium was marked, stained, and sectioned and found to contain three eccentric cells.) For the usual case, a one-to-one relation exists between individual optic nerve fibers in which a discharge of spike action potentials is recorded and corresponding ommatidia, the illumination of which elicits that discharge.

From what has been said above it would seem that of all the nerve fibers emerging from the ommatidia, only those of eccentric cells transmit the trains of impulses that are recorded in the optic nerve as large spike action potentials. This is borne out by microelectrode studies (HARTLINE et al., 1952): successful penetration of an ommatidium yields, on illumination, a generator potential on which are superimposed action potential spikes which are regular, uniform, and coincident with the impulses recorded in the bundle of optic nerve fibers from that ommatidium.

Sometimes, as was noted above, the eccentric cell can be seen in an ommatidium, and penetrated under visual control; these (if alive) invariably yield very large spike action potentials (up to 60 mv) — clearly the responses of the eccentric cell that has been penetrated. Numerous investigations carried out by others (WATERMAN and WIERSMA, 1954; TOMITA et al., 1960; KIKUCHI and UEKI, 1965; BEHRENS and WULFF, 1965) also point directly to the eccentric cell as the source of the spike action potentials. It is for these reasons that we can say that, to the extent that we are concerned with impulses recorded in optic nerve fibers, or recorded by microelectrodes, the ommatidium acts as a functional receptor unit.

This leaves unresolved the question of the role of the retinular cells and their nerve fibers. It seems inescapable to attribute to the retinular cells, which bear the rhabdom, the initial steps in the photoexcitation of the receptor unit. But in spite of numerous careful studies it has been impossible to produce unequivocal evidence of trains of nerve impulses in the dozen or so retinular cell nerve fibers that emerge from the ommatidium and course all the way to the brain — 10 cm or more in large adults. Perhaps these fibers never survive the dissection; perhaps their spikes are too small to be detected, or are swamped by the eccentric cell spikes with which they may be synchronized. Perhaps the retinular cells discharge no impulses at all, but act in other ways. There is no point in discussing this unresolved question any further in this article.

Receptor Properties

The properties of the visual receptor units in the eye of *Limulus*, as measured by the discharge of impulses in their optic nerve fibers, have been described in previous articles (for reviews, cf. HARTLINE, 1940 and 1941—1942; WOLBARSHT and YEANDLE, 1967). Here we need only summarize the salient points as background necessary to the detailed discussion of the interactions of these units.

Illumination of a single ommatidium at moderate intensities elicits a discharge of impulses which begins briskly; the frequency of discharge usually rises slightly to maximum, but quickly subsides in about a second to a considerably lower level — the familiar phenomenon of sensory adaptation. The decline in frequency, after the initial maximum, is monotonic and roughly exponential if the preparation is slightly light adapted. If well dark adapted, however, it goes through a distinct minimum and then rises to its final level. The discharge is maintained with only slight, slow diminution as long as the light continues to shine steadily on the receptor (Fig. 10). The higher the intensity of the stimulating light, the higher is the frequency of the discharge. The frequency of discharge varies approximately linearly with the logarithm of the intensity; the range of a single receptor embracing 5 or 6 factors of 10. The initial maximum can exceed 120 impulses per sec (room temperature); "steady" discharges as high as 60 or 70 per sec can be elicited without producing obvious irreversible changes. At low intensities, the slow maintained discharge tapers off and stops after several seconds. Near threshold only a few impulses are elicited, appearing rather sporadically in the first second or so of illumination.

The discharge of impulses from a photoreceptor unit in response to the onset of steady illumination begins abruptly after a latency that is shorter the higher

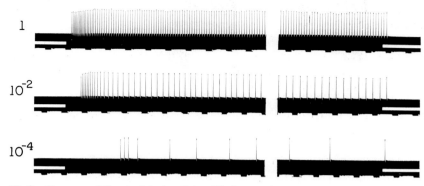

Fig. 10. Oscillograms of the electrical activity (discharge of nerve impulses) in a single eccentric cell axon of the optic nerve of the lateral eye of *Limulus*, stimulated by illumination of the facet of the ommatidium from which that axon arises. Relative values of light intensity given at the left. Time marked in 1/5 seconds in trace at bottom of each record; signal marking period of steady illumination blacks out the white band just above the time marks. From HARTLINE (1941—1942)

the intensity of the light; the latency can be as short as 0.05 sec for very intense light and can be as long as one or two seconds near threshold. When light that has been shining steadily is turned off, the optic nerve discharge stops abruptly (after a delay of approximately 0.1 sec). If the light was very intense, the discharge may be resumed, usually after a silence of a second or so; this "after-discharge" starts at a frequency which is lower than that of the foregoing discharge, and which ordinarily subsides in several seconds (Fig. 11).

Preparations from freshly collected animals rarely show "spontaneous" discharge of impulses even during prolonged stay in absolute darkness. This is true, however, only provided the animals or the excised eyes have not been subjected to heat or other injurious agents. When spontaneous activity develops, it usually begins slowly, with sporadic impulses once every few seconds and is often unnoticed unless the preparation is allowed to become completely dark adapted. For recent studies using the unexcised lateral eye see BARLOW and KAPLAN (1971).

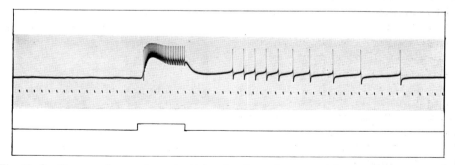

Fig. 11. Oscillogram of generator potential and spike action potentials recorded from a micropipette electrode inserted into an eccentric cell body. Time marked in 0.1 seconds. Period of illumination indicated by signal at the bottom. Note afterdischarge. From RATLIFF et al. (1963)

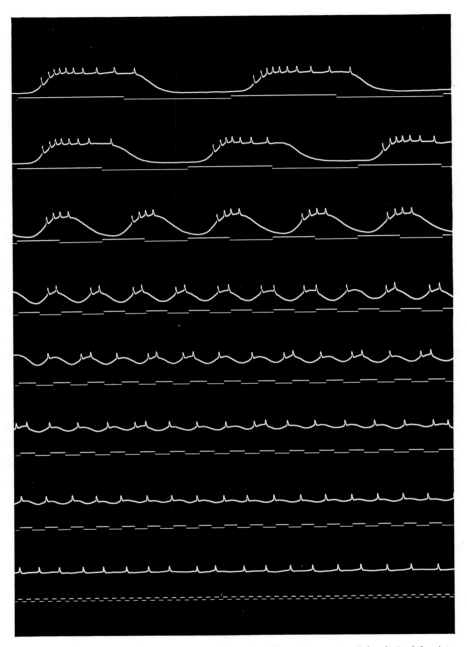

Fig. 12. Oscillographs of generator potentials and spike action potentials elicited by intermittent illumination. Light and dark periods signalled by upward and downward steps in line below each record. Duration of each flash in upper record was 0.5 seconds. From Miller et al. (1961)

The receptors of the *Limulus* lateral eye discharge impulses with considerable regularity. After the first few seconds of activity, however, irregularities in the spacing of impulses do become apparent, to a degree which varies somewhat from preparation to preparation. Irregularity is less pronounced at high levels of frequency. It is less pronounced in the light adapted receptor than when the receptor has become dark adapted. The light-induced discharge is less regular than is the discharge elicited by injecting electric current into the ommatidium (RATLIFF et al., 1968; SHAPLEY, 1969).

A short flash of light, several hundredths of a second long and of moderate intensity, elicits a brief burst of impulses. The latency, average frequency of discharge, and number of impulses discharged depend on stimulus energy, wavelength, and degree of dark adaptation.

Modulation of the light shining on an ommatidium gives rise to fluctuations in the receptor's discharge. A small step-wise increment of intensity elicits (after a brief latency) a rapid rise of frequency which peaks and then subsides to a slightly higher steady level; return to the original intensity is followed by a transient minimum which mirrors the "on" transient (MACNICHOL and HARTLINE, 1948; RATLIFF et al., 1963). Interrupted light (succession of square-wave flashes) causes the frequency of firing to be modulated, up to about 10–15 flashes per sec (Fig. 12). Above about 2–3 flashes per sec, the depth of the modulation is less the faster the flash rate; "fusion" (imperceptible modulation) occurs at higher flash rates the brighter the flashes. The average frequency of impulse discharge in response to rapidly interrupted light is equal to that elicited by steady light of equal average intensity (Talbot's law). This is true even if some modulation of impulse frequency is perceptible (RIGGS and HARTLINE, unpublished observations).

Sinusoidally modulated light has come into use in recent years for studying the *Limulus* receptor, making possible the application of powerful Fourier methods of analysis (PINTER, 1966; DODGE et al., 1968). By keeping the mean intensity well above threshold and the depth of modulation small (10–20 %), the fluctuations in the receptor discharge frequency are strictly sinusoidal, with the period of the stimulus modulation. On-line computer acquisition and monitoring of experimental data, and subsequent computer analysis have greatly advanced the study of receptor properties and mechanisms (cf. DODGE, 1969; and DODGE et al., 1970). The same can be said of the study of the dynamics of the inhibitory interaction, as will be seen later in this article.

Lateral Inhibition: Basic Properties

A discharge of impulses in a single optic nerve fiber can be elicited, we have noted, only by excitation of the one ommatidium in the eye from which it arises. However, activity thus elicited in a fiber can be profoundly affected by excitation of the neighbors of its own ommatidium; the effect they exert is inhibitory (HARTLINE, 1949; HARTLINE et al., 1952). An ommatidium, steadily illuminated, discharges impulses in its fiber at a steady rate; if neighboring regions of the eye are suddenly illuminated, the frequency of discharge in the fiber falls abruptly to a lower level. On extinguishing the light on the neighboring ommatidia, the discharge rises again to its former rate. The brighter the light on the neighbors, the greater is the decrease in frequency; the greater the number of neighbors illuminated, and — broadly speaking — the closer they are to the ommatidium under observation, the greater is the inhibitory effect that they exert. Any ommatidium selected for observation is subject to inhibition by its neighbors; being a neighbor of its neighbors, it inhibits them. Individual ommatidia close to one another usually inhibit each other mutually, but the influences may be unequal in the two directions (HARTLINE et al., 1956).

The discharge of impulses by an ommatidium can also be inhibited by artificial stimulation, electrically, of optic nerve fibers coming from neighboring ommatidia. The inhibition thus produced by antidromic volleys of impulses sent into the eye, first shown by TOMITA (1958), apparently has all the properties that we know for inhibition produced naturally by illumination of neighboring receptor units. It affords a powerful method for the precise analysis of the inhibitory mechanisms; we will discuss it later.

Activity of an ommatidium that takes place when it is in darkness is also subject to inhibition. Thus the afterdischarge which follows strong illumination of a receptor is reduced in frequency by illumination of the receptor's neighbors or by antidromic stimulation of their optic nerve fibers, to the same degree as the discharge during the period of illumination. Spontaneous activity, in darkness, can also be inhibited, as can activity artificially induced by injecting electric current into the eccentric cell of the receptor unit. Thus the inhibition is not the result of some retino-motor change, such as pigment migration, that interferes with the entrance of light into the ommatidium (HARTLINE et al., 1961).

The discharge of a receptor subject to inhibition from its neighbors is approximately as regular as is the uninhibited discharge. Differences have been noted — and they are of interest — but it is important to note that the decrease in frequency is not brought about by the dropping out of impulses in an otherwise regular train. Moreover, there is a one-to-one correspondence between impulses recorded by a micropipette in an eccentric cell, and those recorded in the optic nerve fiber from that receptor unit, even during strong inhibition (RATLIFF et al., 1963).

The inhibitory interaction in the *Limulus* eye is mediated by the lateral branches of the plexus, which interconnect the receptor units. There is ample evidence for this: in fresh preparations that have been sectioned, the fiber bundles of the plexus can be seen under the dissecting microscope, and transverse bundles associated with the bundles of optic nerve fibers emerging from an ommatidium can be cut with fine-pointed scissors. Cutting the branches on one side eliminates the inhibition from neighboring ommatidia on that side; careful sectioning of all of the branches eliminates all inhibition exerted on the ommatidium under observation (HARTLINE et al., 1956). Lateral inhibition in the *Limulus* retina is a synaptic process, as will be discussed later in this article. The synapses are undoubtedly located in the clumps of neuropile that are scattered so copiously throughout the plexus.

The primary function of each receptor unit in the retinal mosaic is to signal the intensity of light reaching it from that small portion of the visual field which it sees. The inhibitory influences which units exert mutually on one another are weak, taken individually, but because each unit is subjected to the summed influences from all its neighbors, substantial alterations in its responses are effected. The overall patterns of activity of the assemblage of all the receptor units thus may differ widely from a faithful representation of the patterns of light and shade in the external visual field. These alterations constitute a first step in the integration of visual information by the nervous system.

The basic experiment for studying the lateral inhibition in *Limulus* consists of singling out, arbitrarily, an ommatidium with its optic nerve fiber, illuminating it steadily, and then testing the effect on its impulse discharge of activating neighboring receptor units. As shown in Fig. 13, illumination of a small group of ommatidia

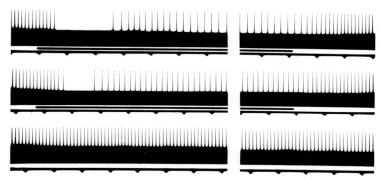

Fig. 13. Oscillograms of nerve action potentials, showing inhibition of the impulses in a single optic nerve fiber of the compound lateral eye of *Limulus*. The ommatidium from which the fiber arose was illuminated steadily, at a fixed intensity, beginning 3 seconds before the start of each of the records; adjacent ommatidia were illuminated during the interval signalled by the blackening out of the white line above the time marks, in the upper two records. For the top record, the intensity of illumination of the adjacent receptors was ten times that used for the middle record. Bottom record is a control (no adjacent illumination). Time marked in 1/5 seconds. From HARTLINE (1959)

neighboring a steadily illuminated test receptor resulted in a slowing of the discharge. The higher the intensity of illumination on the group of neighbors, the greater was the decrease in frequency of impulse discharge by the test receptor (HARTLINE, 1959).

Similar experiments show that, within limits, the greater the number of neighboring ommatidia illuminated, the greater is the inhibition exerted on a test receptor (spatial summation of inhibitory influence). Also, roughly speaking, near neighbors exert a stronger inhibition on a test receptor than distant ones. More exactly, somewhat stronger effects are exerted by receptors at a slight distance ($\approx 1/2$ mm) than by the very nearest neighbors. Beyond 5 mm (eyes of large adults), inhibitory influences generally become imperceptible.

Over a wide range, the level of activity to which a test receptor is excited makes only a small difference in the decrease in frequency that is produced by a given group of neighbors, illuminated at a fixed intensity. Thus we may write, to a good approximation

$$i \approx (e - r) \qquad (1)$$

where the magnitude of the inhibition, i, is to be measured by the difference between the frequency of discharge of the test receptor when it is excited alone (e) and when it is responding to the same illumination but subject to the inhibition exerted upon it (r).

The inhibition exerted on its neighbors by a receptor unit depends on its activity. But its activity is affected by the inhibition exerted on it by those very neighbors whose activity it affects. The mutually exerted inhibitory influences act recurrently — as illustrated diagrammatically in Fig. 14a.

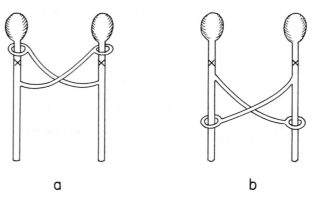

a   b

Fig. 14. Schematic diagram of (a) "recurrent" and (b) "non recurrent" inhibitory systems. In both systems the magnitude of the inhibitory influence exerted by each unit on the other depends upon the level of activity generated at the site x. In the recurrent system each unit exerts influences back on the other at or near the site of impulse generation. In the non-recurrent system (b) each unit exerts influences on the other at some point distant from the site of impulse generation and below the lateral branches. In the Limulus eye, the inhibition acts recurrently. From Ratliff et al. (1963)

Quantitative Studies: The Steady State

To analyze the inhibitory interaction quantitatively, we begin by measuring the optic nerve fiber discharges from two interacting ommatidia, illuminated independently, and restrict our measurements, for the present, to the steady discharges elicited by steady illumination, after all transients have subsided (Hartline and Ratliff, 1957). Oscillograms from such an experiment are shown in Fig. 15. These records show that the discharge of receptors A and B were both lower when the two were illuminated together than when each was illuminated by itself. Since the inhibition of each depended on the activity, r, of the other, we may use Eq. (1) to write a pair of simultaneous equations

$$r_A = e_A - i_{AB}(r_B) \tag{2}$$
$$r_B = e_B - i_{BA}(r_A)$$

where $i_{AB}(r_B)$ expresses the inhibition exerted on receptor unit A by unit B, expressed as a function of the activity, r_B, of the latter (and correspondingly for i_{BA}). To establish the quantitative form of the function i, experimental "runs" similar to those of Fig. 15 are required, using various intensities on A and B, in

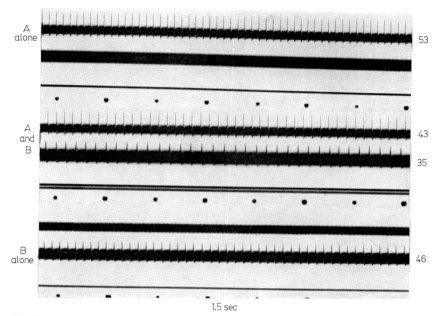

A
alone

53

A
and
B

43

35

B
alone

46

1.5 sec

Fig. 15. Mutual inhibition of two ommatidia close to one another in the eye of *Limulus* steadily illuminated at fixed intensities. Two separate adjacent small spots of light were focussed on the eye, one centered on ommatidium *A*, the other on ommatidium *B*. Each spot illuminated about 5 ommatidia in addition to the ommatidium on which it was centered. Note slowing of discharge from both ommatidia (middle record) when both spots of illumination were on simultaneously. From HARTLINE and RATLIFF (1957)

various combinations. The results of such an experiment are plotted in Fig. 16. Evidently a piece-wise linear relation provides a good approximation for describing the results. Below a fairly distinct threshold — different for each receptor unit — no appreciable inhibition was exerted. Above that threshold, a linear relation held. We re-write (2):

$$r_A = e_A - K_{AB}(r_B - r^0_{AB}) \tag{3}$$
$$r_B = e_B - K_{BA}(r_A - r^0_{BA})$$

where $K_{AB} (\geqq 0)$ is the "inhibitory coefficient" (slope of the line) expressing the action of receptor B on A, and r^0_{AB} the threshold for its action (correspondingly for K_{BA} and r^0_{BA}). It is to be understood that negative values in the parenthesis are to be excluded (replaced by 0) for there is no "negative inhibition" (also, of course, negative values of e and r, which are frequencies, are meaningless).

For any two interacting receptor units, the parameters K and r^0 of inhibitory influence are not necessarily equal in the two directions — indeed more often than not they differ considerably (Fig. 16 is somewhat exceptional in this regard). In magnitude, the values of the coefficients K are usually less than 0.1; we have never observed any that exceeded 0.2. The values of K vary with the separation of the receptors. Roughly speaking, they are greater the closer the interacting receptors (but see below for a more exact description), and fall to zero for receptors

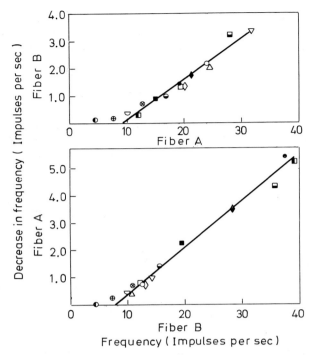

Fig. 16. Mutual inhibition of two ommatidia close to one another in the eye of *Limulus*, illuminated independently at various levels of intensity in various combinations. Amount of inhibition (decrease in frequency) plotted as a function of concurrent level of response (frequency) of the other. The illumination was restricted to these two ommatidia by coating the remainder of the eye with an opaque wax. From Hartline and Ratliff (1957)

separated by about 5 mm or more on the eye. The thresholds, r^0, tend to be smaller the closer the interacting receptors.

When more than two receptor units interact, the set of simultaneous equations describing their responses must be extended, n equations being required for n interacting units. For each receptor unit, the corresponding equation must contain terms representing the contributions to the inhibition of that unit from all the other units; the total inhibition exerted on any one unit is made up of these individual contributions combined according to some law. The law of combination of inhibitory influences turns out to be simple arithmetic addition; this was established empirically by experiment designed to test it (Hartline and Ratliff, 1958).

Thus we may write

$$r_p = e_p - \sum_{j=1}^{n} K_{pj} (r_j - r_{pj}^0) .$$

$$p = 1, 2, \ldots, n$$
$$j = 1, 2, \ldots, n \qquad\qquad (4)$$
$$j \neq p .$$

The same restrictions apply to these equations that were specified for (3) — negative values of the quantities represented are excluded. Also, no negative inhibitory terms (in parenthesis) are allowed; a solution must be sought in which only non-negative terms appear. The conditions that must be met for a unique solution to exist have been set forth by MELZAK (1962). A more subtle restriction concerns the conditions for stability of the responses of a real set of receptors which may satisfy the equations (cf. REICHARDT and MACGINITIE, 1962). The nature of this problem is readily appreciated by considering the pair of Eq. (3) for large values of K. If the determinant $1 - K_{AB}K_{BA}$ should be negative, a valid solution would exist, but the gain around the loop would be greater than unity. Any slight displacement of the responses from the values thus given would then cause the response of one receptor to be depressed to 0 (or to below its threshold) and the response of the other to rise to its e. With time delays in the development of the inhibition, such a "flip-flop" could oscillate between two such states. For more than two receptors, more complex situations are conceivable. In the *Limulus* eye, instabilities that can be related to inhibitory interactions have not been observed.

The restriction $j \neq p$ is written to exclude, for the present, considerations of any possible inhibitory influence of a receptor on itself. As written in (4), with this restriction, we describe only the inhibition of a receptor by all of the $n - 1$ others. However, there is a process in the *Limulus* eye which may properly be termed "self-inhibition", and as we will show later it is important in the dynamics of this system. Even so, it is desirable to preserve this restriction $j \neq p$ in considering, as we do here, the steady-state equation. Indeed, measurements such as we have described, taken during steady conditions, permit no estimate of any self inhibition that there may be. To show this, consider the restrictions removed, and a K'_{pp} admitted, unknown in magnitude. Write e' and K' to identify this condition. By collecting the terms in r_p and dividing each equation through by $1 + K'_{pp}$, the form of (4) is restored, with $e'_p/(1 + K'_{pp})$ in place of the e_p of (4), and $K'_{pq}/(1 + K'_{pp})$ in place of each K_{pq} of (4). It is the unprimed e_p that is measured as "the (steady-state) response of the pth receptor illuminated by itself", and the unprimed K_{pq} that appears as the coefficient of the inhibition exerted by the qth receptor on the pth, in the presence of whatever self inhibitory feed-back the latter may have. We will show later, in the section on dynamics, how the coefficients of self inhibition, K'_{pp}, can be measured, and from them the primed quantities e'_p and K'_{pq} (cf. HARTLINE et al., 1961).

Many experiments in visual physiology concern extended areas in the visual field, in which a number of receptors are uniformly illuminated. If such patches of light are small, the receptors illuminated in each patch may be considered uniform in their properties and interactions, and the system of Eq. (4) may be reduced in number. This is done by collecting terms in r, and partitioning the matrix of their coefficients. Under the assumption that all the e's, K's and r^0's in each group thus partitioned off are the same, the set of Eq. (4) reduce to a fewer number expressing the interactions of receptors within each group on one another, and the action of each group on the other groups.

As an illustration, consider two groups of receptors, each group containing, respectively, n_A and n_B units, each receptor in group A excited at the level e_A

and each responding with the frequency r_A (e_B and r_B for each receptor in group B). Then (4) reduces to

$$r_A = e_A - (n_A - 1) K_{AA} (r_A - r^0_{AA}) - n_B K_{AB} (r_B - r^0_{AB}) \qquad (5)$$
$$r_B = e_B - (n_B - 1) K_{BB} (r_B - r^0_{BB}) - n_A K_{BA} (r_A - r^0_{BA})$$

where K_{AA} is the coefficient of the mutual inhibition of individual receptors in group A, K_{AB} that of each individual receptor in group B acting on each individual of group A (analogously for K_{BB} and K_{BA}). To simplify the illustration, we will neglect the thresholds. We then have

$$r_A = \frac{e_A}{1 + (n_A - 1) K_{AA}} - \frac{n_B K_{AB}}{1 + (n_A - 1) K_{AA}} r_B$$
$$r_B = \frac{e_B}{1 + (n_B - 1) K_{BB}} - \frac{n_A K_{BA}}{1 + (n_B - 1) K_{BB}} r_A \,. \qquad (6)$$

These are in the form $\bar{r}_A = \bar{e}_A - \bar{K}_{AB} r_B$ (correspondingly for $\bar{r}_B$). Here $\bar{e}_A$ is the frequency at which each receptor in group A fires if that entire group is illuminated by itself, and $\bar{K}_{AB}$ the coefficient of the inhibitory action of all the group B receptor units together on each receptor of group A. The quantity $\bar{e}_A$ is less than the "individual" e_A by the factor $\dfrac{1}{1 + (n_A - 1) K_{AA}}$ which expresses the negative feed-back exerted by the "group self inhibition" acting on each receptor of group A by all its companions in that group. The "group coefficient" $\bar{K}_{AB}$ differs from the individual K_{AB}. It is increased by the factor n_B which expresses the summation of the inhibitory influences from all of the units in group B acting together on each unit of A; it is decreased by the factor $1/(1 + (n_A - 1) K_{AA})$ which again expresses the negative feed-back of the "group self inhibition" (Hartline et al., 1961).

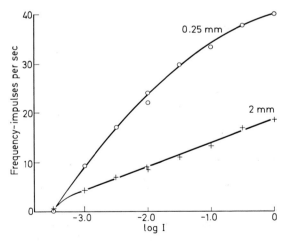

Fig. 17. Relations between intensity of light (log I, abscissae) and frequency of discharge (ordinates) of a single ommatidium when illuminated alone (upper curve, 0.25 mm spot of light centered on its facet), and when illuminated together with a large number (approximately 40) of neighboring ommatidia surrounding it (lower curve, 2.0 mm spot of light centered on its facet). From Hartline et al. (1961)

Experiments that illustrate these points are instructive. In Fig. 17 the frequency of discharge of a receptor illuminated steadily is plotted under two conditions — when it alone was illuminated (upper curve) and when it was illuminated together with about 40 of its neighbors. The smaller slope implies a coefficient of "group self inhibition" equal to approximately 2.5. Note the approach to "saturation" at the higher frequencies for the upper curve is not reached by the lower — the intensity range over which the group can function without reaching saturation is extended by at least 3 factors of 10. Note also the intersection of the curves at small positive value of frequency — indicating the threshold of the interaction.

A simple but important experiment on spatial summation of inhibitory influences is illustrated in Fig. 18. A steadily illuminated test receptor (indicated by X) could be subjected to inhibition by 3 patches of light projected on neighboring regions of the eye, singly or in pairs as indicated. Each patch was adjusted in intensity to yield the same decrement in frequency of the test receptor: when two patches were illuminated together, the decrement produced by their combined actions was almost equal to the sum of their separate actions if they were widely separated, and therefore presumably not interacting. But when the two patches were close together, their combined actions produced a substantially smaller decre-

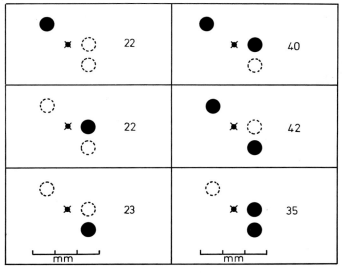

Fig. 18. The summation of inhibitory influences exerted by two widely separated groups of receptors and by two groups of receptors close together. Each panel in the figure is a map of the same small portion of the eye. The test receptor, location indicated by the symbol X, was illuminated steadily by a small spot of light confined to its facet. Larger spots of light were placed singly in one of three locations, as shown in the three panels on the left side of the figure, and in pairs, as shown in the three panels on the right. The filled circles indicate the spots illuminated in each case; the other locations (not illuminated) are indicated in dotted outline merely for purposes of orientation. The number of impulses discharged from the test receptor in a period of 8 sec was decreased upon illumination of the neighboring spot or spots by the amount shown at the right in each panel. From HARTLINE and RATLIFF (1958)

ment. The closely spaced groups evidently inhibited one another, and as a consequence each contributed less inhibition to their combined effect on the test receptor. It was this experiment and similar ones that originally led to the realization that inhibition in the *Limulus* eye is recurrent.

An even more striking instance is illustrated in Fig. 19. In this experiment, a test receptor was subjected to inhibition from a small group of receptors in its neighborhood. Illumination of a more distant group while the nearer was also being illuminated resulted in an *increase* in activity of the test receptor, as it was released from some of the inhibition that was being exerted on it by the receptors in the nearer group. Such "disinhibition" can be quite strong, when large groups of receptors at different distances are brought into action.

Additional quantitative experiments outline the salient features of the steady-state interaction of two or three receptors or small groups of receptors. In Fig. 20 a single test receptor was subjected to inhibition from two groups of neighbors, located 2 mm on either side of it — close enough to it to inhibit it, but widely enough separated so as not to inhibit one another. The combined action of the two when illuminated together is plotted as a function of the sum of their separate inhibitory effects. Arithmetic addition of the separate influences was complete (dotted line at 45°). In the several experiments of Fig. 21, the two receptor groups were close together, and inhibited one another mutually to various degrees. In these experiments, the illumination on one of the groups was fixed, the second

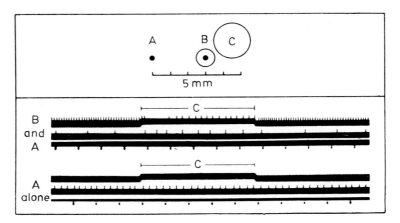

Fig. 19. Oscillograms of the electrical activity of two optic nerve fibers showing disinhibition. The configuration of the pattern of illumination on the eye is sketched above the records. The lower record illustrates the activity of receptor unit *A* in the absence of illumination on *B*, showing that illumination of the large area *C* (signaled by the upward offset of the upper trace) had no perceptible effect. The upper record demonstrates the activity of receptor units *A* and *B* when both were illuminated, showing (1) lower frequency of discharge of *A* (as compared with lower record) resulting from activity of *B*, and (2) effect of illumination of *C* which cause a reduction in the frequency of discharge of *B* and concomitantly an increase in the frequency of discharge of *A*, as *A* was partially released from the inhibition exerted by *B*. Time marked in 1/5 seconds. The black band above the time-marks signals illumination on *A* and *B*, thin when *A* was illuminated alone and thick when *A* an *B* were illuminated together.
Records from HARTLINE and RATLIFF (1957)

group then illuminated to various degrees. The amount of inhibition that the second group added when acting together with the first is plotted against the amount that it exerted when it alone was illuminated. The linearity of the plots reflects the linearity of the inhibitory interaction. The slope of each line — less than unity — reflects the inhibition of the first group by the second. The displacement of the lines to the right is to be attributed to the action of the first group in reducing the effect of the second.

The test receptor itself also takes part in interactions such as those just described. But its effect is small since it is but a "group" of one. By enlarging the spot of light illuminating it, so as to include several of its nearest neighbors, the effects of the receptor group are augmented. In Fig. 22 two widely separated groups were chosen on either side of a third group of several ommatidia, including the test receptor. When illuminated together the inhibition the two groups exerted on the test receptor was actually greater than the arithmetic sum of their separate effects. This may be interpreted by considering that the test receptor's group, being more strongly inhibited when both groups on either side were active, necessarily exerted less inhibition back on them, than when only one of them was illuminated. A trace of this effect may be noted in Fig. 19 in which some of the points lie slightly above the dotted line, even though the test receptor was only a "group" of one.

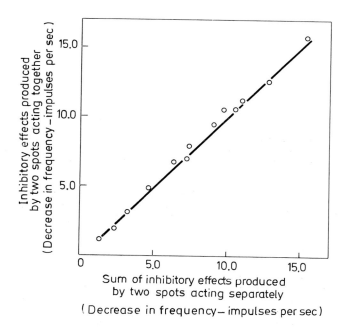

Fig. 20. The summation of inhibitory effects produced by steady illumination, at various intensities, of two widely separated groups of receptors. The sum of the inhibitory effects on a test receptor (steadily illuminated at a fixed intensity) produced by each group acting separately is plotted as abscissa; the effect produced by the two groups of receptors acting simultaneously is plotted as ordinate. From HARTLINE and RATLIFF (1958)

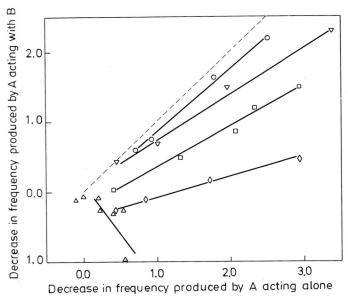

Fig 21 (Legend see p. 407)

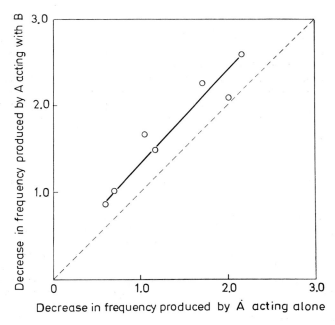

Fig. 22 (Legend see p. 407)

The experiments just described are satisfactorily accounted for by the empirical theory, as expressed in Eqs. (4) and (5). The tedious algebra (cf. HARTLINE and RATLIFF, 1958) can be bypassed by the displays furnished by a simple analog computer, Fig. 23.

A more complete quantitative test of the theory is provided by an experiment (Fig. 24) in which the response of 3 interacting receptor units were recorded when illuminated singly, pairwise, and all together. From the responses to exposures of the single units by themselves the e's are obtained: the pairwise exposures give the K's and r^0's. With these and with the law of spatial summation, the responses of each receptor when all three were illuminated were calculated. Fig. 23 shows that the observed responses agree well with the calculated values.

The success of the piecewise-linear theory in accounting quantitatively for the steady state interactions of 3 receptor units or small groups of receptor units argues for its extension to larger arrays. It has its limitations, however. As long as one considers all elements to be activated above their thresholds, linear theory applies elegantly and simply. At lower levels of response, the non-linearities complicate the analysis, but such complications may comprise many interesting phenomena in vision. Thresholds of individual elements are not perfectly "sharp", and thresholds for group action necessarily blurred by the statistical distribution of individual thresholds. Finally, the inhibitory coefficients, K, are not perfectly constant over wide ranges of activity. Perturbation theory can extend the useful range of analytic formulae; for more extreme ranges, when necessity arises, computer simulations are useful.

Fig. 21. Summation of inhibitory influences exerted on a test receptor (X) by two groups of receptors (A and B) at various distances from one another and from X. Each of the graphs was obtained from an experiment on a different preparation. In each case B refers to a group which was illuminated at a fixed intensity, A to a group illuminated at various intensities. (X was always illuminated at a fixed intensity.) As abscissa is plotted the magnitude of the inhibition (decrease in frequency of the discharge of X) resulting from illumination of A alone. As ordinate is plotted the change in frequency produced by A when it acted with B; that is, the decrease in frequency produced by A and B together less the decrease produced by B alone. In the upper graph A and B were on opposite sides of X; in the others they were on the same side, in various configurations, the lowest being a case showing disinhibition. From HARTLINE and RATLIFF (1958)

Fig. 22. The summation of inhibitory influences exerted by two widely separated groups of receptors upon a test receptor within a third active group of receptors. Spots A and B were located on either side of the test receptor. They were each approximately 1.0 mm in diameter and were centered about 2.0 mm from the test receptor. Unlike the previous experiments, the illumination on the test receptor was not confined to its facet: the spot of light used was about 1.0 mm in diameter and illuminated some 8 or 9 receptors in addition to the one in the center of the group from which the discharge of impulses was recorded. Abscissae and ordinates as in Fig. 21. The positions of the points above the dotted diagonal reflect the influences of the test receptor group, as discussed in the text. Because of the scatter of the points in this experiment the slope of the line that should be drawn through them cannot be determined with precision. The line that has been drawn is in accordance with plausible assumptions concerning the constants of the interacting system. Average $Ix(B) = 1.55$. The equation of this line is:
$$y = 1.13\,Ix(A) + 0.20.$$ From HARTLINE and RATLIFF (1958)

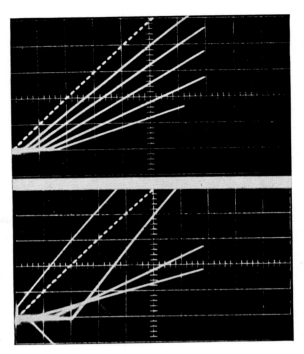

Fig. 23. Solutions generated by an analog computer (constructed by C. C. Yang) imitating the response of three interacting receptor groups. The traces are analogous to the experimental plots of Fig. 20. The decrease in response of a test element (X) inhibited by two interacting elements, A and B, in combination minus the decrease produced by B alone is traced (ordinate) as a function of the decrease in response of X when inhibited by A alone (abscissa). In the upper figure, A and B were caused to inhibit one another to varying degrees (increasing from top to bottom). In the lower figure, various degrees of interaction between A and B are portrayed. The lowest trace (negative slope) illustrates disinhibition. The topmost trace is the only one for which X was caused to inhibit A and B. In this latter case A and B did not interact: this illustrates how their combined effect can sometimes exceed the sum of their separate effects, as in the points above the line in Fig. 22. In both figures the dotted line represents the case of equality of the combined and separate effects of A and B (solid line of Fig. 20). From HARTLINE et al. (1961)

As we have noted above, the inhibitory interaction of receptor units is, in the main, weaker the more widely separated the units are in the receptor mosaic. An experimental study of this dependence has been performed by R. B. BARLOW, JR. (1969). He measured the inhibition exerted by a small group (usually 4 units) on their nearby neighbors and on the more distant ones in a sector of the surrounding mosaic. He found that roughly speaking, the inhibitory coefficients K diminished with increasing distance of the receptors from the group chosen to act as the source of inhibition, and that the threshold rose. As noted above, there was considerable

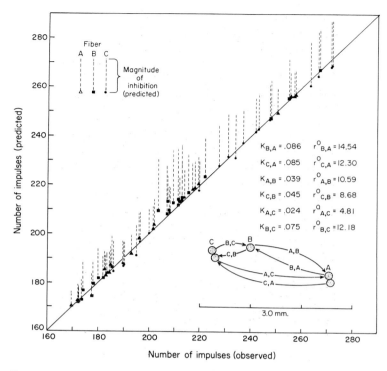

Fig. 24. Comparison of observed and predicted responses of three interacting receptor units (number of impulses generated in a steady-state period of 10 sec). The values of K and r^0 obtained by illuminating the elements in pairs are shown in the table. The configuration of the illuminated ommatidia is shown in the inset. The spots of light A and C each illuminated two ommatidia; the response of only one member of each pair was observed, however, and this response was taken as a measure of the activity of both members of the pair. The spot of light B illuminated only one ommatidium. The uninhibited responses of the three receptor units are indicated — on the ordinate — by the upper end of the dashed line (A, B, and C illuminated alone). The filled symbols plot the responses of A, B, and C illuminated together; predicted responses as ordinates, observed responses as abscissae. From RATLIFF et al. (1963)

variation among receptors, even at comparable distances. The thresholds especially were irregularly distributed.

BARLOW's most striking finding was that the strongest inhibitory action was exerted not on the very nearest neighbors, but on receptors somewhat farther removed — by some 3 to 5 ommatidia — from the source of the inhibition. Beyond this maximum, the K's fell off in value to become imperceptible at distances of about 5 mm. Fig. 25 is one of his figures, illustrating the findings. Fig. 26 is a more idealized "map" representing the distribution of K about a source of inhibition as an elliptical depression with a small hill in the bottom. The detailed configuration of the inhibitory field surrounding each receptor unit is important in considering the role of the inhibitory interaction in vision.

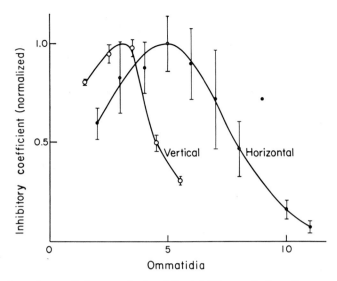

Fig. 25. The dependence of the magnitude of the inhibitory effect on the separation of ommatidia in the retinal mosaic. The magnitude of the effect (measured by the "normalized"
inhibitory coefficient) is plotted on the ordinate as the function of the distance from the source
of inhibition in ommatidial diameters on the abscissa. The coefficients measured in the dorsal
and ventral directions from the source of inhibition are nearly identical and are plotted
together on the "vertical" curve; the same is true for the anteroposterior or "horizontal"
direction. Each point on the vertical curve is the average of four to five experiments with one
exception: the point above the word "Horizontal" represents the only measurement made at
the ninth position in the anteroposterior direction. Standard deviations indicated by vertical bars. The data are normalized by assigning the maximum inhibitory coefficient in each
experiment a value of one and adjusting the other coefficients proportionately. From Barlow
(1969)

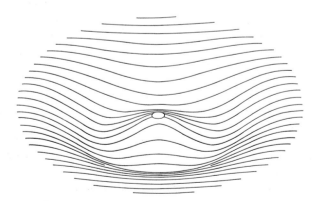

Fig. 26. A three-dimensional map of the inhibitory field in parallel perspective. The map was
constructed using the methods of cartography outlined in Jenks and Brown (1966). The
major axis (anteroposterior) of the inhibitory field lies horizontally. The open circle corresponds to the area occupied by a single ommatidium. The curvature of the lines immediately
surrounding the open circle is based on data extrapolated from Fig. 24. From Barlow (1969)

Cellular Mechanisms of the Inhibition

We have already noted some of the salient features of the inhibitory interaction in the *Limulus* lateral eye: 1) It depends on the integrity of the lateral nerve fibers in the plexus — cutting them abolishes the inhibition exerted on a receptor unit. 2) It does not act by interfering with the entrance of light into the receptor, but lowers the frequency of any discharge of a unit in darkness, such as spontaneous activity shown by some units, after-discharge following intense illumination, or a discharge elicited by passing electric current through a receptor unit. 3) During inhibition, the discharge has essentially the same degree of regularity as that in a discharge of equal frequency in response to lower light intensity on the receptor. At least, there are no gaps in the discharge — the lower frequency is not the consequence of impulses dropped out of a sequence of more closely spaced impulses. Also the recurrent nature of the mutual effects suggest that the inhibition is exerted in the neighborhood of the pacemaker, where the impulses originate, and not at some more proximal point downstream. 4) The amount of inhibition exerted on a receptor depends on the amount of activity in the optic fibers from the neighboring receptors; as we have seen, the quantitative theory associates the inhibition with the response r (not the excitation, e) of each inhibiting receptor. Moreover, antidromic volleys of impulses elicited by electric shocks applied to the optic nerve fibers are equally efficacious in inhibiting a receptor as are orthodromic trains of impulses elicited either naturally by illumination or artificially by the passage of current into the receptor unit. We now turn to the cellular mechanisms underlying the inhibitory interaction.

Inhibition in the *Limulus* eye is a simple synaptic process. The histology of the plexus, described in our earlier section, speaks convincingly for this; the electrophysiological evidence is conclusive.

Penetration of an ommatidium in the *Limulus* eye by a micropipette electrode usually encounters locations in which the potential is 60—80 mv, negative with respect to an indifferent electrode in the bathing medium. This typical "resting potential" in viable preparations undergoes changes in response to influences exerted on the ommatidium. Illumination results in a positive-going, i.e., depolarizing, "generator potential", which appears to be made of summed relatively slow potential fluctuations — the "quantum bumps" discovered by YEANDLE (1958) and studied by FUORTES and YEANDLE (1964) and ADOLPH (1964). After a transient peak the generator potential retains a steady level as long as light shines steadily, fluctuates as the light stimulus fluctuates, and subsides to the resting level when the light is turned off. Superimposed on the relatively slow, graded electrical responses are typical spike-like action potentials, each of which is synchronous with an action potential spike in the bundle of nerve fibers emerging from the ommatidium (HARTLINE et al., 1952).

If the spikes that are thus recorded by the intracellular electrode are small (a few millivolts), and the generator potential relatively prominent, the interpretation is usually made that the pipette has penetrated a retinular cell; if the spikes are large (say, 10 to 40 or more millivolts), it is assumed that an eccentric cell has been penetrated. This conclusion is based largely on the excellent study by BEHRENS and WULFF (1965 and 1967) who injected a marker dye through the recording electrode to identify the cell from which the responses were being recorded. The largest spikes recorded from the eccentric cells rarely if indeed ever show a reversal of membrane potential (electrode becoming positive with respect to the outside medium). From this it is concluded that one usually enters the soma of the cell, and that the "trigger zone", where the spikes originate in the axon, is at an appreciable distance (several hundred microns) proximal to the point of emergence of the axon from the eccentric cell soma. The small spikes, recorded when the pipette is in a retinular cell probably represent only the action potentials of the discharging eccentric cell, attenuated by the impedance of the two cell membranes. Whether nerve impulses are in fact discharged by retinular cells themselves is still open to question (cf. Fuortes' article in this Handbook).

A discharge of nerve impulses by a receptor unit can also be elicited artificially by electric current (Hartline et al. 1952; Macnichol, 1956; Fuortes and Mantegazzini, 1962; Fuortes and Poggio, 1963). Macnichol (1956) showed that injection of current directly into an eccentric cell through a microelectrode, which thus depolarized the cell artificially, initiated a discharge of impulses. If the cell was already discharging impulses (from injury, or when illuminated), the discharge was speeded up when the current passed in the depolarizing direction, slowed down when passed in the opposite (hyperpolarizing) direction. A step of depolarizing current elicited a abrupt transient increase with exponential decay to a steady level. The frequency of the discharge was quite closely linear with the magnitude of the depolarization, i.e., with the magnitude of the injected current.

Concomitant with the generator potential elicited normally by light is an increase in the electrical conductance between the intracellular electrode and the outside medium. This was first shown by Fuortes (1959), further analyzed by Rushton (1959), and confirmed by Purple (1964) and by Purple and Dodge (1965). The accepted interpretation is that excitation by light results in an increased permeability of the membrane of the eccentric cell (and retinular cells), probably to all ions participating in the maintenance of the cell's polarization. The conductance increase is localized in the region of the rhabdom; depolarization spreads electrotonically as the short-circuit drains the polarized membrane, reaching the trigger zone in the eccentric cell's axon, where the local depolarization initiates the rhythmic discharge of nerve impulses.

Inhibition of the *Limulus* photoreceptor unit is also accompanied by electrical events. Illumination of neighboring receptor units, or activation of their optic nerve fibers by antidromic volleys of impulses, which slows the discharge of a unit penetrated by a microelectrode, produces a concomitant increase in membrane potential (Tomita et al., 1960; Hartline et al., 1961; Ratliff et al., 1963). That is, the negativity of the intracellular electrode is increased: the cell becomes repolarized, if it had been already depolarized by light, or hyperpolarized if at rest. The stronger the inhibitory influence, the greater the hyperpolarization. Concomitantly, an increase in the electrical conductance of the cell membrane occurs.

These electrical changes accompanying inhibition of a photoreceptor unit fit the classical picture of an "inhibitory post-synaptic potential". The synapses in the plexus have already been described; the presumed inhibitory transmitter is not known, although Adolph (1966) has presented evidence implicating gamma-amino butyric acid.

The potential changes accompanying inhibition and associated changes in the electrical conductance of the cell are in agreement with the present day concept of synaptic mechanisms. It is generally believed that the effect of the release of inhibitory transmitter at the synapse is to increase the membrane conductance principally for K^+ (possibly Cl^-) ions. This has as a result the increased polarization of the cell membrane, its consequence in turn being the slowing of the rate of discharge of nerve impulses. Direct measurements of the conductance increase during inhibition and the quantitative treatment of the measurements are given by PURPLE (1964) and PURPLE and DODGE (1965).

PURPLE and DODGE's observations, summarized here, are illustrated in Fig. 27. The upper record (a) shows hyperpolarzing, inhibitory post synaptic potential (IPSP) developed in an eccentric cell in response to antidromic activation of optic nerve fibers from receptors neighboring the penetrated ommatidium. In the lower record (b), the separation of the dots shows the conductance increase registered by the bridge circuit. The onset of the IPSP takes place after a delay of ca. 0.2 sec (with a corresponding delay at the cessation of the train of antidromic volleys). Development at onset (and decay at cessation) follow a course that is roughly exponential with a time constant of ca. 0.5 sec (average preparation at ca. 20° C). Note the slight positive deflection at onset, just before the negative-going deflection sets in — likewise a slight increase in the negative direction at cessation, just before recovery begins. The conductance increase parallels the potential change; indeed, there is a linear relation between change in membrane potential V and

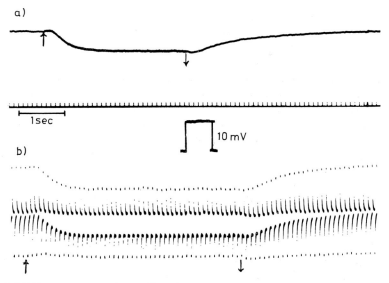

Fig. 27. Inhibitory responses to antidromic stimulation. In (a) antidromic stimulation of 30 impulses/sec was applied between the arrows. In (b) the same stimulus was applied with the bridge circuit operating. The steady-state portions of the lower pulses trace the potential change, and the separation of the traces chart the conductance increase. The pulse transients result from the charging and discharging of the cellular membrane capacitance. Calibration records are for both (a) and (b). From PURPLE and DODGE (1965)

fractional decrease in membrane resistance, $\Delta R/R_{rest}$ in relation to the resting values (Fig. 28). Extrapolation of this line to the membrane voltage at which the fractional decrease is complete ($\Delta R = R_{rest}$) yields the inhibitory equilibrium potential (but see below for a correction to be applied). It is usually some $10-17$ mv below (i.e., negative with respect to) the resting potential.

An important feature of Purple and Dodge's analysis recognizes the fact that the initiation of the optic nerve impulses occurs in a region of the axon some distance (a few tenths of a mm) proximal to its point of emergence from the soma of the eccentric cell. It is the degree of depolarization of this "trigger" zone that determines, by a linear relationship, the rate of impulse discharge. Evidently the inhibitory synapses are close to the trigger zone, and in a favorable position to affect its level of polarization. Because this region of impulse initiation is some distance electrically ($\sim 1/2$ a "cable" length) from the soma, the light-induced depolarization of the cell is somewhat attenuated in its electrotonic spread to the critical region. By the same token, an electrode in the cell body "sees" the full effect of the excitatory depolarization, but sees only to an attentuated degree the inhibitory hyperpolarization, which however is less attenuated at the trigger zone. Hence the observed inhibitory potentials seem to be unduly small in proportion to the marked changes in firing rate with which they are associated. This feature also necessitates the correction to the estimate of the inhibitory equilibrium potential which was noted above.

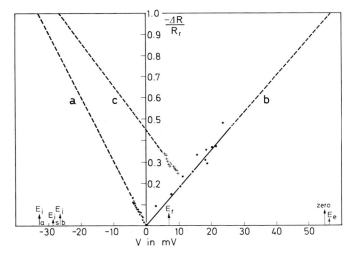

Fig. 28. Relations between the decrease in resistance and the change in membrane potential of an eccentric cell under different conditions of excitation and inhibition. The ordinate is the fractional decrease in the cell resistance, and the abscissa is the change in membrane potential relative to the resting level. (The cell was hyperpolarized during measurements to allow for measurements of the excitatory conductance changes without interference of excessive spiking. E_r is the true resting potential. Zero is the ground potential relative to the V scale and E_e is the extrapolated equilibrium potential for excitation. E_a is the apparent equilibrium potential from curve a and E_b is the apparent equilibrium potential from curve c). E_s is the apparent equilibrium potential for "self inhibition" obtained from data which were not plotted in this figure. From Purple and Dodge (1965)

These considerations are summarized in the schema of Fig. 29, in which the site of impulse initiation is represented as being located at an appreciable distance proximal to the soma of the eccentric cell. The photoexcitatory mechanism is shown at the distal end — in the rhabdom surrounding the dendrite of the eccentric cell. Lateral inhibitory branches from neighboring receptor units are shown as terminating in clumps of neuropile close to the axon, and synapsing on short branches of the axon in the vicinity of the trigger zone. The axon sends branches to its neighbors, which come off in the more proximal stretches of the axon on its way to the optic nerve. No attempt is made to represent the retinular cells, since their role is uncertain.

Corresponding to this schema is the equivalent electrical circuit also shown in Fig. 29. The excitatory mechanism is represented as a battery E_e and variable resistance R_e located in the dendrite. The value of R_e is controlled by the action of light, becoming smaller the brighter the light. A short stretch of distributed R sections is interposed as an electrical cable between soma and "trigger zone" (capacitance is omitted, since the effects being considered here are all too slow to require consideration of it). The inhibitory synapses are represented as a battery E_i and variable resistance R_i the value of which is controlled by the release of inhibitory transmitter at the synapses. From this circuit is derived the expression used by PURPLE and DODGE in their analysis (E_r and R_r are resting values):

$$V = \frac{-\Delta R}{R_r} [E_i - E_r]$$

where

$$\frac{-\Delta R}{R_r} = \frac{R_r}{R_r + R_i}.$$

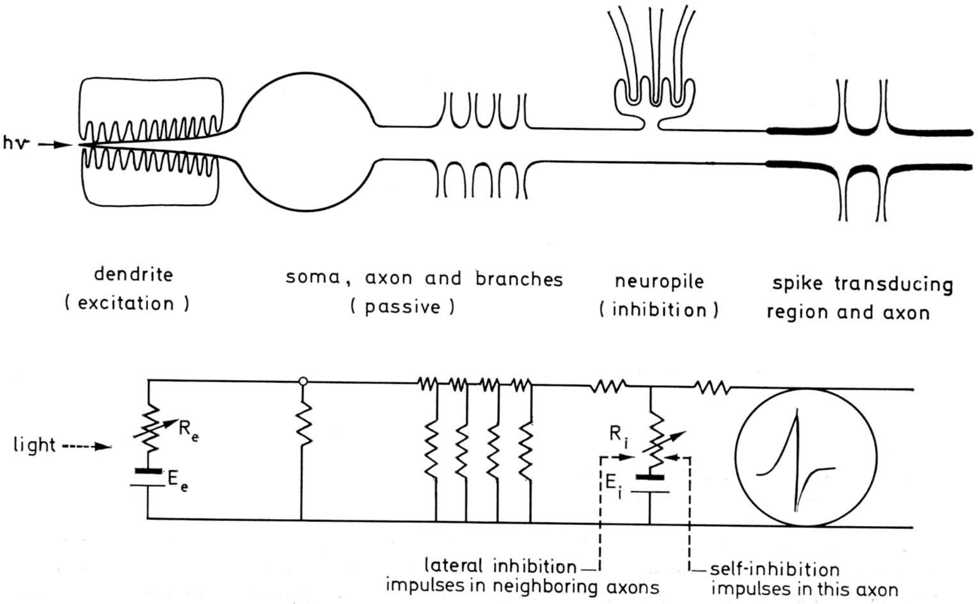

Fig. 29. Two abstractions of the eccentric cell. Upper diagram is a scheme of essential structural elements, lower diagram is the equivalent circuit based upon the structure. After PURPLE and DODGE (1965)

The mechanism whereby transmitter substance released at a synapse effects an increase in ionic conductance is as yet unknown; it is one of the central problems in neurophysiology. It is generally assumed — on good evidence — that each nerve impulse on reaching a synapse liberates a certian aliquot of transmitter substance, which acts to increase the electrical conductance of the post synaptic membrane. In the case of inhibition, only those ions are involved that serve to increase the membrane potential. The membrane is thus hyperpolarized if originally at rest, repolarized if already depolarized. After its release, transmitter substance is removed or inactivated, and the electrical resistance and membrane potential are restored along an approximately exponential course.

In *Limulus*, the rise and fall of the "unitary" pulses of membrane potential elicited at the inhibitory synapses by the arrival of synchronized volleys of impulses have been directly observed by Knight et al. (1970). Groups (2—4) of closely spaced ($\sim$ 10 msec separation) antidromic volleys were repeated at 1 sec intervals and the individual IPSP's averaged (Fig. 30). The records show that the principal response was a *hyperpolarizing* fluctuation; it was preceded by a brief, small *depolarizing* fluctuation. The hyperpolarization developed moderately rapidly, reached a maximum in about 100 msec., and then decayed exponentially with a time constant of about 300 msec. The responses were graded, as shown in the figure; there appeared to be a threshold — a single antidromic volley failed to give any IPSP, and two volleys elicited only a barely perceptible deflection. Above threshold, the peak amplitude of the IPSP varied linearly with number of volleys in the closely spaced group.

The time course of the unitary IPSP's determines the features of the dynamics of the inhibitory influence. The principal, hyperpolarizing component is relatively

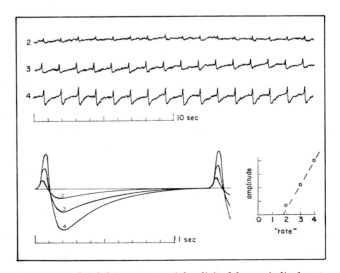

Fig. 30. Computer averaged inhibitory potentials elicited by periodic bursts of antidromic impulses, sample records at top, curves and records labeled according to number of impulses in burst. Insert right: plot of amplitude of inhibitory potential against number of impulses per burst. From Knight et al. (1970)

sluggish (0.3 sec), hence rapid fluctuations of receptor activity will not make themselves felt in the inhibitory influences they exert — only slow changes, and steady levels of activity, will be subject to the effects of inhibitory interaction. We will discuss this fully in the next section. The initial depolarizing component of the unitary IPSP is responsible for the slight positive deflection noted in Fig. 27. It is fast and small compared to the hyperpolarizing component; its principal effect is to slow the development of the hyperpolarization — an effect that has important consequences in the dynamics of the inhibitory action, as we shall see in the next section.

At this point it is possible to discuss the insight that has been achieved by the analysis of the mechanism of the lateral inhibition. To begin with, the "law of spatial summation" of inhibitory influences is seen as a consequence of the addition of conductances in parallel pathways, as additional synapses are activated. The linearity of inhibitory effect, above threshold, is to be attributed to the summation of conductance increments caused by aliquots of transmitter. Aliquots delivered in close succession add linearly, as in the small groups of Fig. 30. In the steady state the balance between delivery of transmitter and its inactivation yields average levels of conductance that are linearly related to frequency of impulse arrivals at the synapses. Increments of conductance at the synapses add linearly; they are the same regardless of the state of polarization of the cell membrane. Hence the magnitude of the inhibitory effect is, to a first approximation, independent of the level of excitation of the receptor — as is written into the equations we have developed.

The basic membrane parameter is the conductance change — conductance given by the movement across the subsynaptic membrane of the specific ions (K^+ and possibly Cl^-) concerned with inhibition and determining the electromotive force represented in the equivalent circuit by the battery E_i. The parameter that is significant in determining the rate of discharge of impulses by the affected eccentric cells, however, is the membrane potential at the trigger zone. Conductance changes brought by the inhibitory synapses will be less effective in changing this potential the closer it is to the equilibrium potential E_i. Hence nonlinearities are to be expected and indeed have been observed. (See PURPLE, 1964; PURPLE and DODGE, 1965; BARLOW, 1967.) As long as the membrane potential differs substantially from E_i, as is the case when the excitatory mechanism depolarizes it, the potential changes follow the conductance changes with fair fidelity, and the observed inhibitory "constants" K are indeed constant to a good approximation.

The mechanism that accounts for a threshold of the inhibition is not yet understood. Evidently one or even two aliquots at each synapse are insufficient to activate any conductance change — even when delivered simultaneously at many synapses.

All of these quantitative considerations are based on the purely empirical finding mentioned in the previous section, that firing rate is quite closely linear with membrane potential. The basic mechanism underlying this fundamental property is as yet unknown.

We now turn to the final topic of this section — self inhibition. In the previous section we have shown that the inhibition of a receptor unit by its own activity cannot be detected in steady state measurements. We have also shown how it is to be included in the set of equations, as the coefficient K'_{pp}, when there arises evidence for its existence and means for estimating its magnitude. STEVENS (1963, 1964) provided that evidence and developed the experimental means for measuring it. Following observations by MACNICHOL (1956) (cf. previous section), STEVENS analysed the changes in firing rate when a step-wise increment of depolarizing current was injected into an eccentric cell through a microelectrode. At the onset of the step, there was an immediate increase in frequency; from this peak the frequency declined exponentially to its final steady level (Fig. 31). This is a form of "neural adaptation" which STEVENS showed was entirely consistent with the concept of self-inhibition: each impulse discharged acted recurrently to delay the discharge of the next. The time constant of the exponential decay — approximately 1/2 sec — implied a transient inhibition after each impulse which decayed with a time constant of the same order of magnitude. The amplitude of the "adaptation" is set by this time constant and the frequency of the discharge; the steady level, being about 1/3 of the peak amplitude, implied a coefficient of self inhibition, K'_{pp}, equal to about 3.

STEVENS' evidence for self inhibition included a demonstration that the decline in frequency was not consistent with the concept of "accomodation" of the impulse-producing mechanism as it is presently understood in terms of work on the squid axon (HODGKIN and HUXLEY, 1952). A brief interruption in the step of current, after a steady level of firing had been reached, elicited a pause in the discharge, followed by a small transient peak which grew exponentially as the duration of the interruption was increased. The interruption of the current needed only to be deep enough to stop the firing in order to permit the growth of the peak along a fixed time course: a deeper interruption, even a strong (hyperpolarizing) reversal, did no more. From this STEVENS argued that the state of the membrane, or possible

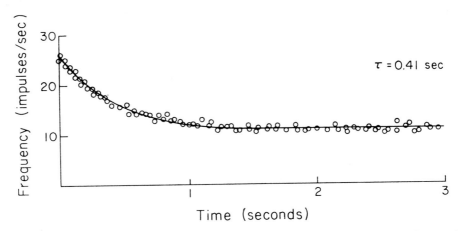

Fig. 31. Response of *Limulus* eccentric cell to rectangular stimulus. The reciprocal of intervals between impulses is plotted as a function of time after the onset of rectangular stimulus. Responses to two successive stimulus presentations are superimposed. The smooth curve through the points is an exponential decay from the initial value to the steady frequency. The time constant of the curve is 0.41 seconds. From STEVENS (1964)

"accommodation" of the firing mechanism, was not the basis for the self inhibition, but rather that a true synaptic process was involved.

PURPLE and DODGE's electrophysiological studies have further elucidated the process of "self inhibition" in the *Limulus* eye. Following each nerve impulse discharged by a receptor unit, a microelectrode in the eccentric cell detects a brief fluctuation in membrane potential in the hyperpolarizing direction, accompanied by an increase in electrical conductance (Fig. 32). The fluctuation starts abruptly with the falling phase of the nerve impulse; the potential and conductance then return to their original level exponentially with a time constant of approximately 1/2 sec. Frequent impulses thus drive the average potential in the hyperpolarizing direction and thus tend to diminish their own frequency of occurence.

Measurements of potential and conductance across the cell membrane during these fluctuations showed them to resemble closely the IPSP's of the lateral inhibitory mechanisms, the extrapolated equilibrium potentials in the two cases being in reasonable agreement. PURPLE and DODGE (1965) therefore concluded, in agreement with STEVENS (1964) that self inhibition in the *Limulus* eye is a synaptic process.

While self inhibition and lateral inhibition are similar in many respects, the two do differ in some fundamental details. In the first place, the self inhibition is always strong: $K'_{pp} \approx 3$. In contrast, lateral inhibitions of any two individual units, p and q, measured under steady condition, are weak: K'_{pq}'s rarely exceed 0.1, and are often much less (cf. BARLOW, 1969). Of course, an experimenter usually searches for the strongest interactions he can find, and these therefore cannot be considered representative. Indeed REICHARDT's analyses of contrast effects (to be reviewed later) implied mean effective K's of the order of 0.02. Reference to our earlier mathematical section will recall that any observed K_{pq} must be multiplied by the factor of $1 + K'_{pp}$ to obtain the value of the "true" coefficient, K'_{pq}. Even so, the large value of K'_{pp} assures that many receptor units can interact

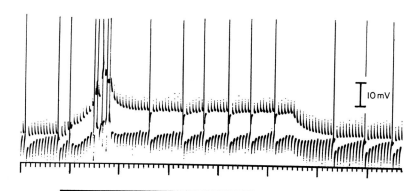

Fig. 32. Bridge records of self inhibitory conductance increase following spike action potentials. Steady state portions (upper ends) of the lower pulses trace the depolarization of the cell induced by the stimulus. The separation of the upper and lower traces indicates the conductance. Note the increase in conductance during the time that the light is on and the transient increases in conductance following each spike action potential. From PURPLE and DODGE (1965)

without endangering the stability of the system (cf. the section on steady state theory).

A second difference between self and lateral inhibition is in the absence of a threshold for the former. After every impulse discharged, no matter how infrequently, a hyperpolarizing potential fluctuation appears, at full value. Thirdly, there is no time delay in the onset of the self inhibition, as can be seen in Purple and Dodge's records, and in Stevens' finding that the very first intervals in the peak transient of the response to onset of injected current fall on the common exponential decay curve of the entire discharge. No trace of an initial depolarizing component precedes the hyperpolarizing fluctuation. Finally, the time constants of decay of IPSP's for self as compared with lateral inhibition have been found to differ significantly when measured under the same conditions in the same preparation: the self inhibitory decay constant is always approximately twice that of the lateral inhibition. This has the consequence that transient lateral influences can act strongly, before they are "clamped" by the more sluggish self inhibitory feed-back. This will be discussed in detail in the next section.

The Dynamics of Excitation and Inhibition:
(1) Transient Responses

Pronounced transient responses to temporal changes in the pattern of illumination on the retina are among the most outstanding features of the neural activity of the lateral eye of *Limulus* — indeed, of all well-developed visual systems. These transient responses result from the combined effects of many and diverse processes — including the photochemical processes in the receptors themselves, the various electrochemical processes underlying the generation of nerve impulses, and the interplay of excitatory and inhibitory influences among neighboring units in the retina. But however many and diverse the causes may be, it is evident that inhibitory influences play a major role in the accentuation of transient responses.

The steady discharge of a single receptor unit in response to constant illumination is altered in a characteristic way by step increments or decrements in the intensity of that illumination (Macnichol and Hartline, 1948). A relatively small increase in intensity produces a large transient increase in the frequency of response which gradually subsides to a steady level only slightly greater than that preceding the change in illumination. Similarly, a small decrease in intensity produces a large momentary decrease in the frequency of discharge, which then returns to a level slightly below that preceding the change in illumination. (As shown above in Fig. 31, self inhibition contributes much to such effects, but — as we shall see later — it is not the sole cause of them.)

For moderate changes in illumination, as in the experiment illustrated in Fig. 33, the transient responses to a step increment (a) and an equal decrement (b) are almost exactly equal but opposite in sign — one being more or less the mirror image of the other. Note also that the transients at the beginning and the end of an increment (or decrement) are also nearly identical. These identities are an indication that the transient response of the system is linear, over this small range at least, which is essential for the quantitative analyses to follow.

Because of the lateral inhibition, these transients in the activity of any one receptor unit, although generated in response to local changes in the stimulus on it, produce similar but opposite transients in the frequency of impulses discharged by neighboring units — even though these neighbors may be steadily illuminated. In the experiment illustrated in Fig. 34 (RATLIFF, 1961) the frequency of impulses was recorded simultaneously from two units near one another. One (solid circles) was illuminated steadily with a small spot of light throughout the period shown. The other unit (open circles), along with several of its immediate neighbors, was illuminated with a larger spot. At $t = 0.0$, a step increment was added to this spot, and at $t = 2.0$ seconds, the illumination was reduced to the former level — as indicated at the bottom of the graph. Accompanying the marked excitatory

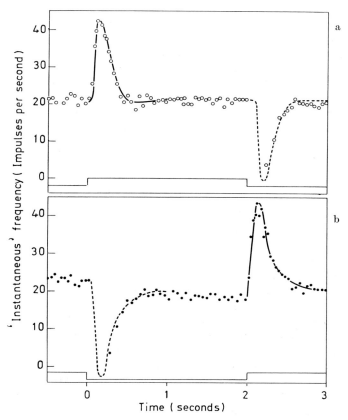

Fig. 33. Changes in optic nerve discharge frequency for incremental and decremental stimuli. Light adapted steadily discharging *Limulus* single optic nerve preparation stimulated by an incremental 2-second flash (a) or 2-second decrement (b). Open and closed circles represent experimental determinations of "instantaneous" frequency of response (reciprocal of time between successive impulses). The bottom lines in each half of the figure indicate onset and termination of the stimulus. Solid lines fitted by inspection to changes in frequency of response during upwardstep stimuli; broken lines are mirror images of the solid lines (for each half of the figure, respectively). Log adapting $I = -0.26$; log I during increment $= 0.0$; log I during decrement $= 0.50$. From RATLIFF et al. (1963)

transients produced by the increment and decrement are slightly delayed and somewhat reduced opposite effects in the neighboring unit — first an inhibitory transient, then a lesser steady inhibition, and finally a release from inhibition with a significant post-inhibitory rebound. Considering the activity of the retina as a whole, an excitatory transient in the response of any one unit is accentuated by the concomitant and opposite inhibitory transient it produces in neighboring units.

To express the dynamics of the inhibitory interaction in a quantitative form similar to our steady state equations we make the following assumptions: Uniform amounts of an inhibitory transmitter substance are produced in a particular receptor unit by each nerve impulse in its neighbors. The transmitter decays, or becomes inactivated, along an exponential time course, and the inhibitory potential is linearly related to the transmitter concentration that exceeds a threshold value. (This threshold is imposed locally — that is, there is no subliminal interaction among units or between self inhibition and lateral inhibition.) And finally, although the time constants for the self inhibition and the lateral inhibition differ significantly, we will neglect this for the moment.

On the above assumptions:

$$r_p(t) = e_p(t) - \sum_{j}^{n} K_{p,j} \left(\frac{1}{T} \int_{0}^{t} \exp\left[(t' - t)/T\right] r_j(t') dt' - r_{j,p}^0 \right), \tag{7}$$

where T is the time constant of the exponential decay of the hypothetical trans-

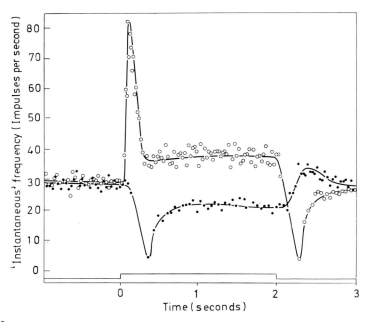

Fig. 34. Simultaneous excitatory and inhibitory transients in the responses of neighboring receptor units. The illumination was constant on one receptor unit (black filled circles). On a small group of neighboring receptor units (the response of one indicated by open circles) the illumination was first increased and then decreased abruptly, as indicated by the signal at the bottom of the graph. From Ratliff (1961)

mitter and $K_{p,j}$ depends on the amplitude of the corresponding inhibitory potential in the steady state. Note that p is not excluded from the summation; that is, "self inhibition" is included. These dynamic integral equations reduce to steady state equations similar to (4) when t approaches infinity and all frequencies become constant.

The transient excitatory and inhibitory actions and reactions, even in response to simple steps in illumination, can be very complex (cf. RATLIFF et al., 1966). To obtain simpler patterns of activity in the optic nerve we have resorted to Tomita's technique for antidromic stimulation that we mentioned earlier. Although the inhibition produced by antidromic stimulation of the optic nerve is admittedly unphysiological (in that the impulses in the nerve travel towards the eye instead of toward the optic ganglion), it seems to have essentially the same properties as the inhibition resulting from impulses that are elicited by stimulating the eye with light and which travel in the normal direction in the same optic nerve fibers (Fig. 35). The antidromic technique has the great advantage that it allows precise control of the excitation of the optic nerve. Over a wide range the frequency of discharge of volleys of antidromic impulses follows exactly the frequency of the electric shocks to the nerve. Furthermore, the group of units, driven antidromically in synchrony, can be treated as a single large unit which exerts strong inhibition on the test unit. The test unit itself exerts no effective inhibition back on units of the group, however, for the externally controlled electrical stimulation is the sole determinant of their frequency of discharge. Under these conditions the lateral inhibition is, in effect, all one way: from the group to the test receptor. Thus, both

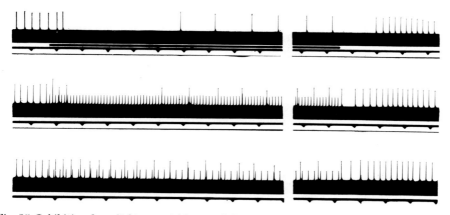

Fig. 35. Inhibition from light on neighbors and from antidromic stimulation. The upper trace is an oscillogram recording of nerve impulses from a single eccentric cell axon. The pulse frequency was relatively constant until a light (designated by the black bar beneath the spikes) was shone on the neighboring ommatidia. In the two lower traces inhibition was produced by antidromic volleys in the other optic nerve fibers. Small spikes were due to spread of the compound action potential to the recording electrode. They do not indicate antidromic firing of the single test fiber. Notice that the amount of inhibition was increased (middle trace) at higher antidromic pulse frequency. Notice also the similarities in the transients of the responses to light and to antidromic inhibition, including: delay, initial undershoot (at the higher antidromic frequency), and post inhibitory rebound. Time marks are at 0.2 second intervals.
(Figure taken from TOMITA, 1958)

No Threshold

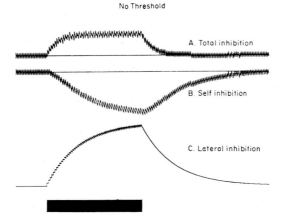

Fig. 36. Model inhibitory pools with no threshold. Trace C shows the lateral inhibitory pool as it builds up during a train of inhibitory pulses designated by the black bar. The sawtooth build-up is essentially exponential. Trace B shows the self-inhibitory pool decreasing as a result of the decreased firing of the inhibited receptor. The rate of decrease and increase of self-inhibition is such that it matches the lateral inhibition so that their sum (Trace A) has no overshoot. (Increase in total inhibition corresponds to decrease in the firing frequency of the test unit.) From LANGE et al. (1966)

With Threshold

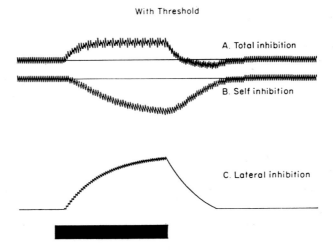

Fig. 37. Model inhibitory pools with a threshold. Trace C shows the lateral inhibitory pool. Notice the delay due to the time to achieve threshold; notice also the shortened decay. Because of this shortened decay the self inhibition (Trace B) cannot compensate exactly for the lateral inhibition (Trace C), and therefore their sum (Trace A) exhibits an undershoot which leads to a post inhibitory overshoot in the frequency of the test unit. The apparent noise in the total inhibition is a consequence of beats between the sawteeth of the lateral and self inhibitory pools. From LANGE et al. (1966)

the experiments and the corresponding theoretical formulation are greatly
simplified, which is most helpful in the initial stages of our analysis.

To test the theory, the dynamic equations were written as a digital computer program. The basis of the program is a loop, which represents one "clock
cycle" — usually taken as one millisecond of time. A representation of the generator potential (g) in the test unit was formed from an excitatory term (e) minus
inhibitory terms, representing pools of self inhibition (I_s) and lateral inhibition (I_L). At each clock cycle of the program the generator (g) was added to
a running sum. When this sum reached a critical value an "impulse" was recorded
for the test unit and the summation was begun again. In short, this amounted
to an integration of the generator and the production of an impulse frequency
proportional to it.

The pool of self inhibition (I_s) was set to zero at the beginning of the program.
When an impulse was produced in the test unit, a fixed quantity of self inhibition
(A_s) was added to this pool. At each cycle of the program a portion of the pool,
proportional to its current value and to the reciprocal of the self inhibitory time
constant (T_s), was substracted from it. Thus there was formed an exponentially
decaying "inhibitory potential" following each impulse generated. (This results in
an exponential approach to the steady state frequency as in Fig. 31.)

In the same way the pool of lateral inhibition (I_L) was formed with a quantity
(A_L) added each time an antidromic pulse occured. At each cycle of the clock the
pool was decremented, the decay being governed by the lateral inhibitory time
constant (T_L). Thus, the impulse frequency of the test unit was governed by the

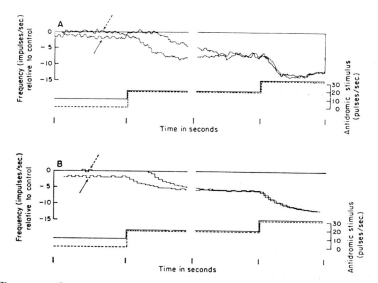

Fig. 38. Responses of an ommatidium (A) and of the model (B) to steps of antidromic frequency,
showing changes in delay. The average frequency of controls in this experiment was 20 impulses/
sec. In B the same inhibitory input was introduced into the model as described in the text.
The difference in the time delay due to previous inhibition is evident. The constants used in
the calculations were: $A_L = 1.3$; $A_s = 2.0$; $T_s = T_i$; and $C = 7.2$. From LANGE et al.,
(1966)

generator that resulted from the excitation minus the self inhibition and the
lateral inhibition:

$$g = e - I_s - I_L.$$

According to the above theory, the relative amounts of the lateral and self
inhibition resulting from a uniform train of antidromic impulses would be that
shown in Fig. 36. Note that the theoretical curve for total inhibition has no latent
period and that the onset and cessation are symmetrical. In these respects, it does
not conform exactly to the results of the experiments. Addition of a threshold to
the lateral inhibitory pool however, improves the representation. The threshold
is entered in the program simply by substracting a constant (C) from the lateral
inhibition (I_L) leaving a remainder (I_r). If this resulted in a negative, I_r was set
equal to zero. Now,

$$g = e - I_s - I_r.$$

Fig. 37 shows the self and lateral inhibition and the total inhibition resulting
from a uniform train of antidromic impulses in this non-linear version of the
model. Note the substantial time delay to the onset of the lateral inhibition and
the asymmetry of the onset and cessation of the total inhibition. Note also the
small post-inhibitory rebound.

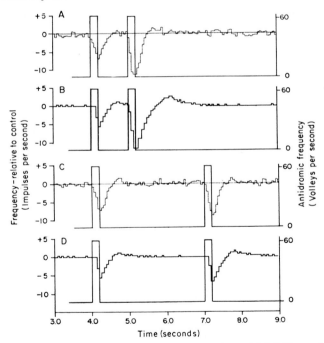

Fig. 39. Facilitation. Fig a and c illustrate the facilitation of inhibition when a burst of
inhibitory pulses (at 4 sec) is followed by another (at 5 sec in A and 7 sec in C). It is evident
that the facilitation decreases with increasing time between bursts. According to the theory
the facilitation of inhibition is the result of superposition of the inhibitory influences produced
by the second burst upon residual subthreshold influences remaining from the first burst.
Figs. B and D illustrate the model simulation of this experiment, as described in the text.
From Lange et al. (1966)

Comparison of actual experimental results with these theoretical predictions are made in the next two figures. In Fig. 38, both the experimental curves (A) and the curves from the theoretical model (B) agree in showing that the onset of inhibition is delayed when the antidromic stimulus was stepped from zero (dotted line), as compared to its onset when the step began from a steady rate (solid line). In the latter case the inhibitory pool was already above threshold, and an increase in antidromic firing was effective almost immediately. In the former case, starting from zero, it took time to fill the pool to threshold. In Fig. 39 the second of two flashes in succession produced a larger inhibitory transient when the interval between the flashes was 1 sec (A, experiment; B, model) than when the separation was 3 sec (C and D). In A and B, the inhibitory pool was partly filled, and the effects of the second flash were facilitated; in C and D, when the inhibition had further decayed, the two transients were more nearly equal. Note the post inhibitory rebound, evident in all traces. This rebound resulted in part from the fact that the lateral inhibition, although decaying asymptotically to zero, reached threshold quickly at a time when the self inhibition was still recovering.

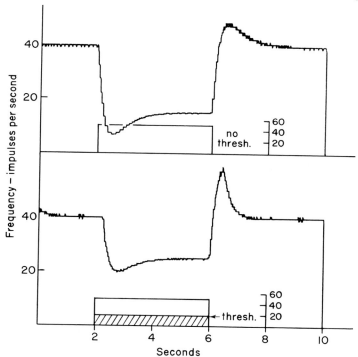

Fig. 40. Simulation by means of a computer program of the responses of a steadily excited receptor subjected to a period of constant inhibition from neighboring receptors. The decay constants assigned to the self inhibitory and lateral inhibitory influences were respectively 1 sec and 0.4 sec. For the upper trace, a threshold of zero was assigned to the lateral inhibition; for the lower tracing, a threshold was introduced that was unrealistically large, considering the strong lateral influence that was assigned. This served to exaggerate, for illustrative purposes, the asymmetries of onset and cessation of inhibition, especially the "post inhibitory rebound". From HARTLINE (1967)

In effect, the threshold of the lateral inhibition acts to give its decay a short time constant. In addition, it is an experimental fact that the actual time constant of decay of lateral inhibition is almost half that of self inhibition. Threshold simply enhances this difference. The computer model, programmed for these differences in time constants, shows responses to a step of antidromic stimulation that are symmetric at on and off (Fig. 40, upper trace, no threshold). Undershoot and overshoot are characteristic of such a linear system with two time constants. Note that the introduction of a threshold for lateral inhibition (Fig. 40, lower

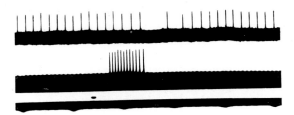

Fig. 41. Transient inhibition of the discharge from a steadily illuminated ommatidium (upper trace) by a burst of impulses discharged by a second ommatidium nearby (lower trace) in response to a 0.01 second flash of light (signalled by the black dot in the white band above the time marks). Time marked in 1/5 seconds. From Hartline et al. (1961)

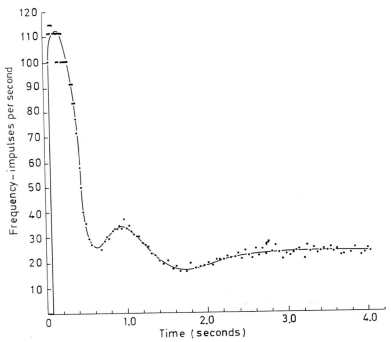

Fig. 42. Plot of the time course of the frequency of discharge of an ommatidium illuminated together with several nearby ommatidia, showing oscillations resulting from the time delay in the action of the mutual inhibitory influences. From Ratliff et al. (1963)

trace) diminishes the "on" undershoot and enhances the rebound at "off". Such asymmetries are usually quite evident in actual experiments.

In general, the quantitative agreement between experiment and theory thus far developed is very good, but there are still discrepancies. These have served to guide further experiments which have led to an improved theory and to further insights into the nature of the underlying mechanisms. In Fig. 38, for example, the experiment (A) shows a considerable delay before the step in antidromic frequency from 10 to 20 volleys per second (dashed line) results in any appreciable inhibition (lower trace). The model (B), however, shows an immediate inhibitory effect. This and other evidence (for example, the long delay that can be seen directly in Fig. 41 and that can be inferred from oscillations in Fig. 42) indicated that there must be some delay in the system in addition to that caused by the time required to reach threshold. It now appears that this additional delay results almost entirely from the peculiar time course of the lateral inhibitory potential described in the previous section: an initial depolarization which serves to delay the hyperpolarization so that it does not reach its maximum for 150 milliseconds or so. It is evident that our simple theory, while quite good as a first approximation, is not adequate to express, in all details, the dynamics of excitation and inhibition in the retina of *Limulus*. In the next section we show how some of the defects in this first approximation have been corrected and the theory expressed in a more exact form.

The Dynamics of Excitation and Inhibition:
(2) Linear Systems Analysis

The following analysis is based on three unitary processes underlying excitation and inhibition in the *Limulus* retina which can now be directly observed: (1) The photoexcitatory "shots" (the so-called quantum bumps); (2) the self inhibitory potential produced in an eccentric cell following each impulse it discharges; and (3) the lateral inhibitory potential following each impulse discharged in neighboring eccentric cell axons. We have discussed the cellular mechanisms of these basic processes in a previous section. (For reviews see DODGE (1969) and DODGE et al., 1970). In this analysis we will only be concerned with the time course of each process. To recapitulate, the discrete photoexcitatory "shots" have a fairly abrupt onset and a decay that is approximately exponential. Their average amplitude changes markedly with the state of light and dark adaptation. The self inhibitory potential has a very abrupt onset and an approximately exponential decay. The lateral inhibitory potential is more complex; a slight excitatory depolarization precedes the inhibitory hyperpolarization. As a result, the inhibitory phase does not begin at once, and its maximum is delayed some 150—200 milliseconds.

The integration of these several processes, and the resulting generation of a train of nerve impulses, may be represented in terms of the block diagram in Fig. 43. Light incident on the ommatidium produces quantum bumps which sum to yield a generator potential. Through a "voltage to frequency" converter (the impulse generating mechanism), a train of spike action potentials is discharged in the eccentric cell axon. This positive excitatory influence is integrated with self inhibition, represented by the negative feedback loop, and with lateral inhibition, represented by the negative input from neighbors. (Note, as mentioned in the section on cellular mechanisms, that similar positive and negative effects may be produced by passing current, in the appropriate direction, through a microelectrode inserted in the eccentric cell body).

Since light incident on an ommatidium, current injected into its eccentric cell, and lateral inhibition exerted on it are all subject to independent control, the behavior of the system may be examined not only as a whole, but in the several parts that correspond to the components of the block diagram. A convenient way to characterize these components is in terms of linear systems analysis — that is, in terms of their responses to sinusoidal inputs.

For a translationally invariant linear system, the response to a sinusoidal input is a sinusoidal output of the same frequency, generally differing in amplitude and phase. The relationship of the amplitude and phase of the output to that of the input, as a function of frequency of the input, completely characterizes the system and is commonly called the "transfer function" or "frequency response" of that system. The *Limulus* retina is sufficiently linear to be analyzed in this way. Such analyses show that the properties of the three underlying processes in the *Limulus* retina (the photoexcitatory "shots", the self inhibitory potentials, and the lateral inhibitory potentials) account almost entirely for the behavior of the entire network of interacting elements.

Let us begin with the two component processes and the one composite process that occur entirely within a single ommatidium: light to generator potential,

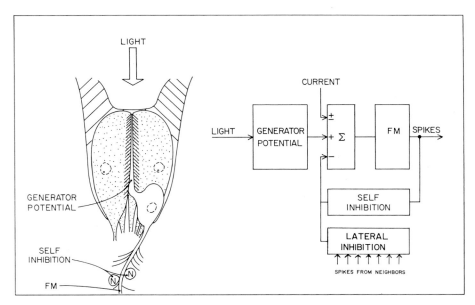

Fig. 43. Block diagram (right) of *Limulus* ommatidium (left). Light incident on the photo-receptors produces a generator potential. This excitatory influence (represented as +) sums with the self and lateral inhibitory influences (—) exerted in the clumps of neuropile (*N*) to drive the voltage-to-frequency converter (*FM*) which determines the rate of discharge of spike action potentials. The same effect can be produced by injecting current into the eccentric cell.
From DODGE et al. (1970)

generator to nerve impulses, and light to nerve impulses. Sample records of data are shown in Fig. 44. The first component, the generator potential elicited in response to sinusoidally modulated light, was recorded after blocking the impulse generating mechanism with tetrodotoxin. This was done to eliminate the self inhibitory potentials generated by the impulses and to avoid obscuring the record by the spike action potentials of the nerve impulses that would be conducted passively to the site of the recording electrode. The second component, generator to spike frequency, was measured by sinusoidal modulation of current passed

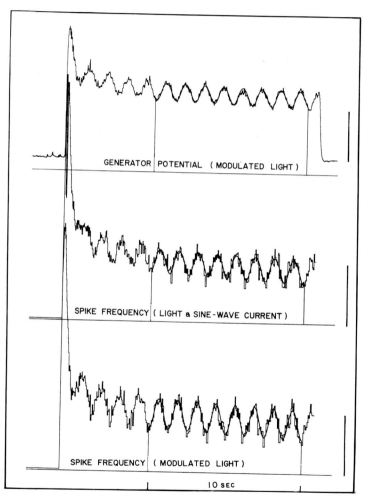

Fig. 44. Sample data records for the component processes in a single ommatidium. The upper record shows modulation of the generator potential by modulated light, the middle record shows modulation of spike rate by modulated current superimposed on steady light, and the lower record shows modulation of the spike rate by modulated light. Smooth curves between the two vertical lines show curves fit to a 10-second span of the data by the processing scheme described in the text. From KNIGHT et al. (1970)

through the recording electrode while the ommatidium was steadily illuminated. The effects of self inhibition appear in this component. The composite response, light to spike frequency, was obtained by measuring the impulse discharge in response to sinusoidally modulated light.

To a span of each record was fitted, by least squares, a combination of six time functions: a constant, a linear ramp, a cosine and a sine of the input modulation frequency, and a cosine and a sine of twice that frequency. (In Fig. 44, the fitted curves are superimposed on a ten second span of the raw data.) The constant measures the mean level of the output and the ramp measures the drift of the response. Cosine and sine of twice the input frequency measure "harmonic content" — a check on the linearity of the system. The coefficients of cosine and sine at the input frequency are easily converted into amplitude gain and phase which — as a function of modulation frequency — is the "frequency response" we seek. For details see Knight et al. (1970).

Frequency responses obtained for the two separate components are shown on the left side of Fig. 45, and the corresponding composite frequency response on the right. Combining the empirical frequency response for the generator potential and self inhibitory feedback (left) predicts fairly accurately the composite frequency response (right). Note the prominent maximum in the overall frequency response at about two cycles per second. Part of this peak results from the self inhibitory feedback, which attenuates the amplitudes of response at low modulation frequencies, but which, because of the long duration of the inhibitory potential, has a

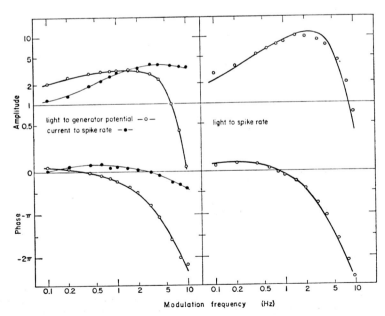

Fig. 45. Left frame: amplitude and phase of component processes, light to generator potential (O) and current to spike rate (●). Right frame: amplitude and phase of composite process, light to spike rate. Data points from direct measurement, solid curves predicted from curves fitted to data of component processes in left frame. From Knight et al. (1970)

diminishing effect as the modulation frequency increases. The remainder of the peaking at about two cycles per second depends on the generator potential itself — in particular, on the shapes and mean amplitudes of the "quantum bumps".

When the illumination is dim and the generator potential is below the threshold of impulse generation, the small quantum bumps can be seen as slow shot-like fluctuations. As the intensity of the illumination increases, the mean rate of the fluctuations increase, and presumably they superimpose to yield an increasingly larger generator potential. Sample records of responses to steady light are shown in Fig. 46 (left). Note that the average amplitude of the fluctuations is not fixed. In general, they appear to become smaller the greater the intensity of the light and the longer the exposure to it. Especially important in this discussion of the dynamics are two effects of the fluctuations on the frequency response for light to generator potential (Fig. 46, right). First, the shape and duration of the fluctuations — whatever their amplitudes may be — impose a steep cut off on the response at high modulation frequency. Second, at high mean intensities the variation in amplitude of the fluctuations with changes in light intensity are sufficiently great, during a single cycle of relatively slow modulation, to attenuate the fluctuations of the response at low modulation frequencies. The effect is somewhat analogous to the self inhibition described above. Indeed, both may be regarded as forms of sensory adaptation.

The lateral inhibition produces a similar low frequency cut off, as expected, because of the long duration of the individual lateral inhibitory potentials. It is difficult to make direct measurements of these potentials because the inhibitory synapses are usually distant from the site of the recording electrode in the eccentric

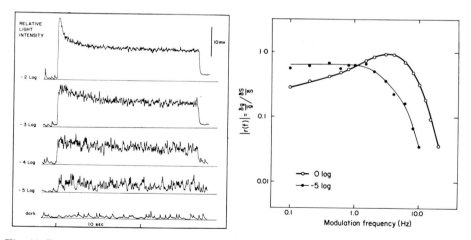

Fig. 46. Representative records of generator potentials (left) elicited by various intensities of steady illumination. Records obtained from an eccentric cell in which the nerve impulses had been blocked by application of tetrodotoxin. Frequency response (right) for the modulation of the generator potential in response to sinusoidal modulation of the light intensity. The mean amplitude of generator potential was 25 mv for 0 log and 4 mv for −5 log. The frequency response amplitude $|r(f)|$ was normalized so that the ordinate is the ratio of the fractional variation $\delta g/\bar{g}$ in excitatory conductance to the fractional variation $\delta s/\bar{s}$ in light intensity.
From DODGE et al. (1968)

cell (see Fig. 29). But the technique of antidromic stimulation, already described above, yields good results occasionally. Sample records are shown on the left in Fig. 47. The lower graph shows the modulated antidromic stimulus rate, the middle graph the modulation of the summed inhibitory potential in the eccentric cell of the test receptor (in the dark). The upper graph shows the resulting modulation about the mean rate of the discharge of a steadily illuminated test receptor. The corresponding frequency responses are shown on the right. The insert shows two estimates of the form of the unitary lateral inhibitory potential computed by taking the Fourier transform of the frequency response of the summed inhibitory potential (light curve) and the Fourier transform of the frequency response of the inhibited spike discharge (heavy curve). Both estimates agree qualitatively with one another and with the direct observations of the unitary potentials that were recorded in the experiment illustrated in Fig. 30 above. Both the calculated and the directly observed unitary inhibitory potentials show a brief small initial depolarization preceding the inhibitory hyperpolarization. (This brief depolari-

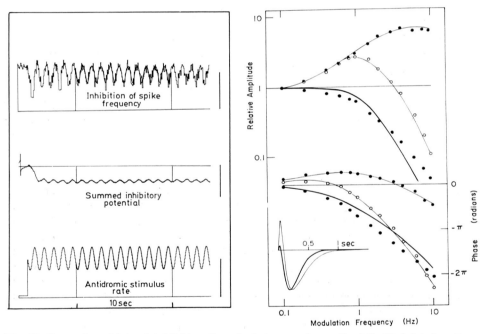

Fig. 47. Dynamics of lateral inhibition. Sample data records on the left. Modulated antidromic stimulus rate at bottom. Top record is modulated spike rate produced by this time-varying inhibition exerted on a steadily illuminated test receptor. Middle record is inhibitory potential produced by modulated antidromic inhibition measured by a microelectrode in the test receptor. On the right, frequency responses for the summed inhibitory potential (●), for inhibition of spike frequency (○), and for self inhibition (●). Insert: Waveform of the unit inhibitory potentials computed by taking Fourier transform of the frequency response of summed inhibitory potential (light curve) and frequency response from spike data (heavy curve). Calibration: 10 millivolts for record of inhibitory potential, 10 impulses/second for records of antidromic stimulus rate and spike discharge frequency. Curves fitted to data between vertical lines by least squares method. From Dodge et al. (1970)

zation is the only evidence that has been seen — to date — for a lateral excitatory influence in the compound eye of *Limulus*). Most striking, however, is the long delay of about 150 milliseconds to the peak of the inhibitory potential. There are several significant consequences of this delay: the delayed response to a step or burst, that we discussed earlier (see Figs. 38 and 39); the damped oscillations illustrated in the response to a large step (Fig. 42); and an amplification of the response to certain frequencies of sinusoidal stimuli. This latter effect we were able to predict from the frequency dependent generalization of our original steady state formulation. Omitting the small threshold corrections and including the self inhibition, the steady state Eq. (4) may be expressed as follows:

$$r_p = E_p - \overline{K} r_p - \sum_{j \neq p} K'_{p,j} r_j. \tag{8}$$

Here the three component processes resulting finally in the firing rate r of the pth ommatidium are represented separately. E_p is the excitatory generator potential, $\overline{K}$ the self inhibitory coefficient ($= K'_{pp}$), and K'_{pq} the lateral inhibitory coefficients linking the pth ommatidium to the others. Now all that is required to express each of the terms representing the component processes in a frequency dependent form is to multiply the time varying input to each process by the appropriate transfer function. Thus, if I is the modulation in light intensity, the modulation of the generator potential is given by

$$E_p = G(f) I$$

where the transfer function $G(f)$ is the complex number whose amplitude and phase, at any particular modulation frequency f, are given by the light to generator potential curve in Fig. 45. The generalization of the entire equation is similarly:

$$r_p = E_p - \overline{K} T_s(f) r_p - T_L(f) \sum_{j \neq p} K'_{pj} r_j. \tag{9}$$

Here T_s and T_L are the self and lateral inhibitory transfer functions scaled so that they are unity when $f = 0$. The amplitude and phase of the self inhibition is given in Fig. 45, of the lateral inhibition in Fig. 47. Now given $E_p = G(f)I$, the set of simultaneous equations can be solved for r_p at each modulation frequency.

In a test of the above formulation the equations were used to predict the outcome of an experiment (Fig. 48) in which frequency responses were measured for a single ommatidium (no lateral inhibition) and for a circular cluster consisting of that same ommatidium and 19 others surrounding it (strong lateral inhibition). The discrepancies between the theory and experiment are small.

An interesting feature of this experiment is the *amplification* of the response by lateral inhibition from about two to five cycles per second. (See RATLIFF et al., 1967 and RATLIFF et al., 1969.) The expectation has generally been that the amplitude of the response of an inhibited system cannot exceed the comparable response without inhibition. Indeed, as the frequency increases, the amplitude of the inhibited response to a sine wave increases and must eventually become the same as the uninhibited. This happens because the duration of each cycle of the stimulus at the higher frequencies is short compared to the duration of the inhibitory influence. If the modulation is very slow, the temporal characteristics of the inhibitory influences may be neglected. Under these conditions any output,

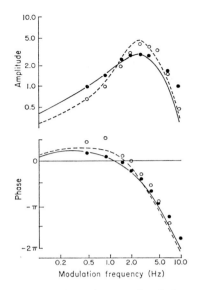

Fig. 48. Amplification by lateral inhibition. Theoretical and observed transfer functions (light to spike rate) for a single ommatidium illuminated alone filled circles, and for the same ommatidium when it is the central one of a cluster of 19 interacting ommatidia illuminated together, open circles. Amplitude is in impulses per second, peak to peak. Mean rate was 25/second for the small spot (a single ommatidium) and 20/second for the large spot (19 ommatidia). Note amplification of inhibited response from about 2.0 to 5.0 cycles per second. Data from experiments by Ratliff et al. (1967). Theoretical curves from Knight et al. (1970)

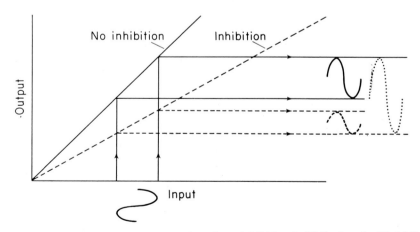

Fig. 49. Input-output curves for a network with no inhibition (solid line) and with inhibition (dashed line). In general the two outputs for the sinusoidal input given would not exceed the limits indicated by the pair of solid lines (no inhibition) and the pair of dashed lines (inhibition). If the inhibition could be turned completely off at the crest of the input and completely on at the trough, the amplitude of the output (peak to peak) would be amplified as indicated by the dotted curve. From Ratliff et al. (1969)

inhibited or uninhibited, would be given simply by drawing a reflection of the input from the appropriate steady state input-output curve, as illustrated in Fig. 49.

This is not necessarily what must happen, however. Possible limits of the output are shown by the dotted curve. That is, if there were some way to turn the inhibition off at the right time, the response could reach the upper limit (i.e., the uninhibited curve) and if the inhibition could be turned on at the right time, the response would then be suppressed to the lower (inhibited) limit. The delay to the maximum of the inhibitory potential causes such an effect at intermediate frequencies. This delay is about 150 milliseconds; therefore the opposed excitatory and inhibitory influences are about 1/2 cycle out of phase at stimulus frequencies of about 3 cycle/sec (period of 333 millisec). The frequency response thus passes through its maximum at the frequency at which the opposed excitatory and inhibitory influences are most out of phase, for that is the frequency at which the

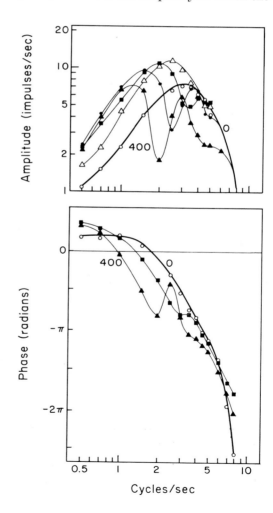

Fig. 50. The effect of delayed inhibition. The frequency responses were obtained by simultaneous sinusoidal modulation of the intensity of a spot of light on the ommatidium whose discharge was recorded and of a larger spot on a neighboring group of ommatidia. Delays of lateral inhibition (modulation of light on neighbors) with respect to direct excitation (modulation of light on test ommatidium) were 0 msec (open circles), 100 msec (open triangles), 300 msec (filled circles), and 400 msec (filled triangles). Phase data are shown for 0, 200, and 400 msec only.
From Ratliff et al. (1970)

greatest inhibitory influence coincides with the smallest excitatory influence and the greatest excitatory influence with the smallest inhibitory influence.

In short, the system is not only "tuned" to transmit particular frequencies, but because of the time delay in the lateral inhibition the network is also an "amplifier". Notice that it is an "AC amplifier" and that the amplification of these variations in the response is at the expense of a reduction in the mean "DC level" of response.

Further delays between excitation and inhibition such as might be caused by different conduction times or different latencies of receptor responses shift the maximum amplitude of the response toward lower frequencies. If the delay is great enough it will cause the transfer function to become multimodal, because with fixed delays the excitation and inhibition go in and out of phase several times over the range of frequencies to which the eye responds. Fig. 50 shows an experimental example of this effect in which very long delays have been produced artificially by introducing a fixed time lag between the sinusoidal stimulus to the test receptor and the sinusoidal stimulus to the neighboring units that exert inhibition on it.

Analogous effects may occur also in the spatial domain. The field of the inhibition about any receptor is bimodal along any diameter, and can be closely approximated by the difference between a broad Gaussian surface and a narrow one (cf. Fig. 26). The spatial frequency responses for both a bimodal distribution and a simple unimodal Gaussian distribution (calculated in one dimension) are shown in Fig. 51. As expected, the low spatial frequencies are always depressed by

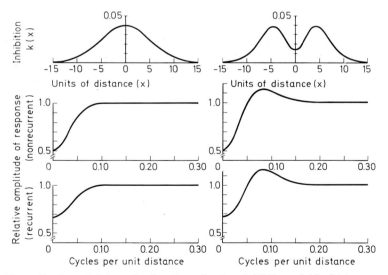

Fig. 51. Normalized spatial transfer functions for two inhibitory fields $k(x)$: upper left, a simple unimodal (Gaussian) inhibitory field. (Calculated in one dimension only.) Upper right, more complex bimodal (broad Gaussian minus a narrow Gaussian) inhibitory field. Below are the corresponding transfer functions for the two distributions in a non-recurrent network and in a recurrent network. Note amplification (amplitude greater than 1.0) for the bimodal inhibitory field. From Ratliff et al. (1969)

inhibition, and for very high frequencies the inhibited response approaches that of the uninhibited, that is, becomes equal to 1.0 in these graphs. (The high frequency cut off resulting from limitations of the optical system is not shown.) For intermediate frequencies, however, there is a marked difference in the behavior of the two networks. One amplifies the response (i.e., the ratio goes above unity), and the other does not. The explanation is essentially the same as for the amplification of responses to temporally varying stimuli. It should be noted that these curves are theoretical. Spatial amplification has not yet been demonstrated by experiment. Furthermore, all of the above spatial considerations are for steady (temporal) state interactions; the analysis of the combined effects of spatial *and* temporal variations of the inhibitory influences has yet to be undertaken.

The Functional Significance of Inhibitory Interaction

Some of the principal effects of lateral inhibition can be seen directly in the familiar phenomena of brightness contrast, especially border contrast. As a matter of fact, many fundamental principles of inhibitory interaction were originally deduced from such phenomena, rather than being based on direct electrophysiological observations. For example, the so-called Mach bands – the bright and dark lines seen at the bright and dark edges of a penumbra (Fig. 52) – are simply a special case of brightness contrast. It was upon these simple phenomena that ERNST MACH – the well known physicist, philosopher, psychologist – based his ideas about inhibitory interactions in the retina, over a hundred years ago (see RATLIFF, 1965).

It is easy to see now how such spatial contrast effects result from lateral inhibition. As MACH postulated, a field of inhibitory influence that extends for some distance laterally is all that is required. Because of the extent of such a field, a unit in the retina within the dimly illuminated area but near the boundary will

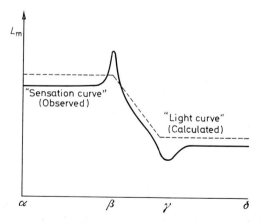

Fig. 52. Actual mean luminance or "light curve" (dashed line) along a uniform bright area (α to β), a uniform dark area (γ to δ) and the gradient (β to γ) between these two areas. The apparent mean luminance or "sensation curve" is represented by the dashed line. From RATLIFF (1965)

be inhibited not only by its dimly illuminated neighbors, but also by some brightly illuminated ones. The total inhibition on it will therefore be greater (and its response less) than that of dimly illuminated neighbors farther from the boundary. Similarly, a unit within the brightly illuminated area, but near the boundary and therefore within the range of inhibitory influences from dimly illuminated neighbors will have less inhibition on it (and a greater response) than elements far from the boundary that are completely surrounded by brightly illuminated neighbors. Because of these differential effects maxima and minima appear in the response of the network, even though there are no maxima and minima in the distribution of the stimulus itself. Such effects have been observed directly in the eye of *Limulus* (RATLIFF and HARTLINE, 1959), and in the vertebrate retina (BAUMGARTNER, 1961; GORDON, 1969).

One supposed function of lateral inhibition in neural networks is to sharpen information. The term "sharpen", although seldom defined exactly when used in this sense, generally means that responses of a network to changes in stimuli — in both space and time — are somehow enhanced by inhibition relative to responses

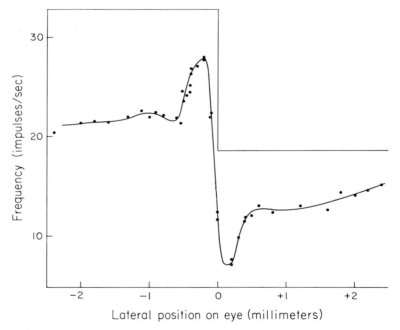

Fig. 53. The discharge of impulses from a single ommatidium in response to a step pattern of illumination in various positions on the retinal mosaic. The upper (rectilinear) graph shows the step pattern of illumination expressed in frequency of discharge which that level of illumination produced when the light was restricted to a single ommatidium by means of a small aperture in the optical system. The lower (curvilinear) graph is the frequency of discharge from the same ommatidium when the aperture was enlarged so that the entire pattern of illumination reached the eye. The effects of the curvature of the eye on the effective intensity at the extreme edges of the pattern are not taken into account here. From BARLOW (1967). See also RATLIFF and HARTLINE (1959)

to uniform stimuli. As MACH put it (see RATLIFF, 1965, p. 306): "Since every retinal point perceives itself so to speak, as above or below the average of its neighbors, there results a characteristic type of perception. Whatever is near the mean of the surroundings becomes effaced, whatever is above or below is disproportionately brought into prominence. One could say that the retina schematizes and caricatures. The teleological significance of this process is clear in itself. It is an analog of abstraction and of the formation of concepts."

An electrophysiological example, from the retina of *Limulus*, is illustrated in Fig. 53. Ordinarily, the magnitude of the response at or near the step itself is no greater, or even smaller, with inhibition than without. It comes into prominence, however, because the responses to the uniform conditions adjacent to the step are strongly affected by the inhibition. Note, in the spatial example shown here, and in the temporal example in Fig. 34 above, how large the differences are between the maxima and minima of responses to steps in stimuli, relative to the corresponding differences between the responses to uniform conditions.

As we have seen, the effect of inhibition in a linear system, such as the retina of the lateral eye of *Limulus*, may be expressed more clearly and in a more general form in terms of its contribution to the composite transfer function of the eye. Knowing this function, the response to any spatial and temporal distribution of illumination within the linear range of the eye may be predicted. As was shown above, lateral inhibition is a "filter" which reduces the amplitudes of responses to low frequencies — both spatial and temporal. This low frequency cut off, in combination with the high frequency cut off that results mainly from optical limitations in the spatial case, and from limitations of impulse generator mechanisms in the temporal case, results in a "tuning" of the system to particular intermediate frequencies. It is this selective attenuation of low frequencies, with a resultant shift of the maximum in the frequency response toward the higher frequencies (which contribute most to sharp curvatures in a Fourier synthesis) that is the basis of the concept of "sharpening". Stated in these terms, the only effect of inhibition is to attenuate selectively certain low frequencies, thus *relatively* increasing the higher frequencies. This is all that is meant by the term "sharpen" — no more and no less. It is important to note that the inhibition cannot, by some magic, restore information once lost. Indeed, even the apparently exceptional cases of amplification (see Figs. 48 and 51 above) also merely result from a selective suppression of responses at particular times and locations. All of these effects are achieved at considerable expense: information about slow changes and absolute levels is diminished. But the cost is easy to bear; such information is usually of little significance to the organism.

Border contrast, the Mach bands, and the like are very striking phenomena — easy to see and easy to understand in terms of lateral inhibition. But that which is obvious often obscures that which is more subtle. Inhibition plays equally important and equally fundamental roles in all aspects of vision. And it is not too much to expect that the enormous complexity of a wide variety of special visual functions may ultimately be reduced to, and expressed in terms of, a few general laws governing the interplay and integration of opposed excitatory and inhibitory influences. Trends in this direction are indicated by the few examples that follow (see also, RATLIFF, 1965, 1968).

Transient Responses. The most striking features of the response of any visual system are the vigorous discharges produced by changes in the intensity of illumination. The "on", "on-off", and "off" types of discharges, first observed to occur separately in separate ganglion cells of the frog retina (Hartline, 1938), are common features of the neural activity, at one level or another, of all visual systems — both vertebrate and invertebrate.

These diverse types of responses have long been attributed to the particular distribution and timing of the excitatory and inhibitory influences acting on particular cells (cf. Adrian and Matthews, 1928; Granit, 1933; Hartline, 1938). In the lateral eye of *Limulus*, for example, by properly balancing these opposed influences, in both space and time, one can synthesize an "on-off" or an "off" response where normally only a maintained "on" discharge would appear (Ratliff and Mueller, 1957). Recent investigations by Bicking (1965), Lange et al. (1965) Bishop and Rodieck (1965), and Rodieck (1965) lend additional support to the view that transient responses result from the opponent processes of excitation and inhibition. Furthermore, these studies state the necessary conditions for such results explicitly and in quantitative form.

Center-Surround Organization of Receptive Fields. Kuffler (1953) found that the transient responses of the cat retina were similar to those in the frog, but all three basic types of discharge could often be observed in a single ganglion cell. For example, stimulation of the center of the receptive field might yield a pure "on" response, the periphery a pure "off" response, and between the two, an "on-off" response (or the reverse, where the configuration of the response types is reversed — as is often the case). This complex organization was first attributed to the slight overlap of a central excitatory field and a peripheral inhibitory field. More recent evidence indicates that both the excitatory and the inhibitory influences extend over the whole of the receptive field (see Wagner et al., 1963). An imbalance of the influences, rather than total absence of one or the other, appears to determine the character of the response elicited by stimulating a given point of the receptive field. This interpretation is based mainly on experiments on color coding in the vertebrate retina.

Color Coding. Wagner et al. (1960) found in their experiments on the gold fish retina that the responses of some ganglion cells depend strongly on the wavelength of the illumination. An "on" center "off" surround type of cell when illuminated in the intermediate zone of its receptive field, for example, will yield nearly equal "on" and "off" responses to light with a wavelength near the middle of the visible spectrum. But stimulation of the same cells in the same locations of their receptive fields by short wavelengths may yield maximal excitation, and long wavelengths maximal inhibition (or the reverse, depending upon the type of cell). Thus the excitatory influences are not confined mainly to the center of the receptive field and the inhibitory influences to the surround, with just a slight overlap. Instead both influences evidently extend over the whole of the receptive field — the character of the response at any point being governed by the relative amounts of the two. This was clearly demonstrated by mapping the field separately with short (excitatory) wavelengths and long (inhibitory) wavelengths.

Motion, and Directional Sensitivity. Many cells in the visual system that respond to a change in the intensity of illumination, e.g., "on-off" cells, respond equally well to a change brought about by motion (cf. HARTLINE, 1938). Some such cells are highly specialized. For example, BARLOW and HILL (1963) have observed ganglion cells in the retina of the rabbit that respond either to changes in intensity of stationary patterns of illumination or to motion of that pattern. But in these experiments on the rabbit the response to motion is selective; it is elicited by movement of the stimulus in one direction and not the other. BARLOW and LEVICK (1965) attribute this directional sensitivity to asymmetrical lateral inhibitory influences in the retina. Experiments indicate that as the stimulus moves in one direction a wave of inhibition moves along with or even ahead of it, cancelling out any excitatory influence that might be elicited. When the stimulus moves in the other direction, the inhibition trails behind and has no immediate effect on the excitatory influence — and a response is generated in the ganglion cell. Thus the preferred direction and velocity of the sensitivity to movement are determined by the spatial and temporal distributions of the opposed excitatory and inhibitory influences.

Spatial and Temporal Filters. All of the special functions described above depend in one way or another upon certain spatial and temporal distributions of inhibition. Although these inhibitory mechanisms may have been shaped into particular forms by evolutionary pressures in order to achieve certain special ends, general characteristics that they all share in common will appear in measures of the overall behavior of the visual system. Following the theoretical work of IVES (1922) and the experimental work of DE LANGE (1952, 1957, 1958) numerous investigators have begun to consider the visual system as a series of optical "filters" in cascade. Analysis in these terms is facilitated by the use of sinusoidal stimuli, as in the techniques of linear systems analysis used above. For analysis of spatial properties, the so-called sine wave gratings are used (for reviews see CAMPBELL, 1968; and FRY, 1969); for temporal properties, sinusoidal variations in time (for a review see LEVINSON, 1968).

All visual systems analyzed in this way thus far are similar in that they transmit low spatial and temporal frequencies poorly, intermediate frequencies relatively well, and high frequencies poorly or not at all. The major determinants of the low frequency cut off in both the spatial and temporal domain appears to be the spatial and temporal distributions of the inhibitory influences. Having a considerable extent and duration, the isolated inhibitory influences act as low-pass filters. But since their ultimate effects are *inhibitory* (i.e., negative) this means that they diminish responses to low frequencies and do not significantly affect responses to high frequencies. In the end, therefore, the typical inhibitory network, no matter what its special functions may be, is simply a high pass filter when considered in these more general terms.

The preparation of this review was supported in part by research grants (EY 188) from the National Eye Institute, (GM 1789) from the National Institute of General Medical Sciences, and (GB-6540) from the National Science Foundation.

References

Adolph, A. R.: Spontaneous slow potential fluctuations in the *Limulus* photoreceptor. J. Gen. Physiol. **48**, 297—322 (1964).

— Excitation and inhibition of electrical activity in the *Limulus* eye by neuropharmacological agents. In: Bernhard, C. G. (Ed.): The Functional Organization of the Compound Eye, pp. 465—482. Oxford: Pergamon Press 1966.

Adrian, E. D., Bronk, D. W.: The discharge of impulses in motor nerve fibers. J. Physiol. (Lond.) **66**, 81—101 (1928).

— Matthews, R.: The action of light on the eye. Part I. The discharge of impulses in the optic nerve and its relation to the electric change in the retina. J. Physiol. (Lond.) **63**, 378—414 (1927).

Barlow, H. B., Hill, R. M.: Selective sensitivity to direction of movement in ganglion cells of the rabbit retina. Science **139**, 412—414 (1963).

— Levick, W. R.: The mechanism of directionally selective units in the rabbit's retina. J. Physiol. (Lond.) **178**, 477—504 (1965).

Barlow, R. B., Jr.: Inhibitory fields in the *Limulus* lateral eye. Thesis. The Rockefeller University, New York, N. Y. (1967).

— Kaplan, Ehud: *Limulus* lateral eye: Properties of Receptor Units in the Unexcised Eye, Science **174**, 1027—1029, 1971.

— Inhibitory fields in the *Limulus* lateral eye. J. Gen. Physiol. **54**, 383—396 (1969).

Baumgartner, G.: Kontrastlichteffekte an retinalen Ganglienzellen: Ableitungen vom Tractus opticus der Katze. In: Jung, R., Kornhuber, H. (Hrsg.): Neurophysiologie und Psychophysik des visuellen Systems, S. 45—55. Berlin-Göttingen-Heidelberg: Springer 1961.

Behrens, M. E., Wulff, V. J.: Light-initiated responses of retinula and eccentric cells in the *Limulus* lateral eye. J. Gen. Physiol. **48**, 1081—1093 (1965).

— — Functional autonomy in the lateral eye of the horseshoe crab, *Limulus polyphemus*. Vision Res. **7**, 191—196 (1967).

Bicking, L. A.: Some quantitative studies on retinal ganglion cells. Thesis. The John Hopkins University (1965).

Bishop, P. O., Rodieck, R. W.: Discharge patterns of cat retina ganglion cells. In: Nye, P. W. (Ed.): Proceedings of the Symposium on Information Processing in Sight Sensory Systems. pp. 116—127. California Institute of Technology, Pasadena, Calif. 1965.

Campbell, F. W.: The human eye as an optical filter. Proc. IEEE. **56**, 1009—1014 (1968).

De Lange, H.: Relationship between critical flicker-frequency and a set of low frequency characteristics of the eye. J. opt. Soc. Amer. **44**, 380—389 (1952).

— Attenuation characteristics and phase-shift characteristics of the human fovea-cortex systems in relation to flicker-fusion phenomena. Thesis, Technical University, Delft, Holland (1957).

— Research into the dynamic nature of the human fovea — cortex systems with intermittent and modulated light. I. Attenuation characteristics with white and colored light. II. Phase-shift in brightness and delay in color perception. J. opt. Soc. Amer. **48**, 777—789 (1958).

Dodge, F. A., Jr.: Inhibition and excitation in the *Limulus* eye. In: Reichardt, W. (Ed.): Processings of Optical Data by Organisms and by Machines. Proceedings of the International School of Physics "Enrico Fermi" Course XLIII, pp. 341—365, 1969.

— Knight, B. W., Toyoda, J.: Voltage noise in *Limulus* visual cells. Science **160**, 88—90 (1968).

— Shapley, R. M., Knight, B W.: Linear systems analysis of the *Limulus* retina. Behav. Sci. **15**, 24—36 (1970).

Fahrenbach, W. H.: The morphology of the eyes of *Limulus*. 1. Cornea and epidermis of the compound eye. Z. Zellforsch. **87**, 278—291 (1968).

— The morphology of the eyes of *Limulus*. II. Ommatidia of the compound eye. Z. Zellforsch. **93**, 451—483 (1969).

Fry, G. A.: Visibility of sine-wave gratings. J. opt. Soc. Amer. **59**, 610—617 (1969).

Fuortes, M. G. F.: Initiation of impulses in visual cells of *Limulus*. J. Physiol. (Lond.) **148**, 14—28 (1959).

FUORTES, M. G. F., MANTEGAZZINI, F.: Interpretation of the repetitive firing of nerve cells. J. Gen. Physiol. **45**, 1163—1179 (1962).
— POGGIO, G. F.: Transient responses to sudden illumination in cells of the eye in *Limulus*. J. Gen. Physiol. **46**, 435—452 (1963).
— YEANDLE, S.: Probability of occurrence of discrete potential waves in the eye of *Limulus*. J. Gen. Physiol. **47**, 443—463 (1964).
GORDON, J.: Edge accentuation in the frog retina. Thesis, Brown University (1969).
GRANIT, R.: The components of the retinal action potential and their relation to the discharge in the optic nerve. J. Physiol. (Lond.) **77**, 207—240 (1933).
— Charles Scott Sherrington — An Appraisal. London: Thomas Nelson and Sons Ltd. 1966.
GRENACHER, H.: Untersuchungen über das Sehorgan der Arthropoden. Göttingen: Vandenhoeck und Ruprecht 1879.
HARTLINE, H. K.: The response of single optic nerve fibers of the vertebrate eye to illumination of the retina. Amer. J. Physiol. **121**, 400—415 (1938).
— The nerve messages in the fibers of the visual pathway. J. opt. Soc. Amer. **30**, 239—247 (1940).
— The neural mechanisms of vision. *The Harvey Lectures*. Series XXXVII. 39—68 (1941 to 1942).
— Inhibition of activity of visual receptors by illuminating nearby retinal elements in the *Limulus* eye. Fed. Proc. **8**, 69 (1949).
— Receptor mechanisms and the integration of sensory information in the eye. Rev. Mod. Phys. **31**, 515—523 (1959).
— Inhibitory interaction in the retina. In: STRAATSMA, B. R., HALL, M. O., ALLEN, R. A., CRESCITELLI, F. (Eds.): The Retina: Morphology, Function and Clinical Characteristics. UCLA Forum in Medical Sciences, pp. 297—317. 1967.
— Visual receptors and retinal interaction. Science **164**, 270—278 (1969).
— GRAHAM, C. H.: Nerve impulses from single receptors in the eye. J. cell. comp. Physiol. **1**, 277—295 (1932).
— RATLIFF, F.: Inhibitory interaction of receptor units in the eye of *Limulus*. J. Gen. Physiol. **40**, 357—376 (1957).
— Spatial summation of inhibitory influences in the eye of *Limulus*, and the mutual interaction of receptor units. J. Gen. Physiol. **41**, 1049—1066 (1958).
— — MILLER, W. H.: Inhibitory interaction in the retina and its significance in vision. In FLOREY, E. (Ed.): Nervous Inhibition, pp. 241—284. New York: Pergamon Press 1961.
— WAGNER, H. G., MACNICHOL, E. F., JR.: The peripheral origin of nervous activity in the visual system. Cold Spr. Harb. Symp. quant. Biol. **17**, 125—141 (1952).
— — RATLIFF, F.: Inhibition in the eye of *Limulus*. J. Gen. Physiol. **39**, 651—673 (1056).
HODGKIN, A. L., HUXLEY, A. F.: A quantitative description of membrane current and its application to conduction and excitation in nerve. J. Physiol. (Lond.) **117**, 500—544 (1952).
HUBBARD, R., WALD, G.: Visual pigment of the horseshoe crab, *Limulus polyphemus*. Nature (Lond.) **186**, 212—215 (1960).
IVES, H. E.: A theory of intermittent vision. J. opt. Soc. Amer. **6**, 343—361 (1922).
JENKS, G. F., BROWN, D. A.: Three-dimensional map construction. Science **154**, 857—864 (1966).
KIKUCHI, R., UEKI, K.: Double-discharges recorded from single ommatidia of horseshoe crab *Tachypleus tridentatus*. Naturwissenschaften **52**, 458—459 (1965).
KIRSCHFELD, K., REICHARDT, W.: Die Verarbeitung stationärer optischer Nachrichten im Komplexauge von *Limulus*. Kybernetik **2**, 43—61 (1964).
KNIGHT, B. W., TOYODA, J., DODGE, F. A., JR.: A quantitative description of the dynamics of excitation and inhibition in the eye of *Limulus*. J. Gen. Physiol. **56** (1970).
KUFFLER, S. W.: Discharge patterns and functional organization of mammalian retina. J. Neurophysiol. (Lond.) **16**, 37—68 (1953).
LANGE, D., HARTLINE, H. K., RATLIFF, F.: The dynamics of lateral inhibition in the compound eye of Limulus. II. In: BERNHARD, C. G. (Ed.): The Functional Organization of the Compound Eye, pp. 425—449. Oxford: Pergamon Press 1966.
LASANSKY, A.: Cell junctions in ommatidia of *Limulus*. J. Cell Biol. **33**, 365—383 (1967).
LEVINSON, J.: Flicker fusion phenomena. Science **160**, 21—28 (1968).

Macnichol,E.F.,Jr.: Visual receptors as biological transducers. In: Grenell,R.G., Mullins, L.J. (Eds.): Molecular Structure and Functional Activity of Nerve Cells, pp. 34—62. Washington, D. C.: Amer. Inst. Biol. Sci. 1965.
— Hartline, H.K.: Responses to small changes of light intensity by the light-adapted photoreceptor. Fed. Proc. **7**, 76 (1948).
Melzak, Z.A.: On a uniqueness theorem and its application to a neurophysiological control mechanism. Information and Control **5**, 163—172 (1962).
Miller, W.H.: Morphology of the ommatidia of the compound eye of *Limulus*. J. biophys. biochem. Cytol. **3**, 421—428 (1957).
— Fine structure of some invertebrate photoreceptors. Ann. N. Y. Acad. Sci. **74**, 204—209 (1958).
— The anatomy of the neuropile in the compound eye of *Limulus*. In: Rohen,J.W. (Ed.): Eye Structure, II. Symp., pp. 159—169. Stuttgart: Schattauer 1965.
— Ratliff,F., Hartline, H.K.: How cells receive stimuli. Sci. Amer. **205**, 222—238 (1961).
Pinter, R.B.: Sinusoidal and delta function responses of visual cells of the Limulus eye. J. Gen. Physiol. **49**, 565—593 (1966).
Purple, R.L.: The integration of excitatory and inhibitory influences in the eccentric cell in the eye of Limulus. Thesis. The Rockefeller Institute. (1964).
— Dodge, F.A.: Interaction of excitation and inhibition in the eccentric cell in the eye of Limulus. Cold Spr. Harb. Symp. quant. Biol. **30**, 529—537 (1965).
Ratliff, F.: Inhibitory interaction and the detection and enhancement of contours. In: Rosenblith,W.A. (Ed.): Sensory Communication, pp. 183—203. Cambridge: M. I. T. Press: New York: John Wiley and Sons 1961.
— Mach Bands: Quantitative Studies on Neural Networks in the Retina. San Francisco, Calif.: Holden-Day 1965.
— On fields of inhibitory influence in a neural network. In: Caianiello,E.R. (Ed.): Neural Networks, pp. 6—23. Berlin-Heidelberg-New York: Springer 1968.
— Hartline, H.K.: The responses of *Limulus* optic nerve fibers to patterns of illumination on the receptor mosaic. J. Gen. Physiol. **42**, 1241—1255 (1959).
— — Lange, D.: The dynamics of lateral inhibition in the compound eye of *Limulus*. I. In: Bernhard,C.G. (Ed.): The Functional Organization of the Compound Eye, pp. 399—424. Oxford: Pergamon Press 1966.
— — — Variability of interspike intervals in optic nerve fibers of *Limulus*. Effects of light and dark adaptation. Proc. nat. Acad. Sci. (Wash.) **60**, 464—469 (1968).
— — Miller, W.H.: Spatial and temporal aspects of retinal inhibitory interaction. J. opt. Soc. Amer. **53**, 110—120 (1963).
— Knight, B.W., Graham, N.: On tuning and amplification by lateral inhibition. Proc. nat. Acad. Sci. (Wash.) **62**, 733—740 (1969).
— — Milkman, N.: Superposition of excitatory and inhibitory influences in the retina of *Limulus*. The effect of delayed inhibition. Proc. nat. Acad. Sci. (Wash.) **67**, 1558—1564 (1970).
— — Toyoda,J., Hartline, H.K.: Enhancement of flicker by lateral inhibition. Science **158**, 392—393 (1967).
— Mueller,C.G.: Synthesis of "On-Off" and "Off" responses in a visual-neural system. Science **126**, 840—841 (1957).
Reichardt,W., Macginitie,G.: Zur Theorie der lateralen Inhibition. Kybernetik **1 /4**, 155—165 (1962).
Rodieck,R.W.: Quantitative analysis of cat retinal ganglion cell response to visual stimuli. Vision Res. **5**, 583—601 (1965).
Rushton,W.A.H.: A theoretical treatment of Fuortes' observations upon eccentric cell activity in *Limulus*. J. Physiol. (Lond.) **148**, 29—38 (1959).
Schwartz, E.A.: Retinular and eccentric cell morphology in the neural plexus of *Limulus* lateral eye. J. Neurobiol. In press (1970).
Shapley, R.M.: Fluctuations in the response to light of visual neurons in *Limulus*. Nature (Lond.) **221**, 437—440 (1969).
Sherrington,C.S.: On reciprocal action in the retina as studied by means of some rotating discs. J. Physiol. (Lond.) **21**, 33—54 (1897).

STEVENS, C. F.: Input-output relation for *Limulus* receptor cells. Biophysical Society, Abst. 7th ann. meeting, New York City, item WF6 (1963).
— A quantitative theory of neural interactions: theoretical and experimental investigations. Thesis. The Rockefeller Institute, New York (1964).
TOMITA, T.: Peripheral mechanisms of nervous activity in lateral eye of horseshoe crab. J. Neurophysiol. **20**, 245—254 (1957).
— Mechanism of lateral inhibition in the eye of *Limulus*. J. Neurophysiol. **21**, 419—429 (1958).
— KIKUCHI, R., TANAKA, I.: Excitation and inhibition in lateral eye of horseshoe crab. In: KATSUKI, Y. (Ed.): Electrical Activity of Single Cells, pp. 11—23. Tokyo: Igaku Shoin Ltd. 1960.
WAGNER, H. G., MACNICHOL, E. F., JR., WOLBARSHT, M. L.: The response properties of single ganglion cells in the goldfish retina. J. Gen. Physiol. **43**, 45—62 (1960).
— — — Functional basis for "On"-center and "Off"-center receptive fields in the retina. J. opt. Soc. Amer. **53**, 66—70 (1963).
WATERMAN, T. H.: Polarized light and angle of stimulus incidence in the compound eye of *Limulus*. Proc. nat. Acad. Sci. (Wash.) **40**, 4, 258—262 (1954a).
— Directional sensitivity of single ommatidia in the compound eye of *Limulus*. Proc. nat. Acad. Sci. (Wash.) **40**, 252—257 (1954b).
— WIERSMA, C. A. G.: The functional relation between retinal cells and optic nerve in *Limulus*. J. exp. Zool. **126**, 59—86 (1954).
WHITEHEAD, R., PURPLE, R. L.: Synaptic organization of the neuropile of the lateral eye of *Limulus*. J. Vision Res. **10**, 129—133 (1970).
WOLBARSHT, M. L., YEANDLE, S.: Visual processes in the *Limulus* eye. Ann. Rev. Physiol. **29**, 513—542 (1967).
YEANDLE, S.: Evidence of quantized slow potentials in the eye of *Limulus*. Amer. J. Ophthal. **46**, No. 3, Part II, 82—87 (1968).

Chapter 11

Optical Properties of Vertebrate Eyes

By

GERALD WESTHEIMER, Berkeley, California (USA)

With 18 Figures

Contents

I. Introduction

Starting with a simple photodetector in protozoa and some of the earliest multicellular invertebrates, evolution has fashioned very elaborate visual organs. For an understanding of these organs and of the way they tie in with the rest of the organism, it needs to be considered at the outset what kind of information they extract from the photic energy impinging on them.

What information is there in the electromagnetic energy coming in from the environment ? In attempting to answer this question, we must remember that almost all that we know about the subject is derived from what was originally funnelled through man's collective visual system. Yet, edifices such as the electro-magnetic theory have found so all-pervading confirmation and application that their sole dependance on properties of the human visual system seems a remote possibility. We will, therefore, attempt to describe and measure the effectiveness of the eye's performance in terms of standards of physical theory, with only the passing thought that we may have been confined, epistemologically, to a circular argument.

In simple terms, the organism can possibly be informed about

(1) the distribution relative to the organism of light-emitting or reflecting points in the environment;
(2) the intensity of such light;
(3) the wave length distribution of such light;
(4) the state of polarization of such light; and
(5) the phase coherence of such light.

Much of this can be discussed in terms of nineteenth century electromagnetic theory, but quantum theory does occasionally enter the picture, e.g., in describing the limits of intensity and wavelength resolution. The equivalent limits of spatial resolution were already part of the wave theory of light, but only when the wave theory of the particle was established in the 1920's did it become apparent that the determination of the energy contents of electromagnetic radiation was subject to limitations similar to those concerning the direction of incidence.

We shall not refer any further in this chapter to mechanisms that have evolved to analyze the variables 2—5 in the above list. Intensity discrimination and wave-length discrimination form the subject of lengthy discourses in these volumes. Polarization discrimination exists in certain invertebrates. Phase discrimination is rarely thought of in vision. In theory its value would be to detect (1) whether the light coming in from various directions is phase coherent, i.e., has issued from the same original source, and, if so, (2) distance and velocity relationships between the origin of two or more such beams. The method of analysis would be some kind of interferometer, and there is no reason in principle why an organism could not have evolved one. The localization by ultrasonic means in bats and electric pulses in electric fish bears witness to what evolution can achieve.

The primary purpose of the optical apparatus of eyes is to provide efficiency in the localization of the sources of the light reaching the organism. An exposed sac of rhodopsin would do as well as an eye in catching quanta, but it would not yield information about their direction of entry unless it were separated into compartments each shielded from all but a narrow range of entry directions. This is, in fact, the way the apposition eyes of insects are built. There is, however, a difficulty here which is apparent at a first glance at an insect's head. (Fig. 1).

The uncertainty about the direction of entry of a quantum of light can only be reduced by enlarging the aperture admitting the light. This concept had been well understood in terms of the wave theory of light and was taken over into quantum theory where it found expression in the uncertainty principle. The basic statement in our context is that there is a reciprocity between locating the position of a quantum in the aperture of an optical system and in the image plane. If one reduces the aperture of an optical system to a infinitesimal pinhole, one conco-

Fig. 1. Head of an insect, showing the large area covered by the many facets (corneas) of the eye compartments (ommatidia)

mitantly loses information about the direction of origin of a quantum observed in any point in the image plane. In symbols, for an aperture of diameter a and wavelength of light λ, the uncertainty of direction of origin of the incident quantum is given by the relation

$$\varDelta \theta \cdot a \sim \lambda \qquad (1)$$

where $\varDelta \theta$ is a measure, expressed in radians, of the spatial spread of quanta arriving from a single point source such as a star. It follows that if an animal needs accurate spatial localization of incident light, it has to have very fine compartmentalization of photoreceptive elements as well as a collecting aperture that is large in absolute (and not relative!) terms. Eq. (1), apart from constants involving the shape of the aperture and the desired kind of measure of spread, states that the directional uncertainty in radians is equal to wavelength divided by the aperture diameter, both being in the same units of length. Since there is not much choice about the wavelength, and since in any case the wavelength constitutes another dimension along which measurement may be wished, physical size of the aperture is the variable directly responsible for the angular resolution of an eye. If each compartment has to have its own aperture, as it does in insects' apposition eyes, the eye of necessity becomes very large (Fig. 1).

A solution is to make all receptor compartments share the one aperture and the resultant organ we call a simple eye (Fig. 2). Vertebrates, and some invertebrates, such as cephalopods and arachnids, feature such an eye. In discussing the optical properties of this kind of eye, we shall refer mainly to the example that has been most thoroughly studied, the human eye. Almost all the optical principles, but almost none of the values of the parameters, apply to other simple eyes. A

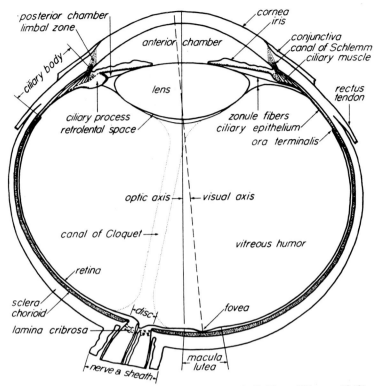

Fig. 2. Horizontal section of right human eyeball (From Walls, 1942)

few species, notably the cat (Vakkur and Bishop, 1963) and the rabbit (Hughes, 1972), have been treated in detail, but size and shapes of vertebrates eyes vary widely (Walls, 1942) so that the numerical values given in this chapter should not be accepted as applying to other species.

II. Optical Parameters of Ocular Structures

It is fortunate that to a first approximation the simple eye may be regarded as an assembly of spherical refracting surfaces whose centers of curvature lie on a straight line. The parameters relevant to the image-forming properties of such a system are the indices of refraction of the media and the positions and radii of curvature of the surfaces. Table 1 gives this information for a hypothetical normal human eye and Table 2 for a cat.

A great deal of effort has gone into the collection of the information in these tables. The problem, of course, is that *in vivo* measurements, desirable to obviate artifacts, require considerable ingenuity of method and analysis. The refractive indices of the cornea and lens are almost impossible to secure *in vivo*, but other data can become almost routinely available, particularly when the technique of measurement is a standard eye-examination procedure.

Table 1. Optical parameters of a typical normal human eye (From WESTHEIMER, 1968)

Surface	Radius of curvature (mm)	Refractive index		Distance from anterior surface of cornea (mm)	Refractive power (D)
		Anterior	Posterior		
Anterior cornea	7.8	1.000 (air)	1.376	0	+ 48.2
Posterior cornea	6.8	1.376	1.336 (aqueous)	0.5	− 5.9
Anterior lens	10.0[a]	1.336	1.386[b]	3.6[a]	+ 5.0[b]
Posterior lens	− 6.0	1.386[b]	1.336 (vitreous)	7.2	+ 8.3[b]
Retina				24.0	

[a] During maximum accommodation the anterior surface of the lens has a radius of curvature of 5 mm and its anterior surface is moved forward to be nearly 3 mm behind the anterior surface of the cornea. Partial accommodation will produce values between these values and those given in the table.

[b] The index of refraction of the lens varies from 1.386 near each surface to 1.406 in the center. The indicated refractive power is for the lens surfaces only. The gradient of refractive index within the lens produces additional refractive power.

Table 2. Optical parameters of typical cat eye (From VAKKUR and BISHOP, 1963)

Surface	Radius of curvature (mm)	Refractive Index		Distance from anterior surface of cornea (mm)	Refractive power (D)
		Anterior	Posterior		
Anterior cornea	8.57	1.000	1.376	0	43.9
Posterior cornea	7.89	1.376	1.336	0.68	− 5.1
Anterior lens	7.2	1.336	1.5544	5.2	30.3
Posterior lens	−8.05	1.5544	1.336	13.7	27.1
Retina				22.3	

A. Index of Refraction

The concept of refractive index is easily applied to such relatively homogeneous substances as the aqueous and vitreous humors and the pre-corneal layer of lacrimal fluid. In their chemical constitution, they all closely resemble blood plasma which is equivalent in many physico-chemical ways to a 0.9% NaCl solution. The refractive index consequently is 1.336 for sodium light at body temperature, and the dispersion is about 60. The refractive index of the pre-corneal tear layer may be somewhat higher because of evaporation and admixture of oily secretion from the glands of the lid margin.

The cornea, lens and prereceptor retinal layers are cellular and cannot strictly be regarded as having a uniform enough constitution to assign to them a single refractive index. Unstained material from these tissues has structure that can be seen with a phase-contrast microscope — obviously an indication of refractive index differences. In addition, the corneal stroma, for one, has a regular ultramicroscope structure that may be a significant factor in making the cornea trans parent; the sclera and scarred portions of the cornea are merely translucent yet differ from the normal cornea only in regularity of disposition of collagen fibrils.

For the purposes of defining the refracting power of corneal or lenticular surfaces, it suffices to assign an average refractive index to the cornea and to the material near the surface of the lens: 1.376 and 1.386 respectively for Na light.

This simple approach fails in two important ways. The first failure is at a level where the refractive index is taken to mean the square root of the specific inductive capacity and used in this sense in more sophisticated applications of Maxwell's electromagnetic equations, e.g., reflections, scattering, etc. To give an example: the assumption of homogeneity of refractive index for a lens with anti-reflection coating is justified when calculating the focal length of such a lens, but not surface reflections. In the eye, several thin layers — pre-corneal tear layer, corneal epithelium, Bowman's membrane — precede the corneal stroma and, while they may not materially modify the refractive power of the anterior surface of the cornea calculated from the formula $(1.376-1.000)/r$, they may influence an estimate of, say, the amount of light lost by reflection at the anterior surface of the eye. The same applies to the crystalline lens which has a capsule and, anteriorly, a layer of epithelium.

The second failure of the assumption of uniform refractive index is in the manifest inhomogeneity of the crystalline lens. The lens is an encapsulated, avascular structure which, however, continuously grows new fibers. The total volume of the lens changes little with age, but there is a compacting of the nuclear region whose solid content, and hence refractive index, in the adult is higher than that of the outer, or cortical, regions. Numerical values for the refractive index are 1.386 just under the capsule, gradually increasing at the center to 1.406 or, according to some authors, even 1.44 or higher. In designing his schematic eye, Gullstrand (1909) made one model in which there is a lens nucleus, surrounded by a lens cortex, the two having refractive index values of 1.386 and 1.406 respectively. Actually, the graded distribution has an additional refractive effect. The passage of light in such a nonuniform medium is not rectilinear, but follows curved trajectories demanded by the basic principle of geometrical optics that the optical path (i.e. the product of refractive index and path element) be stationary (Fermat's principle). The net effect is that the crystalline lens produces more refractive power than that predicted from its surface powers, even when thickness is taken into account. It is possible to estimate what the refractive index of a hypothetical medium would have to be which, when uniformly filling a lens of standard geometrical shape, would yield its demonstrated refractive power. It is about 1.42.

There is probably a great inter-species difference in the optical properties of the lens. Depending on the level of development of the faculty to accommodate, the capsule thickness varies (Fincham, 1937). In view of the progressive changes in lens composition and consistency with age, one would be led to expect more homogeneity in adult animals of species with a very much shorter life-span than man. On the other hand, aquatic animals have less refractive power in their corneas (owing to the drastically reduced refractive index difference when the object-sided medium is water) and need to make it up by more powerful crystalline lenses. As a consequence fish lenses are more spherical, and have higher refractive indices and refractive index gradients than other vertebrate lenses.

B. Radii of Curvature

The anterior and posterior boundaries of the cornea and crystalline lens constitute the four major refracting surfaces of the eye. The description of the geometrical characteristics of such surfaces is simplified by approximating them in the first instance to a section of a sphere, i.e. the surface generated by the revolution of a circular arc around a radius. The easiest way of specifying a sphere is by the position of its center and its radius.

When a light ray reaches such a boundary, its direction of propagation is deviated by an angle that depends on the ratio of the refractive indices of the two media and the angle the incident ray makes with the normal to the surface at the point of incidence (Snell's Law). When a surface is spherical, the normal is, of course, the line from the point of incidence to the center of curvature. Any surface, even a somewhat irregular one, can in a given meridian[1] be approximated by a circular arc over a small region. The procedure of choice is to find the best fitting circular arc in the region covering the central portion of the incident bundle of light, say 2—4 mm. If this arc does not fit more peripheral portions of the surface, a method of specification may be attempted which gives the deviations from this circular shape as a function of, for example, height of incidence.

Sometimes an analytical expression for the surface, e.g., ellipsoid or paraboloid, may be successfully applied. The specification of the evolute, i.e., the locus of the instantaneous centers of curvature of the surface, has not been widely used.

It cannot be generally supposed that ocular boundaries are surfaces of revolution. In this connection, DUPIN's theorem may be of help. It states that any surface, no matter what its characteristics, can over a small enough region be regarded as having a meridian of maximum curvature and a meridian of minimum curvature, and that, moreover, these two meridians are always at right angles to each other.

The experimental determination of the curvature of the ocular surfaces is based on the above theoretical framework. The principal bundle of light making up the image traverses each surface in a defined area. Because there is a prominent change of refractive index, the surface acts as reflector for a proportion of the incident beam. An object of given size placed at a given distance in front of this mirror will give rise to an image whose position and size are governed by the object and by the radius of curvature of the mirror. Considerable ingenuity has gone into building instruments for this purpose and they can achieve an accuracy ($\sim 1/4\%$) unexcelled in comparable routine examination procedures.

The most widely applied measurement technique is the method of *doubling* (Fig. 3). Here, the rays in the reflected image are passed through an optical device which splits the bundle coming from each object point into two bundles with exactly known differences in direction. The result is two images with a precisely known separation. If the separation is now adjusted so that the two opposite edges of the image pairs just touch each other, the separation is equal to the image size. The end point of this determination can be made even in the presence of image movements.

---

[1] Meridian: A plane section of a sphere containing a diameter.

Circular target placed in front of a surface will give rise to a circular image if the curvature in all meridians is the same, but in the more general case the image formed by reflection will exhibit a meridian of maximum and one of minimum size, corresponding to the surface's so-called principal meridians, i.e., meridians of minimum and maximum curvature. In most cases, the region of the surface covered by the entering light beam is small enough to allow Dupin's theorem to be applied so that the specification of the orientation of the principal meridians and their radii of curvature suffices. When the dioptric system is used, it is common to express the power of the surface in one principal meridian as if it were a spherical surface, on which is superimposed a cylindrical surface that has zero power in that meridian, and, in the other principal meridian, a power equal to the difference between the powers in the two meridians — i.e.,

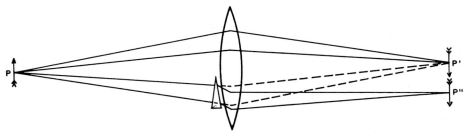

Fig. 3. Doubling principle used to measure size of reflected image. P is an inaccessible image formed by reflection of a target in an ocular surface. Lens forms a further image in plane $P' P''$. Prism covers a portion of lens and the rays are uniformly deviated to form a displaced image P'' as well as a normal image P'. By varying amount of prismatic power, a separation of P' and P'' can be achieved such that opposite edges of P' and P'' just touch. This separation is equal to the image size in that meridian

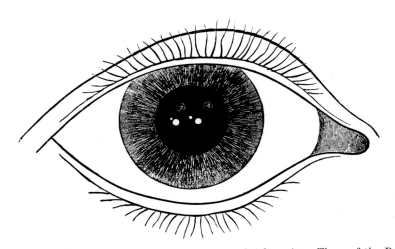

Fig. 4. Catoptric images of a target consisting of two bright points. Three of the Purkinje-Sanson images are seen, the large bright one is reflected from the anterior corneal surface, the small bright one from the posterior lenticular surface, the large dim one from the anterior lenticular surface. Partly schematic from Tscherning (1904)

the surface is a sphero-cylindrical (or toric) surface. This kind of surface produces astigmatic imagery (*vide infra*).

A bright target placed before an eye will be reflected at each of the major optical boundaries. A photograph will show multiple reflected (catoptric) images, so-called Purkinje-Sanson images (Fig. 4). If the refractive indices of the ocular images and the surface positions are known, the radii of curvature of each of the internal surfaces can be calculated. Most of the results in Table 1 were obtained in this way. If the artifacts introduced in a histological preparation are known, a geometrical analysis of a cross-section of an eye is probably the method of choice in most instances.

C. Geometrical Lay-out of the Eye

Again it is best to proceed via the assumption of the utmost simplicity and then see how the eye may best be fitted into such a framework.

The important idealization of optical systems, and one that is almost never departed from in practice, is to have the axis of symmetry common to all surfaces. In the case of spherical surfaces, this means that the centers of curvature of all surfaces lie on one line, and this concept is easily extended to cover an array of paraboloidal or elliptoidal surfaces, and even toric surfaces. The consequence is that all surfaces have at least one common normal. Because a ray of light is not deviated if it is incident normal to a surface, there is at least one, and usually only one, ray that passes undeviated through the system. The line thus described is called the *optical axis*. Its intersection with the cornea and crystalline lens would define the anterior and posterior poles of these structures. The separation of these poles give thickness values for the cornea, anterior chamber, and lens.

The geometrical lay-out of the eye is best obtained from postmortem sections, but there exist techniques for obtaining important locations in the living eye. One such technique involves the analysis of light reflected from the various surfaces when a narrow sheet of light is directed somewhat obliquely into the eye (slit-lamp beam). The thickness of the cornea and lens and their separation can then be measured either from diffusely scattered light, or from bundles of specularly reflected light which permit precise measurement (Ophthalmophakometry, see TSCHERNING, 1904).

More recently, ultrasound has been used to map the relative positions of the ocular structures. High-frequency sonic energy is reflected at the various interfaces and precise time-interval measurements of the echoes permit localization of the interfaces. The velocity of the ultrasound in biological structures varies in much the same, though inverse, manner as that of light, so that inaccuracies may be introduced by an inhomogeneity of the crystalline lens.

The ultrasonic technique (JANSSON, 1963) is one of only two ways of ascertaining the position of the retina in an intact eye. The other is based on the property of the retina of signalling light when stimulated directly by X rays. A sheet of X rays sent in through the bones of the orbit at right angles to the eye's optic axis will intercept the retina in a circle and, as it is moved posteriorly, eventually in only one small patch at the posterior pole of the eye. This method, (RUSHTON, 1938), however, needs the co-operation of the subject to report the visualization of the X radiation and indicate when the sheet of X rays has been moved past the posterior pole. A related technique (GOLDMAN and HAGEN, 1942) makes use of the fact that X rays are not refracted and hence permit stimulation of retinal locations with exactly known differences in position. Comparison of the subjective impression thus conveyed with an optically created image of an outside target permits an absolute measure of the eye's object/image magnification and hence computation of the posterior nodal distance. Both ultrasonic and X ray measuring techniques contain an element of risk of injury to the eye.

In any description of the optical characteristics of an instrument, one further item of information is essential: the structure responsible for the limitation of the width of the bundle making up the image-sided rays, the so-called aperture stop. In the eye it is the pupil, i.e., the opening of the iris, usually situated just in front of the anterior surface of the crystalline lens. The position of the center of the pupil is an important parameter in imagery calculation. Ideally, again, the pupil would be centered on the optical axis of the instrument. In the human eye there is a small but significant displacement of the center of the pupil — typically about 1/2 mm nasally from the optic axis of the eye.

III. Dioptrics of the Eye

The refracting power of an optical surface, i.e. a boundary separating two media of refractive index n_1 and n_0, is given by the expression

$$F_1 = \frac{n_1 - n_0}{r_1} \tag{2}$$

where r_1 is the radius of curvature of the boundary. If r_1 is expressed in meters F_1 is in a unit called diopter whose dimension is meters^{-1}. In the present discussion a sign convention is adopted where the origin of a cartesian coordinate system is placed at the pole of the surface. When the center of curvature is to the right of the surface (a convex surface for light incident from the left), r_1 is positive and F_1 becomes positive if $n_1 > n_0$. When an object point is placed at a distance s_1 from the surface (measured along the direction of the axis) and at a distance y_1 from the

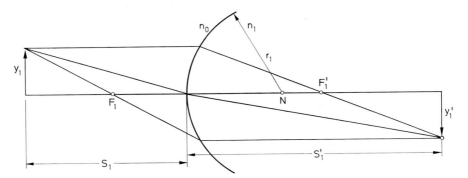

Fig. 5. Object-image relationship for refraction at single surface. A ray passing through F_1, the anterior focal point of the surface, becomes parallel to the axis after refraction. An incident ray that is parallel to the axis emerges after refraction to pass through the posterior focal point F_1'

axis (Fig. 5), first order optical theory yields the following expressions for the corresponding image coordinates s_1' and y_1'

$$\frac{n_1}{s_1'} = \frac{n_0}{s_1} + F_1 \tag{3}$$

and

$$\frac{y_1'}{y_1} = -\frac{n_0\,s_1'}{n_1\,s_1} = 1 + \frac{F_1\,n_0}{s_1}. \tag{4}$$

(Radial symmetry of the system demands that the third coordinate of the object and image be the same, i.e., that the image lie in the plane determined by the object and the axis of the instrument.)

These equations are obtained by application of the most basic laws of geometrical optics but using only the first term in the trigonometric expansions

$$\sin\theta = \theta - \theta^3/3! + \theta^5/5! \ldots$$

and

$$\cos\theta = 1 - \theta^2/2! + \theta^4/4! \ldots \tag{5}$$

The implication of Eqs. (3) and (4) is that corresponding to every object point there is one and only one image point, i.e. that there is a unique conjugacy relationship between object and image spaces. This conjugacy applies not only to points, but also lines and angles.

Of special interest are the points conjugate to an object or image on the axis at infinity, i.e., parallel incident or emergent bundles. These are called the first and second primary focal points and their positions are obtained by substituting $s_1 = \infty$ or $s_1' = \infty$ in Eq. (3).

Object-image calculations by serial applications of Eqs. 3 and 4 become cumbersome when a system is composed of several surfaces. Algebraic manipulation of the equations obtained by sequential stepwise application of Eqs. (3) and (4) to a series of k surfaces $1, 2 \ldots i \ldots k - 1, k$ has yielded the following rules:

Let the system be an assembly of k spherical surfaces whose centers of curvature lie on one line. The ith surface has power $F_i = (n_i - n_{i-1})/r_i$ and is separated from the next following neighboring surface by a distance t_i in a medium of refractive index n_i. Define $d_i = t_i/n_i$. An object placed with coordinates s_1, y_1 (with respect to the first surface) will have its conjugate point (image) with coordinates s_k', y_k'

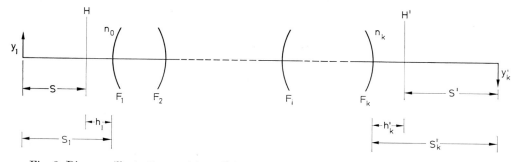

Fig. 6. Diagram illustrating position of object, image, first and second principal planes and various surfaces of a centered system of k spherical surfaces

(with respect to the kth surface) where the following relation holds

$$\frac{s'_k}{n_k} = \frac{\mathfrak{C}\,\dfrac{s_1}{n_0} + \mathfrak{D}}{\mathfrak{A}\,\dfrac{s_1}{n_0} + \mathfrak{B}} \tag{6}$$

and

$$\frac{y'_k}{y_1} = \frac{1}{\mathfrak{A}\,\dfrac{s_1}{n_0} + \mathfrak{B}}\ . \tag{7}$$

The four constants $\mathfrak{A}$, $\mathfrak{B}$, $\mathfrak{C}$, $\mathfrak{D}$ are characteristics of a given system. They can be evaluated from the following matrix formulation

$$\begin{vmatrix} \mathfrak{B} & \mathfrak{D} \\ \mathfrak{A} & \mathfrak{C} \end{vmatrix} = \begin{vmatrix} 1 & 0 \\ F_1 & 1 \end{vmatrix} \cdot \begin{vmatrix} 1 & -d_1 \\ 0 & 1 \end{vmatrix} \cdot \begin{vmatrix} 1 & 0 \\ F_2 & 1 \end{vmatrix} \cdots \begin{vmatrix} 1 & -d_{k-1} \\ 0 & 1 \end{vmatrix} \cdot \begin{vmatrix} 1 & 0 \\ F_k & 1 \end{vmatrix}, \tag{8}$$

i.e. multiplication of the matrices on the right side of the equation will give a matrix whose members are the four characteristic or Gaussian constants for the system. The four constants are related to each other by the equations

$$\mathfrak{B}\mathfrak{C} - \mathfrak{A}\mathfrak{D} = 1$$

and

$$\mathfrak{B} = \frac{\partial \mathfrak{A}}{\partial F_1}, \quad \mathfrak{C} = \frac{\partial \mathfrak{A}}{\partial F_k}, \quad \mathfrak{D} = \frac{\partial_2 \mathfrak{A}}{\partial F_1\, \partial F_k}\ . \tag{9}$$

The following interesting interpretation of Eqs. (6) and (7) is due to GAUSS (1843). Consider the conjugate planes for which $y'_k/y_1 = 1$, i.e., the planes of unit magnification where objects and images are always equal in size. Let the coordinates with respect to the first and last surfaces (i.e. the s_1 and s'_k values) for these two conjugate planes be denoted by h_1 and h'_k (Fig. 6). Substitution of $y'_k/y_1 = 1$ in Eq. (7) and then going to Eq. (6) gives

$$\frac{h_1}{n_0} = \frac{\mathfrak{B} - 1}{\mathfrak{A}}\ ,$$
$$\frac{h'_k}{n_k} = \frac{\mathfrak{C} - 1}{\mathfrak{A}}\ . \tag{10}$$

The planes of unit magnification defined by the coordinates (10) are called principal planes and denoted by H and H'. Their importance lies in the fact that the single surface object/image relationships embodied in Eqs. (3) and (4) may be used for the compound system, provided that the object distance be measured from the first principal plane and the image distance from the second principal plane. As may be verified by substituting in Eq. (6), the value to be used for the focal power of the whole system is the Gaussian constant $\mathfrak{A}$ of Eq. (8). It is called the principal point or equivalent power of the system, for it is the power of the single surface (or thin lens) that can be substituted for the whole assembly of surfaces, provided always that object and image distances are measured, respectively, from the first and second principal planes.

This theoretical framework finds immediate application in the case of the eye. For the many purposes in which the passage of light rays towards the retina has to be calculated, it becomes helpful to put together a schematic eye by using typical values for the parameters of the optical media and their boundaries. This was done by LISTING (1853), Helmholtz (1867) and GULLSTRAND (1909) many years ago and could conceivably be done better now that larger samples of the relevant data are becoming available (STENSTROM, 1947; SORSBY et al., 1957). The difficulty of the inhomogeneity of the crystalline lens is usually overcome by substituting a simplified lens model, e.g. one with a medium of fictitiously high refractive index uniformly filling the intracapsular space, or the hypothetical nucleus/core model of GULLSTRAND. Once these typical values have been assembled, calculations by means of Eq. (8) and (9) give the positions of the principal, nodal and focal planes. They are given in Table 3 for GULLSTRAND's schematic eye.

Table 3. Cardinal points of Gullstrands schematic human eye. All positions are in mm relative to the anterior surface of the cornea

Anterior focal plane	—15.2
First principal plane	1.5
Second principal plane	1.6
First Nodal plane	7.3
Second Nodal plane	7.4
Posterior focal plane	24.0

It turns out that the eyes that have been studied this way have their principal points exceedingly close together. In view of the kind of schematization that has formed the basis of the developments described so far, it is a reasonable step to make the further approximation of allowing both principal points to coalesce in a point. This happens to be an enormous simplification in that the whole compound system has now been reduced to a *single surface* separating the object-sided medium (air) from the image-sided medium (vitreous humor). In this *reduced eye*, the power of the surface is equal to the principal point power. A corresponding fictitious radius of curvature may be calculated from Eq. (2).

IV. Object-Image Calculations with Reduced Eye

For purposes of simple object image calculations the reduced eye stands un-excelled, because only Eqs. (3) and (4) are needed, and the object and image distances are measured from the position of the hypothetical single surface sub-stituting for the whole eye. But in wondering at the simplicity of the reduced eye, we should not forget that it was arrived at from typical real-eye values by the application of Eqs. (6) through (10). The circumstance that the two principal points of the schematic eye virtually coincide makes the approximation of coa-lescing them no more of an approximation than what had preceded.

The position and size of the ocular image corresponding to any object can be found by applying Eqs. (3) and (4) to the reduced eye. Here F is 60 D, and s and s' have to be measured from the principal point which is 1 2/3 mm behind the corneal vertex. When the target is at optical infinity, the image will be formed in the posterior focal plane. An eye whose retina is in the posterior focal plane of its optics is called emmetropic. However, targets not at infinity will now no longer be imaged on the retina. Most eyes have a mechanism, called accommodation, which adds refracting power to its optical system to take care of the situation when

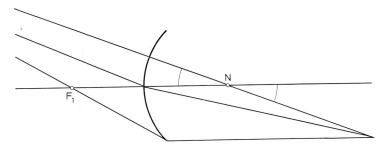

Fig. 7. Use of nodal rays to find image size. Of all the incident rays, the one directed to the nodal point of a surface goes to the image undeviated

the target is close up. When the retina is not in the geometrical image plane of the eye's optics, there is said to be a refractive error. The usual way of expressing refractive errors is in diopters of power excess or deficit; this gives directly the power of the correcting spectacle lens. Target location for clear retinal imagery in corrected eyes with refractive errors follows from the conjugacy relation of Eq. (3).

There are several ways of calculating the size of the retinal image corresponding to a given object. Eq. (4) is applicable here, but there are some simplifications in using nodal rays.

Any ray directed to the center of curvature of the surface of the reduced eye passes straight on to the retina without deviation (Fig. 7). This point, situated in man 5.55 mm behind the mean principal points, or 16 2/3 mm in front of the retina, is the nodal point of the reduced eye.

Its virtue is that the angle subtended at it by an object is equal to that subtended by the image, i.e. it is the point of unit angular magnification. When the eye is in focus for a target plane, the nodal point may be used very effectively for simple calculations of retinal image size. Example: in a normal emmetropic eye, the nodal point is 16 2/3 mm in front of the retina. The moon subtends 31.1 min of arc at the eye, what is the size of its retinal image ?

$$y' = 16.67 \cdot \tan 31.1' = .15 \text{ mm .}$$

Or conversely: a given retinal receptor measures 3 μ in cross section. What does this represent in angular measure in object space ?

$$\theta = \tan^{-1} \left(\frac{.003}{16.67} \right) = 37 \text{ sec of arc .}$$

V. Entrance and Exit Pupil System

When the eye is not focused on the target, however, there are difficulties with the above calculation caused by the fact that the nodal ray is not necessarily in the center of the image-forming bundle and may consequently intersect the retina away from the center of the blurred image.

To pursue this question further, consider the bundle of rays from a point object finally converging on the image point. Not all rays emitted by an object make it

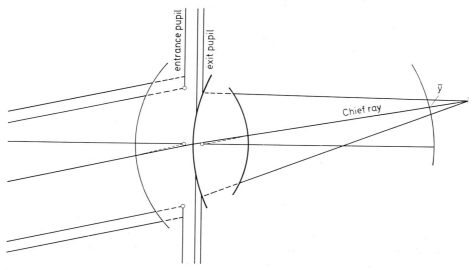

Fig. 8. Diagram illustrating position of entrance pupil, actual pupil and exit pupil of simplified schematic eye. Chief ray passes from object to center of entrance pupil and from center of exit pupil to image. When retina is not in geometrical image plane, a point image will give rise to a blur patch of diameter $\bar{y}$ and given by Eq. (11). Chief ray intersects retina in center of blurred light distribution

to the image on the retina. The structure responsible for the limitation of the bundle is the eye's pupil, but this is situated neither in object space nor in image space, but at a place where the rays from the object have already been subject to refraction (by the cornea) and are subsequently further refracted by the crystalline lens. The concept of entrance pupil has been developed to obviate the necessity of tracing each object ray through the cornea to ascertain whether it is passed or blocked by the iris. The entrance pupil is the image of the real pupil formed by the cornea. It is situated in object space (in man typically 3 mm behind the cornea and 13% larger than the real pupil). Every ray which in its passage through the eye is not blocked by the iris passes through the entrance pupil in object space (Fig. 8). Similarly for the exit pupil and image space. The exit pupil is the image formed of the pupil by the crystalline lens. In man it is typically 3.7 mm behind the corneal vertex and 3% larger than the real pupil, and its significance lies in the fact that the bundle of rays making up an image of a point object is a cone based on the exit pupil and having the Gaussian geometrical image as its apex. The central ray of this bundle, i.e., the ray going from the center of the exit pupil to the image, is called the *chief ray*. The whole configuration has its homologue in object space where the chief ray goes from the object to the center of the entrance pupil, and the entrance pupil intercepts that bundle of all the rays issuing from a point object that ends up in the image.

This geometrical construction is helpful because it forms the basis both in geometrical and wave optics of the retinal light distribution.

First of all, it shows the quantity of light that may be expected in the image. For a given target this depends in the ideal case only on the pupil area. In the real, as distinct from the schematic, eye there are transmission losses (Ruddock, 1972) as well as the complication that some retinal receptors, the cones, are directionally sensitive, i.e., do not equally accept radiation impinging on them from all directions (they demonstrate the Stiles-Crawford effect).

A second application of the geometrical construction in Fig. 8 is in the consideration of what happens when there is a focus error. The retina will intercept the light cone leaving the exit pupil not at its apex but elsewhere. As a first approximation, the resulting retinal light distribution will be a patch (top-hat distribution) proportional in size to the pupil area and the distance between the retina and the geometrical image. In symbols

$$\bar{y} = a \cdot \Delta F \tag{11}$$

where $\bar{y}$ is the blur-patch diameter, ΔF is the focus error in diopters and a is the exit pupil diameter. Assuming that the eye's resolution performance is reduced by a given amount whenever the blur patch has a given diameter, we see from Eq. (11) that, in the blurring of the retinal image, there is reciprocity between pupil diameter and out-of-focusness in diopters. Since the human eye suffers no decrement of performance so long as $\bar{y}$ remains below a certain amount, about 12 μ, the allowable focus error, the depth of focus, is inversely proportional to the pupil size when the focus error is expressed in diopters. The hyperfocal distance is the nearest point of clear vision when the eye's focus is fixed for infinity. Except for a factor probably related to the Stiles-Crawford effect, it is a linear function of pupil diameter (Campbell, 1957).

Regardless of whether image calculations are approximated by a simple geometrical optical model as above, or whether the more accurate model of wave optics is employed, the reference ray is always the chief ray, for it is the central ray in the cone based on the exit pupil. In general it will define the center of the light patch making up the image, blurred or clear. When there are two object points, the linear separation on the retina of their light-distributions will be given by the angular separation at the center of the exit pupil of the chief rays from the two objects. The geometrical relationship between the exit pupil of an eye and the retina is almost constant, even in accommodated states, so that a given chief ray angle will define with a high degree of invariance the separation of two retinal image distributions, clear or blurred. It remains to point out there corresponds for each exit pupil angle a fixed entrance pupil angle and it is now apparent that the method of choice of defining retinal images is the angle subtended by an object at the center of the entrance pupil. This will always be uniquely related to lateral image separation on the retina, no matter whether the retina is in the geometrical image plane or not. In the typical normal human eye, the angular magnification of the entrance/exit pupil system is 0.82, and the position of the exit pupil is 20.3 mm in front of the emmetropic retina. An object subtending a small angle θ at the center of the entrance pupil, will have its image-sided chief rays subtend an angle 0.82θ at the center of the exit pupil, and wherever the retina is, the image will be bounded by these chief rays.

VI. Diffraction

The wave theory of light offers the following model of image formation. Let us restrict the discussion at the outset to a point object of light. Of the wave front centered on the object point, only that portion intercepted by the entrance pupil of the eye will be of consequence. The process of refraction or imagery is now looked on as a change in shape of the wavefront: in an ideal case there now emerges from the exit pupil a spherical wavefront centered on the geometrical image point. All effects in image space may now be deduced from wave-theoretical considerations: by Huygens' principle, each point on the wave front acts as a secondary source of waves. The amplitude of the resulting disturbance is obtained by summing at each point on the receiving surface the radiation received from all points of the exposed wavefront, distance, phase and obliquity relationships being duly allowed for. Diffraction, for example, is now merely the phenomenon resulting from the circumscription of the entering wavefront by the exit pupil. All monochromatic aberrations are now viewed as deviations from sphericity of that wavefront. Obliquity effects and nonuniformity of transmission of the cornea or lens now become local differences in amplitude of the wavefront. Finally, the intensity of illumination at the receiving point is the square of the amplitude of the disturbance at this point resulting from the combined effect from all points of the wavefront corresponding to the object point. In quantum mechanics, the light intensity distribution is viewed as a probability distribution of photon events.

The effect of pure diffraction may be observed in the simplest case of a spherical wavefront uniformly filling the exit pupil and an image screen in the center of curvature of the wavefront and normal to the chief ray. This is physically realized, for example, in the best constructed telescopes, and to a satisfactory approximation also in the human eye for a narrow pupil ($<$ 2 mm diameter). For a circular aperture, there is radial symmetry, and the distribution of light intensity in the image plane is given by the equation

$$I(\varrho) = \left[\frac{J_1(\pi a \varrho/\lambda)}{\pi a \varrho/\lambda} \right]^2 \tag{12}$$

where ϱ is radial distance in image plane, a the diameter of the pupil, and λ the wavelength of light, and J_1 refers to a first-order Bessel function. The intensity distribution, illustrated in Fig. 9, comes to the first zero at 1.22 λ/a from the center, and successive maxima become smaller. The central zone included within the first minimum is called Airy's disk and its diameter is equal to 2.3 min of arc for $\lambda = 555$ nm and $a = 2$ mm. When the pupil is rectangular in shape, the intensity distribution in the image plane along a line parallel to the direction in which the pupil rectangle has side length a, is given by the equation

$$I(\alpha) = \left[\frac{\sin(2\pi a \alpha/\lambda)}{2\pi a \alpha/\lambda} \right]^2 . \tag{13}$$

This has its first zero at a distance $\alpha = \lambda/a$ from the central maximum.

Because the variables determining Airy's disk when expressed in angular measure in an eye's object space are independent of the eye's characteristics except the pupil aperture and the wavelength of light, the above figures hold for all eyes.

Irregular pupils, annular pupils, and pupils with nonuniform transmission can be handled in a similar manner, but the actual distribution is difficult to evaluate analytically and may have to be done by numerical integration on a computer.

Eqs. 12 and 13 emphasize the point made in the introduction to this chapter, viz. that there is an inverse relationship between width of pupil aperture and spread of light on the retina: the more extended the window for the acceptance of the wave, the more concentrated the image distribution.

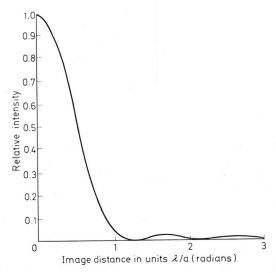

Fig. 9. FRAUNHOFER diffraction image for a perfect optical system with aperture diameter a and for light of wavelength λ (see Table 4). Light distribution comes to zero at 1.22 λ/a, 2.23 λ/a etc. from center, and has maxima at 1.64 λ/a, 2.68 λ/a etc. First bright ring has relative intensity 0.0175, second bright ring has relative intensity 0.0042

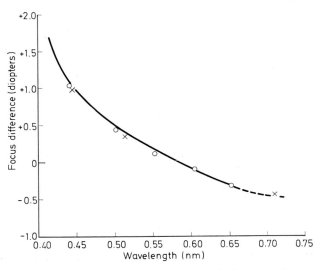

Fig. 10. Axial chromatic aberration of the human eye (x) and of an equivalent system composed of water (solid line) (From HARTRIDGE, 1950)

VII. Aberrations

There are two broad sources of inadequacy of the analysis so far developed. One is its assumption of monochromatic light. The velocity of light varies with the wavelength, and all quantities depending on it — refractive index, angle of refraction, trajectory of light ray, focal length, etc. — also vary correspondingly. The wavelength changes in refractive index of water, and focal power of the eye, are shown in Fig. 10. The close parallel between the two changes suggest that for practical purposes one may regard the ocular media as having the dispersion of water.

A. Axial Chromatic Aberration

When a point object emits light of a wide range of wavelengths, the eye will create images in a range of positions along the chief ray. Only one of these images can ever fall on the retina; the bundle for other wavelengths will be intercepted by the retina in a blur patch, as defined by Eq. (11). In the present connection, the focus error for light of a given wavelength will depend on the refractive index difference between light of the wavelength which is focused on the retina and light of the particular wavelength considered. All blur patches will be directly proportional to the pupil size, so that the deleterious effect of axial chromatic aberration will be enhanced by a large pupil. The analysis is manageable if the source contains just a few spectral bands, for then the retinal image, so far as axial chromatic aberration is concerned, consists of the concentric superimposition of a few blur patches of size as defined by Eq. (11). The situation as far as it pertains to visual stimulation can be exceedingly complicated because the light in each blur patch has a different absorption coefficient in each visual pigment. Spatial distribution of visual pigment bleaching in a broad wavelength band visual stimulus is therefore not easily specified.

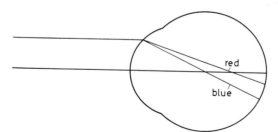

Fig. 11. Diagram to illustrate lateral chromatic aberration. Rays of different wavelengths entering the eye off-center will reach the retina in different locations

B. Lateral Chromatic Aberration

When the chief ray from a heterochromatic point object does not enter along a normal, there results also a lateral displacement of the blur patches for light of the various wavelengths with respect to one another. This is most simply seen in Fig. 11. For a constant angle of incidence, the angle of refraction of the chief ray (which defines the center of the blur patch on the retina) will vary with the refractive index of the vitreous. The effect can easily become several minutes of arc in magnitude. It is the basis of the phenomenon of chromastereopsis, which has its origin in symmetrically opposite lateral chromatic aberration in the two eyes producing a binocular disparity for superimposed heterochromatic stimuli.

C. Correction of Chromatic Aberration

It is possible to construct a lens system with axial chromatic aberration equal and opposite to that of the eye (WRIGHT, 1946). It will eliminate differences in focus due to wavelength changes but it will not, in general, correct for lateral chromatic aberration.

D. The Monochromatic Aberrations

The simple imaging Eqs. (3) and (4) and the equivalent wave-theoretical formulation (spherical wavefront centered on geometrical image point) are idealizations that optical systems only rarely satisfy. They apply best when a narrow bundle enters along the optical axis and they are often referred to as *paraxial* approaches. In general, all image rays from a point object do not intersect in the geometrical image point. An analytical expression for the image-sided ray trajectories has been developed for spherical surfaces by SEIDEL in 1856 using not only the first order of the expansions of Eq. (5) but also the next higher order. The equations worked out by SEIDEL have various additional terms on the right hand side of Eq. (3), the so-called Seidel Sums, which characterize five different kinds of defects, the so-called five third-order aberrations: spherical aberration, coma, astigmatism, distortion and curvature of field. Other ways of determining the image defects in the passage of monochromatic light rays through a system of centered spherical surfaces have been devised (HERZBERGER, 1958), and computation has helped a great deal in lens design.

Light rays are defined as normals to the wave front. Lack of convergence of all rays on to a point image, i.e., lack of stigmatic imagery, has its exact counterpart in lack of sphericity of the wavefront from a point object as it emerges into the image space through the exit pupil. The approach of describing the nature of the wavefront and then computing the intensity distribution in the image plane has the advantage of simultaneously including the effect of diffraction. Mathematically, the situation is handled in the same way as the simpler examples of diffraction. For each point of the receiving surface one sums the effects at this point of the disturbance emanating from each point of the wavefront. An exact knowledge of the nature of the latter is, therefore, a prerequisite to successful computation of light distribution. For a few rather simple defects, the wavefront deviations have been described and the computations attempted (HOPKINS, 1950). They are:

(i) Simple Spherical Focus Error (Refractive Error, Ametropia): when the image receiving plane is not in the geometrical image, first order theory gives a top-hat blur patch light spread [Eq. (11)]. The accurate description is that the wavefront is not centered on the image plane but on a plane in front or behind it. Concerning the light distribution centered on the intersection of the chief ray with the image plane, the actual entering wavefront deviates from the reference wavefront, i.e., the wavefront which would define ideal imagery for the particular position of the image plane. This deviation is a progressively increasing phase error for the light entering the image space as one goes peripherally from the chief ray towards the edge of the pupil. Integration in the image plane must take this phase error into consideration; and this is in fact the subject of FRESNEL diffraction which deals with the general case, whereas FRAUNHOFER diffraction [which leads to Eqs. (12) and (13)] has the restriction of a spherical wavefront centered on the image point (no focus error, no other aberrations).

(ii) A detailed theoretical analysis of the monochromatic aberrations of eyes is not a rewarding study because individual variations are large enough to render any general conclusion invalid. In the human eye, off-axis aberrations are significant only for peripheral vision, where, however, the retinal resolution is not high. Spherical aberration may become important for foveal vision when the pupil is large; it has been investigated in some detail (IVANOFF, 1947; KOOMEN, SKOLNIK and TOUSEY, 1949; WESTHEIMER, 1955; CORNSWEET and CRANE, 1970).

E. Astigmatism

There is a widely prevalent kind of image defect which manifests itself in the development of the image-sided bundle illustrated in Fig. 12 when there is a point object. The light never comes to a point image, or even an approximation of a point image (i.e., it is *astigmatic*), but the rays leaving the exit pupil have directions that take them all through one focal line and then, further on, through another line perpendicular to the first one. In between there is one position in which the bundle is circular in cross-section (the circle of least confusion) while all other cross sections are elliptical with a long axis parallel to the first line at the outset, and beyond the circle of least confusion, parallel to the second focal line.

This kind of image defect occurs under several circumstances. First of all, it occurs whenever a bundle of rays is obliquely incident, even on a spherical surface. Because the fovea is 5° away from the optical axis in the human eye, the foveal bundle of rays from a point object has a small amount of astigmatism induced, even if the cornea is spherical. On the other hand, it is common for ocular surfaces to be not spherical, nor even surfaces of revolution, but to exhibit a different radius of curvature in a pair of meridians, at right angles, usually the vertical and horizontal. Since the anterior surface of the cornea has the highest difference in refractive indices of all the ocular surfaces, toroidal surface characteristics on it are most likely to show up as astigmatic image defects. Finally, as discussed earlier, any irregular surface can, by DUPIN's theorem, be regarded as toroidal over a small enough region and hence engendering astigmatic imagery.

The calculation of the position and length of the focal lines in astigmatic imagery follows straight-forward application of Eqs. (3) and (11), but is must be remembered that a sheet of rays coming in along one of the principal meridians of the surface will be brought to a focus at the focal point corresponding to its power, but it will there help to produce a focal line at right angles to itself, i.e., in the meridian normal to it. Special lenses, so-called cylindrical or sphero-cylindrical lenses, are made to correct for astigmatism. They are ground to have astigmatic imagery themselves, but of power equal and opposite to that they are designed to correct.

VIII. Image Quality

A. Point-Spread, Line-Spread and Modulation Transfer Functions

The factors that contribute to a spreading of light in the image receiving plane have thus far been considered separately, and this is an acceptable approach in so far as something may possibly be done to ameliorate the one or other individually. However, in the final analysis of the performance of an optical instrument, the substantive information rests in the actual spread of light in the image

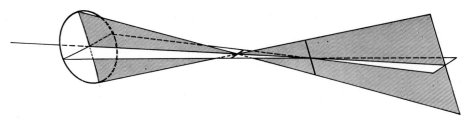

Fig. 12. Astigmatic image formation. A bundle from a point object is not brought to an image in a point but in two lines, perpendicular to each other, at different image distances. Depending on location of receiving screen, image light distribution will be a line, or ellipses of various eccentricities with their long axes parallel to either the first or the second focal line

of a point object, the point-spread function. According to the ideal of geometrical optics, this is a delta function, but the wave theory of light demands a spread according to Eq. (12) for an instrument with a circular aperture. The general concept of point-spread function is the most elementary way of describing the performance of an optical instrument because the light distribution in images of other targets can be deduced from it. All targets may be regarded as made up of an assembly of points. The image of each point is diffused into the shape of a point-spread function and the final image distribution at each point in the image plane is the sum of the height of the point-spread functions of all object points. This procedure of summing the contributions of all target points weighted by a factor depending on distance, is called *convolution*. In symbols: if the light distribution of the object in the two space dimensions x, y, is $O(x, y)$ and the instrument's point-spread function is $s(x, y)$, the image distribution $I(x, y)$ is given by the equation

$$I(x, y) = O(x, y) * s(x, y) \tag{14}$$

where the symbol $*$ denotes convolution.

Written out in full, the convolution integral reads:

$$I(x, y) = \int_{-\infty}^{+\infty} \int_{-\infty}^{+\infty} O(x - x_0, y - y_0)\, s(x_0, y_0)\, dx_0 dy_0 .$$

It states that the intensity at a given image point x, y is made up of contributions weighted by the distance-dependent factor $s(x_0, y_0)$ of all object points removed by a distance x_0, y_0 from the image point, x_0, y_0 going from zero to plus and minus infinity. The dimensions x, y in these equations are the two dimensions of the object and image planes. In the case of the eye, it is usually advantageous to let the dimensions be angles, as subtended by the objects and images at, say, the entrance and exit pupil centers respectively. Strictly speaking, a magnification factor should be introduced but this can be omitted in writing these equations. In this notation, then, Eq. (14) means that $O(\alpha, \beta)$ is the light distribution of the image in Gaussian optical theory, and $I(\alpha, \beta)$, that given by the fact that each object point is spread into the point-spread function $s(\alpha, \beta)$.

In general, the solution of Eq. (14) is quite difficult unless the spread can be expressed as a simple analytical function. Thus if it can be assumed that the point-spread is a Gaussian error function, $e^{-k\alpha^2}$, (FRY and COBB, 1935) or a simple exponential, $e^{-k\alpha}$, (FLAMANT, 1955; WESTHEIMER and CAMPBELL, 1962) the problem becomes tractable.

A simple example will illustrate the concept embodied in Eq. (14). Suppose that the point-spread function is radially symmetrical and has the form $s(\varrho)$ where $\varrho^2 = \alpha^2 + \beta^2$. What is the light distribution in the image of a long line of infinitesimal width? Let the line be given, in a cartesian coordinate system, by the equation

$$\alpha = 0 .$$

The task is to calculate the contribution at each point α along the axis $\beta = 0$ from all the points $\beta = \pm \infty$ of the line object. An element of line length $d\beta$ at a distance β from the α axis will contribute an element of intensity $s(\varrho)d\beta$ where $\varrho^2 = \alpha^2 + \beta^2$. Integration yields

$$A(\alpha) = 2 \int_{\alpha}^{\infty} s(\varrho) \ (\varrho^2 - \alpha^2)^{-1/2} \ \varrho \, d\varrho . \tag{15}$$

The function $A(\alpha)$ is now the distribution of light across the image of an infinitely long, infinitesimally wide line object. It is called the line-spread function and Eq. (15) gives the relationship between it and the point-spread function, assumed radially symmetrical, and forms the basis of image computation by convolution when the object is a target varying in only one dimension, such as a grating, bar or straight edge. Eq. (14) then simplifies to

$$I(\alpha) = A(\alpha) * O(\alpha) .$$

A special case of such an object is a grating whose intensity varies sinusoidally with distance

$$O(\alpha) = 1 + \cos 2 \pi \alpha \omega . \tag{16}$$

This pattern has maxima of 2 intensity units at points

$$\alpha = 0, \ \pm \frac{1}{\omega} , \ \pm \frac{2}{\omega} , \ . \ . \ ..$$

It thus has a spatial period of $1/\omega$ and a spatial frequency of ω cycles/unit image distance.

It turns out that, regardless of the actual shape of the line-spread function $s(\alpha)$, any sinusoidal grating of Eq. (16) has an image that is also a sinusoidal grating satisfying the relationship

$$I(\alpha) = 1 + \tau_\omega \cos\left(2\pi\omega\alpha + \phi\right)$$

i.e., that has maxima of height $1 + \tau_\omega$ and minima $1 - \tau_\omega$ and that may in addition be phase shifted through a fraction ϕ of a spatial period. One defines the modulation of a sinusoidal pattern by Michelson's expression (1927)

$$\frac{I_{max} - I_{min}}{I_{max} + I_{min}} = \tau_\omega . \tag{17}$$

τ_ω is the modulation coefficient for the sinusoidal grating of spatial frequency ω. The function showing the modulation coefficient for all spatial frequencies is an important describing function for any optical instrument. It is called the *modulation transfer function* and its relationship to the line-spread function is given by the equation

$$\tau(\omega) = 2 \int_0^\infty A(\alpha) \cos\left(2\pi\alpha\omega\right) \cdot d\alpha ,$$

i.e., the modulation transfer function is the Fourier cosine transform of the line-spread function. Because no phase shifts occur in the imaging of a sinusoidal grating by a symmetrical optical system, the cosine transform suffices in our discussion. For completeness we also give the relationship between the point-spread function $s(\varrho)$ and the modulation transfer function

$$\tau(\omega) = \int_0^\infty s(\varrho) J_0(2\pi\omega\varrho) \varrho d\varrho$$

which is a Hankel transform, i.e., a transform with the Bessel function J_0 as the kernel function.

The application of the foregoing mathematical formulation is the following. By the Fourier theorem, any light distribution giving the quantity of light as a function of distances α, β in the object or image planes can be equally given as the sum of sinusoidal gratings of various amplitudes, phases and frequencies. The relationship between $O(\alpha, \beta)$, a light distribution (intensity *versus* distance), and the parameters of the equivalent sinusoidal patterns (functions of spatial frequency) is fixed for each case and given by the Fourier transformation equation

$$O(\omega_\gamma, \omega_\beta) = \int\limits_{-\infty}^{+\infty} \int O(\alpha, \beta) \cdot e^{-2\pi i (\alpha\omega_\alpha + \beta\omega_\beta)} \, d\alpha \, d\beta .$$

The simplest case, of course, is when $O(\alpha, \beta) = 1 + \cos 2\pi\alpha\omega$ for which $O(\omega_\alpha, \omega_\beta)$ is a delta function situated at $\omega\alpha$ and is everywhere else zero. The variables ω_α, ω_β are the spatial frequencies of sinusoidal intensity patterns in the α and β direction of object or image planes. Once the Fourier content, or spatial frequency distribution, of the object function has been found, one can obtain the Fourier content or spatial frequency distribution of the image by merely multiplying by the height of the modulation transfer function.
In symbols

$$i(\omega_\alpha, \omega_\beta) = o(\omega_\alpha, \omega_\beta) \cdot \tau(\omega_\gamma, \omega_\beta) . \tag{18}$$

Comparison of Eq. (14) with Eq. (18) indicates that the basis of the whole sequence of operations is Parseval's theorem, which states that the operation of convolution becomes, in the Fourier domain, the operation of multiplication. The two processes, that embodied in Eq. (14) and that embodied in Eq. (18), are in all respects equivalent: to go from object to image one either convolutes the object light distribution with the point-spread function, or one multiplies the Fourier transform of the object light distribution with the modulation transfer function.

The central place of the point-spread function, or its transform, the modulation transfer function, are hereby emphasized, for they epitomize the ultimate capability of the optical system in its role as an image-forming mechanism. In describing how well a system performs, they are equivalent to the impulse function and the frequency response function of electric or mechanical systems. Their range of applicability is almost, but not quite the same. Except in the case of oblique aberrations, the point-spread function is circularly symmetrical about the chief ray intersect and the specification of the cosine transform generally suffices, i.e., there are generally no phase terms in the modulation transfer function. The laws of linearity hold widely here because, in the framework presented, the equations deal with the spatial spread of light, and amplitude saturation effects do not enter (although such limitations may apply to the transduction process, e.g. in the retina). In optical systems there is, however, the problem of changes in point-spread function with image position, even in a fixed image plane, so that inhomogeneity has to be considered, but within these limits the laws of superposition hold. Finally it must be pointed out that the theory so far developed holds strictly only when the object is incoherently lit, i.e., consists of an assembly of independent self-luminous points. If the object is lit by perfectly coherent light (e.g. by a laser) the arguments can be transferred completely into the domain of amplitude (rather than intensity) of electromagnetic disturbance. A reduction case of this type of problem is the so-called Maxwellian View, which has been treated in detail elsewhere (WESTHEIMER, 1966).

Before proceeding to summarize our present knowledge of the shape of the functions $s(\varrho)$, $A(\alpha)$, and $\tau(\omega)$, i.e., the point-spread, the line-spread, and the modulation transfer functions of eyes, certain theoretically determined functions will be valuable.

B. Modulation Transfer Function for Diffraction Limited Systems

The diffraction pattern of Eq. (12) is, in fact, the point-spread function for a round pupil. Its Fourier transform, which is the modulation transfer function, is given in Fig. 13. The ordinates are simply the ratios of image-sided to object-sided modulation of sinusoidal gratings of the spatial frequencies indicated on the abscissae. The latter are given in normalized coordinates, so that the same figure can be used for combinations of different apertures and wavelengths. Also shown is the modulation transfer function for imagery with a rectangular pupil along a direction parallel to a side of the rectangle, with side length a. Both curves come to zero at the cut-off spatial frequency which always has the value a/λ cycles per radian object or image angle in air. When the image is formed at a distance f (in air) from the principal point, the cut-off spatial frequency has period

$$y = f \tan \lambda/a \ .$$

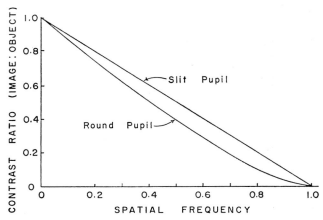

Fig. 13. Modulation transfer function for perfect optical system with a round pupil and a rectangular pupil. Ordinates: Modulation in the image of a sinusoidal object with contrast 1. Abscissae: Spatial frequency in normalized coordinates. Cut-off spatial frequency is λ/a cycles/radian where λ is wavelength and a is pupil diameter (see Table 4)

Any target features of spatial frequencies higher than the cut-off spatial frequency are not represented in the image; those at lower frequencies are attenuated as shown. Fig. 13 is yet another statement of the fact that the performance of an instrument is inversely proportional to the pupil aperture, and directly proportional to the wavelength of light [Eq. (1)]. For the representative situation of a human eye of 2 mm pupil diameter and wavelength 555 nm, the cut-off spatial frequency is 1.02 cycle/min of arc (in object space in air) and the period of the narrowest sinusoidal pattern that has any modulation at all is 4.8 μ on the retina. For other pupil diameters, proportional values apply (Table 4).

Table 4. Image limitation due to diffraction (circular aperture, wavelength 0.56 μ).

Diameter of aperture stop (mm)	Angular subtense of diameter of Airy's disk (in air)		Cut-off spatial frequency cycles/min of arc.
	Milliradians	min of arc.	
1	1.36	4.70	0.51
2	0.68	2.35	1.02
3	0.45	1.57	1.53
4	0.34	1.18	2.04
6	0.23	0.78	3.06
8	0.17	0.59	4.08

C. Modulation Transfer Functions for Defocused Imagery

As Eq. (11) has shown, the two parameters of concern in out-of-focus imagery are the pupil diameter and the dioptral value of the defocus. In all situations in which the wavefront errors at the edge of the exit pupil (see Section VII, Di) are the same, the modulation transfer functions normalized to the cut-off spatial

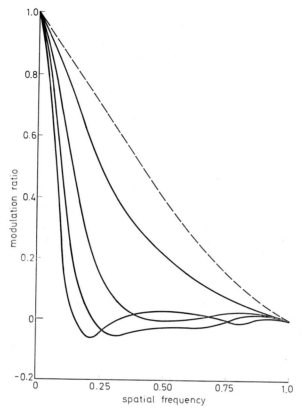

Fig. 14. Modulation transfer function for defocused optical systems. For coordinates, see Fig. 13. Parameter of defocusing is wavelengths of deviation of actual wavefront from reference wave front at edge of pupil. For wavelength 0.56 μ, pupil diameter a millimeters, and defocusing ΔF diopters, error is 0.225 $a^2 \Delta F$ wavelengths in human eye. For a human eye with a 3 mm round pupil, the interrupted line corresponds to perfect imagery and the four solid curves to 0.23, 0.31, 0.62 and $1 D$ defocus.

frequency have the same shape (Fig. 14). It can be shown (HOPKINS, 1949) that the phase error in number of wavelengths at the edge of the exit pupil, which is the parameter in the family of curves in Fig. 14, is given by the expression

$$2 n a^2 \Delta X / \lambda l^2$$

where n is the refractive index of the image space, a is the diameter of the exit pupil, ΔX is the distance between the receiving plane and the Gaussian image plane, λ the wavelength of light in vacuum, and l the distance between the exit pupil plane and the Gaussian image plane. Applied to the typical human eye for light of wavelength 0.56 nm, this reduces to

$$0.225 \, a^2 \, \Delta F$$

where a is the diameter of the eye's pupil — in millimeters — and ΔF is the focus error in diopters. An interesting feature of Fig. 14 is the fact that after an initial

Fig. 15. Pattern to illustrate spurious resolution. Figure should be viewed close to the eye, but with accommodation relaxed. It can be observed that a white spoke becomes dark and then white again, sometimes going through several reversals in contrast as it is followed from the edge of the figure to the center

drop to zero, the modulation coefficient may become negative over a range of spatial frequencies. This negativity, when checked against Eq. (17), implies that $I_{min} > I_{max}$, i.e., that the dark and light regions in the object grating come out to be reversed in the image. We are dealing here with a phenomenon called ,,spurious resolution''; it can be observed by the reader when viewing Fig. 15 out-of-focus. It is of interest to compare the situation of a zero crossing of an out-of-focus modulation transfer function with the concept of cut-off spatial frequency of Fig. 13. No information is present beyond the latter, but in the former, except at zero crossing frequencies, information is available in the image from which, in theory, the object can be reconstructed. The idea of reconstruction of objects from impoverished but not destitute images does not fall within the ambit of this chapter. Suffice it to say that a knowledge of the modulation transfer function (or its equivalent) is a prerequisite to any accurate reconstruction.

D. Modulation Transfer Function in Other Cases

Modulation transfer functions have been computed for a variety of defects of optical systems: spherical aberrations, coma, astigmatism, partial obscuring of the pupil and many more (Born and Wolf, 1959; Hopkins, 1962; O'Neil, 1963). They can occasionally be used with advantage in problems of ocular image formation.

There is a general theorem here that has its origin in diffraction theory. It states that the modulation transfer function is the self-convolution or auto-correlation of the pupil aperture function. This can be checked readily in Fig. 13 where the modulation transfer function of the rectangular aperture is a triangular function which is indeed the autocorrelogram of a rectangular function. The requisite units and scales can be deduced from the fact that when the pupil diameter is a, the cut-off spatial frequency is a/λ cycles/radian. The autocorrelogram of a circular pupil is not easy to compute but it does have the shape given for the modulation transfer function for a circular pupil in Fig. 13. More generally, any pupil function, e.g. an annular or otherwise partially obscured pupil, will lead to the associated modulation transfer function by autocorrelation. Finally, all aberrations can be regarded as phase deviations of the wavefront leaving the exit pupil and the appropriate modulation transfer functions then are approached via complex autocorrelation, i.e., correlation which uses both the amplitude and phase of the pupil disturbance, or which uses the sine and cosine components of the disturbance in quadrature.

IX. Performance Data on Eyes

Hard data on the point-spread or modulation transfer function of eyes has only recently become available.

Earlier attempts at measuring the light-spread in the eye were mainly directed at the long tail of the light distribution. This has considerable practical importance, for the scattered light from an automobile headlight might readily swamp the light reaching the retina from a pedestrian several degrees away. Indirect methods, such as psychophysical brightness matching in the presence of peripheral glare sources, have been used (LeGrand, 1937; Schouten and Orstein, 1939; Fry and Alpern, 1951). Typical results are shown in Fig. 16. Another procedure involves the measurements through a hole in the sclera of the actual light reaching a point at the back of an enucleated eye, (Boynton, Enoch, and Bush, 1954; De Mott and Boynton, 1958), but this may be seriously marred by post-mortem artifacts.

Acceptable determinations of the actual shape of the point-spread function of the eye have their origin in Flamant's procedure (1955). She presented a bright line as an object to the eye. An image is produced at the back of the eye; the information desired is its line-spread function. The fundus of the eye acts as a reflector for a small proportion of the incident light, with reflecting properties that introduce no complications. This light then traverses the eye in the opposite direction and the optics of the eye form an areal image, which may be separated from the original source by a beam splitter. Various techniques have been employed for the analysis of the areal image; Flamant herself used a photographic technique, but more recent replications of her work employed photoelectric means. However it is measured, the light in the areal image has suffered the spreading effects of the eye's optics twice and the data have to be processed to retrieve the line-spread function of the eye. The most recent and most careful replication of Flamant's experiment is that of Campbell and Gubisch (1966) and is shown in Fig. 17. An alternative procedure is to use grating targets instead of the line and measure the modulation in the grating areal image. Since this has been demodulated twice, once during each passage through the eye, the modulation during a single passage, which gives the height of the modulation transfer function at that spatial frequency, is the square root of the observed modulation. The results are compatible with the deductions from the line-spread measurements (Westheimer,

1963). Since all this is an objective procedure, it can be carried out in animals (cat, Westheimer, 1962).

There is some doubt whether the spatial distribution of light reflected from the fundus corresponds to the spatial distribution of intrareceptoral quantum absorption, which is the basic visual stimulus. In principle it could be obtained through measurement of the spatial spread of bleaching on the retina by extending Rushton's rhodopsinometry (Campbell and Rushton, 1955) to the space domain, but this has not been pursued since Kühne's original observation (1879) of a bleached pattern on a frog retina.

The subjective visual response is, of course, dependent on the intrareceptoral quantal absorption distribution and if it can be utilized by a suitable experimental arrangement that leaves aside the photochemical and neural factors that superimpose themselves on the optical ones, a confirmation may be obtained that the Flamant method does indeed yield results meaningful in the actual operation of the visual apparatus.

Such an approach may be made by the interference fringe technique introduced by Westheimer (1959). By one of several arrangements, a system of Young's interference fringes can be created on the retina. It becomes a retinal image that

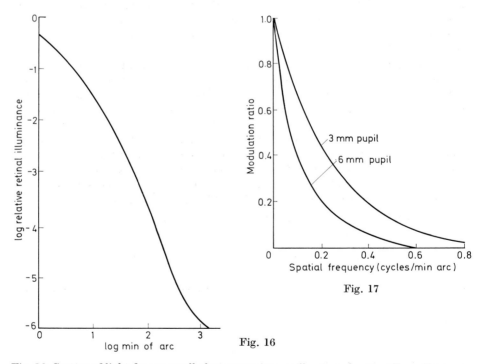

Fig. 16. Scatter of light from a small glare target into unilluminated retina (From Flamant, 1955). Ordinates: log relative intensity; abscissae: log distance from source, in min of arc

Fig. 17. Best current estimate of modulation transfer function of human eye for white light with a 3 mm pupil (Gubisch, 1967) and with a 6 mm pupil (Westheimer, 1963), based on the application of Flamant's method

is in fact a sinusoidal grating of fixed spatial frequency with a modulation that closely approaches 100%. By mixing it in varying proportions with a uniform field, i.e., demodulating it, the modulation threshold of the visual system can be determined. Its value depends on retinal and neural factors that need not concern us here, but the fact that YOUNG's interference fringes start off with practically 100% modulation at the retina allows us to obtain an absolute value for the modulation at threshold. A sinusoidal grating is then presented to the eye as an object and demodulated until the visual modulation threshold is again reached. The knowledge of the absolute modulation threshold allows one to compute the demodulation introduced by the optics of the eye, i.e., the eye's modulation transfer coefficient at that spatial frequency. This technique, which uses the rest of an observer's visual system as a null indicator, was employed by ARNULF and DUPUY (1960) and by CAMPBELL and GREEN (1965) in the human eye. The results show only fair concordance with the most careful replication of the Flamant experiments for reasons that are as yet obscure but may be purely methodological.

X. Image Light Distributions

The final application of the material in this chapter in most instances is the specification of the light distribution on the receiving layer of the retina for a target configuration that is given by the exigencies of a particular stimulus situation.

The principle is that of Eq. (14) or (18), i.e., convolution of the object with the point-spread function, or multiplication of the object's Fourier transform by the modulation transfer function and subsequent retransformation from the Fourier to the space domain. Apart of the computational difficulty of the former approach, there is the additional point that the point-spread function always has a long tail, deduced not only observationally but also theoretically from the fact that a finite aperture leads by diffraction theory to an infinite point-spread function. For these practical considerations, it is preferable to carry out the computation of image light distributions in the Fourier domain, utilizing the modulation transfer function.

The following are examples of light distributions treated in this manner.

A circular disk of radius ϱ_0 has a FOURIER transform

$$o(\omega) = 2\pi \int_0^{\varrho_0} J_0(2\pi\varrho\omega)\varrho\,d\varrho = \frac{\varrho_0}{\omega} \cdot J_1(2\pi\varrho_0\omega)$$

Multiplying this by the modulation transfer function $\tau(\omega)$ and retransforming yields

$$I(\varrho) = 2\pi\varrho_0 \int_0^\infty J_1(2\pi\varrho_0\omega) \cdot J_0(2\pi\varrho\omega)\,\tau(\omega) \cdot d\omega$$

as the rotationally symmetrical light-intensity as a function of image distance ϱ from the center of the image. Fig. 18 gives the resulting light distribution for target disks of several diameters using the modulation transfer function for a human eye in best focus for a 6 mm pupil, i.e., the kind of situation one finds in experiments on rod vision.

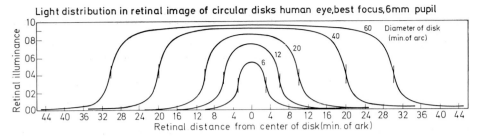

Fig. 18. Light distribution in retinal image of uniformly illuminated disks of various diameters. Light spread is that for human eye with a 6 mm pupil (Westheimer, 1963).

Similarly a long bar of width a_0 has as its Fourier transform the function

$$o(\omega) = 2 \int_0^{\frac{a_0}{2}} \cos(2\pi\omega a)\,da = \frac{\sin(\pi\omega\alpha_0)}{\pi\omega}.$$

Multiplying by the modulation transfer function $\tau(\omega)$ and retransforming, we obtain

$$I(\alpha) = 2 \int_0^\infty \frac{\sin(\pi\omega a_0)}{\pi\omega} \cdot \tau(\omega)\cos(2\pi\omega\alpha)\,d\omega$$

as the intensity function at right angles to the bar as a function of distance α from the center of the bar.

Repetitive targets such as gratings are easier to handle because they have a series of discrete Fourier terms. For example, a square wave grating of 100% modulation of spatial frequency ω has Fourier terms of height $1/n$ at all frequencies $n\omega$ where n is odd and goes from 1 to ∞. Each term has to be multiplied by the height of the modulation transfer function at that frequency and the sinusoidal components thus created have to be superimposed.

Finally, the intensity distribution at a semi-infinite half plane, i.e., at the border of a dark and light region, is given by the equation

$$I(\bar{\alpha}) = \int_{-\infty}^{\bar{\alpha}} A(\alpha)\,d\alpha$$

where $A(\alpha)$ is the line spread function and the border is taken to be at $\bar{\alpha} = 0$.

Preparation of this chapter was aided by grant EY 00220 from the National Eye Institute, U. S. Public Health Service.

References

Arnulf, A., Dupuy, O.: La transmission des contrasts par le système optique de l'oeil et les seuils de contrastes rétiniens. C. R. Acad. Sci. (Paris) **250**, 2757—59 (1960).

Born, M., Wolf, E.: Principles of Optics. New York: Pergamon Press 1959.

Boynton, R. M., Enoch, J. M., Bush, W. R.: Physical measures of stray light in excised eyes. J. Opt. Soc. Amer. **44**, 879—886 (1954).

CAMPBELL, F. W.: The depth of field of the human eye. Optica Acta **4**, 157—164 (1957).

— GREEN, D. G.: Optical and retinal factors affecting visual resolution. J. Physiol. (Lond.) **176**, 576—593 (1965).

— GUBISCH, R. W.: Optical quality of the human eye. J. Physiol. (Lond.) **186**, 558—578 (1966).

— RUSHTON, W. A. H.: Measurement of the scotopic pigment in the living human eye. J. Physiol. (Lond.) **130**, 131—147 (1955).

CORNSWEET, T. N., CRANE, H. D.: Servo-controlled infrared optometer. J. Opt. Soc. Amer. **60**, 548—554 (1970).

DEMOTT, D. W., BOYNTON, R. M.: Retinal distribution of entoptic stray light. J. Opt. Soc. Amer. **48**, 13—21 (1958).

FLAMANT, F.: Etude de la repartition de lumière dans l'image rétinienne d'une fente. Rev. Opt. **34**, 433—459 (1955).

FRY, G. A., ALPERN, M.: Effect of a peripheral glare source upon the apparent brightness of an object. J. Opt. Soc. Amer. **43**, 189 —195 (1953).

— COBB, P. W.: A new method for determining the blurredness of the retinal image. Trans. Amer. Acad. Ophthal. Otolaryng. **40**, 423—438 (1935).

GAUSS, C. F.: Dioptrische Untersuchungen. Abh. Kgl. Gesellsch. f. Wissensch. Göttingen, Vol. 1. 1843.

GOLDMAN, H., HAGEN, R.: Zur direkten Messung der Totalbrechkraft des lebenden menschlichen Auges. Ophthalmologica (Basel) **104**, 15—22 (1942).

GUBISCH, R. W.: Optical performance of the human eye. J. Opt. Soc. Amer. **57**, 407—415 (1967).

GULLSTRAND, A.: In: HELMHOLTZ, H. v.: Handbuch der Physiol. Optik. Bd. 1, Ed. 3. Hamburg: L. Voss 1909.

HARTRIDGE, H.: Recent Advances in the Physiology of Vision. London: J. and A. Churchill Ltd. 1950.

HELMHOLTZ, H. v.: Handbuch der Physiologischen Optik. Leipzig: L. Voss 1867.

HERZBERGER, M.: Modern Geometrical Optics. New York: Interscience Publ. Inc. 1958.

HOPKINS, H. H.: Disturbance near the focus of a spherical wave showing a radial inhomogeneity of amplitude. In: FLEURY, P. (Ed.): La Theorie des Images Optiques. pp. 209—223. Paris: Editions de la Revue d'Optique 1949.

— Wave Theory of Aberrations. Oxford: Clarendon Press 1950.

HUGHES, A.: A schematic eye for the rabbit. Vision Res. **12**, 123—138 (1972).

IVANOFF, A.: Les aberrations de chromatisme et de sphericité de l'oeil. Rev. Opt. **26**, 145—171 (1947).

JANSSON, F.: Measurements of Intraocular Distances by Ultra Sound. Acta Ophthal. (Kbh) Suppl. **74** (1963).

KOOMEN, M., TOUSEY, R., SCOLNIK, R.: Spherical aberration of the eye. J. Opt. Soc. Amer. **39**, 370—376 (1949).

KÜHNE, W.: Chemische Vorgänge in der Netzhaut. In: HERMANN, L. (Ed.): Handbuch der Physiologie, 3. Bd., 1. Teil. Leipzig: F. C. W. Vogel 1879.

LEGRAND, Y.: Diffusion de la lumière dans l'oeil. Rev. Opt. **16**, 201—241 (1937).

LISTING, J. B.: Mathematische Diskussion des Ganges der Lichtstrahlen im Auge. In: WAGNER, R.: Handwörterbuch der Physiologie, Bd. IV, S. 451—504. Braunschweig: Vieweg 1853.

MICHELSON, A. A.: Studies in Optics. Chicago: University of Chicago Press 1927.

O'NEIL, E. L.: Introduction to Statistical Optics. Reading, Mass: Addison-Wesley 1963.

RUDDOCK, K. H.: Light transmission through the Ocular media and macular pigment and its significance for psychophysical investigation — Chapter 17, in Vol. VII/4 of this Handbook.

RUSHTON, R. H.: The clinical measurement of the axial length of the living eye. Trans. Ophthal. Soc. U. K., **58**, 136 (1938).

SCHOUTEN, J. F., ORNSTEIN, L. S.: Measurements on direct and indirect adaptation by means of a binocular method. J. Opt. Soc. Amer. **29**, 168—182 (1939).

SEIDEL, L.: Zur Dioptrik: Über die Entwicklung der Glieder dritter Ordnung. Ast. Nachr. **43**, 289—332 (1856).

Sorsby, A., Benjamin, B., Davey, J. B., Sheridan, M., Tanner, J. M.: Emmetropia and its aberrations. Med. Res. Council (Great Britain) Special Report Series No. 293. London: H. M. Stationery Office 1957.

Stenstrom, S.: Untersuchungen über die Variation and Kovariation der optischen Elemente des menschlichen Auges. Acta Ophthal. (Kbh.) Suppl. XXVI 1946.

Tscherning, M.: Physiologic Optics. Philadelphia: The Keystone 1904.

Vakkur, G. J., Bishop, P. O.: The Schematic Eye in the Cat. Vision Res. 3, 357—381 (1963).

Walls, G. L.: The vertebrate eye. Bloomfield Hills, Mich.: Cranbrook Institute of Science 1942.

Westheimer, G.: Spherical aberration of the eye. Optica Acta 2, 151 (1955).

— Line-spread function of living cat eye. J. Opt. Soc. Amer. 52, 1326 (1962).

— Optical and motor factors in the formation of the retinal image. J. Opt. Soc. Amer. 53, 86—93 (1963).

— The Maxwellian View. Vision Res. 6, 669—682 (1966).

— The Eye. In: Mountcastle, Vernon B. (Ed.): Medical Physiology, 12th Ed. pp. 1532—1553. St. Louis: C. V. Mosby Co. 1968.

— Campbell, F. W.: Light distribution in the image in the living human eye. J. Opt. Soc. Amer. 52, 1040—1044 (1962).

Wright, W. D.: Researches in Normal and Defective Colour Vision. London: H. Kimpton 1946.

Chapter 12

Light-Induced Potential and Resistance Changes in Vertebrate Photoreceptors

By

Tsuneo Tomita, Tokyo (Japan) and New Haven, Connecticut (USA)

With 18 Figures

Contents

1. Introduction

It has been established from ERG analyses that the cornea-negative component, the PIII of Granit, originates at least in part in the receptors themselves. (Concerning the background and present status of our knowledge, see Chapter 17.) Until just recently, however, a few points still remained to be explained. One of these was the unusual sign of the PIII, first pointed out by Granit (1947). While it is general that a receptor, when excited, forms an electric field of such sign as to make its distal end negative relative to the proximal end, and while this generality applies also to most invertebrate photoreceptors (see Section 7 of this Chapter), the polarity of the PIII is just opposite and is such as to shift the distal margin of the receptors positive, instead of negative. Another point requiring a reasonable account in assigning the PIII to the receptors was the parallel relation between the PIII and inhibition, well established also by Granit (1947). There is a building up of inhibition along with the PIII during photic stimulation, and the sudden destruction of the PIII at the termination of light results in a release of excitation. Does this then mean that the response to light of vertebrate photoreceptors is "inhibitory" in nature?

As a step towards explaining these points, it was important to know whether the response to light of the rods and cones is depolarizing, as in any receptor studied so far, or hyperpolarizing. This could be determined by means of intracellular recording from single vertebrate photoreceptors. Several such attempts have been reported (Svaetichin, 1953; Oikawa et al., 1959; Bortoff, 1964; Svaetichin et al., 1965; Bortoff and Norton, 1965a, 1965b; Werblin and Dowling, 1969; Baylor and Fuortes, 1970), although in some of them the identification of the cell types, from which the recordings were made, was not always conclusive, particularly in differentiating a receptor potential from an S potential. Nevertheless, the results agreed in that the cells were hyperpolarized during light.

This Chapter intends to discuss the development of research along this line, which has led to the conclusion that both rods and cones are hyperpolarized by light and that the hyperpolarizing response is associated with an increase in membrane resistance. Readers may also refer to two recent review articles (Tomita, 1970; Witkovsky, 1971). The description in the following sections will begin with the study on single carp cones, since the work is considered to best illustrate the technique of intracellular recording from small cells such as the vertebrate photoreceptors, as well as the technique of cell identification.

Readers who are interested in the early receptor potential (e.r.p.) will refer to Chapter 8.

2. Potential Changes in Single Carp Cones

a) Recording Technique

Jolter Device. In our work using the carp (Tomita, 1965; Tomita et al., 1967), intracellular recording from single receptors was made possible only when the penetration was aided by jolting the retina at a high acceleration against a vertically held, slowly advancing minute pipette. This was identical in principle with the old pipette-hammering technique, but the difficulty in the old technique, that of giving pipettes a purely longitudinal movement, was overcome in this jolting technique, to some extent at least. The jolter device presently being used is illustrated in Fig. 1. The retina, detached from the pigment epithelium, is mounted receptor side up on the glass plate (GP) which serves as the jolting table. The glass plate also permits illumination from underneath, so that the direction of light incident upon the retina is physiological. The natural frequency of the jolting table is about 10 kc, and the vibration in response to each current pulse is damped out in a few milliseconds. The calculated acceleration is of the order of several hundred g's (g = gravity acceleration). This acceleration was found to be too large and had to be reduced to an empirically determined optimal value by a variable resistance connected in parallel with the coil.

The function of the jolter is apparently to facilitate penetration of a micropipette into a cell whose membrane is depressed by, but does not quite yield to,
the tip of the micropipette.

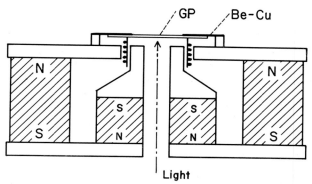

Fig. 1. Schematic diagram of the jolter so designed as to facilitate the penetration of micropipette electrodes into small cells (TOMITA et al., 1967)

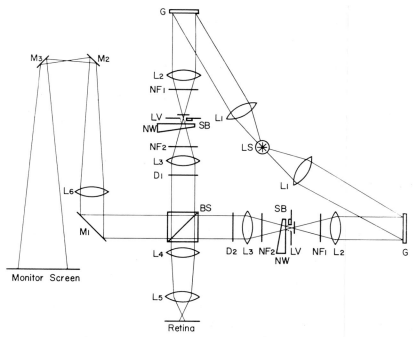

Fig. 2. Schematic diagram of the two-channel photostimulator. BS, beam splitter; D_1—D_2,
diaphragms with changeable apertures, the images of which are focused on the retina; G,
grating; L_1, collimeter lens; L_2—L_6, lenses; LS, xenon lamp of 150 W as the light source;
LV, light valve; M_1—M_3, surface mirrors; NF_1—NF_2, neutral density filters to permit attenuation as a whole from 0 to 4.75 log-units in steps of 0.25 log-unit; NW, neutral density wedge
operated by a servo mechanism for automatic equalization of the quantum flux of monochromatic light available at 18 wavelengths in steps of 20 nm between 400 and 740 nm; SB,
solar battery to signal the light stimulus (TOMITA et al., 1967)

Photostimulator. Fig. 2 is a schematic diagram of our two-channel photo-stimulator with a xenon lamp as the light source (Tomita et al., 1967). The two channels are arranged symmetrically on both sides of a line connecting the light source (LS) and the beam splitter (BS). From each channel, monochromatic light of 18 different wavelengths, ranging between 400 nm and 740 nm in steps of 20 nm, are available. White light can also be obtained, whenever necessary, by rotating the grating to an angle suitable for surface reflection. The spectral scan, which is accomplished in steps of 20 nm, proceeds automatically with repetitive cycles of a given duration of illumination followed by an intermission. Each inter-mission is utilized not only for a 20 nm shift of the wavelength by a step of rotation of the grating but also for automatic equalization of the flux of light quanta impinging on the retina. The adjustment of the flux is achieved by means of a servo mechanism which sets the neutral density wedge (NW) immediately in back of the light valve (LV) to a position predetermined for each wavelength by an absolute measurement of the quantum flux. The full intensity of monochromatic light at the retinal level has been adjusted to a value of 2×10^5 quanta per $\mu^2 \cdot$ sec for all wavelengths of monochromatic light. Two series of neutral-density filters (NF$_1$ and NF$_2$) serve for attenuation from this full intensity down to $- 4.75$ log-units in steps of 0.25 log-unit.

Procedure. A micropipette, filled with 2M-KCl and having a tip diameter of less than 0.1 μ, is advanced vertically into the retina by aid of a micromanipulator. During the advance of the micropipette the jolter on which the retina is mounted is periodically activated at the beginning of each sweep of the oscilloscope beam. Whenever a resting potential and a response to light are observed within the receptor layer, the jolting is stopped and the response to light is studied.

To obtain spectral response curves, the spectrum is scanned at least twice so as to provide an adequate control against deterioration of the response during recording. An ascending and descending series, covering the spectrum from 400 to 740 nm, could be obtained within 40 sec.

b) Spectral Response Curves

Three sample records of single cone spectral responses that peak at different wavelengths are shown in Fig. 3. As is clear from these records, the responses are always hyperpolarizing, irrespective of the wavelength of light. No response curve resembling the chromaticity type of S potential has been found. The records in Fig. 3 represent those with the greatest signal-to-noise ratio obtained from hundreds of records. The typical recordings contained more noise, as would be expected upon consideration of the relatively small amplitude of response (less than 5 mV) and the high resistance (usually over 200 MΩ) of the extremely fine micropipettes. In order to determine whether or not the responses could be classified into sub-types, it was necessary to select those records which were adequate for quantitative treatment. As a result of the screening process, based solely upon the greatest signal-to-noise ratio, 142 recordings were utilized in the statistical analysis. From each of the selected records the response amplitude at individual wavelengths was measured. The data were then normalized with the maximal response set at 100 %, and were plotted on amplitude-wavelength coordinates. The adjacent points were connected by a straight line.

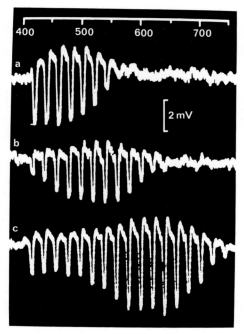

Fig. 3. Three sample records of spectral response curves from single carp cones. The spectrum was scanned in steps of 20 nm with monochromatic light of equalized quantum flux (2×10^5 photons per $\mu^2 \cdot$sec), and with a duration of light of 0.3 sec at each wavelength followed by an intermission of 0.6 sec. A downward deflection indicates negativity. Recording was made with a C-R coupled amplifier having a time constant of 0.5 sec. The spectral scale is given in nm at the top of the figure (TOMITA et al., 1967)

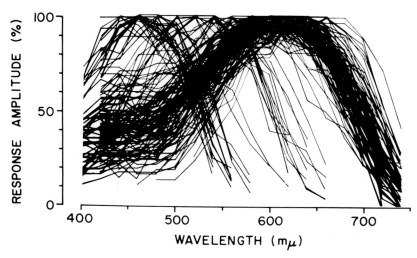

Fig. 4. The result of superimposition of 142 spectral response curves, each plotted from the original record with the maximal response set at 100% (TOMITA et al., 1967)

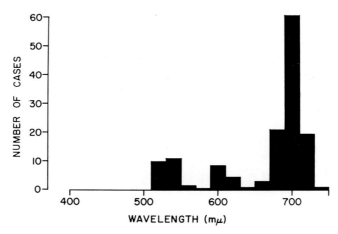

Fig. 5. Histogram from the data in Fig. 4, showing the number of cases within 20 nm intervals at the 50% amplitude level occurring on the phase of the curves falling toward the longer wavelengths. Three groupings are evident (Tomita et al., 1967)

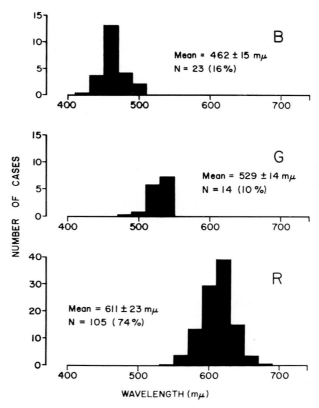

Fig. 6. Histograms of the peaking wavelengths of the three groups derived from Fig. 5 (Tomita et al., 1967)

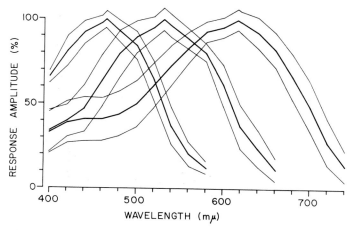

Fig. 7. Mean spectral response curves along with standard deviation curves of the three groups
(TOMITA et al., 1967)

The 142 curves thus obtained were superimposed photographically, with the result shown in Fig. 4. Although the selected recordings were still contaminated with considerable noise, the result of superimposing the response curves suggested three definite groups; one peaking in the red region, another in the green, and still another in the blue. Fig. 5 shows three groupings, in a histogram, of the wavelengths at the 50 % amplitude level occurring on the phase of the curves falling toward the longer wavelengths. On the basis of this histogram, 105 curves out of 142 (74 %) were included in the category of R-type, 14 curves (10 %) in G-type, and 23 (16 %) in B-type. From the curves of each category a histogram of the peaking wavelengths was constructed and the mean value along with the standard deviation was calculated. The result of this analysis is presented in Fig. 6, and the mean spectral response curves in Fig. 7.

The table below shows that the three maximally sensitive wavelengths of the single cone responses are in close agreement with those of difference spectra obtained by MARKS (1965) in single goldfish cones by means of a microspectrophotometer.

B-type	G-type	R-type	
462 ± 15 nm	529 ± 14 nm	611 ± 23 nm	TOMITA et al. (1967) (carp)
455 ± 15	530 ± 5	625 ± 5	MARKS (1965) (goldfish)

One cannot directly compare the three pairs of spectral curves obtained by the electrophysiological and microspectrophotometric methods. To make a comparison valid, the curves obtained electrophysiologically, which represent a kind of response spectra, would necessarily have to be transformed into action spectra.

An example of the procedure and result of such a transformation has been given by Tomita et al. (1967).

c) Cell Identification

Criteria. In intracellular studies of single photoreceptor responses, one needs to be particularly careful in differentiating a receptor potential from an S potential, because both of these are hyperpolarizing responses to white light. While the final identification is made by electrode marking, as is to be described later, there are several other criteria that could be used. The following are those used by Tomita et al. (1967) for the identification of single carp cones. (1) Histology. The rods in the carp retina are too slender to be penetrated. The only possible sites for intracellular recording within the receptor layer are the cone inner segments, which lie at a depth of $50-70\,\mu$ from the distal margin of the receptor layer. (2) Resting potential. Before recording a response to light, the pipette should record a negative d.c. potential shift, as a sign of being intracellular. (3) Electrode depth. The pipette electrode should be at the depth corresponding to the layer of the cone inner segments. (4) Lability of the recorded response in comparison with the S potential. This indicates that the recording is made from a small structure easily damaged or inactivated by impalement with the electrode. (5) Absence of an area effect. The

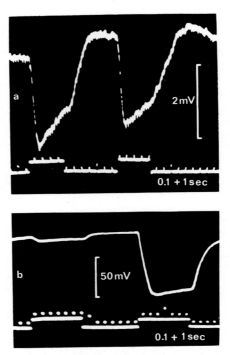

Fig. 8. Responses to two successive lights, first focal (a light spot of 0.2 mm in diameter centered onto the recording site) and second diffuse. *a*, responses of a carp cone, recorded with a C-R coupled amplifier; *b*, d.c. recording of responses of an S cell, showing a distinct area effect (Tomita, 1965)

amplitude of the receptor response depends solely on the light intensity at the site of recording but is independent of the retinal area illuminated (see Fig. 8a).[1] Contrary to this, the S potential is known to have a distinct area effect, as shown in Fig. 8b. (6) Recording of an S potential. Once data are obtained from a tentatively identified receptor, the electrode is penetrated $50-70\,\mu$ further into the retinal tissue so as to record an S potential. If no S potential is found, the data are discarded.

Electrode Marking. While the above were the criteria used for the tentative identification of single carp cones, the final identification was made by means of electrode marking (KANEKO and HASHIMOTO, 1967), using micropipettes filled with a $4-8\%$ aqueous solution of Niagara Sky Blue 6B, the dye previously used by POTTER et al. (1966). Since this dye dissolves and dissociates well in water (CONN, 1961), the resistances of the dye filled microelectrodes are not very much higher than those of pipettes filled with 2M KCl solution. In addition, the extremely low solubility of the dye in organic solvents is advantageous in retaining the dye during the course of dehydration and embedding of the tissue.

After a potential is recorded and tentatively identified as a cone response by the previously outlined criteria, the dye is ejected from the electrode electrophoretically by the arrangement shown in Fig. 9. The lead-off wire of the recording electrode is connected to the cathode of a 135 V battery (B), through a 40 MΩ

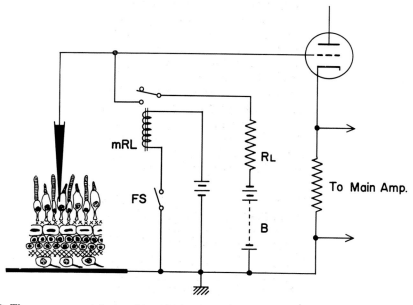

Fig. 9. The arrangement for marking single cells in the retina with intracellular micropipette electrodes filled with an aqueous solution of Niagara Sky Blue 6B. B, 135 V battery; FS, footswitch; mRL, microrelay; RL, current-limiting resistor of 40 MΩ (KANEKO and HASHIMOTO 1967)

---

[1] Lateral effects upon the receptor potential were recently observed in the turtle retina by BAYLOR and FUORTES (1970) and BAYLOR, FUORTES and O'BRYAN (1971). See Section 3c.

current-limiting resistor (RL), by means of a micro-relay (mRL) which is activated by a foot-switch (FS). A small portion of the electrode tip, a micron or two in length, is electrically broken at the instant of current application. This makes the injection of the dye into the cell easier, while the rate of the injection now is limited by the 40 MΩ resistor in series. The dye injected into the impaled cell is observed and controlled under the dissection microscope. A duration of current of less than 1 sec is usually sufficient for successful injection. The stained cell is discerned under the microscope as a tiny blue dot which could be traced through-out the subsequent histological procedures. (A test, similar to the above but with the electrode placed extracellularly, revealed that a blue dot appears also at the electrode tip but it diffuses away and is completely lost within a few seconds after the injection.) The retina is fixed overnight at about $2°$ C in glutaraldehyde acidified to pH 4 by acetate buffer, as suggested by Potter, Furshpan, and Lennox (1966). After dehydration in graded acetone-water mixtures, the retina is cleared in propylene oxide and embedded in Epon (Luft, 1961). Serial sections are cut, stained with 0.5 % basic fuchsine alcohol solution, and examined under the light microscope. By this means, Kaneko and Hashimoto (1967) were successful in distinguishing single cones stained blue at their inner segments.

More recently, Stretton and Kravitz (1968) developed a dye injection method using micropipettes filled with Procion Yellow M4RS. When injected in the cell body this fluorescent dye diffused well down to small branches, revealing the entirety of the injected cell. This method was successfully applied for the identifi-cation of the retinal cells in the goldfish (Kaneko, 1970) and dogfish (Kaneko, 1971). This was again followed by Baylor, Fuortes, and O'Bryan (1971) for the identification of single turtle cones.

d) Intensity-Amplitude Relation of the Cone Response

It remains to be studied as to how the light-induced hyperpolarization in the carp cones is related to the flow of information from the outer segment, where light is absorbed by photopigment molecules, to the proximal end, where the transmission to secondary neurons takes place by way of synaptic contacts. The flow of information within the vertebrate photoreceptors may be by means of electrotonic spread of the hyperpolarization, or by some other means which is nonelectrical. Whatever the case may be, let us make a rough estimation of the potential change per photon absorbed by a single cone. Fig. 10 shows the relation between the log-intensity of light and the response amplitude. As expected, this is almost linear. In Fig. 11 the same data are replotted in linear intensity co-ordinates, showing that at low intensities of light the amplitude is directly pro-portional to the intensity, or to the amount of light per unit area when tested by flashes of light of durations within which the Bunsen-Roscoe law applies. Since the quantum flux of light from the present photostimulator is known, one can estimate the number of photons impinging on single cones for intensities in the region where the amplitude-intensity relation is approximately linear. It was found from several experiments that the approximate linear relation holds at intensities eliciting responses smaller than one third (or 1 mV) of the saturation amplitude (3 mV on the average), and that a flash of white light (10 msec duration and band-passed between 400 nm and 700 nm) at an intensity of 10^3 photons per $μ^2$ could produce

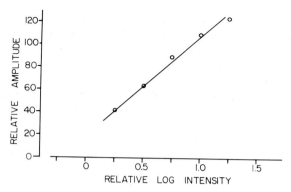

Fig. 10. The amplitude-intensity relation in a carp cone, with a log intensity scale along the abscissa (Tomita, 1968)

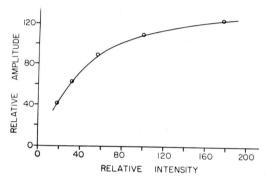

Fig. 11. The same relation as in Fig. 10, replotted with a linear intensity scale along the abscissa (Tomita, 1968)

a hyperpolarization of about 1 mV. This yields a value of 1 μV per photon per μ^2. Since the area of the tangential section of the carp cone outer segment is about 10 μ^2 at the base, 10 photons impinge on a single carp cone to produce 1 μV of hyperpolarization. The coefficient of absorption of the light by single cones is unknown, but it might be safe to assume a value between 0.1 and 0.01. Then, the amplitude of the cone response to a single absorbed photon falls between 1 and 10 μV. A few microvolts might be a reasonable value.

It remains to be seen whether or not the value estimated here for the cones applies to the rods.[2] It should also be admitted that in making the above estimate no consideration was taken of the effect of adaptation which is also probably great in the cones, even if it is not the same as that demonstrated for the rods by WEIN-STEIN, HOBSON, and DOWLING (1967). For instance, using the isolated rod retina of the rat, and measuring both the ERG response and the rhodopsin content,

---

[2] HAGINS, PENN, and YOSHIKAMI (1970) recently estimated in rat rods a value of about 6 μV per photon absorbed at the outer segment or about 3.6 μV at the proximal rod terminal. The value is comparable with that for carp cones.

these authors showed that a 50 % bleaching of rhodopsin causes a 3 log-unit decrease in sensitivity in the ERG, as compared with the fully dark-adapted retina, the rhodopsin content of which was set at 100 %. It would be desirable to know the response amplitude per photon absorbed for a fully dark-adapted rod. If the flow of information is electrical, the potential change per photon absorbed should be sufficiently high to yield the extraordinary sensitivity of vision, as described by Hecht, Shlaer, and Pirenne (1942), according to whom single photons can excite dark-adapted human rods.

3. Single Photoreceptor Responses in Other Species

The mudpuppy, *Necturus maculosus*, and the nocturnal gecko, *Gekko gekko*, were utilized successfully for the intracellular recording of single photoreceptor responses. Both species are known to have large photoreceptor cells. The responses were also hyperpolarizations. Similar results have recently been obtained also in the cones of the turtle, *Pseudemys elegans* (Baylor and Fuortes, 1970), and in the rods of the bullfrog, *Rana catesbeiana* (Toyoda et al., 1970).

a) Necturus maculosus

Bortoff (1964) was the first to observe the activity of individual cells in the retina of *Necturus*. He used micropipette electrodes filled with a saturated aqueous solution of Trypan Blue for both recording and subsequent electrode marking. The electrode was inserted into the retina from the vitreal side of the opened eye which remained in the head. To obtain intracellular recording from such large cells, there was no need of auxiliary devices such as a jolter to facilitate the penetration. During the insertion, the microelectrode recorded the activity of ganglion cells and bipolar cells, but an advancement of the electrode beyond these layers brought it into the region of the retina from which sustained, negative potentials from 5 to 30 mV in amplitude were obtained in response to light. The onset of the response corresponded to that of the a-wave of the ERG, while its termination temporally related to the d-wave. A test with monochromatic light showed that the polarity of the response was negative for all wavelengths. From the results of electrode marking, Bortoff came to the conclusion that the responses were all from single photoreceptors. Horizontal cells of the type found in the Teleost retina were not apparent in *Necturus*, and, for this reason, Bortoff did not distinguish the receptor potential from the S potential. Also in his later work in collaboration with Norton (Bortoff and Norton, 1965a, 1965b, 1967) all the responses from this region were described as photoreceptor potentials. It appears, however, that many of the responses illustrated as photoreceptor potentials in Bortoff and Norton's papers are S potentials, judging from their magnitude which is as large as 20 mV. According to Werblin (1968) and Toyoda, Nosaki, and Tomita (1969), who later worked on the same material but made much more use of variations in method, the difference between the receptor potential and the S potential was distinct. The receptor potential was faster in its rising phase, had no area effect, and was small in response size, typically 5 mV or less, while the S potential was usually some 20 mV in magnitude. Histologically also, Werblin (1968) was able to differentiate horizontal cells from photoreceptor cells, the former being the most variable in position and lying above and below the outer plexiform layer as well as within it.

For the identification of the recording site, both WERBLIN (1968) and TOYODA et al. (1969) used micropipettes filled with aqueous solution of Niagara Sky Blue 6B, and the procedure after KANEKO and HASHIMOTO (1967). Responses typical of S potentials were all localized in horizontal cells, while those of photoreceptors were localized consistently in the inner segments of receptors.

b) Gekko gekko

The response of single *Gekko* photoreceptors was studied by TOYODA, NOSAKI, and TOMITA (1969), using the retina which was detached from the pigment epithelium and mounted receptor side up on the vibrating table of a jolter. In this material, the jolter proved useful for the penetration, but it often occurred that the micropipette already was intracellular at the moment of its contact with the distal surface of the retina. Possibly, two factors may be serving to facilitate the penetration. One is the action of the micropipette electrode as an electromechanical transducer. When the micropipette is held in the air before contacting the retinal surface, the potential of the pipette is shifted considerably from the ground, and hence the pipette undergoes a sudden potential change at the moment of contact with the retinal surface. This probably causes a jolt of the pipette against the retina. The other is the surface tension effect developed around the very tip of the micropipette during contact, causing an agitation similar to a jolt to the tissue just below the tip.

When the pipette failed to be intracellular at the moment of contact with the tissue, it had to be advanced further and aided by jolting for a successful penetration. The depth at which single receptor potentials were obtained was usually shallow, being less than $20\,\mu$ from the surface. Since the photoreceptor outer segment of *Gekko* is $40-50\,\mu$ in length and some $10\,\mu$ in diameter, which is a size large enough to permit intracellular placement of the micropipette, it was highly probable that the pipette was within the outer segment of a receptor. This was confirmed by the electrode marking method. In contrast to *Necturus*, all marks were found in the outer segments of receptors and not in the inner segments.

The resting potential recorded from the outer segment was small (about 10 mV), and the response to light was a hyperpolarization of varied amplitudes ($3-15$ mV). The response was similar to that of carp cones and *Necturus* photoreceptors, except that the response in *Gekko* often was larger than in the other species. It is not clear whether this was because of recording from the outer segment or because of the larger size of photoreceptors in *Gekko*. The peaking wavelength of the spectral response curves was at about 520 nm (KANEKO, HASHIMOTO, and TOMITA, unpublished), showing good agreement with the 521 nm pigment which CRESCITELLI (1963) reports as the principal photopigment in *Gekko gekko*.

c) Turtle Cones

BAYLOR and FUORTES (1970), working on single cones of the turtle, *Pseudemys elegans*, with micropipettes introduced from the vitreal side of the eye cup preparation, also observed a resting potential (about -30 mV) and a hyperpolarization ($15-20$ mV) upon illumination. Their results confirm most of those on single carp cones (Tomita et al., 1967); (1) the small size of the response as compared to the S potential, (2) narrow dynamic range of less than 2.5 logarithmic units, (3) three groupings in terms of spectral response maxima (about 450, 530 and 650 nm), (4) increase in membrane resistance during the hyperpolarizing response for which certain ions having the equilibrium near zero potential level are involved, and (5) small receptive fields. One difference was that they observed interaction among cones. In a succeeding paper, Baylor, Fuortes, and O'Bryan (1971) further studied this point, with the result that the hyperpolarizing response of a cone to a given intensity of light increased in size when the retinal area illuminated was increased toward the surround, but a further increase in the area beyond 70 μ in diameter resulted in a curtailment of response by a potential of the opposite polarity (depolarization). In other words, interaction from the near surround (less than 40 μ distance) was summative, while interaction from the far surround was counteractive. Their two micropipette experiment, one for electrically polarizing a retinal cell and the other for recording from another, led to the conclusion that the summative interaction from the near surround occurs via some kind of connections between cones, while the counteractive interaction from the far surround is mediated by the hyperpolarizing response of the horizontal cells (S-potential).

d) Frog Rods

As afore mentioned (see 3b), intracellular recording from rods has been successful in the nocturnal gecko, *Gekko gekko*. However, the rods of this reptile are considered by Walls (1963) as transmuted from cones, and therefore their activity may be different from other rods, though there is indirect evidence that true rods also respond to light with a hyperpolarization (Penn and Hagins, 1969; Hagins, Penn, and Yoshikami, 1970). Intracellular recording from rods of the bullfrog, *Rana catesbeiana*, was achieved by Toyoda et al. (1970), with the identification of cell type by electrode marking. The response proved to be a hyperpolarization, with the maximum at about 502 nm, corresponding to the absorption maximum of the red rod pigment. The response was characteristic by saturation of the response at relatively low light intensities, and by its slow time course, especially in the decaying phase following termination of stimulation with strong light. Recording from green rods has not been successful.

4. Resistance Changes

The measurement of the resistance of single photoreceptors was made by Toyoda, Nosaki, and Tomita (1969), using the mudpuppy *(Necturus maculosus)* and the nocturnal gecko *(Gekko gekko)*. The recording electrodes were either single or double-barrelled, depending on the purpose of the experiment. The circuit for the resistance measurement was similar to that described by Tomita and Kaneko (1965), and is schematically illustrated in Fig. 12. Square pulses are applied through a high resistance (200 MΩ) with the recording pipette as the cathode. (A control experiment with square pulses applied with the pipette as the anode gave the same result.) The potential drop across the combined resistances of the pipette and the cell membrane is brought to balance by another train of square pulses of the opposite polarity applied from the indifferent electrode. A change in the resistance

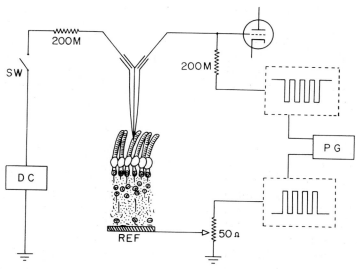

Fig. 12. Schematic diagram of the bridge circuit for measuring the cell membrane resistance (shown in the right-hand part of the figure, and for this purpose alone single pipette electrodes are adequate), and the arrangement for applying extrinsic currents (left-hand part). DC, d.c. source; PG, square pulse generator; REF, reference or indifferent electrode. Explanation in the text (Toyoda et al., 1969)

of the cell membrane results in a bridge imbalance. In experiments with double-barrelled electrodes, one barrel is used to apply an extrinsic current to change the membrane potential, while the other barrel serves to measure the resistance.

Fig. 13 shows three sample records from a *Necturus* photoreceptor. In these bridge records, the onset of each current pulse is signalled by a positive transient, and the cessation by a negative transient. These transients are the result of inter-action of the two trains of pulses of opposite polarities, whose wave forms are not identical enough for a complete balance. While the train of positive pulses applied from the indifferent electrode better maintains the original square wave form, the train of negative pulses applied from the pipette is distorted, owing to the cell membrane capacitance in series with the high resistance (200 MΩ resistor in Fig. 12), to give a rounded shoulder at both the rising and falling phases of each pulse. Since the current pulses applied with the electrode as the cathode are, in this and all subsequent records, in the direction to hyperpolarize the cell membrane, a resistance increase of the membrane displaces the pulse-on portion of the oscil-loscope tracing downward. In Fig. 13a, for example, the pulse-on portion is dis-placed upward in the dark but comes to balance upon illumination, indicating a resistance increase during illumination. Fig. 13b and 13c are d.c. recordings from the same unit at a slower sweep speed, with the bridge balanced in the dark (13c) or balanced during illumination (13b). All these records indicate that the mem-brane resistance is increased during illumination. The effective resistance of *Necturus* receptors was 10–20 MΩ in the dark, and a resistance increase of some 5 MΩ was observed upon exposure to light. Both the potential change and the

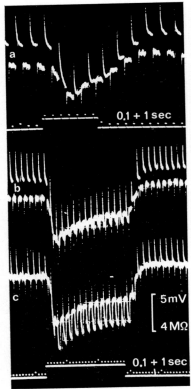

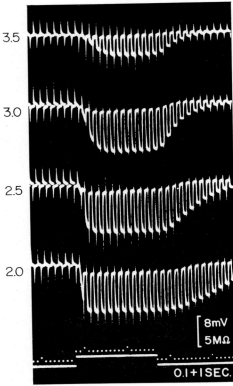

Fig. 13   Fig. 14

Fig. 13. Bridge records of *Necturus* photoreceptor response to light recorded with a single pipette electrode. *a*, C-R recording with a fast sweep speed; *b* and *c*, d.c. recordings from the same photoreceptor with a slower sweep speed and at different levels of bridge balance. All three records reveal a resistance increase by light, as seen in the text (Toyoda et al., 1969)

Fig. 14. Bridge records of *Gekko* photoreceptor response at various light intensities, all showing a resistance increase during response to light. Numbers on the left indicate attenuations in log-units of the stimulus intensity (Toyoda et al., 1969)

resistance change depended on the light intensity, becoming larger as the intensity was increased.

A membrane resistance change similar to *Necturus* photoreceptors was observed also in *Gekko* (Fig. 14). The effective resistance in darkness measured about 10 MΩ and the resistance increase by light was often more than 5 MΩ. Both the potential and the resistance changes became larger as the light intensity was increased, but were finally saturated at an intensity about 1.5 log units above threshold. Further increase of the light intensity only prolonged the decaying phase of the response after the termination of stimulus. Although the amplitude and the time course of the response varied according to the stimulus conditions, the resistance change seemed to be approximately linear with the potential change.

BAYLOR and FUORTES (1970) obtained similar results from turtle cones.

5. Effect of Extrinsic Current

It now is clear that the vertebrate photoreceptors are hyperpolarized by light and the hyperpolarization is accompanied by an increase of the membrane resistance (or a decrease of the membrane permeability). There are two possibilities to account for the observed resistance change. The resistance increase could be either a secondary effect of the hyperpolarization, or a phenomenon directly related to the membrane process that causes the hyperpolarization. The first possibility was proposed by BORTOFF and NORTON (1967) who previously observed in *Necturus* a resistance increase, during light, in what they believed to be photoreceptors but which actually was a mixture of cells consisting of receptors and S cells, or horizontal cells, which also exhibit a resistance increase during their hyperpolarizing response (TOYODA, NOSAKI, and TOMITA, 1969). BORTOFF and NORTON ascribed the resistance increase to the rectification property commonly observed in excitable membranes. If the resistance change is due solely to the rectification of the membrane, however, a passive hyperpolarization of the membrane by extrinsic current should result in a resistance increase equivalent to that induced by light. This was tested in *Gekko*, using the double-barrelled microelectrode with the arrangement shown in Fig. 12. Fig. 15 shows the result. A hyperpolarizing current of the order of na was passed through one barrel of the double-barrelled electrode for about 1 sec in the absence of light. The effect of such current on the membrane resistance was compared with the resistance increase due to illumination. It is clear in the figure that the passive change in the membrane potential does not change the bridge balance, but a considerable imbalance is caused by light even when the potential change is smaller. This seems to exclude the first possibility, and therefore, makes the alternative more likely. A similar experiment in *Necturus* photoreceptors gave the same result.

The extrinsic-current experiment permitted another important observation. As illustrated in Fig. 16, the response to a given intensity of light becomes larger

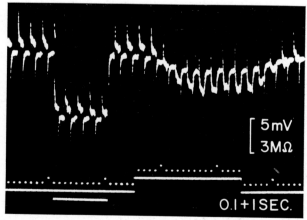

Fig. 15. Comparison of the effects of hyperpolarization of a *Gekko* photoreceptor by extrinsic current (left), and by light (right). The membrane resistance is changed little by extrinsic current, but clearly is increased by light (TOYODA et al., 1969)

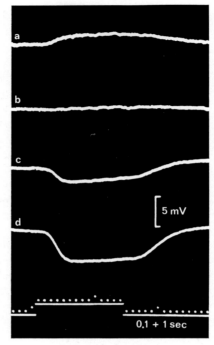

Fig. 16. Effect of extrinsic current on *Gekko* photoreceptor response. Hyperpolarization by current makes the response larger (*d*) compared to the control (*c*), while depolarizing current makes the response smaller (*b*) or even reversed (*a*). Intensity of extrinsic current; + 4 na (*a*), + 2 na (*b*), 0 (*c*), and − 2 na (*d*) (Toyoda et al., 1969)

when the membrane is hyperpolarized by extrinsic current (16d), but becomes smaller when depolarized (16b). A strong depolarizing current can even invert the response (16a). Attempts to determine the potential level for the reversal in polarity of the response have not been very successful, but this was roughly estimated in *Gekko* photoreceptors to be slightly above zero membrane potential. The results were varied, mainly because of technical difficulties arising from the use of high-resistance double-barrelled micropipettes. The relative position of the electrode tip to the electrogenic or active site of the receptor also might have influenced the result.

Similar effects of currents on photoreceptor responses were observed by Baylor and Fuortes (1970) on turtle cones.

6. Receptor Potential and PIII

We are now in a better position to answer the questions raised at the beginning of this Chapter. One of those was the unusual polarity of the electric field which is formed around the receptors and which is manifested as a fraction of the ERG. This component of the ERG, ascribed to the receptors, was isolated in mammals by the method of clamping the retinal circulation (BROWN and WATANABE, 1962). The technique abolished the activity of all cells supported by this circulation, but left the receptors active because they are supported by the choroidal circulation. Thus the ERG component in question was identified as the PIII. A successful isolation of the ERG fraction ascribed to the receptors was accomplished also by HANITZSCH and TRIFONOW (1968) by means of intraretinal microelectrodes applied to the rabbit retina which was isolated from the eye bulb and bathed in an artificial solution. In cold-blooded vertebrates, a fraction of the PIII component termed the distal PIII was identified as being of receptor origin (TOMITA, 1963; MURAKAMI and KANEKO, 1966). In both warm- and cold-blooded retinas, the potential thus isolated is of such a direction as to make the distal tips of the receptors positive, which is just the opposite to the polarity of most receptor potentials. However, in light of the evidence that the vertebrate photoreceptors are depolarized in the dark and repolarized by illumination, the unusual polarity is interpreted as follows: The receptor outer segments, depolarized in the dark, act as a sink to produce an interstitial "dark" current in such a direction as to make their distal tips negative. Upon illumination the sink diminishes in a graded manner to cause a shift of the potential at the distal tips in the positive direction.

A recent experiment of PENN and HAGINS (1969) lends support to the above interpretation. They worked on living sections of the rat retina, similar to those of the squid retina previously used by HAGINS (1965). Multiple extracellular microelectrodes were placed radially along the rod cells, and the electric field due to the rod cell membrane current was measured. The experiment showed that in the dark a steady current of approximately 20 μa per cm^2 flows into the rod outer segments from the inner segments as well as from more proximal parts of the cells, and upon illumination the dark current is reduced with a time course resembling the PIII of the ERG.

It should be noted that the electrical model proposed by BORTOFF and NORTON (1967) also substantiates such dark current, and its reduction during light accompanied by a resistance increase in the membrane. In contrast to TOYODA, NOSAKI, and TOMITA (1969), however, they conclude that the receptor potential is not the result of permeability changes, but of alterations, by light, of some electrogenic mechanism. Worthy of attention in this connection are some previous observations on the effect of various substances on the ERG, particularly on the PIII. FURUKAWA and HANAWA (1955) found that the ERG was completely abolished when the toad retina was bathed in a solution similar to Ringer, but where sodium chloride was replaced by isotonic glucose, and that a sodium free, lithium-Ringer solution was unable to maintain the ERG. HAMASAKI (1963, 1964) further observed that the ERG of the isolated frog retina was completely abolished, but in an almost wholly reversible way, when 95 % of the sodium in Ringer was replaced by choline, sucrose or lithium. The reduction in the ERG with partial replacement of sodium by choline was linearly related to the logarithm of the concentration of sodium in the test solution, suggesting that the ERG amplitude is mainly determined by the equilibrium potential of sodium. FRANK and GOLDSMITH (1967) showed that ouabain added to Ringer solution bathing the isolated frog retina irreversibly abolishes the ERG in several minutes. They consider that the abolition of the ERG by ouabain is due principally to inhibition of the active transport of sodium, and that, in a standard sodium environment, essentially constant activity of the sodium pump is required to prevent rapid and irreversible change.

The studies cited above all agree in that sodium ions are essential for the electrical activity of vertebrate photoreceptors. Their hyperpolarizing response to light thus appears to be a consequence either of permeability decrease for sodium ions in the photoreceptor cell membrane as suggested by Toyoda, Nosaki, and Tomita (1969), or of enhanced activity of an electrogenic pump, such as described by Bortoff and Norton (1967). The observation on the mass receptor potential by Sillman, Ito, and Tomita (1969b) supports Toyoda, Nosaki, and Tomita (see Section 8 of this Chapter).

7. Receptor Potential in Vertebrates and Invertebrates

We have just seen that two hypotheses have been proposed to account for the vertebrate photoreceptor potential. In invertebrates also, two similar hypotheses have been proposed. According to one of them, light causes an increase in the conductance or permeability of the cell membrane to one or more ions, particularly sodium (Fuortes, 1959; Benolken, 1961; Hagins and Adams, 1962; Kikuchi, Naito, and Tanaka, 1962), and as a result the cell is depolarized in essentially the same way as that proposed to account for the endplate potential (Katz, 1962) or for the excitatory postsynaptic potential (Eccles, 1964). While this hypothesis represents the majority of opinions, an entirely different conclusion was drawn from work on *Limulus* photoreceptors by Smith et al. (1968a, b), according to whom the receptor potential is not the result of permeability changes, but rather of alterations in a voltage-sensitive, electrogenic sodium pump.[3]

Thus, the comparison of vertebrate and invertebrate photoreceptor potentials reveals both similarities and differences between them. They are similar in that the generation of the potentials appears to be controlled by some common mechanisms where sodium ions play an important role, the nature of which can be explained by one or the other of the two hypotheses just mentioned. They are different, on the other hand, in that the response to light of the vertebrate photoreceptor is a hyperpolarization accompanied by a resistance increase, whereas the response of the invertebrate photoreceptor is a depolarization accompanied by a resistance decrease.

Fig. 17 summarizes properties of the invertebrate and vertebrate photoreceptor potentials. The curves in the middle of the figure represent controls, one showing the depolarizing response of invertebrates (left), and the other the hyperpolarizing response of vertebrates (right). In invertebrates, the depolarizing response becomes larger if the membrane is hyperpolarized by extrinsic current (lower left). Conversely, it becomes smaller if depolarized, and even reverses (top left) if the depolarization by extrinsic current exceeds a certain potential level, which is at about zero (Fuortes, 1959; Rushton, 1959; Eccles, 1964; Purple and Dodge, 1965). In vertebrates, the response polarity is just the opposite, but nevertheless, the

---

[3] Membrane property studies by voltage clamp on ventral photoreceptor cells of *Limbus* by Millecchia and Mauro (1969) and on photoreceptors in the lateral ocelli of barnacles (1970) provide strong evidence against the sodium-pump theory.

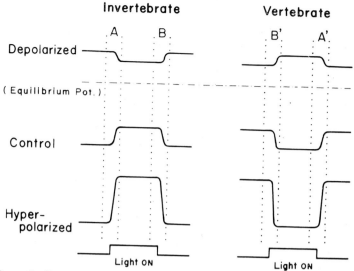

Fig. 17. Schematic diagram for comparison of the responses of invertebrate and vertebrate photoreceptors. Note that the on and off responses in invertebrates (regions A and B in the figure) are similar to the off and on responses in vertebrates (A' and B') (TOYODA et al., 1969)

effect of extrinsic current upon the response size is similar. As in the depolarizing response of invertebrates, the hyperpolarizing response of vertebrates is also enhanced by a hyperpolarizing current, but suppressed, or even reversed, by a depolarizing current. This was shown in Fig. 16 and is reproduced schematically in the right-hand part of Fig. 17.

Fig. 17 shows that the on-response in invertebrates (region A in the figure) is qualitatively the same as the off-response in vertebrates (A'), and that the off-response in invertebrates (B) is the same as the on-response in vertebrates (B'). It thus is concluded that the vertebrate photoreceptors are depolarized in the dark and hyperpolarized by illumination. In other words, the response of the vertebrate photoreceptors to light is "inhibitory" in nature. And this is the answer to the second question in the Introduction. It sounds unusual that a receptor responds to a stimulus with inhibition. However, perhaps this should not be surprising. The complex neural network intervening between the receptors and the ganglion cells could be the seat not only of information transmission but also of conversion of the signal carrying the information from inhibition to excitation, or vise versa. The conversion might require a certain time, and hence a longer latency for excitation at the level of ganglion cells. Indeed, the earliest response to light observed at the ganglion cell level is inhibition. This was described and termed the pre-excitatory inhibition by GRANIT (1947). Studies in the frog retina with intracellular micropipettes also revealed that the on-off type ganglion cells respond with an initial hyperpolarization followed by depolarization superimposed with spike impulses, and that the hyperpolarization of pure-off type ganglion cells starts with a latency shorter than the depolarization of pure-on type ganglion cells (TOMITA et al., 1961).

In the invertebrate retina where the first response to light is a depolarization (excitation) of photoreceptive units, the order of sequence observed is just the opposite. As clearly demonstrated by Hartline, Ratliff, and Miller (1961) in the lateral compound eye of *Limulus*, which possesses also a network for interaction between adjacent ommatidia, the first response to light of the optic nerve fibers is a discharge of impulses, upon which an inhibitory influence is exerted only after a delay.

8. Ionic Mechanism of the Receptor Potential

The technique of intracellular micropipette recording, which has been applied successfully to the study of single vertebrate photoreceptor cells, is not practical for studying the ionic mechanism of the receptor potential. While it is necessary, for this type of work, to observe the effect upon the receptor potential of alterations of the chemical environment, the application of solutions to the tissue almost invariably results in the dislodgement of the penetrating pipette. Therefore, it is desirable to establish a technique which would serve to isolate the receptor potential so as to permit the recording of the mass response by means of an external electrode. Furukawa and Hanawa (1955) report that sodium aspartate is useful for the isolation of the PIII from the other ERG components. Recently, Sillman, Ito, and Tomita (1970a) observed that the PIII isolated by this reagent does not comprise the proximal PIII but only the distal PIII or the mass receptor potential (Murakami and Kaneko, 1966). Using this reagent, Sillman, Ito, and Tomita (1970b) studied the effect of changes in the chemical environment upon the receptor potential.

a) Effect of Sodium and Potassium

As previously referred to (Section 6), the importance of sodium ions for the generation of the vertebrate retinal response was shown by Furukawa and Hanawa (1955) and Hamasaki (1963) who demonstrated that the ERG of the toad and frog was completely suppressed by the absence of sodium in the external medium. Furukawa and Hanawa, using the retina treated with aspartate, also showed qualitatively that the a-wave, which is the leading edge of the PIII, increased in amplitude when sodium concentration was increased.

In the experiments of Sillman, Ito, and Tomita (1969b), aspartate solutions containing sodium in amounts of 100 mM (Sol. B in Table 1), 50 mM (Sol. C),

Table 1

Solution	Ouabain	Na-Aspart.	NaCl	LiCl	KCl	NaHCO$_3$	KHCO$_3$	CaCl$_2$	Glucose
A			95		2	5		1.8	50
B		10	85		2	5		1.8	50
C		10	35	50	2	5		1.8	50
D		10	10	75	2	5		1.8	50
E		10		85	2	5		1.8	50
Q			113				4	1.8	20
R	0.1		113				4	1.8	20
S	0.1	113					4	1.8	20

Composition of the experimental solutions, with all concentrations in mM.

25 mM (Sol. D) and 15 mM (Sol. E) were employed. To obtain such solutions, the sodium in the Ringer solution (Sol. A) was replaced with appropriate amounts of lithium, which was known not to substitute for sodium in the generation of the receptor potential (FURUKAWA and HANAWA, 1955). The complete replacement of the external sodium with lithium resulted in the disappearance of the response, in a readily reversible fashion. Apparently, lithium has no toxic effect.

The result showed clearly that, except at very low concentrations where there was some deviation, the amplitude of the receptor potential increased in direct proportion to the logarithm of the external sodium concentration. With respect to the exponential aspect of the relationship, the data for the receptor response agree with that for the b-wave of the frog retina obtained by HAMASAKI (1963) and HANAWA, KUGE, and MATSUMURA (1967).

Using much the same procedure as for the sodium experiment, the mass receptor potential was studied in relation to variations in the external potassium concentration. Aspartate solutions containing potassium in various amounts, obtained by adjusting the relative amounts of lithium and potassium to the desired level, were used. As was observed with sodium, except at very low concentrations, the relationship between the amplitude of the receptor potential and the external potassium concentration on a logarithmic scale was linear, but it was *inversely* linear in the case of potassium.

b) Effect of Ouabain

As already established, the response of the vertebrate photoreceptor to light is a hyperpolarization. As to how this hyperpolarization is brought about, two possibilities were suggested in Section 6. One of these was that light serves to stimulate a metabolic pump so as to increase the efflux of cations. Against this, TOYODA, NOSAKI, and TOMITA (1969), who found that the hyperpolarization is related to an increase in the membrane resistance, concluded that the action of light is to suppress the influx of cations which occurs in darkness. These two possibilities were tested by SILLMAN, ITO, and TOMITA (1969b) using ouabain, an inhibitor of the metabolic sodium-potassium pump (SKOU, 1965). The frog retina was incubated in a large volume of sodium-free solution (Sol. Q in Table 1) for 2 min to remove a major portion of sodium from the tissue. Then the retina was transferred to a large volume of solution (Sol. R) similar to the previous sodium-free medium, but containing 0.1 mM ouabain, and was left in this solution for 7 min, a duration found by FRANK and GOLDSMITH (1967) to be sufficient to allow complete suppression of the ERG by 0.1 mM ouabain. After the allotted time the preparation was placed on the indifferent electrode where the response to light was measured.

Exposure of the retina to light of duration and intensity great enough to elicit the normal receptor response resulted in no sign of the potential. Three drops of a solution (Sol. S) containing a normal complement of sodium in the form of sodium aspartate, but still containing 0.1 mM ouabain, were then added to the retina. Within seconds a normal receptor potential was observed in response to the same light stimulus. Thus, replacing the sodium in the external medium restored immediately the receptor's ability to generate the potential despite the fact that the metabolic pump was suppressed by the presence of ouabain. If a fresh retina was allowed to remain immersed in the sodium/ouabain-Ringer (Sol. S) for 7 min, no response to light could be observed. Identical results were obtained using solutions containing 1 mM ouabain.

c) Electric Model of the Electrogenesis

It is apparent from the above experiment that sodium ions are of vital importance in the generation of the vertebrate photoreceptor potential. Most likely, sodium plays its role in relation to the ionic flux of the photoreceptor, and an alteration at the outer segment of the sodium flux upon exposure to light is manifested as the receptor potential which, in the vertebrate, is a hyperpolarization. Furthermore, the observation with ouabain seems to provide strong evidence that the primary effect of light in generating the receptor potential is not to activate the metabolic pump, but to decrease the membrane permeability to sodium.

An electric model of the outer segment membrane is proposed in Fig. 18b (H-type), which explains the relationship between the membrane resistance and the receptor potential. Calculation by Sillman, Ito, and Tomita (1969b) on this simple model resulted in the relation below between the membrane potential change (ΔE_m) and the resistance change in the sodium channel (ΔR_{Na}).

$$\Delta E_m = \frac{\left(\dfrac{RT}{F} \ln \dfrac{[Na^+]_o [K^+]_i}{[Na^+]_i [K^+]_o} + i R_K \right) R_K}{(R_{Na} + R_K)^2} \Delta R_{Na} ,$$

where i denotes the intensity of extrinsically applied hyperpolarizing current through the recording electrode.

The relationships represented by the above expression are fully supported by existing evidence. Firstly, ΔE_m, which is the hyperpolarizing receptor potential, should be directly proportional to a change in R_{Na}. This relationship actually was demonstrated by Toyoda, Nosaki, and Tomita (1969). Secondly, the receptor potential which results from a given light stimulus should vary in direct linear

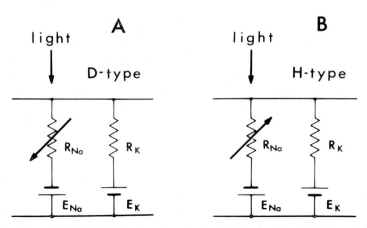

Fig. 18. Electric model of the photoreceptor membrane in invertebrates (A) and vertebrates (B). The arrow pointing left down in A, and that pointing right up in B indicate respectively a decrease and increase in R_{Na} by illumination

proportion to the logarithm of the external sodium concentration, and in inversely linear proportion to the logarithm of the external potassium concentration. These relationships are consistent with the observations just described. Thirdly, we see in the expression that the receptor potential is in linear relation to the intensity of extrinsic hyperpolarizing current i. In other words, if the extrinsic current is in the direction to increase the membrane potential (i.e. if i is positive) then the receptor potential should become larger, while if i is negative then the receptor potential should be decreased. This, in fact, is a quantitative expression of the relationship that actually was observed by TOYODA, NOSAKI, and TOMITA (1969) and illustrated in Fig. 16. Finally, similar reasoning can be used in relation to the effects of trans-retinal current on the mass receptor response. Vitreous-negative transretinal current is in the direction to increase the membrane potential at the distal region of the photoreceptor cell and, therefore, is comparable to hyperpolarizing extrinsic current (i.e. i is positive). Therefore, application of such a current should increase the amplitude of the receptor potential while, conversely, vitreous-positive current should decrease the response. Exactly such a relationship was observed by SILLMAN, ITO, and TOMITA (1969a). Similar observations on the ERG waves were reported by GRANIT and HELME (1939).

We can conclude that the primary effect of light in eliciting the receptor potential is to decrease the permeability of the photoreceptor membrane to sodium. This is made clear by the fact that the receptor potential can still be generated even in the absence of the metabolic pump, so long as the sodium concentration gradient is maintained. The metabolic pump is important to the generation of the potential only with respect to its role in maintaining the sodium concentration gradient necessary for the hyperpolarization response. The proposed model and the equation derived from it show that this conclusion is consistent with the data obtained.

In good agreement with ours are the recent observations of PENN and HAGINS (1969) and HAGINS, PENN, and YOSHIKAMI (1970) on the rods in slices of the albino rat retina. The technique was similar to the one previously utilized in the squid retina by HAGINS (1965) but required far more precautions against deterioration of the warm-blooded retinal tissue. One of the slices, prepared with the aid of infrared radiation visualized with an image converter, was mounted in a chamber on the stage of an inverted infrared microscope, so that the layers of the retina and the applied microelectrodes could be seen in profile through the infrared microscope. The receptor potential was isolated effectively by perfusion of the retinal slice with Ringer of low (1 mM) PO_4. Mapping the electric field in darkness, a steady "dark" voltage gradient was demonstrated in the interstitial spaces along the whole length of rods, indicating a radial flow of interstitial current which emerges from the proximal regions of the rods and disappears into the distal regions across the membrane of the rod outer segments. Upon illumination, another flow of current, that just opposed but never exceeded the dark current, was generated. This photocurrent of PENN and HAGINS (1969) may be interpreted as a reduction of the dark current by the same amount. By so interpreting, their result is identical with that proposed by TOYODA, NOSAKI, and TOMITA (1969), who state: "The receptor outer segments are depolarized in the dark and act as a sink of current, producing an electric field around the receptors such as to make their distal tips negative. Upon illumination the sink disappears or diminishes in a graded manner and as the result the potential at the distal tips is shifted in the positive direction." Indeed, a report that fully supports the above interpretation recently appeared (Yoshikami and Hagins, 1970).

Controversial to the above, on the other hand, are the recent observations of Arden and Ernst (1969, 1970), who, working on the mass receptor potential in the preparation of the isolated rat and pigeon retinas, came to the following conclusion which differs between the cones and rods. (1) The cone outer segments have cation pores which are open in the dark to cause a dark current and closed by light. The cation pores allow passage of not only sodium ions but also potassium and ammonium ions as well. These ions move through the pores along the concentration gradient. Potassium ions are thus synergic to sodium ions, according to Arden and Ernst. However, the result of Sillman, Ito, and Tomita suggests that the role of potassium ions is to maintain the inside-negative membrane potential in the manner principally the same as in the other cell types including invertebrate photoreceptors (cf. Fig. 18a, b). Most controversial is the relationship between the external potassium concentration and the cone response, which according to Arden and Ernst are in a parallel relationship, while according to Sillman, Ito, and Tomita they are in an inverse relationship. Arden and Ernst also stress the importance of chloride ions on the electrogenesis, while Sillman, Ito, and Tomita do not, who observed that substitution of more than 95 % chloride ions by aspartate ions had little effect upon the response size. (2) Arden and Ernst saw that the rod response was more resistant than the cone response against depletion of sodium ions in the external medium and more susceptible to ouabain. The result suggested that the rod response is associated with an active ion transport, contrary to the cone response which is attributed to passive ion movements along the concentration gradient. However, the observation of Yoshikami and Hagins (1970) on the effects of changes in the external sodium concentration and of ouabain on the response of rat rods does not agree with Arden and Ernst, but supports the view held by Toyoda, Nosaki, and Tomita (1969) that in both cones and rods the ionic mechanism for the generation of the receptor potential is analogous and attributed to passive transport of primarily sodium ions through pores in the outer segment membrane along the concentration gradient. The pores are open in the dark and closed by light.

The reason of the above-mentioned discrepancy remains to be clarified. Arden and Ernst used EDTA for the purpose of isolating the receptor potential from the other ERG components. For the isolation of the receptor potential they needed to remove any trace of calcium ions in the media by EDTA. This ingredient seems to be responsible, partly at least, for the discrepancy. Arden and Ernst (1970) themselves admit that EDTA has an important effect on the nature of the results either directly or indirectly. For example, in the absence of EDTA ouabain-treated cones produced responses only to changes in sodium concentration and not to changes in potassium or ammonium concentrations.

In connection with the use of EDTA by Arden and Ernst for chelating calcium ions as a method to isolate the receptor potential, Yoshikami and Hagins (1970) report that both dark current and photocurrent in rat rods are largest in Ca^{++} free solution, and that a trace of Ca^{++} (0.1 mM) applied externally in the medium immediately reduces the dark current in a way analogous to the action of light. In other words, calcium ions mimic the effect of light. A possible role of Ca^{++} upon the rod and cone responses was discussed on this basis by YOSHIKAMI and HAGINS (1971).

Acknowledgement

Work in the author's laboratory in Tokyo was supported in part by grants from the Education Ministry of Japan, U.S. Public Health Service Grants NB06421 and EY00017, and U.S. Air Force Office of Scientific Research Grants through the U.S. Army Research and Development Group (Far East).

References

ARDEN, G. B., ERNST, W.: Mechanism of current production found in pigeon cones but not in pigeon or rat rods. Nature (Lond.) **223**, 528—531 (1969).
— — The effect of ions on the photoresponses of pigeon cones. J. Physiol. (Lond.) **211**, 311—339 (1970).
BAYLOR, D. A., FUORTES, M. G. F.: Electrical responses of single cones in the retina of the turtle. J. Physiol. (Lond.) **207**, 77—92 (1970).
— — O'BRYAN, P. M.: Receptive fields of cones in the retina of the turtle. J. Physiol. (Lond.) **214**, 265—294 (1971).
BENOLKEN, R. M.: Reversal of photoreceptor polarity recorded during the graded receptor potential response to light in the eye of *Limulus*. Biophys. J. **1**, 551—564 (1961).
BORTOFF, A.: Localization of slow potential responses in the *Necturus* retina. Vision Res. **4**, 627—635 (1964).
— NORTON, A. L.: Simultaneous recording of photoreceptor potentials and the PIII component of the ERG. Vision Res. **5**, 527—533 (1965a).
— — Positive and negative potential responses associated with vertebrate photoreceptor cells. Nature (Lond.) **206**, 626—627 (1965b).
— — An electrical model of the vertebrate photoreceptor cell. Vision Res. **7**, 253—263 (1967).
BROWN, H. M., HAGIWARA, S., KOIKE, H., MEECH, R. M.: Membrane properties of a barnacle photoreceptor examined by the voltage clamp technique. J. Physiol. **208**, 385—413 (1970).
BROWN, K. T., WATANABE, K.: Isolation and identification of a receptor potential from the pure cone fovea of the monkey retina. Nature (Lond.) **193**, 958—960 (1962).
CONN, H. J.: Biological Stains. 7th edition. Baltimore: Waverly 1961.
CRESCITELLI, F.: The photosensitive retinal pigment system of *Gekko gekko*. J. gen. Physiol. **47**, 33—52 (1963).
ECCLES, J. C.: The Physiology of Synapses. Berlin-Heidelberg-New York: Springer 1964.
FRANK, R. N., GOLDSMITH, T. H.: Effects of cardiac glycosides on electrical activity in the isolated retina of the frog. J. gen. Physiol. **50**, 1585—1606 (1967).
FUORTES, M. G. F.: Initiation of impulses in visual cells of *Limulus*. J. Physiol. (Lond.) **148**, 14—28 (1959).
FURUKAWA, T., HANAWA, I.: Effects of some common cations on electroretinogram of the toad. Jap. J. Physiol. **5**, 289—300 (1955).
GRANIT, R.: Sensory Mechanisms of the Retina. London: Oxford Univ. Press 1947.
— HELME, T.: Changes in retinal excitability due to polarization and some observations on the relation between the processes in retina and nerve. J. Neurophysiol. **2**, 556—565 (1939).
HAGINS, W. A.: Electrical signs of information flow in photoreceptors. Cold Spr. Harb. Symp. quant. Biol. **30**, 403—418 (1965).
— ADAMS, R. G.: The ionic basis of the receptor current of squid photoreceptors. Proc. 22nd Intern. Congr. Physiol. Sci. **2**, 970 (1962).
— PENN, R. D., YOSHIKAMI, S.: Dark current and photocurrent in retinal rods. Biophys. J. **10**, 380—412 (1970).
HAMASAKI, D. I.: The effect of sodium ion concentration on the electroretinogram of the isolated retina of the frog. J. Physiol. (Lond.) **167**, 156—168 (1963).
— The electroretinogram after application of various substances to the isolated retina. J. Physiol. (Lond.) **173**, 449—458 (1964).

HANAWA, I., KUGE, K., MATSUMURA, K.: Effects of some common ions on the transretinal dc potential and the electroretinogram of the isolated frog retina. Jap, J. Physiol. **17**, 1—20 (1967).

HANITZSCH, R., TRIFONOW, J.: Intraretinal abgeleitete ERG-Komponenten der isolierten Kaninchennetzhaut. Vision Res. **8**, 1445—1455 (1968).

HARTLINE, H. K., RATLIFF, F., MILLER, W. H.: Inhibitory interaction in the retina and its significance in vision. In: FLOREY, E. (Ed.). Nervous Inhibition. Oxford-London-New York-Paris: Pergamon Press 1961.

HECHT, S., SHLAER, S., PIRENNE, M.: Energy, quanta, and vision. J. gen. Physiol. **25**, 819—840 (1942).

KATZ, B.: The transmission of impulses from nerve to muscle, and the subcellular unit of synaptic action. Proc. roy. Soc. B **155**, 455—479 (1962).

KANEKO, A.: Physiological and morphological identification of horizontal, bipolar and amacrine cells in goldfish retina. J. Physiol. (Lond.) **207**, 623—633 (1970).

— Electrical connexions between horizontal cells in the dogfish retina. J. Physiol. (Lond.) **213**, 95—105 (1971).

— HASHIMOTO, H.: Recording site of the single cone response determined by an electrode marking technique. Vision Res. **7**, 847—851 (1967).

KIKUCHI, R., NAITO, K., TANAKA, I.: Effect of sodium and potassium ions on the electrical activity of single cells in the lateral eye of the horseshoe crab. J. Physiol. (Lond.) **161**, 319—343 (1962).

LUFT, J. H.: Improvements in epoxy resin embedding methods. J. biophys. biochem. Cytol. **9**, 409—414 (1961).

MARKS, W. B.: Visual pigments of single goldfish cones. J. Physiol. (Lond.) **178**, 14—32 (1965).

MILLECCHIA, R., MAURO, A.: The ventral photoreceptor cells of *Limulus*. III. A voltage-clamp study. J. gen. Physiol. **54**, 331—351 (1969).

MURAKAMI, M., KANEKO, A.: Differentiation of PIII subcomponents in cold-blooded vertebrate retinas. Vision Res. **6**, 627—636 (1966).

OIKAWA, T., OGAWA, T., MOTOKAWA, K.: Origin of so-called cone action potential. J. Neurophysiol. **22**, 102—111 (1959).

PENN, R. D., HAGINS, W. A.: Signal transmission along vertebrate photoreceptors and the a-wave of the ERG. Biophys. J., **9**, 13th Ann. Meeting Abstr., A-244 (1969).

POTTER, D. D., FURSHPAN, E. J., LENNOX, E. S.: Connections between cells of the developing squid as revealed by electrophysiological methods. Proc. nat. Acad. Sci. (Wash.) **55**, 328—336 (1966).

PURPLE, R. E., DODGE, F. A.: Interaction of excitation and inhibition in the eccentric cell in the eye of *Limulus*. Cold Spr. Harb. Symp. quant. Biol. **30**, 529—537 (1965).

RUSHTON, W. A. H.: A theoretical treatment of FUORTES's observations upon eccentric cell activity in *Limulus*. J. Physiol. (Lond.) **148**, 29—38 (1959).

SILLMAN, A. J., ITO, H., TOMITA, T.: Studies on the mass receptor potential of the isolated frog retina. I. General properties of the response. Vision Res. **9**, 1435—1442 (1969a).

— — Studies on the mass receptor potential of the isolated frog retina. II. On the basis of the ionic mechanism. Vision Res. **9**, 1443—1451 (1969b).

SKOU, J. C.: Enzymic basis for active transport of Na^+ and K^+ across cell membrane. Physiol. Rev. **45**, 596—617 (1965).

SMITH, T. G., STELL, W. K., BROWN, J. E.: Conductance changes associated with receptor potentials in *Limulus* photoreceptors. Science **162**, 454—456 (1968a).

— — — FREEMAN, J. A., MURRAY, G. C.: A role for the sodium pump in photoreception in *Limulus*. Science **162**, 456—458 (1968b).

STRETTON, A. O. W., KRAVITZ, E. A.: Neuronal geometry: Determination with a technique of intracellular dye injection. Science **162**, 132—134 (1968).

SVAETICHIN, G.: The cone action potential. Acta physiol. scand. **29**, Suppl. 106, 565—600 (1953).

— NEGISHI, K., FATEHCHAND, R.: Cellular mechanisms of a Young-Hering visual system. In: DE REUCK, A. V. S., KNIGHT, J. (Eds.): Colour Vision. Boston: Little, Brown Comp. 1965.

TOMITA, T.: Electrical activity in the vertebrate retina. J. Opt. Soc. Amer. **53**, 49—57 (1963).

TOMITA, T., Electrical response of single photoreceptors. Proc. IEEE **56**, 1015—1023 (1968).
— Electrophysiological study of the mechanisms subserving color coding in the fish retina. Cold Spr. Harb. Symp. quant. Biol. **30**, 559—566 (1965).
— Electrical activity of vertebrate photoreceptors. Quart. Rev. Biophys. **3**, 179—222 (1970).
— KANEKO, A.: An intracellular coaxial microelectrode. Its construction and application. Med. Electron. Biol. Eng. **3**. 367—376 (1965).
— — MURAKAMI, M., PAUTLER, E. L.: Spectral response curves of single cones in the carp. Vision Res. **7**, 519—531 (1967).
— MURAKAMI, M., HASHIMOTO, Y., SASAKI, Y.: Electrical activity of single neurons in the frog's retina. In: JUNG, R., KORNHUBER, H. (Eds.): The Visual System: Neurophysiology and Psychophysics. Berlin-Heidelberg-New York: Springer 1961.
TOYODA, J., HASHIMOTO, H., ANNO, H., TOMITA, T.: The rod response in the frog as studied by intracellular recording. Vision Res. **10**, 1093—1100 (1970).
— NOSAKI, H., TOMITA, T.: Light-induced resistance changes in single photoreceptors of *Necturus* and *Gekko*. Vision Res. **9**, 453—463 (1969).
WALLS, G. L.: The Vertebrate Eye and its Adaptive Radiation. New York: Hafner Publ. Co. 1963.
WEINSTEIN, G. W., HOBSON, R. R., DOWLING, J. E.: Light and dark adaptation in the isolated rat retina. Nature (Lond.) **215**, 134—138 (1967).
WERBLIN, F. S.: Functional Organization of the Vertebrate Retina Studied by Intracellular Recording from the Retina of the Mudpuppy, *Necturus maculosus*. Doctoral Dissertation. The Johns Hopkins University, Baltimore 1968.
— DOWLING, J. E.: Organization of the retina of the mudpuppy, *Necturus maculosus*. II. Intracellular recording. J. Neurophysiol. **32**, 339—355 (1969).
WITKOVSKY, P.: Peripheral mechanisms of vision. Ann. Rev. Physiol. **33**, 257—280 (1971).
YOSHIKAMI, S., HAGINS, W. A.: Ionic basis of dark current and photocurrent of retinal rods. 1969 Biophys. Soc. Abstr., 60a, WPM-13 (1970).
— — Light, calcium, and the photocurrent of rods and cones. 1971 Biophys. Soc. Abstr. 47a, TPM-E16 (1971).

Chapter 13

S-Potentials

By

Peter Gouras, Bethesda, Maryland (USA)

With 12 Figures

Contents

Introduction

S-potential is a term that describes a unique response to light which can be detected when relatively fine micro-pipette electrodes are inserted into certain regions of the inner nuclear layer of the retina. This response is of large amplitude (10 to 50 mV), usually negative, superimposed on a negative resting potential (10 to 50 mV) and *maintained* for the duration of light stimulation (Fig. 1). The

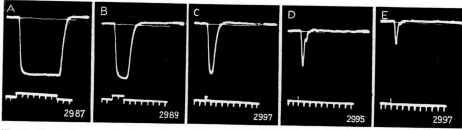

Fig. 1. Recordings of S-potentials elicited at different durations (*A* 500, *B* 150, *C* 20, *D* 5, *E* 0.5 msec) of light stimuli (constant intensity and amplification). Time marks: 70 msec. Light stimulus marked on time scale (Svaetichin, 1953)

response was discovered by Svaetichin (1953) in fish retina and subsequently named "S-potential" by Motokowa (see Tomita et al., 1959) not only to acknowledge the discoverer but also to introduce a non-committal term that did not assume any specific cellular origin for S-potentials since this remained in doubt for some time after their discovery.

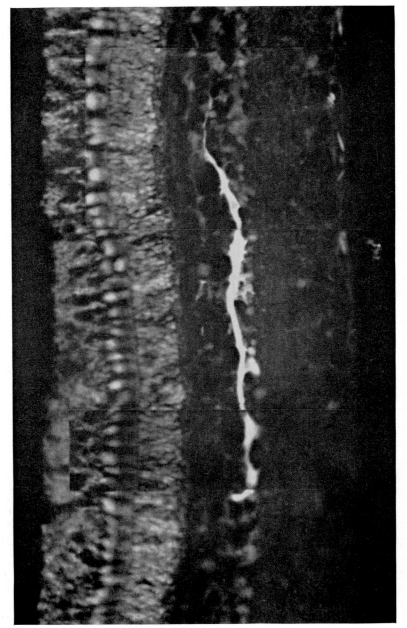

Fig. 2. A montage photograph of an internal horizontal cell in which an S-potential, located vitread, was recorded (Kaneko, 1970)

SVAETICHIN (1953, 1956) originally thought that the response originated within single cones which in fish retina are relatively large (8 micra diameter). Numerous attempts were made to test this hypothesis by iontophoresing dye through a micro-electrode recording an S-potential in order to subsequently identify the site of this response in the retina (MACNICHOL and SVAETICHIN, 1958; MITARAI, 1958, 1960 TOMITA et al., 1959; OIKAWA et al., 1959; GOURAS, 1960; BROWN and TASAKI, 1961). General agreement emerged that the responses which SVAETICHIN (1953, 1956) had found arose somewhere between the external and internal plexiform layers, most likely within horizontal cells which tend to be large in fish retina. Direct confirmation of this impression was not forthcoming, however, until only recently. This was made possible by the development of intracellular marking techniques which delineate the entire profile of a cell (STRETTON and KRAVITZ, 1968). Using this technique KANEKO (1970) demonstrated beyond doubt that S-potentials are produced by horizontal cells (Fig. 2).

The discovery of S-potentials was an important step in retinal physiology since it provided the first opportunity to record from single cells within the inner nuclear layer of the retina. S-potentials are not peculiar to fish but appear to be a typical response of all vertebrate horizontal cells since they have been found in the retinas of cat (GRÜSSER, 1957, 1961; BROWN and WIESEL, 1959, 1961; BROWN and TASAKI, 1961; BROWN and MURAKAMI, 1968; STEINBERG, 1969a, b, c, d; GOURAS and HOFF, 1970), monkey (SVAETICHIN, 1961; BROWN and MURAKAMI, 1968) turtle (TRIFONOV, 1968; BAYLOR, FUORTES, and O'BRYAN, 1971), necturus (WERBLIN and DOWLING, 1969) and lamprey (GOURAS, unpublished).

Receptive Field of S-Potentials

One of the remarkable properties of S-potentials is the extraordinarily large size of their receptive field. This was first brought out by TOMITA et al. (1958) who shined concentric spots of light of different sizes around a micro-electrode recording S-potentials in fish retina and demonstrated that these responses are influenced by retinal stimulation several millimeters away from the recording micro-electrode. These results have been confirmed by OIKAWA et al. (1959), GOURAS (1960), NAKA and RUSHTON (1967), and NORTON et al. (1968). WATANABE and TOSAKA (1959) provided unique evidence that what was spreading from one region of retina to another was current and not scattered light by showing that an S-potential was reduced when the retina was cut at a point between the light stimulus and the recording micro-electrode. In a less direct but more analytical approach NAKA and RUSHTON (1967) reached a similar conclusion. Their evidence is shown in Fig. 3 in which the amplitude of an S-potential is plotted against the energy of spots of white light placed in different parts of the S-potential's receptive field. In one condition spots of light of different energy are used to elicit S-potentials of different amplitude by stimulating the retina at the recording site (E_2). In a second condition ($E + E_2$) a spot of light of *fixed* intensity is superimposed on the E_2 spot of *variable* intensity. The function $E + E_2$ represents a horizontal shift of the E_2 curve which is what is expected if a fixed light is added to the E_2 stimulus. In a third condition another spot (F_2) 0.5 mm away from the micro-electrode is used to elicit S-potentials of different amplitude. In a fourth condition ($E + F_2$) a spot of

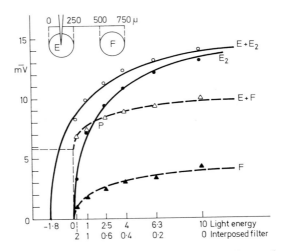

Fig. 3. Inset indicates spatial distribution of light. Curve E_2 is drawn through black circles which plot S-potential when stimulus was a flash at E of energy indicated on the linear horizontal scale. White circles show potential when a fixed light E was added to each E_2. Curve $(E + E_2)$ is curve E_2 displaced to the left. Black triangles show potential at E when the same set of light flashes are presented at E centred 500 μ away. Curve F is curve E_2 scaled down to 2/7. White triangles give potential when the fixed light E was added (at site E). Curve $(E + F)$ is curve F displaced upwards (Naka and Rushton, 1967)

fixed intensity is flashed at E while the variable spot is flashed at F. The function $E + F_2$ represents a vertical shift of the curve F_2 which could occur if the current produced by F_2 added to that produced at E but not if light were scattering from F_2 to E.

The size of a single horizontal cell does not seem to be sufficient to explain the large area of the receptive field of S-potentials. These large receptive fields may depend in the fact that horizontal cells can be electrically coupled to one another (Kaneko, in press; Baylor, Fortes, and O'Bryan, unpublished). This electrical coupling goes along with electronmicroscopic evidence that "tight junctions", which are characteristic of electrically coupled membranes, are found between different horizontal cells (Yamada and Ishikawa, 1965; Stell, 1967). It also agrees with the relatively low total input resistance (10—300 kohms) which has been measured within regions of retina producing large S-potentials (Watanabe et al., 1960; Gouras, 1960; Tomita, 1965).

Another explanation for the large receptive field of S-potentials is the possibility that horizontal cells make synapses with other cells which permit a spread of activity from one horizontal cell to another. There is electrophysiological evidence that horizontal cells synapse on cones (Baylor et al., 1971) which could in turn relay signals to other horizontal cells. There is electronmicroscopic evidence that horizontal cells also form synapses with bipolar cells (Dowling et al., 1966; Dowling and Werblin, 1969) but it is not clear whether such a signal could spread to other horizontal cells.

S-Potentials and Light Energy

SVAETICHIN (1953) demonstrated that the amplitude of an S-potential increases approximately linearly with the logarithm of light energy up to a saturation point. Fig. 4 shows the sigmoidal relation that is usually obtained between S-potential amplitude and the logarithm of retinal illumination for five S-potentials in bream retina. NAKA and RUSHTON'S (1967) results indicate that this relationship is sometimes more complex than is immediately apparent because it seems to be composed of relatively independent contributions from several cone mechanisms. Their evidence for this is shown in Fig. 5 which illustrates S-potential amplitude versus light energy plotted linearly. In Fig. 5 *A* the stimulus is a red light; in *B* a fixed *red* light is added to the variable red light and the resulting relationship represents a horizontal shift of curve *A* as expected if light energy is increased by a fixed ratio. In Fig. 5 *C* a fixed *blue* light is added to the red light but this surpris-

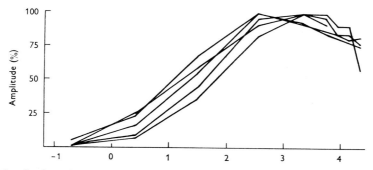

Fig. 4. Graph of amplitude of S-potentials plotted against retinal illumination. Semi-log, scale (GOURAS, 1960)

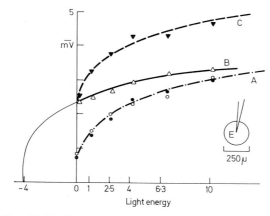

Fig. 5. All lights fall on *E*. Black and white circles are two runs each like black circles (Fig. 3), except that the light was now deep red. Curve *A* is drawn through the points. White triangles (like white circles, Fig. 3) are where a fixed red light was added to each of those in curve *A*. Black triangles are when the fixed light was not red but blue. Curve *B* is curve *A* displaced sideways; curve *C* is *A* displaced upwards (NAKA and RUSHTON, 1967)

ingly shifts curve A not horizontally but vertically which would occur if current but not light were being added. This result implies that different cone mechanisms contribute their signals to S-potentials in a relatively independent way (see also Orlov and Maksimova, 1965).

By proper choice of stimulus parameters, Naka and Rushton (1966a, b, c) were able to isolate the response of only one cone mechanism in a S-potential. The relationship between light energy and S-potential amplitude which empirically fit their results for a single cone mechanism was:

$$V = V_\infty \frac{I}{I + \sigma}$$

where V = voltage of the S-potential, V_∞ = voltage of the S-potential at saturation, I = the light energy of the stimulus and σ = the light energy at half saturation (Rushton, 1969). Fig. 6 shows the sigmoidal relationship which this function determines over 4 logarithmic units when V is plotted against log I.

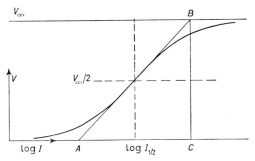

Fig. 6. Theoretical curve representing the expression $V = V_\infty I/I + \sigma$ when V is plotted against log I (Rushton, 1969)

S-Potentials and the Wavelength of Light

By changing the wavelength of stimulation S-potentials can be separated into two distinct classes (Svaetichin, 1956). One class produces a negative or hyperpolarizing response to all wavelengths; the other produces a hyperpolarizing response to some wavelengths and a depolarizing one to other wavelengths.

L-Units

The purely hyperpolarizing S-potentials have been called luminosity or L-units because they have been thought to play a role in brightness as distinct from color perception (Svaetichin, 1956; MacNichol et al., 1957; Svaetichin and Mac Nichol, 1958; MacNichol and Svaetichin, 1958). Based on their responses to monochromatic lights L-units can be subdivided into a variety of subtypes. Many of such studies have been done with stimuli of constant energy or quanta which generate spectral response functions which cannot be directly compared with photopigment absorption curves because of non-linearities in the S-potential response. Some of the results indicate that all L-units in a particular species of fish have approximately the same spectral response (MacNichol and Svaetichin, 1958); others show that *L*-units in the same retina may have different spectral responses (Svaetichin, 1953; Motokowa et al., 1957; Svaetichin, 1961; Tamura and Niwa, 1967). The spectral responses of *L*-units may also have several sub-maxima (Svaetichin, 1956; Svaetichin and MacNichol, 1958; Tomita et al., 1958) indicating that at least two or more photoreceptor systems must contribute to the response.

Spectral sensitivity curves based on constant response criteria have also been determined for *L*-units and although such functions ought to resemble photo-pigment absorption curves more closely than those based on constant input criteria they have not entirely clarified the problem. Motokowa et al. (1957) found four different spectral sensitivity curves with peaks at 470, 550, 600 and 650 nm in *L*-units of carp retina which when studied by Witkowsky (1967) revealed only one action spectrum with two peaks, one at 613 and another at 665 nm. In *L*-units

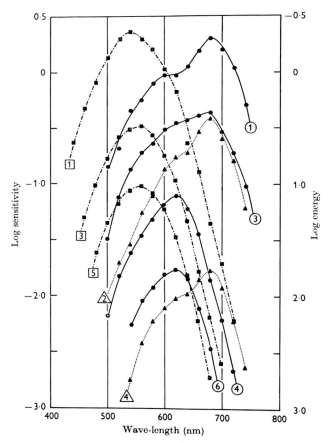

Fig. 7. Action spectra from an *L*-unit. Numbers at end of each curve give the level of hyper-polarization used as criterion of action. Circles when flashes fell upon dark retina, triangles when flashes fell upon steady green adapting light, squares when adapting light was red (Naka an Rushton, 1966 c)

of another Cyprinidae fish (the tench Tinca) NAKA and RUSHTON (1966c) also found two peaks, one at 620 and another at 680 nm but when action spectra were determined for these units in the presence of chromatic adapting lights an additional peak at 540 nm (Fig. 7) and possibly another in the blue region of the spectrum were uncovered. They concluded that L-units receive signals from four different cone mechanisms with peak sensitivities at 680, 620, 540 and also in the blue region of the spectrum. The latter three mechanisms are probably related to photopigments identified in single cones of Cyprinidae fishes by microspectrophotometry (MARKS, 1965; LIEBMAN and ENTINE, 1964). No 680 nm cone pigment has been found microspectrophotometrically.

LAUFER and MILLAN (1970) have determined both action spectra of L-units and the absorption spectra of cone pigments by microspectrophotometry in the same type of fish (Eugerres plumieri) and their results show good agreement between these two variables. L-units were found with spectral peaks at 476, 568, and 606 nm which corresponded to cone absorption peaks at 471, 568, and 604 nm. An important difference between these results and most of the previous ones is that here each cone mechanism is isolated in a single L-unit whereas in the others several cone mechanisms converge on one L-unit. It would have been useful for LAUFER and MILLAN (1970) to have determined whether selective chromatic adaptation affected the action spectra of their L-units.

Both MAKSIMOVA et al. (1966) and NAKA and RUSHTON (1966c) discovered a remarkable effect when L-units were examined in the presence of selective chromatic adaptation. A red adapting light not only depressed the contribution of a red sensitive mechanism but also enhanced considerably the contribution of a green sensitive one to the same L-unit (Fig. 8). This indicates that the red sensitive cone mechanism not only hyperpolarizes L-units but also antagonizes the contributions of green sensitive cones to these same L-units.

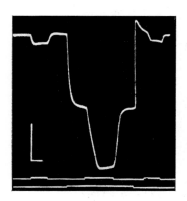

Fig. 8. Amplification of a response to short wavelength stimulation by a red background light in an L-unit of the pike. The middle record shows the duration of the short wavelength stimulus; the lower one shows the duration of the red adapting light (MAKSIMOVA et al., 1966)

C-Units

S-potentials which are hyperpolarized by some wavelengths but depolarized by others (Fig. 9) have been called chromaticity or C-units because they have been thought to play a role in color vision (SVAETICHIN, 1956; MacNICHOL et al., 1957; SVAETICHIN and MacNICHOL, 1958; MacNICHOL and SVAETICHIN, 1958). There

seem to be three general classes of C-units. One is depolarized by red and hyperpolarized by green light (R/G) or vice versa (G/R); another class is depolarized by
green and hyperpolarized by blue light (G/B) or vice versa (B/G); the third is
depolarized by red and blue and hyperpolarized by green light (R, B/G); the
reverse has not been reported. There are other possible combinations of three or
possibly four (MOTOKAWA et al., 1957; NAKA and RUSHTON, 1966a, b) cone
mechanisms that exist in fish retinas but they have never been detected.

There have been two attempts to identify specific cone mechanisms in C-units
by determining action spectra based on constant response criteria in the presence
of selective chromatic adaptation (NAKA and RUSHTON, 1966a, b; WITKOWSKY,
1967). In the first study R/G and G/B type C-units were examined in Cyprinidae
fishes. Selective adaptation successfully isolated a 540 nm or green sensitive cone
mechanism in both the R/G and G/B units which corresponded well with the green

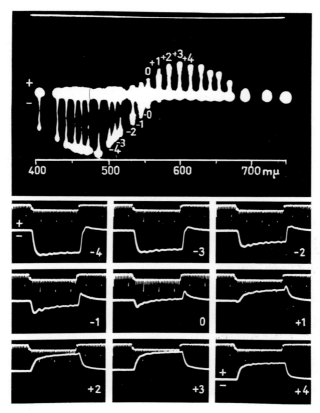

Fig. 9. Changes in time course of responses from a C-unit of Mugil retina as a function of
wavelength. (Top record). Recording of response amplitude as a function of wavelength.
(Lower records.) Responses recorded as function of time. Each numbered record was taken
simultaneously with the response peak bearing the same number in the top record. Top trace
in each record indicates time in tenths and hundredths of a second. Deflection of top trace is
due to output of photocell circuit used to monitor light flash (duration 0.3 sec). Records taken
at approximately three-second intervals (MACNICHOL and SVAETICHIN, 1958)

sensitive pigment identified in goldfish cones (Marks, 1965). The action spectrum of the blue sensitive cone mechanism could not be identified but it was suspected that it also resembled Marks blue sensitive cone pigment. The red sensitive cone mechanism responsible for depolarizing R/G units had an action spectrum which peaked at 680 nm and did not match any photopigment detected in fish cones. Naka and Rushton (1966a, b) acknowledged the possibility that the action spectrum which they detected in the deep red could be produced by the interaction of two pigments but developed a strong argument for the existence of a single 680 nm pigment. This theory has received some support from a later study by Witkowsky (1967) who employed selective chromatic adaptation to

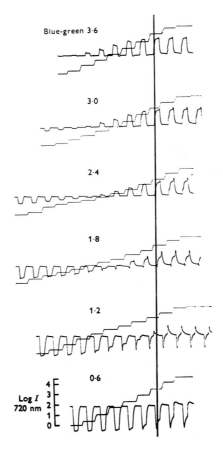

Fig. 10. Time records of the responses of a R/G unit to flashes of a mixture of blue-green light and deep red (720 mm). In each row the blue-green component was fixed in intensity by interposition of a neutral filter whose density is indicated by the number above the row. Throughout each row the red component increased in intensity at each flash, as monitored by the staircase. The level of each step is related to log intensity by the scale shown at bottom left of figure. The vertical line shows the place where the red light is 10% of the maximum intensity available (Naka and Rushton, 1966b)

isolate cone mechanisms in an R, B/G unit and an R/G unit. In both cases the red sensitive mechanism had its peak sensitivity in the far red (665 nm).

The chromatic adaptation experiments of NAKA and RUSHTON (1966a, b) have uncovered an interesting property of the depolarizing mechanism of C-units. Figure 10 illustrates this in an R/G unit in which mixtures of red and green light are used to elicit responses; in each row of responses the red light in the mixture increases from left to right while the green light remains constant; the green light in each mixture is made progressively stronger starting from a weak green light above to the strongest below. Regardless of the stimulus to the green sensitive mechanism (hyperpolarizing), the red sensitive mechanism (depolarizing) saturates at the same point (vertical line), i.e., saturation depends only on the stimulus to red and not green sensitive cones. Saturation of the depolarizing mechanism must occur before the red and green sensitive cone signals interact. The hyperpolarizing mechanism does not appear to saturate as easily as the depolarizing one so that stronger stimuli tend to favor hyperpolarizing responses in C-units (see also ORLOV and MAKSIMOVA, 1965). This is undoubtedly why GOURAS (1960) observed that some S-potentials in bream retina depolarized with weak and hyperpolarized with strong white light stimuli.

There is another important difference between depolarizing and hyperpolarizing S-potentials. Depolarizing potentials have a longer delay and a slower time course than hyperpolarizing ones (MacNICHOL and SVAETICHIN, 1958; GOURAS, 1960) so that when these two opposing potentials are approximately balanced transient hyperpolarizing on − and depolarizing − off responses occur (Fig. 9). This suggests that the depolarizing response may be mediated by one or more additional synapses.

S-Potentials and Cone Potentials

Cones also generate maintained hyperpolarizing responses to light (BORTOFF, 1964; TOMITA, 1965; TOMITA et al., 1967; OIKAWA et al., 1959) and it is interesting to consider how they could be related to S-potentials of horizontal cells. BYZOV and TRIFONOV (1968) have provided an important clue to this problem by stimulating cones with both light and electric current while recording an S-potential (Fig. 11). When cone pedicles are depolarized by current a depolarization is registered in the S-potential implying that cone terminals release a depolarizing transmitter on horizontal cells just as the terminals of other axons release transmitter when depolarized (KATZ, 1969). When light hyperpolarizes an S-potential (Fig. 11) the depolarization produced by current is increased indicating that either more transmitter is released or that the same amount of transmitter has more effect. These investigators provide some evidence to indicate that the former occurs since they occasionally detect S-potentials with small hyperpolarizing responses to light in which the depolarizing response to current is similarly enhanced.

Another important clue to the relation of cones to S-potentials originates from experiments of BAYLOR et al. (1971). A micro-electrode within a cone detected a depolarizing potential whenever a neighboring horizontal cell was hyperpolarized by current passed through a second micro-electrode. These results sug-

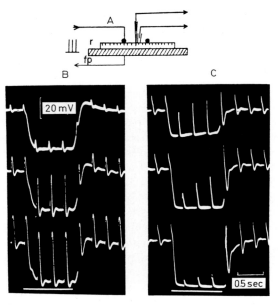

Fig. 11. The responses of S-cells (*L*-type) to repetitive electrical pulses (receptor side positive) in darkness and during the light. *A* Experimental arrangement for passing electrical pulses through the retina and of recording S-potentials. r — isolated retina (receptors up); fp — filter paper. *B* The effect of different strengths of electrical pulses. In the middle curve the strength was twice, and in the lower one 7 times, that of the upper curve. The intensity of light was the same throughout (∼ 30 lux). *C* The effect of different intensities of light on the electrically-evoked responses. In the middle curve the intensity of light was 3 times and in the lower curve 10 times that of the upper curve. The strength of pulse was the same throughout (twice threshold) (Byzov and Trifonov, 1968)

gest that horizontal cell terminals continuously release a hyperpolarizing transmitter on cones which can be reduced by hyperpolarizing the horizontal cell.

When a cone is hyperpolarized by light its impedance increases (Bortoff and Norton, 1967; Toyoda et al., 1969; Baylor and Fuortes, 1970) and following the results of Byzov and Trifonov (1968) the depolarizing transmitter it releases should decrease. As a consequence the impedance of horizontal cells should also increase. Kaneko and Tomita (Tomita, 1965) have found that the impedance of hyperpolarizing horizontals cells does increase in the presence of light but slower than the hyperpolarizing response, itself. This delay has not yet been explained. It has been difficult to measure impedance changes in horizontal cells because their relatively low resistance demands the passage of large amounts of current through fine micro-electrodes (Watanabe et al., 1960; Tasaki, 1960; Gouras, 1960).

Together these results suggest a relatively simple circuit (Fig. 12) which can explain many of the phenomena associated with S-potentials. In this arrangement cones release a depolarizing transmitter on horizontal cells and horizontal cells release a hyperpolarizing transmitter on cones. Horizontal cells are hyperpolarized by light because a depolarizing transmitter being continuously liberated by cones in darkness is reduced when cones are hyperpolarized by light. Hyperpolarizing horizontal cells tend to depolarize cones and this in turn depolarizes horizontal cells. *C*-units receive a hyperpolarizing input from only one cone mechanism and therefore tend to be depolarized whenever other cone mechanisms are stimulated by light. This depolarizing signal goes through two additional synapses and is consequently more delayed than the hyperpolarizing one. *L*-units receive hyperpolarizing signals from several different cone mechanisms and this tends to conceal any depolarizing signals they receive. That such antagonistic responses are actually present is revealed by chromatic adaptation in which the response of one cone mechanism is enhanced by light-adapting the other (MAKSIMOVA et al., 1966; NAKA and RUSHTON, 1966c).

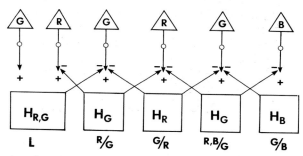

Fig. 12. A scheme based on the idea that S-potentials originate in horizontal (*H*) cells (KANEKO, 1970) which receive a depolarizing (+) transmitter (BYZOV and TRIFONOV, 1968) from three possible cone mechanisms (R, B, G) and release a hyperpolarizing (—) transmitter (BAYLOR et al., 1971) on cones. *C*-units are considered to receive direct signals from only on cone mechanism and *L*-units from more than one. The lower row of symbols signifies the type of S-potentials obtained (as described in the text)

S-Potentials of Rods

Most of the early studies on S-potentials have been done on light-adapted, excised retinas in which rod adaptation is undoubtedly abnormal so that there was little chance for finding rod signals. Rod function did not appear to be essential for S-potentials since they were recorded from retinas in which rod outer segments had been stripped away (MITARAI and YAGASAKI, 1955; MACNICHOL and SVAETICHIN, 1958). MITARAI et al. (1961) first demonstrated that rods contributed to S-potentials by detecting a Purkinje shift together with lower threshold, slower responses in S-potentials of carp retina which had been removed from eyes dark-adapted for 12 h. On the other hand WATANABE and HASHIMOTO (1965) failed to detect a Purkinje shift in S-potentials of carp retina with an intact blood supply where rod adaptation was presumably adequate. WITKOWSKY (1967) using prolonged dark-adaptation to enhance rod activity in *L*-units of excised carp retina obtained an increased sensitivity to short wavelengths which he could not conclusively identify as a rod mechanism. NAKA and RUSHTON (1968) were also unable to detect rod activity in *L*-units of dark-adapted tench retina. LAUFER and MILLAN (1970) claimed to have found rod S-potentials in dark-adapted retina of the fish,

Eugerres plumieri, and implied that these horizontal cells were not receiving inputs from cones. Although these results on fish retina are not in total agreement, it seems possible to conclude that rod signals may not always be present in fish horizontal cells and when they are they are not easy to detect.

There is better evidence that rods contribute to S-potentials in the intact mammalian retina. Brown and Murakami (1968) used the time course of decay of S-potentials to demonstrate a shift from rod (slow decay) to cone (fast decay) function. They also claimed that horizontal cells mediated antagonistic interaction between rods and cones. Steinberg (1969 b, c, d) confirmed their responses but failed to detect any rod — cone antagonism. Rod and cone signals appeared to add their effects independently in cat S-potentials. Boycott (unpublished) has found that cat horizontal cells appear to send dendrites to cones and a long (0.5 nm) axon to rods. Kolb (1970) has observed a similar arrangement in monkey retina. In order to reconcile these findings with the physiological evidence that both rods and cones send signals into horizontal cells it may be necessary to consider that both terminals of such a cell may have dendritic properties. This interpretation is supported by the fact that propagating action potentials have not been found in S-potentials.

S-Potentials and Ganglion Cells

Witkowsky (1967) compared the spectral properties, light- and dark-adaptation and increment thresholds of S-potentials and ganglion cells in carp retina and concluded that the responses of these two different retinal elements were unrelated. Steinberg's (1969d) results in cat retina tend to support this hypothesis. Naka and Kishida (1967) on the other hand found that the amplitude of an S-potential was reflected in the latency of a ganglion cell response and suggested that these two variables might be related. Maksimova (1969) points out that correlations made in experiments in which light is the only stimulus are not conclusive because they may simply be due to the fact that both of these elements receive signals from the same source, i.e., the receptors. In her investigation interconnections between horizontal cells and ganglion cells were more directly examined by passing current through a horizontal cell (L-units) with one micro-electrode while recording from a ganglion cell with another. Her results show that polarization of a horizontal cell of 5 to 10 mV has effects on the impulses generated by ganglion cells. Hyperpolarization of an L-unit produces the surround response in a ganglion cell, i.e., depolarizes an off-center, hyperpolarizes an on-center ganglion cell. Naka (1971) has confirmed these findings. S-potential activity must therefore influence the output of retinal ganglion cells.

S-Potentials and Vision

The evidence is sufficient to conclude that S-potentials represent the activity of horizontal cells in probably all vertebrate retinas. If one accepts two simplifying assumptions: 1) a depolarizing transmitter goes from cones to horizontal cells (Byzov and Trifonov, 1968); 2) a hyperpolarizing transmitter goes from horizontal cells to cones (Baylor et al., in press), then there seems to be two classes of horizontal cells, those that receive a direct input from only one class of cones (*C*-units) and those that receive a direct input from more than one class of cones and probably rods (*L*-units). This arrangement allows horizontal cells to exert negative feedback on photoreceptors. When photoreceptors are put in the light (receptor membrane hyperpolarizes) the horizontal cells tend to put them back in the dark (receptor membrane depolarizes). This circuit could increase the dynamic range over which photoreceptors operate.

In addition the arrangement could also enhance spatial contrast since when one photoreceptor is illuminated horizontal cells will tend to reduce the effects of illumination on neighboring receptors somewhat analogous to the spatial interaction described in Limulus (HARTLINE et al., 1961). *C*-units could also enhance color because whenever one cone mechanism is stimulated the responses of cone mechanisms with different spectral sensitivities will tend to be suppressed. Such horizontal cells could thereby confer color opponent properties on single cones.

References

BAYLOR, D. A., FUORTES, M. G. F. Electrical responses of single cones in the retina of the turtle. J. Physiol. (Lond.) **207**, 77—92 (1970).

— — O'BRYAN, P.: Receptive fields of cones in the retina of the turtle. J. Physiol. (Lond.) **214**, 265—294 (1971).

BORTOFF, A.: Localization of slow potential responses in the Necturus retina. Vision Res. **4**, 626—627 (1964).

— NORTON, A. L.: An electrical model of the vertebrate photoreceptor cell. Vision Res. **7**, 253—263 (1967).

BROWN, K. T., MURAKAMI, M.: Rapid effects of light and dark adaptation upon the receptive field organization of S-potentials and late receptor potentials. Vision Res. **8**, 1145—1171 (1968).

— TASAKI, K.: Localization of electrical activity in the cat retina by an electrode marking method. J. Physiol. (Lond.) **158**, 281—295 (1961).

— WIESEL, T. N.: Intra-retinal recording with micropipette electrodes in the intact cat eye. J. Physiol. (Lond.) **149**, 537—562 (1951).

— — Localization of origins of electroretinogram components by intra-retinal recording in the intact cat eye. J. Physiol. (Lond.) **158**, 257—280 (1961).

BYZOV, A. L., TRIFONOV, YU. A.: The response to electric stimulation of horizontal cells in the carp retina. Vision Res. **8**, 817—822 (1968).

DOWLING, J. E., BROWN, J. E., MAJOR, D.: Synapses of horizontal cells in rabbit and cat retinas. Science **153**, 1639—1641 (1966).

— WERBLIN, F. S.: Organization of retina of the mudpuppy, Necturus maculosus. I Synaptic structure. J. Neurophysiol. **32**, 315—338 (1969).

GOURAS, P.: Graded potentials of bream retina. J. Physiol. (Lond.) **152**, 487—505 (1960).

— HOFF, M.: Retinal function in an isolated, perfused mammalian eye. Invest. Ophthal. **9**, 388—399 (1970).

GRÜSSER, O. J.: Receptorpotentiale einzelner retinaler Zapfen der Katze. Naturwissenschaften **44**, 522 (1957).

— Receptorabhängige R-potentiale der Katzenretina. In: JUNG, R., KORNHUBER, H. (Eds.): The Visual System. Neurophysiology and Psychophysics, pp. 56—61. Berlin-Göttingen-Heidelberg: Springer 1961.

HARTLINE, H. K., RATLIFF, F., MILLER, W. H.: Inhibitory interaction in the retina and its significance in vision in Nervous Inhibition, pp. 241—284. FLOREY, E. (ed.) New York: Pergamon Press 1961.

KANEKO, A.: Physiological and morphological identification of horizontal, bipolar and amacrine cells in goldfish retina. J. Physiol. (Lond.) **207**, 623—633 (1970).

KATZ, B.: The Release of Neural Transmitter Substance. Liverpool: University Press 1969.

Kolb, H.: Organization of the outer plexiform layer of the primate retina: electron microscopy of Golgi — impregnated cells. Phil. Trans. B **258**, 261—283 (1970).

Laufer, M., Millan, E.: Spectral analysis of L-type S-potentials and their relation to photopigment absorption in a fish (Eugerres plumieri) retina. Vision Res. **10**, 237—251 (1970).

Liebman, P. A., Entine, G.: Sensitive low-light level microspectrophotometer: detection of photo-sensitive pigments of retinal cones. J. Opt. Soc. Amer. **54**, 1451—1459 (1964).

MacNichol, E. F., Jr., MacFerson, L., Svaetichin, G.: Studies of spectral response curves from fish retina. Visual problems of colour. Nat. Phys. Lab. Symp. No. 8, **2**, 531—536 (1957).

— Svaetichin, G.: Electric responses from the isolated retinas of fishes. Amer. J. Ophthal. **46**, 26—40 (1958).

Marks, W. B.: Visual pigments of single goldfish cones. J. Physiol. (Lond.) **178**, 14—32 (1965).

Maksimova, E. M.: Effect of intracellular polarization of horizontal cells on ganglion cell activity in the fish retina. Biofizika **14**, 537—544 (1969).

— Maksimov, V. V., Orlov, O. Yu.: Intensified interaction between signals of receptors in cells that are sources of S-potentials. Biofizika **11**, 472—477 (1966).

Mitarai, G.: The origin of the so-called cone action potential Proc. Jap. Acad. **34**, 299—304 (1958).

— Determination of ultra-microelectrode tip position in the retina in relation to S-potential. J. gen. Physiol. **43**, part **2**, 94—99 (1960).

— Svaetichin, G., Vallecalle, E., Fatehchand, R., Villegas, J., Laufer, M.: Glia-neuron interactions and adaptational mechanisms of the retina. In: Jung, R., Kornhuber, H. (Eds.): The Visual System Neurophysiology and Psychophysics, pp. 463—481. Berlin-Göttingen-Heidelberg: Springer 1961.

— Yagasaki, Y.: Resting and action potentials of single cone. Ann. Rep. Res. Inst. environm. Med. Nogoya Uni **1955**, 54—64.

Naka, K.: Receptive field mechanism in the vertebrate retina. Science **171**, 691—693 (1971).

— Kishida, K.: Recording of S- and spike potentials from the fish retina. Nature (Lond.) **214**, 1117—1118 (1967).

— Rushton, W. A. H.: S-potentials from color units in the retina of fish (Cyprinidae). J. Physiol. (Lond.) **185**, 536—555 (1966a).

— — An attempt to analyse colour reception by electrophysiology. J. Physiol. (Lond.) **185**, 556—586 (1966b).

— — S-potentials from luminosity units in the retina of fish (Cyprinidae). J. Physiol. (Lond.) **185**, 587—599 (1966c).

— — The generation and spread of S-potentials in fish (Cyprinidae). J. Physiol. (Lond.) **192**, 437—561 (1967).

Norton, A. L., Spekreyse, H., Wolbarsht, M., Wagner, H. G.: Receptive field organization of the S-potential. Science **160**, 1021—1022 (1968).

Oikawa, T., Ogawa, T., Motokawa, K.: The origin of the so-called cone action potential. J. Neurophysiol. **22**, 102—111 (1959).

Orlov, O. Yu, Maksimova, E. M.: S-potential sources as excitation pools. Vision Res. **5**, 573—582 (1965).

Rushton, W. A. H.: S-potentials in Proc. International School of Physics "Enrico Fermi", pp. 256—269. Reichardt, W. (Ed.). New York: Academic Press 1969.

Steinberg, R. H.: High-intensity effects on slow potentials and ganglion cell activity in the area centralis of cat retina. Vision Res. **9**, 333—350 (1969a).

— Rod and cone contributions to S-potentials from the cat retina. Vision Res. **9**, 1319—1329 (1969b).

— Rod-cone interaction in S-potentials from cat retina. Vision Res. **9**, 1331—1344 (1969c).

— The rod after — effect in S-potentials from cat retina. Vision Res. **9**, 1345—1355 (1969d).

Stell, W. K.: The structure and relationships of horizontal cells and photoreceptor — bipolar synaptic complexes in goldfish retina. Amer. J. Anat. **121**, 401—424 (1967).

Stretton, A. O. W., Kravitz, E. A.: Neuronal geometry: determination with a technique of intracellular dye injection. Science **162**, 132—134 (1968).

SVAETICHIN, G.: The cone action potential. Acta physiol. scand. **29** (Suppl. 106); 565—599 (1953).

— Spectral response curves of single cones. Acta physiol. scand. **39** (Suppl. 134), 18—46 (1956).

— Origin of the R-potential in the mammalian retina, in the Visual System. In: JUNG, R., KORNHUBER, H. (Eds.): Neurophysiology and Psychophysics, pp. 61—64. Berlin-Göttingen-Heidelberg: Springer 1961.

— MACNICHOL, E. F., Jr.: Retinal mechanisms for chromatic and achromatic vision. Ann. N.Y. Acad. Sci. **74**, 385—404 (1958).

TAMURA, T., NIWA, H.: Spectral sensitivity and color vision of fish as indicated by S-potential. Comp. Biochem. Physiol. **22**, 745—754 (1967).

TASAKI, K.: Some observations on the retinal potentials of the fish. Arch. Ital. Biol. **98**, 81—91 (1960).

TOMITA, T.: Electrophysiological study of the mechanisms subserving color coding in the fish retina. Cold Spr. Harb. Symp. quant. Biol. **30**, 559—566 (1965).

— KANEKO, T., MURAKAMI, M., PAUTLER, E. L.: Spectral response curves of single cones in the carp. Vision Res. **7**, 519—531 (1967).

— MURAKAMI, M., SATO, Y., HASHIMOTO, Y.: Further study on the origin of the so-called cone action potential (S-potential). Its histological determination. Jap. J. Physiol. **9**, 63—68 (1959).

— TOSAKA, T., WATANABE, K., SATO, Y.: The fish ERG in response to different types of illumination. Jap. J. Physiol. **8**, 41—50 (1958).

TOYODA, J., NOSAKI, H., TOMITA, T.: Light-induced resistance changes in single photoreceptors of Necturus and Gekko. Vision Res. **9**, 453—463 (1969).

TRIFONOV, YU. A.: Study of synaptic transmission between photoreceptors and horizontal cells by means of electric stimulation of the retina. Biophysics, Moscow **13**, N5 (1968).

WATANABE, K., HASHIMOTO, Y.: S-potential in light and dark adaptation in the live carp. Abstr. XXIII. Intern. Cong. Physiol. Sc. Tokyo 361 (1965).

— TOSAKA, T.: Functional organization of the Cyprinid fish retina as revealed by discriminative responses to spectral illumination. Jap. J. Physiol. **9**, 84—93 (1959).

— — YOKOTA, T.: Effects of extrinsic electric current on the cyprinid fish EIRG (S-potential). Jap. J. Physiol. **10**, 132—141 (1960).

WERBLIN, F.S., DOWLING, J. E.: Organization of the retina of the mudpuppy, Necturus maculosus. II. Intracellular recording. J. Neurophysiol. **32**, 339—355 (1969).

WITKOWSKY, P.: A comparison of ganglion cell and S-potential response properties in carp retina. J. Neurophysiol. **30**, 546—560 (1967).

YAMADA, E., ISHIKAWA, T.: The fine structure of the horizontal cells in some vertebrate retinae. Cold Spr. Harb. Symp. quant. Biol. **30**, 583—591 (1965).

Chapter 14

Receptive Fields of Retinal Ganglion Cells

By

W. R. LEVICK, Canberra (Australia)

With 14 Figures

Contents

Introduction

The concept of *receptive field* is important because it draws attention to the fact that the sensitivity of a neurone to photic stimuli is spread out over a region. The distribution of sensitivity is a functional sign that neural analysis of the retinal image has begun. The changing visual scene is not like "snow" on a television screen: it contains objects which are characterized by properties of local extension and local connectedness of various kinds. In order to detect their presence and progression, neurones must reflect at least some of the properties in their spatial distribution of sensitivity (referred to the receptor mosaic or to the external visual field).

It is therefore not surprising that the interaction of spatially distinct stimuli is a theme which unites a large proportion of the work which has been done on retinal ganglion cells. The story begins with the pioneering work of ADRIAN and

Matthews (1927, 1928a, b) on the excised eye and optic nerve of the eel. They recorded from the optic nerve which contained only 10000 fibres and hoped to produce physiological isolation of a single unit by reducing the size of the spot of light which stimulated the retina. Even with a retinal image of $10\,\mu$ diameter, the recording revealed the activity of several nerve fibres. Furthermore, they were able to show that increasing the area of the stimulus spot reduced the latency of the shower of spikes in the optic nerve. An increase in the frequency of the discharge might simply mean more fibres activated, but an effect on latency implies spatial interaction provided an optical explanation can be excluded. For example, if the image-forming components of the eye were so disturbed by the experimental procedures that the image of a "point" source was relatively spread out on the retina as a blurred disc, then small changes in the spatial dimensions of the "point" would cause changes mainly in the illumination of the disc, rather than in its size. In this case, latency decrease would be attributable to intensity changes without implying spatial interactions. Adrian and Matthews went to considerable trouble to ensure and check the optical quality of the retinal image and there seems no reason to doubt that the effect of stimulus area on latency was a physiological interaction, linking the effects produced by stimuli acting on spatially separated receptors. By stimulating the exposed retina with four symmetrically disposed spots of light, a separation could be found at which simultaneous flashing gave a latency no shorter than that for any spot flashed on its own; under such circumstances the application of strychnine to the retina could cause the latency for simultaneous flashing to become shorter than that for any individual spot. Since strychnine could hardly affect the amount of scattered light but would more likely act on the neural synapses, Adrian and Matthews found support for their thesis of physiological spatial interaction within the retina.

A direct demonstration of the nature and organization of spatial interactions followed when Hartline (1938, 1940a, b) succeeded in obtaining recordings from the axons of single ganglion cells in the frog's excised eye. The procedure involved removal of cornea, lens and vitreous so that an external optical system had to be substituted for the natural optics. The technique had the advantage that the quality of the retinal image could be directly checked through a dissecting microscope and the light scattered beyond the boundary of the geometrical image could be shown to be less than 1 % of the focal illumination of the image. Hartline confirmed the observations of Adrian and Matthews that the optic nerve response to light consisted of an outburst of impulses at the onset of light (on-response), followed by a subsiding discharge during the continued illumination and a further outburst when the light was turned off (off-response). He made the fundamental discovery that individual ganglion cells produced sharply different contributions to the total discharge (Fig. 1): 20 % of the fibres showed a high-frequency burst of impulses at light-on followed by a lower-frequency maintained discharge which stopped at light-off; 50 % showed bursts only at on and off; 30 % showed only a vigorous and prolonged discharge after the light was turned off. This discharge could be promptly suppressed by reillumination. Subsequently the epithets: "on units", "on-off units", "off-units" have been applied to these ganglion cell types. The responses of each type were shown to be graduated over very large ranges of stimulus intensity.

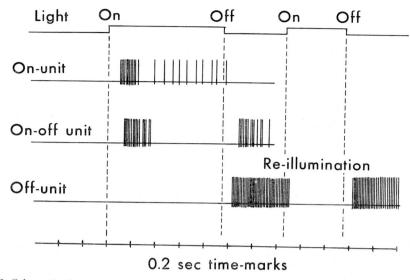

Fig. 1. Schematic diagram of three characteristic response types found in optic nerve fibres of the frog (based on Figs. 1 and 5 E of HARTLINE, 1938). The top line signals the light stimulus, the next three show the responses of an on-unit, on-off unit and off-unit, and the last line is the time-base. In the case of the off-unit, reillumination during the off-response produces prompt inhibition of the discharge

By exploring the retinal surface with a small spot of light HARTLINE could map the area from which responses were obtainable in a single fibre when the spot was turned on or off. This region he termed the *receptive field* of the fibre. The location was a permanent property of each fibre, but the spatial extent depended on the intensity and size of the exploring spot and the state of adaptation. Some of the dependence of receptive field size on spot luminance is attributable to stray light. With very bright spots the light which excites the ganglion cell may not be the focal light absorbed within the geometrically defined retinal image of the target, but the non-focal light incident on retina even well beyond the margin of the geometric image. There are many sources of such extraneous light: multiple reflections at surfaces of any optical system including the retina and choroid (especially if there is a tapetum), scatter in optical components (retina is a segment of a sphere, each part having an uninterrupted view of all other parts), optical aberrations and diffraction. For moderate intensities and small spots the receptive fields of most fibres were roughly circular with a diameter of the order of 1 mm. This dimension is very large compared with the size of receptors: frog rods are about 6 μ in cross-section (WOLKEN, 1961) and cones are smaller according to WALLS (1942); however, it roughly agrees with the spread of the dendritic trees of some ganglion cells (MATURANA et al., 1960). HARTLINE found that the response-type (on, on-off or off) remained unchanged at all points in the receptive field of a particular fibre. The sensitivity of the receptive field (defined as the reciprocal of the threshold for a small test spot) was greatest in its central portion and fell off steadily with

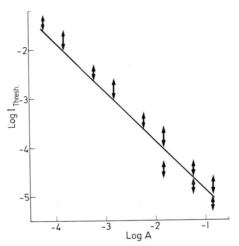

Fig. 2. Relation between area of a centred stimulus spot and the threshold intensity for activation of a single optic nerve fibre of the frog. For each arrow, upper point gives lowest intensity which elicited one or two impulses; lower point gives highest intensity which failed to elicit any response (determinations made to nearest 0.3 or 0.4 log unit). Where duplicate determinations coincided, arrows are drawn more heavily. Line drawn through points has slope of —1; this would imply that threshold is determined by a constant product of area and intensity. Log $I = 0$ equivalent to 3×10^5 lm/m²; area in mm². From Hartline (1940b)

increasing distance from the centre. Furthermore, over the central portion he was able to show that threshold intensity was inversely proportional to area of illumination (Fig. 2). Thus the ganglion cell was capable of combining subthreshold effects distributed over its receptive field.

Microdissection of the intraretinal bundles of axons was a difficult technique. Therefore Granit and Svaetichin (1939) made a major advance by showing that it was possible to record from individual units (later shown to be large ganglion cells, — Rushton, 1949) by means of glass-insulated, silver pins advanced onto the retinal surface after removal of cornea and lens. Since Granit (summarized in 1947, 1955) applied the technique mainly in the area of spectral sensitivity of ganglion cells, this work is considered elsewhere in this volume (see Abramov's article).

Modifications of the Receptive Field Concept

Silent Inhibitory Surround. Using Granit's technique of placing a microelectrode directly on the retina, Barlow (1953a, b) recorded from frog retinal ganglion cells and was able to confirm and extend Hartline's results. He repeated area-threshold measurements of the kind shown in Fig. 2 but with larger stimulus spots. When the stimuli exceeded the receptive field size, the threshold for off-units reached a constant level, whereas that of on-off units actually rose. It seemed as though regions outside the receptive field had inhibitory effects on activity elicited from within the receptive field. This was directly demonstrated by a two-

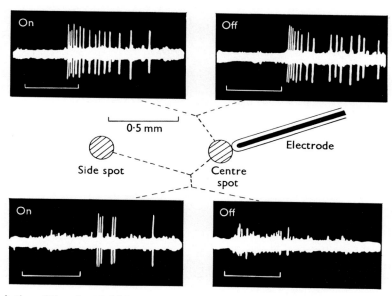

Fig. 3. Action of the silent inhibitory surround of an on-off ganglion cell in the frog's retina. Above: on- and off-responses from illumination of a centred spot alone. Below: simultaneous illumination of the centred spot and a side spot 1 mm distant in the inhibitory surround. The on-response is considerably reduced and the off-response completely abolished. Time marked on records, 0.2 sec. From BARLOW (1953b)

spot experiment: one spot was in the centre, the other was in the surround and both were turned on and off together (Fig. 3). Light-on in the surround reduced the response to light-on in the centre, and light-off in the surround abolished the response to light-off in the centre. The surround spot flashed on its own evoked no response from the ganglion cell. Silent inhibitory surrounds have also been observed in the frog by LETTVIN et al. (1959), GRÜSSER et al. (1964) and by BARLOW and LEVICK (1965) and LEVICK (1967) in connection with direction-selective and local-edge-detecting units in the rabbit.

Antagonistic Surround. HARTLINE's picture of a uniform response type throughout the receptive field did not hold for ganglion cells in the cat's retina. Preserving the cat's own natural optics, KUFFLER (1952, 1953) succeeded in recording from ganglion cells by introducing glass-ensheathed platinum-iridium wire (5–15 μ diameter) through a fine cannula puncturing the eyeball behind the limbus. He discovered that different response-types could be obtained in the same ganglion cell by moving the small testing spot to different positions. The receptive field could be subdivided into concentric zones: there was an approximately circular central zone of high sensitivity which yielded either on-responses or off-responses depending upon the particular cell; outside this was a concentric annular zone of lower sensitivity to the small spot which yielded the opposite type of response — off-responses if the centre had on-responses, on-responses if the centre had off-responses (Fig. 4). There was also a zone of overlap between the two regions in

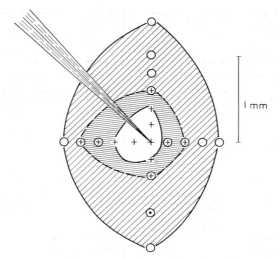

Fig. 4. Receptive field map of an on-centre ganglion cell in the cat's retina. Exploring spot was 0.2 mm in diameter, about 100 times threshold at the centre of the field. Background retinal illumination approximately 25 lm/m², uncorrected for transmission of the ocular media. Crosses indicate positions giving responses at "on", circles at "off". The field is composed of a central zone giving only on-responses, a peripheral annular zone giving only off-responses and an intermediate region of overlap yielding responses both at "on" and at "off". From Kuffler (1953)

which on-off-responses could be obtained. The arrangement of these zones has resulted in the descriptive term "concentric units" for such ganglion cells. Since the response-type of the central zone is a fixed property of the neurone whereas the surround may disappear at low levels of adaptation (Barlow et al., 1957; Barlow and Levick, 1969), the neurones have been designated "on-centre concentric" and "off-centre concentric" units. They are also described as belonging to the B (bright-activated)-system and D (dark-activated)-system respectively (Jung, 1964).

Kuffler demonstrated a striking interaction by simultaneous stimulation of the central zone and annular surround (Fig. 5). It appeared that if a spot yielded impulses at light-on, then it also evoked inhibition at light-off, because the response elicited by stimulation with another spot was diminished or abolished. In the same way, it was evident that a spot yielding impulses at light-off also evoked inhibition at light-on. The generalization that emerged was that the centre and surround are functionally antagonistic: whatever the effect of stimulation of the centre, the same stimulus applied to the surround has an opposing effect. This is the point of similarity with the silent inhibitory surrounds: both oppose activity elicited by stimulation of the central zone; however, the antagonistic surrounds can be made to generate impulses from the neurone, whereas the silent surrounds ordinarily cannot.

Silent and Inhibitory Receptive Fields. There are some ganglion cells which cannot be *activated* by a small spot turned on or off anywhere over the retina. In

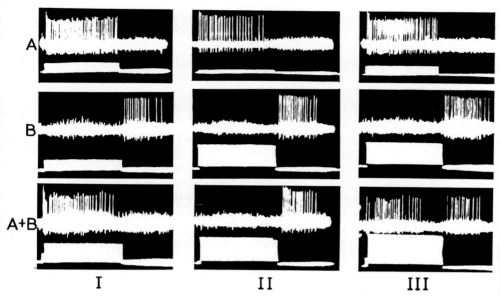

Fig. 5. Mutual antagonism between centre and surround of cat retinal ganglion cell. Spot A, 0.2 mm in diameter, was located in centre of the receptive field; spot B, 0.4 mm in diameter, was 0.6 mm distant in the surround. Flashed separately, they set up on-responses (A) and off-responses (B). With simultaneous flashing, $A + B$ in column I, the off-response was suppressed and at the same time the number of on-discharges in $A + B$ was slightly reduced as compared with A. In column II, the intensity of spot A was reduced, while that of spot B was increased (note flash strength indication on the second beam) resulting in elimination of the on-response. In column III, the intensities were adjusted to yield individually strong responses; simultaneous flashing caused reduction of both on- and off-responses. Flash durations were 0.33 sec, potentials 0.3 mV. From KUFFLER (1953)

this group are some of the "movement-gated, dark-convex-edge detectors" of LETTVIN et al. (1959) which respond only when the image of a small, round target, darker than the background, moves across a restricted area of the retina, but not when a small light spot is flashed (LETTVIN et al., 1961). It is nevertheless possible to map out the sensitive region of the retina by moving small dark targets. In the same category are the "uniformity detectors" present in the rabbit retina (LEVICK, 1967) and the "suppressed-by-contrast" ganglion cells found rarely in the cat retina (RODIECK, 1967). These units have a strong maintained discharge and all forms of stimulation including small spots of light turned on or off produce a suppression or decrease of the discharge, but never a true activation. Once the inhibitory property is recognized, it is a simple matter to map out the receptive field.

Periphery Effect. Another observation that must be accommodated in the definition of the receptive field is that cat retinal ganglion cells can be excited by special stimuli applied well beyond the limits of both the centre and the antagonistic surround (McILWAIN, 1964, 1966). The effect is typically elicited by a large target of coarsely patterned light and dark areas moved to and fro briskly, as distant as 45° from the centre of the conventional receptive field. The response consists of a slow, unsynchronized build-up of the maintained discharge, quite different from the crisp response to contrast targets moved through the centre of the conventional receptive field. The behaviour cannot be explained as an artifact from stray light (Levick et al., 1965). Whatever the significance of the periphery effect, it is certainly far less specific than is the conventional centre or antagonistic surround.

How should Hartline's definition of the receptive field be rephrased? An ideal statement would have to recognize the nature of the operation which the ganglion cell is doing upon the retinal image, give appropriate emphasis to the permanent as compared to the labile or non-specific features of the spatial distribution and also provide a suitable correspondence with an underlying structural basis. At the current state of knowledge, it is still necessary to proceed operationally: the receptive field is *that region of retina within which appropriate stimulation modifies the activity of the neurone, due emphasis being given to the specificity of the effects.* The usefulness of the concept of receptive field is not diminished by allowing it to occupy the whole retina (e.g. by including the periphery effect) because the sharpness of localization is preserved by the maximally sensitive centre and specificity is retained by the centre in cooperation with the surround.

The difficulties of providing a completely satisfactory definition of receptive field are evident from the variety of terms that have been employed in recent years: "responsive receptive field" (Maturana et al., 1960); "excitatory receptive field" (includes both centre and antagonistic surround) and "inhibitory receptive field" (includes only the silent surround) of Grüsser et al. (1964). Higher up the visual system, the problems are greater and the term "minimum response field" has been used (Barlow et al., 1967; Nikara et al., 1968).

Another subtle modification of the term receptive field has taken place with the development of experiments in which the animal's natural optics are preserved and maps are made on an external screen of the regions within which appropriate stimulation modifies the neurone's activity (e.g. Lettvin et al., 1959; Hubel and Wiesel, 1960, 1961; Barlow and Hill, 1963a; Grüsser et al., 1964; Rodieck and Stone, 1965a, and many others). Such maps are then regarded as the receptive fields. In making this identification, it should be recognized (and often is) that in general, the retinal image is a more or less degraded copy of the external target used to make the map. Whether this matters depends upon the nature of the result. For instance, it would be misleading to speak of areal summation if this were attributable to blurring of the images of small spots by optical imperfections, natural or induced. Receptive fields may be spuriously enlarged or incorrectly localized (discussed by Thomson, 1953) by light scattered from excessively bright images falling on intraocular structures, such as the myelinated fibre bands in the rabbit or the tapetum in the cat. Optical problems are discussed in detail elsewhere in this volume (see Westheimer's article).

Here, it is worth mentioning that overall refractive errors may be corrected by external spectacle lenses, spherical aberration may be controlled by avoiding pupillary dilatation or by using an artificial pupil, and the effects of scattered light may be minimized by supplying steady background illumination to the region involved (cf. Levick et al., 1965). Refractive errors are often assessed by retinoscopy (e.g. Hubel and Wiesel, 1960) or ophthalmoscopy (e.g., TALBOT and KUFFLER, 1952; BURNS et al., 1962; McILWAIN, 1964), but these methods all determine the plane conjugate to the scattering surface behind the receptors. If the receptors funnel light (RUSHTON, 1956), the effective image plane may lie in front by a distance equal to the length of the outer segments. The difference is negligible under ordinary circumstances, but may be important for eyes as small as the frog's or in the primate fovea where tiny receptive field centres are found (HUBEL and WIESEL, 1960) and the cone outer segments are relatively long (40 μ, POLYAK, 1941). BARLOW and LEVICK (1965) described an objective method for correcting the focus, based on the ability of receptive fields to resolve fine grating patterns. Useful approximate formulas for the out-of-focus error (E, dioptres) and angular diameter of an out-of-focus blur-circle (B, degrees) are:

$$E \simeq x/f^2$$
$$B \simeq 0.057 \, d\,E$$

where: $x =$ length of outer segments in microns,
               $f =$ posterior nodal distance in mm,
               $d =$ diameter of pupil in mm.

Since the cat has been widely used in receptive field studies the following background information is helpful: the cat's posterior nodal distance is 12.5 mm (VAKKUR et al., 1963); the retinal image of a thin line target has a full-width at half-maximum ranging from 8—16 min of arc when the eye is in best focus with a 6 mm round pupil (WESTHEIMER, 1962).

Variety of Receptive Fields

It has become abundantly clear that there are many distinct functional types of ganglion cell. The types vary from one species to another, some species having many, others few. Some types occur in several species. At present there is no satisfactory explanation of the diversity, just as there is no explanation why the number of optic nerve fibres per retina varies so much (e.g. a particular cat had 85926 – DONOVAN, 1967; rabbit averaged 265000 – BRUESCH and AREY, 1942; frog 440000 – MATURANA, 1959). A suggested generalization is that animals with substantial binocular visual fields have mainly or only ganglion cells which signal local contrast, the idea being that further elaboration ought to be postponed until information from the two eyes can be brought together (BISHOP, 1965). Another possibility is that the acquisition of a well-developed visual area in the neo-cortex relieves the need for extensive retinal pre-processing that a less flexible tectum might demand.

Frog. In a series of three papers LETTVIN et al. (1959, 1961; MATURANA et al., 1960) developed an experimentally-based interpretation of the behaviour of frog retinal ganglion cells that went far beyond the first simple picture described by HARTLINE (1938) or BARLOW (1953b). It is likely that they discovered a new class of unit (convex-edge-detectors), but the principal point of departure was the way in which they interpreted their results in terms of *operations* by which the neurones tested stimuli applied to their receptive fields. Thus they describe "edge-detectors", rather than "on-off response types". Their results, particularly in relation to the convex-edge-detectors, have been substantially confirmed by GRÜSSER-CORNEHLS et al. (1963) and GRÜSSER et al. (1964), with only minor differences.

HARTLINE's on-units became "sustained-edge-detectors" in the scheme of LETTVIN et al. The correspondence was established by observing (as HARTLINE did) that the unit responded to flashing a small light spot with a burst of impulses at "on" followed by a lower frequency maintained discharge for the duration of illumination. The justification for edge-detection lay in the further observations: no response to flashing the background (maximum flux increment on receptive field, but no edge present); target darker than the background shifted into the receptive field elicited the burst plus maintained discharge (decrement of flux but increment of edge, sustained). Some of these units were also selective for the direction of movement of targets. The weaker responses to large targets suggested that there may be a silent inhibitory surround. Keating and Gaze (1970) criticized the identification of this class with Hartline's on-units because of the work of Muntz (1962) who recorded units (probably optic tract endings) in the diencephalic nuclei which behaved just as did Hartline's on-units but which had none of the edge-detecting features described above. Instead, the units were differentially responsive to light in the blue part of the spectrum. It is not excluded that the on-units may be a heterogeneous collection.

Lettvin's second type, the "movement-gated, dark-convex-edge-detectors" had no obvious counterpart amongst Hartline's types. In seeking an explanation for the discrepancy, one may note that the retina was separated from the brain in the work of Hartline and Barlow, thus excluding activity in possibly existing centrifugal efferent fibres (Maturana, 1958). Alternatively, the range of stimuli employed may not have encompassed the very specific requirements discovered by Lettvin et al.; most of the units responded only to a small target, darker than the background, moved into the receptive field in a direction towards its centre; furthermore, the edge presented within the receptive field had to be curved with the dark part of the target on the inside of the curve. These are not off-units because turning off the background illumination gave no response. The description "movement-gated" is necessary because if the proper target was placed in the receptive field in darkness and then revealed by turning on the background, there was no response, even though the same target, moved into the same position with the light on, evoked a strong and sustained discharge. This response also had a related property called "erasability": the sustained response could be abolished by turning off the background light and did not reappear with reillumination; thus, when the frog blinks his eye, he wipes clean not only the cornea but also the current picture transmitted by members of this class of unit.

The above designations have not gone unchallenged. Keating and Gaze (1970) questioned the validity of criteria used to distinguish between "dark-convex-edge-detectors" and "sustained-edge-detectors" mainly on the basis of an experiment reported by Gaze and Jacobson (1963). The latter suggested that the lack of response to the long, straight border of a dark target could be attributed to the fact that the continuation of the border passed over the silent inhibitory surround of the receptive field. A mask with a hole was placed close to the object plane so as to screen off the view of the parts of the target traversing the surround; the response to the straight-bordered target was restored (Fig. 6). However, the demonstration as it stands is not entirely convincing: consider the scene that the receptive field observes; does it lack convex borders? No; where the straight edge meets the edges of the hole, the dark region is formed into two very sharp corners which move around the boundary of the receptive field as the straight edge advances. Clearly, additional controls are required. Maturana et al. (1960) had already covered the objection indirectly: a response to a long straight border appeared

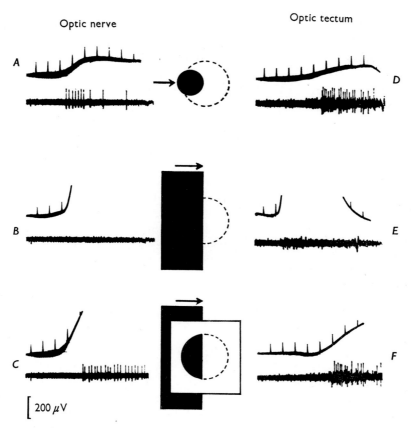

Optic nerve　　　　Optic tectum

Fig. 6. Test of "convexity-detection" by a class of retinal ganglion cells in the frog. Responses of single unit recorded either in the optic nerve (left) or optic tectum (right) on lower trace and output of a phototransistor monitoring the scene on the upper trace with 100 msec time marks. *A* and *D:* moderate response to small, black 2° disc moved through receptive field (dashed circle 5° diameter). *B* and *E:* no response to movement of a large, black, straight-edged figure across the receptive field. *C* and *F:* stationary mask with a 5° hole placed so as to hide events occurring in the inhibitory surround restored response to straight-edged figure. For further discussion see text. From GAZE and JACOBSON (1963)

only if there was a small dark angle projecting from it. GRÜSSER-CORNEHLS et al. (1963) made out a case that "change-of-position" rather than movement was the true operation carried out by the convex-edge-detectors as well as the sustained-edge-detectors above and the changing-contrast-detectors described below. It is hard to see the value of this distinction since the first two types show continued responses if the target is left *stationary* in the centre of the receptive field. Furthermore, LETTVIN et al. only claimed that the second class was gated (i.e. turned on) by movement.

These units have a silent inhibitory surround (GRÜSSER et al., 1964) which may be five times the diameter of the central, excitatory part of the receptive field.

Inhibition is evoked by dot patterns moved in the surround, but the direction relative to that of a target moved through the centre did not significantly modify the inhibitory effect. The same number of dots distributed uniformly in the surround were more effective than when grouped in restricted parts of the surround. By using much larger targets Keating and Gaze (1970) showed that the inhibitory surround was even more widespread, extending out to 45° from the centre of the receptive field. Also, the inhibitory effects in different cells could be sustained (persisting while a large target remained in the surround) or transient (occurring both at the moment of appearance of the target as well as at its withdrawal).

The on-off units of Hartline and Barlow became "moving-edge-detectors" in the scheme of Lettvin et al. (1959), later revised to "changing-contrast-detectors" (Maturana et al., 1960). A transient burst of impulses occurred whenever a constrasting border appeared, disappeared or was moved in the receptive field. Hartline (1940a) had already noted the great sensitivity of these units to movement of targets in the receptive field. Barlow (1953b) showed that this result was not merely the effect of moving the spot from a region of one sensitivity to a region of different sensitivity because even within a zone of constant sensitivity, movement elicited impulses. Thus, the behaviour could not be interpreted simply as a response to change of total effective flux, since the total remained unchanged throughout the movement. Wagner and Wolbarsht (1958) reached a similar conclusion from a slightly different experiment. As mentioned earlier, these units also have a silent inhibitory surround (Barlow, 1953b).

The off-units of Hartline and Barlow are the "dimming-detectors" of Lettvin et al. They simply respond to a weighted summation of the amount of darkening over the rather large receptive field with a shower of impulses, the rate of which gradually declines with a time constant measured in minutes or longer. The initial part of the discharge may show grouping of impulses into rhythmic bursts at 20/sec which are synchronous over the retina for the first few seconds. Hartline (1938) was struck by the speed and completeness with which the off-discharge could be suppressed by reillumination; indeed the latency of the suppression was shorter than the latency of the on-discharges in other fibres simultaneously recorded (Fig. 1). The observations of Keating and Gaze (1970) have raised the possibility that the dimming detectors are also a mixed collection. They found that many could be excited by stimuli applied over very wide areas which considerably exceeded the region yielding responses to moving black discs and which sometimes included the entire visual field. There appeared to be discrete regions of on-responses as well as off-responses arranged in different patterns from unit to unit. While stray light might account for the widespread distribution of responses, it could hardly be responsible for the presence of distinct on-responses.

Maturana et al. (1960) added a rare fifth class of unit called "dark-detectors" which differed from the off-units by having a very large vague receptive field, a maintained discharge at all levels of illumination inversely related to luminance, and sluggish responses to sharp changes of illumination or target movement.

Cat. In contrast with the frog, feline ganglion cells are of only two main types: on-centre and off-centre, as already described. Within each type, however, there is considerable quantitative variation with respect to the size of the central zone. By

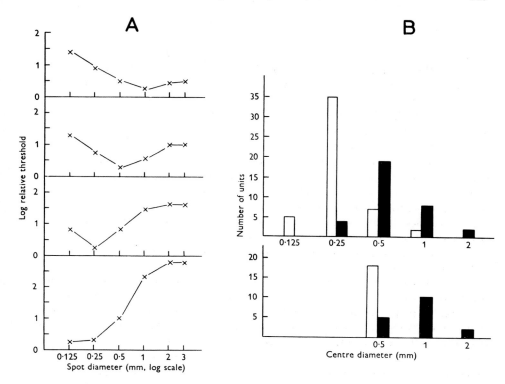

Fig. 7. *A Area-threshold curves* for four different on-centre ganglion cells in the cat's retina at a background retinal illumination of 2.2 lm/m² (uncorrected for transmission of the ocular media). For centred spots of different diameters superimposed on the background and flashed for 0.8 sec, the threshold was measured as the lowest intensity which produced a clear and repeatable change in the maintained discharge. Threshold intensities are plotted on a logarithmic vertical scale such that 0.0 corresponds to 2.2×10^{-2} lm/m². The spot size at which threshold was minimum was taken as the size of the receptive field's centre. *B Size-distribution of receptive field centres*. White columns indicate cells recorded in the area centralis; black columns, cells from the retinal periphery. Upper: eighty-two ganglion cells recorded with micropipettes, 49 from area centralis and 33 from periphery; lower: thirty-five cells recorded with platinum electrodes, 18 from area centralis and 17 from periphery. Note that the area centralis has smaller field centres than the periphery; also, the coarser platinum electrodes missed cells having the smallest centres. From WIESEL (1960)

taking the size of a centred spot of light which gave the minimum increment threshold (Fig. 7a), WIESEL (1960) found the receptive field centres varied from 0.5—8° (Fig. 7b) or approximately 0.125 to 2 mm on the retina. In the region of greatest ganglion cell density (STONE, 1965), or *area centralis*, the smaller receptive field centres preponderated. Fig. 7b also shows that the nature of the result depends upon the type of recording electrode used: the coarser platinum electrodes which do not penetrate the inner limiting membrane miss the cells with small receptive field centres, presumably because the neurones are smaller if we accept the argument of BROWN and MAJOR (1966).

The area-threshold curves of Fig. 7a provide additional information. Note how the threshold rose when the spot was enlarged beyond the receptive field centre; this is another way of demonstrating the antagonistic surround (cf. Fig. 5). If it is accepted that the approximate outer limit of the surround is reached when threshold shows little further change, then the surrounds have relatively much the same peripheral extent (6—12°, 1.5—3 mm on the retina approximately) regardless of the centre size. This arrangement of the antagonstic surround has a major influence on the relative sensitivities of different ganglion cells to the size of targets. Fig. 7a shows that the threshold *luminance* for targets of optimum size is much the same regardless of the particular centre size. However, the small-centred cell is more than 10 times more sensitive to a 0.5° target than the large-centred cell. Under the same circumstances, the latter is 100 times more sensitive to a 4° target than the former, a remarkable selectivity.

The centre-surround type of organization depends upon the general level of illumination and its recent past history. During adaptation to darkness, there comes a time (as much as 2 h delay if the preceding illumination was very bright) when the threshold for the surround response becomes high or unobtainable (Barlow et al., 1957). Under these circumstances, the receptive field organization is simplified to a single response type (on or off, depending upon the original centre) spread over a patch as large as or slightly larger than the original centre. This change is not associated with the changeover from cone-dominated to rod-dominated performance which also occurs during dark-adaptation but at a rather earlier time. The disappearance of the inhibitory surround occurs in both on-centre and off-centre cells. If, starting from darkness, the background is increased allowing time for retinal equilibrium to be reached, the antagonistic surround emerges at about 3×10^{-2} cat trolands (i.e. field luminance in cd/m² times pupil area in sq. mm) (Barlow and Levick, 1969) whereas the changeover from rods to cones occurs at a much higher level: about 25 scotopic cat trolands according to Barlow and Levick (1968), 125 photopic cat trolands (Daw and Pearlman, 1969) or even higher (Andrews and Hammond, 1970).

The antagonistic surround organization is a sufficient basis to account for the perceptual effect known as simultaneous contrast (Barlow et al., 1957; Baumgartner and Hakas, 1962). A grey disc located on a dark background looks distinctly brighter than an identical grey disc on a white background. The receptive field organization causes the ganglion cells to respond to the *difference* in the illuminations falling on the centre and on the surround, rather than merely to the absolute flux on the centre. For an on-centre cell, an increase in surround illumination with a constant intensity on the centre decreases firing. An off-centre cell would be activated. On decreasing surround intensity, the on-centre cell would be activated, the off-centre cell depressed. If it is assumed that activation of on-centre units and depression of off-centre units give rise to the sensation "lighter", and *vice versa* for "darker" and if it is further assumed that these sensations are referred to the place in the visual field corresponding to the centres of the receptive fields, then indeed, one grey disc will look lighter than the other. Corresponding responses are found in behavioural experiments on cats (Grüsser and Snigula, 1968).

There is another class of perceptual experiences known as Mach bands (extensively discussed by Ratliff, 1965) which seem to be closely related to the

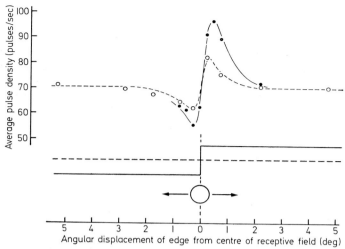

Fig. 8. Neurophysiological parallel of Mach bands. Starting from a uniformly illuminated field, a contrast-edge was suddenly established by raising the luminance on one side of the field and simultaneously lowering it on the other side, the two regions meeting at a straight border which could be located at various positions across the receptive field (centre only shown) of a cat retinal ganglion cell (on-centre X-type, cell No. 68 of Fig. 13). The average discharge frequency ("pulse density") over the period 10—20 sec after introduction of the edge was measured for each edge position. Filled circles are for a stimulus of contrast 0.4 (bright side 1.4 times the luminance of uniform field, dark side 0.6 times that luminance); open circles for a contrast of 0.2. Note that when the centre lies just within the darker side of the edge the discharge is below the frequency corresponding to a position further into the darker area. Just within the lighter side, the frequency is elevated above the level it would be well within the lighter area. From ENROTH-CUGELL and ROBSON (1966)

antagonistic surround organization. The classical viewing situation (MACH, 1865) consists of a field of lower luminance joined to a field of higher luminance by a zone where luminance increases at a constant rate with distance. Where the ramp begins, there is a perceptual impression of a band darker than regions on either side; at the end of the ramp a band lighter than adjacent regions appears. MACH attributed these effects to retinal interactions which produced a sensitivity to the negative of the second derivative of luminance with respect to distance on the retina. RATLIFF discusses the various models proposed to explain the sensitivity; they all have in effect an antagonistic surround organization. Neurophysiologically, most work has been done in a rather special limiting condition where the ramp is very steep and there is a sharp step from lower luminance to higher (BAUMGARTNER, 1961; BAUM-GARTNER and HAKAS, 1962; BAUMGARTNER et al., 1965). In this case, the bands are very close to each other, the brighter band being on the side of higher lumin-ance, the darker on the darker side. The corresponding behaviour of a retinal ganglion cell is shown in Fig. 8 (ENROTH-CUGELL and ROBSON, 1966). The Her-mann illusion is also considered to arise from the centre-surround organization (BAUMGARTNER, 1960).

   Quite apart from the range of centre-sizes, the two classes: on-centre and off-centre are inhomogeneous in another way. It has been known for some time that

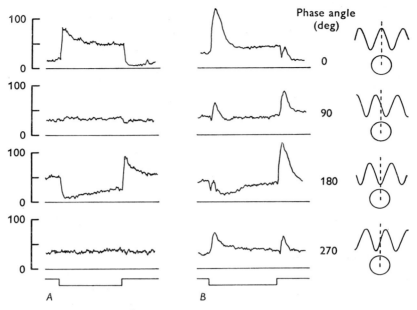

Fig. 9. Differentiation of X-type ganglion cells from Y-type in the cat's retina. The stimulus is a stationary pattern of lighter and darker stripes such that the luminance is a sinusoidal function of distance across the field. The picture is switched periodically (0.45 c/s) between the above stimulus and a uniformly illuminated field of the same mean luminance. The peak-to-peak modulation depth of the pattern was 64% of the mean luminance (contrast = 0.32). Responses of an off-centre X-cell are shown in column A, those of an off-centre Y-cell in B. The upper trace of each record is the pulse density of ganglion cell discharge (scale at left: pulses/sec); the length of the zero line represents a duration of 2 sec. The lowest trace in A and B gives stimulus timing: upward deflexion indicates introduction of the pattern. The parameter varied is the position of the pattern expressed as a "phase angle": 0° and 360° correspond with a peak of luminance centred on the field, and each 90° increase in phase angle represents a displacement of the pattern (to the left in the diagram) by 1/4 of a cycle. With the X-cell, positions can be found (90°, 270°) where there is essentially no change with introduction or withdrawal of the pattern; no such positions exist for Y-cells. From ENROTH-CUGELL and ROBSON (1966)

there are two subdivisions of off-centre units. BAUMGARTNER (1961) showed that most of the neurones which responded to diffuse illumination both at on and off had off-centres. The on-response, late and preceded by inhibition, arose by illumination of the surround. Thus, BAUMGARTNER subdivided off-centre neurones into pure off-neurones and on-off-neurones with pre-excitatory inhibition at on. Similarly, on-centre neurones included pure on-neurones and a rarely encountered group of on-off-neurones with pre-excitatory inhibition at off. Although detailed comparisons are lacking, it seems possible that the on-off neurones are the Y-cells of ENROTH-CUGELL and ROBSON (1966) and the pure on- and pure off-neurones are the X-cells. The essential feature of X-cell performance is that it was always possible to find an axis of symmetry dividing the receptive field such that the effects of application of a stimulus pattern on one side could be exactly balanced (no change in maintained discharge) by simultaneously presenting the spatially complementary pattern on the other side. It was never possible to find such an axis with Y-cells (Fig. 9). The patterns included spatial sinusoidal modulations of the mean luminance level as well as spatial rectangular step modulation, and stimulation consisted of abrupt transitions between the pattern and a uniform field of the same mean luminance. The implication is that processing (transduction plus combination) of light signals is effectively linear within the X-type organization, but obviously non-linear within the Y-type. Further discussion is deferred until later. X-cells may be either on-centre or off-centre, and the same holds for Y-cells.

Other types of receptive field of rare occurrence have been reported in the cat retina. When the instances are few, it is always difficult to be certain that the behaviour is not simply an unusual example of the more commonly occurring concentric organization, recorded under unfavourable conditions. Nevertheless, there seems little doubt about the on-off direction-selective unit described by STONE and FABIAN (1966) and the "suppressed-by-contrast" type of RODIECK (1967). The discovery by DAW and PEARLMAN (1970) of rarely occurring geniculate cells with differential sensitivity to blue light at photopic background levels also implies segregation of different cone inputs at least to some retinal ganglion cells which should result in functional differences in their receptive fields as well.

Monkey. In the spider monkey, receptive fields of retinal ganglion cells have the concentric organization found in the cat (HUBEL and WIESEL, 1960): mapped with spots of white light, the fields had sharply demarcated on-centres with antagonistic off-surrounds, or the reverse. The smallest centres (down to 4 min arc) occurred near the fovea and with considerable scatter, the centre sizes increased with eccentricity. Colour-specific responses were found in only 3 of 100 cells.

The situation is different in the *rhesus* monkey retina. WIESEL and HUBEL (1966) showed that lateral geniculate neurones in this species generally possessed concentrically organized receptive fields with on-centres and antagonistic off-surrounds or the reverse; in the majority there were sharp differences in the spectral sensitivity of centre and surround components. Opponent-colour responses had been earlier described by DEVALOIS (1960). It would be expected that the same basic organizational patterns would be found in the receptive fields at the retinal ganglion cell level. Confirmatory evidence may be found in a series of papers by GOURAS who recorded directly from the cell bodies in the retina. Thus, increasing the size of a stimulus spot beyond a critical value led to reduction of response (GOURAS and LINK, 1966) or elevation of threshold (GOURAS, 1967). The majority of on-centre ganglion cells received excitatory signals from only one cone mechanism, either blue-, green- or red-sensitive, in the centre, and inhibition from another cone mechanism in the surround of their receptive fields (GOURAS, 1968). Their responses were clearly coding colour. The balance of on-centre cells received signals from both green- and red-sensitive cone mechanisms, both of which excited in the centre and inhibited in the surround. The responses of these neurones did not appear to encode colour specifically. No information was given for off-centre cells. However, it was shown that the colour-coded cells had well-sustained responses to standing stimuli (tonic type) whereas the others responded only transiently under the same conditions (phasic type). Tonic cells were relatively more frequent toward the fovea and their axons conducted at slower speeds than those of the phasic cells (GOURAS, 1969). Responses to coloured light are discussed in more detail elsewhere in this volume (see ABRAMOV's article).

Rabbit. A majority of the ganglion cells in this animal are concentrically organized like those of the cat, with on-centres and antagonistic off-surrounds or the reverse (BARLOW et al., 1964). Sometimes, the surround response could not be evoked unless a light spot was left steadily on the centre. It has not yet been established whether this method of bringing out the surround works by desensitizing the centre to light scattered from an annular stimulus or whether there is actually a genuine sensitization of the surround.

In addition to concentric units, there is a sizeable group of cells specifically sensitive to the direction in which targets move across their receptive fields (BARLOW and HILL, 1963a). Typical behaviour is shown in Fig. 10 (BARLOW et al., 1964). Mapped with a small light spot, the receptive field yielded on- and off-responses at all points without any obvious and consistent suggestion of asymmetry. Nevertheless, there was a path through the receptive field which yielded a powerful response for movement of a light spot in one direction ("preferred") and zero response for movement in the opposite direction ("null"). Intermediate direc-

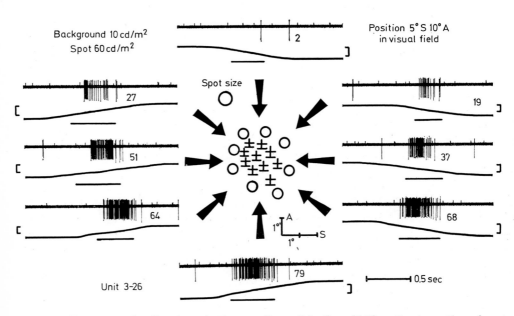

Fig. 10. Responses of a direction-selective ganglion cell in the rabbit's retina to motion of a small spot of light in different directions. A map of the receptive field is shown in the centre. Symbols: ±, response to stationary flashed spot at both "on" and "off"; O, no response; there were no responses outside the ring of O's. Anterior (*A*) and superior (*S*) directions in the visual field are shown together with 1° calibration marks. Each pair of records shows (lower trace) movement of a spot of light (size indicated near map) right through the receptive field in the direction of the adjacent arrow, and (upper trace) the response elicited; the number of spikes is shown immediately after each response. For the lower trace, the vertical calibration bar shows 5° displacement of light spot; horizontal bar indicates approximately when the spot was within the receptive field. From BARLOW et al. (1964)

tions yielded responses of intermediate strength. Spots darker than the background also yielded the same pattern showing that the selectivity for direction was independent of contrast. This selectivity persisted (through response magnitudes changed) despite substantial variations of background illumination as well as size, shape and velocity of the target. The same selectivity for direction could be obtained for small displacements of a target anywhere in the receptive field and the minimum amount of receptive field required for a selective response was a small fraction of the size of the whole receptive field (BARLOW and LEVICK, 1965). It is as though the selective operation is elaborated locally and extensively replicated throughout the receptive field.

Further evidence suggested that the operation was achieved by an inhibition spreading locally in the null direction from any stimulated point. Morphological elements which could fulfil the requirements of the operation are horizontal cells of the type which have a dendritic tree contacting local receptors and a long axon-like process making connections with a laterally displaced region (CAJAL, 1955; Figs. 192—194). To account for the fact that a target crossing the receptive field in the null direction elicited no impulses, the distant connections would have to be on bipolar cells. Such connections have been observed by DOWLING et al. (1966) in the rabbit and cat. However, horizontal cells are not the only elements that could perform the task. The multiplicity of form of amacrine cells (BOYCOTT and DOWLING, 1969) together with the profusion of their synaptic contacts (DOWLING and BOYCOTT, 1965, 1966; RAVIOLA and RAVIOLA, 1967) would suggest that the operation might be achieved in the inner plexiform layer. Direction-selective neurones also have a silent inhibitory surround, but the nature and mechanism have not been examined in detail.

The preferred directions of different on-off direction-selective neurones fall into 4 non-overlapping groups (OYSTER and BARLOW, 1967; OYSTER, 1968) which seem to correspond to the directions of apparent object displacement produced by contractions of the four rectus muscles. This led to the suggestion that the outputs of each group of neurones could, without further processing, provide the error signal for a visual servo-system minimizing retinal image motion. Units selecting all four directions may be found in all parts of the retina; thus, unambiguous signalling of the direction of motion of a target is possible throughout the visual field. If units of this kind are the basis for the visual perception of direction of movement, the question arises as to whether they show after-effects of stimulation of a type which might account for the subjective after-effect known as the "waterfall illusion" (WOHLGEMUTH, 1911). If one gazes fixedly at a scene showing sustained localized movement of contours in one direction (e.g. distant waterfall) for about 1/2 min, and then abruptly diverts the gaze to a non-moving part of the scene, there is a definite impression of apparent movement in the opposite direction at a much slower speed within the new region of gaze; it lasts for several seconds at least. With direction-selective neurones, sustained stimulation in the preferred direction caused sustained activation which was followed at the end of stimulation by a period of depressed firing below the maintained level in the absence of stimulation. Similar stimulation in the null direction had no effects on the maintained discharge either during or afterwards. Thus the after-effect would be identified with an imbalance in the maintained discharges of neurons signalling opposite directions of motion (Barlow and Hill, 1963b). The evidence is rather against a retinal location of such an effect in man (Barlow and Brindley, 1963).

Other less commonly occurring classes are found in the rabbit retina. These include a distinctly different type of direction-selective unit, with special sensitivity to slowly moving targets lighter than the background and on-type receptive field. Only three preferred directions seemed to be represented, two of which were rather different from those of the on-off type. "Large-field" units, characterized by brisk responses to rapid movement of targets brighter or darker than the background and by vigorous bursts to abrupt dimming of ambient illumination might be considered to be specializations of certain off-centre and on-centre concentric units with exaggerated Y-type behaviour (see discussion on the *cat*).

Three other classes have been investigated in detail (Levick, 1967): orientation-selective units, local-edge-detectors and uniformity-detectors. Units of the first class responded to small targets in much the same way as concentric units;

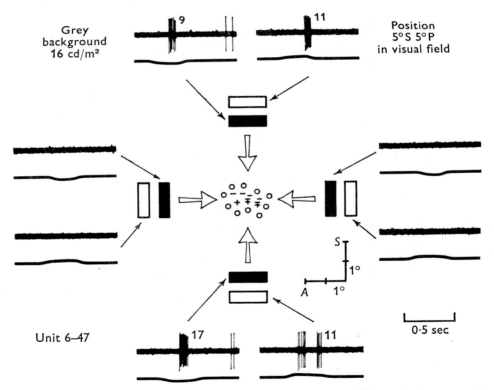

Fig. 11. Responses of an orientation-selective ganglion cell of the rabbit's retina to movement of a rectangular target either darker or lighter than the background across the receptive field in different orientations. Map of receptive field in centre. Symbols: +, response to stationary spot at "on"; —, at "off"; ±, at both "on" and "off"; ○, no response; other conventions as in Fig. 10. Each pair of records shows (upper trace) the response elicited by target movement signalled (lower trace) by the output of a photomultiplier focussed on the receptive field (upward deflexion, increased light). Only horizontally orientated targets yielded responses.
From Levick (1967)

however, testing with extended rectangular targets revealed sharply different behaviour (Fig. 11): maximum responses required a particular orientation of the rectangle and no response arose from an orientation at right-angles. Analysis showed that these units were organized like the concentric units except that the antagonistic surround was missing at opposite ends of a diameter of the receptive field corresponding to the optimum orientation.

Optimum orientations of different units were either horizontal or vertical. If the rabbit used only the output of this class of unit to determine the orientation of extended targets, then it should be confused by targets orientated at 45°, and 135° to the horizontal. In fact, the rabbit's ability to discriminate orientation is much better (VAN HOF, 1966); hence the function of the retinal orientation-detectors is obscure. The local-edge-detectors have much in common with the "changing-contrast-detectors" (MATURANA et al., 1960) of the frog. The argument for edge-detection was based on the behaviour of the unit to patterns which presented the same length of edge but very different quantities of light flux or very different lengths of edge but much the same quantity of flux. The responses more closely followed the amount of edge present, weighted by the contrast across it. These units had a silent inhibitory surround which was brought into action by the presence or movement of an edge; hence, the qualification "local" was required. Uniformity-detectors had the remarkable property that all forms of stimulation produced only a diminution or cessation of their brisk maintained discharge. This was the case for spots of all sizes flashed on or off, targets lighter or darker than the background and coarse or fine grating patterns moved into or out of the receptive field. Units with similar properties in the cat retina have been called "suppressed-by-contrast-types" (RODIECK, 1967).

Other Species. In the *pigeon*, MATURANA (1962) and MATURANA and FRENK (1963) differentiated six types of ganglion cells: verticality-detectors, horizontality-detectors, general edge-detectors, directional moving-edge-detectors, convex-edge-detectors and luminosity-detectors. Some of these units have points in common with corresponding units in the frog and rabbit, but the orientation-sensitive units may well be modified edge-detectors rather than modified concentric units as in the rabbit. Additional observations have been reported by HOLDEN (1969).

In the *rat*, BROWN and ROJAS (1965) found concentrically organized receptive fields (on-centre with off-surround and vice versa) which closely resembled those of the cat. In addition, there were others which lacked surrounds. Detailed search did not reveal any more sophisticated types.

The mexican ground *squirrel (Citellus mexicanus)* is exceptional in possessing a retina containing only cones (DOWLING, 1964). MICHAEL (1966a, b, 1968a, b, c) showed that it had concentrically organized on-centre and off-centre ganglion cells like the cat, direction-selective cells like the rabbit and colour-coded cells which are discussed elsewhere in this volume (see ABRAMOV's article).

According to JACOBSON and GAZE (1964) the receptive fields of retinal ganglion cells in the *goldfish* may be classified into on-centre units with inhibitory surround, on-off units, on-units and off-units. The first three of these classes were occasionally direction-selective. In the on-off units, regions of on-and off-responses were mostly separate from each other and arranged side by side rather than concentrically. Since the fish was out of water in these experiments, there is some uncertainty about the quality of the retinal image. It seems that some of the on-off units are coded for colour (Wagner et al., 1960, 1963) and this aspect is discussed elsewhere in this volume (see Abramov's article).

Some Quantitative Analyses of Retinal Ganglion Cell Function

The stage is now set for discussion of quantitative aspects of ganglion cell performance. The analysis of colour-coding (see Abramov's article) and the problems of adaptation are taken up in other chapters. In this section some analyses of spatial properties are discussed.

Preliminaries. The concept of a linear system is helpful in understanding some of the work to be described. Without delving into the many subtleties, a linear system is one for which the *principle of superposition* holds. That is to say, if one arbitrary stimulus I_1 leads to a response R_1 and another stimulus I_2 leads to R_2, then the response to $I_1 + I_2$ is the same as $R_1 + R_2$ for all I_1 and I_2. There is no interaction or interference between responses to different stimuli. The significance of linear systems is that exact solutions of their behaviour can readily be calculated by standard techniques whereas each non-linear system requires special individual treatment. Linear analysis may often be used to treat non-linear systems over small ranges of input.

In electrical engineering usage, stimuli and responses are commonly encountered as functions of time, but they may just as easily be functions of space as for example in discussing image formation by an optical system such as the eye (Westheimer, 1964, 1965), or they may be functions of both time and space. The relations are best discussed with the aid of a diagram (Fig. 12). The schema in the middle represents a hypothetical system consisting of a spatially distributed light stimulus at the top (point of light superimposed on a uniform background of illumination) which is transduced and operated on by the retinal network to produce the output ("impulse response") shown at the bottom. The output can equivalently be considered as (i) the spatial distribution of output activity in a set of identical ganglion cells spread over the field or (ii) the output of a single ganglion cell as a function of the position of the stimulus in relation to the field. It has to be mentioned that whereas the stimulus can be represented as a continuous function in space, the retinal operation is handled by discrete elements, so the spatially distributed response at the bottom should properly appear as a set of discrete samples of the continuous curve. An arbitrary stimulus may be represented by a set of impulses whose weights correspond with the amplitude of the stimulus at each point. By superposition, the response may be synthesized by adding up the set of corresponding impulse responses, appropriately scaled, signed and located. This process is known mathematically as a "convolution" of the input waveform with the impulse response.

In the general case, responses are functions of time as well so that there is a different $R(x, y)$ for each instant after the application of a stimulus. The left-hand schema illustrates the relations in the time domain at a particular position. The response to a rectangular step stimulus is shown as well as that to an impulse because it permits some short cuts when considering bar-shaped stimuli, but the general principle for calculating the response to an arbitrary stimulus remains unchanged: the stimulus is represented by a temporal succession of impulses at each place in the field and the appropriately timed, scaled, signed and located impulse responses are all added up to determine the time course of ganglion cell output. For further details, Rodieck's (1965) paper should be consulted.

If the parameters of a linear system remain constant, the response to sinusoidal stimulation is also sinusoidal of the same frequency; no new frequencies are generated. Only relative amplitude and phase are changed. Since an arbitrary stimulus waveform may be represented by the sum of a number of sinusoids of appropriate frequencies, amplitudes and phases (Fourier transformation), the principle of superposition enables us to synthesize the response from a knowledge of the relative amplitude $R(f)$ and phase changes $\varphi(f)$ imposed by the system on the individual sinusoidal components (lower two diagrams on right of Fig. 12). The resultant

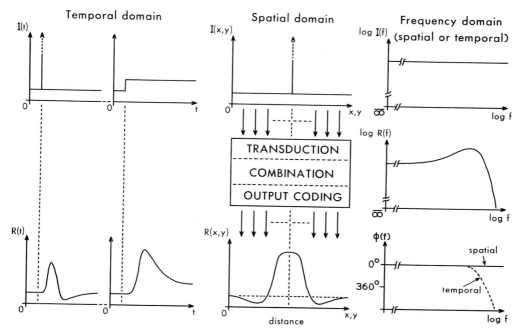

Fig. 12. Schema for the application of linear-system analysis to the behaviour of a retinal ganglion cell. An input stimulus (top) is represented as a spatial distribution of light intensity, $I(x, y)$ in two spatial coordinates, x, y (middle column); at each point, the intensity is a function $I(t)$ of time, t (left-hand column); alternative representations in terms of spatial frequency components and temporal frequency components $I(f)$ as functions of spatial and temporal frequency, f are shown on the right. For purposes of analysis, the retinal operation on the input is often separated into three stages (centre); non-linearities may involve one or more stages, but there are examples of how linear analysis may circumvent them by good experimental design. Assuming all the processes are linear, the responses $R(x, y)$ and $R(t)$ (bottom) to point function inputs are important elements in the calculation of output from a general input. In the frequency domain, the spatial and temporal modulation transfer functions $R(f)$ in conjunction with the phase shift functions $\Phi(f)$ play an alternative equivalent role in the calculation

sinusoids are reassembled (inverse Fourier transformation) to obtain the output response. Although only one pair of diagrams is shown there is a different R and φ for both the spatial and temporal domains of frequency. The $R(f)$ diagram is commonly known as a "modulation-transfer function" or "contrast-sensitivity function". Some mathematical complexities concerned with the frequency representation of patterns varying in both spatial dimensions have been omitted in the interests of simplicity.

Spatial Analyses of Retinal Ganglion Cell Performance

In a series of three papers RODIECK and STONE (1965a, b) and RODIECK (1965) developed an interesting picture to account for the responses of cat retinal ganglion cells to targets moved across their receptive fields. The analysis was based upon measurements of the responses of ganglion cells to a small stationary spot of light flashed on and off at different positions over the receptive field, and a demonstration that the principle of superposition was applicable to the transformations connecting stimulus and ganglion cell response. The demonstration was not entirely satisfactory because the response could only follow a phase of depression down to zero impulses/sec and the further time course was inaccessible. Nevertheless, RODIECK (1965) was able to obtain at least qualitative agreement between calculated and observed responses for a variety of targets and modes of movement.

However, it has become abundantly clear that the responses of retinal ganglion cells are non-linear functions of stimulus intensity (HUGHES and MAFFEI, 1966; CLELAND and ENROTH-CUGELL, 1966; ENROTH-CUGELL and ROBSON, 1966; BÜTT-NER and GRÜSSER, 1968; STONE and FABIAN, 1968) and therefore the principle of superposition cannot be validly employed to calculate ganglion cell responses.

It might therefore have been thought that the considerable simplifications of linear analysis were unavailable as an aid to interpretation. ENROTH-CUGELL and ROBSON (1966) restored the situation to some extent by using the technique of constant response from the ganglion cell in order to determine relations between stimulus parameters. This separates the processes of transduction and combination from the process of output coding (see Fig. 12) and enabled them to discover a class of ganglion cells (the X-type, see. p. 546) in which transduction and combination ("summation") were approximately linear even though ganglion cell output was clearly non-linear. In the other class (Y-type) the summation was very non-linear.

ENROTH-CUGELL and ROBSON went on to measure the one-dimensional spatial contrast-sensitivity function of a number of X-cells with the results shown in Fig. 13. From the earlier discussion, it is clear that each curve provides a basis for predicting the sensitivity of that cell to other stimulus patterns varying only along one spatial dimension. The prediction was checked successfully in the case of a single contrasting edge presented at various positions across the receptive field. It is not easy to predict the spatial impulse response (cf. Fig. 12) or "receptive field sensitivity profile" so the authors used an indirect method to arrive at the size of the receptive field centres. The sizes lay within the range found by others: $0.5-8°$ (Fig. 7b from WIESEL, 1960), $0.5-4°$ (RODIECK and STONE, 1965b).

The curves of Fig. 13 are also a basis for discussions of visual acuity. From the curve at top right it can be seen that detectable modulation was present in the output of that ganglion cell up to a spatial frequency of 3 cycles/degree. After making allowances for differences in pupil size and mean level of illumination, this agrees fairly well with a behaviourally determined acuity of at least 5.5 cycles/degree (SMITH, 1936).

Linearity of the processes preceding the ganglion cell output has been tested in other ways. CLELAND and ENROTH-CUGELL (1968) measured the spatial sensitivity profile of the receptive field and used it to calculate the sensitivity of the cell to centred spots of various diameters. As before, the sensitivities were estimated as the reciprocal of the added luminance required to produce a constant, threshold response. At least in some cases the agreement was good. Since the sensitivity profiles were generally flat-topped over the centres, the area-sensitivity relation plotted on log-log coordinates had a slope of 1 over this part of the centre, a relation known as RICCO's law (1877, cited in BRINDLEY, 1960) from human psychophysical studies. Further support for linear transduction and combination

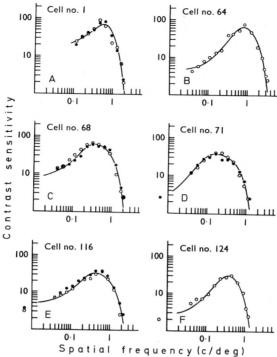

Fig. 13. Spatial contrast sensitivity functions (one-dimensional) for five on-centre X-cells $(A - E)$ and one off-centre X-cell (F). Sinusoidal grating patterns were drifted across the receptive fields at one bar per second and the contrast determined for a just-audible modulation of discharge frequency. Two sets of symbols indicate duplicate measurements separated by 40 min or longer. Points to the left of the vertical axis in $D - F$ indicate the contrast sensitivity at "zero spatial frequency" (sinusoidal temporal modulation of a field of uniform luminance at 1 c/s). Continuous lines are theoretical curves. From ENROTH-CUGELL and ROBSON (1966)

came from experiments in which stimulation was applied at four approximately equisensitive spots arranged symmetrically within the receptive field centre. The measured sensitivities for the spots stimulated in pairs or all together was very close to the calculated sensitivities based on measurements with each spot individually.

It might have been thought that the combination of RICCO's law with the demonstration of a flat-topped sensitivity profile would have been sufficient to establish linear summation. It is not. EASTER (1968) obtained precisely this combination of results from goldfish retinal ganglion cells but then showed convincingly that the operations preceding the ganglion cell output were non-linear. The demonstration involved a two-spot experiment in which the intensities required to produce a constant response were compared when stimulating with either spot alone or both together. The locations were equisensitive. If the processing were linear, the intensity for a single spot should be twice that for two spots together. In fact, a single spot required about *four* times the intensity. This relation held for a wide range of stimulus intensities.

To interpret the result, Easter divided the processing into an initial stage in which the physical stimulus is translated neurally into an intermediate quantity, E, which he called the "excitation", followed by a stage in which the excitations are summed and the total directly determines the ganglion cell's response. The two stages are indicated in Fig. 12 as "transduction" and "combination". The results of two-spot experiments with unequal intensities made non-linear combination most unlikely, so that non-linearity involved the generation of excitation only. In fact, the results implied that excitation was approximately proportional to the square root of intensity.

Ricco's law was not explained in Easter's treatment. If one makes the simple assumption that the excitation is being generated not by flux-detecting subunits but by edge-detecting subunits, then the number of subunits active for spots larger than some minimum size is proportional to the square root of the area of the spot. With Easter's formulation (E = total excitation; $e_j(I)$ = excitation of j th subunit; n = number of subunits activated; A = area of spot):

$$E = \sum_n e_j(I)$$

$$= k \cdot A^{1/2} \cdot I^{1/2}.$$

For constant E, $I \propto 1/A$ ... Ricco's law.

Two points can be made from the above discussion. Firstly, Easter's approach is a sound start towards the analysis of the types of non-linear system found in the retina. Secondly, Ricco's law is seen to arise in quite different contexts, so that terms such as "complete summation", implying a sensitivity to flux independent of its distribution, should not necessarily be taken at face value.

Temporal Analyses of Retinal Ganglion Cell Performance

In other parts of this Handbook, detailed accounts are given of the responses to flickering light and the neurophysiology of movement perception (see Grüs-ser's articles in volume VII/3), so they will not be specifically dealt with here.

Problems and Points of View

Although it has not been emphasized, there is considerable variation in detail of the behaviour and function of ganglion cells within an ostensibly uniform class. Indeed, there can be as many classes as there are ganglion cells; the key to their study is the recognition of what is essential, durable and invariant in the jumble of accidental variation.

There are also persistent difficulties with off-responses. At low levels of background illumination, an off-centre X-type concentric unit responds to a weak light stimulus to its centre with a well-sustained depression of the maintained discharge; when the stimulus is turned off, the discharge accelerates back to the preceding level with little or no overshoot. Should this acceleration be called an off-response even though it is merely a direct return to the prestimulus discharge rate? At higher background levels this problem of definition tends to disappear because the depression with illumination is largely transient and the discharge may already have returned to near the prestimulus level when the light is turned off. Then the discharge is transiently accelerated well above the prestimulus level. The excess is easily identified as an off-response. But now the difficulty is to explain how such a response comes about. Anodal-break excitation of nerve is the familiar model for such rebound phenomena, but neither real nerve, nor the theoretical formulation of it (HODGKIN and HUXLEY, 1952) appear to be sufficient to account for more than the first impulse or two of a slowly subsiding off-response, wherever it may be generated.

The question of the existence and effects of possible centrifugal efferents to the retina is a problem of long standing which has been reviewed from time to time (BRINDLEY, 1960; OGDEN, 1968). Recent anatomical evidence supports their existence in the frog (MATURANA, 1958) and pigeon (COWAN and POWELL, 1963; MATURANA and FRENK, 1965; DOWLING and COWAN, 1966) but is against it in the cat (BRINDLEY and HAMASAKI, 1966) and is conflicting in the primate (DOWLING and BOYCOTT, 1966; HONRUBIA and ELLIOTT, 1968; BOYCOTT and DOWLING, 1969). If centrifugal efferents are nothing more than the autonomic nerves ending on retinal blood vessels (WOLTER, 1965) they would be much less significant for the organization of vision. The existence of "associational ganglion cells" (GALLEGO and CRUZ, 1965) the axons of which end within the retina, complicates the interpretation of morphological studies not employing degeneration methods. The physiological evidence for centrifugal efferents on the cat (GRANIT, 1955) and rabbit (DODT, 1956) is open to various interpretations (see BRINDLEY, 1960). Since then, OGDEN and BROWN (1964) discovered a slow late positive potential in the inner plexiform layer of the primate ("P wave") which they interpreted as a postsynaptic potential of ganglion cell dendrites elicited by centrifugal efferents via amacrine cells. It now seems possible (OGDEN, 1966) to explain the potential by passage of antidromic activity of ganglion cells backwards through "tight junctions" (DOWLING and BOYCOTT, 1965) into bipolar cells. SPINELLI et al. (1965) and SPINELLI and WEINGARTEN (1966) claimed to have recorded centrifugal efferents in the cat's optic nerve, but additional controls are required. WEINGARTEN and SPINELLI (1966) claimed to have produced reliable dimensional changes in receptive fields by auditory and somatic stimuli but the changes were small and may have been caused by eye-movements which can occur with "paralyzed" preparations (RODIECK et al., 1967).

From the survey in a previous section it is clear that a map of the receptive field or a verbal description of the distribution of on- and off-responses fails to convey the essential behaviour of the ganglion cells in several cases. In the rabbit for example, ganglion cells with on-off receptive fields may be local-edge-detectors or direction-selective units; the horizontal-edge-detectors in the pigeon do not even have a receptive field in the sense of HARTLINE since they do not respond to a light spot turned on or off. For cells like these, some epithet is required to supplement the concept of receptive field; indeed one may need to know the behaviour before a proper concept of the receptive field can be developed. MATURANA et al. (1960) employed the term "contextual optimum" to refer to such an epithet, but "trigger feature" of BARLOW and LEVICK (1965) is a more explicit generic. The trigger feature is that constellation of stimulus attributes for which the neurone is specially selective; the receptive field remains the region of retina (or visual field, with reservations) over which the neurone may be influenced by appropriate stimulation.

The mechanism that formed the basis of the trigger feature of direction-selective ganglion cells suggested an interesting generalization (Barlow and Levick, 1965). As one passes centrally through successive layers of synapses, it appears that the flow of information is modified by successive alternation of two distinct types of process. The first is a stage of summation or pooling of selected excitatory inputs, the second a stage of inhibitory interaction. A bipolar cell connects with a group of receptors and thereby gains in sensitivity, but information is lost because the bipolar output no longer reflects exactly where in the group the light quanta were absorbed. A laterally conducting element exerts an inhibitory influence and thus makes the bipolar cell specifically sensitive to the sequence of activation of groups of receptors. The pooling operation repeats at the ganglion cell which picks up from a field of sequence-selective bipolars and again, laterally conducting elements bring in inhibitory influences to make the output sensitive to local events. It remains to be seen whether neural pattern recognition, which is the purpose of sensory coding, is based generally on a hierarchical organization of alternating, oppositely directed processes. Recent developments in cat visual cortex (Hubel and Wiesel, 1962, 1965) appear to support the idea.

Another point concerns the angular size of receptive fields. At first, it might seem that large receptive fields would imply that the animal has lost a considerable amount of the spatial resolution that was available from the fine receptor mosaic. In some cases there is such a loss, but it tends to be offset by advantages along other dimensions. For instance, large receptive fields are more sensitive to the centring of large fuzzy objects than are small fields; large fields have lower absolute luminance thresholds than individual receptors could possibly have. In other cases the loss of spatial resolution is slight: if the cell is pooling the outputs of a spatially distributed set of subunits each of which is developing a signal based on fine local differences in illumination of the receptor mosaic, then resolution of detail is still present. All that is lost is the precise location of the detail (Maturana et al., 1960; Barlow and Levick, 1965).

The relation between structure and function is an intriguing problem in the retina because the geometrical aspects of the retinal image that are functionally significant are expected to have their counterparts in the horizontal structure and arrangement of the neural components. Thus Lettvin et al. (1961) related each of the particular types of receptive field with a ganglion cell having a particular type of dendritic tree (Cajal, 1893) by correspondence of receptive field sizes, conduction velocities (and thus axon diameters and ganglion cell body diameters) and histological spread of dendritic fields. The inner plexiform layer of the frog has several distinct strata to each of which particular types of bipolar cells make specific contributions. Maturana et al. (1960) suggested that three types of bipolar cells deliver three types of information: on-information, off-information and edge-information, to specific strata. The specificity of ganglion cell function therefore depends upon which layers the dendrites contact and the way in which the dendrites sample them. The edge-function is generated by arranging for the particular bipolar to be driven by two groups of receptors having opposite actions on it, thus making it report local difference of illumination. This whole speculative scheme cannot be said to have been proved by the observations, but it certainly made sense of much of the structural detail and provided a powerful call for further quantitative morphology and exploration of the connectivity of the retina. LETTVIN et al. (1961) hinted that the geometry of the dendritic tree generated the nonlinear combination of inputs sampled from the layers in a specific way ("E-sha-ped" and "H-shaped" arbors). DOWLING (1968) and BOYCOTT and DOWLING (1969) pointed out the great variety of size, form and connectivity of amacrines in the frog and pigeon and would place the operations with these elements. All could be correct in some sense.

All are agreed that the range of horizontal spreads of dendritic trees of ganglion cells approximates the range of sizes of receptive field *centres* in the cat (GALLEGO, 1965; BROWN and MAJOR, 1966; LEICESTER and STONE, 1967) and probably in the rat (BROWN, 1965). To account for the presence of the antagonistic surround additional laterally spreading elements are needed and detailed functional require-ments for such connections were listed by BARLOW et al. (1957). Both DOWLING and BOYCOTT (1966) and RODIECK (1967) have argued persuasive cases implicating amacrine cells as mediators of the antagonistic surround. The processes of at least some amacrine cells are spread almost as widely as the dendritic trees of the largest ganglion cells (LEICESTER and STONE, 1967) and the appropriate synapses (bipolar to amacrine/ganglion cell — the "dyad" junction; amacrine to bipolar; amacrine to ganglion cell; amacrine to amacrine) have been described by DOWLING and BOY-COTT (1965), who offered the scheme of Fig. 14. To make the model work bipolars are assumed to excite amacrine cells, which then exert a weaker throttling effect back on the bipolar cells over a wider area. An on-centre ganglion cell is assumed to be excited by the bipolars and inhibited by the amacrines which contact it over its dendritic field. A spot of light excites directly via the bipolars if it is turned on within the field but beyond the bounds of the dendritic tree, the effect is indirect: amacrine cells centred on the spot are excited and throttle the excitatory bipolars within the field, as well as exerting a direct inhibitory effect on the ganglion cell by virtue of the overlap of their respective fields. Off-centre neurones are assumed to be inhibited by bipolar cells and excited by the amacrines. Whether this simple scheme is adequate remains to be seen, but one weakness is immediately apparent: the maintained discharge of off-centre ganglion cells evidently would depend upon the amacrine cells; how is the maintained discharge accounted for if the surround drops out in dark adaptation?

More recently, some elegant experiments on the retina of the mudpuppy, *Necturus maculosus*, (DOWLING and WERBLIN, 1969; WERBLIN and DOWLING, 1969) have raised the possibility that surround effects might be mediated in the outer plexiform layer by horizontal cells. The argument was based upon intracellular recordings from receptors, horizontal cells, bipolar cells, amacrine cells and ganglion cells, all identified by injection of stain from the recording electrode and subsequent microscopic examination. Single bipolar cells responded to photic stimulation within a region about $100\,\mu$ wide by displacement of membrane potential in one direction; stimulation with a narrow concentric annulus $250\,\mu$ in diameter yielded an oppositely directed potential displacement. The spread of bipolar cell dendrites can account for the central zone of responsiveness, but only the processes of horizontal cells are extensive enough to mediate the surround behaviour in the

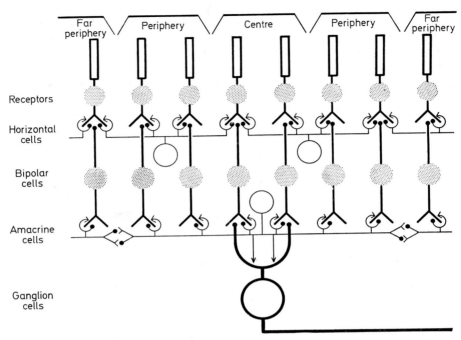

Fig. 14. A "wiring diagram" for a ganglion cell receptive field based on the anatomy and physiology of the retina. λ, excitatory synapse; ↓, inhibitory synapse. The synaptic contacts between processes of horizontal cells and receptor terminals, and also between amacrine processes and bipolar terminals are represented as reciprocal contacts; the others as one-way junctions. From Dowling and Boycott (1966) with the permission of the Royal Society

outer plexiform layer. If a ganglion cell was excited exclusively by such a bipolar cell or a small group of them, it would show many of the features of the antagonistic centre-surround organization described earlier. At least some of the *Necturus* ganglion cells behaved in this way.

The same diagram as before (Fig. 14) and a similar discussion can be used to explain the mechanism, but now the three elements participating are receptors, horizontal cells and bipolars, instead of bipolars, amacrines and ganglion cells. There are, however, some subtle differences: Dowling and Boycott (1966) originally pictured horizontal cells as making reciprocal contacts back onto receptors but this cannot be correct in *Necturus* because it would induce a centre-surround organization in the receptors which they do not have (Werblin and Dowling, 1969). Instead, Dowling and Werblin (1969) found the morphology to be more distinctive in *Necturus* than in the primate and observed conventional synapses from horizontal processes upon bipolar dendrites. Horizontal cells are thought to be driven by receptors over a wide field and to throttle bipolar cells gathering activity from a smaller group of receptors.

It is worth noting that the unconventional nature of the recorded neural activity admits unconventional interpretations of its origin. Although bipolar cells must transmit centripetally, the origin of the potentials recorded in the cell body could be proximal as easily as distal. WERBLIN and DOWLING did not discuss the possibility that the oppositely directed polarization from surround stimulation might arise from the synapses of amacrine cells onto the bipolar processes in the inner plexiform layer. A host of new problems has arisen by virtue of the concept of non-impulsive conduction of activity. How they may be resolved in the retina remains to be seen.

The secretarial assistance of Mrs. L. COWAN and Miss B. FERGUSON was much appreciated.

References

ADRIAN, E. D., MATTHEWS, R.: The action of light on the eye. Part I. The discharge of impulses in the optic nerve and its relation to the electric changes in the retina. J. Physiol. (Lond.) **63**, 378—414 (1927).

— — The action of light on the eye. Part II. The processes involved in retinal excitation. J. Physiol. (Lond.) **64**, 279—301 (1928a).

— — The action of light on the eye. Part III. The interaction of retinal neurones. J. Physiol. (Lond.) **65**, 273—298 (1928b).

ANDREWS, D. P., HAMMOND, P.: Mesopic increment threshold spectral sensitivity of single optic tract fibres in the cat: cone-rod interaction. J. Physiol. (Lond.) **209**, 65—81 (1970).

BARLOW, H. B.: Action potentials from the frog's retina. J. Physiol. (Lond.) **119**, 58—68 (1953a).

— Summation and inhibition in the frog's retina. J. Physiol. (Lond.) **119**, 69—88 (1953b).

— BLAKEMORE, C., PETTIGREW, J. D.: The neural mechanism of binocular depth discrimination. J. Physiol. (Lond.) **193**, 327—342 (1967).

— BRINDLEY, G. S.: Inter-ocular transfer of movement after-effects during pressure blinding of the stimulated eye. Nature (Lond.) **200**, 1347 (1963).

— FITZHUGH, R., KUFFLER, S. W.: Change of organization in the receptive fields of the cat's retina during dark adaptation. J. Physiol. (Lond.) **137**, 338—354 (1957).

— HILL, R. M.: Selective sensitivity to direction of movement in ganglion cells of the rabbit retina. Science **139**, 412—414 (1963a).

— — Evidence for a physiological explanation of the waterfall phenomenon and figural aftereffects. Nature (Lond.) **200**, 1345—1347 (1963b).

— — LEVICK, W. R.: Retinal ganglion cells responding selectively to direction and speed of image motion in the rabbit. J. Physiol. (Lond.) **173**, 377—407 (1964).

— LEVICK, W. R.: The mechanism of directionally selective units in the rabbit's retina. J. Physiol. (Lond.) **178**, 477—504 (1965).

— — The Purkinje shift in the cat retina. J. Physiol. (Lond.) **196**, 2—3P (1968).

— — Changes in the maintained discharge with adaptation level in the cat retina. J. Physiol. (Lond.) **202**, 699—718 (1969).

BAUMGARTNER, G.: Indirekte Größenbestimmung der receptiven Felder der Retina beim Menschen mittels der Hermannschen Gittertäuschung. Pflügers Arch. ges. Physiol. **272**, R 21 (1960).

— Kontrastlichteffekte an retinalen Ganglienzellen: Ableitungen vom Tractus opticus der Katze. In: Neurophysiologie und Psychophysik des visuellen Systems. Berlin-Göttingen-Heidelberg: Springer 1961.

— BROWN, J. L., SCHULZ, A.: Responses of single units of the cat visual system to rectangular stimulus patterns. J. Neurophysiol. **28**, 1—18 (1965).

— HAKAS, P.: Die Neurophysiologie des simultanen Helligkeitskontrastes. Reziproke Reaktionen antagonistischer Neuronengruppen des visuellen Systems. Pflügers Arch. ges. Physiol. **274**, 489—510 (1962).

BISHOP, P. O.: Discussion of paper, p. 133. In: Information Processing in Sight Sensory Systems. Pasadena: California Institute of Technology 1965.

Boycott, B. B., Dowling, J. E.: Organization of the primate retina: light microscopy. Phil. Trans. Roy. Soc. B **255**, 109—184 (1969).

Brindley, G. S.: Physiology of the Retina and Visual Pathway. London: Arnold 1960.

— Hamasaki, D. I.: Histological evidence against the view that the cat's optic nerve contains centrifugal fibres. J. Physiol. (Lond.) **184**, 444—449 (1966).

Brown, J. E.: Dendritic fields of retinal ganglion cells of the rat. J. Neurophysiol. **28**, 1091 to 1100 (1965).

— Major, D.: Cat retinal ganglion cell dendritic fields. Exp. Neurol. **15**, 70—78 (1966).

— Rojas, J. A.: Rat retinal ganglion cells: Receptive field organization and maintained activity. J. Neurophysiol. **28**, 1073—1090 (1965).

Bruesch, S. R., Arey, L. B.: The number of myelinated and unmyelinated fibers in the optic nerve of vertebrates. J. comp. Neurol. **77**, 631—665 (1942).

Burns, B. D., Heron, W., Pritchard, R.: Physiological excitation of visual cortex in cat's unanaesthetized isolated forebrain. J. Neurophysiol. **25**, 165—181 (1962).

Büttner, U., Grüsser, O. J.: Quantitative Untersuchungen der räumlichen Erregungssummation im rezeptiven Feld retinaler Neurone der Katze. Kybernetik **4**, 81—94 (1968).

Cajal, S. R. y.: La rétine des vertébrés. La Cellule **9**, 119—257. Lierre: J. van In and Co.; Louvain: A. Uystpruyst, Libraire 1893. Reprinted as an appendix in Travaux du Laboratoire de Recherches biologiques de l'Université de Madrid **28**, 1—144.

— Histologie du Système nerveux. Vol. II. French edition. Madrid: Cosejo superior de Investigaciones cientificas, Instituto Ramon y Cajal 1955.

Cleland, B., Enroth-Cugell, C.: Cat retinal ganglion cell responses to changing light intensities: sinusoidal modulation in the time domain. Acta. physiol. scand. **68**, 365—381 (1966).

Cleland, B. G., Enroth-Cugell, C.: Quantitative aspects of sensitivity and summation in the cat retina. J. Physiol. (Lond.) **198**, 17—38 (1968).

Cowan, W. M., Powell, T. P. S.: Centrifugal fibres in the avian visual system. Proc. Roy. Soc. B **158**, 232—252 (1963).

Daw, N. W., Pearlman, A. L.: Cat colour vision: one cone process or several? J. Physiol. (Lond.) **201**, 745—764 (1969).

— — Cat colour vision: evidence for more than one cone process. J. Physiol. (Lond.) **211**, 125—137 (1970).

DeValois, R. L.: Color vision mechanisms in the monkey. J. gen. Physiol. **43**, (6) Suppl. 2, 115—128 (1960).

Dodt, E.: Centrifugal impulses in rabbit's retina. J. Neurophysiol. **19**, 301—307 (1956).

Donovan, A.: The nerve fibre composition of the cat optic nerve. J. Anat. (Lond.) **101**, 1—11 (1967).

Dowling, J. E.: Structure and function in the all-cone retina of the ground squirrel. In: The Physiological Basis of form Discrimination. Providence: Walter S. Hunter Laboratory of Psychology, Brown University 1964.

— Synaptic organization of the frog retina: an electron microscopic analysis comparing the retinas of frogs and primates. Proc. Roy. Soc. B **170**, 205—228 (1968).

— Boycott, B. B.: Neural connections of the retina: fine structure of the inner plexiform layer. Cold Spr. Harb. Symp. quant. Biol. **30**, 393—402 (1965).

— — Organization of the primate retina: electronmicroscopy. Proc. Roy. Soc. B **166**, 80—111 (1966).

— Brown, J. E., Major, D.: Synapses of horizontal cells in rabbit and cat retinas. Science **153**, 1639—1641 (1966).

— Cowan, W. M.: An electron microscope study of normal and degenerating centrifugal fibre terminals in the pigeon retina. Z. Zellforsch. mikr. Anat. **71**, 14—28 (1966).

— Werblin, F. S.: Organization of retina of the mudpuppy, *Necturus maculosus*. I. Synaptic structure. J. Neurophysiol. **32**, 315—338 (1969).

Easter, S. S.: Excitation in the goldfish retina: evidence for a non-linear intensity code. J. Physiol. (Lond.) **195**, 253—271 (1968).

ENROTH-CUGELL, C., ROBSON, J. G.: The contrast sensitivity of retinal ganglion cells of the cat. J. Physiol. (Lond.) **187**, 517—552 (1966).

GALLEGO, A.: Connexions transversales au niveau des couches plexiformes de la rétine. Ann. Inst. Farm. Esp. **14**, 181—204 (1965).

— CRUZ, J.: Mammalian retina: associational nerve cells in ganglion cell layer. Science **150**, 1313—1314 (1965).

GAZE, R. M., JACOBSON, M.: Convexity detectors in the frog's visual system. J. Physiol. (Lond.) **169**, 1—3P (1963).

GOURAS, P., LINK, K.: Rod and cone interaction in dark-adapted monkey ganglion cells. J. Physiol. (Lond.) **184**, 499—510 (1966).

— The effects of light-adaptation on rod and cone receptive field organization of monkey ganglion cells. J. Physiol. (Lond.) **192**, 747—760 (1967).

— Identification of cone mechanisms in monkey ganglion cells. J. Physiol. (Lond.) **199**, 533—547 (1968).

— Antidromic responses of orthodromically identified ganglion cells in monkey retina. J. Physiol. (Lond.) **204**, 407—419 (1969).

GRANIT, R.: Sensory Mechanisms of the Retina. London: Oxford University Press 1947.

— Receptors and Sensory Perception. New Haven: Yale University Press 1955.

— SVAETICHIN, G.: Principles and technique of the electrophysiological analysis of colour reception with the aid of microelectrodes. Upsala Läk.-Fören. Förh. **45**, 1—4, 161—177 (1939).

GRÜSSER, O.-J., GRÜSSER-CORNEHLS, U., BULLOCK, T. H.: Functional organization of receptive fields of movement detecting neurons in the frog's retina. Pflügers Arch. ges. Physiol. **279**, 88—93 (1964).

— SNIGULA, F.: Vergleichende verhaltensphysiologische und neurophysiologische Untersuchungen am visuellen System von Katzen. II. Simultankontrast. Psych. Forsch. **32**, 43—63 (1968).

GRÜSSER-CORNEHLS, U., GRÜSSER, O.-J., BULLOCK, T. H.: Unit responses in the frog's tectum to moving and non-moving visual stimuli. Science **141**, 820—822 (1963).

HARTLINE, H. K.: The response of single optic nerve fibers of the vertebrate eye to illumination of the retina. Amer. J. Physiol. **121**, 400—415 (1938).

— The receptive fields of optic nerve fibers. Amer. J. Physiol. **130**, 690—699 (1940a).

— The effects of spatial summation in the retina on the excitation of the fibers of the optic nerve. Amer. J. Physiol. **130**, 700—711 (1940b).

HODGKIN, A. L., HUXLEY, A. F.: A quantitative description of membrane current and its application to conduction and excitation in nerve. J. Physiol. (Lond.) **177**, 500—544 (1952).

HOLDEN, A. L.: Receptive properties of retinal cells and tectal cells in the pigeon. J. Physiol. (Lond.) **201**, 56—57P (1969).

HONRUBIA, F. M., ELLIOTT, J. H.: Efferent innervation of the retina. Arch. Ophthal. **80**, 98—103 (1968).

HUBEL, D. H., WIESEL, T. N.: Receptive fields of optic nerve fibres in the spider monkey. J. Physiol. (Lond.) **154**, 572—580 (1960).

— — Integrative action in the cat's lateral geniculate body. J. Physiol. (Lond.) **155**, 385—398 (1961).

— — Receptive fields, binocular interaction and functional architecture in the cat's visual cortex. J. Physiol. (Lond.) **160**, 106—154 (1962).

— — Receptive fields and functional architecture in two non-striate visual areas (18 and 19) of the cat. J. Neurophysiol. **28**, 229—289 (1965).

HUGHES, G. W., MAFFEI, L.: Retinal ganglion cell response to sinusoidal light stimulation. J. Neurophysiol. **29**, 333—352 (1966).

JACOBSON, M., GAZE, R. M.: Types of visual response from single units in the optic tectum and optic nerve of the goldfish. Quart. J. exp. Physiol. **49**, 199—209 (1964).

JUNG, R.: Neuronale Grundlagen des Hell-Dunkelsehens und der Farbwahrnehmung. Ber. dtsch. ophthal. Ges. Heidelberg **66**, 69—111 (1964).

KEATING, M. J., GAZE, R. M.: Observations on the surround properties of the receptive fields of frog retinal ganglion cells. Quart. J. exp. Physiol. **55**, 129—142 (1970).

KUFFLER, S. W.: Neurons in the retina: organization, inhibition and excitation problems. Cold Spr. Harb. Symp. quant. Biol. **17**, 281—292 (1952).

— Discharge patterns and functional organization of mammalian retina. J. Neurophysiol. **16**, 37—68 (1953).

LEICESTER, J., STONE, J.: Ganglion, amacrine and horizontal cells of the cat's retina. Vision Res. **7**, 695—705 (1967).

LETTVIN, J. Y., MATURANA, H. R., McCULLOCH, W. S., PITTS, W. H.: What the frog's eye tells the frog's brain. Proc. Inst. Radio. Eng. (N. Y.) **47**, 1940—1951 (1959).

— — PITTS, W. H., McCULLOCH, W. S.: Two remarks on the visual system of the frog. In: Sensory Communication. New York-London: M.I.T. Press and John Wiley and Sons Inc. 1961.

LEVICK, W. R.: Receptive fields and trigger features of ganglion cells in the visual streak of the rabbit's retina. J. Physiol. (Lond.) **188**, 285—307 (1967).

— OYSTER, C. W., DAVIS, D. L.: Evidence that McIlwain's periphery effect is not a stray light artifact. J. Neurophysiol. **28**, 555—559 (1965).

MACH, E.: Über die Wirkung der räumlichen Verteilung des Lichtreizes auf die Netzhaut. Sitzungsberichte der mathematisch-naturwissenschaftlichen Classe der kaiserlichen Akademie der Wissenschaften, Wien **52**, (2), 303—322 (1865). Translated in RATLIFF, F.: Mach bands. San Francisco: Holden-Day 1965.

McILWAIN, J. T.: Receptive fields of optic tract axons and lateral geniculate cells: peripheral extent and barbiturate sensitivity. J. Neurophysiol. **27**, 1154—1173 (1964).

— Some evidence concerning the physiological basis of the periphery effect in the cat's retina. Exp. Brain Res. **1**, 265—271 (1966).

MATURANA, H. R.: Efferent fibres in the optic nerve of the toad *(Bufo bufo)*. J. Anat. (Lond.) **92**, 21—27 (1958).

— Number of fibres in the optic nerve and the number of ganglion cells in the retina of Anurans. Nature (Lond.) **183**, 1406—1407 (1959).

— Functional organization of the pigeon retina. P. 170—178 in Information Processing in the Nervous System. International Congress Series No. 49, Vol. III. Amsterdam-New York-London-Milan-Tokyo: Excerpta Medica Foundation 1962.

— FRENK, S.: Directional movement and horizontal edge detectors in the pigeon retina. Science **142**, 977—979 (1963).

— LETTVIN, J. Y., McCULLOCH, W. S., PITTS, W. H.: Anatomy and physiology of vision in the frog *(Rana pipiens)*. J. gen. Physiol. **43**, Suppl. 2, 129—175 (1960).

MICHAEL, C. R.: Receptive fields of directionally selective units in the optic nerve of the ground squirrel. Science **152**, 1092—1095 (1966a).

— Receptive fields of opponent color units in the optic nerve of the ground squirrel. Science **152**, 1095—1097 (1966b).

— Receptive fields of single optic nerve fibers in a mammal with an all-cone retina. I. Contrast-sensitive units. J. Neurophysiol. **31**, 249—256 (1968a).

— Receptive fields of single optic nerve fibers in a mammal with an all-cone retina. II. Directional selective units. J. Neurophysiol. **31**, 257—267 (1968b).

— Receptive fields of single optic nerve fibers in a mammal with an all-cone retina. III. Opponent color units. J. Neurophysiol. **31**, 268—282 (1968c).

MUNTZ, W. R. A.: Microelectrode recordings from the diencephalon of the frog *(Rana pipiens)* and a blue-sensitive system. J. Neurophysiol. **25**, 699—711 (1962).

NIKARA, T., BISHOP, P. O., PETTIGREW, J. D.: Analysis of retinal correspondence by studying receptive fields of binocular single units in cat striate cortex. Exp. Brain Res. **6**, 353—372 (1968).

OGDEN, T. E.: Intraretinal slow potentials evoked by brain stimulation in the primate. J. Neurophysiol. **29**, 898—908 (1966).

— On the function of efferent retinal fibers. P. 89—109 in Structure and Function of Inhibitory Neuronal Mechanisms. New York: Pergamon Press 1968.

OGDEN, T., BROWN, K. T.: Intraretinal responses of the cynamolgus monkey to electrical stimulation of the optic nerve and retina. J. Neurophysiol. **27**, 682—705 (1964).

OYSTER, C. W.: The analysis of image motion by the rabbit retina. J. Physiol. (Lond.) **199**, 613—635 (1968).

— BARLOW, H. B.: Direction-selective units in the rabbit retina: distribution of preferred directions. Science **155**, 841—842 (1967).

POLYAK, S. L.: The Retina. Chicago: University of Chicago Press 1941.

RATLIFF, F.: Mach bands: Quantitative Studies on Neural Networks in the Retina. San Francisco: Holden-Day 1965.

RAVIOLA, G., RAVIOLA, E.: Light and electron microscopic observations on the inner plexiform layer of the rabbit retina. Amer. J. Anat. **120**, 403—426 (1967).

RODIECK, R. W.: Quantitative analysis of cat retinal ganglion cell response to visual stimuli. Vision Res. **5**, 583—601 (1965).

— Maintained activity of cat retinal ganglion cells. J. Neurophysiol. **30**, 1043—1071 (1967).

— PETTIGREW, J. D., BISHOP, P. O., NIKARA, T.: Residual eye movements in receptive-field studies of paralyzed cats. Vision Res. **7**, 107—110 (1967).

— STONE, J.: Response of cat retinal ganglion cells to moving visual patterns. J. Neurophysiol. **28**, 819—832 (1965 a).

— — Analysis of receptive fields of cat retinal ganglion cells. J. Neurophysiol. **28**, 833—849 (1965 b).

RUSHTON, W. A. H.: The structure responsible for action potential spikes in the cat's retina. Nature (Lond.) **164**, 743—744 (1949).

— The difference spectrum and the photosensitivity of rhodopsin in the living human eye. J. Physiol. (Lond.) **134**, 11—29 (1956).

SMITH, K. U.: Visual discrimination in the cat. IV. The visual acuity of the cat in relation to stimulus distance. J. gen. Psychol. **49**, 297—313 (1936).

SPINELLI, D. N., PRIBRAM, K. H., WEINGARTEN, M.: Centrifugal optic nerve responses evoked by auditory and somatic stimulation. Exp. Neurol. **12**, 303—319 (1965).

— WEINGARTEN, M.: Afferent and efferent activity in single units of the cat's optic nerve. Exp. Neurol. **15**, 347—362 (1966).

STONE, J.: A quantitative analysis of the distribution of ganglion cells in the cat's retina. J. comp. Neurol. **124**, 337—352 (1965).

— FABIAN, M.: Summing properties of the cat's retinal ganglion cell. Vision Res. **8**, 1023—1040 (1968).

TALBOT, S. A., KUFFLER, S. W.: A multibeam ophthalmoscope for the study of retinal physiology. J. opt. Soc. Amer. **42**, 931—936 (1952).

THOMSON, L. C.: The localization of function in the rabbit retina. J. Physiol. (Lond.) **119**, 191—209 (1953).

VAKKUR, G. J., BISHOP, P. O., KOZAK, W.: Visual optics in the cat, including posterior nodal distance and retinal landmarks. Vision Res. **3**, 289—314 (1963).

VAN HOF, M. W.: Discrimination between striated patterns of different orientation in the rabbit. Vision Res. **6**, 89—94 (1966).

WAGNER, H. G., MacNICHOL, E. F., WOLBARSHT, M. L.: The response properties of single ganglion cells in the goldfish retina. J. gen. Physiol. **43**, Suppl. 2, 45—62 (1960).

— — — Functional basis for "on"-center and "off"-center receptive fields in the retina. J. opt. Soc. Amer. **53**, 66—70 (1963).

— WOLBARSHT, M. L.: Studies on the functional organization of the vertebrate retina. Amer. J. Ophthal. **46**, 46—59 (1958).

WALLS, G. L.: The Vertebrate Eye and its Adaptive Radiation. Michigan: Cranbrook Institute of Science 1942. Reprinted, New York: Hafner 1963.

WEINGARTEN, M., SPINELLI, D. N.: Retinal receptive field changes produced by auditory and somatic stimulation. Exp. Neurol. **15**, 363—376 (1966).

WERBLIN, F. S., DOWLING, J. E.: Organization of the retina of the mudpuppy, *Necturus maculosus*. II. Intracellular recording. J. Neurophysiol. **32**, 339—355 (1969).

WESTHEIMER, G.: Line-spread function of living cat eye. J. opt. Soc. Amer. **52**, 1326 (1962).
— Pupil size and visual resolution. Vision Res. **4**, 39—45 (1964).
— Applications of Fourier methods to the human visual system. In: Information Processing in Sight Sensory Systems. Pasadena: California Institute of Technology 1965.
WIESEL, T. N.: Receptive fields of ganglion cells in the cat's retina. J. Physiol. (Lond.) **153**, 583—594 (1960).
— HUBEL, D. H.: Spatial and chromatic interactions in the lateral geniculate body of the rhesus monkey. J. Neurophysiol. **29**, 1115—1156 (1966).
WOHLGEMUTH, A.: On the after-effect of seen movement. Brit. J. Psychol. Monograph supplement **1**, 1—117 (1911).
WOLKEN, J. J.: A structural model for a retinal rod. In: The Structure of the Eye, pp. 173—192. New York-London: Academic Press 1961.
WOLTER, J. R.: The centrifugal nerves in the human optic tract, chiasm, optic nerve and retina. Trans. Amer. opthal. Soc. **63**, 678—707 (1965).

Chapter 15

Retinal Mechanisms of Colour Vision*

By

Israel Abramov, New York, New York (USA)

With 11 Figures

Contents

One of the goals in analyzing a sensory system is a clearer understanding of how physical aspects of a stimulus are changed into a nervous message. This chapter will deal with those retinal responses for which wavelength of light is the important aspect of a stimulus. More generally, it will deal with the role played by various retinal mechanisms in determining the messages which pass along the optic nerve and allow an organism to distinguish colours. The subject matter will be divided between two broad areas: the first treats the spectral characteristics of the light transducers or visual receptors, since these set the primary limitations on colour discrimination; the second part is concerned with the ways in which the responses of the receptors are treated, analyzed, and transmitted by the later neural elements. Although most of the data has been obtained from animal studies,

* This work was partially supported by the following grants: Research Grant NB-00864 from the National Institute of Neurological Diseases and Blindness, U.S. Public Health Service; Research Grant GB 6540X from the National Science Foundation; and Training Grant GM 01789 from the National Institute of General Medicine, U.S. Public Health Service. The author thanks Dr. James Gordon for critically reading the manuscript and offering many helpful suggestions.

the aim will be to understand something about the mechanisms underlying human colour vision.

Colour vision can be defined as the ability to discriminate between light stimuli which differ only in wavelength without any additional differences in their brightness; the latter condition is necessary because a totally colour-blind organism might still distinguish between the stimuli on the basis of a difference in their apparent intensity rather than colour. Even the crudest form of colour vision requires that there be at least two types of receptor whose spectral sensitivities are not the same. Moreover, at some level of the visual system, the outputs of the two receptors must be compared with each other to obtain information about wavelength. The fineness of wavelength discrimination might be due to having many different receptor types each maximally sensitive to a narrow region of the spectrum; but in this case small, physically identical stimuli, placed side by side might not appear to have the same colour. Alternatively, the eye might contain only a few receptor types of different but broadly overlapping spectral sensitivities so that all would be stimulated to some degree by most spectral lights; the ratios, for example, of their responses could serve to identify each spectral region.

The latter scheme is essentially the one described by THOMAS YOUNG at the beginning of the last century and is the one which still appears correct today. Young, moreover, was the first to state clearly that the number of receptor types was probably three. Since that time psychophysicists have accumulated a wealth of evidence supporting the trichromacy of colour vision. (For a historical review of this topic see BALARAMAN, 1962.) Briefly, given any four lights it is possible to choose one of them as a standard and exactly match its appearance with the other three "primaries". Generally this can be done by mixing the three primaries and appropriately adjusting their intensities; but in some cases two of the primaries must be mixed and the third added to the light being matched. Provided the intensities of the primaries can be continuously and separately adjusted the match will be exact. Thus, three independent, continuous parameters are sufficient to specify our colour discrimination.

However, the three primary processes, each of different spectral sensitivity, required for a trichromatic visual system need not be limited necessarily to the receptors or even the retina; information about colour could be filtered and confined to three channels at some later stage (see BRINDLEY, 1960; MARRIOTT, 1962, for more on this). Nonetheless, it has generally been assumed that the limitation to three parameters is imposed at the earliest level in the retina. Since the first step in a light-sensitive system must be the absorption of light by a pigment, a simple explanation of colour-mixing data would be that there are, in the retinal receptors, three photopigments of different spectral sensitivities.

A. Primary Processes and Photopigments
1. Psychophysical Measures

The retinae of most species which have colour vision contain two anatomically distinct receptors — rods and cones (POLYAK, 1957; BOYCOTT and DOWLING, 1969). It is the outer segments of the rods and cones which contain the light transducing pigments, one of which (rhodopsin) is found in the rods of many vertebrate eyes, including those of man. The pigment is photolabile and its absorption spectrum agrees very closely with the spectral sensitivity (scotopic) of the fully dark-adapted eye. However under such adaptation conditions lights appear colourless. From this, and other observations (see GRAHAM, 1965, for a summary) the Duplicity Theory of vision was formulated; this theory says that the rods mediate scotopic vision, while light-adapted (photopic) vision and colour discriminations are determined by the cones of the retina. Most of the observations on which the theory is based do not preclude rods from contributing to colour perception, but it seems true that they play at best a small part. More will be said later about possible rod contributions to colour vision; however, for the moment it can be accepted that it is the cones which are of major interest. The rest of this section, then, will deal with the spectral sensitivities of the primary cone mechanisms.

Unfortunately, from psychophysical experiments alone one can derive a wide variety of possible spectral functions for the primary chromatic systems. However it can now be said, more by virtue of hind-sight than anything else, that a pair of related approaches have come the closest to identifying the basic mechanisms: determining the spectral sensitivities (a) of colour-blind subjects, and (b) normal subjects under conditions of chromatic adaptation. (The stimuli are limited to the foveal — central — region of the retina which is largely rod free.). The approach stems from the observation that only two parameters are needed to describe the colour matches of colour-blind (dichromatic) subjects (e.g., PITT, 1944). Assuming that these subjects lack one or other of the normal subject's three mechanisms, the spectral sensitivities of dichromats should be determined only by two mechanisms. Furthermore, since the fovea appears partly deficient in the blue-sensitive mechanism (e.g., WILLMER, 1955), a dichromat's sensitivity in the fovea should be determined largely by one mechanism; for the protanopic (red-deficient) dichromat the result will be due to the green-sensitive mechanism, and for the deuteranope (green-deficient) the curve will be of the red-sensitive mechanism. Results from such experiments (WILLMER, 1955; HSIA and GRAHAM, 1957) and from analyses of dichromats' colour matching functions (PITT, 1944) showed two systems of broad, overlapping sensitivities peaking at about 540 and 580 nm. More recently BLACK-WELL and BLACKWELL (1961) were able to measure the sensitivity of mono-chromats who had apparently only a blue-sensitive system maximally receptive at 440 nm.

The other line of the above experimental approach argues that if the sensitivity of one or other of a normal trichromat's mechanisms can be depressed, his remaining spectral sensitivity will follow, as before, the function of the remaining mechanisms. Appropriately chosen intense chromatic backgrounds can be used to adapt selectively one of the foveal mechanisms and so create an artificial state of colour blindness. In this way colour mechanisms peaking at 540 and 560—580 nm were isolated and found to agree very closely with the functions from dichromats (BRINDLEY, 1953; WILLMER, 1955). A closely related technique devised by STILES (1949) — see MARRIOTT (1962) for a good description — measured the thresholds for spectral lights added to monochromatic backgrounds of relatively low intensities and isolated mechanisms maximally sensitive at about 440, 540, and 580 nm. WALD (1964) using more intense backgrounds repeated and largely confirmed STILES' results.

The problem is whether the sensitivities of the primary colour mechanisms described above are each based on a separate cone pigment, or are the results of neural processing. In deciding if a response function directly reflects the absorption of light by a single visual photopigment, it is instructive to examine the shape of the curve. All visual photopigments which have been extracted have a quantum absorption function of the same shape as rhodopsin when plotted against wave number (Dartnall, 1953). This is substantially correct for rhodopsin-like pigments based on the retinene$_1$ chromophore, but for those derived from retinene$_2$ — as in some fish — a slightly different curve applies (Munz and Schwanzara, 1967). Thus if a sensitivity function is fitted by this standard curve or nomogram, it would indicate that it is determined by one pigment, although this is not strictly a necessary conclusion (see, e.g., Naka and Rushton, 1966b). In addition, if a response function is supposedly due to only one photopigment, chromatic adapting lights should merely depress the function and not change its shape; if more than one mechanism is involved a given chromatic background will preferentially desensitize the one which is initially more sensitive to that spectral region and so produce a change in the response function. However in applying these tests to psychophysical measures of sensitivity various precautions must be observed (Wyszecki and Stiles, 1967, pp. 582—587). The techniques using one form or other of chromatic adaptation carry with them the implication that the threshold for a test stimulus depends only on whichever mechanism is most sensitive to it in the given situation. If there are interactions between chromatic mechanisms in threshold situations the data will not be derived just from one mechanism. There is evidence that considerable interactions can occur in some situations (Boynton et al., 1964; Guth et al., 1969). When comparing results with some pigment standard like Dartnall's nomogram, the sensitivity function must be in quantal and not energy terms, and the comparison must be for quantal flux at the receptor layer. But the various refractive tissues in the eye absorb violet and blue lights quite heavily, the central or macular portion of the retina is unevenly covered with a blue-absorbing pigment, and both effects increase with the subjects' age (e.g., Wald, 1964; Ruddock, 1965). All this necessitates considerable correction of the psychophysical data with attendant uncertainty in the final function.

Nonetheless, with the above caveats in mind, it seems as if each of the primary mechanisms described earlier does directly depend on light absorption by a single photopigment, at least over the greater part of its spectral range. Stiles (1959) and Willmer (1955) showed that the sensitivity functions they obtained agreed closely, except in the very short wavelengths, with standard rhodopsin-like curves. Wald's (1964) curves, when corrected for pre-receptor filtering, fit the nomogram quite well and the shape of the mechanisms he isolates stays the same even with additional chromatic adaptation.

Obviously psychophysical experiments by themselves will not provide complete verification of the three-pigment hypothesis. Direct observations of the pigments themselves are needed, but, unlike rhodopsin, it has proven very difficult to extract cone pigments. Wald et al. (1955) distinguished a cone pigment — iodopsin — in extracts from the chicken eye, and Bridges (1962) has extracted another from the pigeon retina; but so far none has been obtained from the eyes of primates. As a result, some attempts have been made to explain the spectral functions described earlier by assuming that all three cone types contain the same photopigment but each type has a different spectral filter placed in front of the photopigment. This could provide a basis for a form of colour vision, and a possible mechanism does exist in the coloured oil droplets found in the inner segments of cones in certain avian and reptilian retinae; as yet, though, the effects of these coloured droplets have not been fully investigated. In any case, fixed filters of this sort can not account for human colour vision, as was nicely shown by BRINDLEY and RUSHTON (1959); they observed that a spectral light reaching the receptors in the normal fashion appeared to be of the same colour as one shone through the sclera, thereby arriving at the receptors from the reverse direction and avoiding any putative filters.

2. Retinal Spectrophotometry

In recent years various people have developed methods for demonstrating directly that photolabile pigments other than rhodopsin are present in the retina, and for examining their spectral characteristics. One technique, reflection densitometry, was developed by several people to provide information on the pigments in the living human eye (CAMPBELL and RUSHTON, 1955; RUSHTON, 1958; WEALE, 1959). Briefly, lights of different wavelengths are shone into the eye; the amounts reflected back out after a double traverse through the retina, and hence not absorbed, are measured. Any photolabile pigments are then bleached away with intense light and the spectral reflection again measured. The ratio of the amount of incident light absorbed before bleaching to that absorbed after bleaching is used to define the spectral absorption of the photolabile pigments existing prior to the bleach in the irradiated portion of the retina. (The form of difference spectrum just described specifies only the absorption of a pigment bleached away by light — absorptions by photostable elements are cancelled out. See DARTNALL, 1962, for more details.) This method permits one to study the visual pigments *in situ* and observe the kinetics of bleaching and regeneration. However, without additional steps the difference spectrum obtained would be a function of all the photopigments present. To isolate single cone pigments, observations were confined to the fovea, and, as with earlier psychophysical experiments, colour-blind subjects were used as well as normal observers after exposure to chromatic lights intense enough to bleach out one or other of the cone pigments.

Although the early work on reflection densitometry showed that cone pigments existed, there were problems with the spectral functions. The difficulties included low levels of light reflected back out of the eye, absorption due to by-products of bleaching, and the interfering effects of stray light; also the subjects had to maintain very good fixation throughout the experiment otherwise the data would include measures from previously non-irradiated parts of the retina, and so on. RIPPS and WEALE (1964), WEALE (1965), and RUSHTON (1965a) provide detailed discussions of these problems.

RUSHTON'S more recent experiments correct appropriately for many of the artifacts in the earlier work and the results, showing one pigment with peak absorption at 570 nm and another at 540 nm, agree quite well with the pigments previously inferred from psychophysical work (RUSHTON, 1963; RUSHTON, 1965 b; BAKER and RUSHTON, 1965). In the earlier experiments, a pigment peaking at about 440 nm had also been postulated but no detailed evidence could be obtained through reflection densitometry, due to measurement difficulties in the short wavelengths and to foveal insensitivity to blue light.

BROWN and WALD (1963) used a related technique to measure the light transmitted through (and thus not absorbed) small areas of retinae excised from eyes of humans and macaque monkeys. (The macaque is of interest since psychophysical observations show it to have colour vision essentially identical to human — DE VALOIS, 1965 a; DE VALOIS and JACOBS, 1968.) In both species pigments were found with peaks at about 565 nm, 525—540 nm, and some evidence for one around 440 nm.

One still can not tell, just from the above experiments, which structures contain the photopigments, nor if they are segregated in separate receptors. Several

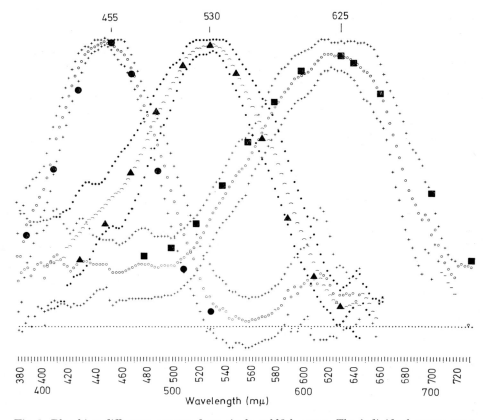

Fig. 1. Bleaching difference spectra from single goldfish cones. The individual curves were normalized to the same height and the averages are shown here bracketed by the standard deviations. The dotted horizontal line represents zero difference on bleaching. The large symbols show the shapes of DARTNALL (1953) nomogram pigments with peak wavelengths indicated. (From MARKS, 1965a)

laboratories have developed microspectrophotometric techniques for measuring light absorption by single receptors in excised retinae (BROWN, 1961; LIEBMAN, 1962; MARKS, 1963, 1965a). The approach is essentially the same as that used in retinal densitometry, but with the prime difference that now the light is confined to a single photoreceptor identified under a microscope. MARKS (1963, 1965a) measured difference spectra from the outer segments of cones from goldfish retinae and found three separate pigments confined one to a cone. (For summaries of behavioural studies of colour vision in fish see: WALLS, 1942; SVAETICHIN and MacNICHOL, 1958. Specifically on goldfish see: YAGER, 1967, who also shows that the species is almost certainly trichromatic.) A major problem was that the lights used to obtain measurements were intense enough to bleach large amounts of pigment thus distorting the absorption functions. MARKS developed a sophisticated procedure to correct for bleaching distortion and found the wavelengths of peak absorption of the pigments to be respectively 455, 530, and 625 nm; the shapes of the function corresponded nicely to the standard DARTNALL (1953) nomogram (see Fig. 1). The results were confirmed by LIEBMAN and ENTINE (1964) using low light levels such that insignificant amounts of pigment were bleached and no corrections of the data were required.

Single-cone spectrophotometry has been extended to primate cones, technically a much more difficult procedure since primate cones are very small. MARKS et al. (1964) reported, for both humans and macaque monkeys, cone pigments with absorption maxima at 445, 535, and 570 nm, while WALD and BROWN (1965) reported values of about 440, 530, and 565 nm for human cones. As with goldfish, each cone appears to contain only one pigment, although the latter authors emphasize the possibility that a few cones may contain mixtures of red- and green-absorbing pigments. The curves are fitted reasonably well by a standard nomogram photopigment (MacNICHOL, 1964).

It would seem, therefore, that the simple three-pigment explanation of the trichromacy of colour vision is correct. But even after the publication of the single-cone data, attempts were made to explain the results on the basis of one pigment together with a form of fixed light filter. The diameters of cones are not much greater than the wavelengths of light, so that they transmit energy as wave-guides (ENOCH, 1963). Since one can construct wave-guides which will transmit energy only in selected spectral regions, it has been suggested that the cones might be so divided into different categories, each allowing a limited wave-band to pass to the photopigment. (MACHOL, 1967, reports a comprehensive debate of these possibilities.) In general, explanations of this sort are obviated by the use of a difference spectrum which takes some ratio of transmissions before and after bleaching; the wave-guide properties of a receptor depend on its geometry which is not affected by bleaching, and so the wave-guide terms cancel. However, MURRAY (1968) demonstrated that the above argument can be used to reject a wave-guide basis for colour only in the following case: the spectrophotometer is so arranged that only light which has passed through the cone's photopigment is sensed while light passing between photoreceptors is ignored. Since in practice this is very difficult to achieve, MURRAY prepared suspensions of the outer segments of cones from macaque foveae, broke up the cones with an ultrasonifier, and then examined the resulting homogenate in a spectrophotometer. Since the anatomical structures

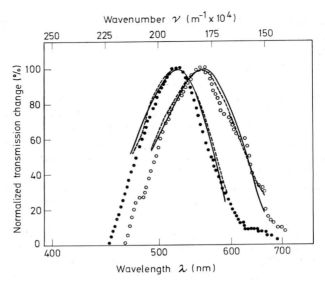

Fig. 2. Symbols show the mean difference spectra obtained by chromatic bleaching of a homogenate of the outer segments of cones from the primate fovea. The curves are "best-fit" functions derived from a standard photopigment; the two curves represent alternative ways of fitting the data. (From Murray, 1968)

were destroyed, no wave-guide mechanisms could be involved. Using chromatic bleaching he obtained difference spectra for two photolabile pigments whose curves agreed with the previous microspectrophotometric results from macaque cones (Fig. 2).

In summary, one can say that the three primary processes of normal colour vision are based on three separate photolabile pigments. These pigments, found in the cones of the retina, are of the same general class of photopigments as rhodopsin, the rod pigment. Thus, the limitation of colour information to three channels is imposed as the very first stage of the visual system.

B. Electrophysiological Experiments

A knowledge of the spectral characteristics of the elements that transduce the incident radiant energy does not directly provide information about how the receptors respond to light, how the responses of the receptors are processed, and so on. The answers are probably best sought by direct investigation of the nervous system — by recording the electrical activity which is its response to stimulation. But before doing so a very brief digression on the anatomy of the retina may be useful. More detailed descriptions are given by Boycott and Dowling (1969), Dowling and Boycott (1966), Missotten et al. (1963), Polyak (1957), and Walls (1942). The receptors make synaptic contacts with bipolar cells which then make connections with the ganglion cells, whose axons leave the retina as the optic nerve. In the receptor-bipolar synaptic complexes, the bipolar endings are flanked by processes from horizontal cells; further, in fish there are generally two layers of horizontal cells and in goldfish the external ones terminate in cone synaptic regions while the inner ones make contact with the rods (STELL, 1967). Amacrine cells provide lateral interconnections at bipolar-ganglion cell synapses in primates, but in other vertebrates several successive amacrines may be interposed between bipolars and ganglion cells (DOWLING, 1968).

For almost a century visual physiologists have been recording one of the electrical concomitants of vision – the electroretinogram, or ERG, which appears to arise largely from the region of the bipolar cells (TOMITA, 1963). But the classical ERG is the sum of many different responses in the eye, and as such is too gross a measure for detailed investigations of colour vision; as will be seen, many of the units respond with differential patterns of excitation and inhibition, and this aspect may be lost in the ERG; furthermore, in the ERG, the activities of different stages of processing are superimposed and mutually mask each other (TOMITA, 1963; MACNICHOL, 1966). It should be noted, though, that with appropriate (and rather complex) techniques some aspects of colour vision can be examined with the ERG. One of the basic problems is eliminating rod responses which otherwise predominate. RIGGS et al. (1966) isolated a photopic(cone-related) human ERG using rapid shifts from one colour to another within a bright field of alternately coloured bars; from the data they were able to compute relative response functions for three independent chromatic mechanisms. But the shapes of their curves do not resemble those of the primary photopigment mechanisms described earlier, possibly because they obtained their functions on the assumption that the ERG response is a simple weighted sum of three separate inputs; BURKHARDT (1966) showed that the goldfish's photopic ERG is not a simple sum of responses of units whose spectral sensitivities are those of that species' photopigments, suggesting interactions between the chromatic mechanisms. Using Riggs' method, larger differences in the wavelengths of the alternating bars are needed to evoke a change in the ERG as compared with a cortical potential (RIGGS and STERNHEIM, 1969); this was interpreted as evidence for more central sharpening of discrimination, but it may merely show that the ERG is not a very sensitive measure of colour discrimination. To add to the complexity, it seems that the different parts of the ERG waveform do not have the same spectral sensitivity; in carp the a-wave is largely green-sensitive while the b-wave receives inputs largely from red-absorbing cones (WITKOVSKY, 1968), and in goldfish the locally recorded ERG has a green-sensitive slow component and a red-sensitive oscillatory process (SPEKREIJSE et al., 1967).

The ERG has the advantage that it can be measured both from human and animal subjects but at present its complexities are sufficient to make interpretation difficult. The more profitable approach for colour vision is the recording of responses from single cells in the visual system; the method suffers from colossal sampling problems, but has the advantage of examining the responses of the elements from which the system is constructed. This approach goes back at least to 1927 when ADRIAN and MATTHEWS recorded the simultaneous discharge of several nerve fibres in the optic nerve of the Conger eel and showed that information is transmitted from the retina by changes in frequency of spike responses. HARTLINE (1938) was the first to succeed in recording from a single cell in a

vertebrate eye when he obtained responses from individual fibres dissected out of
the frog's optic nerve. The vast growth in recent years of single-unit recording
stems from that pioneering work and the development of microelectrodes by
GRANIT and SVAETICHIN (1939), and WILSKA (1940). These techniques have been
widely applied and responses have been obtained from most structures in the
retina.

1. Cone Potentials

The spectrophotometric results cited in the previous section indicate what the
spectral sensitivities of individual cones should be; to complete the picture one
would want to record physiological responses from single cones and show that
their action spectra correspond to the photopigment measures. But recording from
single cones is technically very difficult. TOMITA and his associates were the first to
record electrical responses from single cones in the retinae of carp (TOMITA, 1965;
TOMITA et al., 1967). They used the technique of mounting the retina on a coil

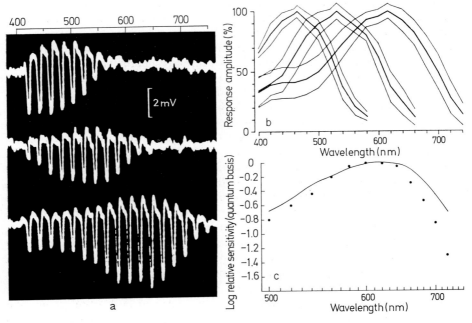

Fig. 3. (a) Potentials recorded from single carp cones. Each deflection is a response to a flash of
light of wavelength indicated by the scale at the top. Lights were adjusted to equal quantum
flux, and lasted 0.3 sec at each wavelength followed by 0.6 sec of darkness. Downward deflections are negative. Traces are from three different receptors with maximal responses respectively
to blue, green, and red lights (from TOMITA et al., 1967). (b) Averaged spectral response functions and standard deviation curves of the three types of cone in carp (from TOMITA et al.,
1967). (c) Points define the spectral sensitivity of a single red-sensitive carp cone (redrawn
from TOMITA et al., 1967). The line is the mean absorption function, from goldfish cones, of
the photopigment peaking at 625 nm. (The curve is redrawn from Fig. 1, which is from
MARKS, 1965a)

which, when electrically pulsed, jolted the retina with very high acceleration against an extremely fine micro-electrode and facilitated the penetration of the cone. At first the responses were only tentatively identified as coming from cones; however by electrophoretically injecting a dye through the recording micro-pipette it was conclusively shown that the responses were intracellularly recorded from the inner segments of the cones (KANEKO and HASHIMOTO, 1967).

The cones respond to light with a graded potential change; the resting trans-membrane potential is negative inside and light increases the negativity. This apparently hyperpolarizing response to stimulation is unusual; invertebrate photo-receptors typically depolarize when stimulated (HARTLINE et al., 1952; HAGINS, 1965). However, the same sort of hyperpolarizing response has been obtained from single receptors in the eye of the mudpuppy, *Necturus* (WERBLIN and DOWLING, 1969). For present purposes, the more interesting aspects of TOMITA's results are the spectral characteristics of the carp's cone responses (Fig. 3a). Individual cones could only be recorded from for very short periods and there is considerable uncertainty in the data. Nonetheless, the spectral response curves can be divided into three distinct classes with mean response maxima respectively at 462, 529, and 611 nm, plus or minus some 15 nm (Fig. 3b). These peaks compare favourably with MARKS' (1965a) spectrophotometric determinations of cone pigments (maxima at 455, 530, and 625 nm) in the goldfish; the two species are closely related, belong-ing to the same family, *Cyprinidae*. (Note that it is more difficult to specify long-wavelength maxima since the absorption spectra of visual pigments become broa-der and flatter when plotted against wavelength – DARTNALL, 1953.)

It is difficult to make more detailed comparisons between the two sets of data because the spectral response curves in TOMITA et al. (1967) are not spectral sensi-tivity curves. The latter define the relative amounts of energy required to elicit some constant criterion response; thus, if the function relating response amplitude and light intensity were known, the response curves could be translated into sensitivity functions. The requisite information has been published for an indi-vidual red-sensitive cone and the appropriate comparison with MARKS' data is shown in Fig. 3c. The sensitivity function of this particular cone is narrower than the photopigment function, which is itself slightly narrower than the nomogram curve (see Fig. 1). It has been said (WITKOVSKY, 1968) that the other two cone types described by TOMITA and his colleagues were also substantially narrower than the functions of nomogram pigments. But a simple comparison of the cone responses in Fig. 3b with the pigment functions in Fig. 1 does not support this. For example, for the blue-sensitive cones lights of 420 and 505 nm elicit responses of the same amplitude. This implies that the photopigment involved absorbs equally at these two wavelengths; however, from Fig. 1 it can be seen that the photopigment absorbs equally at 420 and 480 nm, indicating quite strongly that the sensitivity function derived from the cones' electrical responses would be, if anything, broader than the nomogram function. A similar comparison of the response and absorption functions of the green-sensitive cones suggests they have about the same spectral sensitivity.

There is a discrepancy between the electrical and spectrophotometric data in the proportions in which the three cone types were observed. Green-absorbing cones were the majority in MARKS' (1965a) sample, with the blue-absorbing the fewest, whereas TOMITA et al. (1967) found a large preponderance of red-sensitive cones, and relatively few green-sensitive. Perhaps there are minor differences in the shapes of the cone types making some easier to penetrate than others; a possible difference is that in goldfish many cones are members of joined double-cone structures where one is usually red-absorbing and the other green (MARKS, 1965a).

Responses of single receptors have not been recorded from higher vertebrates, although responses of populations of receptors have been observed in gross recordings from the eye. K.T. BROWN and his associates first isolated early and late receptor potentials (ERP and LRP) from the primate eye by recording the ERG through small electrodes inserted into the retina to the level of the receptors; separate rod and cone components were identified (see BROWN et al., 1965, for a review of their work). The LRP is the leading edge of the a-wave of the normal ERG and appears to be related to activity in the synaptic regions of the receptors, while the ERP seems to be very directly related to the initial light-evoked photochemical events (CONE, 1967). As yet neither response has been used in studies of the different colour mechanisms, despite the finding that the ERP from the mixed rod and cone retina of the frog, macaque, and man are generated almost entirely by the cones (GOLDSTEIN, 1967, 1969; GOLDSTEIN and BERSON, 1969).

To summarize, the action spectra of cones in trichromatic species fall into three groups each corresponding closely to the sensitivities of one of the three photopigments. All the cones respond to the light they absorb in the same fashion. The next sections deal with the ways in which later units in the visual pathway treat the outputs of the cones.

2. Spectrally Opponent and Non-Opponent Responses

a) S-Potentials

SVAETICHIN (1953) was the first to apply microelectrodes to the retinae of shallow-water marine fish, and identified a new, apparently intracellular, potential. He originally thought he had recorded cone action potentials, but careful dye-injection experiments showed that the potentials came from more proximal structures at about the level of the bipolar cells (MACNICHOL and SVAETICHIN, 1958; TOMITA et al., 1959). This potential is a slow, graded potential which is maintained for the duration of a stimulus light. (For reports and reviews of work on S-potentials see MACNICHOL, 1966; SVAETICHIN and MACNICHOL, 1958; SVAETICHIN et al., 1965; TOMITA, 1965.)

The role of the S-potential in the chain of transmission to the brain was, and is, unclear but it is found in most vertebrate eyes (TOMITA, 1965), and does possess certain features of interest for colour vision. Two categories of response have been noted. Certain units respond to all wavelengths with some degree of hyperpolarization, while others exhibit opponent spectral responses, responding with hyperpolarization to one portion of the spectrum and depolarization to other wavelengths (Fig. 4).

The spectrally non-opponent units have broad spectral sensitivities and have been termed L-type (luminosity) units, although direct comparisons have not been made with the subjects' spectral sensitivities. However, SVAETICHIN and MACNICHOL (1958) pointed out that the spectral maxima of L-units in various

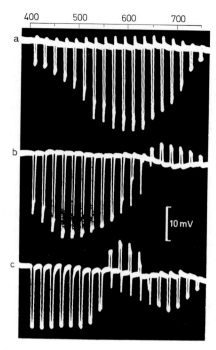

Fig. 4. S-potentials recorded from the carp retina. The deflections (negative downward) are the responses to flashes at wavelengths indicated by the scale. Trace (a) shows a luminosity (L-type) response; (b) and (c) are respectively biphasic and triphasic chromaticity (C-type) responses. (From Tomita, 1963)

Venezuelan marine fish are appropriate to their respective habitation depths; and during dark-adaptation the sensitivity of L-units in carp does shift towards shorter wavelengths, though not to the point of matching the rod sensitivity function (Witkovsky, 1967). There is some evidence that all L-units in a given species do not have the same spectral characteristics; there may be three types, each maximally sensitive to red, green, or blue (Svaetichin et al., 1965; MacNichol, 1966; Svaetichin, 1967). But in carp (Witkovsky, 1967) and tench (Naka and Rushton, 1966c) the shapes of the spectral sensitivity functions and the maxima change with the level of hyperpolarization chosen as the criterion response. The variations are presumably due to different ways in which the S-units are connected to the receptors — this whole topic is dealt with in detail later.

The spectrally opponent units are generally termed C-type since they are presumed to convey colour information. As with the L-type units, no direct comparisons have been made between C-type responses and colour discrimination functions, but they (C-units) do represent the first stage at which wavelength discrimination can take place. An L-unit can not by itself provide a basis for such discrimination. Given a pair of stimuli differing in wavelength it is possible so to adjust their intensities that they elicit the same response from an L-unit; however if the stimuli are such that, from a C-unit, one evokes hyperpolarization and the other depolarization, then, in general, that C-unit will respond differentially to the stimuli regardless of their intensities.

Several types of response have been recorded from C-units and these can be divided into biphasic and triphasic responses (Fig. 4). Usually the biphasic units show a hyperpolarizing response to the shorter wavelengths and depolarization to long wavelengths, with loci of maximal responses respectively in the red and green or yellow and blue regions of the spectrum (Svaetichin and MacNichol, 1958; Naka and Rushton, 1966a). Triphasic units typically respond with depolarization to green lights and hyperpolarization to the ends of the spectrum. The reciprocals of these responses (e.g., biphasic units depolarized by short wavelength stimuli) are also found (Svaetichin et al., 1965; Svaetichin, 1967), though they may be rarer, or possibly absent, in some species such as the carp (Norton et al., 1968; Witkovsky, 1967). Since the incidence of the various C-type responses differs considerably from species to species both in shallow-water marine fish (Svaetichin and MacNichol, 1958) and fresh-water fish (Orlov and Maksimova, 1965), care should be exercised in generalizing from one species to another. The combinations of inputs from the receptors which underlie the above responses are considered later.

It is not entirely clear which anatomical units are the sources of the S-potentials. L-type responses are the first encountered in penetrations from the receptor side and are usually ascribed to horizontal cells (Lipetz, 1963; Svaetichin et al., 1965). However L-units are sometimes found at levels much further from the receptors (Orlov and Maksimova, 1965). It is uncertain what is the site of the C-potentials; they are typically observed at levels further from the receptors than most L-units, but both have been found at the same level, and occasionally a C-type changes smoothly and gradually to an L-type (Tomita, 1965). Svaetichin and his associates (1965; Svaetichin, 1967) implicate the amacrine cells as the source of C-potentials. Recently the use of techniques for injecting dye into a cell through the intracellular recording micropipette have permitted positive identification of the cell types involved. Using these methods on cells in the retina of *Necturus* Werblin and Dowling (1969) observed that the responses from the horizontal cells do appear to have many of the attributes of L-potentials; C-type responses were not found. Responses of amacrine cells in no way resembled S-potentials. From similar experiments on goldfish Kaneko (1970) ascribed both L- and C-type responses to the horizontal cells. In both these species S-potentials could not be from bipolar cells since these cells had spatially antagonistic receptive fields, while S-units have very large, uniform fields (see later). However this still does not preclude the possibility that S-potentials (especially in other species) may also be recorded from cells other than horizontal cells; in general, S-potentials are identified as such from their responses to diffuse illumination without investigating spatial aspects or identifying the cell by dye-injection.

There are also problems about the role played by the S-units. S-units have many properties which appear to set them apart from other nerve cells; for example, by passing current through an intracellular electrode, the resting potential of an S-cell can be changed very considerably without affecting the amplitude or sign of the response to light (see Tomita, 1965, for a summary). Based on such findings, and the apparent sensitivity of S-cells, as compared with other nerve cells, to metabolic poisons it has been suggested that S-cells are not neurons and that they exert their influence in ways other than through classical synaptic paths (see in particular SVAETICHIN et al., 1965; SVAETICHIN, 1967). It is not even certain that S-cells "drive" later units in the pathway. SVAETICHIN and his colleagues and LIPETZ (1963) show that in fish and turtle, respectively, there are definite qualitative similarities between magnitudes and time courses of C-potentials and the variations in frequency of firing of ganglion cell spikes. However, WITKOVSKY (1967) reports that, at least in the carp, this correspondence (and the postulated connections between these units) is more apparent than real. Some of the discrepancies between the response patterns of S-units and ganglion cell responses will be considered in the following sections.

b) Ganglion Cells

Although the early responses of the visual system are graded potential changes, the responses which are transmitted from the retina and along the optic nerve are spike potentials. (Spikes have also been recorded from the region of the bipolar cells in frog retinae — KANEKO and HASHIMOTO, 1968; from the amacrine cells of *Necturus* — WERBLIN and DOWLING, 1969; and from amacrines in goldfish — KANEKO, 1970.) Thus, at the level of the retinal ganglion cells it is variations in the firing rate of spikes which signal stimulus events.

There are several ways of describing spike responses. Ganglion cells in the retinae of many vertebrates generally maintain some firing rate even in the absence of specifiable stimulation, and so responses can be either increases ("excitation") or decreases ("inhibition") of the spontaneous rate; there are no very compelling reasons why only increases of firing rate should be considered responses — a decrease can be just as much of a "response" provided there is a spontaneous rate which permits observation of such a response. The more traditional classification (first used by HARTLINE, 1938, to describe frog optic-nerve responses) is into "on", "off", and "on-off" responses, where the responses are an increase in firing and the names refer to the relationship of the response to stimulus onset and offset; this approach is especially useful when there is no spontaneous rate, as is the case in many of the experiments on excised fish retinae. Most authors make the tacit assumption that the "off" responses reflect the same mechanisms as "on-inhibition" (decrease in firing during the stimulus). In general this may be acceptable for comparing results from different experiments but some caution is needed; WAGNER et al. (1960) recorded from a few goldfish ganglion cells which were spontaneously active and found that although the spectral sensitivity functions for "off" and for "on-inhibition" peaked at the same wavelength, the former was appreciably broader. With these comments in mind, one can turn to a detailed examination of some of the results. The discussion at first will be limited to spectral responses; differences in receptive field organization and variations in combinations of inputs from receptors will be taken up afterwards.

GRANIT was one of the first to use very fine electrodes to isolate responses from single ganglion cells; experiments were done on an assortment of animals, but mostly on the cat (see GRANIT, 1947, for a full statement). His major findings are well known and will be touched on only briefly. Some units ("dominators") showed broad spectral sensitivity with maxima at either 500 or about 560 nm, depending on whether the animal was dark- or light-adapted; others ("modulators") had narrow spectral sensitivities with maxima distributed across the spectrum. Modulator functions were obtained in two ways: some units characteristically responded only to limited spectral bands, while in other cases the functions were calculated from the differences in dominator responses before and after chromatic adaptation.

From these data Granit developed a theory in which the dominators determined the spectral sensitivity of the species, while the modulators constituted a colour channel. The two types of dominator function (scotopic or photopic) were shown to agree very well with the known or presumed spectral sensitivities of the subjects; these cells also readily detected changes in stimulus energy. All this served to confirm the hypothesis that dominators subserved luminosity. Granit did not attempt any detailed correlation between his modulators and psychophysical functions, except to point out that they could form the basis of a colour discriminating mechanism.

One of the difficulties with using these results as a general model of colour vision systems is that much of it is based on the cat; this animal does have cones in its retina (Walls, 1942) but nonetheless possesses only the most rudimentary colour sense (see, e.g., Sechzer and Brown, 1964). Other problems are raised by the ways in which modulator functions were obtained. The functions are too narrow to be derived directly from receptor mechanisms, although it should be noted that Granit did not preclude modulator curves from being the result of interactions. But, as will be discussed later, the use of chromatic adaptation to isolate chromatic mechanisms is not simple. In cases where modulators were observed directly (and from subjects possessing colour vision) the narrow-band responses were probably due to ignoring inhibitory responses to other parts of the spectrum.

It is now generally accepted that Granit's conception is not entirely correct. Rather, in species with well developed colour vision there are two categories of ganglion cell responses to wavelength, and these are essentially the same as those noted earlier for S-potentials: (i) spectrally non-opponent cells which respond with either excitation or inhibition to all wavelengths; these can also be described as pure "on" or pure "off" cells. (ii) Spectrally opponent cells for which the response depends on stimulus wavelength. These two classes of response have been recorded from retinal ganglion cells in goldfish (Wagner et al., 1960; MacNichol et al., 1961) and carp (Motokawa et al., 1960; Witkovsky, 1965); from the macaque monkey (Gouras, 1968) and from the optic nerves of the spider monkey (Hubel and Wiesel, 1960), and the ground squirrel (Michael, 1966, 1968). Extensive studies have also been made of similar responses at the level of the lateral geniculate nucleus (where optic nerve fibres terminate) in macaque (De Valois et al., 1958, 1966; De Valois, 1965b; Wiesel and Hubel, 1966) and squirrel monkeys (Jacobs, 1964; Jacobs and De Valois, 1965).

The majority of cells in the goldfish retina are spectrally opponent, and of these most are of the green-red type; that is, they respond with excitation to green light and inhibition to red light, or vice versa, and the transition from one response to the other is usually quite sharp (Wagner et al., 1960; MacNichol et al., 1961;

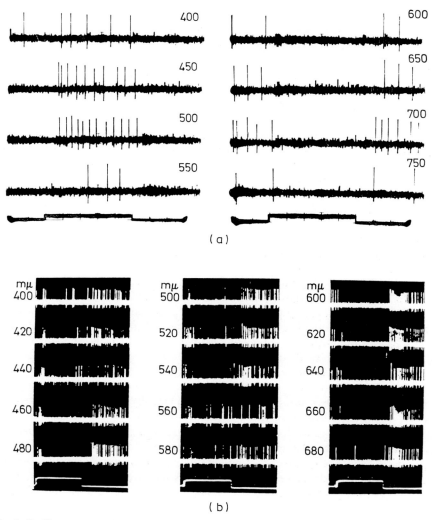

Fig. 5. Oscilloscope records of the responses of single, spectrally opponent, ganglion cells to the monochromatic lights indicated next to each record; stimulus interval is shown by the displacement in the traces at the bottom of each figure. (a) Records from goldfish. Stimulus duration was 0.5 sec. Stimuli were not equated for energy, shorter wavelengths having less energy. Spikes occurring before the onset of a stimulus are not "spontaneous", but are "off" responses from a preceding stimulus. (From WAGNER et al., 1960.) (b) Records from carp. Stimuli equated for energy; stimulus duration about 0.4 sec. (From MOTOKAWA et al., 1960)

and see Fig. 5a). However, some of the cells are better described as blue-yellow or blue-green in terms of the maxima of the opposed responses (BICKING, 1965). A similar situation is found in the carp, though most of the opponent cells appear to be blue-orange (WITKOVSKY, 1965; and see Fig. 5b). The foregoing description can be somewhat misleading. For instance, both the cells in Fig. 5 can be said to respond to blue lights (in the region of 450 nm) with excitation or an "on" response, and yet the temporal distributions of spikes elicited by the stimuli are clearly very different. The responses to blue in Fig. 5b could be very transient (phasic) "on" responses, or the responses might include an inhibitory component of longer latency than the excitation; the records suggest the second case but this would have to be tested — for example, by investigating if threshold is raised immediately following the "on" response. Spekreijse (1969) has used time-varying stimuli to classify goldfish responses as either phasic or tonic. A phasic unit maintains its firing rate during either the rising or falling phase of a low-frequency triangular stimulus, depending on cell type and wavelength; whereas, the responses of tonic units vary in time with the stimulus. In the macaque most spectrally opponent cells show tonic responses, maintaining some firing for the duration of the stimulus (Gouras, 1968). Spectrally non-opponent cells typically respond with either excitation or inhibition, but, in goldfish, some may respond with "on" and "off" responses, both responses having the same spectral sensitivity (Wagner et al. 1960).

A detailed description has not been made of the spectral responses of ganglion cells in the monkey retina. However, the results from cells in the lateral geniculate nucleus probably also describe the situation in the retina, at least as far as colour is concerned; responses from optic nerve fibres and geniculate neurons in monkeys appear to be much the same (Hubel and Wiesel, 1960; Wiesel and Hubel, 1966). DeValois et al. (1966) presented a detailed classification of the responses of units in the lateral geniculate of the macaque to diffuse, monochromatic lights. The two classes of spectrally opponent and non-opponent cells were readily distinguished. The non-opponent cells constituted about one-third of a large, and presumably random, sample; they were about equally divided between cells responding with excitation to all parts of the spectrum and cells showing inhibition to all wavelengths. Determining the types of spectrally opponent cells was not, however, quite so straight forward. From statistical analyses of various response parameters (e.g., wavelengths of peak excitation and inhibition, wavelength of change from excitation to inhibition), the authors concluded that spectrally opponent cells could be divided most parsimoniously into four types on the basis of their responses to diffuse coloured lights: +R−G (red excitation and green inhibition), +G−R, +Y−B, and +B−Y. These names described the spectral loci of peak excitation and inhibition but did not necessarily say how the inputs from the various cones were combined to give the different response types. There were no overwhelming differences in the relative numbers of the four types, though +G−R and +R−G were the most numerous and +B−Y the least common.

In species with less well developed colour vision similar responses have been observed, but with the interesting differences that spectrally opponent cells are less prevalent and of fewer types. Squirrel monkeys are protanomalous trichromats (Jacobs, 1963) and in recordings from the lateral geniculate nucleus only some 10−15 % of the cells are spectrally opponent (Jacobs and DeValois, 1965); however all four of the types observed in the macaque were found. Behavioural experiments have shown that the ground squirrel is clearly a dichromat and has red-deficient colour vision (Jacobs and Yolton, 1969); for this animal Michael (1966) reported that about 20 % of the cells in the optic nerve are spectrally opponent, and all are of the blue-green or green-blue types.

3. Receptor Connections

a) General Considerations

It was suggested above that the variations in spectral response patterns were due to different combinations of inputs from the cones. This must now be examined in greater detail. A spectrally opponent cell must receive inputs from at least two types of cone, the input from one eliciting excitation and from the other inhibition. However, the spectral sensitivities of the cones are very broad (Fig. 1) so that most lights stimulate more than one cone type (this is even more marked in the macaque whose long-wavelength pigment peaks at a much shorter wavelength than that of the goldfish); in these cases the neural response that is recorded is some function of the difference between opposed inputs. Even if only two photopigments are involved, it requires only small variations in the relative magnitudes of the antagonistic inputs to produce a wide range of spectral response functions. In general it is therefore impossible to decide which photopigments determine a unit's responses just by examining the spectral response function. Similar arguments apply to those spectrally non-opponent cells which receive inputs from more than one cone type and whose responses are some additive function of the inputs. It should also be noted that a triphasic spectral response (e.g., Fig. 4) does not necessarily imply inputs from three cone types. If, for example, green- and red-sensitive cones provide respectively the excitatory and inhibitory inputs, then at the long wavelengths the net response is inhibition, changing to net excitation in the middle of the spectrum; inhibition evoked by short wavelengths may again be due to the red-sensitive cones (which still absorb appreciably at these wavelengths) if the input from the green-sensitive cones falls below some threshold.

The usual approach to identifying the cone inputs to a cell is to use a relatively intense adapting light of a wavelength chosen to desensitize preferentially one or other of the cone types. As mentioned earlier, essentially the same technique is used in psychophysical experiments to isolate the primary chromatic mechanisms. For simplicity the discussion will deal with spectrally opponent cells where chro-

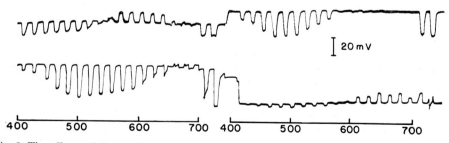

Fig. 6. The effects of chromatic adapting lights on two C-type S-cells. The deflections are the responses (hyperpolarizing downward) to stimuli shown by the scale. To the left are the spectral responses in the absence of the adapting lights, followed by two responses to white lights. Onset of the adapting lights immediately follow the white light and the responses to the right are to stimuli presented on the adapting field. The top records show the effects of a red adapting light. The long-wavelength depolarizing responses are abolished and the short-wavelength responses increase in amplitude and spectral range. The bottom records show similar results using a blue adapting light. (From WITKOVSKY, 1967)

matic adaptation is used to desensitize either the inhibitory or excitatory chromatic input, allowing examination of the relatively non-adapted one without much interference from the antagonist. An example of such an experiment (from Wit-kovsky, 1967) on a spectrally opponent, or C-type, S-potential is given in Fig. 6. The records on the left are the spectral responses of two cells to flashes of different wavelengths in the absence of any adapting light. On the right are shown the responses to spectral stimuli superimposed on red (upper record) or blue (lower record) adapting lights. In the first cell adaptation to red reduces the depolarization elicited by long-wavelength stimuli and increases the amplitude and spectral range of the hyperpolarization to shorter wavelengths. Similarly (lower record) a blue adapting light isolates the long-wavelength mechanism.

The problem is to decide if a single response mechanism determined by one photopigment has been isolated and, if so, which of the cone types is involved. The simplest way is to compare (using appropriate units) the spectral sensitivity of the responses isolated by adaptation with the sensitivities of the various cone types in the given species. Alternatively, if the unit's spectral function can be fitted by the Dartnall (1953) nomogram this would indicate that only a single photopigment was contributing to the observed response and the spectral locus of maximal sensitivity would identify the cone type.

Care must be used in relating the spectral sensitivity of a unit to a single cone type. Typically, one finds for each wavelength the energy required to elicit a threshold response, but this is difficult to do if the unit is spontaneously active. More generally, the responses to spectral stimuli are recorded at several intensities, some amplitude of response is chosen as the criterion, and the energies required at each wavelength to reach the criterion are found by interpolation on the amplitude-intensity functions. (For comparison with a photopigment's sensitivity these measures must be converted from energy to number-of-quanta for the criterion response). When a single photopigment determines the responses, the shape of the amplitude-intensity function should be the same for all wavelengths. However, if the chromatic adaptation was not completely effective, the amplitude-intensity functions will vary with wavelength; in spectrally opponent cells increasing intensity in one portion of the spectrum produces greater response in one direction (positive slope) while in another region the result is a greater response in the opposite direction (negative slope). Thus, over a given spectral range, an opponent unit's responses can not be determined only by inputs from a single cone type if the amplitude-intensity functions do not all fit the same "template" curve. But the converse may not necessarily be true; if, for a series of wavelengths, the responses all follow the same amplitude-intensity function they may still be determined by more than one photopigment; the applicability of the "fixed-template" criterion for identifying a single input will depend on the details of how the system is constructed. In general, though, it has been widely accepted that if all amplitude-intensity curves are the same then for that spectral region this is presumptive evidence for a single input. This, incidentally, means that chromatic adaptation is not indispensable for observing responses derived from one cone type; in some cells the balance between the opposed inputs is so heavily tipped in favour of one that the "fixed-template" condition is met over an appreciable spectral range, but appropriate chromatic adaptation will extend this range further. Note that if the above conditions are adhered to one can not immediately compare the spectral sensitivity of a chromatically isolated mechanism measured just at absolute threshold with a photopigment function — there is no guarantee that the responses at each wavelength would follow the same amplitude-intensity relation indicating a single underlying photopigment.

The effects of chromatic adaptation on spectrally opponent cells are usually quite dramatic (e.g., Fig. 6), but the same is not true of non-opponent cells. If a non-opponent cell receives inputs only from one receptor type, coloured adapting lights will merely depress the responses to all wavelengths uniformly. Only if there are several inputs can chromatic adaptation of one or more of these change the shape of the cell's spectral response function; however the changes will only be of relative amplitudes of response and not of sign as well. Moreover, if the separate inputs all respond to light increases at the same rate, the responses of the non-opponent cell will show the same amplitude-intensity functions at all wavelengths; thus the "fixed-template" amplitude-intensity criterion (discussed above) does not always signify that the input from a single photopigment has been isolated. One can, however, observe if the spectral sensitivity of the responses during chromatic adaptation follows the absorption function of some photopigment.

b) S-Potentials

Usually experiments using chromatic adaptation allow a clear identification of the cone inputs and the data agree quite well with the spectrophotometric measures. However, recently some rather puzzling results have been reported. Although the results were from S-units, the problems raised are important for other responses as well; the comments apply to all attempts at identifying the inputs underlying any cell's responses. NAKA and RUSHTON (1966a, b) recorded spectrally opponent potentials from the retina of the tench and observed the two types (blue-yellow, and green-red) previously described by SVAETICHIN and MacNICHOL (1958); for each of these C-units the response to long wavelengths was depolarization. By using chromatic adaptation to suppress one or other of the inputs to the cells they were able to obtain the action spectra of the cones contributing to the responses observed. The responses were taken to be due to a single cone type only over the spectral range for which the amplitude-intensity functions could be fitted by the same "template" curve. As expected, since the tench, like carp and goldfish, is a Cyprinid, the spectral sensitivity of the input underlying the responses to the green region of the spectrum agreed tolerably well with the absorption spectra of the goldfish cones measured by MARKS (1963, 1965a); however the long-wavelength system was very different. MARKS found that the long-wavelength pigment absorbed maximally at 625 nm while NAKA and RUSHTON's data placed peak sensitivity at 680 nm; moreover the spectral range of this 680 nm system was very much narrower than that of any previously measured cone pigment. None of the C-type units appeared to receive inputs from cones with a 625 pigment, but spectrally non-opponent (L-type) S-units did show long-wavelength inputs from 625 nm cones as well as from a 680 nm system (NAKA and RUSHTON, 1966c).

The criteria, discussed earlier, for identifying responses from a single pigment are very persuasive. According to these standards the very long-wavelength input to S-units is derived from a single pigment. Naka and Rushton (1966b) did consider the possibility that their 680 nm "pigment" was result of interactions between two other pigments more sensitive in other spectral regions. They showed that it is generally possible to find a system of interaction whereby the outputs of two pigments can be combined to give a resultant identical with the output of some third pigment; but this interaction "pigment" will not allow one to match responses elicited by monochromatic lights with responses elicited by mixtures of two standard lights in the way which is possible with a true single pigment. In this way they confirmed that their 680 nm system was based on a single pigment.

One of the problems with the above findings is that specific attempts have been made to obtain spectrophotometric measures of such very long-wavelength cone pigments, but none have been found (Marks, 1965a, b; Svaetichin et al., 1965) and the same is true of recordings from single cones (Tomita et al., 1967). As a rule it is easier to use chromatic adaptation to isolate short-wavelength inputs because adapting lights of sufficiently long wavelength can be used so that only one cone pigment is affected; in these cases the isolated short-wavelength inputs agree well with spectrophotometric measures (De Valois and Abramov, 1972; Naka and Rushton, 1966a, b; Wagner et al., 1960). Thus, it might be possible that the chromatic adaptation used by Naka and Rushton was insufficient to isolate fully the long-wavelength system and that the responses were in fact due to the 625 nm cones. For example, Svaetichin et al. (1965) recorded from a spectrally opponent S-unit (depolarized by long wavelengths) whose peak depolarization was at 690 nm, but use of a blue adapting light shifted peak depolarization to 590 nm, which corresponded with the absorption maximum of one of the cones in the species used. However, Witkovsky (1967) found that even with such adaptation opponent S-units in the carp retina appear to have a long-wavelength input (peak at about 660 nm) similar to the 680 nm pigment. Witkovsky's records, though, indicate that the short-wavelength mechanism was by no means completely suppressed by the blue adapting light (see lower trace in Fig. 6). Furthermore, Orlov and Maksimova (1965) derived the spectral sensitivities of the inputs to opponent red-green and non-opponent S-units in carp from colour mixing experiments. They found that the short-wavelength input was maximally sensitive at 530 nm while the long-wavelength system had its maximum between 600 and 630 nm — not at 660—680 nm. The spectral sensitivities of the inputs were obtained directly from the relative intensities of standard red and green lights in a mixture required to match monochromatic lights of equal quantal content; a given mixture was considered to match a single light when the one could be substituted for the other without detectable response. The spectral sensitivity function of their long-wavelength input is redrawn in Fig. 7a together with the function for the cones containing the 625 nm pigment.

Continuing the argument, one can compare the spectral sensitivity of the 680 nm pigment with the sensitivity of the 625 nm cones. Naka and Rushton (1966a, b, c) presented four separate and quite similar estimates of the spectral sensitivity of the 680 nm pigment; three were obtained by fitting a fixed template curve to amplitude-intensity functions and the fourth from colour matching. Their data have been averaged together and redrawn for comparison with the mean spectral absorption function of the 625 nm cones. The comparison is shown in

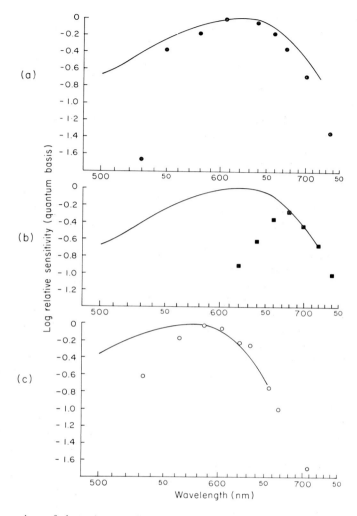

Fig. 7. Comparison of photopigment absorption functions (lines) with the spectral sensitivities (symbols) of the inputs to various cells. (a) Sensitivity of the long-wavelength input to spectrally opponent and non-opponent S-units in the carp retina, as obtained from colour mixing experiments, compared with the 625 nm pigment from goldfish cones. (Points replotted from ORLOV and MAKSIMOVA, 1965. Photopigment function redrawn from mean curve in Fig. 1, which is taken from MARKS, 1965a.) (b) Sensitivity of the "680 nm" input to spectrally opponent and non-opponent S-units in tench. Points are the averages of the four separate measurements of this input given by NAKA and RUSHTON (1966a, b, c); the measures were obtained from chromatic adaptation and colour mixing experiments. The sensitivities were converted to a quantal basis for comparison with the same photopigment curve as in (a). The mean S-unit sensitivity was equated with the 625 nm photopigment curve at 680 nm. (c) Mean sensitivity of the long-wavelength input to $+R-G$ and $+G-R$ spectrally opponent cells in the lateral geniculate nucleus of the macaque; this input was isolated by chromatic adaptation. (Data from DE VALOIS and ABRAMOV, 1972.) The line represents the absorption of the long-wavelength cones in the macaque, as measured by MARKS et al. (1964). The mean function is redrawn from MACNICHOL (1964)

Fig. 7b where the mean spectral sensitivity function of the 680 nm system has been equated with the spectrophotometric results at 680 nm. Clearly the agreement is good for wavelengths longer than 680 nm, suggesting that the antagonistic input was only partially suppressed, thereby shifting the apparent peak of the long-wavelength input. A similar phenomenon has been observed in recordings of spikes from cells in the lateral geniculate nucleus of the macaque. De Valois and Abramov (1972) used short-wavelength adapting lights to isolate the long-wavelength systems of +R−G and +G−R cells and observed that the peak response to long wavelengths was shifted from 640 nm to about 590 nm in both cell types. Spectral sensitivity functions were obtained over the spectral range where the response (spike frequency)-intensity functions could be fitted by curves of a fixed shape. Fig. 7c shows a comparison of these results with the macaque's long-wavelength cones as measured by Marks et al. (1964) − mean absorption function taken from MacNichol (1964). Like the results in Fig. 7a and b, the physiological results deviate from the spectrophotometric at wavelengths shorter than the peak. Yet there is other evidence that the geniculate cells in question (+R−G, and +G−R) receive inputs only from the cones with peak absorptions at 535 and 570 nm and that the inputs are not as narrow as implied by the function from adaptation experiments. The evidence comes from experiments in which it was assumed that these cells receive inputs only from 535 and 570 nm cones whose absorption spectra follow the Dartnall (1953) nomogram. One can then predict which mixtures of various spectral lights will elicit the same responses from these cells (Abramov, 1968).

The isolated long-wavelength inputs to cells in tench, carp, and macaque are obviously very different. Yet in all these cases the sensitivities of the inputs agree closely with the respective photopigment functions at long wavelengths. It is suggested that the deviation at shorter wavelengths is caused by the incompletely suppressed antagonist. Unfortunately, though, this does not explain why the points in Fig. 7 which show this deviation all obeyed the criteria for coming from a single pigment. However it seems preferable to relegate the mystery to the physiology of the cells rather than to invoke otherwise undetected pigments − especially since the absorption spectra of the cones agree so well with a wide range of phenomena including recordings from single receptors, the short-wavelength inputs to S-units, and, as will be seen, responses of ganglion cells.

Spectrally non-opponent S-units (L-type) also present some problems. The shapes of the spectral response curves of L-units in certain shallow-water marine fish (teleosts) can not be changed with chromatic adaptation (Svaetichin and MacNichol, 1958). The curves also fall into three categories which appear to be the same as the types of photopigment measured in these species (Svaetichin et al., 1965; Svaetichin, 1967). It seems that in these fish a given L-unit receives inputs from a single cone type. However in many fresh-water species (including Cyprinids) this is not the case; the spectral response curves of L-units differ from species to species and sometimes within a species, but none are connected to a single receptor type. Orlov and Maksimova (1965) proved this for a variety of species, including carp. They showed that it was impossible to equate different pairs of spectral lights by adjusting their intensities so that no response was evoked when they were alternated; however appropriate mixtures of two standard spectral lights could be matched with single lights — that is, over the range investigated there were two separate chromatic inputs. From the colour mixing data they derived the spectral characteristics of the inputs, which were from photopigments peaking respectively at 530 nm and 600—630 nm (see Fig. 7a); the responses appeared to be the simple sum of the inputs. NAKA and RUSHTON (1966c), however, concluded that L-units in tench receive inputs from the three cone pigments identified previously together with their 680 nm system (see above), and that the units' responses reflected very complex interactions between the inputs. WITKOVSKY (1967) also found a complicated input from a very long-wavelength system in the carp. It was suggested earlier that these very long-wavelength inputs were in fact due to opponent interactions between cone types; if this is so, it is possible that some L-units also receive inputs from a spectrally opponent unit.

c) Ganglion Cells

Chromatic adaptation has also been used to identify the inputs to spectrally opponent and non-opponent ganglion cells. WAGNER et al. (1960) and MACNICHOL et al. (1961) showed that, in goldfish, opponent cells of the green-red type probably received inputs from only two cone types, whose peaks agree quite well with the spectrophotometric measures of MARKS (1963, 1965a) and LIEBMAN and ENTINE (1964). An example of their results is shown in Fig. 8 which gives the sensitivity of a green-"on", and red-"off" cell before and during adaptation by red light. The adaptation greatly extends the spectral range of the "on" response and also increases its absolute sensitivity. Since the spectral region of maximal "on"

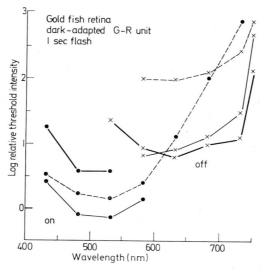

Fig. 8. Intensity necessary to elicit various types of threshold responses from a single, spectrally opponent, ganglion cell at different wavelengths before, during, and after exposure to a red adapting light. Heavy solid line indicates thresholds before adaptation, dashed lines thresholds during red adapting light, thin solid line indicates thresholds approximately 10 min after extinction of the adapting light. Duration of adaptation = 27 min. (From WAGNER et al., 1960)

response was not shifted, the lowered threshold during adaptation suggests that the responses are not simply the difference between the two inputs; rather, adaptation appears to release the input underlying the "on" response from some inhibition by the input which also produces the "off" response. It was generally observed that although chromatic adaptation broadened the spectral range of the non-adapted system, the region of maximal sensitivity was not changed. This is understandable because the goldfish cone absorption functions do not overlap very much. It also means that for this species one can probably infer from the spectral loci of maximal "on"- and "off"-responses the cone types associated with an opponent ganglion cell. By this means Bicking (1965) showed that all possible pairings of the three cone types can be found in responses of ganglion cells in the goldfish retina.

Comparable detailed studies have not been made of ganglion cells in primates; however, as already pointed out, responses from the lateral geniculate nucleus probably also provide a good indication of events at the ganglion cell level. From chromatic adaptation experiments De Valois (1965 b) and De Valois and Abramov (1972) concluded that 570 nm cones determine the responses to long wavelengths in all the four types of spectrally opponent cells they identified in the macaque geniculate; the different types are formed by having either the 445 or 535 nm cones as the antagonistic input. Unlike the situation in goldfish, all the possible pairings of cone inputs were not found. Abramov (1968) reached the same conclusions from experiments based on colour mixtures rather than chromatic adaptation. He was able to predict which mixtures of spectral lights would be equally effective in eliciting some fixed response from cells of the different types; since the stimuli were specifically equated for absorption by the above pairs of inputs only they would not have been equated if the cells also received an input from the remaining cone type. Wiesel and Hubel (1966), in mapping the receptive fields of spectrally opponent cells in the macaque geniculate (discussed later), also specified the cone inputs. In general their findings agree with the conclusions just stated, but they did find some evidence for blue-green cells with opposed inputs from 445 and 535 nm cones. Gouras (1968) recorded from retinal ganglion cells in the macaque and also concluded that spectrally opponent cells had only two inputs; in a few cases he too observed units with inputs from 445 and 535 nm cones.

There is no clear indication of which cones are associated with spectrally non-opponent cells in primates. These cells, at the level of the geniculate, have inputs from more than one cone type, since their spectral sensitivities are typically too broad to be determined by one photopigment, and chromatic adaptation does shift their spectral response functions. However the changes are usually too small to allow one to estimate the spectral sensitivities of the inputs (Jacobs, 1964; De Valois, 1965 b). Recording directly from retinal ganglion cells (macaque) which were apparently spectrally non-opponent, and using chromatic adaptation, Gouras (1968) was able to identify quite nicely inputs from two cone types – probably 535 and 570 nm cones.

The identification of the cone types underlying the spectral responses of ganglion cells (and more central units) is less problematic than is the case with S-units. But this in itself raises serious questions. In carp, for example, very long-wavelength inputs to S-units have been reported, but there is no evidence for such

inputs to the ganglion cells. WITKOVSKY (1967) argued, from this and other observations, that the ganglion cells function independently of the S-cells. RIPPS and
WEALE (1969) have even suggested that S-cells may merely be passively "looking"
at certain aspects of retinal processing of colour but themselves play no part in
transmitting the messages to the central nervous system; but this is not very likely
if S-cells are the horizontal cells (KANEKO, 1970).

d) Rod-Cone Interactions

In general, the transmission of signals in the visual system involves interactions
between the responses due to the different receptors. Certainly by the level of the
ganglion cells there no longer exist pathways connected to only one type of
receptor. For example, spectrally opponent cells receive inputs from two cone
types and appear in some fashion to subtract the inputs of one type from those of
the other. One form of interaction which has been tacitly ignored in the preceding
discussion is the possibility of rods influencing cones and vice versa.

As light adaptation is increased to the point where sensitivity shifts from the
rods to the cones, the rods are still very far from being "saturated" (AGUILAR and
STILES, 1954). Also, for very small stimulus spots the dark-adapted absolute
threshold is very similar for rods and cones; with larger spots the rod threshold
drops much lower implying that the rod system sums over much greater areas
(ARDEN and WEALE, 1954). Thus psychophysical experiments show that both
classes of receptors can and do function simultaneously over a wide range of
intensities. And yet, under conditions of dark-adaptation cones do not contribute
to detecting weak lights, while at higher photopic light levels rods do not contribute
appreciably (see GRAHAM, 1965). It seems as if at some point in the pathway the
responses from one or other class are blocked as a function of adaptation level. The
problems are where this occurs and whether the transition is such that rod responses do not affect cone functions and colour vision.

There is psychophysical evidence that under some conditions rods do contribute
to colour vision. The relative amounts of standard lights in colour mixtures
required to match single spectral lights presented peripherally are not the same as
those required for matches in the fovea, which is free of rods. However, the discrepancies can be accounted for on the assumption of additional inputs from rods
in the periphery (CLARKE, 1963). In a very comprehensive study, LIE (1963) found
that during dark-adaptation the threshold for identifying the colour of a spectral
light rises steeply before absolute threshold for detecting the presence of the light
shifts from cones to rods. Similarly, peripherally viewed test flashes of photopic
intensity become much less saturated in appearance than foveal ones during the
rod phase of dark adaptation. However, rods and cones do not appear to interact
in the "flash after-effect" — that is, where the threshold of a test flash in a small
retinal area is raised by a second (after-) flash falling only upon a surrounding
region and presented a short time after the test flash. ALPERN (1965) found that
after-flashes equated for their effects on the rod system all had the same effect on a
test flash activating only rods; this was so even though the after-flashes were
sufficiently intense to stimulate cones.

Many units in the retinal ganglion layers and lateral geniculate nuclei of
macaques receive inputs from both rods and cones (DE VALOIS and JONES, 1961;

WIESEL and HUBEL, 1966). GOURAS and LINK (1966) have investigated these inputs, and particularly the interactions between them. Recording from ganglion cells in the dark-adapted macaque eye, they found that very dim lights elicited long-latency responses and in these cases the units' spectral sensitivities fitted the rod function; more intense stimuli elicited responses of much shorter latency whose spectral sensitivities were photopic, and hence these responses were due to cones. Using pairs of flashes, of which one stimulated mainly the rods and the other the cones, they varied the timing between the flashes, thus varying the temporal separations with which the rod and cone messages arrived. If the signals from the cones arrived first, they appeared to block the effectiveness of the rod signals and none were seen; similarly, if rod signals were earlier they blocked the cone inputs. Thus, after single flashes which stimulated both rods and cones, only cone responses were seen because they had a shorter latency and pre-empted the channel. GOURAS (1966) argued that the point where this blocking occurred was somewhere close to the ganglion cells since the interaction is not seen in the b-wave of the ERG which seems to arise near the bipolar cells. However there is evidence that this rod-cone interaction may occur at the generator of the late receptor potential (LRP); the LRP has been identified with the synaptic regions of the receptors (BROWN et al., 1965). The decay phase of a pure-rod LRP is much slower than from cones; using a fairly large stimulus spot BROWN and MURAKAMI (1968) found that the macaque LRP showed both slowly and rapidly decaying components when light adapted, but only the slow rod decay when dark adapted; and yet the stimuli were sufficiently intense to evoke both decay components when their area was small. This argues strongly that the rods tend to block cones from contributing to at least part of the LRP when the stimuli are large enough to favour the spatial summation of the rods. However, it should not be concluded that when stimulus conditions favour the cones they always block messages from the rods. GOURAS and LINK's work suggests this, but the cells they observed were probably spectrally non-opponent. One can infer this from the photopic spectral sensitivity functions they obtained with more intense light because one can not obtain such functions from spectrally opponent cells. Since the psychophysical evidence for joint rod and cone contributions deals with the appearance of coloured lights it would seem that these effects should be looked for in spectrally opponent cells, some of which do receive rod inputs (WIESEL and HUBEL, 1966).

4. Receptive Fields

The previous sections have described the responses of cells to coloured lights, but the shape, retinal extent and position of the stimuli were largely ignored. The opponent responses described were opponent with respect to wavelength, but there are also opponent responses related to locus of stimulation in a unit's receptive field. The classical examples of this are the receptive fields of retinal ganglion cells in the cat: a central area responds with excitation in some cells and inhibition in others, while stimulation of an area around the centre elicits the opposite response (KUFFLER, 1953; see also Chapter 14, this volume). The questions for colour vision are whether the spectrally opponent units also show spatial antagonisms of this sort, and how the chromatic mechanisms are distributed over the receptive field.

Most of the work has been done on ganglion cells, but a few things can be said about the receptive fields of more peripheral units. Single-receptor records from carp and *Necturus* indicate that a receptor responds independently of its neighbours; neither TOMITA (1965) nor WERBLIN and DOWLING (1969) found any change in response with increasing stimulus area. However, even their smallest stimuli were large compared to a single cone and so they might already have exceeded a single receptor's interaction field. Using much smaller stimuli, BAYLOR et al. (1971) have demonstrated that cones in the turtle retina do interact; the response of a given cone will be increased by also illuminating its immediate neighbors, but simultaneously stimulating more distant regions causes the response to decay more rapidly; the former interaction seems to be mediated by direct cone-cone connections while the latter is mediated by the horizontal cells. Some interactions have also been observed in the extracellularly recorded late receptor potential (BROWN and MURAKAMI, 1968).

Only cursory attention has been paid to the receptive fields of S-units, but they do appear to be very large (GOURAS, 1960). NORTON et al. (1968), recording from carp, found that stimulus area and intensity were reciprocally related over areas covering virtually the entire retina; no spatial antagonisms were observed. Chromatic adaptation was used to isolate the cone inputs to spectrally opponent S-cells, but still no differences were found in the spatial extents of the inputs. S-cells show summation over very wide areas and no opponent regions in their receptive fields. This provides strong support for the argument that S-cells may not participate directly in transmitting messages to the ganglion cells which, as will be seen do have opponent receptive fields.

Spectrally opponent ganglion cells in the goldfish retina appear to have concentrically organised receptive fields with spatial antagonisms. Thus, a given stimulus in the centre may evoke an "on" response but elicit an "off" response when projected on the surround of the receptive field (WAGNER et al., 1960). Since the responses of these units are also spectrally opponent it is possible, using chromatic adaptation, to examine the spatial distributions of the cone inputs. WAGNER et al. (1963) concluded that both subsystems were maximally sensitive in the centre of the field, but differed in the rates at which their sensitivities dropped as the stimulus spot was moved away from the centre of the receptive field; as a result, the excitatory process predominated in one part of the field, while in the other the inhibitory was more sensitive (Fig. 9a). A similar situation exists in the ground squirrel (MICHAEL, 1966, 1968). With such cells it is possible, using appropriate stimuli, to elicit spectrally opponent responses over much of the receptive field. Spectrally non-opponent cells in goldfish have concentric opponent fields, but both centre and surround have the same spectral sensitivity (WAGNER et al., 1963).

DAW (1967, 1968) extended the previous results from goldfish by stimulating the receptive fields not just with small spots but also annuli. He concluded that for the majority of spectrally opponent units the centres corresponded to what WAGNER et al. (1960, 1963) had described as the entire field; this area was surrounded by a wide peripheral region whose presence was only detectable with annular stimuli — presumably considerable areal summation was needed for the response.

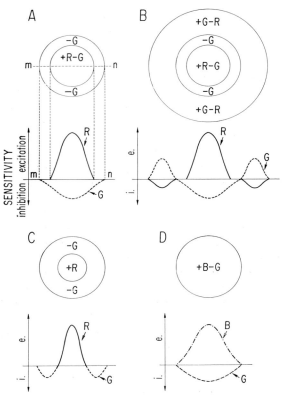

Fig. 9. Receptive fields of various spectrally opponent cells. The upper drawing in each section is a top view of the receptive field; the letters represent the maximally effective wavelengths while the plus and minus signs indicate respectively excitatory and inhibitory responses. The graphs represent the relative sensitivities (along the diameter of each receptive field) of the chromatic systems contributing to each cell's response; the shapes of the curves are not exact— the graphs are intended only as aids to understanding the organization of the receptive fields. (A) and (B): Receptive fields of ganglion cells in goldfish retina; if the fields are examined with small spots of light they seem to be organized as in (A), but if, in addition to spots, annular stimuli are used a large peripheral zone can be demonstrated (B). (C) and (D) are typical receptive fields of cells in the lateral geniculate nucleus of the macaque monkey; there are, however, some questions about the organization as shown here and these are discussed in the text. (A and B are derived from WAGNER et al., 1963; and DAW, 1968; C and D are based on WIESEL and HUBEL, 1966. The entire figure is from ABRAMOV, 1973)

The same pair of chromatic mechanisms associated with the centre were also present in the periphery but with an inversion of sign; if the centre was red-excitatory and green-inhibitory, the periphery would be red-inhibitory and green-excitatory (Fig. 9b). This amounts to an antagonistic centre-surround arrangement for the red-sensitive mechanism superimposed on a reverse centre-surround arrangement for the green; however, the boundaries between centre and surround for red are not the same as for green, which gives rise to the narrow surround zone described by WAGNER et al. (1963). In these cells there are both spatial and chromatic opponent relationships but the chromatic mechanisms do not occupy different parts of the field — a spectrally opponent response can be elicited from most regions of the receptive field. An interesting feature of DAW's findings is the size of the receptive fields which, including the newly found periphery, subtend about 60° at the retina; even the centres subtend some 20°.

Since spectral response patterns are much the same in different species with good colour vision, one might expect similarities in receptive field organisation as well. GOURAS (1968) has reported that ganglion cells in the macaque retina have concentric opponent receptive fields. Cells which apparently are spectrally non-opponent have excitatory centres and inhibitory surrounds but the spectral characteristics of both regions are the same. Cells with spectral antagonisms receive an excitatory input from one cone type in the centre and inhibition from another type in the surround; however, chromatic adaptation indicated that the different cone inputs were not completely restricted to centre or surround. WIESEL and HUBEL (1966) have completed a comprehensive analysis of the receptive fields of cells in the lateral geniculate nucleus of the macaque and have indicated that their findings probably apply to the retina as well. Four types of field were identified. Spectrally opponent cells generally had receptive fields organised in one of two ways: the vast majority (Type I) had concentrically arranged fields with either excitatory or inhibitory centres and opponent surrounds; the spectral sensitivities of centre and surround differed such that there appeared to be a complete spatial separation of the chromatic inputs — the one determined the centre's response, and the other that of the surround (Fig. 9c). Chromatic adaptation confirmed the spatial separation of the opponent spectral inputs so that these units are different from the types found in goldfish. Most Type I units were of the red-green variety. The fields were about 5—6° in size with very small centres (2 min of arc to 1°), especially in the central retina. Type II cells had no centre-surround arrangement but gave spectrally opponent (blue-green) responses over the whole field (Fig. 9d). Spectrally non-opponent cells (Type III) had opponent centre-surround receptive fields but the spectral sensitivities of both parts were the same and were determined by inputs from more than one cone type. Type IV cells were found only in the ventral layers of the nucleus; these were the only cells in which the surround (red-sensitive and inhibitory) predominated when the whole field was illuminated. However, small stimuli in the centre evoked excitatory responses both with red and green light; in this way Type IV cells are spectrally opponent, but would not appear so with diffuse light.

The results just described would indicate that receptive fields in the macaque may not be the same as in the goldfish, although the responses to diffuse light seem comparable. WIESEL and HUBEL's (1966) Type I fields differ from goldfish in that the centre and surround each receive inputs only from one cone type. Moreover, this is in conflict with MEAD's (1967) results from very similar experiments on the macaque. In the absence of chromatic adaptation he found much the same type of receptive field as WIESEL and HUBEL; but, by using chromatic adaptation both cone inputs were found to be distributed over most of the field in much the same manner as in the earlier goldfish experiments. Receptive fields in the macaque also differ in not having the wider peripheral region found in goldfish by DAW (1968). WIESEL and HUBEL did use annular stimuli but failed to find such a periphery. It should be noted that they often found it difficult to obtain a response from the surround alone; in some cases it was only demonstrated by steadily illuminating the centre and turning on and off an annulus on the surround. It remains to be seen if a wider peripheral zone will yet be found for cells in the macaque. (A very few cells have been found in the cortex of the macaque with receptive fields very like those found by Daw, 1968, at the ganglion cell level in the goldfish — Hubel and Wiesel, 1968.)

5. Neural Responses and Psychophysical Data

The general observation that in species with colour vision some cells are spectrally opponent while others are not suggests a two-channel basis for colour vision. In the one channel (colour), wavelengths are differentiated on the basis of excitation and inhibition (spectrally opponent system). The other channel (spectrally non-opponent) can not distinguish between coloured stimuli of appropriately scaled energies, but it can indicate differences in intensity. The added aspect of spatial differences in the inputs from the various cones would, presumably, be related to simultaneous colour and brightness contrasts. But the mere observation that the responses of a given cell type are wavelength selective would not permit one to say that an organism's colour vision is in fact based on the messages from those cells. Some closer correlation between physiological and psychophysical observations is needed. The idea is not novel: Granit (1947) showed that his dominators probably subserved the luminosity function by comparing physiological and behavioural measures of spectral sensitivity.

Unfortunately relatively little has been done to compare in detail the retinal data dealt with in the previous sections with various colour vision functions from the species used. A series of such comparisons has, however, been made by De Valois and his associates who have tried to show how the macaque's performance in psychophysical tests of colour vision might be determined by the responses of units in the lateral geniculate nucleus. One of the reasons for choosing the macaque was that its spectral sensitivity, wavelength discrimination, and colour matching functions, obtained in psychophysical experiments, were found to be essentially identical to those from normal humans (De Valois, 1965a, b; De Valois and Jacobs, 1968). The comparisons have been presented in detail elsewhere (De Valois et al., 1966, 1967; Abramov, 1973; and in particular De Valois, Chapter 3, Vol. VII/3 of this handbook). Even though the comparisons are not strictly based on retinal recordings they probably reflect the state of affairs at the retinal ganglion level and so a few examples will be given together with some work from the goldfish retina.

The luminosity function defines the relative visual effectiveness of different wavelengths. If it is correct that spectrally non-opponent cells are a luminosity channel, then their spectral sensitivity function should fit the luminosity function of the given species. These units cannot form a colour channel since all wavelengths can be adjusted in intensity so as to elicit the same response from these cells; nevertheless they alone may not form the luminosity channel.

The photopic, or light-adapted, luminosity function of the macaque has been obtained psychophysically both by flicker-fusion (De Valois, 1965a) and from threshold measures (SCHRIER and BLOUGH, 1966). The results agree with each other and show that the macaque photopic function is very similar to the curve from humans. The relative spectral sensitivity of non-opponent excitatory cells in the macaque geniculate has been obtained for photopic conditions and agrees well

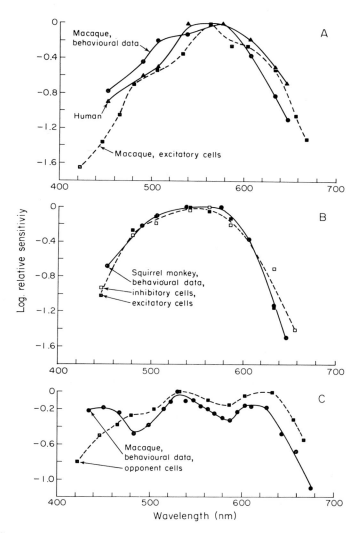

Fig. 10. Comparison of psychophysical determinations of photopic spectral sensitivity with the sensitivities of various cells in the lateral geniculate nucleus. (*A*) compares behavioural data from humans and macaques with spectrally non-opponent excitatory cells in the macaque. (*B*) shows similar functions obtained from squirrel monkeys. In (*C*) the psychophysical function was obtained under conditions of high light-adaptation; this function is compared with the sensitivity of the entire class of spectrally opponent cells in the macaque. (*A* and *B* are redrawn from DE VALOIS and JACOBS, 1968, and JACOBS, 1964. Behavioural curve in *C* is an average of data given by SPERLING et al., 1968; physiological data from DE VALOIS et al., 1966. The entire figure is from ABRAMOV, 1973)

with the psychophysical results (De Valois et al., 1966, and see Fig. 10a). Similarly, Jacobs (1964) has shown that in the squirrel monkey both inhibitory and excitatory non-opponent cells have the same relative spectral sensitivity, and both functions fit the animal's spectral sensitivity very closely (Fig. 10b); this species is protanomalous and less sensitive to long wavelengths than the macaque or man (Jacobs, 1963; De Valois and Jacobs, 1968). Spectrally non-opponent units are also highly sensitive to changes in level of illumination, even with white light for stimuli (De Valois et al., 1962; Jacobs, 1965); spectrally opponent cells generally show little response to white light since both antagonistic mechanisms are activated and tend to cancel each other. However, spectrally non-opponent cells are obviously not the only units whose responses change with intensity of illumination; under some conditions spectrally opponent cells appear to provide not only messages about colour but also signal luminosity. The macaque photopic function obtained under some conditions of high light adaptation is markedly different from the more usual curves (Sperling et al., 1968); but it is quite similar to the spectral sensitivity of a system related to all the spectrally opponent cells in the macaque geniculate (Fig. 10c); the curve was obtained by summing arithmetically the responses of the various types of opponent cells, weighted according to the frequency of occurence of each type.

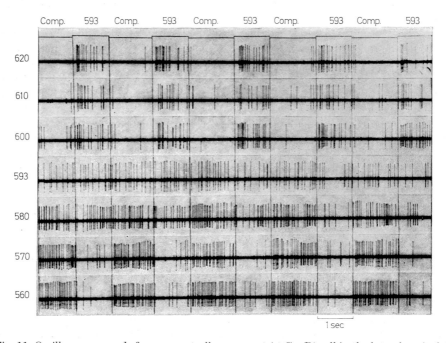

Fig. 11. Oscilloscope records from a spectrally opponent ($+$G$-$R) cell in the lateral geniculate nucleus of the macaque. The responses are to wavelength shifts between a 593 nm standard and various comparison wavelengths whose values are on the ordinate. All stimuli equated for luminance. Top line is the signal marker: upward deflection indicates standard wavelength, downward deflection comparison wavelength. (From De Valois et al., 1967)

In short, spectrally non-opponent cells probably do subserve luminosity under many photopic conditions, but they are not the exclusive source of messages about intensity (see also GUTH et al., 1969). Under conditions of dark-adaptation the sensitivity of many of these cells shifts to the rod function (macaque ganglion cells: GOURAS and LINK, 1966; macaque geniculate: WIESEL and HUBEL, 1966) however, again, they are not the sole luminosity channel since some spectrally opponent cells change to non-opponent units with rod inputs in the dark adapted state (WIESEL and HUBEL, 1966).

The defining characteristic of colour vision is the ability to discriminate between different spectral lights. The behavioural wavelength discrimination function is obtained by measuring the smallest difference which can be detected (at some constant criterion) between a standard wavelength and a comparison light of different wavelength; the standard wavelengths are spaced across the spectrum; and the lights are all equated for luminosity so that detection can only be based on wavelength differences. In such experiments both humans and macaques can detect very fine wavelength differences in the regions about 490 and 590 nm (DE VALOIS and JACOBS, 1968). DE VALOIS et al. (1967) essentially repeated these procedures in experiments on spectrally opponent cells in the macaque geniculate: standard and comparison wavelengths were alternated and the wavelength differences for constant change in spike rate obtained. To maintain comparibility with the psychophysical work, all stimuli were equated for luminance. All these equal-luminance stimuli evoke the same responses from the spectrally non-opponent cells, which therefore contribute nothing to the decision of whether two spectral lights are different. The opponent cells, on the other hand, readily detect the shifts in wavelength (Fig. 11). Wavelength discrimination functions were obtained for each of the four spectrally opponent cell types described by DE VALOIS et al. (1966); and, as might be expected, it was found that they were not all equally good in all spectral regions. The +R−G and +G−R cells discriminated best around 590 nm, while the +B−Y and +Y−B cells did better in the shorter wavelengths, particularly around 490 nm. It was also shown that the psychophysical curve could be matched by the lower bound of the superimposed functions from the four cell types; that is, at any given spectral point the subject's discriminative capabilities are determined by whichever cell type is the most sensitive to wavelength differences in that spectral region. It should be noted that the regions of best discrimination are about the points where the cells' responses cross over from excitation to inhibition; that is, where the opposed inputs from the cones are about in balance. Around this point a small wavelength shift in one direction tips the balance in favour of the excitatory mechanism, while a shift in the other direction favours inhibition.

So far the comparisons have been with psychophysical experiments where spatial aspects were not specifically part of the stimulus. These aspects are relevant to situations in which simultaneous colour contrast is observed. It is well known that if an achromatic area is surrounded by a chromatic area, the achromatic area now appears to be of a hue that is complementary to the surround; for example, a red surround about a grey area "induces" an appearance of green in the grey area. Various schemes have been proposed to account for this; some depend on relative sensitivity changes between cone systems and inhibitory interactions between them (e.g., Alpern, 1964; Alpern and Rushton, 1965), and some demand changes in response mechanisms (e.g., Jameson and Hurvich, 1964).

This is not the place to deal in great detail with this very complex topic, but the following points can be noted. For simultaneous colour contrast, the contrast is probably encoded before the signals from the two eyes meet (Land and Daw, 1962; De Valois and Walraven, 1967); thus, one should find units in the peripheral levels of the visual system with receptive fields capable of encoding contrast. Since these contrasts appear even in the absence of luminance differences, these units are probably spectrally opponent. By themselves none of the receptive fields of spectrally opponent cells in the macaque geniculate (Wiesel and Hubel, 1966) or retina (Gouras, 1968) can play this role; the opposed chromatic mechanisms are spatially separated into centre and surround so that stimulating the centre with one colour and the surround with its complement would cause the responses of the two systems to cancel each other – and yet psychophysically such a stimulus configuration would induce considerable contrast. Daw (1967, 1968) showed that the receptive field organisation he found in the goldfish retina yielded responses which were consistent with the phenomena of simultaneous colour contrast. For example, in a unit with red-"on" and green-"off" centre and the reverse for the surround, neither a red circle in the centre nor a green annulus on the surround gave as much "on" response as the two together.

It is a little dangerous to infer just from such qualitative agreements that these units alone signal all of simultaneous contrast, but the results are very suggestive. There is also the problem that stimuli neatly centred on the receptive field do not correspond to what happens in a behavioural situation. In the latter case a given receptive field will either be entirely covered with one uniform part of the stimulus, or will have its field transected by some boundary of the stimulus figure. For closer comparison of the two situations one really needs to known how these cells respond to a boundary between two fields of complementary colour projected in various positions across the field. Daw (1967, 1968) used such stimuli as well, and the responses to them gave some of his clearest evidence for colour contrast.

From the foregoing comparisons it seems that the broad aspects of the two-channel hypothesis for colour vision hold up under more detailed scrutiny. Spectrally opponent cells are concerned largely with wavelength and colour, while non-opponent ones deal more with intensity. Obviously this hypothesis is limited to situations in which colour and brightness are the dimensions of interest; the same units doubtless also subserve all the other attributes of vision, but these were not the primary concern here.

References

Abramov, I.: Further analysis of the responses of LGN cells. J. Opt. Soc. Amer. **58**, 574—579 (1968).
— Physiology of the visual system: mechanisms of colour vision. In: Cole, A. (Ed.): Theoretical and Experimental Biophysics, Vol. 3. New York: Marcel Dekker 1973.
Adrian, E. D., Matthews, R.: The action of light on the eye. Part I. The discharge of impulses in the optic nerve and its relation to the electrical change in the retina. J. Physiol. (Lond.) **63**, 378—414 (1927).
Aguilar, M., Stiles, W. S.: Saturation of the rod mechanism of the retina at high levels of stimulation. Optica Acta **1**, 59—65 (1954).

ALPERN, M.: Relation between brightness and color contrast. J. Opt. Soc. Amer. **54**, 1491—1492 (1964).
— Rod-cone independence in the after-flash effect. J. Physiol. (Lond.) **176**, 462—472 (1965).
— RUSHTON, W.A.H.: The specificity of the cone interaction in the after-flash effect. J. Physiol. (Lond.) **176**, 473—482 (1965).
ARDEN, G.B., WEALE, R.A.: Nervous mechanisms and dark-adaptation. J. Physiol. (Lond.) **125**, 417—426 (1954).
BAKER, H.D., RUSHTON, W.A.H.: The red sensitive pigment in in normal cones. J. Physiol. (Lond.) **176**, 56—72 (1965).
BALARAMAN, S.: Color vision research and the trichromatic theory: a historical review. Psychol. Bull. **59**, 434—448 (1962).
BAYLOR, D.A., FUORTES, M.G.F., O'BRYAN, P.M.: Receptive fields of cones in the retina of the turtle. J. Physiol. (Lond.) **214**, 265—294 (1971).
BICKING, L.A.: Some quantitative studies on retinal ganglion cells. Doctoral thesis. Baltimore: Johns Hopkins University, 1965.
BLACKWELL, H.R., BLACKWELL, O.M.: Rod and cone receptor mechanisms in typical and atypical congenital achromatopsia. Vision Res. **1**, 62—107 (1961).
BOYCOTT, B.B., DOWLING, J.E.: Organization of the primate retina: light microscopy. Phil. Trans. B **255**, 109—184 (1969).
BOYNTON, R.M., IKEDA, M., STILES, W.S.: Interactions among chromatic mechanisms as inferred from positive and negative increment thresholds. Vision Res. **4**, 87—117 (1964).
BRIDGES, C.D.B.: Visual pigment 544, a presumptive cone pigment from the retina of the pigeon. Nature (Lond.) **195**, 40—42 (1962).
BRINDLEY, G.S.: The effects on colour vision of adaptation to very bright lights. J. Physiol. (Lond.) **122**, 332—350 (1953).
— Physiology of the Retina and the Visual Pathway. London: Edward Arnold 1960.
— RUSHTON, W.A.H.: The colour of monochromatic light when passed into the human retina from behind. J. Physiol. (Lond.) **147**, 204—208 (1959).
BROWN, K.T., MURAKAMI, M.: Rapid effects of light and dark adaptation upon the receptive field organization of S-potentials and late receptor potentials. Vision Res. **8**, 1145—1171 (1968).
— WATANABE, K., MURAKAMI, M.: The early and late receptor potentials of monkey cones and rods. Cold Spr. Harb. Symp. quant. Biol. **30**, 457—482 (1965).
BROWN, P.K.: A system for microspectrophotometry employing a commercial recording microspectrophotometer. J. Opt. Soc. Amer. **51**, 1000—1008 (1961).
— WALD, G.: Visual pigments in human and monkey retinas. Nature (Lond.) **200**, 37—43 (1963).
BURKHARDT, D.A.: The goldfish electroretinogram: relation between photopic spectral sensitivity functions and cone absorption spectra. Vision Res. **6**, 517—532 (1966).
CAMPBELL, F.W., RUSHTON, W.A.H.: Measurement of the scotopic pigment in the living human eye. J. Physiol. (Lond.) **130**, 131—147 (1955).
CLARKE, F.J.J.: Further studies of extra-foveal colour metrics. Optica Acta **10**, 257—284 (1963).
CONE, R.A.: Early receptor potential: photoreversible charge displacement in rhodopsin. Science **155**, 1128—1131 (1967).
DARTNALL, H.J.A.: The interpretation of spectral sensitivity curves. Brit. med. Bull. **9**, 24—30 (1953).
— The photobiology of visual processes. In: DAVSON, H. (Ed.): The Eye, Vol. 2, pp. 321—533. New York: Academic Press 1962.
DAW, N.W.: Goldfish retina: organization for simultaneous color contrast. Science **158**, 942—944 (1967).
— Colour-coded ganglion cells in the goldfish retina: extensions of their receptive fields by means of new stimuli. J. Physiol. (Lond.) **197**, 567—592 (1968).
DE VALOIS, R.L.: Behavioral and electrophysiological studies of primate vision. In: NEFF, W. D. (Ed.): Contributions to Sensory Physiology, Vol. 1, pp. 137—178. New York: Academic Press 1965a.

De Valois, R.L.: Analysis and coding of color vision in the primate visual system. Cold Spr. Harb. Symp. quant. Biol. **30**, 567—579 (1965b).
— Abramov, I.: Effects of chromatic adaptation on the responses of single cells in the lateral geniculate nucleus of the macaque. In preparation (1972).
— — Jacobs, G. H.: Analysis of response patterns of LGN cells. J. Opt. Soc. Amer. **56**, 966—977 (1966).
— — Mead, W. R.: Single cell analysis of wavelength discrimination at the lateral geniculate nucleus in the macaque. J. Neurophysiol. **30**, 415—433 (1967).
— Jacobs, G. H.: Primate color vision. Science **162**, 533—540 (1968).
— — Jones, A. E.: Effects of increments and decrements of light on neural discharge rate. Science **136**, 986—988 (1962).
— Jones, A. E.: Single-cell analysis of the organization of the primate color-vision system. In: Jung, R., Kornhuber, H. (Eds.): The Visual System: Neurophysiology and Psychophysics, pp. 178—191. Berlin-Göttingen-Heidelberg: Springer 1961.
— Smith, C. J., Kitai, S. T., Karoly, A. J.: Response of single cells in monkey lateral geniculate nucleus to monochromatic light. Science **127**, 238—239 (1958).
— Walraven, J.: Monocular and binocular aftereffects of chromatic adaptation. Science **155**, 463—465 (1967).
Dowling, J. E.: Synaptic organization of the frog retina: an electron microscopic analysis comparing the retinas of frogs and primates. Proc. roy. Soc. B **170**, 205—228 (1968).
— Boycott, B. B.: Organization of the primate retina: electron microscopy. Proc. roy. Soc. B **166**, 80—111 (1966).
Enoch, J. M.: Optical properties of the retinal receptors. J. Opt. Soc. Amer. **53**, 71—85 (1963).
Goldstein, E. B.: Early receptor potential of the isolated frog retina. Vision Res. **7**, 837—845 (1967).
— Contribution of cones to the early receptor potential in the rhesus monkey. Nature (Lond.) **222**, 1273—1274 (1969).
— Berson, E. L.: Cone dominance of the human early receptor potential. Nature (Lond.) **222**, 1272—1273 (1969).
Gouras, P.: Graded potentials of bream retina. J. Physiol. (Lond.) **152**, 487—505 (1960).
— Rod and cone independence in the electroretinogram of the dark-adapted monkey's perifovea. J. Physiol. (Lond.) **187**, 455—464 (1966).
— Identification of cone mechanisms in monkey ganglion cells. J. Physiol. (Lond.) **199**, 533—547 (1968).
— Link, K.: Rod and cone interaction in dark-adapted monkey ganglion cells. J. Physiol. (Lond.) **184**, 499—510 (1966).
Graham, C. H.: Some fundamental data. In: Graham, C. H. (Ed.): Vision and Visual Perception, pp. 68—80. New York-London-Sydney: John Wiley & Sons 1965.
Granit, R.: Sensory Mechanisms of the Retina. London: Oxford University Press 1947.
— Svaetichin, G.: Principles and technique of the electrophysiological analysis of colour reception with the aid of microelectrodes. Upsala Läk.-Fören. Förh. **65**, 161—177 (1939).
Guth, S. L., Donley, N. J., Marrocco, R. T.: On luminance additivity and related topics. Vision Res. **9**, 537—575 (1969).
Hagins, W. A.: Electrical signs of information flow in photoreceptors. Cold Spr. Harb. Symp. quant. Biol. **30**, 403—418 (1965).
Hartline, H. K.: The response of single optic nerve fibers of the vertebrate eye to illumination of the retina. Amer. J. Physiol. **121**, 400—415 (1938).
— Wagner, H. G., MacNichol, E. F., Jr.: The peripheral origin of nervous activity in the visual system. Cold Spr. Harb. Symp. quant. Biol. **17**, 125—141 (1952).
Hsia, Y., Graham, C. H.: Spectral luminosity curves of protanopic, deuteranopic, and normal subjects. Proc. nat. Acad. Sci. (Wash.) **43**, 1011—1019 (1957).
Hubel, D. H., Wiesel, T. N.: Receptive fields of optic nerve fibers in the spider monkey. J. Physiol. (Lond.) **155**, 572—580 (1960).
— — Receptive fields and functional architecture of monkey striate cortex. J. Physiol. (Lond.) **195**, 215—243 (1968).
Jacobs, G. H.: Spectral sensitivity and color vision of the squirrel monkey. J. comp. physiol. Psychol. **56**, 616—621 (1963).

JACOBS,G.H.: Single cells in squirrel monkey lateral geniculate nucleus with broad spectral sensitivity. Vision Res. **4**, 221—232 (1964).

— Effects of adaptation on the lateral geniculate response to light increment and decrement. J. Opt. Soc. Amer. **55**, 1535—1540 (1965).

— DE VALOIS, R. L.: Chromatic opponent cells in squirrel monkey lateral geniculate nucleus. Nature (Lond.) **206**, 487—489 (1965).

— YOLTON, R. L.: Dichromacy in the ground squirrel. Nature (Lond.) **223**, 414—415 (1969).

JAMESON, D., HURVICH, L. M.: Theory of brightness and color contrast in human vision. Vision Res. **4**, 135—154 (1964).

KANEKO, A.: Physiological and morphological identification of horizontal, bipolar and amacrine cells in goldfish retina. J. Physiol. (Lond.) **207**, 623—633 (1970).

— HASHIMOTO, H.: Recording site of the single cone response determined by an electrode marking technique. Vision Res. **7**, 847—851 (1967).

KUFFLER, S. W.: Discharge patterns and functional organization of mammalian retina. J. Neurophysiol. **16**, 37—68 (1953).

LAND, E. H., DAW, N. W.: Binocular combination of projected images. Science **138**, 589—590 (1962).

LIE, I.: Dark adaptation and the photochromatic interval. Docum. ophthal. (Den Haag) **17**, 411—510 (1963).

LIEBMAN, P. A.: In situ microspectrophotometric studies on the pigments of single retinal rods. Biophys. J. **2**, 161—178 (1962).

— ENTINE, G.: Sensitive low light level microspectrophotometer — Detection of photosensitive pigments of retinal cones. J. Opt. Soc. Amer. **54**, 1451—1459 (1964).

LIPETZ, L. E.: Glial control of neuronal activity. IEEE Trans. Milit. Electron MIL-7, 144—155 (1963).

MACHOL, R. E.: Chairman and Ed. Debate and panel discussion: How to apply system analysis properly to biological systems. IEEE Internal. Convention Record, Part **9**, 210—252 (1967).

MACNICHOL, E. F., JR.: Three-pigment color vision. Sci. Amer. **211**, 48—56 (1964).

— Retinal mechanisms of color discrimination. In: BURIAN, H. M., JACOBSON, J. H. (Eds.): Clinical Electroretinography, pp. 55—73. London: Pergamon Press 1966.

— SVAETICHIN, G.: Electric responses from the isolated retinas of fishes. Amer. J. Ophthal. **46**, 26—39 (1958).

— WOLBARSHT, M. L., WAGNER, H. G.: Electrophysiological evidence for a mechanism of color vision in the goldfish. In: MCELROY, W. D., GLASS, B. (Eds.): Light and Life, pp. 795—814. Baltimore: Johns Hopkins Press 1961.

MARKS, W. B.: Difference Spectra of the Visual Pigments in Single Goldfish Cones. Doctoral Thesis. Baltimore: Johns Hopkins University 1963.

— Visual pigments of single goldfish cones. J. Physiol. (Lond.) **178**, 14—32 (1965a).

— Discussion. In: DE REUCK, A. V. S., KNIGHT, J. (Eds.): Ciba Foundation Symposium: Colour Vision, Physiology and Experimental Psychology, pp. 280—285. Boston: Little-Brown 1965b.

— DOBELLE, W. H., MACNICHOL, E. F., JR.: Visual pigments of single primate cones. Science **143**, 1181—1183 (1964).

MARRIOTT, F. H. C.: Color vision. In: DAVSON, H. (Ed.): The Eye, Vol. 2, pp. 219—320. New York: Academic Press 1962.

MEAD, W. R.: Analysis of the Receptive Field Organization of Macaque Lateral Geniculate Nucleus Cells. Doctoral Thesis. Bloomington, Indiana: Indiana University 1967.

MICHAEL, C. R.: Receptive fields of opponent color units in the optic nerve of the ground squirrel. Science **152**, 1095—1097 (1966).

— Receptive fields of single optic nerve fibers in a mammal with an all-cone retina. III: Opponent color units. J. Neurophysiol. **31**, 268—282 (1968).

MISSOTTEN, L., APPELMANS, M., MICHIELS, J.: L'ultrastructure des synapses des cellules visuelles de la retine humaine. Bull. Soc. Franc. Ophthal. **76**, 59—82 (1963).

MOTOKAWA, K., YAMASHITA, E., OGAWA, T.: Studies on receptive fields of single units with colored lights. Tohoku J. exp. Med. **71**, 261—272 (1960).

Munz, F. W., Schwanzara, S. A.: A nomogram for retinene$_2$-based visual pigments. Vision Res. 7, 111—120 (1967).

Murray, G. C.: Visual pigment multiplicity in cones of the primate fovea. Doctoral thesis. Baltimore: Johns Hopkins University 1968.

Naka, K. I., Rushton, W. A. H.: S-potentials from colour units in the retina of fish (Cyprinidae). J. Physiol. (Lond.) 185, 536—555 (1966a).

— — An attempt to analyse colour perception by electrophysiology. J. Physiol. (Lond.) 185, 556—586 (1966b).

— — S-potentials from luminosity units in the retina of fish (Cyprinidae). J. Physiol. (Lond.), 185, 587—599 (1966c).

Norton, A. L., Spekreijse, H., Wolbarsht, M. L., Wagner, H. G.: Receptive field organization of the S-potential. Science 160, 1021—1022 (1968).

Orlov, O. Yu., Maksimova, E. M.: S-potential sources as excitation pools. Vision Res. 5, 573—582 (1965).

Polyak, S.: The Vertebrate Visual System. Chicago: University of Chicago Press 1957.

Pitt, F. H. G.: The nature of normal trichromatic and dichromatic vision. Proc. roy. Soc. B 132, 101—117 (1944).

Riggs, L. A., Johnson, E. P., Schick, A. M. L.: Electrical responses of the human eye to changes in wavelength of the stimulating light. J. Opt. Soc. Amer. 56, 1621—1627 (1966).

— Sternheim, C. E.: Human retinal and occipital potentials evoked by changes of the wavelength of the stimulating light. J. Opt. Soc. Amer. 59, 635—640 (1969).

Ripps, H., Weale, R. A.: On seeing red. J. Opt. Soc. Amer. 54, 272—273 (1964).

— — Color vision. Ann. Rev. Psychol. 20, 193—216 (1969).

Ruddock, K. H.: The effect of age upon colour vision. II. Changes with age in light transmission of the ocular media. Vision Res. 5, 47—59 (1965).

Rushton, W. A. H.: Human cone pigments. In: Visual Problems of Colour, pp. 71—105. Natl. Phys. Lab. Symp. No. 8. London: H.M.S.O. 1958.

— A cone pigment in the protanope. J. Physiol. (Lond.) 168, 345—359 (1963).

— Stray light and the measurement of mixed pigments in the retina. J. Physiol. (Lond.) 176, 46—55 (1965a).

— A foveal pigment in the deuteranope. J. Physiol. (Lond.) 176, 24—37 (1965b).

Sechzer, J. A., Brown, J. L.: Color discrimination in the cat. Science 144, 427—429 (1964).

Schrier, A. M., Blough, D. S.: Photopic spectral sensitivity of macaque monkeys. J. comp. physiol. Psychol. 62, 457—458 (1966).

Spekreijse, H., Wagner, H. G., Wolbarsht, M. L., Heffner, D. K.: Color coded components of the intraretinal action potential. Physiologist 10, 311 (1967).

— Rectification in the goldfish retina: analysis by sinusoidal and auxiliary stimulation. Vision Res. 9, 1461—1472 (1969).

Sperling, H. D., Sidley, N. A., Dockens, W. S., Jolliffe, C. L.: Increment-threshold and spectral sensitivity of the rhesus monkey as a function of the spectral composition of the background field. J. Opt. Soc. Amer. 58, 263—268 (1968).

Stell, W. K.: The structure and relationships of horizontal cells and photoreceptor-bipolar synaptic complexes in goldfish retina. Amer. J. Anat. 120, 401—424 (1967).

Stiles, W. S.: Increment thresholds and the mechanisms of color vision. Docum. Ophthal. (Den Haag) 3, 138—165 (1949).

— Colour vision: the approach through increment threshold sensitivity. Proc. nat. Acad. Sci. (Wash.) 45, 100—114 (1959).

Svaetichin, G.: The cone action potential. Acta physiol. scand. 29, Suppl. 106, 565—600 (1953).

— Celulas horizontales y amacrinas de la retina: propriedades y mecanismos de control sobre las bipolares y ganglionares. Acta Cient. Venezolana. Suppl. 3, 254—276 (1967).

— MacNichol, E. F., Jr.: Retinal mechanisms for chromatic and achromatic vision. Ann. N. Y. Acad. Sci. 74, 385—404 (1958).

— Negishi, K., Fatehchand, R.: Cellular mechanisms of a Young-Hering visual system. In: de Reuck, A. V. S., Knight, J. (Eds.): Ciba Foundation Symposium: Colour Vision, Physiology and Experimental Psychology, pp. 178—203. Boston: Little-Brown 1965.

TOMITA, T.: Electrical activity in the vertebrate retina. J. Opt. Soc. Amer. **53**, 49—57 (1963).
— Electrophysiological study of the mechanisms subserving color coding in the fish retina. Cold Spr. Harb. Symp. quant. Biol. **30**, 559—566 (1965).
— KANEKO, A., MURAKAMI, M., PAUTLER, E. L.: Spectral response curves of single cones in the carp. Vision Res. **7**, 519—531 (1967).
— MURAKAMI, M., SATO, Y., HASHIMOTO, Y.: Further study of the origin of the so-called cone action potential (S-potential). Its histological determination. Jap. J. Physiol. **9**, 63—69 (1959).
WAGNER, H. G., MacNICHOL, E. F. JR., WOLBARSHT, M. L.: The response properties of single ganglion cells in the goldfish retina. J. gen. Physiol. **43**, Suppl. 6, part 2, 45—62 (1960).
— — — Functional basis for "on"-center and "off"-center receptive fields in the retina. J. Opt. Soc. Amer. **53**, 66—70 (1963).
WALD, G.: The receptors of human color vision. Science **145**, 1007—1017 (1964).
— BROWN, P. K.: Human color vision and color blindness. Cold Spr. Harb. Symp. quant. Biol. **30**, 345—359 (1965).
— — SMITH, P. H.: Iodopsin. J. gen. Physiol. **38**, 623—681 (1955).
WALLS, G. L.: The Vertebrate Eye and Its Adaptive Radiation. New York: Hafner 1942, reprinted 1963.
WEALE, R. A.: Photo-sensitive reactions in fovea of normal and cone-monochromatic observers. Optica Acta **6**, 158—174 (1959).
— Vision and fundus reflectometry: a review. Photochem. Photobiol. **4**, 67—87 (1965).
WERBLIN, F. S., DOWLING, J. E.: Organization of the retina of the mudpuppy, *Necturus maculosus*. II. Intra-cellular recording. J. Neurophysiol. **32**, 315—355 (1969).
WIESEL, T. N., HUBEL, D. H.: Spatial and chromatic interaction in the lateral geniculate body of the rhesus monkey. J. Neurophysiol. **29**, 1115—1156 (1966).
WILLMER, E. N.: A physiological basis for human colour vision in the central fovea. Docum. Ophthal. (Den Haag) **9**, 235—313 (1955).
WILSKA, A.: Aktionspotentialentladungen einzelner Netzhautelemente des Frosches. Acta Soc. Med. „Duodecim" **22**A, 63—75 (1940).
WITKOVSKY, P.: The spectral sensitivity of retinal ganglion cells in the carp. Vision Res. **5**, 603—614 (1965).
— A comparison of ganglion cell and S-potential response properties in carp retina. J. Neurophysiol. **30**, 546—561 (1967).
— The effect of chromatic adaptation on color sensitivity of the carp electroretinogram. Vision Res. **8**, 823—837 (1968).
WYSZECKI, G., STILES, W. S.: Color Science: Concepts and Methods, Quantitative Data and Formulas. New York-London-Sydney: John Wiley & Sons 1967.
YAGER, D.: Behavioral measures and theoretical analysis of spectral sensitivity and spectral saturation in the goldfish, *Carassius auratus*. Vision Res. **7**, 707—727 (1967).

Chapter 16

Light and Dark Adaptation

By

Peter Gouras, Bethesda, Maryland (USA)

With 13 Figures

Contents

Introduction

Living photoreceptor systems modify their behavior to light stimulation. After exposure to light, their sensitivity decreases and this is accompanied by a speeding up of responses and an increase in space-time resolution; in darkness the entire process is reversed. These changes, called light — and dark — adaptation have evolved to optimize the function of visual receptor systems in an ever-changing external world so that they can enhance sensitivity at the expense of spatial and temporal accuracy when light energy is scarce or do the converse when energy is abundant.

Evolution has discovered many mechanisms, most of which are used conjointly, to change the sensitivity of vision. All of them reside almost entirely within the eye. These include reflex changes in pupillary size, migration of pigment epithelial cells, extensions and contractions of photoreceptor cells (Walls, 1942), the bleaching of photopigments as well as neural changes within the retina. The mechanisms of greatest importance occur within the retina and involve neural and chemical (bleaching) alterations within individual photoreceptor cells as well as groups of similar photoreceptors. When all the mechanisms acting within one photoreceptor system are exhausted, nature is capable of introducing an entirely different class of photoreceptors designed to operate over another range of light energy.

Duplex-Vision

The retinas of almost all vertebrates have two such photoreceptor systems. One is extremely sensitive, operates best in dim illumination and employs a class of photoreceptor cells, called *rods*, all of which in man contain the same photopigment, rhodopsin. The other is relatively insensitive, acts most effectively in broad daylight and uses a class of photoreceptors, called *cones*, which in man are comprised of three distinct types based on the photopigments they contain. When retinal illumination increases sufficiently, vision changes over completely from rod to cone function and this is accompanied by a change in the wavelength of maximal visual sensitivity (Koenig and Ritter, 1891), the so called Purkinje (1825) shift.

This duplex theory of vision began with Müller's (1857) microscopic observations of two distinct classes of photoreceptor organelles in vertebrate retina, rods and cones. Max Schultze (1866) first realized that rods and cones were most common in nocturnal and diurnal animals, respectively, and proposed that the former were the receptors for dim and the latter for bright light. This idea had little impact at the time and it was only revived years later by Parinaud (1885), an ophthalmologist, who associated clinical abnormalities of night and day vision with rods and cones, respectively, and independently by von Kries (1895), a physiologist, who had been studying various manifestations of the Purkinje shift. Selig Hecht (1937) has since formulated how duplex function can be identified in so many aspects of visual function that today this theory stands as one of the most fundamental links between structure and function in vertebrate vision.

Responses of Dark-Adapted Rods and Cones

In man the dark-adapted rod mechanism is more sensitive to light than the cone retina except in the long wavelength region of the visible spectrum (Wald, 1945). The cone retina, on the other hand, has better spatial and temporal resolution. These differences are due to many factors. Rod outer segments are long and contain large amounts of photopigment in order to catch as large a fraction of the light that reaches them. This trend is especially exaggerated in deep sea fishes (Denton and Warren, 1957). Except in the foveal region of some vertebrates cone outer segments tend to be short and consequently have less photopigment available to absorb light. Rods respond to light rays reaching them from a large angle; cones only respond to a narrow angle of incident light (Stiles-Crawford, 1933), which in man is approximately the size of the pupillary aperture in daylight. Because of this directional selectivity cones will not be excited as easily as rods by light scattering in the retina and as a result lose in sensitivity but gain in spatial accuracy.

The total spatial summation (intensity × area = constant) of the dark-adapted rod system is as large as 1° of visual angle in man; the cone system has a much smaller area of total spatial summation (Wald, 1959). These differences are also reflected in the duplex receptive field organization of vertebrate ganglion cells when tested with concentric spot stimuli (Gouras, 1967; Andrews and Hammond, 1970a, b). One field is determined by the rods, the other by the cones. The rod receptive field is larger than that of the cones so that changes in the spatial parameters of stimulation can shift the threshold response of a ganglion cell from rod to cone function even in the dark-adapted state. Similarly dark-adapted human vision undergoes a Purkinje shift when the spatial parameters of stimulation change sufficiently (BROWN, KUHNS, and ADLER, 1957).

In the dark-adapted human retina spatial summation of vision also depends upon the duration of light stimulation, being greater for stimuli of short than for those of long duration (BOUMAN, 1950, 1952; BARLOW, 1958). A similar decrease in spatial summation within the receptive field of cat ganglion cells occurs when the duration of stimulation is increased and this had been considered to reflect a longer integration time for antagonistic influences originating in the more peripheral regions of the cell's receptive field (BARLOW, FITZHUGH, and KUFFLER, 1957).

The total temporal summation time (intensity $\times$ time = constant) of the rod system is about 0.1 sec; the cone system has a shorter temporal summation time (GRAHAM and MARGARIA, 1935; BAUMGARDT, 1959). The temporal integration time of both rods and cones is not constant, however, but decreases as the area of retinal illumination increases (BOUMAN, 1950, 1952; BARLOW, 1958) indicating that groups of receptors acting together influence the responses of each other. The entire time course of the rod response is slower than that of the cones and this is reflected in the lower flicker fusion frequencies (HECHT, 1937) of rod vision, the slower time course (ADRIAN, 1946; ARMINGTON, JOHNSON and RIGGS, 1952) and lower flicker fusion frequency (DODT, 1957; JOHNSON and CORNSWEET, 1954) of the rod ERG and the longer latency of rod signals arriving at S-potentials (STEINBERG, 1970) and ganglion cells (GOURAS, 1967; ANDREWS and HAMMOND, 1970a, b) in vertebrate retina. This greater delay of the rod system must begin within the receptor cell, itself, since it is evident in the earliest components of the ERG. Recordings of single rod receptors have not yet been performed although the mass response of many rod receptors isolated by clamping the central retinal artery of the monkey shows a slower decay phase than that of cones (BROWN, WATANABE and MURAKAMI, 1965). A possible criticism of the latter approach is that destruction of cells in the inner nuclear layer of the retina may deprive rods of a synaptic input which contributes to their normal response.

Light-Adaptation of Rods

When rods are exposed to light their thresholds rise. The traditional way of demonstrating this is to measure rod thresholds to a spot of light appearing upon a steady background. The relationship between threshold and the brightness of the background field is called an *increment threshold* curve. Fig. 1 shows such a relationship over the entire range of human rod vision. In this case rods have been isolated from cones by examining the retina 9° from the fovea where rod sensitivity is maximal with test stimuli of large size (9° diameter) and long duration (0.2 sec) and by appropriately selecting the wavelength and the point of entry in the pupil of both the test and the adapting beam (AGUILAR and STILES, 1954). Dim adapting lights have a weak effect on rod threshold; stronger adapting lights have more effect as thresholds begin to rise in almost direct proportion to the luminance of the adapting field. At this point the behavior of rod vision resembles a well known generalization in sensory psychology, the Weber-Fechner relationship. At higher

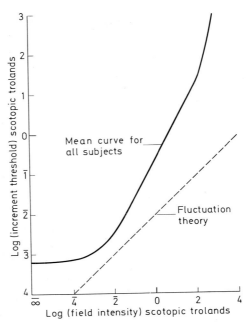

Fig. 1. Comparison of the observed variation of log T (threshold) with log F (adapting field)
and the variation derived by fluctuation theory (broken line) (Aguilar and Stiles, 1965)

levels of retinal illumination this resemblance ceases as human rod vision saturates
so that any further increase in stimulation produces no concomitant increase in
rod response.

The mechanisms responsible for all phases of the rod desensitization by light
illustrated in Fig. 1 are not entirely understood. The line marked, *fluctuation
theory*, indicates the physical limits set on the performance of an ideal photodetector. The quantal fluctuations reaching any photodetector increase with the
square root of the mean number of quanta in the field and stimuli must at least
exceed this value if they are to be detected (Rose, 1948; DeVries, 1943; Pirenne,
1944). Rod vision does not follow this relationship under most conditions and
therefore rod desensitization by light cannot be explained by fluctuation theory
alone (Aguilar and Stiles, 1954; Barlow, 1958; Jones, 1959).

Calculations on the amount of light being absorbed by the rods under the conditions of Fig. 1 indicate that the rod desensitization cannot be due to an exhaustion of the rhodopsin available, even at saturation, (Aguilar and Stiles,
1954) and this has been confirmed by direct measurements of rhodopsin in the
human eye (Campbell and Rushton, 1955). Similar conclusions have been obtained in the rat (Dodt and Echte, 1961; Dowling, 1963) and skate (Dowling and
Ripps, 1970) by comparing increment thresholds of the ERG with rhodopsin
detected in the retina.

An important point that Stiles (1959) has made using increment threshold
curves to isolate receptor mechanisms in human vision is that the elevation of

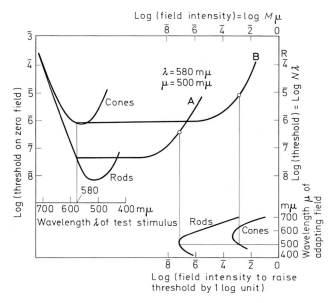

Fig. 2. Diagram illustrating how the composite increment threshold curve of two component mechanisms, A (rod) and B (cone), whose positions with respect to the axes of log $N\lambda$ and log $M\mu$ for any test and field wave lengths λ and μ is determined by the respective spectral sensitivity curves of the mechanisms shown in the auxiliary diagrams (STILES, 1959). The test wavelength, 580 mμ, (vertical line) and the adapting wavelength, 500 Mμ, (horizontal line) affect the rods at a lower intensity than they effect the cones. As the intensity of the adapting field increases (along the horizontal line), the thresholds of the rods (A) rise above those of the cones (B) and threshold for the 580 mμ stimulus is now determined by the cones entirely. At this point the increment threshold runs parallel to the abscissa until the adapting wavelength (500 Mμ) becomes intense enough to affect the cones. At this point the increment threshold begins to rise again due to adaptation of the cones. In all situations the rod and cone mechanisms seem to operate independently of one another

threshold of any receptor mechanism by a steady background light is due to light acting on that particular receptor mechanism alone and this is especially true for rods (Fig. 2). Even rod saturation is not due to cones inhibiting rods but resides entirely within the rod system (FUORTES, GUNKEL, and RUSHTON, 1961; GOURAS, 1965). Therefore it is the action of light on rhodopsin alone which leads to a reduction in rod sensitivity and ultimately eliminates rod vision entirely at saturation. This effect could occur within the rod receptor cell or at some later stage in vision but it must take place at a point where rod signals remain independent of cones.

This desensitization must also involve structures which link the responses of groups of rods. The evidence for this is the following. PIRENNE (1959) found that rod thresholds in a small area (0.1°) of retina, 10° from the fovea, were elevated by a steady annulus of light whose inner edge was about 2° from the test area. Since the stray light falling into the test area from the annulus was insufficient to produce an equivalent change in sensitivity, it was concluded that inhibitory

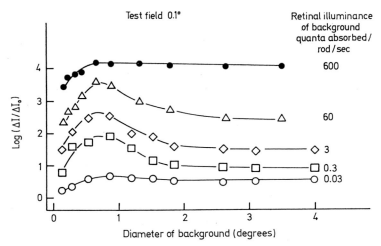

Fig. 3. Incremental threshold for a small blue-green test flash seen in the middle of red backgrounds of various diameters and retinal illuminance levels. Zero on the axis of ordinates denotes the absolute threshold for the test flash. Peripheral vision for dark-adapted eye (WESTHEIMER, 1965)

signals were traveling from rods excited by the annulus to rods determining threshold for the test stimulus. Subsequently RUSHTON (1965b) demonstrated that light-adapting only 1 % of the rods in a 2° area of retina, 10° from the fovea, elevated the thresholds of all the other rods in the neighborhood to the same extent. BARLOW and ANDREWS (1967) disagree with RUSHTON on the extent over which this rod desensitization spreads. ALPERN (1965) also found that rod thresholds in one area of retina were increased by a light stimulus presented 50 msec *later* to another area of retina about a millimeter away from the first. This desensitization was shown to be exerted exclusively by rods upon rods. Therefore a mechanism must exist which spreads rod desensitization by light over adjacent areas of retina to involve the responses of other rods.

Light-adaptation not only desensitizes but under suitable conditions also sensitizes rod vision. CRAWFORD (1940) observed this effect while determining how the size of a steady adapting field influenced threshold in a small area (0.5°) of retina, 8° from the fovea. Increasing the size of the adapting field decreased the sensitivity of the test area until a size was reached where any further increase began to *reduce* this desensitization. WESTHEIMER (1965, 1970; see also TELLER et al., 1970) has re-examined this sensitizing effect of light more specifically for the rod system (Fig. 3). In retina, 10° from the fovea, rod sensitization was best produced by a concentric zone surrounding the test area at a radius of about 300 micra. Sensitization increased with the amount of light used up to a saturation point but the distance for maximum effect did not appear to change. Stimuli that sensitized rods were incapable of sensitizing cones. WESTHEIMER and WYLIE (1970) have also found additional regions beyond the sensitizing annular zone which counteracted this rod sensitization. Light-adaptation of rods must involve the interaction of both excitatory and inhibitory signals which act over relatively large distances to modulate the sensitivity of rod pathways in any one region of retina. This arrangement is reminiscent of "disinhibition" demonstrated in the lateral eye of Limulus (HARTLINE, RATLIFF, and MILLER, 1961). The specificity of these effects for rods alone suggests that they are mediated by neurons with specific connections and not by chemicals diffusing across the retina.

An explanation of these effects in terms of the underlying anatomy and physiology of the rod system is still difficult. Rod and cone signals converge on single ganglion cells in all vertebrate retinas examined so far, so that rod desensitization must take place at an earlier stage in the retina where rod signals are still independent of cones. In primates rod and cone ERGs seem to be independent of one another (GOURAS, 1966) implying that rod and cone receptor cells do not act directly upon each other. On the other hand, BROWN and MURAKAMI (1968) report antagonistic rod-cone interaction in both receptor and S-potentials in primates but STEINBERG (1970b) failed to confirm this in cat retina. KANEKO (1970) found antagonistic lateral interaction in rod bipolar cells in goldfish retina but did not determine whether this involved both rods and cones. CAJAL (1894) believed that there were separate bipolar and horizontal cells for rods and cones in vertebrate retina and PARTHE (unpublished) has confirmed this for fish retina (see also STELL, 1967). MISSOTTEN (1965) and BOYCOTT and DOWLING (1969) have also found separate rod and cone bipolar cells in primates, correcting an error that POLYAK (1941) made using Golgi technique without the aid of electronmicroscopy and without the intuition of CAJAL. KOLB (1970) finds that horizontal cells of primates send dendrites to cones and axons to rods. BOYCOTT (unpublished) has been successful in completely tracing a cat horizontal cell impregnated with silver and demonstrating by electronmicroscopy that its dendrites go to cones while its axon goes to rods. These terminals ending on rods and cones are connected by axon of enormous length (500 micra) and small diameter (0.1 micra). If horizontal cells do not produce impulses as all evidence indicates then perhaps the rod and cone terminals of such horizontal cells receive and send signals to either rods or cones but not to both being kept electrically isolated by this long, thin "axonal" process. STEINBERG's (1969a) finding of both rod and cone signals in cat S-potentials seems to be evidence against this hypothesis. Whatever the outcome of this problem human vision seems to require that rods have some means of exciting and inhibiting one another specifically and this must presumably be through either receptor-receptor contacts or by a special set of rod horizontal cells as Cajal originally proposed.

Can light desensitize a portion of a rod or a single rod independently of others ? RUSHTON (1965a) has concluded that the desensitization is much less in single human rods than in the pool into which they feed. Implicit in the analysis is the assumption that rod receptors do not directly influence one another. In squid retina, HAGINS, ZONANA and ADAMS (1962) have detected that light stimulation has a smaller effect on light-adapted than on dark-adapted regions of the *same* photoreceptor. In other invertebrates where the responses of single photoreceptors are more attainable, considerable adaptation also appears to occur at an early stage within or near the receptor membrane (FUORTES and HODGKIN, 1964). RATLIFF's (1965) results indicate that this desensitization involves only local portions of the ommatidium (photoreceptors) of Limulus. It would be surprising if some local adaptation did not also occur within the outer segments of vertebrate rods.

Dark-Adaptation of Rods

When the eye is placed in darkness following exposure to bright light, the sensitivity of rod vision begins to recover. In man this recovery process can be observed by measuring the threshold for light detection at different times after an adapting light has been turned off (Fig. 4). Because cones recover more quickly than rods, rod thresholds can only be examined during the later stages of rod recovery, a problem that confronts similar studies in most other vertebrates. The brighter the light exposure, the longer will the recovery time be up to a point. It is important to realize that it is the light exposure to the rods alone and not the cones that determines entirely the subsequent course of rod dark-adaptation. This has been shown by comparing the time course of rod dark-adaptation after exposures to lights of different wavelengths (Hecht and Hsia, 1945; Smith, Morris, and Dimmick, 1955) and also by the coincidence of the dark-adaptation curve of a rod monochromat (a person with normal rod but little or no cone vision) with the rod component of normal dark-adaptation curves (Rushton, 1961). Rod dark-adaptation must take place entirely within the rod system and presumably within the retina where rod signals remain independent of those of cones.

In certain regions of the human retina where the ratio of cones to rods is relatively high, a slight retardation of rod dark-adaptation can be produced by the action of dark-adapting cones (Rushton, 1968). This rod-cone rivalry appears to depend upon competition for a common substrate, presumably 11-*cis* vitamin A. The absence of this competition in most other areas of retina where rods and cones also co-exist indicates that this phenomenon plays no significant role in the usual dark-adaptation of rods.

The parallelism between the slow return of rod sensitivity in the dark and the regeneration of rhodopsin was recognized early in vision research but any quanti-

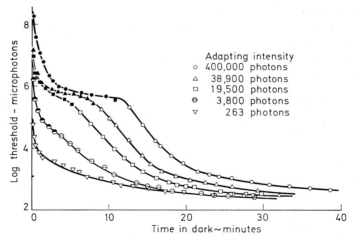

Fig. 4. The course of dark adaptation as measured with violet light following different degrees of light adaptation. The filled-in symbols indicate that a violet color was apparent at threshold, while the empty symbols indicate that the threshold was colorless (Hecht, Haig, and Chase, 1937)

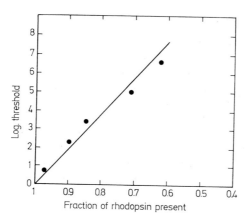

Fig. 5. Relation between log threshold and fraction of rhodopsin present in a rod monochromat (RUSHTON, 1961)

tative relationship between these two variables was difficult to establish (LYTHGOE, 1940; HECHT, 1942). It was realized that the bleaching of rhodopsin did not make rods insensitive merely by reducing their ability to catch quanta because small changes in rhodopsin produced disproportionately large changes in rod sensitivity (LYTHGOE, 1940; DEVRIES, 1943; BAUMGARDT, 1949; WALD, 1954; RUSHTON and COHEN, 1954). A more quantitative description of this amplified effect of rhodopsin bleaching on rod sensitivity has emerged during the past decade. DOWLING (1960) found that in the rat there was a linear relationship between the amount of rhodopsin bleached and the logarithm of the threshold required to produce a ERG response of constant amplitude. This relationship was strengthened when it was also observed to hold not only when the concentration of rhodopsin was reduced by light but also when it was reduced by vitamin A deficiency (DOWLING and WALD, 1960). RUSHTON (1961) uncovered a similar function between subjective visual threshold and bleached rhodopsin in a rod monochromat (Fig. 5). This simple logarithmic relationship is not without criticism (WEALE, 1964).

Several explanations have been proposed for why the bleaching of small amounts of rhodopsin produce such large changes in rod sensitivity. GRANIT et al. (1938) suggested that perhaps only rhodopsin at the surface of the rod outer segment was functional and the rhodopsin in the interior remained inert until it perhaps migrated to the outer membrane. WALD (1954) proposed another hypothesis in which the rod outer segment was considered to be composed of compartments, presumably provoked by the then recent discovery of disc-like structures in vertebrate photoreceptors (SJÖSTRAND, 1953), and that when a rhodopsin molecule in one compartment was inactivated by light the effectiveness of all the other molecules within this compartment was also lost. WALD'S compartment theory, though somewhat less wasteful of rhodopsin than GRANIT'S has conceptual difficulties. It seems more reasonable to expect that whenever rhodopsin catches a quantum of light this event is not wasted when rods are desensitized but is put to use in some other way to make the visual process more and not less

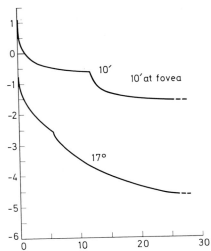

Fig. 6. Dark adaptation curves taken with 10′ and 17° test fields (Craik and Vernon, 1941)

effective. This concept is perhaps best illustrated by the model which Fuortes and Hodgkin (1964) have proposed to be operating within the photoreceptors of Limulus in which the sensitivity and the time scale of the receptor response are linked by a feedback mechanism which utilizes all the quanta caught in the photoreceptor to automatically optimize time resolution and sensitivity depending upon the availability of light. What is a loss in sensitivity is a gain in temporal resolution and vice versa.

Another difficulty with a theory like Wald's is that it places desensitization entirely within the rod receptor cell and takes little cognizance of the fact that dark-adaptation not only involves changes in sensitivity but also in the spatial integration of rod vision. This idea is best brought out by the pioneering experiments of Craik and Vernon (1941) in which dark adaptation curves obtained with small (10′) and large (17°) test stimuli were compared (Fig. 6). Rod adaptation begins sooner, continues for a longer time and reaches a lower threshold with large stimuli than it does with small stimuli, an effect that has since been confirmed by others (Arden and Weale, 1954; Rushton and Cohen, 1954). The amount of rhodopsin bleached in each rod and its subsequent regeneration are identical for these two conditions; the behavior of rod thresholds is quite different. Therefore at any one time bleached rhodopsin cannot be related to a *unique* rod threshold but to a family of thresholds which depend upon the spatial characteristics of the stimulus. Although such a change could be interpreted to reflect a greater probability for spatial summation when rods contain more rhodopsin (Wald, 1959), it seems more reasonable to associate it with neural modifications occuring in the dark-adapting retina that increase spatial summation of rod signals as well as rod sensitivity.

An important insight into the nature of this problem has come from the early research of Stiles and Crawford on the desensitizing effects of glare (Stiles and

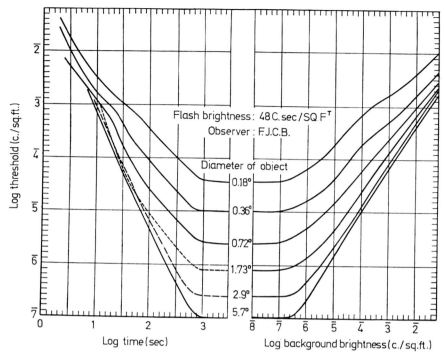

Flash brightness: 48 C. sec / SQ. F^T
Observer : F.J.C.B.

Diameter of object
0.18°
0.36°
0.72°
1.73°
2.9°
5.7°

Log threshold (c./sq.ft.)

Log time (sec)   Log background brightness (c./sq.ft.)

Fig. 7. (CRAWFORD, 1946)

CRAWFORD, 1932; CRAWFORD, 1947). They developed the concept of an "equivalent background" which expresses by *one parameter* the state to which the retina can be brought by either background light or by its past history of bleaching. This principle emerges from comparisons of increment threshold and dark-adaptation curves obtained with stimuli of different sizes (Fig. 7). Increment thresholds for small fields increase more gradually than those for large fields as background luminance increases; similarly dark-adaptation thresholds obtained with small test stimuli decrease more gradually with time in the dark than those obtained with large test stimuli. What seems to decay during dark-adaptation is an "equivalent background" which elevates thresholds to all field sizes just as it does when it is a real light illuminating the retina.

Examination of this concept in a rod monochromat (BLAKEMORE and RUSHTON, 1965a, b; RUSHTON, 1965b, c, d) has revealed how well the principle applies over a large range of rod vision (Fig. 8). Replotting the data of Fig. 8 to eliminate threshold demonstrates how this parameter, equivalent background, changes with time during rod dark-adaptation (Fig. 9). The equivalent background must uniquely determine threshold and bleached rhodopsin presumably generates the equivalent background although it is interesting that rhodopsin regeneration follows a simple exponential function but the disappearance of the equivalent background does not.

This concept suggests that the same mechanisms produce rod desensitization in both light and darkness. If this were so, all of the lateral interactions detectable when rods are being illuminated should also be active during the dark-adaptation

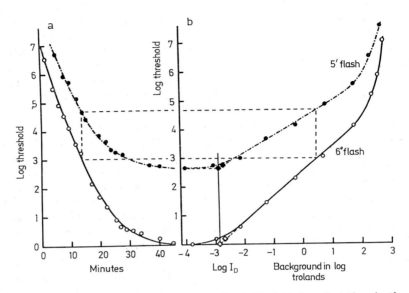

Fig. 8. (a) Left half, dark adaptation curve, log threshold plotted against time in the dark; (b) right half, increment threshold curve, log threshold plotted against log background field. Threshold flash subtending 6°, open circles; subtending 5′, filled circles (BLAKEMORE and RUSHTON, 1965)

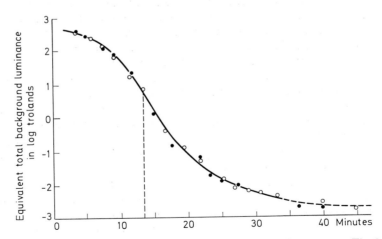

Fig. 9. Replot of Fig. 8 to eliminate "threshold". A horizontal line drawn across Fig. 8 at any level of log threshold cuts the left curve at a "time in the dark" and the right curve at a "background luminance". These two quantities are plotted against each other in Fig. 9; open circles when the test flash subtended 6°, filled circles when it subtended only 5′ (BLAKEMORE and RUSHTON, 1965)

of rods. Experiments supporting this idea have been reported by RUSHTON and WESTHEIMER (1962). They light-adapted a 1° area of retina, 5° from the fovea, with a square wave grating (1/2° light, 1/2° dark) and found that rod adaptation curves for both the illuminated and unilluminated portions of the retina were identical. Physical analysis of the light stimulus indicated that the result could not be attributed to scattered light and must be due to a spread of desensitization from the illuminated to the unilluminated regions of the retina. Similar effects seem to have been detected within the receptive field of frog ganglion cells (LIPETZ, 1961) but in this case the effects of scattered light do not appear to have been fully recognized (see EASTER, 1968). As in the presence of light, rods contraining bleached rhodopsin must send out signals which desensitize other rods or a structure upon which the responses of other rods converge.

ALPERN and CAMPBELL (1963) have demonstrated that light-adapted rods actually do send out continuous signals in the dark. This was done by measuring the size of the pupil of one eye in total darkness with an infra-red system after the other eye had been exposed to a strong bleaching light. The pupil of the unbleached eye, after initially constricting, showed a slow relaxation of constriction that followed the time course of rhodopsin regeneration. This constriction was dependent upon retinal signals coming from the eye containing the bleached rhodopsin because when pressure was applied to this eye sufficient to produce ischemic blindness, the pupil of the unbleached eye dilated; when this pressure was released the pupil of the unbleached eye reconstricted to its position appropriate to the time course of dark-adaptation. By using adapting lights of different wave-lengths these effects were shown to depend upon rod action predominantly. Most of these results have recently been confirmed (NEWSOME, 1971).

Experiments of BARLOW and SPARROCK (1964) suggest that these signals of bleached rods can actually be seen in darkness as positive after-images, but with difficulty because they are stabilized on the retina. They reached this conclusion by matching the brightness of a positive after-image with that of an external light field, *stabilized* on the retina. They found that the brightnesses of these two images faded away with the time course of rod dark-adaptation and that the matching brightness of the external field had the value required for the equivalent background obtained from increment threshold curves. Together these results imply that the positive after-image is the equivalent background and have been considered as support of an hypothesis (BARLOW, 1964) that the elevated thresholds of rods in darkness are due to noise from "spurious light signals" generated by bleached rhodopsin which in themselves prevent the detection of weak stimuli. NEWSOME (1971), however, has observed that positive after-images are reduced or absent in rod monochromates and therefore may not be a good index of rod function. The Barlow-Sparrock experiment considered without this qualification suggests that signals sent out in the dark by bleached rods are identical to those sent out in the presence of light. RUSHTON (1965c, d) objects to this idea on the grounds that real and equivalent backgrounds have different properties of spatial integration. The equivalent background seems to depend upon the average amount of bleached rhodopsin in the rods independent of its distribution in the field; real backgrounds appear to depend upon the average field luminance which undergoes a quasi-logarithmic transformation before it desensitizes rod vision. RUSHTON's

(1965 c, d) scheme envisions one gain reducing or desensitizing mechanism, resembling an automatic gain control (A.G.C.), already proposed for Limulus photoreceptor potentials (Fuortes and Hodgkin, 1964), which in the vertebrate retina can be activated either by real light stimulating the rods directly or by signals from bleached rhodopsin lingering in rods in total darkness. These two signals can simply add their effects in activating the single A.G.C. mechanism to produce an equivalent result but their ability to act reflects their different sources. Rushton doesn't attempt to identify the retinal circuitry of his A.G.C. but stipulates that it depends upon the pooling of signals from many rods and acts not to reduce the ability of a rod to respond to a quantum of light but to reduce the effect which this quantum catch exerts in contributing to rod vision.

Westheimer's (1968) studies with rod sensitization during dark-adaptation poses some problems for such a theory in which the same mechanism desensitizes rods in the presence of light or in darkness following exposure to light. Rod sensitization produced by real light cannot be detected when equivalent light is employed. Equivalent backgrounds neither sensitize the retina to other equivalent backgrounds nor can they be sensitized by real backgrounds or vice versa. The principle of equivalence appears to apply only to the desensitizing and not to the sensitizing effects which real backgrounds exert on rod vision. This implies that the gain reducing mechanisms brought into operation by real light and by bleached rhodopsin are not identical. Westheimer cites that heretofore almost all the comparisons between real and equivalent backgrounds have involved uniform stimulation of large retinal areas. When restricted areas were used the principle of equivalence appears to break down (Rushton, 1965 c, d; Westheimer, 1968). Westheimer has suggested that there are two desensitizing systems, one produced by large uniform backgrounds of real light which includes the summation of spatially different desensitizing and sensitizing influences and another produced by bleached rhodopsin which only involves desensitizing effects. As the size of the adapting field is reduced the sensitizing influences are preferentially reduced and thresholds become proportionately higher in the presence of real backgrounds than in the presence of equivalent ones. He suggests that this difference should be compensated for when estimates of the equivalent background are made from increment threshold curves and claims that when this is done the disappearance of the equivalent background in darkness resembles more closely an exponential function.

A distinction between real and equivalent backgrounds also appears when rod saturation is considered. Large uniform fields of real light saturate human rod vision (Aguilar and Stiles, 1953). In darkness following exposure to light the rod mechanism does not appear to saturate even when more than 40 % of the rhodopsin has been bleached away (Rushton, 1961). The primate ERG also saturates in the presence of light which bleaches little rhodopsin (Gouras, 1965) but does not seem to do so in darkness even after much rhodopsin has been bleached away. Again as Westheimer's idea suggests real light appears to desensitize rod vision more than equivalent light.

The ERG could provide a better test of the equivalence principle since it allows an examination of the entire waveform of the rod response in the presence of light or in a desensitized retina in darkness. The only exploration of this has been that of CONE (1964) who found that the principle of equivalence did not completely apply to the predominantly rod ERG of rat because responses desensitized in the presence of light always had shorter latencies than those obtained from desensitized retinas in darkness.

There is also evidence that at least two distinct mechanisms are involved in rod dark-adaptation. Original data of BLANCHARD (1918) has been used by WALD (1959) to show that there is a rapid component which is unrelated to rhodopsin and considered to be "neural" and a slower component which is presumably related to rhodopsin regeneration. DOWLING (1963, 1967) gives strong support to this conclusion by his studies of the ERG and rhodopsin content of rat retina. In the rat there are also two distinct phases to dark-adaptation, one rapid and virtually unrelated to bleached rhodopsin, which DOWLING also calls neural adaptation, and the other slow and directly related to rhodopsin regeneration, which he calls photochemical adaptation. Additional evidence for this view has recently been obtained in the skate (DOWLING and RIPPS, 1970). On the other hand the dark-adaptation curves obtained from rod monochromats by RUSHTON (1961; 1965a, b) do not obviously show two distinct phases but only one which closely follows rhodopsin regeneration. It is difficult, however, to measure the first few minutes of dark-adaptation especially in rod monochromats who are so dazzled by bright lights.

There are, of course, some extremely rapid alterations of visual sensitivity, lasting fractions of a second, which accompany sudden changes in retinal illumination (CRAWFORD, 1947; BAKER, 1953). These are undoubtedly not related to the somewhat slower processes of light- and dark-adaptation we have been considering so far. Some of these effects can be produced by changing the illumination of one eye while testing the sensitivity of the other (BOUMAN, 1955; WAGMAN and BATTERSBY, 1959) and therefore must take place beyond the retina, the major site of light- and dark-adaptation.

Is the disappearance of bleached rhodopsin alone responsible for the slow recovery of rod vision during dark adaptation or can some of the other photoproducts of bleaching also play a role, perhaps a more significant one, in rod desensitization? DONNER and REUTER (1968) have attempted to demonstrate this by comparing thresholds of ganglion cells with measurements of rhodopsin and its photoproducts in frog retina. One of their conclusions was that metarhodopsin II was closely related to the rapid initial change of rod sensitivity in the dark. On the other hand, FRANK and DOWLING (1968) could find no correlation between the sensitivity of the b-wave of the ERG and the kinetics of any of the long-lived photoproducts of rhodopsin in the isolated rat retina. DONNER and REUTER (1968) believe that the b-wave of the ERG is not as accurate an index of rod sensitivity as the ganglion cell response in the frog and cite BAUMANN and SCHEIBNER's (1967) results as an example but in the skate DOWLING and RIPPS (1970) provide strong evidence that the b-wave and ganglion cell thresholds are closely related.

There are well known effects which the products of rhodopsin photolysis exert on rod dark-adaptation but these only occur after short brilliant flashes of light. Under such circumstances photo-isomerization of the photoproducts of rhodopsin back to either the 11-*cis* or 9-*cis* configuration prevents the complete bleaching of rhodopsin (HAGINS, 1958). This effect delays the time course of regeneration of the remaining rhodopsin and consequently rod dark-adaptation in several ways (Dowling and Hubbard, 1963). Photoproducts in the all trans-configuration must subsequently hydrolyze to completely bleached rhodopsin before being regenerated into rhodopsin and secondly the availability of 11-*cis* retinal so necessary for the regeneration of rhodopsin is also reduced. A retardation of normal rod dark-adaptation in man has also been observed after bleaching with similarly brief brilliant flashes (Wald and Clark, 1937).

The results indicate that dark-adaptation of the rods depends upon the regeneration of rhodopsin, expecially the slower phase following bright pre-adapting lights. The more rapid components presumably do not. Dark-adaptation appears to activate some of the same mechanisms responsible for rod desensitization in the presence of light but these two processes are not exactly equivalent since the latter produces relatively more desensitization and exhibits a more complex interaction. I suspect that one cause of desensitization resides entirely within the rod receptor, itself, as has been found in invertebrates and this component is especially dependent upon the presence of bleached rhodopsin in the outer segment. Another cause of desensitization must depend upon signals which leave the rod and spread to involve neighboring rods, probably by either horizontal cells or rod axons or both and this process is dependent more upon the presence of light than on bleached rhodopsin. When light is turned off both processes disappear but the latter does so more quickly than the former which must wait until all the rhodopsin in the outer segments has been regenerated.

The neurophysiology of rod dark-adaptation is too preliminary to explain the cellular events underlying rod desensitization. A valuable piece of information would be measurements of the membrane potential of single rod throughout the time course of rhodopsin regeneration. Since it is only through a change in its membrane potential that a rod can communicate with other rods any desensitization that spreads to other rods would be reflected in this potential. If the potential recovers more rapidly than rhodopsin regenerates, it would indicate that one stage of rod desensitization precedes the site of membrane electrogenesis within the rod receptor cell.

Light-Adaptation of Cones

Cone systems are usually more complex than those of rods since in most animals there are at least two or more types of cones. Human vision relies on three types of cones, each of which contains a different photopigment within its outer segment (Marks, Dobelle and MacNichol, 1964; Brown and Wald, 1964). The action spectrum of these three cone mechanisms (Fig. 10) have been isolated psychophysically be Stiles (1939, 1949, 1959). One mechanism is most effective in the blue ($\pi 1$), another in the green ($\pi 4$) and third in the red ($\pi 5$) region of the spectrum.

Stiles identified these cone mechanisms by analyzing increment threshold curves in which the wavelength of both the test and the adapting light were systematically altered. If, as Stiles assumed, each cone mechanism can be desensitized by light independently of the others than any point on an increment threshold curve would reflect the function of only cone cone mechanism, the one most sensitive to the particular test conditions (Fig. 11). The transition from a condition

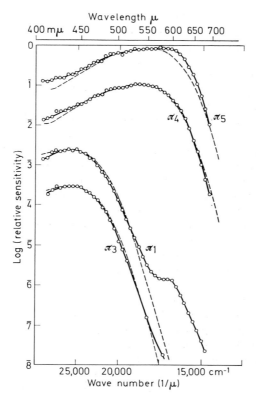

Fig. 10. Field spectra sensitivity curves of mechanisms $\pi 1$, $\pi 3$, $\pi 4$, $\pi 5$ ($\pi 1$ and $\pi 3$, reflect the blue sensitive cone mechanism predominantly) corrected for pre-receptor light losses and compared in each case with the similarly corrected sensitivity curve of rod vision displaced parallel to the two axes to give the best fit (STILES, 1959)

where one cone mechanism determines threshold to one where another cone mechanism does is usually detectable as a discontinuity in the increment threshold curve. By examining how these discontinuities in cone increment threshold curves vary with the wavelength of both the test and the adapting light the action spectrum of each of the three cone mechanisms of human vision can be isolated. The success of this approach confirms the validity of STILE's original assumption that light-adaptation desensitizes each cone mechanism independently of the others. There are circumstances where this principle appears to be violated (STILES, 1949) indicating some degree of interaction between different cone mechanisms but these effects seem to be of such a minor nature as not to weaken the major validity of the idea. So strong is this independence in human vision that each of the three cone mechanisms can be identified by merely determining the spectral sensitivity of vision in the presence of adapting lights which strongly but selectively light-adapt the other cone mechanisms (BRINDLEY, 1953; Wald, 1964).

As with rods, desensitization of cones by light is not usually due to exhaustion of their photopigment (BAKER and RUSHTON, 1965; RIPPS and WEALE, 1965)

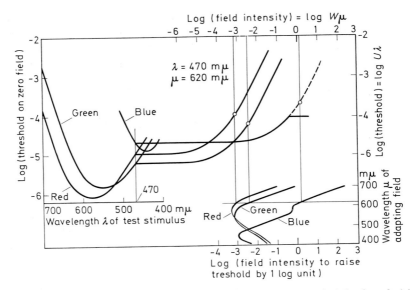

Fig. 11. Relation of the curve of log (threshold) against log (field intensity) for foveal vision to the spectral sensitivity curves of the three cone mechanisms. (STILES, 1949.) The test wavelength (470 mµ) and the adapting wavelength (620 mµ) affect each of the three cone mechanisms in a relatively independent way just as was the case when rods and cones were involved as in Fig. 2

although measurements of cone pigments and their relationship to visual adaptation is less complete than it is for rod vision. What is known suggests that the mechanisms involved in the light- and dark-adaptation of rods have their counterparts in cone adaptation.

For example cones in one part of the retina can be desensitized by cones in another part of the retina. ALPERN and RUSHTON (1965) have demonstrated that cone thresholds tested in a 1° region centered on the fovea can be elevated by an annulus surrounding the test zone and flashed 50 msec *later* than the threshold stimulus. Each of STILES's cone mechanisms appear to act independently in this effect, i.e., if threshold is determined by $\pi 5$, it is only the $\pi 5$ mechanism in the after-flash that desensitizes; the extent to which the $\pi 1$ or $\pi 4$ mechanisms are stimulated by the after-flash is irrelevant.

A similar specifity occurs when human cone vision is sensitized by light. Cone sensitivity like that of rods can be enhanced by appropriately placing a *steady* adapting light in an annular zone surrounding the test area (CRAWFORD, 1940; RATOOSH and GRAHAM, 1941; BATTERSBY and WAGMAN, 1964; WESTHEIMER 1968). This sensitization of cone vision acts over a distance of 20—30 micra which is about tenfold shorter than it does in rod vision (WESTHEIMER, 1964; see p. 9). McKEE and WESTHEIMER (1970) found that this sensitization only operated within a specific cone system, i.e., green sensitive cones ($\pi 4$) could only sensitize other green sensitive cones and the same arrangement was found for the red sensitive cone mechanism ($\pi 5$). The blue sensitive system was not examined.

These sensitizing and desensitizing processes which affect adjacent areas of retina reflect neural mechanisms of inhibition and excitation which are undoubtedly involved in enhancing spatial contrast. When spatial contrast, itself, is examined by measuring the threshold energy gradient necessary to detect a sinusoidal grating a similar independence of cone mechanisms is observed (GREEN, 1968). The effectiveness of a red background for decreasing the contrast of a green grating is determined only by the extent to which it excites the green ($\pi 4$) mechanism. Spatial contrast determined by the other two cone mechanisms shows a similar independence.

Dark-Adaptation of Cones

If an adapting light desensitizes each cone mechanism independently of the others, then the subsequent recovery of cone sensitivity in darkness might also be expected to proceed as independently and such appears to be the case. This was first suggested by AUERBACH and WALD's (1954) finding of two temporal components to human cone dark-adaptation after preadapting with an orange light while testing visual thresholds with a violet light. The action spectrum of the earlier component was that of the blue sensitive cones ($\pi 1$). DuCROZ and RUSHTON (1966) have since demonstrated that each of the three cone mechanisms in man exhibits its own dark-adaptation curve, each of which can be manipulated independently of the others by appropriate selection of the wavelength of the test and the adapting light. They also showed that the principle of equivalent backgrounds holds for the green sensitive cone mechanisms (Fig. 12) just as it does for the rods and assumed that it undoubtedly held for the other cone mechanisms as well. On the other hand RINALDUCCI, HIGGINS, and CRAMER (1970) have not been

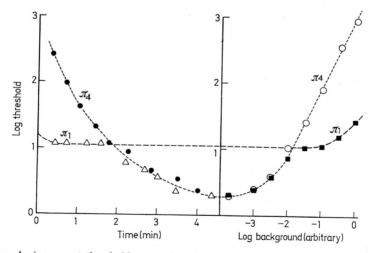

Fig. 12. a An increment threshold curves for a blue test-flash (squares) and a yellow-green test flash (circles) against a yellow-green background; circles displaced vertically so as to coincide with squares at absolute threshold. Dark adaptation curves for a blue test-flash after a white bleach (circles and after an orange bleach, triangles), the bleaching lights being equivalent for $\pi 4$. Ordinate scale the same in a und b (DuCROZ and RUSHTON, 1966)

able to confirm that the principle of equivalence holds for normal cone vision. They only found it to occur in certain color-defective individuals in whom a more complete isolation of cone mechanisms seemed possible. As a result they interpret the breakdown of equivalence which they observed to interaction among different cone mechanisms. Their objection to the principle of equivalence is therefore somewhat less restrictive than that of Westheimer's (1968; see p. 21) which applies to visual adaptation within a single photoreceptor system.

There are other phenomena which indicate that cone dark-adaptation does not always proceed independently for each cone mechanism. Stiles (1949) has noticed that the threshold of the blue sensitive cone mechanism, $\pi 1$, can *rise* several seconds after the eye is put into darkness after pre-adapting with red light and this peculiarity has been confirmed by Das (1964). There is also evidence that suggests interaction between the red and green sensitive cone mechanisms in adaptation (Boynton, Ikeda, and Stiles, 1964).

The electrophysiology of cone vision has progressed somewhat further than that of rod vision because the large size of some vertebrate cones has permitted intracellular recordings (Bortoff, 1964; Tomita, 1965). Recently Baylor, Fuortes, and O'Bryan (1971) have shown that single turtle cones receive sensitizing (hyperpolarizing) signals from neighboring cones and desensitizing (depolarizing) ones from horizontal cells which provides some physiological basis for the psychophysical results upon which much of our present understanding of cone adaptation depends. It should now be possible to determine whether a continuous alteration of cone membrane potential parallel cone dark-adaptation, whether the lateral interactions between cones is actually specific for each cone system and whether the mechanisms responsible for desensitizing cones in the presence of light are the same as those which do so in darkness.

Space-Time Resolution

Light-adaptation not only decreases the sensitivity of rods and cones but also decreases their spatial and temporal integration (Bouman, 1952; Commichau, 1955; Barlow, 1958) and increases temporal (Lythgoe and Tansley, 1929; De Lange, 1954) and spatial (Hecht, 1928; Craik, 1939) resolution. These changes seem to be more pronounced for cone than for rod vision.

The temporal resolution of cone vision can best be described by plotting the frequency of a sinusoidally modulated light against the modulation depth necessary to just perceive flicker (Fig. 13). Light-adaptation decreases the modulation depth and produces a resonant peak in this function. The model of Fuortes and Hodgkin (1964) in which light-adaptation speeds up the time scale of receptor responses (see also Baumann, 1968) can provide some explanation for this improved temporal resolution. Appropriate delays in the feedback of such a model can also be responsible for resonant peaks (Pinter, 1966).

The spatial resolving power of cone vision also increases with light-adaptation. One of the likely explanations of this is that when more light enters the eye more and more neural units are brought into play thereby facilitating spatial discrimination (Hecht, 1928; Pirenne and Denton, 1952). This cannot be the entire explanation because Craik (1938) has observed that acuity is poor when the

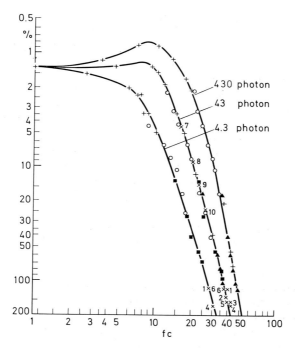

Fig. 13. Low-frequency characteristics of the system fovea — light impression. At three values of retinal illumination, ripple ratio r amplitude of fundamental/average luminance of the stimulus, is plotted against the critical frequency. The observation points ■, ▲, ○, and × refer to sinusoidal time functions obtained in different ways and + to trapezoidal variation (DeLange, 1954). The top of the curves for which $r < 1.35$ % indicates a resonance effect. At critical frequency, if $r < 2\%$, the low-pass filter in the eye reduces the variation in brightness caused by the r ratio of any time function to a brightness variation caused by a sinusoidal variation at very low frequency with $r = 1.35\%$, this being the threshold value of sinusoidal successive contrast

illumination of a test object is much brighter than that to which the eye has been adapted to. The eye does best when the illumination of both test and adapting lights are about equal. Another explanation has been suggested by BARLOW, FITZHUGH, and KUFFLER's (1957) results with cat retinal ganglion cells. Light-adaptation increases the antagonistic surround mechanism in the receptive field of such cells and thereby reducing the effective size of the receptive field center (see GLEZER, 1965). This explanation has not been supported by CLELAND and ENROTH-CUGELL (1968) who found after more quantitative studies that the receptive field centers of cat ganglion cells did not change their size during light-adaptation. Another possibility is that spatial resolution also depends upon the accurate timing of impulses converging on hierarchical systems of cells involved in the detection of spatial patterns. The improved temporal resolution produced by light-adaptation at an early stage, probably within the receptor cells themselves, might facilitate this detection.

References

Adrian, E. D.: Rod and cone components in the electric response of the eye. J. Physiol. (Lond.) **105**, 24—37 (1946).

Aguilar, M., Stiles, W. S.: Saturation of the rod mechanism of the retina at high levels of stimulation. Optica Acta **1**, 59—65 (1964).

Alpern, M.: Rod-cone independence in the after-flash effects. J. Physiol. (Lond.) **176**, 462—471 (1965).

— Campbell, F. W.: The behavior of the pupil during dark adaptation. J. Physiol. (Lond.) **165**, 5—7 (1963).

Andrews, D. P., Hammond, P.: Mesopic increment threshold spectral sensitivity of single optic tract fibres in the cat cone-rod interaction. J. Physiol. (Lond.) **209**, 65—81 (1970a).

— — Suprathreshold spectral properties of single optic tract fibres in cat, under mesopic adaptation, cone-rod interaction. J. Physiol. (Lond.) **209**, 83—103 (1970b).

Arden, G. B., Weale, R. A.: Nervous mechanisms and dark-adaptation. J. Physiol. (Lond.) **125**, 417—426 (1954).

Armington, J. C., Johnson, E. P., Riggs, L. A.: The scotopic a-wave in the electrical response of the human retina. J. Physiol. (Lond.) **118**, 289—298 (1952).

Auerbach, E., Wald, G.: Indentification of a violet receptor in human color vision. Science **120**, 401—405 (1954).

Baker, H. D., Rushton, W. A. H.: The red-sensitive pigment in normal cones. J. Physiol. (Lond.) **176**, 56—72 (1965).

Barlow, H. B.: Temporal and spatial summation in human vision at different background intensities. J. Physiol. (Lond.) **141**, 337—350 (1958).

— Andrews, D. P.: Sensitivity of receptors and "receptor pools". J. Opt. Soc. Amer. **57**, 837—838 (1967).

— Fitzhugh, R., Kuffler, S. W.: Change of organization in the receptive fields of the cat's retina during dark adaptation. J. Physiol. (Lond.) **137**, 338—354 (1957).

— Sparrock, J. M. B.: The role of after images in dark adaptation. Science **144**, 1309—1314 (1964).

Battersby, W. S., Wagman, I. H.: Light adaptation kinetics: the influence of spatial factors. Science **143**, 1029—1031 (1964).

Baumann, Ch., Scheibner, H.: Die Dunkeladaptation einzelner Neurone in der isolierten, umspülten Froschnetzhaut. Pflügers Arch. ges. Physiol. **197**, 85 (1967d).

Baumann, F.: Slow and spike potentials recorded from retinula cells of the Honeybee Drone in response to light. J. gen. Physiol. **52**, 855—875 (1968).

Baumgardt, E.: Les théories photochimiques classique et quantiques de la vision et l'inhibition nerveuse en vision liminaire. Rev. Opt. (théor. instrum.) **28**, 453—478 (1949).

— Visual spatial and temporal summation. Nature (Lond.) **184**, 1951—1952 (1959).

Baylor, D., Fuortes, M. G. F., O'Bryan, P.: J. Physiol. (Lond.) **214**, 265—294 (1971).

Blakemore, C. B., Rushton, W. A. H.: Dark adaptation and increment threshold in a rod monochromat. J. Physiol. (Lond.) **181**, 612—628 (1965a).

— — The rod increment threshold during dark adaptation in normal and rod monochromat. J. Physiol. (Lond.) **181**, 629—640 (1965b).

Blanchard, J.: The brightness sensibility of the retina. Phys. Rev. Ser. 1, **11**, 81—99 (1918).

Bortoff, A.: Localization of slow potential responses in the Necturus retina. Vision Res. **4**, 627—633 (1964).

Bouman, M. A.: Peripheral contrast thresholds of the human eye. J. Opt. Soc. Amer. **40**, 825—832 (1950).

— Peripheral contrast thresholds for various and different wavelengths for adapting field and test stimulus. J. Opt. Soc. Amer. **42**, 820—831 (1952).

— On foveal and peripheral interaction binocular vision. Optica Acta **1**, 177—183 (1955).

Boycott, B. B., Dowling, J. E.: Organization of the primate retina: light microscopy. Phil. trans. B **255**, 109—184 (1969).

Boynton, R. M., Ikeda, M., Stiles, W. S.: Interactions among chromatic mechanisms as inferred from positive and negative increment thresholds. Vision Res. **4**, 87—117 (1964).

BRINDLEY, G. S.: The effects on colour vision of adaptation to very bright lights. J. Physiol. (Lond.) **122**, 332—350 (1953).

BROWN, J. L., KUHNS, M. P., ADLER, H. E.: Relation of threshold criterion to the functional receptors of eye. J. Opt. Soc. Amer. **47**, 198—204 (1957).

BROWN, K. T., MURAKAMI, M.: Rapid effects of light and dark adaptation upon the receptive field organization of S-potentials and late receptor potentials. Vision Res. **8**, 1145—1171 (1968).

— WATANABE, K., MURAKAMI, M.: The early and late receptor potentials of monkey cones and rods. Cold Spr. Harb. Symp. quant. Biol. **30**, 457—482 (1965).

BROWN, P. K., WALD, G.: Visual pigments in single rods and cones of the human retina. Science **144**, 44—51 (1964).

CAJAL, S. RAMÓN: Die Retina der Wirbelthiere. Wiesbaden: Bergmann 1894.

CAMPBELL, F. W., RUSHTON, W. A. H.: Measurement of the scotopic pigment in the living human eye. J. Physiol. (Lond.) **130**, 131—147 (1955).

CLELAND, B. G., ENROTH-CUGELL, C.: Quantitative aspects of sensitivity and summation in the cat retina. J. Physiol. (Lond.) **198**, 17—38 (1968).

COMMICHAU, R.: Adaptationszustand und Unterschiedsschwellenenergie für Lichtblitze. Z. Biol. **108**, 145—160 (1968).

CONE, R. A.: The rat electroretinogram I. Contrasting effects of adaptation on the amplitude and latency of the b-wave. J. gen. Physiol. **47**, 1089—1105 (1964).

CRAIK, K., VERNON, M.: The nature of dark adaptation. Brit. J. Psychol. **32**, 62—81 (1941).

CRAIK, K. J. W.: The effect of adaptation upon acuity. Brit. J. Physiol. **29**, 252—266 (1939).

CRAWFORD, B. H.: The effect of field size and pattern on the change of visual sensitivity with time. Proc. roy. Soc. B **129**, 94—106 (1940).

— Visual adaptation in relation to brief conditioning stimuli. Proc. roy. Soc. B **134**, 283—302 (1947).

DAS, S. R.: Foveal increment thresholds in dark adaptation. J. Opt. Soc. Amer. **54**, 541—546 (1964).

DE LANGE, H.: Research into the dynamic nature of the human fovea cortex system with intermittent and modulated light. J. Opt. Soc. Amer. **48**, 777—789 (1959).

DENTON, E. J., WARREN, F. J.: Study of the photosensitive pigments in the retinas of deep-sea fish. J. Mar. Biol. Ass. UK. **36**, 651—662 (1957).

DE VRIES, H.: The quantum character of light and its bearing upon the threshold of vision, the differential sensitivity and visual acuity of the eye. Physica **19**, 553—564 (1943).

DODT, E.: Cone electroretinogram by flicker. Nature **168**, 783 (1957).

— ECHTE, K.: Dark and light adaptation in pigmented and white rat as measured electroretinogram threshold. J. Neurophysiol. **14**, 427—445 (1961).

DONNER, K. O., REUTER, T.: Visual adaptation of the rhodopsin rods in the frog's retina. J. Physiol. (Lond.) **199**, 49—87 (1968).

DOWLING, J. D., WALD, G.: The biological function of vitamin A acid. Proc. nat. Acad. Sci. (Wash.) **46**, 587—608 (1960).

DOWLING, J. E.: Chemistry of visual adaptation in the rat. Nature (Lond.) **188**, 114—118 (1960).

— Neural and photochemical mechanisms of visual adaptation in the rat. J. Gen. Physiol. **46**, 1287—1301 (1963).

— The site of visual adaptation. Science **155**, 273—279 (1967).

— HUBBARD, R.: Effects of brilliant flashes on light and dark adaptation. Nature (Lond.) **199**, 972—975 (1963).

— RIPPS, H.: Visual adaptation in the retina of the skate. J. gen. Physiol. **56**, 491—520 (1970).

DUCROZ, J. J., RUSHTON, W. A. H.: The separation of cone and mechanisms in dark adaptation. J. Physiol. (Lond.) **183**, 481—496 (1966).

EASTER, S. S., JR.: Adaptation in the goldfish retina. J. Physiol. (Lond.) **195**, 273—281 (1968).

FRANK, R. N., DOWLING, J. E.: Rhodopsin effects on electoretinogram sensitivity in isolated perfused rat retina. Science **161**, 487—489 (1968).

FUORTES, M. G. F., GUNKEL, R. D., RUSHTON, W. A. H.: Increment thresholds in a subject deficient in cone vision. J. Physiol. (Lond.) **156**, 179—192 (1961).

— HODGKIN, A. L.: Changes in time scale and sensitivity in the ommatidia of Limulus. J. Physiol. (Lond.) **172**, 239—263 (1964).

Glezer, V. D.: The receptive fields of the retina. Vision Res. 5, 497—525 (1965).

Gouras, P.: Saturation of the rods in rhesus monkey. J. Opt. Soc. Amer. 55, 86—91 (1965).

— Rod and cone independence in the electroretinogram of the dark-adapted monkey's perifovea. J. Physiol. (Lond.) 187, 455—464 (1966).

— The effects of light-adaptation on rod and cone receptive field organization of monkey ganglion cells. J. Physiol. (Lond.) 192, 747—760 (1967).

Graham, C. H., Margaria, R.: Area and the intensity-time relation in the peripheral retina. Amer. J. Physiol. 113, 299—305 (1935).

Granit, R., Holmberg, T., Zewi, M.: On the mode of action of visual purple on the rod cell. J. Physiol. (Lond.) 94, 430—440 (1938).

Green, D. G.: The contrast sensitivity of the colour mechanisms of the human eye. J. Physiol. (Lond.) 196, 415—429 (1968).

Hagins, W. A.: The quantum efficiency of bleaching rhodopsin in situ. J. Physiol. (Lond.) 129, 22P—23P (1955).

— Zonana, H. V., Adams, R. G.: Local membrane current in the outer segments of squid photoreceptors. Nature (Lond.) 194, 844—847 (1962).

Hartline, H. K., Ratliff, F., Miller, W. H.: Inhibitory interaction in the retina and its significance in vision. In: Florey, E. (Ed.): Nervous Inhibition, pp.-141—184. New York: Pergamon Press 1961.

Hecht, S.: The relation between visual acuity and illumination. J. gen. Physiol. 11, 155—281 (1928).

— Rods, cones and the chemical basis of vision. Physiol. Rev. 17, 239—290 (1937).

— The chemistry of visual substance. Ann. Rev. Biochem. 11, 465—496 (1942).

— Haig, C., Chase, A. M.: The influence of light-adaptation on subsequent dark-adaptation of the eye. J. gen. Physiol. 20, 831—850 (1937).

— Hsia, Y.: Dark adaptation following light adaptation to red and white lights. J. Opt. Soc. Amer. 35, 261—267 (1945).

Johnson, E. P., Cornsweet, T. N.: Electroretinal photopic sensitivity curves. Nature (Lond.) 174, 614—615 (1954).

Jones, R. C.: Quantum efficiency of human vision. J. Opt. Soc. Amer. 49, 645—653 (1959).

Kaneko, A.: Physiological and morphological identification of horizontal, bipolar and amacrine cells in goldfish retina. J. Physiol. (Lond.) 207, 623—633 (1970).

Koenig, A., Ritter, R.: Über den Helligkeitswerthe der Spektralfarben bei verschiedener absoluter Intensität. Arch. Psychol. Physiol. Sinnesorg. 9, 81 (1891).

Kolb, H.: Organization of the outer plexiform layer of the primate retina: electron microscopy of Golgi-impregnated cells. Phil. Trans. B 258, 261—283 (1970).

Lipetz, L.: A mechanism of light adaptation. Science 133, 639—640 (1961).

Lythgoe, R. J.: The mechanism of dark adaptation. A critical resume. Brit. J. Ophthal. 27, 21—43 (1940).

Marks, W. B., Dobelle, W. H., MacNichol, E. F.: Visual pigments of single primate cones. Science 143, 1181—1183 (1964).

McKee, S. P., Westheimer, G.: Specificity of cone mechanisms in lateral interaction. J. Physiol. (Lond.) 206, 117—128 (1970).

Missoten, L.: The Ultrastructure of the Retina. Bruxelles: Editions Arscia S. A. 1965.

Müller, G. E.: Anatomisch-physiologische Untersuchungen über die Retina des Menschen und der Wirbelthiere. Z. wiss. Zool. 8, 1 (1856—1857).

Newsome, D.: After-image and pupillary activity following strong light exposure. Vision Res. 11, 275—288 (1971).

Parinaud, H.: La vision. Paris: Octave Doin 1898.

Pinter, R. B.: Sinusoidal and delta function responses of visual cells of the Limulus eye. J. gen. Physiol. 49, 565—593 (1966).

Pirenne, M. H.: Contribution to the discussion of the paper by Stiles. Proc. Phys. Soc. Lond. 56, 354—355 (1944).

— Some aspects of the sensitivity of the eye. Ann. N. Y. Acad. Sci. 74, 377—384 (1958).

— Denton, E. J.: Accuracy and sensitivity of the human eye. Nature (Lond.) 170, 1039—1042 (1952).

Polyak, S. L.: The Retina. Chicago: Univ. of Chicago Press 1941.

PURKINJE, J.: Beobachtungen und Versuche zur Physiologie der Sinne, Bd. 2, S. 109—110. Berlin: G. Reimer 1825.

RATLIFF, F.: Selective adaptation of local regions of the rhabdom in an ommatidium of the compound eye of Limulus. In: BERNHARD, C. G. (Ed.): The Functional Organization of the Compound Eye, Vol. 7. London: Pergamon Press 1965.

RATOOSH, P., GRAHAM, C. H.: Areal effects in foveal brightness discrimination. J. exp. Psychol. **42**, 367—375 (1951).

RINALDUCCI, E. J., HIGGINS, K. E., CRAMER, J. A.: Nonequivalence of backgrounds during photopic dark adaptation. J. Opt. Soc. Amer. **60**, 1518—1524 (1970).

RIPPS, H., WEALE, R. A.: Analysis of foveal densitometry. Nature (Lond.) **205**, 52—56 (1965).

ROSE, A.: The sensitivity performance of the human eye on an absolute scale. J. Opt. Soc. Amer. **38**, 196—208 (1948).

RUSHTON, W. A. H.: Rhodopsin measurement and dark-adaptation in a subject deficient in cone vision. J. Physiol. (Lond.) **156**, 193—205 (1961a).
— A foveal pigment in the deuteranope. J. Physiol. (Lond.) **176**, 24—37 (1965a).
— The sensitivity of rods under illumination. J. Physiol. (Lond.) **178**, 141—160 (1965b).
— Bleached rhodopsin and visual adaptation. J. Physiol. (Lond.) **181**, 645—655 (1965c).
— The Ferrier Lecture, 1962. Visual adaptation. Proc. roy. Soc. B **162**, 20—46 (1965d).
— Rod/cone rivalry in pigment regeneration. J. Physiol. (Lond.) **198**, 219—236 (1968).
— COHEN, R. D.: Visual purple and the course of dark adaptation. Nature (Lond.) **173**, 301—302 (1954).
— WESTHEIMER, G.: The effect upon the rod threshold of bleaching neighboring rods. J. Physiol. (Lond.) **164**, 318—329 (1962).

SCHULTZE, M.: Zur Anatomie und Physiologie der Retina. Arch. mikr. Anat. **1**, 165—286 (1866).

SJÖSTRAND, F. S.: The ultrastructure of the outer segments of rods and coneds of the eye as revealed by the electron microscope. J. cell. comp. Physiol. **42**, 15—44 (1953).

SMITH, S. W., MORRIS, A., DIMMICK, F. L.: Effects of exposure to various red lights upon subsequent dark adaptation measured by the method of constant stimuli. J. Opt. Soc. Amer. **45**, 502—506 (1955).

STEINBERG, R. H.: Rod and cone contributions to S-potentials from the cat retina. Vision Res. **9**, 1319—1329 (1969a).
— Rod-cone interaction in S-potentials from the cat retina. Vision Res. **9**, 1331—1344 (1969b).

STELL, W. K.: The structure and relationship of horizontal cells and photoreceptor-bipolar synaptic complexes in goldfish retina. Amer. J. Anat. **121**, 401—424 (1967).

STILES, W. S.: The directional sensitivity of the retina and the spectral sensitivities of the rods and cones. Proc. roy. Soc. B **127**, 64—105 (1939).
— Increment thresholds and the mechanisms of colour vision. Docum. Ophthal. (Den Haag) **3**, 138—165 (1949).
— Color vision: the approach through incremental threshold sensitivity. Proc. nat. Acad. Sci. (Wash.) **45**, 100—114 (1959).
— CRAWFORD, B. H.: Equivalent adaptation levels in localized retinal areas. In: Report of Discussion on Vision, pp. 194—211. London: Physical Society 1932.
— — The luminous efficiency of rays entering the eye pupil at different points. Proc. roy. Soc. B **112**, 428—540(1933).

TELLER, D. Y., MATTER, C. F., PHILLIPS, W. D.: Sensitization by annular surrounds: Spatial summation properties. Vision Res. **10**, 549—561 (1970).

TOMITA, T.: Electrophysiological study of the mechanisms subserving color coding in the fish retina. Cold Spr. Harb. Symp. quant. Biol. **30**, 559—566 (1965).

VON KRIES, J.: Über die Funktion der Netzhautstäbchen. Z. Psychol. Physiol. Sinnesorg. **9**, 81—123 (1896).

WALD, G.: Human vision and the spectrum. Science **101**, 653—658 (1945).
— On the mechanism of the visual threshold and visual adaptation. Science **119**, 887—892 (1954).
— Retinal chemistry and the physiology of vision, Selig-Hecht commemorative lecture. In: Visual Problems of Colour, Vol. 1, pp. 7—61. London: H.M.S.O. 1959.
— The receptors of human color vision. Science **145**, 1007—1017 (1964).
— CLARK, A.: Visual adaptation and chemistry of the rods. J. gen. Physiol. **21**, 39—105 (1937).

WAGMAN, I. H., BATTERSBY, W. S.: Neural limitations of visual excitability II. Retrochiasmal interaction. Amer. J. Physiol. **197**, 1237—1242 (1959).

WALLS, G. L.: The Vertebrate Eye and its Adaptive Radiation. Bloomfield Hills, Mich.: Cranbrook Inst. Sci. 1942.

WEALE, R. A.: Relation between dark adaptation and visual pigment regeneration. J. Opt. Soc. Amer. **54**, 128—129 (1964).

WESTHEIMER, G.: Spatial interaction in the human retina during scotopic vision. J. Physiol. (Lond.) **181**, 882—894 (1965).

— Bleached rhodopsin and retinal interaction. J. Physiol. (Lond.) **195**, 97—105 (1968).

— Rod-cone independence for sensitizing interaction in the human retina. J. Physiol. (Lond.) **206**, 109—116 (1970).

— WILEY, R. W.: Distance effects in human scotopic retinal interaction. J. Physiol. (Lond.) **206**, 129—144 (1970).

Addendum

Several publications have recently appeared which bear significantly on the site of visual adaptation. DOWLING and RIPPS (S-potentials in the Skate Retina. J. Gen Physiol. **58**, 163—189, 1971) found that the membrane potential of horizontal cells was independent of the state of light and dark-adaptation supporting earlier observations along these lines by NAKA and RUSHTON (S-Potential and dark adaptation in fish. J. Physiol. (Lond.) **194**, 259—269, 1968). These results imply that horizontal cells can not mediate the spatially dependent sensitizing and desensitizing interactions produced by adapting lights. It would be important to know how accurately the membrane potentials recorded in the soma of this cell describe the transmembrane potential at its dendritic terminals since dendro-dendritic interactions are occuring in horizontal cells. GRABOWSKI, PINTO and PAK (Adaptation in retinal rods of Axolotl: Intracellular Recordings. Science **176**, 1240—1243, 1972) have found that rod receptors are capable of dark-adapting when little (less than 5%) to no rhodopsin regenerates suggesting that there is a neural phase of dark-adaptation within the photoreceptor that is independent of the photopigments. It remains possible, however, that the disappearance of a photoproduct determines the recovery of sensitivity in these rods.

Chapter 17

The Electroretinogram, as Analyzed by Microelectrode Studies

By

Tsuneo Tomita, Tokyo (Japan) and New Haven, Connecticut (USA)

With 20 Figures

Contents

1. Introduction

It was more than a century ago that Holmgren (1865) discovered the electrical response of the retina to light, the electroretinogram (ERG) of the present day. The ERG is usually recorded with a pair of electrodes placed on the opposite sides of the retina. The ERG may differ in shape according to the species and the state of adaptation (Fig. 1), but typically it starts with a cornea-negative deflection termed the *a*-wave, followed by a cornea-positive deflection (*b*-wave). At the termination of light there occurs another deflection (*d*-wave), the polarity of which is either cornea-positive or cornea-negative, depending on the species used. In the dark adapted retina, there is also a very slow cornea-positive deflection (*c*-wave).

Besides these four waves, two other waves have been described. One is the *x*-wave which is an early and fast, cornea-positive deflection, first observed in the human ERG by Motokawa and Mita (1942). They identified it as dependent upon the activity of cones. This was later confirmed by Armington (1952, 1953) and Schubert and Bornschein (1952), who regarded it as a *b*-wave of short latency associated with cone activity. The other is the *e*-wave which is

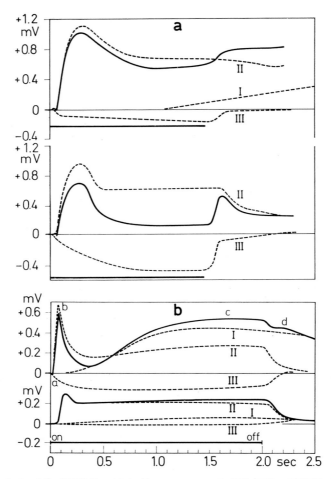

Fig. 1.a. Analysis of the ERG (frog) into three components (PI, PII and PIII) at dark- (upper) and light-adapted state (lower). (Granit and Riddell, 1934) b. Analysis of the ERG (cat) at two intensities of light, 14 ml (upper) and 0.14 ml (lower). (Granit, 1933)

a slow, cornea-positive wave appearing in response to the termination of light but with a considerable delay. This wave was described in the frog by Sickel and Crescitelli (1967) and Crescitelli and Sickel (1968) and also (in the developing retina of the tadpole) by Crescitelli (1970). The most characteristic property of the e-wave is that it responds to the termination of light with a delay that increases with increase in light intensity. Delays up to 20 sec have been observed. The e-wave is easily distinguishable from the d-wave by the long delay, and also from the c-wave by its persistence in the retina detached from the pigment epithelium, from which the c-wave originates.

Many attempts to analyze the ERG into components have been made since its discovery, but the analysis by Granit into three components (PI, PII and PIII) is the one most generally accepted. The background and status as of the 1940s were reviewed extensively by Granit (1947) in his monograph "Sensory mechanisms of the retina".

With the advent of the microelectrode technique, a more direct localization of ERG components by means of depth recording in the retina became possible. Since the start of this type of work (TOMITA, 1950), however, it took more than a decade to reach general, though perhaps not complete, agreement among investigators concerning the localization of the components in retinal layers. Individual observations during this period often were contradictory and confusing. It is intended in this Chapter to review the development of our knowledge concerning the ERG components since the introduction of the microelectrode technique. The description will be confined to the vertebrate ERG. Some early conflicting results also will be discussed in order to elucidate the reasons for confusion in the past. On the other hand, this article does not intend to cover all the papers in this and related fields, nor does it try to review contradictions on minor points such as are inevitable in any type of active research. Readers will find in a review article by WITKOVSKY (1971) good coverage of recent papers on the peripheral mechanisms of vision, including the ERG. Concerning the early receptor potential (e.r.p.) discovered by BROWN and MURAKAMI (1964), see Ch. 12, Vol. 1 of this Hdbk.

It should be admitted that the current viewpoint attained by the aid of the microelectrode technique is about the same as that of GRANIT in 1947, except for some aspects which are now clear but then were discussed merely as possibilities. I would like to give, firstly, a summary of the current viewpoint. This will be followed by a presentation of the actual evidence in the subsequent sections.

a) PI of GRANIT, which represents a slow cornea-positive potential, originates in the pigment epithelium as the result of some interaction between the receptors and the pigment epithelium cells.

b) PII is a fast cornea-positive potential. It originates in the inner nuclear layer, and is related to excitatory processes.

c) PIII is another fast, but cornea-negative, potential. It consists of two subcomponents; one originates in the receptors themselves, and the other in structures in the inner nuclear layer. Both seem to be related to "inhibitory" processes. Evidence was provided in Chapter 12 that the vertebrate photoreceptors remain depolarized in the dark and are polarized by light.

2. Principles of ERG Analysis with Microelectrodes

Let us assume a dipole layer within the retina, as shown in Fig. 2, to represent one component of the ERG. Radially oriented cells like the receptors and bipolar cells, and possibly the Müller cells, should be the most efficient sources for such dipoles. The other cell types (horizontal, amacrine, and ganglion cells) cannot be excluded as ERG sources, but their contribution, if any, is considered small in view of their more lateral orientation.

It should be theoretically easy to localize the dipole layer with a penetrating microelectrode (TOMITA, 1950). When the dipole layer is between the microelectrode E_1 and the indifferent electrode E_2, the recorded potential e is

$$e = e_0 - e_0 R_2/(R_1 + R_2 + R_3) ,$$

where e_0 is the e.m.f. of the dipole or the battery. After the electrode has penetrated

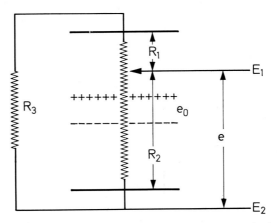

Fig. 2. Equivalent circuit of intraretinal recording. R_1 and R_2, transretinal resistances above and below the intraretinal microelectrode E_1; R_3, external resistance; E_2, indifferent electrode; e_0, potential of a dipole layer representing one ERG component; e, potential recorded between E_1 and E_2. The relation between e_0 and e is shown in the text. (Tomita, 1950)

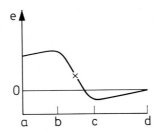

Fig. 3. Showing the amplitude-depth relation of intraretinally recorded potential (e in Fig. 2). a, uppermost retinal surface as the zero reference depth; b and c, depths corresponding to the positive and negative poles of the dipole layer e_0; d, lowermost retinal surface in contact with the indifferent electrode E_2

through the dipole layer to the side of the indifferent electrode, e becomes

$$e = - e_0 R_2/(R_1 + R_2 + R_3) \, .$$

Fig. 3 shows the amplitude-depth relation. As might be clear from the equations, the gradient of curve in the layer between a and b and also in the layer between c and d, each of which constitutes part of the passive pathway of current, becomes smaller with a decrease in current due to an increase in the external resistance R_3. An uneven distribution of the resistances across the tissue also is reflected in the sequence of potentials recorded. If a high resistance membrane exists at a certain depth outside the dipole layer, a potential iRm appears across this membrane, where i is the intensity of current passing the membrane, or $e_0/(R_1 + R_2 + R_3)$, and Rm is the resistance of the membrane. The potential across the membrane becomes larger with a decrease of the external resistance R_3. This potential, which represents a passive e.m.f., differs from the potential recorded

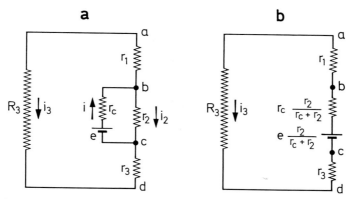

Fig. 4. Two ERG circuits (a and b), which are mutually equivalent, to supplement Fig. 2. See text for explanation. Points a, b, c and d symbolize the same as those in Fig. 3

across the dipole layer in that, due to the internal resistance, the latter becomes smaller with a decrease of R_3 which causes an increase of i. This provides a means of discriminating a passive e.m.f. from an active one: If it is active, the potential recorded is decreased when R_3 is decreased, but if it is passive, the reverse happens. In an experiment using the excised opened eye of the frog, R_3 was made negligibly small by immersing the opened eye in Ringer solution (TOMITA and TORIHAMA, 1956). Another practical way of changing the external resistance is to change the amount of retinal areas illuminated. If the stimulus is a small light spot, the whole retina except the area illuminated acts as a short-circuiting external resistance, resulting in a small R_3. In this case, however, the complication should be noted that the non-illuminated area responds to stray light to produce an ERG to low intensity stimulation (FRY and BARTLEY, 1935; ASHER, 1951; BOYNTON and RIGGS, 1951; BOYNTON, 1953; BROWN and WIESEL, 1961a).

Fig. 4a supplements the ERG circuit in Fig. 2 by showing that the dipole layer consists of a group of cells as the possible source of an ERG component, represented by e and r_c in series, and the extracellular space r_2 connected in parallel with the cell group. The total ERG current i through r_c is

$$i = \frac{e}{r_c + \dfrac{1}{\dfrac{1}{r_2} + \dfrac{1}{r_1 + R_3 + r_3}}} \tag{1}$$

since i is divided into i_2 and i_3 in the proportion of

$$\frac{r_1 + R_3 + r_3}{r_1 + r_2 + r_3 + R_3} \quad \text{and} \quad \frac{r_2}{r_1 + r_2 + r_3 + R_3},$$

$$i_3 = i \frac{r_2}{r_1 + r_2 + r_3 + R_3}. \tag{2}$$

Calculating from Eq. (1) and Eq. (2),

$$i_3 = \frac{e \dfrac{r_2}{r_c + r_2}}{r_c \dfrac{r_2}{r_c + r_2} + (r_1 + R_3 + r_3)} \tag{3}$$

Eq. (3) denotes that the structure of the dipole shown in Fig. 4a is equivalent to a battery, with an e.m.f. of $e \, r_2/(r_c + r_2)$ and an internal resistance of $r_c \, r_2/(r_c + r_2)$, which is connected in series with a resistance of $(r_1 + R_3 + r_3)$. The equivalent circuit is shown in Fig. 4b. Under the condition of $r_c \gg r_2$, which doubtless applies to the retina as does to other tissues, the equivalent internal resistance $r_c \, r_2/(r_c + r_2)$ approximates r_2.

Depending on whether the indifferent electrode is behind or in front of the retina, the sequence of intraretinal potentials recorded as a function of electrode depth might look quite different, but the curve showing the amplitude-depth relation (Fig. 3) should remain the same. It is only the reference level of the curve that changes according to the position of the indifferent electrode. This was emphasized correctly also by Byzov (1965) and by Rodieck and Ford (1969).

3. Observations on the Excised Eye of the Frog

a) Early Conflicting Results

Fig. 5 illustrates schematically the results from a number of depth recordings with a micropipette electrode advanced from the vitreal side into the retina in the frog's eye cup. The indifferent electrode was on the scleral side. The moment the electrode tip touched the retinal surface (the internal limiting membrane) was signalled by a surge of potential produced by the action of the pipette electrode as a sensitive mechanoelectric transducer. The depth at which this potential change

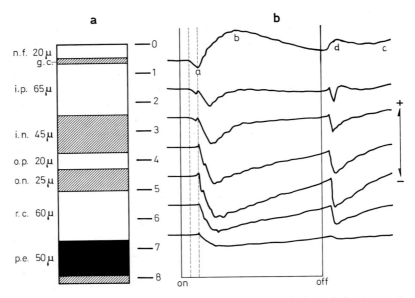

Fig. 5. Schematic diagram showing the depth-configuration relation of the intraretinally recorded ERG (bullfrog), with the indifferent electrode on the scleral side. Left diagram (a) shows the histologically determined dimension of each layer of the bullfrog's retina. Numerals in the middle column indicate distances from the retinal surface, each scale division being 35 μ. n.f., nerve fiber layer; g.c., ganglion cell layer; i.p., inner plexiform layer; i.n., inner nuclear layer; o.p., outer plexiform layer; o.n., outer nuclear layer; r.c., layer of rods and cones; p.e., pigment epithelium. (Tomita, 1950)

appeared was taken as the reference level (depth 0) and the depths therefrom were measured from readings of the micrometer gauge of the micromanipulator. The pipette was moved back and forth in steps of 35 μ, and at each step the response to light was recorded. Fig. 5a shows the histologically determined thickness of the retinal layers, the numbers 0—8 indicating 35 μ intervals. Fig. 5b shows that the normal ERG is recorded at superficial retinal layers, but a gradual inversion of polarity takes place as the micropipette is advanced, and at the depth corresponding to the outer plexiform layer the inversion is virtually complete for the a-, b- and d-wave, but not for the c-wave. The c-wave retained the same polarity throughout, but once the retina was detached from the pigment epithelium, the c-wave was lost from both the retina and the remaining part of the eye still covered by the pigment epithelium (TOMITA, 1950).

Based on the above observation, it was concluded that the a-wave, or the rising phase of the PIII of GRANIT, has its main origin in the inner nuclear layer, and the c-wave is a potential associated with some metabolic interaction between the receptors and the pigment epithelium cells. Concerning the b- and d-waves, the conclusion was withheld until the publication of subsequent papers (TOMITA et al., 1951; TOMITA et al., 1952; TOMITA and FUNAISHI, 1952), in which these waves were localized in the inner nuclear layer. However, in the light of current interpretation, to be given later, this localization for the b- and d-waves was made in an entirely wrong way, though the conclusion as such happened to be approximately true. We now know that the intraretinally recorded negative potentials seen in Fig. 5b are the b- and d-waves inverted after passing through their sites of origin. This possibility was given due consideration at that early time (see, for instance, TOMITA and FUNAISHI, 1952), but it was contradicted by the following three observations. Firstly, exploring the retina with a small light spot, the intraretinal negative potentials fell off as the light spot was moved away from the site of recording, whereas both the b- and d-waves responded irrespective of whether the light spot was on the recording site or far from it (TOMITA et al., 1952). Secondly, the intraretinal negative potentials were more susceptible to chemical agents, and disappeared sooner than the b- and d-waves. Thirdly, after the disappearance of the intraretinal negative potentials, an ERG of normal shape but of smaller size usually remained (TOMITA and FUNAISHI, 1952). These observations led us to the premature conclusion that the intraretinal negative potentials were responses of a type easy to record intraretinally but contributing little to the ERG.

Experiments similar to ours but leading to entirely different conclusions were reported by OTTOSON and SVAETICHIN (1952, 1953). The ERG showed no substantial change until the penetrating micropipette reached the outer plexiform layer. It then decreased in amplitude during the penetration through the receptor layer. No reversal in polarity of the ERG, such as seen in Fig. 5b, was observed in their depth recordings. The conclusion of OTTOSON and SVAETICHIN was that all ERG components originate exclusively in the receptors. This conclusion appeared to be strongly supported by SVAETICHIN's (1953) discovery, in the fish retina, of a large sustained potential in response to light. In the belief that this potential was recorded from within single cones, it was termed the cone action potential.

The conflicting results of SVAETICHIN and TOMITA which resulted from substantially similar experiments, stimulated BRINDLEY (1956a, b, c) to perform another series of microelectrode studies with similar technique. With a micro-electrode advanced in steps from the vitreal side of the frog's opened eye, he first measured the distribution of electric resistance across the retina-pigment epi-thelium-choroid complex. At each step of advance of the micropipette, weak current pulses of intensity I were passed between a pair of silver wire electrodes, one in the vitreous and the other below the eye. The resistance ΔR between suc-cessive steps of advance of the microelectrode, calculated from Ohm's law, is

$$\Delta R = \Delta V/I ,$$

where ΔV is the potential difference between pulses recorded at successive steps. In this experiment, Brindley (1956a) found a layer of high resistance and high capacitance within the eye, and termed it the R membrane. He also found that the largest component of the resting potential is the potential difference across this R membrane.

In the intraretinal recording of the ERG, Brindley (1956c) distinguished two cases. While the eye cup preparation was fresh, a sequence of records resulted which was similar to that of Tomita (1950) shown in Fig. 5b, but in the aged prep-aration, a sequence similar to that of Ottoson and Svaetichin (1953) was ob-served. In the latter case, the ERG continued to be recorded with no substantial change up to a certain electrode depth, to disappear rather rapidly beyond it (cf. Fig. 8). Brindley (1960) called the former type of sequence Tomita's complex pattern and the latter type Svaetichin's simple pattern. Concerning the intra-retinal negative potentials in the complex pattern, however, Brindley mistook them, just as I did, for potentials of a kind easy to record intraretinally but con-tributing little to the ERG, which he held to be masked at the intraretinal site until uncovered in the aging retina. The amplitude-depth relation of the ERG measured by Brindley (1956c) in the simple pattern indicated that the major part of the ERG was produced across the R membrane which at that time he held to be the external limiting membrane. Since the external limiting membrane is penetrated by all receptors, the above observation suggested strongly that the ERG exclusively originated in the receptors. This, and other evidence such as the additivity of locally elicited ERGs (Brindley, 1956b, 1958), led him to the same premature conclusion as Ottoson and Svaetichin, that all the ERG components originate in the receptors.

The current interpretation emerged from the above mentioned conflicting data, when Svaetichin's potential (S potential) which he first thought to be the cone action potential was found to arise in structures within the inner nuclear layer (Tomita, 1957; Tomita et al., 1958; Mitarai, 1958; MacNichol and Svaetichin, 1958; Oikawa et al., 1959; Tomita et al., 1959; Gouras, 1960; Brown and Tasaki, 1961), and when the site of the R membrane was shifted from inside the retina to just back of it (Brown and Wiesel, 1958, 1959; Tomita et al., 1960; Brown and Tasaki, 1961; Brindley and Hamasaki, 1963).

b) Current Interpretation

The current interpretation originated in a repetition of Brindley's experiment using a modification of his method. Fig. 6a shows the arrangement. Weak current pulses (15 μa) are passed across the eye from a wick-electrode in the vitreous humor

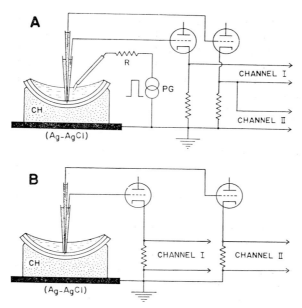

Fig. 6. Arrangements for the measurement of the radial resistance of the bullfrog's opened eye (A), and for the recording of responses to light (B). CH, Ringer-soaked chalk; PG, square pulse generator; R, 1 MΩ resistance as a current limiter. Further explanation in text. (Tomita et al., 1960)

to a chlorided silver plate below the eye, which also serves as the indifferent lead. Potentials developed by the current pulses across the tissue are led off from the internal and external pipettes of a coaxial type microelectrode (Tomita, 1962) through cathode followers to a two-channel amplifier as shown in Fig. 6a. The external pipette is placed at the retinal surface, and the internal pipette is inserted into the retina. The two pipettes, connected differentially to one of the channels (Channel I in Fig. 6a) of the amplifier, record, through this channel, potential pulses whose amplitude is directly proportional to the resistance of the tissue layer intervening between them. For recording the sequence of responses to light as a function of electrode depth, either Channel II in Fig. 6a or the arrangement in Fig. 6b is used. The latter serves for simultaneous recording of the surface ERG (through Channel I) and the intraretinal response (through Channel II). Reconnection from Fig. 6a to 6b, or the reverse, is made by a multipolar switch.

The left column of Fig. 7 illustrates a sequence of potential pulses, showing that the resistance increases gradually as the depth of the internal pipette is increased, but that at a depth slightly over 250 μ there is a sudden marked increase, indicating that the R membrane has been penetrated (bottom record). Recordings from opposite sides of this membrane could be repeated many times by moving the internal pipette up and down within a range of 20 μ. The right column of Fig. 7 shows a sequence of intraretinal records of response to diffuse illumination over the whole retina. The sequence of records is typical of the complex pattern. The response is seen to fall off abruptly when the pipette penetrates to the opposite side

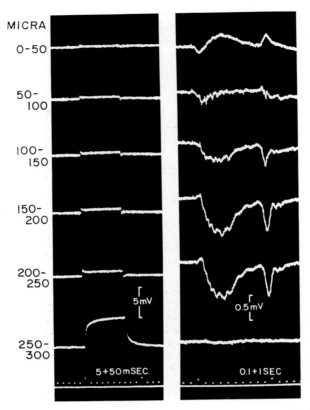

Fig. 7. Depth recording of potential pulses produced by current pulses of 15 μa across the eye (left column), and of responses typical of the complex pattern (right column). Channel I in Fig. 6a was used for recording of the potential pulses whose amplitude at each depth is proportional to the resistance of the tissue layer intervening between the outer and inner pipettes of the coaxial microelectrode. Responses to light were recorded through Channel II of Fig. 6a. (Tomita et al., 1960)

of the R membrane. It should be noted, however, that there is no reversal in polarity after the penetration. A trace of potential similar to that before the penetration remains.

Fig. 8 illustrates the simple pattern of response obtained from a retina which has been excised for some time. In the right column, simultaneous recordings of the surface ERG (upper tracing) and the intraretinal response (lower tracing) are shown. The intraretinal response exhibits no substantial change from the surface ERG as long as the pipette is vitreal to the R membrane, but an abrupt decrease in amplitude occurs after the R membrane has been penetrated. It is important to note that in this case also the polarity of the response remains the same on the opposite sides of the R membrane.

The results so far are in good agreement with Brindley. However, when the experiment was repeated on the eye cup after the removal of the retina, but still

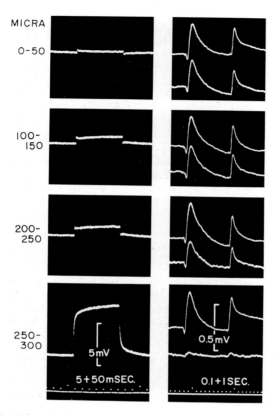

MICRA

0-50

100–150

200–250

250–300

0.5 mV

5 mV

5+50 mSEC.

0.1+1 SEC.

Fig. 8. Similar to Fig. 7, but showing responses typical of the simple pattern in the right column. The surface ERG (upper tracing) and the intraretinal response (lower tracing) were recorded simultaneously by the arrangement in Fig. 6b

covered with the pigment epithelium, a resistance comparable to that of the R membrane was detected near the inner surface of the preparation. If the pigment epithelium happened to be damaged or removed along with the retina, leaving the choroid and sclera, the resistance was lost completely. This observation supported the conclusion of BROWN and WIESEL (1958) that the R membrane is the Bruch's membrane just back of the pigment epithelium.

The R membrane is now identified as the pigment epithelium (BRINDLEY and HAMASAKI, 1963; NOELL, 1954; FABER, 1969) or the complex of pigment epithelium and Bruch's membrane (BROWN, 1968).

The resistance of the R membrane ($R_{R\ \text{memb}}$) and the resistance of the retina (R_{Retina}) are calculated from the following equations:

$$R_{R\ \text{memb}} = \frac{V_2 - V_1}{I}\, S\,,$$

$$R_{\text{Retina}} = \frac{V_1}{I}\, S\,,$$

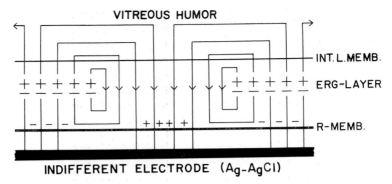

Fig. 9. Schema of the flux of ERG current across an inactivated region (central part of the diagram) and active regions of the retina. Explanation in text. (Tomita et al., 1960)

in which V_1 and V_2 are the amplitudes of potential pulses just before and after the penetration of the R membrane by the internal pipette, and S is the area of the retina across which current pulses of intensity I (15 μa) are passed. The resistance of the R membrane thus calculated was 216 Ω cm² on the average, and was several times greater than the resistance across the whole retinal tissue, which was 43 Ω cm² on the average. Resistance across the choroid after removal of the pigment epithelium was negligibly small.

Fig. 9 is the model proposed by Tomita et al. (1960) to account for the variety of results obtained in similar experiments. The model incorporates a concept of "functional non-uniformity": One region of the retina (central region of this model) is assumed to have been inactivated so that this area, which otherwise contributes to the ERG, does not generate any potential. The ERG current originating in the neighboring, active retinal region then passes this inactivated region which now acts simply as a passive external circuit for the ERG current. As a result, the R membrane under the inactivated region undergoes a polarization, the magnitude of which is

$$V_{R\,\text{memb}} = \frac{R_{R\,\text{memb}}}{R_{\text{Retina}} + R_{R\,\text{memb}}}\, V_{\text{ERG}}\;,$$

in which V_{ERG} is the ERG potential between the vitreous humor and the indifferent electrode. Substituting 43 Ω cm² for R_{Retina} and 216 Ω cm² for $R_{R\,\text{memb}}$,

$$V_{R\,\text{memb}} = 0.84\, V_{\text{ERG}}.$$

It thus is suggested that an electrode located in an inactivated region of the retina records an apparently normal ERG, without much loss of potential, at any depth above the R membrane, resulting in the simple pattern of response, while at an active region of the retina the ERG current polarizes the R membrane underneath in the opposite direction, and accordingly the microelectrode inserted in this region beyond the ERG-producing layer records an inverted ERG, or the complex pattern of response.

The proposed model was tested, using a fresh retina with a limited area around the recording microelectrode inactivated artificially. One way of doing this

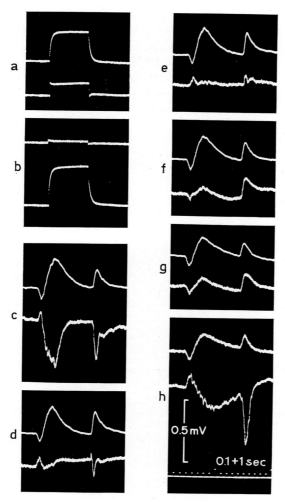

Fig. 10. Effect of local inactivation of the retina at the recording site upon the response to diffuse illumination. a and b, pulse potential recordings by means of which the inner pipette of a coaxial microelectrode was located to a depth some $30\,\mu$ vitreal to the R membrane. The lower tracings in $c - g$ are responses to light at this point, while the upper tracings are surface ERGs recorded simultaneously. c is a control, recorded immediately after the local application of 5% cocaine, showing response typical of the complex pattern. Times of subsequent recordings after cocainization; 30 sec (d), 45 sec (e), 2 min (f), and 5 min (g). The response in g is typical of the simple pattern. After g, recording of h was made at a peripheral region of the same retina. The arrangement in Fig. 6b was used for these recordings. (Tomita et al., 1960)

was by local application of an anesthetic (5% cocaine) to the site of recording. Fig. 10$c-h$ show simultaneous recordings of the surface ERG (upper tracing) and the intraretinal response from a depth just vitreal to the R membrane (lower tracing). Fig. 10c is the control, recorded immediately after local cocainization, showing a response typical of the complex pattern. Later, the intraretinal response is decreased in size (*d* and *e*), reversed in polarity (*f*), and finally acquires a configuration similar to the surface ERG (*g*). This demonstrates clearly that the initial complex pattern has turned to the simple pattern by local inactivation at the recording site. Responses obtained at this stage from neighboring normal retinal regions are of the complex pattern (*h*). In another experiment using a fresh retina, the retinal region around the recording electrode was inactivated locally by exposure to a bright light spot. In the presence of this bright light as the background, the response to diffuse light of lower intensity was found to be typical of the simple pattern, but in the absence of the light spot the response turned out to be of the complex pattern. By turning on and off this intense background light, the response to the diffuse illumination could be alternated between the complex and the simple pattern as many times as desired (Fig. 11).

It should be clear from the above experiments that the proposed model in Fig. 9 accounts satisfactorily for the genesis of both complex and simple patterns

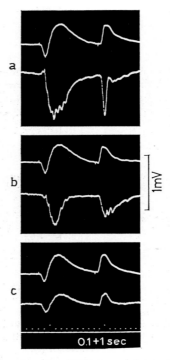

Fig. 11. Simultaneous recordings of surface ERG (upper tracing) and intraretinal ERG at a point just vitreal to the R membrane (lower tracing), obtained with different combinations of two lights, one of which is a light spot of 2 mm in diameter centered onto the recording site (focal light), and the other is a diffuse light but with a shade of 4 mm in diameter centered to the recording site (non-focal light). *a*, response (typical of the complex pattern) elicited by both lights applied at the same time; *b*, response to non-focal light alone, which exhibits still a complex pattern as the result that the retinal area in the shade also was activated by stray light; *c*, response (typical of the simple pattern) elicited by non-focal light in the presence of focal light as a background to eliminate the effect of stray light. The arrangement in Fig. 6b was used for the recordings. (Tomita et al., 1960)

of response. Between these two typical patterns, there should be every intermediate pattern, since the condition of the retina at the recording site could be
varied from entirely normal to completely inactive. Some of these intermediate
patterns which were recorded during the course of local anesthesia by cocaine are
illustreated in Fig. 10$d-f$. Such intermediate patterns were demonstrated also in
experiments similar to that shown in Fig. 11, where the recording site was inactivated locally to various degrees by varying the intensity of the background light
spot (Fig. 12).

So far, I have discussed functional non-uniformity produced by artificial
means, but GOURAS (1958) reports a type of non-uniformity which develops
spontaneously in the excised retina. This is the phenomenon of spreading depression which resembles in many respects the spreading cortical depression of LEÃO
(1944). The depression wave, which is usually observed in aged preparations,
marches periodically across the surface of the retina and lasts for $2-3$ min at any
one point. At these times even the most intense photic stimulation becomes unable
to elicit a response. Mechanical stimulation often triggers the spreading depression,
and hence the site of electrode penetration could be a nucleus for the development
of such depression.

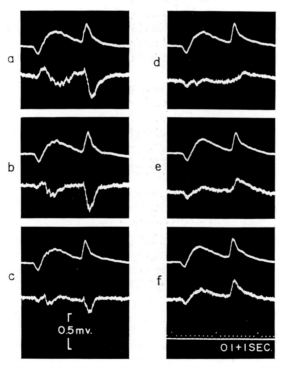

Fig. 12. Responses to non-focal light in the presence of various intensities of focal light as a
background. The arrangement of Fig. 6b was used. a, control without focal light, obtained
under conditions identical with those for Fig. 11b. Intensities in log-unit of focal light in
reference to non-focal light; -1.8 (b), -1.2 (c), -0.6 (d), 0 (e), and 1.0 (f). Note the transition
from the complex pattern to simple pattern. (TOMITA et al., 1960)

The conclusion from the reported results is very simple. For microelectrode localization of ERG components in the excised retina, the preparation should be fresh so that the condition of functional uniformity is better satisfied. Re-examination of the results of depth-recording from this viewpoint confirmed our earlier conclusion that the main site of both PII and PIII is in the inner nuclear layer (Tomita et al., 1960). Concerning the origin of PIII, however, an amendment was later found to be necessary, as described in the subsequent two sections.

4. Observations on Cat and Monkey Retinas *in situ*

A series of ERG analyses with penetrating microelectrodes, using the *in situ* eye of the cat and monkey, was carried out by Brown and his coworkers (Brown and Wiesel, 1958, 1959, 1961a, b; Brown and Tasaki, 1961; Brown and Watanabe, 1962a, b; Arden and Brown, 1965; Brown et al., 1965). The whole work was recently reviewed in detail by Brown (1968). The methods of analysis are in principle the same with warm- as with cold-blooded retinas. Work with warm-blooded animals, however, involves more difficult technical problems, such as the abolishment of eye movements, the insertion of microelectrodes into the unopened eye, the accurate localization of microelectrodes within the retina, and so forth. Once these technical problems were solved, however, the results could be more easily interpreted. Because the retina is supported by blood circulation, there exist no such pitfalls as the functional non-uniformity encountered in the frog's excised eye. The following is the outline of the methods of Brown and his coworkers.

The animal is anesthetized by a barbiturate anesthetic, and skeletal muscle movements are abolished by continuous intravenous infusion of succinylcholine. The animal is then artificially respirated. The normal optics of the eye is maintained by a glass contact lens. A special device attached to the eye incorporates channels for three needles inserted into the vitreous humor. One needle-channel is used either for a micropipette filled with 3 M-KCl, or for a tungsten microelectrode which is insulated except for the very tip. The microelectrode is advanced through the retina by a hydraulic electrode advancer. The second needle-channel is used for a chlorided silver wire which is inserted in the vitreous humor to serve as the reference electrode. The third needle-channel may be used for any of several special devices, such as a rod for applying pressure upon the optic disc so as to clamp the retinal circulation selectively. Positioning of the intraocular devices is made under visual control through the normal optics of the eye using a hand ophthalmoscope. The retina is also viewed directly for focusing and positioning the stimulus spot. Light stimuli from a tungsten lamp are provided by a special optical stimulator.

The difficult problem of knowing the actual depth of the microelectrode in the retina was solved by the success of Brown and Wiesel (1958) in setting three landmarks, utilizing the micropipette electrode capacity to behave as an extremely sensitive mechanoelectric transducer. The pipette records the pulse beat of blood vessels in the tissue through which the electrode passes. The retinal circulation of the cat extends from the inner surface to the outer margin of the inner nuclear layer (Michaelson, 1954); deeper layers of the retina are avascular. Thus a pulse

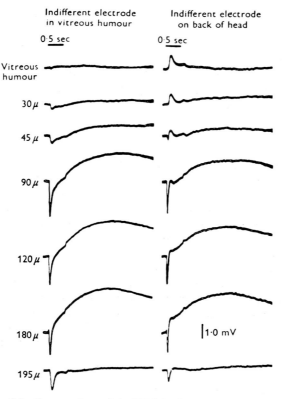

Fig. 13. Two series of depth recorgings of the ERG in the cat, with the indifferent electrode in the vitreous humor (left column) and on the back of the head (right column). (BROWN and WIESEL, 1961a)

beat first appeared in the record when contact was made with the retina, remained strong throughout the surface layers of the retina, and disappeared or was sharply reduced when the electrode passed the outer margin of the inner nuclear layer. But when the electrode reached the pigment epithelium, backed by BRUCH's membrane, a strong pulse beat reappeared which was due to the choroidal circulation. A slight additional advance resulted in the penetration of the high resistance membrane, first described by BRINDLEY (1956a) in the frog as the R membrane, and later identified as the complex of BRUCH's membrane and pigment epithelium. After penetrating the R membrane, a strong pulse beat was recorded throughout the choroid. Thus, recording the pulse beat provided a means for identifying three major electrode locations; the retinal surface, the outer margin of the inner nuclear layer, and the retinal side of the pigment epithelium. Once the relation between the depth and the sequence of intraretinal responses was established utilizing these three landmarks, the micropipette electrodes could be replaced by tungsten electrodes. The tungsten electrodes do not record pulse beats but the depth of the electrode can now be countermeasured from the known sequence of responses.

Fig. 13 illustrates two series of depth recordings in the cat by Brown and Wiesel (1961 a). The indifferent electrode was either in the vitreous humor (records in the left column) or on the back of the head (right column). According to the position of the indifferent electrode, distinct differences are seen between responses at each depth, but as mentioned before (Section 2), the configurations of curves showing the amplitude-depth relation for the a-, b-, and c-wave should be independent of the position of the indifferent electrode. Fig. 14 shows such curves plotted from another series of depth recordings with the indifferent electrode in the vitreous. Curves for the a- and c-wave are similar in shape, with maxima close to the retinal side of the R membrane. On the other hand, the b-wave shows a peak in or near the inner nuclear layer. The correlation between the amplitude maxima and retinal layers has been confirmed by electrode marking (Brown and Tasaki, 1961).

From these and related experiments, the b-wave (PII) was localized in the inner nuclear layer, the a-wave (PIII) in the receptor layer, and the c-wave (PI) in the pigment epithelium. The conclusion that the c-wave arises in cells of the pigment epithelium was based on additional evidence provided by Noell (1953, 1954), who showed that sodium iodate, which selectively destroys the pigment epithelium, also selectively reduces the c-wave.

Noell (1953, 1954) reported that, in the rabbit, PI consists of two separate potentials. One is a cornea-positive deflection, labelled by him as the azide-sensitive potential from the effect of sodium azide upon it, and the other a cornea-negative deflection of similar time course, labelled as the azide-insensitive potential. Sodium iodate, which affects the pigment epithelium and the rods most severely, removes the azide-sensitive potential, unmasking the azide-insensitive potential. Noell concluded from his work with metabolic poisons, that the azide-sensitive potential originates in the pigment epithelium and the azide-insensitive potential in the retina proper. His first conclusion was substantiated by the recent success in intracellular recording of the c-wave from cells of the pigment epithelium in the cat (Steinberg, Schmidt,

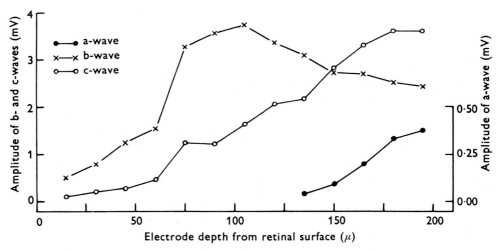

Fig. 14. Amplitudes of a-, b- and c-waves as a function of electrode depth, measured from a series of depth recordings in the cat with the indifferent electrode in the vitreous humor. (Brown and Wiesel, 1961 b)

and BROWN, 1970; SCHMIDT and STEINBERG, 1971), and his second conclusion by the observation of BORNSCHEIN, HANITZSCH and LÜTZOW (1966) of a slow deflection corresponding to
the azide-insensitive potential in the rabbit retina isolated from the pigment epithelium. FABER
(1969), working mainly on these slow potentials, postulates a glial (Müller cell) origin of the
azide-insensitive potential, or the slow PIII according to his terminology.

The localization of PII in the inner nuclear layer is consistent with the
general view held since GRANIT (1947), and with the results of microelectrode
studies in the frog. For the recent work of MILLER and DOWLING (1970) who conclude that the *b*-wave originates in the Müller (glial) cells, see Section 6 b.

BROWN and WIESEL (1961b) consider that PII is made up of two separate components;
the *b*-wave and the d.c. component. The former corresponds to the initial transient phase and
the latter to the steady phase of PII. These authors find the reason for this subdivision in their
observation that, when intensity is reduced the *b*-wave decreases in size, leaving only the d.c.
component, and that the *b*-wave is more susceptible to anesthesia than the d.c. component.
However, I prefer to adhere to GRANIT's classical analysis, until it becomes clear that the
separation of two components promotes actually better understanding of the retinal mechanisms. As discussed by GRANIT (1962), it is rather common that a cell responds to a strong
stimulus with an initial transient phase followed by a steady phase. Furthermore, the site of
b-wave and the site of d.c. component are inseparable by electrode depth studies (STEINBERG,
1969). This question is likely to be answered before long, since intracellular microelectrodes are
being successfully applied to the study of the activity of single cells in the inner nuclear layer
(BORTOFF, 1964; KANEKO and HASHIMOTO, 1968, 1969; WERBLIN, 1968; WERBLIN and DOW-
LING, 1969; KANEKO, 1970).

Concerning the localization of PIII, a distinct difference is noted between the
results in mammals and frogs. In mammals the entire PIII was ascribed solely to
the receptors, while in frogs its major portion was attributed to the inner nuclear
layer. Before discussing this difference, however, let us turn our attention to some
further observations on the PIII of mammals.

BROWN and WATANABE (1962a, b) developed methods, for localizing the ERG
components, that are entirely independent of electrode depth studies. One method
was to record, with a tungsten microelectrode, the local ERG from the central
fovea of the cynomolgus monkey retina, where the ganglion cell layer is absent and
the inner nuclear layer is reduced to a few scattered cells, while the receptor layer,
comprising pure cones, is well developed. Foveal local ERGs were then compared
with those from the peripheral retina. The other method was to clamp the retinal
circulation by pressure (applied by means of a stainless steel rod inserted into the
eye through the third needle-channel) upon the vessels emerging from the optic
disc. The choroidal circulation was unaffected by this procedure because the major
choroidal vessels penetrate the sclera at a distance from the optic disc (POLYAK,
1957). Since the retinal circulation extends only to the outer margin of the inner
nuclear layer and terminates sharply at that level, it is reasonable to assume that
the ganglion cells and the cells of the inner nuclear layer are supported primarily
by the retinal circulation. This was confirmed by the histological effects of clamping the retinal circulation; severe degeneration of the ganglionic and inner nuclear
layers (NOELL, 1954; TANSLEY, 1961; BROWN et al., 1965; STONE, 1969). Since the
receptors and pigment epithelium cells are left almost intact after clamping the
retinal circulation, they must supported primarily by the choroidal circulation, as
has long been thought. It is expected, therefore, that the selective clamping of the
retinal circulation abolishes components generated by cells proximal to the

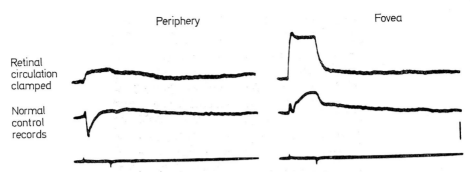

Fig. 15. Records from adjacent to the retinal side of the R membrane, under normal conditions and with the retinal circulation clamped, from both the peripheral retina and fovea of the cynomolgus monkey. The light spot for stimulation was about 0.25 mm in diameter, centered on the electrode and applied for 0.3 sec. Calibration 1.0 mV. (Brown and Watanabe, 1962a)

receptors, but leaves unaffected components generated either by the receptors or by cells in the pigment epithelium.

In Fig. 15 responses from the periphery and fovea of the cynomolgus monkey with the retinal circulation normal, are compared to those obtained when the circulation is clamped. Before clamping, the local ERG from the periphery is normal in shape, but the local ERG from the foveal region is markedly deformed, undoubtedly reflecting the specific structure of the fovea. After clamping the retinal circulation, the components originating in layers supported by the retinal circulation (P II) were completely abolished, leaving those corresponding to the P III of Granit. The P III from the fovea, which is considered to represent the response of cones, is more or less square-shaped, showing a rapid decay to the base line at the termination of the stimulus. The response from the periphery, on the other hand, decays in two steps when the stimulus terminates; a rapid fall followed by a slow decay. The decay in two steps was interpreted as being the result of superimposition of the rapidly decaying cone response and the slowly decaying rod response. This interpretation was confirmed by similar work on the cat, which has a predominantly rod retina, and also on the night monkey, which has an almost exclusively rod retina. After clamping the retinal circulation in these animals, there was no indication of rapid decay at the termination of stimulus.

5. Two Subcomponents of PIII in Cold-Blooded Retinas

Because of the work in mammals, which demonstrated the receptor origin of the P III, the old observation on the frog, in which the major portion of P III was ascribed to the inner nuclear layer, had to be re-examined.

The arrangement in one experiment by Tomita (1963) is the same as in Fig. 6b. A coaxial microelectrode is applied to the frog's opened eye. Its outer pipette is in contact with the inner retinal surface to record the surface ERG through one channel of a two-channel amplifier. The inner, superfine pipette, which is connected to the other channel of the amplifier, is set within the retina at the precise depth where the intraretinally obtained ERG is just reversed in polarity (lower tracing in

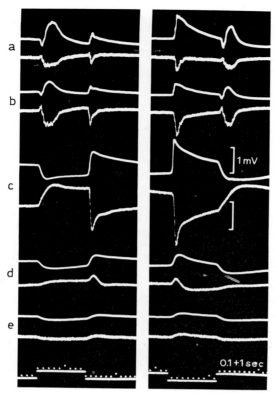

Fig. 16. Effect of sodium azide on both surface and intraretinal ERGs in the frog, obtained simultaneously by the arrangement in Fig. 6b. The inner pipette of the coaxial electrode was located at a depth of 140 μ from the vitreal retinal surface. *a*, a control record before azide, but after draining off the vitreous humor; *b*, another control record after filling the eye cup with Ringer solution. Subsequent records; 2 min (*c*), 4 min (*d*) and 6 min (*e*) after replacement of Ringer in the eye cup with 0.1 % azide-Ringer. With on-off of light for the records in the left column, and with off-on for those in the right column. (TOMITA, 1963)

Fig. 16a). This reversal indicates that the inner pipette has just penetrated through the layers producing the major portion of the ERG to the opposite side (TOMITA et al., 1960). After thus positioning the pipettes, 0.1 % azide-Ringer, which is a PII depressant (NOELL, 1953; MÜLLER-LIMMROTH and BLÜMER, 1957), is applied in the eye cup. Component PIII soon becomes dominant in both the surface and intraretinal ERGs (*c*). Since the intraretinal ERG continues to appear with reversed polarity in this phase after azide, the e.m.f. for this PIII-dominant ERG undoubtedly intervenes between the two pipettes. In the course of time, both surface and intraretinal ERGs become small (*d*), and eventually the intraretinal ERG takes on the same polarity and amplitude as the surface ERG (*e*), indicating clearly that the e.m.f. responsible for this remaining response no longer exists in the layers between the two pipettes, but somewhere else. By further advancement of the inner pipette toward the *R* membrane, the response from the inner pipette

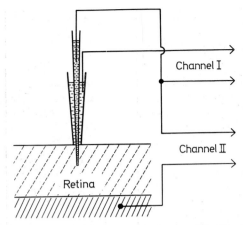

Fig. 17. The arrangement for fractional recording of the ERG in Fig. 18. Channel I records the potential between the inner and outer pipettes of the coaxial microelectrode, and Channel II the potential between the inner pipette and the indifferent electrode (MURAKAMI and KANEKO, 1966)

reversed its polarity again so as to produce a mirror image of that from the outer pipette (not illustrated). It was concluded, therefore, that the fraction of PIII which survived azide originates more distally than all the rest of the ERG in which a large fraction of PIII is involved. It was suggested that the distal fraction of PIII arises from the receptors, while the remaining, large fraction originates from cells in the inner nuclear layer.

A more clear-cut result was obtained by means of fractional recording of the ERG (MURAKAMI and KANEKO, 1966). Fig. 17 shows the arrangement. The frog's isolated retina is mounted receptor side up on the indifferent electrode. The outer pipette of the coaxial electrode is located at the distal margin of the retina, and the inner pipette protrudes from the tip of the outer pipette into the retina. Channel I of the amplifier records the potentials between the two pipettes (upper tracing of each record in Fig. 18), while Channel II records those between the inner pipette and the indifferent electrode (lower tracing). The algebraic sum of the potentials, thus recorded simultaneously, should equal the transretinal ERG. A sequence of simultaneous records are shown in Fig. 18 as a function of the depth of the inner pipette. To a certain depth, Channel I records no potential, but at some 80 μ from the receptor surface, a small positive deflection is discerned in the upper tracing. With further advancement of the inner pipette, the positive deflection becomes apparent in Channel I, and its amplitude reaches a maximum at a depth of 120 μ. Since the thickness from the outer margin of the receptors to the outer plexiform layer is about 105 μ in the frog (TOMITA and TORIHAMA, 1956), it is reasonable to assume that this positive sustained potential is the fraction of PIII arising from the receptors, while the large a-wave still remaining in the lower tracing, at this depth, should come from more proximal retinal layers. MURAKAMI and KANEKO termed these two fractions the distal and proximal PIIIs. As the inner pipette is advanced further, ERG potentials are seen to transfer gradually from Channel II (lower) into Channel I (upper).

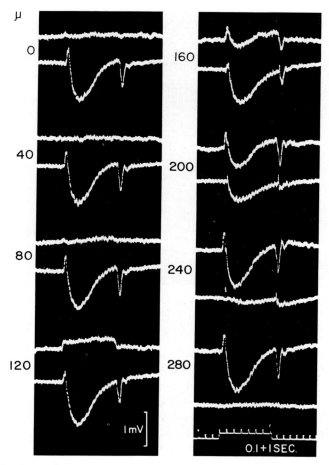

Fig. 18. Fractional recordings of the ERG, with the arrangement shown in Fig. 17, from the frog's retina mounted receptor side up on the indifferent electrode. The upper tracing of each record was obtained through Channel I and the lower tracing through Channel II. The depth of the inner pipette is shown in μ on the left of each column. (MURAKAMI and KANEKO, 1966)

Another approach employed by MURAKAMI and KANEKO to differentiate the distal PIII from the proximal PIII was to compare their latencies. If the distal PIII represents the activity of receptors and the proximal PIII that of cells in the inner nuclear layer, the latency should be shorter in the former. In the record made at 120 μ in Fig. 18, it is considered that the leading edge of the potential on Channel I consists primarily of the rising phase of the distal PIII, while that on Channel II consists primarily of the rising phase of the proximal PIII. The difference in latency between these PIIIs was measured, using an averaging computer. Fig. 19 illustrates a sample record, in which the tracing "d" represents the leading edge of the distal PIII and the tracing "p" that of the proximal PIII. The latency of the distal PIII in this case is about 15 msec, and is 7 msec shorter than that of

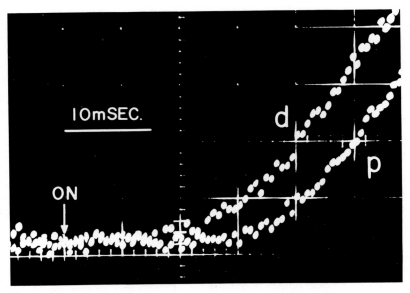

Fig. 19. Record of a data-processing digital computer, showing the difference in latency between the distal PIII (indicated as d) and the proximal PIII (indicated as p). Addition, 70 times; ON, indicating the beginning of photic stimulation. (Murakami and Kaneko, 1966)

the proximal. Like results were obtained from similar experiments on the turtle retina. The separation of the PIII into two components was also supported by differential sensitivity of the distal and proximal PIIIs to chemicals. Results of testing the effect of chemicals on the a-wave of the toad and frog by Brown (1966) and Yonemura and Hatta (1966) confirmed the separation of two components within the cold-blooded PIII. In summary, the PIII component of the cold-blooded retina is of dual origin. This was earlier suspected by Granit (1947) who states in his monograph (p. 113) "It is possible that PIII might turn out to be made up of two components, one localized in the neural structures, the other in the receptors." It is difficult to understand why no proximal PIII is detectable in mammals. It might be that the separation of PIII into two components applies to all the vertebrates, but that the dominant component is species dependent. Hanitzsch and Trifonow (1968), working on the isolated rabbit retina, report some indications, though not conclusive, that the PIII of the rabbit also consists of two components.

6. Cellular Origin and Physiological Significance of the ERG Components

a) PI

It was mentioned previously that the origin of the PI (c-wave) was established by intracellular recording of this component from cells of the pigment epithelium in the cat (Steinberg, Schmidt, and Brown, 1970; Schmidt and Steinberg,

1971). Concerning the mechanisms of its generation, however, the earlier notion of TOMITA (1950) and NOELL (1953) that the PI seems to be a manifestion of some metabolic interactions between receptors and pigment epithelium cells still is a good possibility. The PI is not discerned in pure cone retinas (MESERVEY and CHAFFEE, 1927; BERNHARD, 1941) and hence is considered to represent a process related mainly to the rod system (GRANIT, 1947, 1962). With regard to the biochemical aspects of PI, reference will be made to the article by SICKEL in this Volume (Chapter 18).

b) PII

The opinion of investigators now is consistent with GRANIT's in that the PII is dependent upon the activity of cells in the inner nuclear layer. Indeed, a large *b*-wave is recorded from retinas whose inner nuclear layer is well developed (MOTO-KAWA, 1966). Among the three neuronal cell types in this layer (horizontal, bipolar and amacrine), the horizontal cells were established to be the site of the *S* potential (SVAETICHIN et al., 1961; MITARAI, 1965; MITARAI et al., 1965; KANEKO, 1970; STEINBERG and SCHMIDT, 1970). In view of the cell arrangement which is radial in the bipolar cells but lateral in the horizontal and amacrine cells, it appeared reasonable to suspect the bipolar cells as principal sites of the PII.

Recent intracellular microelectrode studies have demonstrated three response types among cells in the inner nuclear layer (WERBLIN and DOWLING, 1969; KANEKO, 1970); (1) the on type, which represents a depolarization with light, (2) the off type, which represents a hyperpolarization with light, and (3) the on-off type, which exhibits a phasic depolarization at both the turning on and the turning off of the light stimulus. The on-off type has been identified as of amacrine cell origin. It does not change its response pattern with variation in the photic stimuli within the receptive field, and often carries spikes on the depolarizing phases. The other two types, which mostly originate in bipolar cells, are usually dependent upon stimulus patterns. Individual bipolar cells have a concentric organization of the receptive field, which is either an on center-off surround type or vice versa. The receptive field, as measured from the size of the light spot that influences the response size, often extends a few millimeters. Thus, the functional organization of the receptive field of single bipolar cells seems to be much the same as that in single ganglion cells demonstrated by KUFFLER (1953). The large receptive field with concentric organization in cells of the inner nuclear layer was reported previously by BROWN and WIESEL (1959), who used the *in situ* eye of the cat. Their extracellular microelectrode was not adequate for recording slow potentials from these cells, but was suitable for mapping the receptive field in terms of impulse spikes. The impulse activity in cells other than ganglion cells once was doubted by BYZOV (1959) and TOMITA et al. (1961), but it is now evident that at least some amacrine cells respond with impulse spikes.

The large receptive field of cells in the inner nuclear layer leads to the prediction that their electrical activity extends into the surrounds of the area illuminated. The results by MOTOKAWA et al. (1959) on the spatial distribution of the locally recorded ERG are consistent with this prediction. A microelectrode placed on the distal surface of the carp retina, which was detached from the pigment epithelium

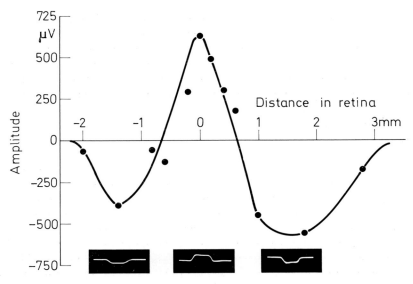

Fig. 20. Spatial distribution of the slow response in the carp, obtained from a pipette electrode on the distal retinal surface at point 0 and by scanning a light spot across the site of recording (Motokawa et al., 1959)

and mounted receptor side up on the indifferent electrode, records a characteristic pattern of slow potential (such as shown in Fig. 20) when the retina is scanned with a light spot across the site of recording. When the light spot is on the recording site, a positive response is obtained, but as the light spot is moved away from the recording site, the response turns negative. On the notion that the recording is made from the receptor surface, and that the polarity of response at one site is determined by the result of a competition between PII and PIII, the positive potential in the center shows a predominance of PIII over PII and the negative potential in the surrounds a predominance of PII over PIII. It was further demonstrated by Motokawa et al. (1961) that the characteristic distribution pattern is reflected in the discharge pattern of on-off type ganglion cells; the on-discharge is enhanced in the surrounds where PII is dominant, but suppressed in the central area where PIII is dominant. The result is consistent with Granit's classical view that the PII is related to excitatory processes and the PIII to inhibitory. It is interesting to note in this connection that all the cornea-positive waves in the ERG which are known to originate in the retina proper (this excludes the c-wave which originates in the pigment epithelium) seem to be related to processes that are excitatory to the ganglion cells. This applies not only to the b- and d-wave but also to the x-wave (Motokawa and Mita, 1942) and the e-wave or the delayed off-response (Sickel and Crescitelli, 1967; Crescitelli and Sickel, 1968).

On these considerations it appeared safe to conclude that the PII is an extracellular or a transretinal manifestation of the depolarizing response of bipolar cells. Recent work on the mudpuppy retina by Miller and Dowling (1970), however, provided evidence that the b-wave (PII) is only an indirect manifestation of neuronal cell activities via Müller cells which are glial. Using a combination of intracellular recording and electrode marking, they obtained slow depolarizing potentials from Müller cells. Such responses were slower than any of the responses of neuronal cells (horizontal, bipolar and amacrine) and were recorded from a wide range of retinal depths which corresponded to the anatomical extent of Müller cells. The Müller cell responses resembled the b-wave in many respects, such as in the latency, the waveform, and the intensity-amplitude relation. The result strongly suggested that the Müller cells are the origin of the b-wave. From the distribution of the b-wave in the retinal layers, as measured by depth recording, cells whose activity results in the depolarization of Müller cells appeared to lie in the distal portion of the inner nuclear layer. The responses of these cells (horizontal and bipolar) had a considerably shorter latency and faster time course than the Müller cell response. MILLER and DOWLING postulated that the Müller cells may generate the b-wave by a K^+-regulated mechanism similar to glial cell potentials reported in the nervous system by ORKAND et al. (1966). Their results are explicit, but nevertheless it would be desirable to clarify the following points. (1) Their result should be confirmed in other animal forms than the mudpuppy. The notion of MILLER and DOWLING that the "slow bipolar" cells in the frog and axolotle retina described by BYZOV (1965) might be Müller cells should be experimentally supported. BYZOV's slow bipolar responses match the b-wave closely, but their origin has not yet been identified by electrode marking. (2) Why the activity of only distal neurons (horizontal and bipolar) but not of proximal neurons (amacrine and ganglion cells) influences Müller cells to generate the b-wave should be explained. Anatomical relationships of the proximal neurons with the Müller cells appear to be as close as those of the distal neurons. (3) Since the bipolar cells are well oriented radially and are electrically as active as the photoreceptors, it should be determined why the electrical activity of the bipolar cells has no direct contribution to the ERG.

c) PIII

One of the two components of the PIII, termed the distal PIII by MURAKAMI and KANEKO (1966), has been localized in the receptors of cold-blooded animals, and identified with the mammalian late receptor potential of BROWN, WATANABE, and MURAKAMI (1965). Its physiological significance was discussed in Chapter 12.

The other component, the proximal PIII, was localized in the inner nuclear layer (MURAKAMI and KANEKO, 1966). Arguments as those used to ascribe the PII to the depolarizing response of bipolar cells suggest that the proximal PIII may be ascribed to the hyperpolarizing response of bipolar cells, though it remains open whether this is a direct manifestation of such response or is mediated by Müller (glial) cells. The same conclusion for both PII and PIII, relating them to the depolarizing and hyperpolarizing responses of bipolar cells, was drawn recently by three Swedish workers (KNABE, MØLLER, and PERSSON, 1972) who analysed the sheep ERG with an averaging technique which enabled detailed studies of the ERG characteristics below the "b-wave threshold".

Besides the hyperpolarizing response of bipolar cells, some other potentials such as the S potential may also contribute to the proximal PIII. The S cells (horizontal cells) might also be responsible for spreading the current generated in the illuminated retinal area to the surround by way of their electrical inter-connections. However, this is not adequate to explain the characteristic spatial distribution of the locally recorded ERG shown in Fig. 20. If the positive and negative potentials distributed spatially (Fig. 20) were merely in a physical source-sink relationship, they should be inseparable, but their separation by chemicals affecting the PII and PIII selectively was demonstrated by Murakami and Sasaki (1968a, b).

Acknow'edgement

Work in the author's laboratory in Tokyo was supported in part by grants from the Education Ministry of Japan, U. S. Public Health Service Grants NB 06421 and EY 00017, and U. S. Air Force Office of Scientific Research Grants through the U. S. Army Research and Development Group (Far East).

References

Arden, G.B., Brown, K.T.: Some properties of components of the cat electroretinogram revealed by local recording under oil. J. Physiol. (Lond.) **176**, 429—461 (1965).

Armington, J.C.: A component of the human electroretinogram associated with red color vision. J. Opt. Soc. Amer. **42**, 393—401 (1952).

— Electrical responses of the light-adapted eye. J. Opt. Soc. Amer. **43**, 450—456 (1953).

Asher, H.: The electroretinogram of the blind spot. J. Physiol. (Lond.) **112**, 40P (1951).

Bernhard, C.G.: The negative component PIII in the retinogram of the tortoise. Acta physiol. scand. **3**, 132—136 (1941).

Bornschein, H., Hanitzsch, R., Lützow, A.v.: Off-Effekt und negative ERG-Komponente des enukleierten Bulbus und der isolierten Retina des Kaninchens. I. Einfluß der Reiz-parameter. Vision Res. **6**, 251—259.

Bortoff, A.: Localization of slow potential responses in the *Necturus* retina. Vision Res. **4**, 627—636 (1964).

Boynton, R.M.: Stray light and the human electroretinogram. J. Opt. Soc. Amer. **43**, 442—449 (1953).

— Riggs, L.A.: The effect of stimulus area and intensity upon the human retinal response. J. exp. Psychol. **42**, 217—226 (1951).

Brindley, G.S.: The passive electrical properties of the frog's retina, choroid and sclera for radial fields and currents. J. Physiol. (Lond.) **134**, 339—352 (1956a).

— The effect on the frog's electroretinogram of varying the amount of retina illuminated. J. Physiol. (Lond.) **134**, 353—359 (1956b).

— Responses to illumination recorded by microelectrodes from the frog's retina. J. Physiol. (Lond.) **134**, 360—384 (1956c).

— The sources of slow electrical activity in the frog's retina. J. Physiol. (Lond.) **140**, 247—261 (1958).

— Physiology of the Retina and Visual Pathway. London: Edward Arnold Publ. Ltd. 1960.

— Hamasaki, D.I.: The properties and nature of the R-membrane of the frog's eye. J. Physiol. (Lond.) **167**, 599—606 (1963).

Brown, K.T.: The analysis of ERG and the origin of its components. Jap. J. Ophthal. **10**, Suppl. (Proc. 4th ISCERG Symp.) 130—140 (1966).

— The electroretinogram: Its components and their origins. Vision Res. 8, 633—677 (1968).

— Murakami, M.: A new receptor potential of the monkey retina with no detectable latency. Nature (Lond.) **201**, 626—628 (1964).

— Tasaki, K.: Localization of electrical activity in the cat retina by an electrode marking method. J. Physiol. (Lond.) **158**, 281—295 (1961).

— Watanabe, K.: Isolation and identification of a receptor potential from the pure cone fovea of the monkey retina. Nature (Lond.) **193**, 958—960 (1962a).

— — Rod receptor potential from the retina of the night monkey. Nature (Lond.) **196**, 547—550 (1962b).

BROWN, K. T., WATANABE, K., MURAKAMI, M.: The early and late receptor potentials of monkey cones and rods. Cold Spr. Harb. Symp. quant. Biol. **30**, 457—482 (1965).

— WIESEL, T. N.: Intraretinal recording in the unopened cat eye. Amer. J. Ophthal. **46**, 91—96 (1958).

— — Intraretinal recording with micropipette electrodes in the intact cat eye. J. Physiol. (Lond.) **149**, 537—562 (1959).

— — Analysis of the intraretinal electroretinogram in the intact cat eye. J. Physiol. (Lond.) **158**, 229—256 (1961 a).

— — Localization of origins of electroretinogram components by intraretinal recording in the intact cat eye. J. Physiol. (Lond.) **158**, 257—280 (1961 b).

BYZOV, A. L.: Sources of the impulses recorded from the inner layers of the frog retina (in Russian). Biofizika **4**, 414—421 (1959).

— Functional properties of different cells in the retina of cold-blooded vertebrates. Cold Spr. Harb. Symp. quant. Biol. **30**, 547—558 (1965).

CRESCITELLI, F.: The e-wave and inhibition in the developing retina of the frog. Vision Res. **10**, 1077—1091 (1970).

— SICKEL, W.: Delayed off-responses recorded from the isolated frog retina. Vision Res. **8**, 801—816 (1968).

FABER, D. S.: Analysis of the Slow Transretinal Potentials in Response to Light (Ph.D. Thesis). University of New York at Buffalo 1969.

FRY, G. A., BARTLEY, S. H.: The relation of strong light in the eye to the retinal action potential. Amer. J. Physiol. **111**, 335—340 (1935).

GOURAS, P.: Spreading depression of activity in amphibian retina. Amer. J. Physiol. **195**, 28—32 (1958).

— Graded potentials of bream retina. J. Physiol. (Lond.) **152**, 487—505 (1960).

GRANIT, R.: The components of the retinal action potential and their relation to the discharge in the optic nerve. J. Physiol. (Lond.) **77**, 207—240 (1933).

— Sensory Mechanisms of the Retina. London-New York-Toronto: Oxford Univ. Press 1947.

— Neurophysiology of the retina. In: DAVSON, H. (Ed.): The Eye, Vol. 2. The Visual Process. New York-London: Academic Press 1962.

— RIDDELL, H. A.: The electrical responses of light- and dark-adapted frog's eyes to rhythmic and continuous stimuli. J. Physiol. (Lond.) **81**, 1—28 (1934).

HANITZSCH, R., TRIFONOW, J.: Intraretinal abgeleitete ERG-Komponenten der isolierten Kaninchennetzhaut. Vision Res. **8**, 1445—1455 (1968).

HOLMGREN, F.: Method att objectivera effecten av ljusintryck pa retina. Upsala Läk.-Fören. Förh. **1**, 177—191 (1865—1866).

KANEKO, A.: Physiological and morphological identification of horizontal, bipolar and amacrine cells in goldfish retina. J. Physiol. (Lond.) **207**, 623—633 (1970).

— HASHIMOTO, H.: Localization of spike-producing cells in the frog retina. Vision Res. **8**, 259—262 (1968).

— — Electrophysiological study of single neurons in the inner nuclear layer of the carp retina. Vision Res. **9**, 37—55 (1969).

KNAVE, B., MØLLER, A., PERSSON, H.: A component analysis of the electroretinogram. Vision Res. (in press).

KUFFLER, S. W.: Discharge patterns and functional organization of mammalian retina. J. Neurophysiol. **16**, 37—68 (1953).

LEÃO, A. A. P.: Spreading depression of activity in the cerebral cortex. J. Neurophysiol. **7**, 359—390 (1944).

MACNICHOL, E. F., SVAETICHIN, G.: Electric responses from isolated retinas of fishes. Amer. J. Ophthal. **46**, Pt. 2, 26—40 (1958).

MESERVEY, A. B., CHAFFEE, E. L.: Electrical response of the retina in different types of cold-blooded animals. J. Opt. Soc. Amer. **15**, 311—330 (1927).

MICHAELSON, I. C.: Retinal Circulation in Man and Animals. Springfield: Charles C. Thomas 1954.

Miller, R. E., Dowling, J. E.: Intracellular responses of the Müller (glial) cells of mudpuppy retina: Their relation to b-wave of the electroretinogram. J. Neurophysiol. **33**, 323—341 (1970).

Mitarai, G.: The origion of the so-called cone action potential. Proc. Japan Acad. **34**, 299—304 (1958).

— Glia-neuron interaction in carp retina, glia potentials revealed by microelectrode with lithium carmine. In: Seno, S., Cowdry, E. V. (Eds.): Intracellular Membraneous Structure. Okayama: Japan Soc. Cell Biol. 1965.

— Watanabe, I., Niimi, K.: Further study on the origin of S-potentials and the function of glia cells in the retina. Proc. 23rd Intn. Congr. Physiol. Sci., Abstr. 838 (1965).

Motokawa, K.: Electrogenesis of ERG and optic nerve discharge. Jap. J. Ophthal. **10**, Suppl. (Proc. 4th ISCERG Symp.) 141—148 (1966).

— Mita, T.: Über einfachere Untersuchungsmethoden und Eigenschaften der Aktionsströme der Netzhaut des Menschen. Tohoku J. exp. Med. **42**, 114—133 (1942).

— Oikawa, T., Tasaki, K., Ogawa, T.: The spatial distribution of electric responses to focal illumination of the carp's retina. Tohoku J. exp. Med. **70**, 151—164 (1959).

— Yamashita, E., Ogawa, T.: The physiological basis of simultaneous contrast in the retina. In: Jung, R., Kornhuber, H. (Eds.): The Visual System: Neurophysiology and Psychophysics. Berlin-Heidelberg-New York: Springer 1961.

Müller-Limmroth, W., Blümer, H.: Über den Einfluß von Monojodessigsäure, Natriumazid und Natriumjodat auf das Ruhepotential und das Electroretinogramm des Froschauges. Z. Biol. **109**, 420—439 (1957).

Murakami, M., Kaneko, A.: Differentiation of PIII subcomponents in cold-blooded vertebrate retinas. Vision Res. **6**, 627—636 (1966).

— Sasaki, Y.: Analysis of spatial distribution of the ERG comp onents in the carp retina. Jap. J. Physiol. **18**, 326—33 6 (1968a).

— — Localization of the ERG components in the carp retina. Jap. J. Physiol. **18**, 337—349 (1968b).

Noell, W. K.: Studies on the Electrophysiology and the Metabolism of the Retina. School of Aviation Med. Rep. No. 1. Randolph Field, Texas 1953.

— The origin of the electroretinogram. Amer. J. Ophthal. **38**, 78—90 (1954).

Oikawa, T., Ogawa, T., Motokawa, K.: Origin of so-called cone action potential. J. Neurophysiol. **22**, 102—111 (1959).

Orkand, R. K., Nicholls, J. G., Kuffler, S. W.: Effect of nerve impulses on the membrane potential of glial cells in the central nervous system of amphibia. J. Neurophysiol. **29**, 788—806 (1966).

Ottoson, D., Svaetichin, G.: Electrophysiological investigations of the frog retina. Cold Spr. Harb. Symp. quant. Biol. **17**, 165—173 (1952).

— — Electrophysiological investigations of the origin of the ERG of the frog retina. Acta physiol. scand. **29**, Suppl. 106, 538—564 (1953).

Polyak, S.: The Vertebrate Visual System. Chicago: University of Chicago Press 1957.

Rodieck, R. W., Ford, R. W.: The cat local electroretinogram to incremental stimuli. Vision Res. **9**, 1—24 (1969).

Schmidt, R., Steinberg, R. H.: Rod-dependent intracellular responses to light recorded from the pigment epithelium of the cat retina. J. Physiol. (Lond.) **217**, 71—91 (1971).

Schubert, G., Bornschein, H.: Beitrag zur Analyse des menschlichen Elektroretinogramms. Ophthalmologica (Basel) **123**, 396—413 (1952).

Sickel, W.: Retinal metabolism in dark and light. This Volume, Chapter 18.

— Crescitelli, F.: Delayed electrical responses from the isolated frog retina. Pflügers Arch. ges. Physiol. **297**, 266—269 (1967).

Steinberg, R. H.: Comparison of the intraretinal b-wave and d.c. component in the area centralis of cat retina. Vision Res. **9**, 317—331 (1969).

— Schmidt, R.: Identification of horizontal cells as S-potential generators in the cat retina by intracellular dye injection. Vision Res. **10**, 817—820 (1970).

— — Brown, K. T.: Intracellular responses to light from cat pigment epithelium: Origin of the electroretinogram c-wave. Nature (Lond.) **227**, 728—730 (1970).

STONE, J.: Structure of the cat's retina after occlusion of the retinal circulation. Vision Res. **9**, 351—356 (1969).

SVAETICHIN, G.: The cone action potential. Acta physiol. scand. **29**, Suppl. 106, 565—600 (1953).

— LAUFER, M., MITARAI, G., FATEHCHAND, G., VALLECALLE, E., VILLEGAS, J.: Glial control of neuronal networks and receptors. In: JUNG, R., KORNHUBER, H. (Eds.): The Visual System: Neurophysiology and Psychophysics. Berlin-Göttingen-Heidelberg: Springer 1961.

TANSLEY, K.: Comparative anatomy of the mammalian retina with respect to the electro-retinographic response to light. In: SMELSER, G.K. (Ed.): The Structure of the Eye. New York: Academic Press 1961.

TOMITA, T.: Studies on the intraretinal action potential. Part I. Relation between the locali-zation of micropipette in the retina and the shape of the intraretinal action potential. Jap. J. Physiol. **1**, 110—117 (1950).

— A study on the origin of intraretinal action potential of the cyprinid fish by means of pencil-type microelectrode. Jap. J. Physiol. **7**, 80—85 (1957).

— A compensation circuit for coaxial and double-barreled microelectrodes. IRE Trans. biomed. Electron. **9**, 138—141 (1962).

— Electrical activity in the vertebrate retina. J. Opt. Soc. Amer. **53**, 49—57 (1963).

— FUNAISHI, A.: Studies on intraretinal action potential with low resistance microelectrode. J. Neurophysiol. **15**, 75—84 (1952).

— — SHINO, H.: Studies on the intraretinal action potential. Part II. Effects of some chemical agents upon it. Jap. J. Physiol. **2**, 147—153 (1951).

— MIZUNO, H., IDA, T.: Studies on the intraretinal action potential. Part III. Intraretinal negative potential as compared with b-wave in the ERG. Jap. J. Physiol. **2**, 171—176 (1952).

— MURAKAMI, M., HASHIMOTO, Y.: On the R membrane in the frog's eye. Its localization, and relation to the retinal action potential. J. gen. Physiol. **43**, Pt. 2, 81—94 (1960).

— — SASAKI, Y.: Electrical activity of single neurons in the frog's retina. In: JUNG, R., KORNHUBER, H. (Eds.): The Visual System: Neurophysiology and Psychophysics. Berlin-Göttingen-Heidelberg: Springer 1961.

— — SATO, Y., HASHIMOTO, Y.: Further study on the origin of the so-called cone action potential (S-potential). Its histological determination. Jap. J. Physiol. **9**, 63—68 (1959).

— TORIHAMA, Y.: Further study on the intraretinal action potentials and on the site of ERG generation. Jap. J. Physiol. **6**, 118—136 (1956).

— TOSAKA, T., WATANABE, K., SATO, Y.: The fish EIRG in response to different types of illumination. Jap. J. Physiol. **8**, 41—50 (1958).

WERBLIN, F.S.: Functional Organization of the Vertebrate Retina Studied by Intracellular Recording from the Retina of the Mudpuppy, *Necturus maculosus*. Doctoral Dissertation. The Johns Hopkins Univ., Baltimore 1968.

— DOWLING, J. E.: Organization of the retina of the mudpuppy, *Necturus maculosus*. II. Intra-cellular recording. J. Neurophysiol. **32**, 339—355 (1969).

WITKOVSKY, P.: Peripheral mechanisms of vision. Ann. Rev. Physiol. **33**, 257—280 (1971).

YONEMURA, D., HATTA, M.: Localization of the minor components of the frog's electro-retinogram. Jap. J. Ophthal. **10**, Suppl. (Proc. 4th ISCERG Symp.) 149—154 (1966).

Chapter 18

Retinal Metabolism in Dark and Light

By

Werner Sickel, Cologne (Germany)

With 30 Figures

Contents

Introduction

One may expect to find in this concluding chapter on metabolism an account for the expenses incurred by the retina in the performance of the activities outlined in the preceding contributions to this volume. Information on its source and its fate — as a building block or fuel — should provide some insight in the inner workings, from which eventually a rationale may be derived for the beneficial intervention in the repair or even improvement of visual function, as perhaps in night vision. The uses of the substrate under specified load may serve to sub-stantiate the powering of amplifier actions invoked in many guises, providing gain, feedback, preventing contamination of the visual signal to be processed, etc. The kinetics of metabolic reactions may help to determine which of a number of processes is the deciding one, e.g. during restoration of sensitivity after exposure to light. Or, the measured rate of metabolism could substitute for "neural work", a quantity not otherwise available at present.

The solution of these problems hinges upon the demonstration of inter-relationships between visual function and metabolism. Linkages are evident in the extremes, e.g. in the experience of blackout under stress, of pain in glare, and from studies of sur- and re-vival of visual functions in impaired metabolic supply (for a recent review: [101, 83a]). Connections may also be inferred from controlling mechanisms governing under operating conditions, which probably involve adjust-ment of the chemical milieu [177], and may be visualized to account for such phenomena of competition as rod-cone rivalry [80, 173], or certain types of periodicity. But the expected correspondence can hardly be extracted from documented measurements of retinal metabolism in dark and light [163], which rather serve to establish for the retina a situation not unlike that found in brain: with high basic metabolic rates, orthodox routes, but little change as a consequence of imposed changes of activity. The difficulties reside not only in the wide range of experimental conditions and in discrepancies of reported data, which for oxygen uptake exceed several hundred percent with two standard techniques [32], but also in the fact that biochemists generally strive for maximal rates to disclose metabolic capacities, whereas minimal rates compatible with function would appear to be the operational ones. To quote an observation: lactate effectively substitutes for glucose, but pyruvate, while increasing oxygen uptake above the rate under glucose, results in diminished responses to light.

A system of simultaneous non-destructive measuring techniques was, therefore, built around the presumed metabolic machinery (Fig. 1) and applied to a retina preparation (Fig. 2) believed to be preserved in a functional state although isolated in order to render it accessible to experimental manipulations. The techniques include measurements of slow transretinal electrical potential changes, which would occupy their place in the metabolic scheme as indicators of available energy and at the same time serve as proven criterion of visual activity, with the especial merit of integrating this activity in a meaningful mode, rather than with undue stress on activities of contributing component parts. It is the design of this ap-proach to contribute to an understanding of the slow electrical phenomena, both ERG and DC-potential, in terms of retinal metabolism by comparing them with optical phenomena of a similar time resolution, which then will be linked to more conventional metabolic measurements, with the simultaneous electrical recordings always authorizing the correlations. Structural inferences would seem warranted to the extent of the resolution of the techniques. Outside the scope are chronic effects of light on retinal development. Photochemical aspects will not be treated in any detail.

Metabolic Outfit

For the purpose of correlating individual data on retinal behavior obtained with the diverse techniques a brief description is given here on the metabolic machinery. The picture is sufficiently general to be supported by numerous investigations in many fields (for a review: [107]), with retinal biochemistry holding a fair share (for reviews: [119, 70, 188, 163, 83, 50a]). The lines of evidence include gasometric studies, the demonstration of chemical intermediates as well as of the necessary

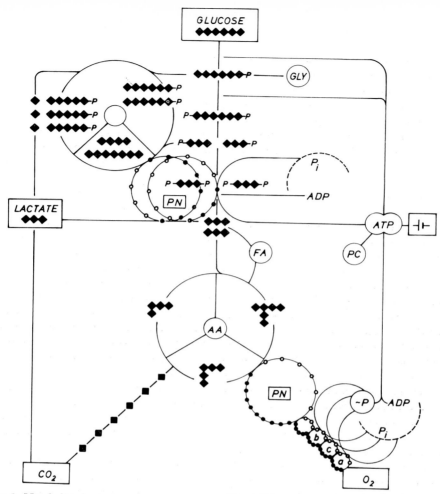

Fig. 1. Metabolic chart.

◆ carbon atoms,
   e.g. ◆◆◆ pyruvate, lactate;
○ oxidized form of carrier,
● reduced form of carrier;
b, c, a cytochromes;
PN pyridine nucleotides e.g. nicotin-amide-
   adenine dinucleotide (NAD, DPN);
ADP, ATP adenine nucleotices
   (adenosine-di-, tri-phosphates);

Pi; $\sim P$ inorganic and high-energy
     phosphates;
◯ pools of
   GLY glycogen,
   PC phosphocreatine,
   FA fatty acids,
   AA amino acids;
⊣⊢ asymmetry potential,
   electroretinogram (ERG)
   DC-component;
▭ test points.

For further explanation see section "Metabolic outfit"

enzymes and cofactors, and also examinations of function by means of the ERG
in connection with judicious choice of blocking agents.

    In the scheme of Fig. 1 glucose is the starting material. It is taken up in brain
in sufficient quantity to cover all energy requirements [100], it will be shown a

sufficient substrate for retina below. The 6-carbon skeleton becomes phosphorylated in two subsequent kinase reactions involving phosphate in a special high-energy form which is supplied from ATP-pools. Only the split triose phosphate accepts inorganic phosphate, a reaction mediated by a redox cofactor, oxidized DPN, which becomes reduced in the process. Its reoxidation occurs in a later reaction, the formation of lactate from pyruvate. Molecular oxygen is not involved, and the sequence of reactions may proceed under anaerobic conditions, as is the case in lower organisms. In the formation of pyruvate phosphate is released in the high-energy form and accepted by ADP. Part of the ATP so formed serves to replenish the store tapped in the initial phosphorylations, but there is a net gain of 2 ATP per one glucose molecule degraded, available for use in a diversity of energy requiring processes.

Instead of being reduced the 3-carbon fragment pyruvate may be oxidatively metabolized. This occurs in a sequence of reactions known as the tricarboxylic acid cycle, or Krebs cycle. The principle of its operation is to make the molecule accessible to the action of several enzymes, which bring about the separation of the carbon and hydrogen atoms. The carbon atoms are released as CO_2, the hydrogen atoms are taken up by pyridine nucleotide-linked dehydrogenases and fed into the respiratory chain. Many more individual steps than shown are involved and material other than pyruvate may find its way into the cycle, e.g. fatty acids. At several sites interconnections, potentially shunting part of the cycle, exist with a pool of amino acids typical in pattern for central nervous tissue [35].

A similar cycle, however drawing substrate from the first phosphorylation step and feeding fragments back into the glycolytic chain at a later step, thereby shunting part of it, also results in the formation of CO_2 and hydrogen. In this case hydrogen transport requires a slightly different cofactor, oxidized TPN. The significance of the "pentose phosphate shunt" is envisaged not so much in the provision of energy, but of various carbon fragments suitable for synthetic reactions as well as the reducing power required in the processes. Its operation in the retina has been widely discussed [83, 147].

The energy to be derived from the carbohydrate molecule resides predominantly in its hydrogen atoms and their reaction with molecular oxygen. But contrary to an open combustion, the energy is not derived in one step — with the consequent loss of most of it as heat — but in a series of oxidation-reduction reactions, three of which are linked to energy-conserving reactions, which transfer inorganic phosphate into the high-energy form suitable for coupling it to ADP. Thus 18 ATP are formed in the respiratory (or: electron-transport) chain by the oxidation of one molecule of glucose. The high energy phosphate may be transferred to creatine, with phosphocreatine buffering the ATP level.

The significance of the oxidative phosphorylation as compared to the glycolytic formation of ATP lies not only in the higher yield, but especially in the fact that through inhibitory mechanisms the flow of electrons along the chain toward oxygen as the final acceptor can occur only if and when the energy thereby released can be utilized. Prerequisite, therefore, is the availability of inorganic phosphate and/or the phosphate acceptor, ADP. Since both substances are available only in limited amounts, they acquire control function. And since they are regenerated from ATP, when split in energy consuming processes, in a cyclic reaction dissipation of energy is tied to its restoration and oxygen uptake [66]. The respiratory control rate, the quotient of oxygen uptake with and without phosphate acceptor [42] is a quantitative expression for the tightness of this coupling, whereas the P/O ratio describes the efficacy of the energy utilization, and the phosphate potential (ATP/ADP + P_i) its result.

Glycolysis and respiration involve numerous enzyme-mediated reactions, all of which have to proceed in pace, and various mechanisms have been demonstrated which effect just that, e.g. product- or feedback-inhibition. Special regulatory mechanism obviously govern at branching sites, such as that of glucose-6-phosphate, which may be metabolized glycolytically (phosphofructokinase), but may also enter the shunt pathway (glucose-6-phosphate dehydrogenase), or add to the glycogen store (phosphorylase), or even end in a reversed hexokinase reaction as glucose. The pH may be deciding [115], perhaps by determining configuration and affinity of the enzymes concerned. The fate of pyruvate is of special significance. In its being reduced the pyridine nucleotide previously used is reoxidized for further use and the product, lactate, may be eliminated as an end product. Thus this segment of reactions is self-sufficient, and although the energy yield is low the ATP produced in the cytoplasmic matrix of the cell is easily available [65] and may subserve specific functions [57]. In the presence of oxygen a major part of pyruvate enters the Krebs cycle to donate hydrogen to the respiratory chain, which is located in the mitochondria. The proportion of carbohydrate glycolysed and oxidized, respectively, is delicately balanced, as can be inferred from two well-documented inhibitory effects: excess glucose leads to decreased oxygen uptake (Crabtree effect, [44]) and increased oxidation impairs utilization of glucose (Pasteur effect, [124]). The control appears to be exerted by inorganic phosphate, which may be trapped for phosphorylations either for the initial glucose phosphorylations or in the respiratory chain, and create a bottleneck at the other site respectively [133].

Pyridine nucleotides are ubiquitous cofactors in hydrogen transfer reactions. On accepting hydrogen they exhibit strong absorption in the near ultraviolet region of the spectrum, which is lost on reoxidation. The spectrophotometric phenomenon is extensively made use of in *in vitro* enzyme assays ("optical test"). In the tissue their role is more complicated, because i) two types of pyridine nucleotides, DPN and TPN, are involved which are not discriminated optically, ii) their location may be cytoplasmic or mitochondrial with restricted exchange between them, iii) their functional significance may be twofold: Thermodynamically they represent ATP-dependent redox equilibria poised by the reduced and oxidized components of substrate pairs, which eventually may be communicated to the extracellular space (lactate-pyruvate quotient). In this sense the higher degree of reduction represents the higher energy level. Second, like the cytochromes pyridine nucleotides mediate the flow of electrons along the respiratory chain, and their instantaneous state of oxidation mirrors the velocity of this flow. There is the added technical difficulty of detecting pyridine nucleotide oxidation-reduction from light absorption in turbid media, with light scatter present and, in the tissue, of potential functional significance [41, 128, 203, 139, 206]. Nevertheless, owing to extensive investigations in purified systems and more complex situations changes in pyridine nucleotide oxidation-reduction are being better understood in living organs [37, 39] and the ease and speed of their assay can be exploited for a deeper insight into the metabolic machinery. In the retina pyridine nucleotides are present in high concentration [191, 8, 186] and fairly evenly distributed [149, 140]. Cyclic oxidation of pyridine nucleotides signalling a spurt of metabolic energy production has been found on and in proportion to light stimulation [178]. The redox change correlates with the electrical process PI (c-wave of the ERG), as there are also correlations, on a longer time scale, between the redox state of pyridine nucleotides and the DC-potential of the retina [182].

Pyridine nucleotides, then, are qualified to furnish information on cellular energy metabolism under functional conditions, provided the inherent ambiguity is overcome through suitable supplementary measurements. Observation of the metabolic in- and outputs is possible without disturbances of function, and oxygen uptake and CO_2-production can be determined in a continuous fashion, thereby obviating scatter of data introduced by sampling techniques. Slow electrical activity, which has been known intimately linked to metabolism [151—153], proves a close function of the phosphate potential. Through adequate probing it holds out the prospect of a localization — both layerwise [131] and type-wise [190] — of the metabolic events, without resort to long-term manifestations, with the well-ordered outer retinal layers usually scoring badly under histological examinations.

The "test points", rectangularly marked in the scheme of Fig. 1, are experimentally accessible in the preparation of the perfused retina [177, 180].

General Techniques[1]

Routinely frog retinas (R. esc., temp., pip.) were used. Other vertebrate species, cold- and warm-blooded including human, have been examined for comparison. The smaller eyes were hemisected by means of a cigar cutter-type slicer, from the larger eyes pieces of suitable size were punched out with a trephine. Isolation of the retinas was done under experimental solution, which requires a minimum of handling, no magnifying devices and no more than dim red illumination. Slight shaking usually suffices to detach the retina proper from the underlying pigment epithelium, the more easily so when the animals have been dark adapted previously for about an hour. From the solution the retina was taken flat on a supporting mesh and transferred to a perfusion chamber inside a light tight housing, where it was continously washed by a nutrient medium on either side. The fellow retina, when not used immediately, was stored floating in the solution in a cold box at $2-4°$ C. It proved quite useful the next day or days.

Special types of chambers were used for different purposes (Fig. 2). An open flow system fed from reservoirs which could be quickly exchanged served to determine the requirements the perfusing solution has to meet for optimal responses to light stimulation, or to study response behavior in impaired supplies.

---

[1] Details will be given at the appropriate sections below. Particulars and experiences have been reported [177, 180, 51, 201, 158]; on photochemical aspects [14, 15]; in human [184, 96, 91]; in rabbit [90, 24, 132, 103]; in rat [207]. A similar approach [5] in an improved version [6] has been used for morphological studies [206].

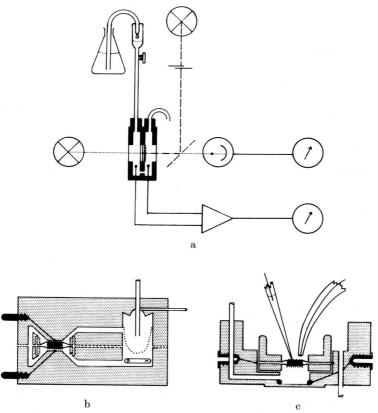

Fig. 2. The perfused retina preparation. Chambers used for measuring transretinal electrical potential together with *a* *(top)* optical absorption; *b* *(lower left)* oxygen consumption, CO_2 output; *c* *(lower right)* intraretinal electrical activity. Details under "General techniques" and respective sections

In the cuvette-type chamber, a brass cage sealed with optical cover glasses and O-rings, a plastic part holds the retina circularly clamped on a grid in a central opening and carries in- and outflow channels and silver ring electrodes contacting the solution in either compartment. Separate perfusion of the two sides [88] was possible, but not used here, instead an overflow offering sufficiently high electrical resistance allowed series perfusion, which was found more convenient and gentle.

The physical dimensions and optical properties of the cuvette had been selected to accommodate it in a spectrophotometer for absorption measurements. The investigations were aimed not at photopigments but at respiratory pigments ("metabolic microscope", [143]). Several arguments will be considered below in differentiating between the two groups of light absorbing material. The simplest, most effective means, and mandatory for preservation of function is to use low light intensities. This puts stringent demands on the photometric techniques, which were met with two types of single-beam apparatus. For fast time resolution a photomultiplier photometer was used, which had originally been designed for bioluminescent work and further stabilized [135], for longer-term observation a modified Eppendorf photometer. Several confirmatory results were obtained with a dual-wavelength technique [36].

In order to detect changes in the medium as a consequence of retinal metabolism the volume of the incubation medium has to be kept small. However, a certain flow rate must be maintained to preserve function (Fig. 9). A recirculating system was, therefore, designed and used in various ways. As shown in Fig. 2b, the chamber consists of two blocks of plexiglass, which are tightly screwed together after the retina, sandwiched between two layers of mesh, has been inserted in its compartment between the electrodes. A known volume of bathing solution, 0.5—1 ml, is then injected from a syringe and continuously circulated through the system of channels by means of a fast rotating magnetic stirring bar (driven by an external magnet underneath; not shown), which serves as a centrifugal pump. In doing so it stirs up the fluid in its compartment to a continuously renewed meniscus, forming a virtually large surface to facilitate the exchange with a gas stream in order to replenish the oxygen used up and to clear the solution of the CO_2 formed. In experiments using isotope-labeled substrate a continuous record of $^{14}CO_2$ production is available, and samples may be taken from the bathing solution at intervals to determine glucose uptake and lactate formation.

In a modification of the recirculating chamber an oxygen electrode was inserted in the exchanger compartment occupying part of its space. It left a total volume of 1 ml for the recirculating system, which was connected to, and filled air-free from an external reservoir of bathing solution. This arrangement could be operated in two ways. During a sufficiently fast fluid flow, supplying dissolved oxygen at a rate not significantly reduced by retinal oxygen uptake, a baseline was established for the PO_2-meter reading. The fluid flow was then discontinued but stirring maintained. The decrease of PO_2 could be observed for a given time, after which the flow was restored to replenish the oxygen, and to recover function if it had decreased. Alternatively the flow rate from the reservoir could be reduced to such an extent that the rate of supply of oxygen approached the rate of uptake by the tissue. Thus the meter reading would eventually reach a steady state immediately reflecting rate of uptake and reveal small changes momentarily.

The advantages of the continuous recording as compared with sample measurements may be put to use for better resolution in a number of analysing problems, e.g. washout from a retina loaded with other radioactive material. The slow-flow technique permits determination of total acid production of about a milligram dry weight of tissue, even in a highly buffered medium, by pH-measurement in the outflow from the chamber with conventional instrumentation.

The composition of the bathing solution is given in Table 1. It was arrived at empirically to meet minimum requirements rather than to copy body fluids. Some of the effects of deviations from the recipe will be discussed below. Striking is the high buffer concentration and the alkaline pH, probably necessitated by a high rate of metabolic production of acid equivalents. Part but not all of the phosphate buffer could be replaced by other buffer systems. CO_2/bicarbonate would increase both speed and amplitudes of the light responses, but could not solely substitute.

It was avoided for simplicity. The oxygen requirement was satisfied by room-air saturation of the solution, even in the case of human retina. No need was found to supplement the medium with substrates other than glucose, in particular no nitrogen source was required. In the case of the human retina an addition of plasma had been found indispensible earlier (at higher concentration of calcium and magnesium). On fractioning a thermostable dialysate has been found similarly effective [24, 132]. No additions were made in the present investigations for the sake of lucidity. Stock solutions were usually made up of the chlorides and buffer, respectively, and ampullized and autoclaved. Sterility was observed in experiments using the recirculating system only. — The temperature varied between 20 and 23° C; in the gas-analysis experiments it was kept constant within a few tenths of a degree.

For exploratory purposes the retina among excitible tissues offers the unique possibility of being easily controlled by its natural stimulus and of producing a conveniently measured electrical response. It ought to be realized, however, that it covers an extraordinary wide range of stimulus parameters. It functions in dim light as well as in a surround many million times brighter, it can detect movements, i.e. resolve stimuli in space and time, it can discriminate wavelength, etc. But it cannot[2] and need not[3] perform of these tasks all at a time, it adapts, i.e. specializes for the performance of one particular set of tasks at the expense of others: good visual acuity is not available when the demand is for high sensitivity, but good spacial discrimination goes with high time resolution. Control of intensity and timing of the exposure to light are, therefore, important and informative.

Stimulation of the retina and control of its adaptational state was effected with the aid of a simple stimulator. Two beams were obtained from a tungsten car lamp run off a constant voltage supply. Each could be controlled independently by neutral density filters. One of them, the stimulus beam, served to deliver flashes, usually of 1 second duration timed by a magnetic shutter, the other, the adapting beam, as a background if needed. The beams were re-combined, collimated and directed onto the preparation. Additionally neutral density or interference filters could be accommodated in the common path. Attenuation of light intensity at this place would affect both stimulus and adapting beam by the same factor and was used to change from photopic to scotopic conditions. "Dark adapted" in the following, therefore, means no or subthreshold background.

Large field illumination of the retina does not explicitly disclose certain discriminatory skills of the retina which in its output are revealed a result of inhibitory activity [94, 141, 109]. Activity of the optic ganglion cells, be it for their comparatively low number, does not contribute much to the ERG [169, 96]. The ERG rather samples activity, without regard of the sign in the output, in a spacially additive fashion [26, 162], as do the other analysing techniques described above. In connection with the recording of gross electrical activity homogeneous illumi-

---

[2] Coding of intensities by sequences of impulses would have to discriminate against single flashes, i.e. start above flicker fusion frequency, and be limited by refractoriness of the ganglion cells. At a ten percent error no more than 25 steps could be expected and have been found [167] to cover a three log unit range of light intensities.

[3] Light intensities reflected from pigments occuring in a natural terrestrial environment never differ by more than 1: 1000 at a time [172].

nation is, therefore, appropriate and more easily accomplished with a flat retina preparation than with the optics of the eye intact and the response determined by the large retinal area hit by scattered light only.

Recording of electrical activity from a retina submerged in a bathing solution is not common practice. Shunting by the electrolyte can be kept small compared with intraretinal current paths by close positioning of the electrodes, as is the case in the chambers, and response amplitudes of customary size be obtained. The advantages are reproducible and unchanging recording conditions. The artificial environment will not be perfect, but certainly better than any atmosphere, and perhaps not much inferior to body fluids flooded with anesthetics.

Localized recording was possible in a chamber built on the same principle but offering access to the insertion of microelectrodes (Fig. 2c). It was placed on the stage of an inverted microscope, with the optics in the reverse direction demagnifying the size of the stimulus if required. Smooth fluid flow, achieved through hydrostatic force and syphoning, ensured sufficiently stable recording. Metal electrodes have been used to advantage in the recording of superficial spike activity [14], but glass capillaries proved preferable for penetrations. Due to their high electrical resistance, they require impedance matching, which was done with a negative-capacity input stage [136].

Amplification was conventional, both direct-recording for true display and capacitor-coupling was employed. Direct-writing recorders, particularly of the ink-jet type, were found of adequate frequency response and convenient for instantaneous intervention during the course of an experiment. Dual traces permitted simultaneous recording of electrical behavior and anyone of the other analysing procedures. In some cases an electromechanical or electronic timer served to program the experiment. In these cases interrupted recordings were obtained, i.e. the recorder was advanced for a few seconds, during which the stimulus was applied and the response recorded, but stopped thereafter for 3 min until the next cycle of operation. In this way original recordings can be presented on a compressed time scale instead of processed data.

Actions of Light and Gross Electrical Response

The electroretinogram will be used as an instantaneous indicator of the functional state of the retina. It is treated more exhaustively elsewhere in this volume [198][4]. The aspects taken up here are to illustrate the electrical behavior of the retina under the experimental conditions, specifically with respect to its detachment from the pigment epithelium and its exposure to an artificial environment. Because heterogeneous retinal structures contribute to the gross electrical behavior, several exploratory procedures have been adapted to disclose its make-up. These include differentiating between stimulating, adapting, and bleaching effects of light.

A series of ERGs of a dark adapted retina to stimuli of various intensities is shown in the left column of Fig. 3. The smallest of the responses was to a stimulus

---

[4] This article should also be consulted for the nomenclature. Further review articles: [82, 144, 27, 29].

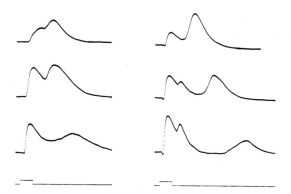

Fig. 3. Electroretinograms of the perfused frog retina. *Left*: dark adapted; *right*: light adapted (10^{-2} lx), stimulus intensities from top to the bottom: (left) 10^{-3}, 10^{-2}, 10^{-1} lx; (right) 0.1, 1, and 10 lx; stimulus duration 1 sec. Deflections of stimulus marks correspond to 20 μV. Notice: growing delay of off-response on increase of stimulus intensity; rod- and cone-responses

of 1 millilux. The gain of the amplifier could easily have been increased and responses be observed to stimuli more than two log units weaker, approaching the threshold, which in the preparation has been determined at 10^{-14} W/cm² of monochromatic light of wavelength 507 nm in frog [14] or 4×10^{-6} lx of white light in rabbit [24]. This is less than one log unit above human threshold.

On increasing the stimulus intensity the amplitudes of the on-reaction grow, but not beyond 10 millilux. No more than a three log units range of light intensities is covered by amplitude grading. The range of graded b-waves may, however, be reversibly shifted to a higher light level through the action of a permanent background light (right column). This shift ensures that stimuli of an intensity comparable to that of the respective background come to lie on the linear part of ampliture-vs.-log intensity characteristics, i.e. that they are well discriminated. Thus adaptation in the preparation exhibits the known feature of a zero-level shift.

A parallel shift of the intensity characteristics may also be effected by changing the pH of the perfusion medium (see below; Fig. 13a), but not by changing the temperature, in which case the slope changes. The family of b-waves-vs.-intensity curves taken at different temperatures converges towards low intensities and intersects the abscissa in one point, which denotes an intensity below which no processes having a $Q_{10} > 1$ take place, i.e. the threshold intensity.

There are conspicuous changes in the shapes of the ERG, most evident in the off-effects. Several of the off-properties have been known [93, 187, 78, 30, 183], but have been more fully investigated in this preparation [49]. Most striking is that the off-reaction, above a certain stimulus intensity, occurs the later the stronger the stimulus, and double in photopic conditions. The scotopic off-effect, the only one in dark adaptation and the second one in light adaptation, has mostly been missed in capacitor-coupled recordings and/or for too intense stimulation. Its time of occurence on equal-energy stimulation is delayed according to a rhodopsin spectrum, and shortened in proportion to the amount of visual purple removed by a previous bleaching, as had been determined by extraction. The time-to-peak of the off-reaction in dark adaptation is a precise measure of the number of light quanta and of visual pigment molecules interacting in the rods, uneffected by other processes which may influence response amplitudes. By that token, i.e. growing off-delays in responses to constant stimuli in a period following a bleach, regeneration of rhodopsin does occur in the preparation devoid of pigment epithelium (Fig. 4).

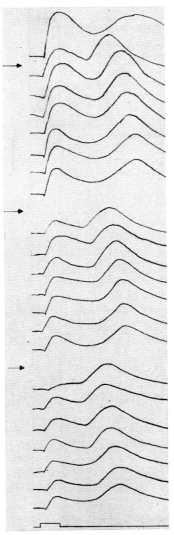

The photopic off-response, the first one in light adaptation, does not exhibit a rhodopsin spectrum, but is more sensitive in the red. It shows the same delay and decay with increasing stimulus intensity, however at a higher light level and at a much reduced time scale, closely locked to the off of the stimulus. It becomes prominent together with the a-wave (Fig. 3; right), and both ERG deflections are held due to onset and decay, respectively, of one process, the sustained "negative" component PIII of GRANIT [82]. PIII has been assumed to reflect the generator potentials of retinal receptors [29]. No indication of receptor activity is seen in dark adapted ERGs with equally high positive potential changes, but the process may be demonstrated in localized recordings (below) and be inferred from double-stimulus explorations (Fig. 5).

As is well known [137, 56, 10, 63, 175] considerable time has to elapse after a conditioning stimulus before full responsiveness is regained. Some of the recovery curves, at suitable stimulus parameters, show a marked discontinuity (comp. Fig. 15). In fact, on narrowing the interval between two stimuli the test response may even grow (Fig. 5a, b), and stimuli too weak or too short to cause a response after longer intervals may become effective in this phase. This "subthreshold facilitation" phenomenon has been observed in *Limulus* light receptors and been ascribed to the outlasting of the generator potential, which in its decay phase is restored to renewed trigger action by the addition of respectively smaller amounts of excitation [95, 134]. In preventing the decay of the generator process its off-trigger action is abolished, and the time course of the decay of the receptor process may be determined from the disappearance of the off-reaction of the first response as it is swallowed within a critical interval by the approaching second response (Fig. 5c). The second response

Fig. 4. ERGs before and after repeated bleaching of approx. 20% of rhodopsin each. Test stimuli (10^{-1} lx, 1 sec) delivered 6, 9, 12, 15, 19, 25, and 30 min after bleachings ($\rightarrow$: 6 sec; 3×10^{-6} W/cm²; 503 nm). Stimulus mark corresponds to 20 µV. Notice: growing delay of off-reaction during regeneration

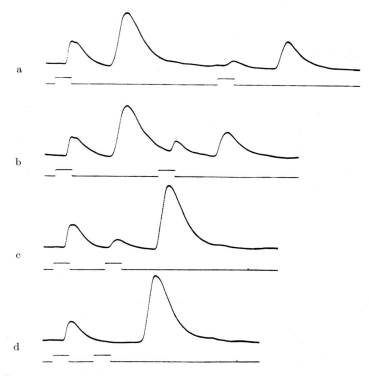

Fig. 5. ERG responses to double stimulation. Stimuli: 0.03 lx, 1 sec; interval decreasing from top to bottom (a ... d). Stimulus marks correspond to 25 μV. Notice: second b-wave facilitated in vicinity of first off-reaction (a; b), first off-reaction "swallowed" by approaching second response (c; d)

in the process takes over more and more of the first one's off-activity, until the dark gap between the two stimuli remains unnoticed (Fig. 5d).

A receptor potential directly recorded from a micro-pipette in response to a scotopically supermaximal stimulus is seen in Fig. 6a. It exhibits a long, sustained hyperpolarization from which it returns to the resting level with the time course just inferred from gross recording and in coincidence with the drawn-out off-reaction of the ERG, which had been recorded simultaneously. The micro-recording proves intracellular by its amplitude, the membrane oscillation seen during its decay, and the comparatively low chance of detecting and keeping it. A minute displacement of the tip of the electrode suffices to yield responses of the same general character, however much reduced in amplitude, but stable enough for more detailed study. The extracellular recordings of Fig. 6b are responses to stimuli of increasing intensity, and show in their maintained negativity the same increase in duration that had been observed in the delays of the scotopic off-reaction of the ERG (comp. Fig. 3). They have the appearance of being caused by some graded chemical process of exponential onset [46] and decay [199], throwing and keeping the membrane in a state of maximum excursion.

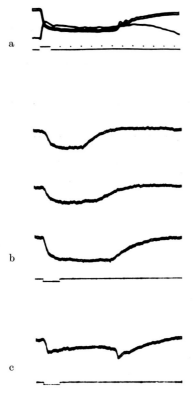

Fig. 6. a Receptor potential and ERG (fainter trace); stimulus 1 lx, 1 sec. b Extracellularly recorded receptor potentials; stimuli 0.01, 0.03, and 0.1 lx; 1 sec. c Intraretinal ERG; 0.1 lx, 1 sec. Stimulus marks correspond to 50 μV (2.5 mV for intracellular recording in a)

A slight further dislocation of the electrode tip from the receptor site toward the inner layer of the retina results in an entirely different type of response (Fig. 6c; same intensity as in last recording of 6b). It is no longer maintained activity, but transients in the same negative direction both at onset and end of the receptor potential. It appears as if only a synapse, having capacitative and rectifying properties, had been traversed in passing from first to second order neuron domain. It should however be noted that no unitary response of the transient-type has ever been found[5], that the recording site is much less critical, and that the second order neuron activity projects to outside electrodes (PII component of Granit [82]) with the polarity opposite to the receptor potential. Because the time course of PII resembles the impulse density of the majority of activated nerve cells [1] it seems to acquire slow potential properties secondarily from the ionic imbalance, which accumulates in a circumscribed extracellular space, as a consequence of time-scattered unit activity and as a cause for outreaching current flow.

Unit activity from the site of PII generation has been frequently observed [27, 196, 31, 179, 110]. It is readily detected in the loudspeaker but much less easily isolated for recording. Off-bursts from a recording site external to the ganglion cell layer have been published, which are triggered with growing delay on more intense stimulation and are accompanied by the slow potential changes [183]. On-activity, recorded from a central retinal layer with a coupling time constant suitable to demonstrate the relations with slow activity, is shown in Fig. 7. It will be seen that impulse firing is gated for the duration of the receptor hyperpolarization, marked by local slow potentials at beginning and end.

Negative and positive potential changes representing receptor and neural function, respectively, are sampled by external electrodes to make up the gross response to light flashes[6], and that both in scotopic and photopic conditions. But while individual rod (Fig. 6a) and cone [197] receptor potentials are of comparable

---

[5] Potential changes of this type may be recorded "from single cells", which may exhibit other kinds of activity of their own. In that case the impaled cells act simply as an extension of the microelectrode.

[6] On the PI-component, the c-wave, and DC-potential see below.

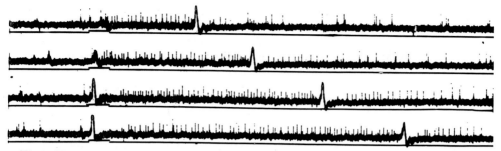

Fig. 7. Microelectrode recording from inner nuclear layer. Stimuli (from top to bottom) 0.1, 1, 10, and 100 lx; 1 sec (time constant 0.03 sec). Notice: growing delay of slow potential change following end of stimulus; increased rate of impulses between slow potentials

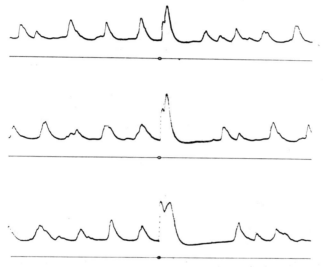

Fig. 8. Spontaneous slow potential changes (PII) of dark adapted retina; silenced by light; stimuli (0.003, 0.01, 0.03 lx; 1 sec); RC-coupling (time constant 1 sec)

size, the scotopic PIII is negligably small. That means that most of the potential difference is lost along current paths which are closed when more parallel generators are active. The proportion of PII and PIII, or the b-wave — a-wave ratio of the ERG, therefore, is a measure of the number of nerve cells fed from one receptor, or the lateral spread of excitation. Lateral summation is the most effective means to achieve the high gain in dark adaptation [18]. The increased tangential traffic would seem to implicate a higher degree of excitability of neural elements in dark adaptation, which may go as far as to amplify spontaneous membrane fluctuations (bumps: [19, 76, 208, 60]), which are silenced by light (Fig. 8).

Environmental Influences

Retinal function is highly susceptible to changes of its chemical environment and easily impaired by adverse conditions. Survival times in ischemia are equally short in retina as they are in brain, longer revival times of the retina will shrink under more rigorous experimental procedures, in particular on excluding residual circulation [102] and diffusion of oxygen through the eye media. "Chemical slicing" [194, 82, 50] has contributed much to the understanding of the ERG. Conversely, the ERG has been put to use as a powerful tool in a systematic study of retinal metabolism [151—153]. Much of this work has been included in a comprehensive review [144], together with numerous observations confirming in frog many of the findings from other vertebrate species.

With the natural environment replaced by an artificial one certain requirements have to be met and may be specified. This task is facilitated in a simplified incubation medium (Table 1), which suggests focussing on ionic constituents, buffer properties, and substrate utilization. Criterion for adequacy is the ERG, because it is available the same *in vitro* and *in vivo*, is instantaneous, and is of a sensitivity

Table 1. Composition of perfusion medium (mM/l)

NaCl	80.0 for frog
	120.0 for human
KCl	2.0
CaCl$_2$	0.1
MgCl$_2$	0.1
NaH$_2$PO$_4$	1.5
Na$_2$HPO$_4$	13.5
Glucose	5.0

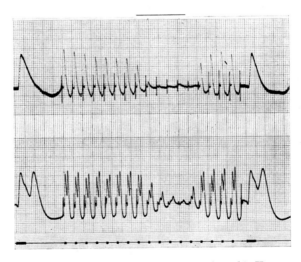

Fig. 9. Decay of ERGs on flow stop of perfusate (horizontal mark). *Upper trace*: dark adapted (stimuli 0.03 lx, 1 sec); *lower trace*: light adapted (0.01 lx; stimuli 30 lx, 1 sec). Calibration 200 μV/cm; RC-coupling (time constant 1 sec). "Interrupted recording": the recorder was advanced during stimulation for 5 sec (at higher paper speed at beginning and end), but was stopped during the 3 min intervals. Impulse-like deflections written during standstill of recorder are the delayed off-reactions (s. Fig. 3)

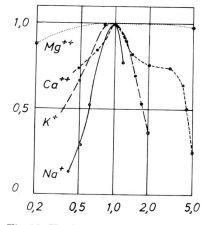
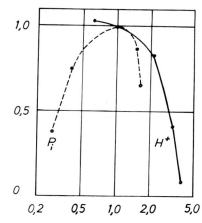

Fig. 10. Heights of ERG-*b*-waves under changed concentrations of ionic constituents of incubation medium. Results are referred to standard concentrations of Tab. 1. Abscissae: multiples of respective concentrations on log scale. Exposure to each of the solutions 30 min; dark adaptation

surpassing ganglion cell discharges [17]. Size, shape and constancy in time may be evaluated.

In a fluid medium exchange of matter is not confined to volatile material, but the exchange of gases is more difficult than in an atmosphere. Of paramount importance is, therefore, that the medium be flowing. In a flow stop the responses decay rapidly (Fig. 9), and time course and extent of recovery depend on the duration of the stop. The effect would have been more drastic still, had the retina been encapsuled in a smaller volume. The experiment was done in a twin version of the cuvette-chamber (Fig. 2a), and the two retinas of one frog were exposed to the same adapting light with 1-sec stimuli superimposed, but one of the chambers had in front of it a neutral density filter which reduced both background and stimuli by a factor of a thousand. The on-reactions are about the same size, but *a*-waves and off-effects mark the responses of the light-adapted retina. There is an indication, that in dark adaptation the retina is more demanding. This result is easily verified by repeating the experiment with the neutral density filter shifted in front of the other retina. Increase of any one of the constituents of the medium could not lower the required flow rate, which was found adequate at 2—5 ml/min depending on the geometry of the chamber.

The significance for retinal function of common ions has long been recognized [20], and their role been discussed under various aspects of the ionic theory of nervous excitation including cation-stimulated splitting of ATP [86, 72, 158, 88]. On the other hand distinct ionic effects on individual enzymatic reactions [120, 165] are known. In fact, most reactions of biological interest involve reactants which may combine with protons or metal ions and implicate a mutual coupling on LE CHATELIER's principle [2]. In integrated function this commends some reserve towards inferences from changes of the ionic mileu, particular if these are

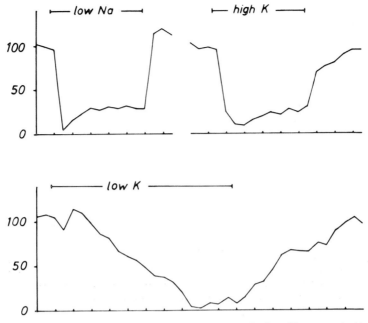

Fig. 11. Heights of *b*-waves (percent of initial values) in low Na-concentration (40 mM/l; sucrose added for isotonicity), high K-concentration (10 mM/l), and low K-concentration (1 mM/l). Dark adaptation (as in Fig. 9); time marks 10 min

large. A descriptive presentation of such changes and their consequences on the ERG of dark adapted frog retinas is given in Fig. 10. Results of a more recent reevaluation [23] are included. The data points are the fractional heights of *b*-waves as resulting after a 30 min exposure of the retina to a medium in which the respective ion concentration had been varied as indicated by the abscissae, while the other constituents remained constant, except in the case of low sodium when sucrose was added for isotonicity. A log plot of multiples of the standard concentrations (Table 1) was used for the abscissae to facilitate references and show more clearly the weight of each change. From the graphs the empirical concentrations appear to be the optimal ones. However, other combinations with more than one constituent changed are imaginable.

Most influential from the slopes of Fig. 10a are sodium and potassium, and equally so. This is suggestive of the involvement in the generation of the *b*-wave of regenerative unit activity, which is probed for excitability, or membrane potential, by increased potassium concentration and for responsiveness, or spike production, by decreased sodium concentration. Unit activity at the site of *b*-wave generation (Fig. 7) is in support of this view. Actions on cell membranes of low Na and high K follow from their kinetics (Fig. 11a). Both changes take immediate effect by establishing new ionic gradients, with the slight delay in high K possibly introduced through preferred access of this ion to the interior of the cell. The small and incomplete recoveries under administration of the altered solutions may

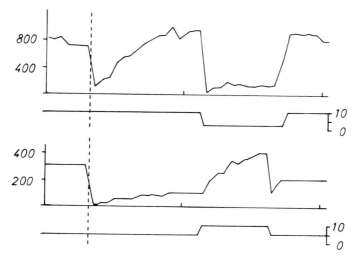

Fig. 12. *b*-waves (μV) in light adaptation followed by dark adaptation under varied concentrations of the phosphate buffer. At the times vertically marked the adapting (0.1 lx) and
stimulating (100 lx) lights had been attenuated by the factor of 1000. Concentrations (mM/l)
of phosphate buffer (pH 7.7) below ERG traces. Time marks in hours

indicate corrective fluxes across the cell membrane. Assuming a NERNST relationship the *b*-wave process could be estimated at 10 mV half-maximum at most,
recovered through external electrodes with a 1 % efficiency or better.

For osmotic reasons the effect of sodium cannot well be studied much above
the standard concentration. In a potassium concentration below the normal level
the *b*-waves are reduced. But K-deprivation proves different, both from the log
concentration plot and from its kinetics. The decay of the responses, as well as
their eventual recovery, is slow and no steady state is reached (Fig. 11 b). Small
transients on withdrawal and readmission may indicate membrane effects, but a
depletion of intracellular K follows and leaves chemical reactions short of a
necessary cofactor.

Calcium and magnesium effects are included in the same graph (Fig. 10a) for
comparison, but the importance of magnesium would be better appreciated in
longer observation periods. Calcium has multiple effects. Its stabilizing effect on
the membrane potential is well known [28], but it has also been accused for
uncoupling of oxidative phosphorylation [37]. These effects may become manifest
at different concentrations and explain the irregular dose-effect curve. Besides
calcium increases the requirements for magnesium [185], which may be observed
in a Ca-Mg antagonism in the ERG.

The influences of the concentration and of the pH of the phosphate buffer are
plotted in Fig. 10b using the same coordinates as in Fig. 10a. The two curves are
very much like mirror images indicating that a major function of the phosphates
is unspecific and serves to neutralize hydrogen ions. This is substantiated by the
fact that other buffer systems may partly substitute. Non-specificity also follows
from similar pH curves in other electrical phenomena [148] or in oxygen uptake

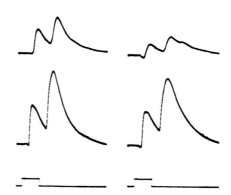

Fig. 13. Electroretinograms to stimuli of red (*left*) and blue (*right*) lights which had been matched for equal responses in alkaline medium (pH 7.7; *below*), but proved of unequal effect in acidic medium (pH 7.4; *above*). Dark adaptation; stimulus duration 1 sec; heights of stimulus marks correspond to 50 μV

[161, 170]. But concentration and pH, i.e. proportion of the primary and secondary component, of phosphate are quite specific in that they affect scotopic and photopic function in a selective manner (Fig. 12, 13 and 13a). In the experiment of Fig. 12 two retinas in different phosphate concentrations, previously light adapted, were tested during dark adaptation. The responses were smaller in low phosphate at the beginning, but the more conspicuous difference is the lack of adaptability to darkness in the low buffer concentration. In Fig. 13 two broad-band stimuli, red and blue, had been matched to yield comparable responses in high pH. In low pH they were clearly different, red being preponderant. The smaller blue response in low pH, furthermore has a large *a*-wave, giving it the appearance of a photopic response, although no background light was present. From the reduced *b*-wave — *a*-wave ratio, according to the foregoing consideration, summative ability has dropped out in low pH.

In addition to the spectral sensitivity (Purkinje effect) and inferred spatial integration, temporal summation and intensity discrimination have been found changed by pH in the sense that hydrogen ions mimic light adaptation (see: Fig. 13a [177]). Observations pointing in the same direction have been made in other studies [126, 45]. The phenomenon is not easily understood, particularly because there is no increased net acid production in light adaptation, as will be shown later. But its significance may be visualized in the fact that the multifacetted change of visual adaptation, including discrimination of intensity, space and time and spectral sensitivity, becomes a single-parameter problem.

In a search for mechanism underlying the pH-phenomenon it may be of importance that calcium in this respect acts much like hydrogen ions, while magnesium corresponds to hydroxyl ions. These are parameters, and the "right" constellations, which determine not only the amount of energy to be derived from ATP-hydrolysis but also its make-up [174, 2]. This may be summarized by saying that in alkaline medium more energy is released, but in acid medium ATP synthesis proceeds faster. That phosphate was not entirely replacable would follow by necessity (comp. scheme of Fig. 1).

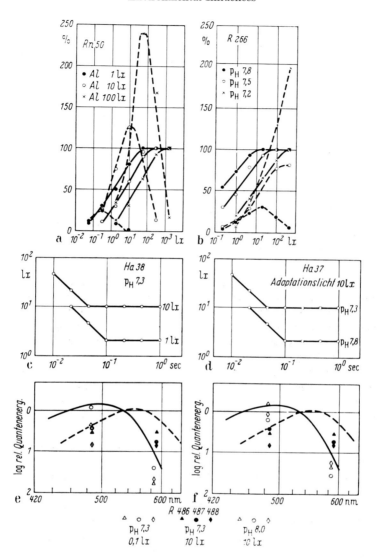

Fig. 13 a. Control of the operational range of the retina through adapting lights (*left: a, c, e*) and environmental pH (*right: b, d, f*) with respect to intensity (*upper: a, b*), temporal (*middle: b, c*) and spectral (*lower: e, f*) characteristics.

a, b: Amplitudes of *b*-waves (full lines) and off-effects (dashed) as a function of stimulus (1 sec) intensity; saturation *b*-waves, which differed but slightly in absolute size, taken as 100%. *c, d*: $I \times t$ — functions for half maximum *b*-waves, showing degeneration of Bunsen-Roscoe law (kinks) for different stimulus durations. *e, f*: In 3 retinas (triangles, squares, circles) the number of quanta of three spectral lights (referred to 534 nm) were determined for equal *b*-waves, first in dark adaptation and low pH (open symbols, left,) then with a white adapting light present (filled symbols, left = right), and finally — with the background light still present — after the spectral shift had been counteracted by raising the pH (open symbols, right). Solid and dashed lines: visual purple absorption and photopic dominator for reference (from Ref. [177])

Drastic effects on metabolism and light response will ensue from withdrawal of the energy source. Controlling the glucose level is no easy task in an intact organism, because it is held constant through special regulatory mechanisms. To probe those mechanisms load tests have been devised employing the ERG as an indicator [189]. To the isolated preparation any glucose concentration may be impressed. On deprivation function eventually suffers [4], but stores of endogenous substrates may fill in temporarily (Fig. 20 below). When function finally fails it proves more severely affected than in comparable reduction from oxygen lack, certain blocking agents etc. No steady state of reduced function will be attained, structural changes soon follow [205], and recovery on restored substrate supply will be impaired. On the other hand glucose is by no means indifferent. High glucose concentration reduces oxygen uptake [44], glucose uptake [113] and light responses [16].

Scotopic and photopic functions of the retina are differently affected by glucose lack. The light response does not simply vanish, but certain parts are more resistant than others (Fig. 14). The b-wave proves most sensitive, but the PIII component persists for some time, as is the case in most any adverse condition. Notably, the off-effect doubles, and whereas the first off soon disappears, the second one, ascribed to scotopic activity, even increases. It would disappear on prolonged exposure to the glucose-free solution, but it also fades on readmission of glucose, when the photopic off reappears.

Less encroachingly, before they result in deformation of the ERG and impeded reversibility, the effects of reduced glucose may be recognized and differentiated through double-stimulus exploration. In Fig. 15 stimuli were regularly applied every three minutes. In addition test stimuli were injected which followed a regular one, respectively, at increasing intervals as indicated by the abscissae

Fig. 14. Electroretinograms (from top to bottom) before, 12 and 27 min after glucose deprivation, and 12 min after restitution of glucose supply (5 mM/l). Light adaptation 0.03 lx; stimuli 30 lx, 1 sec; stimulus mark corresponds to 50 μV

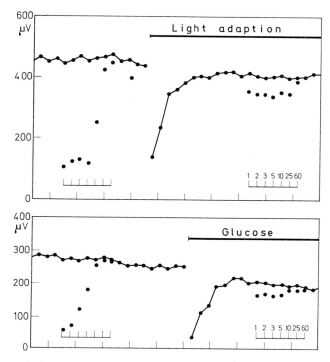

Fig. 15. Effects of adaptation and of glucose revealed from double-stimulations. *Curve traces*: *b*-waves (μV) to stimuli delivered every 3 min. *Individual symbols*: *b*-waves to stimuli injected after a preceding one at intervals indicated by inset abscissae (sec). Experimental time marked at 10 min-intervals. Stimuli as in Fig. 9; *upper part*: 5 mM/l glucose throughout; initially dark adapted, then (horizontal mark) light adapted; *lower part*: light adapted throughout; initially 0.5 mM/l, then (horizontal mark) 5 mM/l glucose

(approx. log scale in seconds). The responses, *b*-waves, to the test stimuli are represented by individual symbols, while the responses to the conditioning stimuli, which are always preceded by an interval of two minutes or more, are connected by a line. Adaptation was changed in Fig. 15a by increasing equally log background and stimulus intensities. Except transitorily this does not affect the control responses, but thoroughly the test responses. In dark adaptation the reaction to a light stimulus leaves the retina in a state of reduced responsiveness to renewed stimulation, from which it recovers in a characteristic time course which consists of two branches. In the early part the maximum of Fig. 5 will be recognized. Complete restitution is halted, perhaps for continuing activity (Fig. 7), and begins only after 5 sec to reach a final state not much before one minute. In light adaptation nearly full responsiveness obtains within one second, and only a minor fraction is contributed by the second process.

In Fig. 15b the same procedure was adopted under continuous light adaptation, but glucose in the perfusate had previously been reduced to one tenth the normal concentration. While this affected the widely spaced responses but little, it deprived the retina of its characteristic ability of fast responding in the

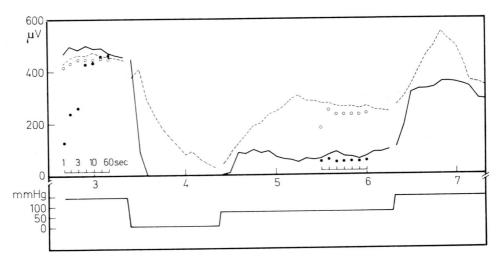

Fig. 16. Heights of *b*-waves in oxygen lack. *Line trace* and *filled symbols*: dark adaptation; *dashed trace* and *open symbols*: light adapted. Equilibrating oxygen atmosphere (mm Hg) in lower part. Abscissae: experimental time in hours; otherwise as in Fig. 15

photopic state. Notice that immediately after restoring the glucose supply the ERGs are smaller, as if the energy to produce them had been trapped for other uses, as in the initial phosphorylation of glucose (comp. Fig. 1).

Different tolerances in dark and light were also found toward oxygen lack (Fig. 16). In a room air-saturated medium the *b*-wave responses of the two retinas of one frog, examined in a twin chamber, were of the same size in dark and light, but differing in their recovery characteristics: fast in the light, delayed in the dark. On withdrawing the oxygen the scotopic response disappears abruptly, but the photopic response is diminished more gradually. Half-restoration of the oxygen supply leads to a partial recovery, which is more pronounced for the light adapted retina. Above the critical partial pressure of oxygen (see below) the responses are sufficiently stable to repeat the double stimulation. Neither the photopic nor the scotopic response exhibit any appreciable change as a function of the interval. It is thus the second branch of the bipartite recovery characteristics, which suffers predominantly in oxygen lack, the one that proved more resistant in glucose lack.

The two-step recovery does not reflect photopigment regeneration, because it is not the amount of light, but its effect that determines the time course [92]; a bipartite recovery may be obtained in responses to "dark flashes". A similar reasoning was used in interpreting "exhaustion" [21], certain types of "inhibition" [192], or "suppression" [10, 63]. Influences on the recovery characteristics are also exerted by excess of both oxygen [177] and glucose [16] in that high oxygen pressure resembles low glucose concentration, and excess of glucose oxygen lack. Because phosphate relieved oxygen toxicity, the abnormal supplies appear to involve actions on phosphate control of energy metabolism corresponding to the Crabtree and Pasteur effects.

From the selective actions on the first — the glucose dependent — and the second — the oxygen dependent — branch of the recovery some autonomy is suggested of the initial and final segments of carbohydrate utilization, possibly on the ground of their cytoplasmatic and mitochondrial location, respectively. Photopic function appears to draw on easily available and rapidly replenishable energy stores, whereas scotopic function is tightly geared to mitochondrial activity. Attempts have been made [52] to further substantiate this contention by measuring the products of oxydative phosphorylation, ATP and phosphocreatine, at different intervals following a conditioning light stimulus in dark adaptation. The test stimulus was replaced, so to speak, by chemical analysis employing a quick-transfer version of perfusion chamber. At the shortest intervals reached for stopping reactions in perchloric acid, about twenty seconds, ATP was reduced by 20 % and PC by 40 %. Both high-energy phosphates were restored to their steady state levels by the time the ERG recovery was complete.

The fact that even ATP, which is buffered from PC by fast transphosphorylations, is measurably reduced by a single stimulus testifies to energetic limitation of retinal functions, e.g. in flicker-fusion, and makes plausible the high susceptibility to metabolic disturbances. On the other hand, ATP subserving diverse functions as a "common currency" implies that induced specific activity may evade coarse measurements of over-all metabolism.

Metabolic Activities of the Retina Proper in Function

Retinal metabolism, having been known exceptional in rate and pattern from the introduction of the tissue slice technique by WARBURG [204], has been investigated extensively for various motivations, e.g. for suspected parallels with tumors or brain, or just methodological reasons. Attempts to integrate it with other retinal activities have been less rewarding [163, 50a], so that relevant data are sought on less direct routes [7]. In a pioneering and penetrating study, doing justice to earlier work, COHEN and NOELL [44, 45] have investigated the metabolism of the young and the adult rabbit's retina, respectively, and have compared it with the development of visual functions in that period. Effects of light on retinal metabolism have been denied in this work, but it is explicitly referred to for other details. The approach here adopted is to preserve and control function under the metabolic measurements.

Three types of metabolic measurements were employed: radiorespirometry, polarographic determination of oxygen uptake, and spectrophotometric assay of pyridine nucleotides, each combined with electrical testing of function. Application of radio isotope-labeled substrates provides a rigorous test of product formation, whereas O_2-uptake measurements show utilization of endogeneous substrates in addition. Supplementary information on intermediate reactions may be extracted from the absorption of respiratory pigments, which also offers the advantage of better time resolution approaching that of the electrical measurements.

The tracer studies furnished data on uptake of glucose, output of lactate and CO_2, and incorporation of labeled material into soluble amino acids. The experiments were done using the recirculating system (Fig. 2b).

The chamber was sterilized by a 30 min treatment with 50% alcohol. This concentration proved adequate to avoid microbial contamination, but would not be deleterious to the plastic parts. The chamber was dried thoroughly and the retina inserted. Exactly 1 ml of sterile experimental solution was taken from an ampoulle and added to a small test tube to the bottom of which 4 μC uniformly ^{14}C-labeled glucose of high specific activity had been dried previously. The test tube was stirred, and 0.5—0.8 ml were transferred to the chamber, the remainder served for reference purposes. The chamber was positioned inside a light-tight air thermostat on top of a rotating magnet in a holder which carried a mirror to direct the light from the stimulator to the retina. Electrical and gas flow connections were made and the box closed. The gas supply came from a cylinder containing water-pumped compressed air or a 95% oxygen 5% CO_2 mixture. In the latter case the medium contained bicarbonate and was pregassed. After reduction of the pressure the gas flow was divided into a main branch, which was humidified and traversed the chamber, and a secondary branch, which bypassed the chamber and served to dilute the humid gas to prevent condensation on its way to and within the counting cell. The counting cell was a steel block with a shallow excavation of 5 ml volume sealed by a thin mylar membrane which was placed immediately beneath a low level β-detector. Accumulated counts were printed periodically (for details on counting equipment, calibration etc. see the original publication of Horowicz and Larrabee [104]).

Production of labeled CO_2 in a typical experiment is shown in Fig. 17 (lower portion). The preparation was done in red light within 10 min, and zero time is the moment of starting the gas flow. The ERGs (not shown) were recorded every three minutes. Their amplitudes showed some adjustment to light level, pH and temperature [180] during the first 20 min but remained of constant amplitude for the rest of the experiment. By contrast the counting rate rose gradually and did not reach a stationary state before three hours. It would remain constant from there on until about ten hours when it started declining. This time course was taken as a safeguard against bacterial contamination, in which case no steady state would be

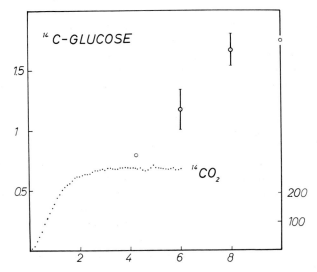

Fig. 17. Glucose uptake and CO_2-production of perfused retina. *Left ordinate* and *open circles* (mean; SD): ^{14}C-glucose (μM) extracted from medium per retina (2.38 $\pm$ 0.38 mg dry weight); *right ordinate* and *dotted data*: $^{14}CO_2$ (nM/h) cleared continuously from medium by carrier gas (accumulated counts/6 min; single experiment). Abscissae: incubation in hours

reached during the experimental period and the counting rate would not, as it did under aseptic conditions, return to the background level after removal of the retina from the chamber.

The slow rise of the $^{14}CO_2$-curve cannot be fitted by a single exponential function and several factors can be isolated that determine its shape. Instrumental lag contributes some but no substantial part. Cutting the main gas flow and using the bypassing branch only is followed by a rapid return of the counting rate to the baseline; and the counts not cleared from the solution during this procedure appear in a burst on reopening the main gas flow. The second of the 6-min printouts is clearly above the background, which indicates that diffusion of glucose in, and of CO_2 out of the preparation are fast. Assuming that the rate of production of CO_2 was constant from the beginning of the experiment, as is suggested by the light responses and the linear uptake of glucose, the "missing counts" may be determined by back-extrapolation of the steady state counting rate to zero time. In another experiment $^{14}CO_2$-counts were pooled during three subsequent 3-h periods, during equilibration, steady state and after removal of the retina from the chamber, respectively:

		Total counts
First	3-h period (equilibration)	34 773
Second	3-h period (steady state)	50 919
Third	3-h period (retina removed)	9 217
	"missing counts"	16 146
Retained in bathing solution		6 252 . . . 9 217
Retained in retina		9 894 . . . 6 929

An amount of ^{14}C-counts equivalent to one fifth of the counts cleared during the steady state period was recovered from the bathing solution after the retina had been removed. Even if the fraction retained in the solution may be higher during equilibration, the amount will be not. Thus a conservative estimate shows that an amount of glucose carbon equivalent to thirty minutes steady state CO_2-production, or 10 nanomoles of glucosyl units per milligram dry weight of tissue, was retained in the retina. It does not find its way to the final product directly, but serves to equilibrate pools of intermediates. A pool of this size does not reflect all the carbohydrate reserves, because lactate production in the absence of an exogenous source may go on using carbon much in excess, unless it limits itself from exhaustion of buffer capacity. Apart from intermediates of the direct path to CO_2 the pool comprises largely [62] soluble amino acids, which are rapidly labeled from ^{14}C-glucose and may easily be used up, as is evident from ammonia formation in glucose lack [171]. In two-dimensional chromatograms of retinal homogenates the amino acids were found to exhibit the "GEIGER pattern" of brain tissue [181], the same as in rat [35] or rabbit [200]. Equilibration of the pool was not appreciably faster in higher temperature, when CO_2 production, but not the ERG, was increased.

In the steady state liberation of labeled CO_2 can be regarded proportional to the rate of its production, irrespective of retention. The factor of proportionality depends on the counting efficiency, rate of gas flow, and the specific activity of the substrate in the medium. It had been determined by various procedures [104] and been verified in the present investigations by injecting into the chamber of a known amount of ^{14}C-labeled bicarbonate. The rate of CO_2-production is included in Table 2 referred to dry weight, which was determined in this series (r. pipiens) at 15.8 % of the wet weight, or 2.38 ± 0.38 mg/retina (mean and standard deviation of 14 retinas). The experiments had originally been designed to differentiate CO_2-production rates in dark and light, and an increased rate was found in the dark (see below). Because not all of the measurements of Table 2 exhibited the light/ dark differences with a sufficient degree of statistical significance, the results have been pooled.

Glucose uptake and lactate production have been determined in the same retinas that furnished the data on CO_2-production (Table 2).

Known amounts of bathing solution, 5 and 25 µl, were taken after different periods of incubation and dried on filter paper. The paper strips, together with those containing the same quantities of unused solution, were chromatographed in ascending manner using butanol-acetic acid-water (80 : 20 : 20). Good separation was achieved within less than one hour, after it had been ensured in more prolonged and two-dimensional runs (second: phenol-water) that no other radioactive material was to be expected in the bathing solution that had been in contact with the retina up to ten hours. The strips were allowed to dry at room temperature and subsequently scanned with a Geiger counter. Radioactivity appeared at about RF 0.15 (glucose) and 0.65 (lactate). The areas under the curves were measured by means of a precision planimeter. Glucose uptake was determined from the differences between the 5 µl sample and control values, lactate production in the 25 µl samples in addition.

No data were taken for incubations shorter than 4 h in order not to disturb the kinetics of the CO_2 evolution, but the data after 4, 6 and 8 h (Fig. 17) show that glucose uptake was linear until that time. The tenth-hour values were correspondingly lower, but by that time the glucose concentration in the medium had fallen by one third, and so had the ERGs, and the experiments were usually discontinued. The data included in Table 2 are those for 6 h incubation, but the accuracy would be improved by incorporating adjacent data from the curve.

The rates of Table 2 on the whole are extraordinarily low compared to any data reported, but they are obviously high enough to support function, i.e. the production of the highly delicate ERG. The smallness of the observed rates is not primarily an expression of a species particularity or experimental temperature, but of a metabolic restraint, which is easily lost, but preserved as long as b-waves are produced. This is demonstrated more directly in measurements of oxygen uptake (Figs. 19 and 30), but follows also from the proportions of the entries.

More than one third of the carbon atoms taken up with glucose appear in lactate. This fraction is low compared to Warburg-determination in rat or rabbit, in which it accounts for more than 90 % of the glucose taken up aerobically, but high from an economical point of view and suspect of possible impairment of

Table 2. U-^{14}C-glucose utilization of perfused frog retina 22° C[µM/mg dry wt · h];means ($\pm$ SD)

Glucose uptake	0.082 ($\pm$ 0.012)
Lactate output	0.063 ($\pm$ 0.011)
CO_2 production	0.115 ($\pm$ 0.019)

oxidative utilization of pyruvate. The existence of an anoxic core can be excluded, because a considerable margin of safety is provided by the partial pressure of oxygen maintained during the experiment (Fig. 18); moreover, the ERG would show such failure immediately (Fig. 16). A possible functional significance is that glycolytic ATP be available in preference to [128], or faster than [53, 142] mitochondrial ATP to certain key functions of the cell. Another interpretation [44] focusses on the potentiality of lactic dehydrogenase to couple reduction of pyruvate with reoxidation of TPNH instead of DPNH and thereby facilitate the operation of the pentose phosphate shunt [166]. Certain of the known LDH isozymes, which channel pyruvate into lactate, were observed predominant when shunt path activity was high [98, 84, 22], and added pyruvate stimulated direct oxidation of glucose [116, 115, 77]. Because shunt path activity is held to subserve syntheses rather than catabolism [111, 168], a defective balance of uptake and immediate outputs of labeled carbon should be expected.

The rate of ^{14}C-glucose uptake less lactate output is high enough to cover more than twice the $^{14}CO_2$-output observed up to 8 h of incubation. Hence, an amount of carbon equivalent to 0.25 micromoles of glucosyl units per mg of dry tissue, or 4.5% of its weight, has not found its final destination during the observation period, while the rate of $^{14}CO_2$-output was constant. In rat sympathetic ganglia, which are largely occupied by nerve cells and synaptic structures, but seem to lack the capacity to form all of the amino acids typical of central nervous tissue [146], a fraction of 12 % of the exogenous glucose has been traced into lipids, proteins and nucleic acids. These groups of compounds were found to be traversed with a time constant of more than ten hours. They were, therefore, considered slowly exchanging intermediates of carbohydrate metabolism (LARRABEE [123]). In retina

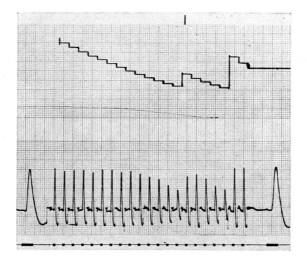

Fig. 18. Oxygen uptake and light responses in varying oxygen supply; *above*: concentration of oxygen (0.260 µM/ml f.s.) in chamber (recirculating volume 1 ml); *below*: ERGs; 200 µV/cm; RC 1 sec; stimuli 0.03 lx, 1 sec; "interrupted recordings" (s. Fig. 9). Prior to reproduced section of experiment the flow of the medium from the external reservoir had been stopped. Vertical mark: admission of 0.1 ml, later of 0.3 ml of medium

turnover of lipids [113], proteins [155] and nucleic acids [200] has been observed in tracer studies. One third of the glucose unaccounted for in the immediate outputs of lactate and CO_2 appears adequate to account for the more remote paths.

The partitioning of glucose carbon demonstrates that the pathways are more involved than is commonly visualized with regard to energy metabolism. Whereas lactate production proceeds on a direct route and tracer results truely reflect total lactate formation [44], the $^{14}CO_2$-rates have to be taken with the reservations dictated from the existence of pools of fast and of slowly exchanging intermediates. An appreciation of the energy resulting from complete degradation of the substrate, not requiring long equilibration, is furnished in the measuring of the oxygen consumed in the process. Simultaneous recording of the ERG shows to what avail.

In order to measure O_2-uptake the recirculating chamber (Fig. 2b) was modified as described under "General Techniques". The polarograph was battery operated and had an output $dc - dc$ converter permitting floating operation without electrical interference to/from other electrical measurements. The solutions were equilibrated with room air; the temperature was kept at 20° C; sterility was observed. Retinas of *Rana esculenta* were used in this series of experiments having an average dry weight of 1.78 mg. As in the tracer studies the experiments were programmed by a timer providing interrupted recordings. Capacitor coupling of one-second time constant was employed to have a stable baseline in the ERG recordings.

In Fig. 18, lower trace, ERGs of a dark adapted retina to 1-second stimuli are recorded periodically at 3 min intervals, initially at 5 mm/sec, but at only 1 mm/sec thereafter except for the very last, which was again written with the higher paper speed for comparison. Altogether 69 min of the experiment are displayed. The upper trace indicates the partial pressure of oxygen in the chamber. It was initially at the upper margin ($0.260 \mu M$ O_2/ml at room air saturation of the solution and fast flow from the reservoir), but the flow of the solution had been stopped prior to the recorded period, while stirring inside the chamber was maintained. The decline of the trace caused by the uptake of oxygen from the medium by the retina appears in a stepwise fashion, because the recorder advanced only during the recording of the ERG, but the PO_2 drop occured continuously and linearly at first. The initial rate was $0.235 \mu M$ O_2/h but it dropped to $0.150 \mu M$/h after the oxygen concentration had fallen below half-saturation. Coincident with the reduction of O_2-uptake the ERG amplitudes decay and are recovered immediately on admitting only a few drops of solution (vertical bar), which carried enough oxygen to restore the partial pressure to near half-saturation. The recovery is not complete and the ERGs decay again as the added oxygen is expended. A final addition of 85 picomoles of oxygen is sufficient to raise the PO_2 above the critical level, to increase the O_2-uptake, and restore the responses completely.

Fig. 18 shows that below a critical level of oxygen pressure the ERG follows oxygen supply and oxygen uptake in a reversibly coordinated manner. The span of PO_2-determined b-wave amplitudes, roughly $150-30 \mu M$ O_2/ml, may be transposed to a higher level by introducing, or rather preserving, a diffusion barrier. In isolated frog eyeballs, both submerged in a well-stirred bathing solution and in an atmosphere, the limits happen to be between 100 and 20 % oxygen. In this conveniently expanded range stationary states of reduced b-waves and reduced oxygen uptake have been observed [13] which showed so close a correlation as to warrant the b-waves to be taken vicariously for oxygen uptake. It would also follow that oxygen utilization was of constant efficiency within that range. This was no longer the case outside that range, neither in high oxygen pressure when the b-waves vanished without a parallel reduction of oxygen uptake, nor in oxygen lack that proved much longer reversible for oxygen uptake than for the b-waves. Hence, the linkage between oxidation and phosphorylation is a likely site of attack in both conditions of inadequate supply. While the higher concentration of oxygen needed in the eyeballs comes close to the toxic limit, a considerable margin of safety obtains in the exposed retina, as in the intact eye, between optimum and toxic pressures. Independency from PO_2 of oxygen uptake and b-waves within that range indicates the operation of a restraining control, protective against varying supplies but accessible to varying demands (next section).

The conditions were different in this series of experiments from those of the tracer studies, but the response behavior of the retinas (*R. pipiens* and *esculenta*, respectively) was quite similar. Therefore, the measured rates of O_2-uptake and $^{14}CO_2$-output cannot be directly related, but it seems safe to say that the oxygen consumed is well covered by exogenous substrate taken up. Some more technical

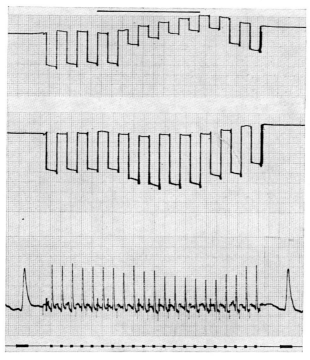

Fig. 19. Oxygen uptake and light responses under cyanide and dinitrophenol; *below*: ERGs; $200\,\mu V/cm$; RC 0.3 sec; stimuli 10^{-2} lx, 1 sec; "interrupted recordings"; *upper traces*: PO_2 (scale: $0.208 \ldots 0.260\,\mu M\ O_2/ml$). Alternatingly, the chamber had been rapidly perfused from the external reservoir and the fluid flow had been stopped, for 3 min each. During horizontally marked periods the following additions had been made to the perfusate: $10^{-4}\ M/l$ NaCN (uppermost; respective ERGs not reproduced), 3×10^{-5} DNP (lower trace together with ERGs).

At the stop-time selected 1 mm corresponds to an uptake of oxygen of $2 \times 10^{-8}\ M/h$

points emerge from Fig. 18. The critical pressure of oxygen is well below room air saturation, and a moderate reduction should be tolerated without serious consequences. Furthermore, the changes of PO_2 on refilling the chamber, and of oxygen uptake on changed PO_2 are fast enough to employ a flow—stop technique for a quasi-continuous follow-up of the oxygen uptake. This is examplified in Fig. 19 in simultaneous recordings of ERGs and oxygen uptake under the influence of two known blocking agents.

A flow of standard solution from a reservoir through the chamber was established initially fast enough for the PO_2 not to be affected much by retinal O_2-uptake (each of the upper two traces). Immediately after a flash was delivered and the response recorded (lowermost trace), the flow was stopped until after the next flash three minutes later. The PO_2 dropped during that time, as the O_2 supply was confined to the 1-ml portion recirculated, and the level reached was recorded as a near horizontal line for 5 sec with the next advancement of the paper. The flow was then opened and the cycle repeated throughout the experiment. In this way the rate of oxygen uptake is represented through the modulation depth of the PO_2 trace. During the stop—periods preceding the onset and end of the horizontal line the reservoir was exchanged and the new solution admitted to the chamber with the next flow—period. Some delay was introduced deliberately through a length of tubing inside the thermostat to ensure equilibration of temperature. A shift of the baseline may result from slight differences of dissolved oxygen in different solutions without serious effect on determined uptake. A ten second response time of the electrode should be taken into account in precise measurements. The sensitivity of the polarograph was increased fivefold with only the upper 20% displayed on the chart.

Cyanide and dinitro-phenol affect respiration and phosphorylation in a selective manner. Accordingly the ERGs are reduced similarly in both cases (not reproduced for cyanide), but with a decrease and an increase of oxygen uptake, respectively. These facts are known [125, 193, 145], but the simultaneous demonstration of metabolic and ERG effects eliminates doubts as to effective concentrations, reversibility etc. As regards the "normalcy" of the oxygen uptake measured with

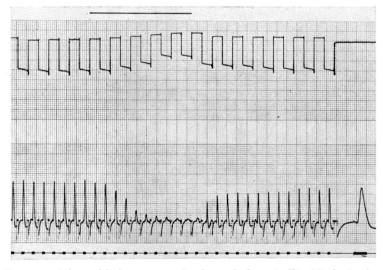

Fig. 20. Oxygen uptake and light responses in glucose lack; as in Fig. 19; during horizontally marked period (30 min) the substrate had been withdrawn; otherwise 5 mM/l glucose

the technique, the experiments of Fig. 18 show that the rates are high enough to be reduced by cyanide to 50 % with only moderate and readily reversible impairment of the ERG, but under sufficient restraint by tight coupling between oxidation and phosphorylation for dinitro-phenol to raise the O_2-uptake through uncoupling by 40 %.

The same flow-stop technique was applied to investigate the relation between O_2-uptake and ERG in glucose lack (Fig. 20). After a few flow—stop cycles the glucose was withdrawn from the medium for thirty minutes (horizontal line) and readmitted thereafter. There is an immediate onset of the effect in the ERGs, which rather precipitously disappear 10 min later. By contrast, the oxygen uptake proceeded nearly undiminished at first and became drastically reduced by one third only with the last two cycles of the glucose lack period. On restitution of the glucose supply the oxygen uptake resumes its initial rate immediately, but there is a more gradual recovery of the ERG, which approaches completion with a further rise only half an hour later. On reduction, instead of complete omission, of glucose the process would be delayed and ERGs of intermediate height would be obtained as long as oxygen uptake was still high. A two-step decay of the ERG may also be observed on offering the "wrong" substrate, 2-deoxyglucose [164].

Failure of function in glucose lack [57], as in oxygen lack [157] with the oxygen uptake unimpaired has been a puzzling observation. It is of interest in the present context in view of the selective effects of glucose and oxygen lack on photopic and scotopic function (Figs. 15 and 16). In the absence of an exogenous substrate some intermediate obviously serves to consume the oxygen. Pyridine nucleotides (PN), which collect hydrogen for use in the respiratory chain, were considered worthwhile investigating.

Spectrophotometric detection of PN should offer no insurmountable difficulties. A coarse estimate shows that 0.3 micromoles of PN per mg of wet tissue weight (rounded figure as determined for total PN in rat [186, 140]) should in the reduced form cause an absorption of the order of 10% at the appropriate wavelength in a layer equal in thickness to retinal tissue. Some of the problems implicated have been outlined above ("Metabolic Outfit"). Low analysing light intensities, mandatory from the presence of light sensitive photopigments, were compensated for by a two-step amplification. In a modified Eppendorf photometer, the signals were selectively amplified and rectified, part of the output was biased, and further dc-amplification was provided from a fast potentiometer recorder. In the cuvette-type chamber (Fig. 2a) an area of 0.3 cm² of homogenous retinal tissue could be used, which helped further to decrease the density of the incident radiant energy to approx. 10^{-9} W/cm² at 365 nm. The transmitted fraction of the analysing light was funneled onto the photosensitive surface of the detector through a polished aluminum tube. Two microscopic cover glasses of known optical densities could be inserted in the light path. One was usually present and could be removed, or the second one added, so that positive and negative calibration signals were available to check the linearity around the operating light level.

The silver ring electrodes were carefully chlorided to monitor the asymmetry potential across the retina. The fluid shunt of the chamber proved of high enough electrical resistance to make separate perfusion of the two sides dispensable. A dc—microvoltmeter served as a preamplifier, the output of which was displayed on the second trace of the dual-recorder. ERGs were observed on a CRT—oscilloscope but not recorded.

In order to determine PNs participating in substrate utilization a cycle of oxidation-reduction was performed by varying the glucose concentration of the perfusion medium (Fig. 21). Changes are seen to occur in both the optical and the

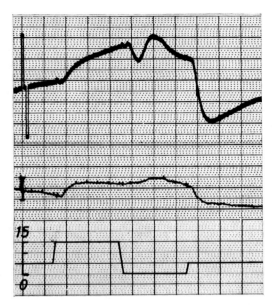

Fig. 21. Pyridine nucleotide oxidation/reduction and asymmetry potential of the retina; *lowermost*: concentration of glucose (mM/l) in perfusate; *center part*: transretinal potential difference; calibration ± 1 mV, vitreous-positive upward; *uppermost*: transmitted UV-light (365 nm); calibration + 14% (upward, "oxidation"); —12% (downward, "reduction"); time divisions: 10 min

electrical trace on each change of the medium, which are similar to each other but not correlated to the concentration of glucose in a simple fashion. An increase of the transmitted light, oxidation of PN, follows the increase from 5—15 mM/l, but a decrease results later from raising the concentration from 0—5 mM/l, and a multiphasic change ensues on complete withdrawal of glucose in the center part of the recording. Unlike strong reducing agents, e.g. dithionite, glucose causes PN reduction only at moderate concentration but PN oxidation at higher, non-physiological concentration, as incidentally does oxygen lack initially.

An interpretation of the optical events, in the order of the experimental periods of Fig. 21, is offered as follows. *Step 5 . . . 15 mM/l:* Initially in 5 mM/l glucose the retina is in an essentially normal state, for ERGs would be obtained under this condition for many hours. Increasing the glucose concentration threefold would neither increase the light responses nor oxygen uptake, on the contrary both would eventually decline (Crabtree effect; [16, 44]). The increased oxidation of PN is, therefore, taken to signal the loss of available energy spent in the additional phosphorylation of glucose. Reduced PN is thus revealed equivalent to ATP. In fact, PN reduction can be brought about at the expense of ATP in mitochondrial preparations (reversed electron flow: [40, 117]), but the functional significance of energy-linked PN reduction has been questioned [112]. *Step 15 . . . 0 mM/l:* On complete withdrawal of glucose there is an immediate stop to the drain of ATP, which is soon followed by a renewed oxidation of PN as a first symptom of glucose lack. A pool of reduced PN substitutes in maintaining oxygen uptake (comp. Fig. 20), but as it becomes exhausted O_2-uptake decreases with PN reduction now reflecting decelerated electron flow along the respiratory chain. By analogy these processes are intramitochondrial events. *Step 0 ... 5 mM/l:* The large reduction of PN on restoring glucose is a known effect in glycolysing systems, as in ascites tumor cells. It is most likely caused by the production of DPNH at the glyceraldehyd-3-phosphate step and of cytoplasmatic ATP. It coincides with the prompt onset of ERG recovery. The strong density increase masks the beginning of mitochondrial PN oxidation, on the arrival of ADP from the hexokinase reaction, which signals resumed normal oxygen uptake.

The optical changes of Fig. 21 are suspiciously large, and changes in light scatter, which are known to accompany PN oxidation-reduction may be contributory. A dual-wavelength technique [36] would eliminate scatter by subtracting from the light transmitted at the wanted wavelength that transmitted at a nearby wavelength, however, at the cost of higher total irradiation. It was applied in several such experiments, which confirmed the directions and proportions of the changes of Fig. 21. The UV absorption measurements are, therefore, taken to reflect predominantly redox changes of PN, and that in their dual significance of representing stored energy and electron flow. Knowledge of oxygen uptake allows to decide between the two meanings.

The entirely independent electrical recordings show slow potential changes which are of striking similarity, though not congruent, to the optical changes in all phases of the experiment with a (vitreous-) positive deflection prevailing during functional impairment both in high and in low glucose concentration. The occurrence in the retina proper of DC-potential changes is not unexpected [121, 160, 195], although slow potential changes are mostly discussed as a function of the pigment epithelium [151, 153, 29], which was absent in the experiments.

From the effective parameter, glucose concentration, and the observed consequences on oxygen uptake and ERG, the slow trans-retinal potential changes are clearly metabolically determined. But the congruity with PN oxidation/reduction was not perfect and would be less so in an altered concentration of the phosphate buffer. Negative deflections were observed on lowering the pH or, more conventional, on increasing the PCO_2 or on lowering the PO_2 in the perfusate. These are parameters that affect the "resting potential" (electrooculogram: EOG) of the intact eye in the same sense and to the same extent [69, 99]. These are also parameters that influence the cortical DC-potential, with the surface becoming positive [34]. Incidentally, an "azide response" (NOELL [150, 151]), a huge (vitreous-) positive slow potential change on intravenous injection of the poison, which has been made a strong case for the role of the pigment epithelium, may be observed the same in the detached retina and, surface negative, in cortex. The retinal asymmetry potential, therefore, does not seem to be unique, but share essential properties with slow brain potential changes. The structural convergence or the predominant direction of signal flow perhaps accounts for the respective polarities.

Another correspondence between retinal and cortical processes has been disclosed by the discovery in the retina [79] of "spreading depression" (LEÃO [127]). It is a state of reduced responsiveness to light stimuli, which may be recognized from a milky wave slowly traversing the retina and a pronounced slow electrical

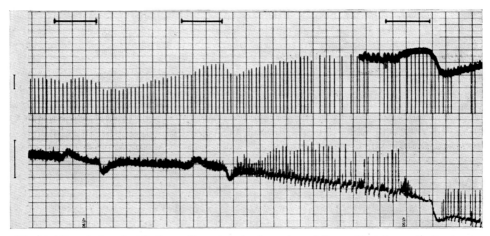

Fig. 22. Electrical and optical manifestations in the retina of "spreading depression"; *above*: transmitted UV-light (365 nm); zero suppressed, calibration 12% change, increase upward; *below*: transretinal potential difference; calibration 2 mV, vitreous-positive upward; time divisions: 10 min. Horizontal markings: glucose deprivation. Up to 20 min prior to last glucose lack the retina had been exposed to the UV-light for 5 sec every 3 min; from then on continuous UV exposure with occasional five-second interruptions

potential change. It is observed in impaired metabolic conditions and preferably in dark adaptation. In a perfused retina it does not normally occur. A chance observation of the development of spreading depression in the preparation is shown in Fig. 22.

A retina which happened to exhibit vigorous spontaneous activity was subjected to repeated periods of glucose lack (horizontal lines). It was stimulated every three minutes by 5-sec flashes from the UV analysing light beam in order to observe at the same time the ERG responses and to record the fraction of light transmitted during each flash (vertical deflections in upper part). The transretinal potential is displayed in the lower trace. After the second exposure to glucose lack huge periodic potential changes are seen to develop. They approach a constant size [89] when they occur spontaneously, but they may be triggered by light flashes shortly before they are due. If they preceded a light flash, then no ERG response would result from the flash. Twenty minutes prior to the last glucose lack period the analysing light was kept on continuously, and only briefly interrupted at intervals to observe the ERG-off effects. The continuous exposure to UV-light leads to a transitory decrease of the slow depressive electrical potentials. The small fluctuations of the continuous record of the transmitted light, perhaps from the light scatter, allow to predict the occurrence of a slow potential change. Both the optical and the electrical oscillations are suppressed during glucose lack, but are resumed thereafter, when the gradual increase of the transmitted light signals resumed mitochondrial function. Optical [139] and electrical [138] localizations point to the inner retinal layers as the site of generation of spreading depression, a dense array of cells, as in cortex, with high and easily endangered metabolic activity.

Effects of Light on Retinal Metabolism

In an evaluation of past work PIRIE and v. HEYNINGEN [163] have come to the conclusion that there is no influence of light on retinal energy metabolism. It is difficult to argue against this view, even from more exhaustive or more recent reviews [32, 50a]. One reason would be close at hand: that special precautions have to be observed in order to enable the retina to react to the imposed light changes. This refers to the environmental conditions [129, 83a], but also to the regimentation of the light exposure. For light is by no means physiologically indifferent. Its aggravating effect in the manifestation of retinal degeneration is well known [61], and detrimental effects on retinal viability have been seen of only moderate light intensities [154]. A study in which the prerequisites have been fulfilled is that of NOYONS, WIERSMA, and of JONGBLOED [156, 108], notably by avoiding any artificial medium, by preserving the retinas in the eyecups, and by the use of low light intensity. The result was the unpopular finding of a reduction of oxygen uptake and CO_2 output in the light. A confirmation (Fig. 23b) and an extension (Fig. 23a) comes from tracer studies in the perfused retina [180] (for technicalities: p. 692).

In Fig. 23a CO_2-production from ^{14}C-labeled glucose is reproduced for a retina which had been exposed to light throughout the experiment. But during the equilibration period and the next two hours it was steady light, with only occasionally stimuli superimposed to monitor responsiveness. After the fifth experimental hour the intensity of the light was doubled, but it was presented in

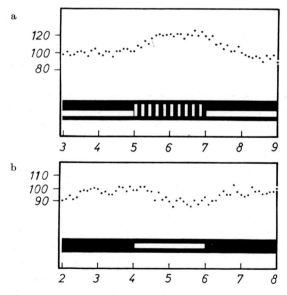

Fig. 23. Effects of light on $^{14}CO_2$ formation of the retina; *upper part*: continuous light (0.1 lx) vs. flickering light (1 sec flashes, 0.2 lx, 1 sec dark); *lower part*: dark adaptation vs. light adaptation (0.1 lx). Ordinates: counts/6 min; 100% = stationary level before light change. Abscissae: experimental time in hours

flashes of one second duration with dark intervals of equal length. Thus the amount of light remained the same, but the retina had to respond to each on and off. The counting rate as a consequence rises by approximately 20 %, and this increase is reversible after the seventh hour. In Fig. 23b the retina had been kept dark, except for occasional exploratory stimuli, up to the fourth hour and following the sixth hour, but exposed to an adapting light in between. Isotopic equilibration was not completed at the beginning of the experimental period reproduced, but a reversible depression of $^{14}CO_2$ production by approximately 10 % will be recognized during the period of exposure to the steady light.

Two opposite effects of light are apparent from Fig. 23. One has, therefore, carefully to distinguish between stimulating and adapting effects of light. Only the former activate metabolism whereas light adaptation depresses it. The magnitudes of the changes in dark and light are small, but about as big as can be obtained by control of the light parameters. One should realize that they do not reflect a rest-activity transition, and effects much larger would rather appear suspicious, especially if they prove not reversible. Still, the possibility that extra-expenses are paid from oxidation of substrates other than ^{14}C-labeled ones suggests measuring oxygen uptake.

A further increase of the sensitivity of the O_2-uptake measurement was achieved in the following way: While fast circulation of the medium within the chamber was maintained (recirculating system: Fig. 2b), the fluid flow through the chamber was much reduced and kept constant at a rate of 6 ml/h by means of a precision piston pump. As a result of the lowered supply of dissolved oxygen the PO_2 in the chamber dropped and reached a steady state level determined by the oxygen consumption of the retina. Thus the meter readings can be directly calibrated in rates of oxygen uptake, an upward deflection, or higher PO_2, corresponding to a lower rate. The amount of oxygen taken up less, or in excess may be determined from the area under the curve, while the time course of the changes will be distorted by the slow flow of the medium. From Fig. 18 no limitation would be expected from the lowered oxygen supply as long as the uptake does not exceed 0.78 μMO_2/retina · h. Less than half this rate was usually found, and was displayed, through compensating most of the oxygen current, in the center of the chart with a span of 25% change of the rate for full scale deflection.

In the experiment of Fig. 24 adaptation was controlled simply by reversing the shutter operation: In dark adaptation ERG—responses were elicited as usual by one-second light flashes every three minutes, in light adaptation the same light was kept on all the time except for brief one-second interruptions every three minutes, which served as "dark flashes" to record the off-responses. Light adaptation was presented several times (see stimulus track at bottom of figure), while during the periods of dark adaptation interference- and neutral density-filters were changed for different wavelengths (top of figure) but equal responses. Responses at beginning and end at expanded time scale were to stimuli of white light for comparison.

During the light periods the oxygen trace climbed with little delay to a level which would not have been far exceeded but stationarily be maintained, had the light period been extended to more than 15 min (comp. Fig. 26). In the interest of accomplishing repeated tests the background light was switched off after the fifth dark flash and normal operation resumed. The oxygen trace returned immediately, but the preexisting level was not quite reached within the time allowed before the next light period. This explains the over-all upward trend of the trace.

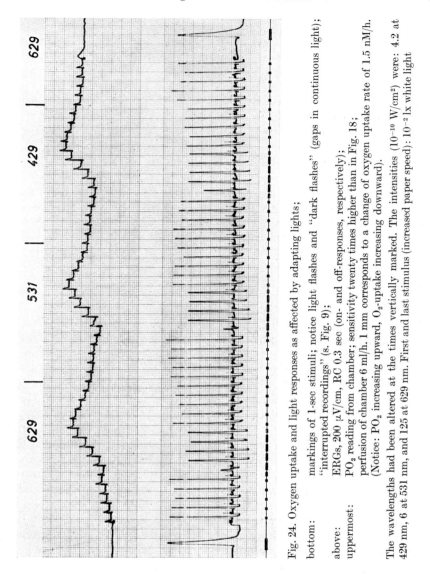

Fig. 24. Oxygen uptake and light responses as affected by adapting lights;

bottom: markings of 1-sec stimuli; notice light flashes and "dark flashes" (gaps in continuous light); "interrupted recordings" (s. Fig. 9);

above: ERGs, 200 μV/cm, RC 0.3 sec (on- and off-responses, respectively);

uppermost: PO₂ reading from chamber; sensitivity twenty times higher than in Fig. 18; perfusion of chamber 6 ml/h. 1 mm corresponds to a change of oxygen uptake rate of 1.5 nM/h. (Notice: PO₂ increasing upward, O₂-uptake increasing downward).

The wavelengths had been altered at the times vertically marked. The intensities (10^{-10} W/cm²) were: 4.2 at 429 nm, 6 at 531 nm, and 125 at 629 nm. First and last stimulus (increased paper speed): 10^{-2} lx white light

The change of the filters (vertical bars) was of little consequence for the oxygen trace. The ERG recordings reveal that the match of the lights was not perfect; not quite enough energy was available at the short wavelength. Still, the effects of light on oxygen uptake, depressing as on CO_2-output, seem to be similar, if the effects on the electrical behavior of the retina are similar.

The light intensities chosen in the experiment of Fig. 24 were such as not to cause substantial bleaching but to produce about half-maximum depression of oxygen uptake. Maximum metabolic change, still completely reversible, was found at about 100 lx of white light. Interestingly, this level of retinal illumination cor-

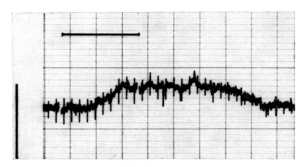

Fig. 25. Acid production of the retina and illumination. The pH (calibration: 0.01 pH; increase upward) had been measured continuously in the outflow from the chamber during slow (6 ml/h) perfusion. Notice lag of meter reading on exposure of the retina to steady light (0.03 lx, 30 min; horizontal mark). From the buffer concentration according to Van Slyke's formula a total acid production results of approx. 0.6 µeq/h. The reduction of acid production under light shown amounts to 8%

responds closely to that (10000 td) causing maximum "light rise" of the EOG [11]. Beyond that level of irradiation oxygen uptake would stay low in the following dark period and the ERGs would not be restored to their previous heights. Within a range of adapting intensities intermediate to those used in Fig. 24 and maximum the oxygen trace would reach its dark level with an undershoot (Fig. 26) indicating extra-oxygen uptake following, and in proportion to, light exposure.

Extra-oxygen uptake following exposure to light is plausible on the assumption of additional requirements arising from the elimination of bleaching products and regeneration of photopigments. A rate-term found necessary to relate photopigment concentration and sensitivity [58] may reveal the drain of energy for regeneration from a store which also powers other functions in the processing of the light response. Or, "rod/cone-rivalry" [173] may as well be a competition for a common source of energy as of photopigment precursors. Regeneration of photopigments would not be expected to take place in a retina devoid of its pigment epithelium if migration back and forth of detached prosthetic groups [59] was integral in the process. But regeneration of rhodopsin may be observed in the preparation, provided bleaching had not been effected in one intense blast and provided the retina is well supplied. Regeneration was absent when the fluid flow had been stopped or cyanide been added to the perfusate. Perhaps leakage of prosthetic groups will then occur, for encasing the detached retina within close confines also has been found to help regeneration in detached rat retina [47].

Reduction of oxygen uptake during exposure to light is corroborated by several related observations. An artificially established oxygen gradient was less steep in light [87]. The requirements are higher in the dark (Fig. 16), and it seems common experience that some light serves to stabilize isolated preparations, probably by preventing them from becoming metabolically exhausted. Administration of CO_2 also has a stabilizing action [190], as has the lowering of the pH of the medium, which was shown (see Fig. 13a) to mimic other effects of light. The production of CO_2 is reduced in light and so is total acid formation (Fig. 25). This result is at variance with older reports to the contrary [9, 130, 55, 118[7]] and with more recent less definite statements [159]. Increased oxygen uptake in the light has been found in insect eyes [12], but those eyes differ not only in structure but also in their metabolic pattern, lacking lactate production [122]. Whether particularities of that nature account for the enormous rise of O_2-uptake in light reported from sectioned outer layers of vertebrate retina [68] wants investigating. Preferential formazan staining, however, on exposure to light of receptor cells, which has been put to use for marking purposes [67], does not indicate a higher rate of energy production. It rather suggests impaired utilization of substrate in light, leaving hydrogen to be trapped by the tetrazolium salt [71], for the blue staining occurs the same at cut edges and in any demaged retinal area.

Considering the inherent time lag of the techniques, the metabolic changes on exposure to light take immediate effect and are sustained for the whole duration of the exposure (for CO_2-production: Fig. 23b; pH-change: Fig. 25; O_2-uptake: Fig. 26). This by far exceeds the time span light quanta are integrated for the immediate response (summation time of b-wave, of spike bursts) but is just the property of light adaptation which may be read from such objective manifestations as pupil size [25, 3]. No measurements were made in this series of experiments of bleaching of visual pigments. But from the recoveries in the dark after different length of light exposure (compare undershoot in Fig. 26 and 24) bleaching and formation of products has occurred in a cumulative fashion, while responsiveness and metabolic activities were unchanged. Hence a non-correlation between concentration of visual pigments or products and the state of adaptation obtains under the experimental situation. Eventually a state of photochemical equilibrium may be reached through onsetting regeneration, but this was not the case here when the "oxygen debt" appears to account for the amount of pigment to be regenerated afterwards. The concentration of bleaching products, therefore, may be indicative of, but hardly causally related to the state of adaptation.

Reduced metabolic activity does not seem to be the result of a decreased demand in the light. In the experiment of Fig. 26 after a dark period a background light was switched on (electrical reaction at missing stimulus mark) and removed (off-reaction without stimulus mark) one and a half hours later. During light adaptation the stimulus intensity was increased to approximate that of the background, and in the center part of the recording the stimuli were presented for 30 min at increased frequency (one every five seconds instead of every three minutes; irregular time track). It will be seen that the increased driving, rather than to run the metabolic blockade toward the dark level of oxygen uptake, results in electrical responses of reduced size. Only after removal of the steady light is increased oxygen uptake resumed, overshooting while the response amplitudes are being restored.

Attempts were made to inquire into the mode of the adaptive light action by comparing the effects on O_2-uptake of different wavelength, of which Fig. 24 is an example. The energy at the long wavelength far exceeded that of the blue light,

---

[7] The electrometric measurements obviously have been done under more physiological conditions. Unfortunately, no indication has been given on the position of the reference electrode. A few millivolts (vitreous-positive) potential change in light could account for the observed meter readings.

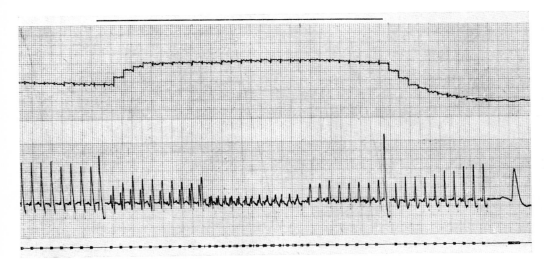

Fig. 26. Oxygen uptake and light responses in dark and light adaptation, and increased rate of stimuli. Techniques and display as in Fig. 24. Stimulus intensity 10^{-2} lx. During period marked horizontally a steady light of 3 lx had been presented and stimulus intensity increased 100 fold. In the center part of the light adaptation period the rate of stimulation had been increased from 1 every 3 min to 1 every 5 sec. Aberrantly large responses to the switching on and off of the background light (no marks)

yet the depression of oxygen uptake did not. Unspecific heat effects, therefore, are excluded as possible cause. The wavelength characterization is not yet good enough to specify an action spectrum for metabolic depression, but is does indicate that more than one pigment is involved, for absorption of rhodopsin is very low at the long wavelength used. This also applies to the light rise of the EOG [81, 11, 64].

In view of the difficulties still existing in present-day interpretations of visual adaptation [75] it is perhaps not out of order to draw attention to the fact that in vertebrate retina the entire neural network with its metabolic outfit is situated in the light path. Several of the respiratory pigments absorb strongly in the visible and near ultraviolet range. Absorption of light by flavins has been shown to result in DPNH oxidation [202, 73]. DPNH itself would be a good target, could light of appropriate wavelength reach it. This is known to be the case in aphakics [211], whose UV-perception has tentatively been attributed to mediation of pyridine nucleotides [38]. Such action of light on respiratory pigments in addition to that on visual pigments would at the same time testify to the significance for retinal metabolism and function of vitamines of the B-group [33]. It would also biochemically pinpoint light-induced metabolic depression on the mitochondrial segments of energy production, which proved of lesser importance in light adaptation (see Fig. 16).

Control of scotopic/photopic behavior by the chemical milieu (Fig. 13a) could accordingly include facilitated accomodation by hydroxyl ions of metabolically produced acids, and the oxidizing property of hydrogen ions. Electrical manifestation, on the other hand, of steady light in a positive shift of the resting potential of the eye, the light rise of the EOG, is to be considered a retinal event resulting from, and signalling reduced metabolic activity. An old notion of FRÖHLICH'S [74], who likened light to harm, appears not entirely unsubstantiated.

In contrast to steady illumination, flashes of light increase retinal metabolism (Fig. 23). The extra energy expenditure accrues from the processing of the response, which under the experimental conditions is believed to occur in a manner not far from normal. It is therefore considered of some interest to exploit the unique possibility offered by the retina of determining how much energy is spent additionally, of precisely specifying the uses the increment energy subserves, and from the kinetics learn about how the stimulated energy production proceeds and correlates with other forms of activity. The results obtained so far cannot be taken for anything but preliminaries.

Assuming additivity a coarse estimate of the contribution of each flash to the steady state increase of the $^{14}CO_2$ counting rate (Fig. 23a) would give a figure of the order of 10 picomoles of CO_2 per milligram dry tissue per flash. In this experiment the associated adapting effect of the light was eliminated, light adaptation served to create a low level of "resting" activity. Attempts were made to determine the extra CO_2-production in dark adaptation by adding a number of counting periods with one time-locked light response in each. But this technique proved not very profitable, measuring O_2-uptake offered better chances.

The procedural description given for the experiments of Fig. 24 and 26 was followed, except that the flow rate of the perfusate was 3 ml/h. With small retinas of about 10 mg wet weight the resulting partial pressure of oxygen would be still above the critical PO_2 (Fig. 18), but wide spacing of the stimuli became necessary. In Fig. 27 stimuli differing in intensity by half a log unit each had been delivered every 21 min. The concentration of oxygen dropped after each response, reaching

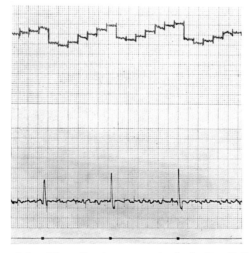

Fig. 27. Extra oxygen uptake of the retina in response to single stimuli. Techniques and display as in Fig. 24. Perfusion of chamber 3 ml/h; stimuli 10^{-10}, 3×10^{-9}, 10^{-9} W/cm² — 503 nm; 1 sec. Intervals between stimuli 21 min ("interrupted recordings"). Calibration: ERG (below) 200 μV/cm; RC 0.3 sec; O_2-uptake (above): 1 mm corresponds to a change of 400 pM/h

minimum not before 6 min, and is slowly restored. The delay of the minimum indicates the slow communication to the fast circulating medium of retinal oxygen consumption, while the slow recovery of the PO_2 is due to the slow replacement of the solution in the chamber. While the ERGs show the expected gradation of their amplitudes, the first of the "oxygen-responses" appears correspondingly bigger. This may indicate that the time integral of the electrical response, which is not faithfully represented here for capacitor coupling, was a better correlate. The last of the responses is of maximum magnitude in dark adaptation. Under these conditions the extra oxygen uptake determined from the areas under the curves averaged at about 3×10^{-10} moles of oxygen per milligram dry retina per flash.

No data exist for the immediate comparison of these results. Neither is, from the reasons given above, a strict comparison warranted between the two sets of experiments. But because the retina is activated from its natural photic stimulus, one may, on a coarse scale, reasonably compare the energy content of the stimulus with the amount of energy expended in producing the response. A gain factor of several millions would result for the example in dark adaptation. From an engineering point of view this should necessitate rigorous and costly measures against contamination of the signal by noise, perhaps one of the uses of the high metabolic rate in dark adaptation.

The amount of oxygen taken up in the payment for one single stimulus is of the order of 0.1 % of that consumed in one hour without stimulation. The extra-oxygen uptake could be calculated to equal the basic rate at a stimulus frequency of one every three seconds. But a hundred percent increase of the metabolic rate is never seen, instead more than twenty seconds must be allowed between stimuli for full responses in dark adaptation, a result which would follow from the measured rates. Thus slowness is imposed upon the retina from its energy metabolism, whereas photochemical processes can hardly be visualized limiting under the conditions. Higher rates of stimuli are processed by the retina, but only at the expense of gain, in light adaptation, and perhaps necessitating renewal of the retinal image on adjacent parts of the retina [54].

The high energy intermediates ATP and phosphocreatine are measurably depleted by one single stimulus [52] and are slowly replenished through oxidations (Fig. 16). The exact stoichiometry still has to be established, but the involvement of oxidative phosphorylation holds out the prospect of recognizing stimulated oxygen utilization from cyclic oxidation of pyridine nucleotides. Thereby limitations from diffusion and from sampling techniques are circumvented and the time resolution of the electrical manifestation of the induced activity may be approached.

The technique of the combined recording of the electrical responses to light flashes and of the changes of the transmitted UV-light resulting from such stimulation is shown schematically in Fig. 2a. Details have been given in the original publication [178]. Special care had been taken to prevent the stimulus light from directly affecting the detector of the analysing light. To this purpose the stimulus light was limited by an interference filter to a narrow band of visible wavelengths, which was rejected by a guard filter in front of the photomultiplier. Alternatively, the beams could be rapidly (1000 c.p.s.) intersected in such a way, that the multiplier surface was covered whenever the stimulus beam was free. Attention had been paid to possible contribution to the optical effects from unspecific light scatter. The contention reached, that it is oxidation of pyridine nucleotides which is reflected by an increase of the transmitted UV-light, would seem further corroborated by the observations above on ATP and oxygen.

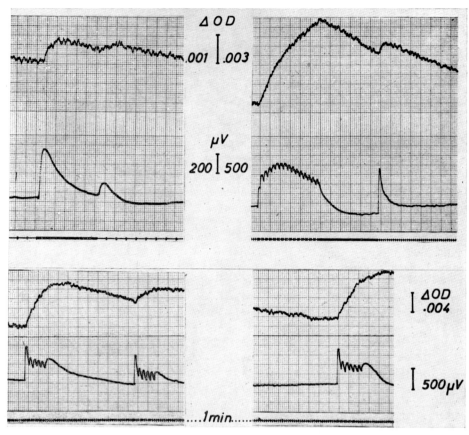

Fig. 28. Respiratory and electrical responses of the retina to light stimulation. In the upper traces the UV-light (350 nm) transmitted through the retina is shown with an increase result- ing in an upward deflection ("oxidation"); beneath are the simultaneously recorded ERG responses (RC 1 sec in *a* and *b*; dc cou ling in *c*). Stimulus marks superimposed on time marks (1 sec). a (*upper left*): single stimulu 10^{-8} W/cm² — 574 nm; b (*bottom*): repeated trains of stimuli; 1 lx; c (*upper right*): 15 stim li of 1 sec, subsequently continuous light, 10^{-8} W/cm² — 574 nm. (Adapted from Ref. [178, 180])

Simultaneously recorded optical and electrical responses to light stimulation are shown in Figs. 28 and 29. In Fig. 28a it was a single flash that resulted in an ERG having a barely discernible *a*-wave, but well developed *b*-wave and off-effect. Oxidation of PN occurs at onset and end of the stimulus. The loss of optical density can, therefore, not be attributed to changes of light absorbing visual pigments, for opposite effects would be expected during and after light exposure, quite apart from the fact that the stimulus intensity was low enough not to cause any measur- able bleaching at all. The proportions of PN oxidations at on and at off of the stimulus are the same as those of the electrical response. But the *b*-wave and off- effect, accentuated through the RC-recording, peak prior to the optical changes.

Given a causal relationship, the *b*-wave and off-effect represent a load, probably an ionic imbalance, which initiates, via cation-stimulated ATPase, its own subsequent oxidative removal. The restorative character of the transient PN oxidation becomes apparent on repeated stimulation (groups of five stimuli, for distinctness, in Fig. 28b): Full responses are not obtained unless the cycle of PN oxidation/reduction is completed, i.e. before accelerated electron flow, or oxygen uptake, has eliminated the "debt" and returned to the normal rate.

In Fig. 28c the retina had been stimulated by a train of one second flashes and then been exposed to a 30 sec steady light. DC-recording was employed for the electrical responses. Only the first flash resulted in a full-sized *b*-wave, but the subsequent smaller flash responses appear superimposed on a slow positive potential change, which decays in steady light. A strong electrical reaction follows the switching off of the light. Different from the *b*-waves, the optical response exhibits summative behavior, approaching saturation only after a prolonged train of stimuli. Steady light does not add to the level of oxidation reached, instead the trace reverses, only to reflect the payment of a renewed load at off of the steady light. Each phase of the experiment is clearly marked both optically and electrically. In fact, by feeding the optical signal through an electrical network having a 10-sec coupling time constant the time course of the optical signal can be made nearly identical with the slow electrical potential change. The slow component of the electrical response is, therefore, associated with restorative processes, whereas

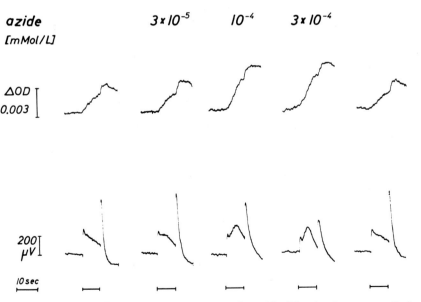

Fig. 29. Respiratory and electrical light responses under azide. The simultaneous optical and electrical recordings (techniques and display as in Fig. 28c) have been selected from the ends of 30-min periods in which NaN₃ had been added to the perfusate at the concentrations indicated. The sequence (from left to right) corresponds to the course of the experiment. Notice: *c*-wave together with larger transmission increment in moderate concentration of azide

the faster components, the b-wave and off-effect, reflect the "debt" incurred in the responding.

More specifically, it is the component PI of GRANIT [82], or the c-wave[8] of the single flash ERG that correlates closest with the redox-process. The background of the UV-analysing beam, however weak, would not be favorable to demonstrate a c-wave; an artifice was therefore employed. In Fig. 29 simultaneous recordings of PN oxidation and ERGs to ten-second stimuli are reproduced from an experiment in which sodium azide had been added to the perfusion medium at the concentration indicated. Each concentration level had been maintained for thirty minutes, and the recordings were selected from the ends of each period after stationary response behavior had been reached. At 10^{-4} M/l NaN$_3$ a c-wave is clearly seen with only minor alterations of the b-wave and off-effect, and the c-wave even persists when the other components of the ERG already suffer in the higher concentration of azide. It is concomitant with the high c-wave that PN oxidation shows correspondingly large deflections. The enhancing effect on the c-wave of azide is well known [151]. Biochemically azide seems to create a bottleneck to the flow of electrons near oxygen [42]. The consequences are correspondingly larger redox changes of the carries of the respiratory chain on disturbances of the steady state. This is the case in stimulated oxidation, which electrically manifests itself in the c-wave, but occurs also on sudden flooding with azide, which causes the electrical "azide response" [150] of the DC-potential (p. 701).

Concluding Remarks

In view of numerous negative results (e.g. [43, 106]) differentiating retinal metabolic activities in dark and light would seem to require i) a study object capable of exhibiting the full set of physiological consequences of an exposure to light and ii) measuring techniques of a certain degree of delicateness hardly applicable unless in isolated preparations. It is with the full realization of the character of a compromise that the perfused retina preparation has been selected for the purpose. Some of the properties reported, e.g. sensitivity, adaptability, would justify the procedures, but the viability of the preparation is of limited duration and normalcy may be questioned on this ground. To facilitate, or to invite just that, technicalities have been reported in some detail and original recordings been presented in most cases. Extending longevity would be an interesting subject of its own.

The results of this study were from a cold-blooded species and want confirming in higher vertebrates. But species differences should perhaps not be overemphasized. It is rather remarkable that the electrical behavior of the eye has many features in common in all vertebrates [114]. Differences that do exist refer in many instances to auxiliary mechanisms of nutritional supply. Some of the inherent problems are apparent from an attempt to use human retina for a comparison of the findings in frog.

Fig. 30 is the protocol of an experiment in which the ERGs and oxygen uptake were recorded under conditions identical to those described, except for a higher concentration of sodium chloride of the perfusion medium (Table 1). The retina

---

[8] The c-wave is frequently held to originate in the pigment epithelium, which was absent in the present studies. For a comment: [180], see also p. 701 regarding the DC-potential.

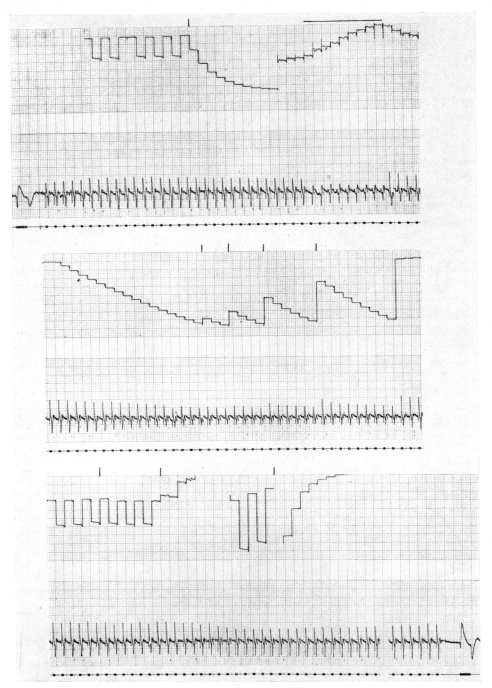

Fig. 30. Oxygen uptake and ERG of an isolated human retina.

Techniques and display: see Figs. 2b, 18, 20, and 24.
Stimuli 10^{-1} lx, 1 sec; temperature $20°$ C.
a, b, and c (from top to bottom) in continuous sequence.

came from a tumor eye, which had been enucleated in dim light[9]. The ERGs were smaller, with the negative *a*-wave predominant, but the light intensity needed to elicit them was about the same as in frog. Hence, the sensitivity was well preserved. The oxygen uptake per unit weight was slightly higher than in frog (a) and so was the critical PO_2 at which the responses started declining (b), but depression of oxygen uptake in steady light was similar as in frog (a; horizontal line). Increasing the glucose concentration (c; 1st vertical bar) did not change oxygen uptake nor ERG. Raising the temperature (c; 2nd vertical bar) increased the oxygen uptake and the *a*-wave, but impaired the *b*-wave—*a*-wave ratio; "uncoupling" occured, and would become worse in higher temperature still. It is therefore doubtful, that oxygen uptake should be very much larger at body temperature, i.e., that the retina should work in man much less economically than in frog.[10] It rather appears that the higher body temperature offers the possibility of faster reactions, but necessitates the more the controlling metabolic restraint. Perhaps it is the lesser evil to work with a species that puts less stringent demands on the preservation of this restraint and preserve it than to dispense with it in a more demanding species. Plasma has been found to normalize human ERGs [184], and fractionation of an effective component [132] holds out a rational supplementation of the perfusate to extend the studies on warm-blooded species.

The view that the well-ordered sequence of chemical reactions termed metabolism is not an alternative but integral part of organ function is not new. But through the manifestation in slow electrical activity the retina offers the unique possibility of immediately recognizing energy dissipating and restoring processes. The classical component analysis (nomenclature after Granit [82]) would cover this use: with PIII and PII reflecting receptor and higher order neuron activity, respectively, and PI, a modulation of the retinal asymmetry potential, oxidative payment of the debts previously incurred.[11] In fact, conditions may be arranged in

---

[9] The operation had been performed by Prof. J. G. H. Schmidt, Universitäts-Augenklinik Köln.

[10] In frog reducing the temperature from 20 °C to 15 °C resulted in a reduction of the oxygen uptake by 63%, but of the steady state ERGs by only 23%. The recovery from a previous responding to a light stimulus, which was 63% complete after 30 seconds at the higher temperature, took twice as long at 15 °C. The time course of b-wave recovery of a human retina nearly coincided with that of frog at room temperature, but was much faster prior to the enucleation (16, 96).

[11] The different temperature dependencies of the b-waves and of the oxygen uptake, respectively (see footnote 10), which may be interpreted in terms of activation energies, indicate that the immediate generation of P II is powered by stored energy, the release of which involves only physical processes of the diffusion type, whereas the restoration of responsiveness requires the completion of the P I-process (comp. Fig. 28), i.e. mitochondrial activity.

---

a O_2-uptake from flow-stop: 0.325 µM/h (dry weight of retina approx. 1 mg);
   vertical mark: flow reduced to 6 ml/h, subsequently PO_2-reading at 4 × gain;
   horizontal mark: steady light with "dark flashes".

b Complete flow stop (PO_2-reading at 1/20 gain).
   Marks: Admission of 0.1, 0.2, 0.4, and 0.8 ml of perfusate; at end fast flow (3 ml/min).

c Flow-stop (as in a):
   1st mark: glucose concentration from 5 to 15 mM/l;
   2nd mark: temperature from 20° C to 25° C (PO_2-zero adjustment);
   3rd mark: O_2 saturation of perfusate from 20 to 100% (PO_2-gain reduced)

such a way that specific functions — for example regeneration of visual pigments — can be recognized electrically — from increasing delay of scotopic off-effects (Fig. 4) — or from metabolic measurements — from the extra oxygen uptake following light exposure (Fig. 26).

It is therefore not surprising to find retinal energy metabolism under strict control with remarkably low over-all rates, compared with reported claims that is. Striking, however, is the high rate of lactate production, which accounts for one third of the glucose-C taken up. This suggests the operation of alternative metabolic pathways subserving special functions. Tentatively the view has been advanced [181] that the pentose-phosphate shunt of carbohydrate break-down may be of particular significance in light adaptation. But this does by no means exhaust the possibilities of explaining qualitative differences in dark and light. Microsomal activities [97, 176] have not yet found much attention. And from the utilization of "end products", of lactate (p. 668, [77]) and of CO_2 [48], and from the involvement in the paths of carbon of pools of different sizes it is clear that a qualitative penetration of retinal metabolic activities is far from complete.

Disregarding chronical and toxic actions, three separate effects of light on retinal metabolism have been demonstrated:

i) a spurt of oxygen uptake caused by a stimulating light flash,

ii) depression of metabolism during steady light and

iii) extra oxygen uptake from regeneration after bleaching.

The magnitude of stimulated oxygen uptake shows that the processing of a single light-response can be a fairly expensive and time-consuming affair, which accounts for certain functional limitations. Depression of oxygen uptake is one of the many facetts of light adaptation, which are so closely associated as to suggest a single-parameter control, but which are not at all easily explained on a unitary basis (Fry [75]). Bringing into the play the respiratory pigments may not add too much to the embarrassment. Extra energy expenditure for visual pigment regeneration seems plausible [210, 105], but the occurrence of regeneration in the preparation raises serious doubts as to the function of the pigment epithelium, as do the presence of the c-wave, of the DC-potential and of the azide response. This is not to deny the significance for retinal function of the pigment epithelium but rather to suggest a reevaluation in the light of more recent results, especially on visual pigment renewal [209, 85] — and the removal of remnants.

Postscript

"Metabolism of the Retina in Dark and Light" in this article is understood to ask

i) are there changes detectable in quantity or pattern of retinal metabolism as a consequence of exposure to light; and if so, can these changes be used to specify the energetic expense accruing from a given kind of "work" performed ? And

ii) given the above correlation, might the respective rate/pattern of metabolism determine the functional properties of the retina, i.e. act as a control in the processing of the visual information ?

It was assumed initially that a beginning toward this aim could be made simply by sorting data from literature into observations made in light and dark, respectively. However this proved impractical, because most studies had been undertaken with questions other than the above in mind and valuable information had to be discarded that way, as, on the other hand, necessary data on e.g. light intensities used, functional state of the retina etc., were missing, making it impossible to appraise for the purpose the results adequately. This attempt was, therefore, dismissed, the more readily so, as an updated review of retinal metabolism along more orthodox lines was due and has appeared since this manuscript has been completed [83a]. Indeed, general biochemistry, apart from photochemistry, has now come to be recognized a legitimate subject of retinal activity to be covered by a chapter in the second edition of the now five-volume treatise of The Eye (DAVSON [50a]). Instead, it was preferred to offer an experimentally uniform approach, the perfused retina technique, which the author has been using for some time, with — it was hoped — sufficient reference to work from other laboratories. Unapologizable omissions will have occurred, as there are also new groups to join the endeavour [103a].

The results reported do little more than hold out the prospect of eventually answering the questions raised, but do seem to indicate some avenues to persue. For one, it is the remarkable demand for oxidative metabolism in the dark, or put the reverse, the fact that light exposure saves rather than costs energy. Preliminary results obtained since show that this is — to a so far undetermined extent — a property of the receptor population. The evidence comes from employing aspartate in the perfusion medium, which with unusual promptness eliminates the higher-order neuron contribution, i.e. the b-wave of the ERG [184a, 89a] together with the stimulated extra-oxygen uptake (Fig. 27), while leaving most of the receptor potential together with the light-depression of oxygen uptake (Fig. 24). Structurally the receptors remain intact [92a]. The effect is very reminiscent of that of ammonia [126] of which aspartate, as does glutamate, constituates a store, releasing NH_3 as the remainder of the molecule is metabolized, particularly if in preference to glucose (p. 693). A reduction of the energy consumption in the light of receptors, moreover, would be postulated, if their reaction was not, as in most other receptor types, a depolarizing but a hyperpolarizing response. This is the case in vertebrate rods (Fig. 6; [198b]) and cones [198a, 190a, 24a, 207a, 18a]. Measurements of currents [84a], permeabilities [184b, 198c] and ion fluxes [200a] support the contention.

Extra-energy expenditure attributable to especial activity has been looked for very hard in brain — with little reward, unless in unphysiological conditions. This may be due to inaccessibility to the measurements of sufficiently homogeneous-reacting structures, but some comfort has also been derived from information theory, which states that for all practical purposes no reasonable relationship exists between the amounts of energy and information processed: computers employing solid state devices perform equally well as their first-generation predecessors — at much reduced energy expenditure. If, now, one envisages technical devices no longer requiring for reliability volt-ranges, but operating in the millivolt range, one would probably approach a level of energy, relevant energy, which stands in reasonable proportion to "information", or neg-entropy for that matter. The "biological device", then, would appear to be close to the ideal one, most economical at the level of single cells, but entering the book-keeping as these co-act. It seems the retina a promising object to study in-unison activity, to stimulate the production of measurable amounts of relevant energy — and measure them.

Extra-energy expenditure, amount and kinetics, resulting from the processing of a brief light stimulus was sought under conditions where the make-up of the response is still simple and can be extracted from gross electrical recording, i.e. inhibitory interactions can be assumed to be of small weight. Accordingly, the additional energy is expended in the processing of the visual information in the neural network of the retina. It seems intriguing, but will require further work, to reconcile the metabolic findings with the much more detailed observations from recent microelectrode probings in diverse species [110a, 142a]. Other uses of extra-energy are more easily separated, because they occur only after the light has exerted its bleaching action. Presumably they serve to restore visual pigments. But little definite is known where exactly the energy comes in in the sequence of dark reactions following the absorption of light quanta. Moreover, assessing photochemical processes from the energy participating would imply, as does the validity of an $i \times t$—law for adapting lights over a considerable span of time, a linear additivity with halted thermal regeneration while the light shines, i.e. to say a non-correlation between pigment concentration and light reaction. The findings reported, thus, raise rather than settle problems.

This is equally true for the second of the questions asked above: We seem to have two eyes in one, one for the day and an entirely different one for the night [80a]. The functional changes involve all the heterogeneous structures of the retina, and that with a remarkably tight coupling (p. 675): in light adaptation spatial resolution (visual acuity) does not improve without improvement of temporal resolution (flicker fusion), and both proceed at the expense of gain (sensitivity). Given metabolic consequences of visual function it is perhaps not too far fetched to inquire into a contributing common control on the functional properties, "fed back" by the chemical "milieu". So far but the pH-phenomenon (Fig. 13a), with an acid environment mimicking light adaptation, an alkaline one dark adaptation, can be offered with only speculations as to the mechanisms involved. Perhaps the principle was better understood as a means for range-shifting, if it was technically utilized, but so far in technical multilayered structures comparable parameters such as temperature etc. seem to have been controlled only to keep the operational properties constant, not to modify them.

Acknowledgement

The author wishes to acknowledge the help in the early stages of this work of his colleagues at the University of Leipzig; the hospitality and support of Drs. E. F. MacNichol, Jr., and M. G. Larrabee of the Johns Hopkins University, Baltimore, and B. Chance of the Johnson Foundation, Philadelphia. Mrs. Sickel assisted throughout this work.

Financial support was obtained during the final period from the Deutsche Forschungsgemeinschaft.

References

1. Adrian, E. D., Matthews, R.: The action of light on the eye. Part I: The discharge of impulses in the optic nerve and its relation to the electric change in the retina. J. Physiol. (Lond.) **63**, 378—414 (1927).

2. ALBERTY, R. A.: Effects of pH and metal ion concentration on the equilibrium hydrolysis of adenosine triphosphate to adenosine diphosphate. J. biol. Chem. **243**, 1337—1343 (1968).

3. ALPERN, M., CAMPBELL, F. W.: The spectral sensitivity of the consensual light reflex. J. Physiol. (Lond.) **164**, 478—507 (1962).

4. AMES, A. III.: Studies of morphology, chemistry and function in isolated retina. In: GRAYMORE, C. N. (Ed.): Biochemistry of the Retina. London-New York: Academic Press 1965.

5. — GURIAN, B. S.: Measurement of function in an in vitro preparation of mammalian central nervous tissue. J. Neurophysiol. **23**, 676—691 (1960).

6. — — Electrical recordings from isolated mammalian retina mounted as a membrane. Arch. Ophthal. **70**, 837—841 (1963).

7. ANDERSON, B., SALTZMAN, H. A.: Retinal oxygen utilization measured by hyperbaric blackout. Arch. Ophthal. **72**, 792—796 (1964).

8. ANFINSEN, C. B.: The distribution of diphosphopyridine nucleotide in the bovine retina. J. biol. Chem. **152**, 279—284 (1944).

9. ANGELUCCI, A.: Physiologie générale de l'oeil. L'Encyclopédie Francaise d'Ophthalmologie **2**, 106—241 (1905) (cit. by KRAUSE).

10. ARDEN, G. B., GRANIT, R., PONTE, F.: Phase of suppression following each retinal b-wave in flicker. J. Neurophysiol. **23**, 305—314 (1960).

11. — KELSEY, J. H.: Changes produced by light in the standing potential of the human eye. J. Physiol. (Lond.) **161**, 189—204 (1962); Some observations in the relationship between the standing potential of the human eye and the bleaching and regeneration of visual purple. J. Physiol. (Lond.) **161**, 205—226 (1962).

12. AUTRUM, H. J., TSCHARNTKE, H.: Der Sauerstoffverbrauch der Insektenretina im Licht und im Dunkeln. Z. vergl. Physiol. **45**, 695—710 (1962).

13. BAUEREISEN, E., LIPPMANN, H.-G., SCHUBERT, E., SICKEL, W.: Bioelektrische Aktivität und Sauerstoffverbrauch isolierter Potentialbildner bei Sauerstoffdrucken zwischen 0 und 10 Atm. Pflügers Arch. ges. Physiol. **267**, 636—648 (1958).

14. BAUMANN, CH.: Die absolute Schwelle der isolierten Froschnetzhaut. Pflügers Arch. ges. ges. Physiol. **280**, 81—88 (1964).

15. — Sehpurpurbleichung und Stäbchenfunktion in der isolierten Netzhaut. Pflügers Arch. ges. Physiol. **298**, 44—81 (1967).

16. — DETTMAR, P., HANITZSCH, R., SICKEL, W.: Untersuchungen an der umströmten Froschnetzhaut zur Analyse des ERG. Acta Ophthalm. (Kph.) Suppl. **70**, 156—163 (1962).

17. — HEISS, W. D.: Elektroretinogramm und retinale Impulsaktivität in Hypothermie. Experientia (Basel) **22**, 184—185 (1966).

18. BAUMGARDT, E.: Absolute Schwelle und Differentialschwellen. In: JUNG, R., KORNHUBER, H. (Eds.): The Visual System: Neurophysiology and Psychophysics, S. 400—409. Berlin-Göttingen-Heidelberg: Springer 1961.

18a. BAYLOR, D. A., FUORTES, M. G. F.: Electrical responses of single cones in the retina of the turtle. J. Physiol. (Lond.) **207**, 77—92 (1970).

19. BENOLKEN, R. M.: Effects of light and dark adaptation processes on the generator potential of the Limulus eye. Vision Res. **2**, 103—124 (1962).

20. BEUCHELT, H.: Die Abhängigkeit der photoelektrischen Reaktion des Froschauges von den ableitenden Medien. Z. Biol. **73**, 205—230 (1921).

21. BÖCK, J., BORNSCHEIN, H., HOMMER, K.: Die Erholungslatenz der Helligkeitsempfindung. und des Elektroretinogramms nach retinaler Ischämie. Albrecht v. Graefes Arch. Ophthal. **167**, 276—283 (1964).

22. BONAVITA, V., GUARNERI, R., PONTE, F.: Neurochemical studies on the inherited retinal degeneration of the rat. II. NAD- and NADP-linked enzymes in the developing retina. Vision Res. **5**, 113—121 (1965).

23. BORCHARD, U.: Untersuchungen über Belichtungspotential und Ionentransport im Retina-Membransystem. Diss. Universität Köln 1969.

24. BORNSCHEIN, H., LÜTZOW, A. V.: Electroretinographic threshold in the isolated rabbit retina. Nature (Lond.) **215**, 1394—1395 (1967).

24a. Bortoff, A., Norton, A. L.: An electrical model of the vertebrate photoreceptor cell. Vision Res. **7**, 253—263 (1967).

25. Bouma, H.: Size of the static pupil as a function of wavelength and luminosity of the light incident on the human eye. Nature (Lond.) **193**, 690—691 (1962).

26. Brindley, G.S.: The effect on the frog ERG of varying the amount of retina illuminated. J. Physiol. (Lond.) **134**, 353—359 (1956).

27. — Physiology of the Retina and Visual Pathways. London: Arnold 1960.

28. Brink, F.: Role of calcium ions in neural processes. Pharmacol. Rev. **6**, 243—298 (1954).

29. Brown, K.T.: The electroretinogram: Its components and their origins. Vision Res. **8**, 633—677 (1968).

30. — Murakami, M.: Delayed decay of the late receptor potential of monkey cones as a function of stimulus intensity. Vision Res. **7**, 179—189 (1967).

31. — Wiesel, T. N.: Intraretinal recording in the unopened cat eye. Amer. J. Ophthal. **46**, 91—96 (1958).

32. Campos, R.: Ricerche sul ricambio della retina. Ann. Ottal. Clin. Ocul. **64**, 456, 538, 577, 594—602 (1936).

33. Capolongo, G.: La lattoflavino nel metabolismo della retina. Ann. Ottal. Clin. Ocul. **75**, 271—278 (1949).

34. Caspers, H.: Relations of steady potential shifts in the cortex to the wakefulness — sleep spectrum. In: Brazier, M. (Ed.): Brain Function: Cortical Excitability and Steady Potentials, pp. 177—213. Berkeley: Univ. of Calif. Press 1963.

35. Catanzaro, R., Chain, E. B., Pocchiari, F., Reading, H.W.: The metabolism of glucose and pyruvate in rat retina. Proc. roy. Soc. B **156**, 139—143 (1962).

36. Chance, B.: Techniques for the assay of respiratory enzymes. In: Colowick, S.P., Kaplan, N.O. (Eds.): Methods of Enzymology, pp. 273—329. New York: Academic Press 1957.

37. — Quantitative Aspects of the Control of Oxygen Utilization. CIBA Found. Symp. on Cell Metabolism 1959, pp. 91—121.

38. — Fluorescence emission of the mitochondrial DPNH as a factor in the ultraviolet sensitivity of visual receptors. Proc. nat. Acad. Sci. (Wash.) **51**, 359—361 (1964).

39. — Biochemical studies of transitions from rest to activity. In: Sleep and Altered States of Conciousness. Association for Research in Nervous and Mental Disease, Vol. 45, pp. 48—63. Baltimore: Williams and Wilkins Comp. 1967.

40. — Hollunger, G.: Energy-linked reduction of mitochondrial pyridine nucleotide. Nature (Lond.) **185**, 666—672 (1960).

41. — Packer, L.: Light scattering and absorption effects caused by addition of adenosine diphosphate to rat-heart-muscle-sarcosoma. Biochem. J. **68**, 295—297 (1958).

42. — Williams, G.R.: The respiratory chain and oxidative phosphorylation. In: Nord, F. F. (Ed.): Advances in Enzymology, Vol. 17, pp. 65—134. New York: Interscience Publ. 1956.

43. Chase, A.M., Smith, E.: Regeneration of visual purple in solution. J. gen. Physiol. **23**, 21—39 (1939).

44. Cohen, L. H., Noell, W. K.: Glucose catabolism of rabbit retina before and after development of visual function. J. Neurochem. **5**, 253—276 (1960).

45. — — Relationship between visual function and metabolism. In: Graymore, C. N. (Ed.): Biochemistry of the Eye, pp. 36—49. London-New York: Academic Press 1965.

46. Cone, R. A.: The rat electroretinogram. II. Bloch's law and the latency mechanism of the b-wave. J. gen. Physiol. **47**, 1107—1116 (1964).

47. — Brown, P.: Spontaneous regeneration of rhodopsin in the isolated rat retina. Nature (Lond.) **221**, 818—822 (1969).

48. Crane, R.K., Ball, E.G.: Relation of $^{14}CO_2$-fixation to carbohydrate metabolism of the retina. J. biol. Chem. **189**, 269—276 (1951).

49. Crescitelli, F., Sickel, E.: Delayed off-responses recorded from the isolated frog retina. Vision Res. **8**, 801—816 (1968).

50. Danis, P.: Contribution à l'etude électrophysiologique de la rétine. Bruxelles: Impr. Med. Sci. 1959.

50a. GRAYMORE, C. N.: General aspects of the metabolism of the retina. In: DAVSON, H. (Ed.): The Eye, 2. Ed., Vol. 1. London-New York: Academic Press 1969.

51. DEMIRCHOGLIAN, G.G.: Physiology and Pathology of the Retina. Moscow: Medical Press 1964. (Russ.).

52. DETTMAR, P.: Energiereiche Phosphorsäureverbindungen und ihr Verhalten unter Lichtreizen in der isolierten, umströmten Froschnetzhaut. Biochem. Symp., S. 23—26. Dresden-Leipzig: Th. Steinkopff 1963.

53. DICKENS, F., GREVILLE, G. D.: Metabolism of normal and tumor tissue. Biochem. J. **27**, 1134—1140 (1933).

54. DITCHBURN, R. W., GINSBORG, B. C.: Vision with a stabilized image. Nature (Lond.) **170**, 36—37 (1952).

55. DITTLER, R.: Über die chemische Reaktion der isolierten Froschnetzhaut. Pflügers Arch. ges. Physiol. **120**, 44—50 (1907).

56. DODT, E.: Beiträge zur Elektrophysiologie des Auges. II. Über Hemmungsvorgänge in der menschlichen Retina. Albrecht v. Graefes Arch. Ophthal. **153**, 152—162 (1952).

57. DOLIVO, M., ROUILLER, C. H.: Changes in ultrastructure and synaptic transmission in the sympathetic ganglion during various metabolic conditions. Progr. Brain Res. **31**, 111—123 (1969).

58. DONNER, K. O., REUTER, T.: Visual adaptation of the rhodopsin rods in the frog's retina. J. Physiol. (Lond.) **199**, 59—87 (1968).

59. DOWLING, J. E.: Chemistry of visual adaptation in the rat. Nature (Lond.) **188**, 114—118 (1960).

60. — Discrete potentials in the dark adapted eye of the crab Limulus. Nature (Lond.) **217**, 28—31 (1968).

61. — SIDMAN, R. L.: Inherited retinal dystrophy in the rat. J. Cell Biol. **14**, 73—109 (1962).

62. EBATA, M.: cit. by KISHIDA, K., NAKA, K. I.: Amino acids and the spikes from the retinal ganglion cells. Science **156**, 648—650 (1967).

63. ELENIUS, V.: Decay of suppression of retinal function after short flashes of light. Docum. ophthalm. (Den Haag) **18**, 529—536 (1964).

64. — KARO, T.: Cone activity in the light-induced response of the human electro-oculogram. Pflügers Arch. ges. Physiol. **291**, 241—248 (1966).

65. ELLIOTT, K. A. C., WOLFE, L. S.: Brain tissue respiration and glycolysis. In: ELLIOTT, K. A. C., PAGE, I. H., QUASTEL, J. H. (Eds.): Neurochemistry, 2 nd. Ed., pp. 177—211. Springfield, Ill.: Charles C. Thomas Publ. 1962.

66. ENGELHARDT, W. A.: Die Beziehungen zwischen Atmung und Pyrophosphatumsatz in Vogelerythrocyten. Biochem. Z. **251**, 343—368 (1932).

67. ENOCH, J. M.: Validation of an indicator of mammalian receptor response: Recovery in the dark following exposure to a luminous stimulus. Invest. Ophthal. **6**, 647—656 (1967).

68. EPSTEIN, M. H., O'CONNOR, J. S.: Enzyme changes in isolated retinal layers in light and darkness. J. Neurochem. **13**, 907—911 (1966).

69. FENN, W. O., GALAMBOS, R., OTIS, A. B., RAHN, H.: Corneoretinal potential in anoxia and acapnia. J. appl. Physiol. **1**, 710—715 (1949).

70. FISCHER, P.: Ernährung und Stoffwechsel der Gewebe des Auges. Ergebn. Physiol. **31**, 507—591 (1931).

71. FOWLKS, W. L., PETERSON, D. E.: Substrate inhibition of tetrazolium salt reduction in dark adapted retinae. Proc. Soc. exp. Biol. (N. Y.) **118**, 491—494 (1965).

72. FRANK, R. N., GOLDSMITH, T. H.: Effects of cardiac glycosides on electrical activity in the isolated retina of the frog. J. gen. Physiol. **50**, 1585—1606 (1967).

73. FRISELL, W. R., MACKENZIE, C. G.: The photochemical oxidation of DPNH with riboflavin phosphate. Proc. nat. Acad. Sci. (Wash.) **45**, 1568—1572 (1959).

74. FRÖHLICH, F. W.: Beiträge zur allgemeinen Physiologie der Sinnesorgane. Z. Sinnesphysiol. **48**, 28—165 (1913).

75. FRY, G. A.: Mechanisms subserving bright and dark adaptation. Amer. J. Optom. **46**, 319—338 (1969).

76. FUORTES, M. G. F.: Electrical activity of the cells in the eye of Limulus. Amer. J. Ophthal. **46**, 210—223 (1958).

77. FUTTERMAN, S., KINOSHITA, J. H.: Metabolism of the retina. I. Respiration of cattle retina. J. biol. Chem. **234**, 723—726 (1959).

78. GOTO, M., TOIDA, N.: On the splitting of off-response in electroretinogram. Jap. J. Physiol. **4**, 123—130 (1954).

79. GOURAS, P.: Spreading depression of activity in amphibian retina. Amer. J. Physiol. **195**, 28—32 (1958).

80. — Rod and cone interaction in dark-adapted monkey ganglion cells. J. Physiol. (Lond.) **184**, 499—510 (1966).

80a. — Electroretinography; some basic problems. Invest. Ophthal. **9**, 557—569 (1970).

81. — CARR, R. E.: Cone activity in the light induced dc response of monkey retina. Invest. Ophthal. **4**, 318—321 (1965).

82. GRANIT, R.: Sensory Mechanisms of the Retina. London-New York-Toronto: Oxford Univ. Press 1947.

83. GRAYMORE, C. N. (Ed.): Biochemistry of the Retina. (Suppl. to Exp. Eye Res.) London-New York: Academic Press 1965.

83a. — Biochemistry of the Retina. In: GRAYMORE, C. N. (Ed.): Biochemistry of the Eye. London-New York: Academic Press 1970.

84. — TOWLSON, M. J.: The metabolism of the retina of the normal and alloxan diabetic rat. The levels of oxidised and reduced pyridine nucleotides and the oxidation of the carbon-1 and carbon-6 of glucose. Vision Res. **5**, 379—389 (1965).

84a. HAGINS, W. A., PENN, R. D., YOSHIKAMI, S.: Dark current and photocurrent in retinal rods. Biophys. J. **10**, 380—412 (1970).

85. HALL, M. O., BOK, D., BACHARACH:, A. D. E.: Visual pigment renewal in the mature frog retina. Science **161**, 787—789 (1968).

86. HAMASAKI, D.: The effect of sodium ion concentration on the electroretinogram of the isolated retina of the frog. J. Physiol. (Lond.) **167**, 156—168 (1963).

87. HANAWA, I.: cit. by NOELL, W. K.: Cellular physiology of the retina. J. Opt. Soc. Amer. **53**, 36—48 (1963).

88. — KUGE, K., MATSUMURA, K.: Effects of some common ions on the transretinal dc potential and the electroretinogram of the isolated frog retina. Jap. J. Physiol. **17**, 1—20 (1967).

89. — — — Mechanisms of the slow depressive potential production in the isolated frog retina. Jap. J. Physiol. **18**, 59—70 (1968).

89a. — TATEISHI, T.: The effect of aspartate on the electroretinogram of the vertebrate retina. Experientia (Basel) **26**, 1311—1312 (1970).

90. HANITZSCH, R., BORNSCHEIN, H.: Spezielle Überlebensbedingungen für isolierte Netzhäute verschiedener Warmblüter. Experientia (Basel) **21**, 484—485 (1964).

91. — BYZOV, A. L.: Methodische Voraussetzungen zur Ableitung mit Mikroelektroden an der isolierten menschlichen Netzhaut. Vision Res. **3**, 207—212 (1963).

92. — SICKEL, W.: Restitutionsprozesse der b-Welle des Elektroretinogramms nach Reizbelastung. Pflügers Arch. ges. Physiol. **274**, 34 (1961).

92a. HANSSON, H. A.: Ultrastructural studies in the long term effects of sodium glutamate on the rat retina. Virchows Arch. Abt. B. Zellpath. **6**, 1—11 (1970).

93. HARTLINE, H. K.: The electrical response to illumination of the eye in intact animals, including the human subject; and in decerebrate preparations. Amer. J. Physiol. **73**, 600—612 (1925).

94. — Visual receptors and retinal interaction. Science **164**, 270—278 (1969).

95. — MACNICHOL, E. F., WAGNER, H. G.: The peripheral origin of nervous activity in the visual system. Cold Spr. Harb. Symp. Quant. Biol. **17**, 125—140 (1952).

96. HASCHKE, W., SICKEL, W.: Das Elektroretinogramm des Menschen bei Ausfall der Ganglienzellen und partieller Schädigung der Bipolaren. Acta Ophthal. (Kph.) Suppl. **70**, 164—167 (1962).

97. HEATH, H., FIDDICK, R.: The ascorbic acid-dependent oxidation of reduced nicotinamide-adenine dinucleotide by ciliary and retinal microsomes. Biochem. J. **94**, 114—119 (1965).

98. — RUTTER, A. C., BECK, T. C.: Reduced and oxidized pyridine nucleotides in the retinae from alloxandiabetic rats. Vision Res. **2**, 333—342 (1962).

99. HECK, J., PAPST, W.: Über den Ursprung des corneo-retinalen Ruhepotentials. Bibl. Ophthal. **48**, 96—107 (1956).

100. HIMWICH, H. E.: Brain Metabolism and Cerebral Disorders. Baltimore: Williams and Wilkins 1951.

101. HIRSCH, H., SCHNEIDER, M.: Zur Durchblutung und O_2-Versorgung der Netzhaut. Proc. XX. Int. Congr. Ophthal. 1966, S. 123—133.

102. — — Durchblutung und Sauerstoffaufnahme des Gehirns. In: OLIVECRONA, H., TÖNNIS, W (Hrsg.): Handb. Neurochirurgie, S. 434—552. Berlin-Heidelberg-New York: Springer 1968.

103. HOMMER, K.: Die Wirkung des Chinins, Chlorochins, Jodacetats und Chlordiazepoxids auf das ERG der isolierten Kaninchennetzhaut. Albrecht. Graefes Arch. Ophthal. **175**, 111—120 (1968).

103a. HONDA, Y.: Studies on electrical activities of the mammalian retina and optic nerve in vitro. Acta Soc. Ophthal. Jap. **73**, 1865—1899 (1969).

104. HOROWICZ, P., LARRABEE, M. G.: Oxidation of glucose in a mammalian sympathetic ganglion at rest and in activity. J. Neurochem. **9**, 1—21 (1962).

105. HOSOYA, Y., OKITA, T., AKUNE, T.: Über die lichtempfindliche Substanz in der Zapfennetzhaut. Tohoku J. exp. Med. **34**, 532—541 (1938).

106. HWANG, T.: Respiration of retina tissue. Jap. J. Physiol. **1**, 169—172 (1951).

107. JÖBSIS, F. F.: Basic processes in cellular respiration. Handb. of Physiol., Sect. 3 Respir. Vol. 1, pp. 63—124. Washington D.C.: Amer. Physiol. Soc. 1964.

108. JONGBLOED, J., NOYONS, A. K.: Sauerstoffverbrauch und Kohlendioxydproduktion der Froschretina bei Dunkelheit und bei Licht. Z. Biol. **97**, 399—408 (1936).

109. JUNG, R.: Neuronale Grundlagen des Hell-Dunkelsehens und der Farbwahrnehmung. Ber. Ophthalm. Ges., 66. Zus., Heidelberg 1964, S. 69—111.

110. KANEKO, A., HASHIMOTO, H.: Electrophysiological study of single neurons in the inner nuclear layer of carp retina. Vision Res. **9**, 37—56 (1969).

110a. — Physiological and morphological identification of horizontal, bipolar and amacrine cells in goldfish retina. J. Physiol. (Lond.) **207**, 623—633 (1970).

111. KAPLAN, N. O., SWARTZ, M. N., FRECH, M. E., CIOTTI, M. M.: Phosphorylative and nonphosphorylative pathways of electron transfer in rat liver mitochondria. Proc. nat. Acad. Sci. (Wash.) **42**, 481—487 (1956).

112. KARLSON, P.: Kurzes Lehrbuch der Biochemie, 6. Aufl., S. 175. Stuttgart: Georg Thieme 1967.

113. KEEN, H., CHLOUVERAKIS, C.: Metabolism of isolated rat retina. The role of non-esterified fatty acids. Biochem. J. **94**, 488—493 (1965).

114. KIKAWADA, N.: Variations in the corneo-retinal standing potential of the vertebrate eye during light and dark adaptation. Jap. J. Physiol. **18**, 687—702 (1968).

115. KINOSHITA, J. H.: The stimulation of phosphogluconate pathway by pyruvate in bovine corneal epithelium. J. biol. Chem. **228**, 247—253 (1957).

116. — FUTTERMAN, S.: Lactic acid dehydrogenase activity with TPNH in ocular tissue. Fed. Proc. **18**, 260 (1950).

117. KLINGENBERG, M., SLENCZKA, W., RITT, E.: Vergleichende Biochemie der Pyridinnucleotid-Systeme in Mitochondrien verschiedener Organe. Biochem. Z. **332**, 47—66 (1959).

118. KOBAKOWA, J. M.: pH-changes of the frog retina. Fiz. Zh. SSSR **32**, 385—394 (1946) (Russ.; Engl. summary).

119. KRAUSE, A. C.: Biochemistry of the Eye. Baltimore: Johns Hopkins Press 1934.

120. KREBS, H. A., EGGLESTON, L. V., TERNER, C.: In vitro measurements of the turnover rate of potassium in brain and retina. Biochem. J. **48**, 530—537 (1951).

121. KÜHNE, W., STEINER, J.: Über das electromotorische Verhalten der Netzhaut. Unters. Physiol. Inst. Heidelberg **3**, 327—377 (1880).

122. LANGER, H.: Die Wirkung von Licht auf den chemischen Grundaufbau des Auges von Calliphora Erythrocephala Meig. J. Insect. Physiol. **4**, 283—303 (1960).

123. LARRABEE, M. G.: Metabolic effects of nerve impulses and nerve-growth factor in sympathetic ganglia. Progr. Brain Res. **31**, 95—110 (1969).

124. LASER, H.: Tissue metabolism under the influence of low oxygen tension. Biochem. J. **31**, 1671—1676 (1937).

125. Laser, H.: Tissue metabolism under the influence of carbon monoxide. Biochem. J. **31**, 1677 — 1682 (1937).
126. Laufer, M., Svaetichin, G., Mitarai, G., Fatehchand, R., Vallecalle, E., Villegas, J.: The effect of temperature, carbon dioxide, and ammonia in the neuron-glia unit. In: Jung, R., Kornhuber, H. (Eds.): The Visual System: Neurophysiology and Psychophysics, pp. 457—463. Berlin-Göttingen-Heidelberg: Springer 1961.
127. Leão, A. A. P.: Spreading depression of activity in cerebral cortex. J. Neurophysiol. **7**, 359—390 (1944).
128. Lehninger, A. L.: Water uptake and extrusion by mitochondria in relation to oxidative phosphorylation. Physiol. Rev. **42**, 467—517 (1962).
129. Le-Van Nham: La consommation d'oxygène de la rétine des boeuf a la lumière du jour et dans l'obscurité. Bull. Soc. Zool. France **80**, 70—74 (1955).
130. Lodato, G.: Imutamenti della retina sotto l'influenza della luce, dei colori e di altri agenti fisici e chimichi, con speciale riguardo alla reazione chimica. Contribuzione alla fisiologia della retina. Arch. Ottal. **7**, 335 (1900). (Cit. by Krause.)
131. Lowry, O. H., Roberts, N. R., Lewis, C. H.: The quantitative histochemistry of the retina. J. biol. Chem. **220**, 879—892 (1956), see also: J. biol. Chem. **236**, 2813—2820 (1961).
132. v. Lützow, A.: Die Bedeutung der Plasmafaktoren für die isolierte umströmte Kaninchennetzhaut. Experientia (Basel) **22**, 215—216 (1966).
133. Lynen, F.: Phosphatkreislauf und Pasteur-Effekt. 8. Coll. Ges. Physiol. Chem. in Mosbach 1957, S. 155—184. Berlin-Göttingen-Heidelberg: Springer 1958.
134. MacNichol, E. F.: Subthreshold excitatory processes in the eye of Limulus. Exp. Cell Res. Suppl. **5**, 411—425 (1958).
135. — Cit. by Chase, A. M. In: Glick, D. (Ed.): Methods of Biochem. Analysis, Vol. 8, p. 61. New York: Interscience Publ. 1960.
136. — Wagner, H. G.: A high impedance input circuit suitable for electrophysiological recording from micropipette electrodes. Nav. Res. Inst. Bethesda, Md. **12**, 97—118. Rept. No. 7 (1954).
137. Mahneke, A.: Electroretinography with double flashes. Acta Ophthal. (Kph.) **35**, 131—141 (1957).
138. Maksimov, V. V., Zenkin, G. M.: Spreading depression in bipolar cells of frog retina. Fiz. Zh. SSSR **51**, 1188—1191 (1965) (Russ.); Fed. Proc. **25**, T 663—664 (1966) (Engl.).
139. Martins-Ferreira, H., de Oliveira Castro, G.: Light scattering changes accompanying spreading depression in isolated retina. J. Neurophysiol. **29**, 715—726 (1966).
140. Matschinsky, F. M.: Quantitative histochemistry of nicotinamide adenine nucleotides in retina of monkey and rabbit. J. Neurochem. **15**, 643—657 (1968).
141. Maturana, H. R., Lettvin, J. Y., McCulloch, W. S., Pitts, W. H.: Anatomy and physiology of vision in the frog (rana pipiens). J. gen. Physiol. **43**, 129—175 (1960).
142. McIlwain, H.: Electrical influences and speed of chemical change in the brain. Physiol. Rev. **36**, 355—375 (1956).
142a. Miller, R. F., Dowling, J. E.: Intracellular responses of the Müller (glial) cells of mudpuppy retina: their relation to b-wave of the electroretinogram. J. Neurophysiol. **33**, 323—341 (1970).
143. Millikan, G. A.: Experiments on muscle haemoglobin in vivo; the instantaneous measurement of muscle metabolism. Proc. Roy. Soc. B **123**, 218—241 (1937).
144. Müller-Limmroth, W.: Elektrophysiologie des Gesichtssinnes. Berlin-Göttingen-Heidelberg: Springer 1959.
145. — Pohlschmidt, W.: Die Wirkungen des 2,4-Dinitrophenol auf das Elektroretinogramm. Naturwissenschaften **47**, 44 (1960).
146. Nagata, Y., Yukoi, Y., Tsukada, Y.: Studies on free amino acid metabolism in excised cervical sympathetic ganglia from the rat. J. Neurochem. **13**, 1421—1431 (1966).
147. Nakajima, A. (Ed.): Retinal Degenerations; ERG and Optic Pathways. Proc. 4th ISCERG Symp. Hakone, Japan 1965. Jap. J. Ophthal. **10**, Suppl. 1966.
148. Negishi, K., Svaetichin, G.: Effects of anoxia, CO_2 and NH_3 on S-potential producing cells and neurons. Pflügers Arch. ges. Physiol. **292**, 177—205 (1966).

149. NIEMI, M., MERENMIES, E.: Cytochemical localization of the oxidative enzyme systems in the retina. I. Diaphorases and dehydrogenases. J. Neurochem. **6**, 200—205 (1961).

150. NOELL, W. K.: Azide-sensitive potential difference across the eye bulb. Amer. J. Physiol. **170**, 217—238 (1952).

151. — Studies on the electrophysiology and the metabolism of the retina. US Air Force, SAM Project 21-1201-0004 (1953).

152. — The origin of the electroretinogram. Amer. J. Ophthal. **38**, 78—90 (1954).

153. — Cellular physiology of the retina. J. Opt. Soc. Amer. **53**, 36—48 (1963).

154. — WALKER, V. S., KANG, B. S., BERMAN, S.: Retinal damage by light in rats. Invest. Ophthal. **5**, 450—473 (1966).

155. NOVER, A., SCHULTZE, B.: Autoradiographische Untersuchung über den Eiweißstoff-wechsel in den Geweben und Zellen des Auges. Albrecht v. Graefes Arch. Ophthal. **161**, 554—578 (1960).

156. NOYONS, A. K. M., WIERSMA, C. A. G.: L'influence de la lumière sur la consommation d'oxygène de la rétine d l'oeil de grenouille. Acta brev. neerl. **3**, 156—157 (1933).

157. OPITZ, E., SCHNEIDER, M.: Über die Sauerstoffversorgung des Gehirns und den Mechanis-mus von Mangelwirkungen. Ergebn. Physiol. **46**, 126—260 (1950).

158. OSTROVSKII, M. A., DETTMAR, P.: Effect of ouabain on the electroretinogram of isolated perfused frog retina. Biofizika **11**, 724—726 (1966) (Russ.).

159. — FEODOROVICH, J. B., POLIAK, S. E.: Change in pH of the medium during illumination of the retina and of a suspension of outer segments of photoreceptors. Biofizika **13**, 338—339 (1968) (Russ.).

160. OTTOSON, D., SVAETICHIN, G.: Electrophysiological investigations of the origin of the ERG of the frog retina. Acta physiol. scand. **29**, 538—563 (1953).

161. OYAMA, N.: Einflüsse der Wasserstoffionenkonzentration auf den Sauerstoff-Verbrauch der Hellnetzhaut von Kaninchen in vitro. Tohoku J. exp. Med. **35**, 567—599 (1939).

162. PILZ, A., SICKEL, W., BIRKE, R.: Lokale Entstehung und Ausbreitung des Elektroretino-gramms der isolierten Froschnetzhaut. Pflügers Arch. ges. Physiol. **265**, 550—562 (1958).

163. PIRIE, A., v. HEYNINGEN, R.: Biochemistry of the Eye. Blackwell Scientific Publ., Oxford 1956.

164. PONTE, F., LAURICELLA, M., BONAVITA, V.: Electroretinographic changes in the rat after injection of 2-deoxyglucose. Vision Res. **4**, 355—359 (1964).

165. QUASTEL, J. H., BICKIS, I. J.: Metabolism of normal tissue and neoplasma in vitro. Nature (Lond.) **183**, 281—286 (1959).

166. RAHMAN, M. A., KERLY, M.: Pathways of glucose metabolism in ox retina. Biochem. J. **78**, 536—540 (1961).

167. RANKE, O. F.: Die optische Simultanschwelle als Gegenbeweis gegen das Fechnersche Gesetz. Z. Biol. **105**, 224—231 (1952).

168. READING, H. W.: Protein biosynthesis and the hexosemonophosphate shunt in the developing normal and dystrophic retina. In: GRAYMORE, C. N. (Ed.): Biochemistry of the Eye, p.. 73—82. London-New York: Academic Press 1965.

169. RIGGS, L. A.: Human retinal response. Ann. N.Y. Acad. Sci. **74**, 372—376 (1958).

170. RÖE, O.: The effect of pH on the oxygen consumption of the retina. Acta Ophthal. (Kph.) **32**, 181—193 (1954).

171. RÖSCH, H., TEKAMP, W.: Über Ammoniakbildung bei der Belichtung der Netzhaut. Z. physiol. Chem. **175**, 158—177 (1928).

172. RÜCHARDT, E.: Sichtbares und unsichtbares Licht. Berlin:J. Springer 1938.

173. RUSHTON, W. A. H.: Rod/cone rivalry in pigment regeneration. J. Physiol. (Lond.) **198**, 219—236 (1968).

174. RUTMAN, R. J., GEORGE, P.: Hydrogen ion effects in high-energy phosphate reactions. Proc. nat. Acad. Sci. (Wash.) **47**, 1094—1109 (1961).

175. SCHWEITZER, N. M. J., TROELSTRA, A.: The recovery of the b-wave in the electroretino-graphy during dark adaptation. Vision Res. **4**, 345—353 (1964).

176. SHICHI, H.: Microsomal electron transfer system of bovine retinal pigment epithelium. Exp. Eye Res. **8**, 60—68 (1969).

177. Sickel, W.: Stoffwechsel und Funktion der isolierten Netzhaut. In: Jung, R., Korn-huber, H. (Eds.): The Visual System: Neurophysiology and Psychophysics, pp. 80—94. Berlin-Göttingen-Heidelberg: Springer 1961.

178. — Respiratory and electrical responses to light stimulation in the retina of the frog. Science **148**, 648—651 (1965).

179. — Microelectrode recording of ERG components and unit activity in the isolated perfused frog's retina. Biophys. J. **5**, 76 (1965).

180. — The isolated retina maintained in a circulating medium: combined optical and electrical investigations of metabolic aspects of generation of the electroretinogram. In: Burian, H.M., Jacobson, J.H. (Eds.): Clinical Electroretinography, pp. 115—124. Oxford Pergamon Press 1966. Suppl. to Vision Res.

181. — Metabolism of the retina in relation to the ERG. Proc. 4th ISCERG Symp. Hakone, Japan 1965. Jap. J. Ophthal. **10**, Suppl. 36—52 (1966).

182. Retinal Oxidation-Reduction States. Proc. 5th ISCERG Symp. Ghent, Belgium 1966, pp. 232—242. Basel-New York: Karger 1968.

183. — Crescitelli, F.: Delayed electrical responses from the isolated frog retina. Pflügers Arch. ges. Physiol. **297**, 266—269 (1967).

184. — Lippmann, H.-G., Haschke, W., Baumann, Ch.: Elektrogramm der umströmten menschlichen Retina. Ber. dtsch. Ophthal. Ges., 63. Zus. Berlin 1960. S. 316—318.

184a. Sillman, A.J., Ito, H., Tomita, T.: Studies on the mass receptor potential of the isolated frog retina; 1. General properties of the response. Vision Res. **9**, 1435—1442 (1969a).

184b. — — — Studies on the mass receptor potential of the isolated frog retina; 2. On the basis of the ionic mechanism. Vision Res. **9**, 1442—1451 (1969b).

185. Skou, J.C.: The influence of some cations on an adenosine triphosphatase from peripheral nerves. Biochim. biophys. Acta (Amst.) **23**, 394—401 (1957).

186. Slater, T.F., Heath, H., Graymore, C.N.: Levels of oxidized and reduced pyridine nucleotides in rat retina. Biochem. J. **84**, 37 P (1962).

187. Smit, J.A.: Over den invloed van intensiteit en golflengte van licht op de electrische verschijnselen van het oog. Diss. Utrecht 1934 (cit. by Granit).

188. Süllmann, H.: Auge und Tränen. In: Flaschenträger, B., Lehnartz, E. (Hrsg.): Handb. d. Physiol. Chemie, Vol. II/2a, S. 864—948. Berlin-Göttingen-Heidelberg: Springer 1956.

189. Sverak, J., Peregrin, J., Hradecky, F.: Electroretinographic study on the metabolism of the retina (glucose test). Ophthalmologica (Basel) **138**, 287—291 (1959).

190. Svaetichin, G., Negishi, K., Fatehchand, R., Drujan, B.D., Selvin, D.E., Testa, A.: Nervous function based on interactions between neuronal and non-neuronal elements. Progr. Brain Res. **15**, 243—266 (1965).

190a. — — — Cellular mechanisms of a Young-Hering visual system. In: De Reuck, A.V.S., Knight, J. (Eds.): Color Vision. Boston: Little, Brown and Co. 1965.

191. Sym, E., Nilsson, R., v. Euler, H.: Co-Zymasegehalt verschiedener tierischer Gewebe. Z. physiol. Chem. **190**, 228—246 (1930).

192. Tansley, K., Copenhaver, R.M., Gunkel, R.D.: Some observations of the mammalian cone ERG. J. Opt. Soc. Amer. **51**. 207—213 (1961).

193. Terner, C.: Anaerobic and aerobic glycolysis in lactating mammary gland and in nervous tissue. Biochem. J. **52**, 229—237 (1952).

194. Therman, P.O.: The neurophysiology of the retina in the light of chemical methods of modifying its excitability. Acta Soc. Sci. Fenn. New Ser. B, Nr. 1, 2—74 (1938).

195. Tomita, T.: Studies on intraretinal action potentials. Part I: Relation between the localization of the micropipette in the retina and the shapes of the intraretinal action potential. Jap. J. Physiol. **1**, 110—117 (1950).

196. — Electrical activity in the vertebrate retina. J. Opt. Soc. Amer. **53**, 49—57 (1963).

197. — Electrical response of single photoreceptors. Proc. IEEE **56**, 1015—1023 (1968).

198. — The electroretinogram. In: Handbook of Sensory Physiology, Vol. VII/2. Berlin-Heidelberg-New York: Springer 635—665

198a. — Electrophysiological study of the mechanisms subserving color coding in the fish retina. Cold Spr. Harb. Symp. Quant. Biol. **30**, 559—566 (1965).

198b. Toyoda, J., Hashimoto, H., Anno, H., Tomita, T.: The rod response in the frog as studied by intracellular recording. Vision Res. **10**, 1093—1100 (1970).

198c. — Nosaki, H., Tomita, T.: Light-induced resistance changes in single photoreceptors of *Necturus* and *Gekko*. Vision Res. **9**, 453—463 (1969).

199. Trappl, R., Bornschein, H.: Ein mathematisches Modell für die Komponente P III des Elektroretinogramms. Kybernetik **4**, 40—43 (1967).

200. Tsukada, Y., Uyemura, K., Matsutani, T.: Metabolism of amino acids and nucleic acids in the isolated rabbit retina. Advan. Neurol. Sci. (Tokyo) **10**, 210—218 (1966) (Jap.).

200a. Turini, S., Sorbi, T., Cavaggioni, A.: The effect of illumination on the efflux of radioactive potassium and rubidium from the isolated frog retina. IUPS-ISCERG Symposium, Brighton 1971.

201. Val'tsev, V. B.: Investigation in flowing liquid of retinal potentials. Biofizika **11**, 1095—1096 (1966) (Russ.).

202. Vernon, L. P.: Photochemical oxidation and reduction reactions catalysed by flavin nucleotides. Biochim. biophys. Acta (Amst.) **36**, 177—185 (1959).

203. Wang, D. Y., Slater, T. F., Dartnall, H. J. A.: Swelling properties of mitochondrial preparations from the retina. Vision Res. **3**, 171—181 (1963).

204. Warburg, O., Posener, K., Negelein, E.: Über den Stoffwechsel der Carcinomzelle. Biochem. Z. **152**, 309—344 (1924).

205. Webster, H. D., Ames III., A.: Reversible and irreversible changes in the fine structure of nervous tissue during oxygen and glucose deprivation. J. Cell Biol. **26**, 885—909 (1965).

206. — — The effects of osmotic changes on the phase and electron microscopic appearance of nervous tissue. J. Neuropath. exp. Neurol. **26**, 160—161 (1967).

207. Weinstein, G. W., Hobson, R. R., Dowling, J. E.: Light and dark adaptation in the isolated rat retina. Nature (Lond.) **215**, 134—138 (1967).

207a. Werblin, F. S., Dowling, J. E.: Organization of the retina of the Mudpuppy *Necturus maculosus*. J. Neurophysiol. **32**, 339—355 (1969).

208. Yeandle, S. S.: Studies on the slow potential and the effects of cations on the electrical responses of the Limulus ommatidium. Ph. Diss., Johns Hopkins Univ. 1957.

209. Young, R. W., Droz, B.: The renewal of protein in retinal rods and cones. J. Cell Biol. **39**, 169—184 (1968).

210. Zewi, M.: On regeneration of visual purple. Acta Soc. Sci. Fenn. S. B, Nr. 2, 1—56 (1939).

211. Ziv, B., Burian, H. M.: Electric response of the phakic and aphakic human eye to stimulation with near ultraviolet. 18. Conc. Ophthalm. 1958, pp. 644—647. Bruxelles: Press Med. Sci. 1959.

Author Index

The numbers shown in square brackets are the numbers of the references in the bibliography. Page numbers in *italics* refer to the bibliography.

Subject Index